2nd International Conference on Materials, Manufacturing and Machining for Industry 4.0 (ICMMM-2020)

IOP Conference Series: Materials Science and Engineering
Volume 1059

Erode, India
9 - 10 October 2020

Editors:

M. Kumar
D. Dinesh
A.D. Pradeep
B. Kalidasan

ISBN: 978-1-7138-4573-7
ISSN: 1757-8981

Printed from e-media with permission by:

Curran Associates, Inc.
57 Morehouse Lane
Red Hook, NY 12571

Some format issues inherent in the e-media version may also appear in this print version.

Institute of Physics
Dirac House, Temple Back
Bristol BS1 6BE UK

Phone: 44 1 17 929 7481
Fax: 44 1 17 920 0979

techtracking@iop.org

Additional copies of this publication are available from:

Curran Associates, Inc.
57 Morehouse Lane
Red Hook, NY 12571 USA
Phone: 845-758-0400
Fax: 845-758-2633
Email: curran@proceedings.com
Web: www.proceedings.com

TABLE OF CONTENTS

IOP Conf. Series: Materials Science and Engineering 1059 (2021) 011001 doi:10.1088/1757-899X/1059/1/011001

2nd International Conference on Materials, Manufacturing, and Machining for Industry 4.0 (ICMMM 2.0) - *A Virtual Conference*

Organized by Bannari Amman Institute of Technology, Sathyamangalam - 638401, Erode, Tamil Nadu, INDIA

9 - 10 October 2020 www.icmmm.ml

PROCEEDINGS

Publishing Partners

IOP Conf. Series: Materials Science and Engineering 1059 (2021) 011001 doi:10.1088/1757-899X/1059/1/011001

PROCEEDINGS

2nd International Conference on Materials, Manufacturing and Machining for Industry 4.0 (ICMMM 2020) – *A Virtual Conference*

October 9 – 10, 2020

Organised by

Department of Mechanical Engineering

Bannari Amman Institute of Technology
Sathyamangalam – TK, Erode – DT,
Tamil Nadu, INDIA -638401

CONVENERS

Dr. M. Kumar | Mr. D. Dinesh

CO - CONVENERS

Dr. P. Sangaravadivel | Mr. A. D. Pradeep
Mr. C. Sathishkumar | Mr. A. Tajdeen
Mr. S. Manojkumar | Mr. K. Kamal Basha
Mr. B. Kalidasan | Mr. G. Sundarraju | Mr. M. Sureshkumar

IOP Conf. Series: Materials Science and Engineering 1059 (2021) 011001 doi:10.1088/1757-899X/1059/1/011001

PROCEEDINGS

2nd International Conference on Materials, Manufacturing and Machining for Industry 4.0 (ICMMM 2020) – *A Virtual Conference*

October 9 – 10, 2020

Editors

Dr. M. Kumar

Mr. D. Dinesh

Mr. A. D. Pradeep

Mr. B. Kalidasan

IOP Conf. Series: Materials Science and Engineering 1059 (2021) 011001 doi:10.1088/1757-899X/1059/1/011001

First Print: December 2020

DISCLAIMER

Due to COVID-19 pandemic situation, the conference meeting was changed to *Virtual Mode* which offered a chance to the students, faculty members, researchers and industrial persons to share their knowledge, experience and expertise virtually that augments the 'Technological Advancements' ensues across countries. We took this conference as an initiative through Virtual Mode as a challenging and innovative way of transferring the technology in the fields of Materials, Manufacturing and Machining. The conference was virtually conducted through google meet platform on 09 & 10th October 2020 by M/s. Bannari Amman Institute of Technology, Department of Mechanical Engineering, Alathukombai (Po), Sathyamangalam & Taluk, Erode District - 638 401 Tamilnadu, India.

The conference was designed to be conducted virtually that consists of three parallel technical paper presentation sessions on each day. Authors were given 12 min for their presentation and 3 min for question and answers session. In addition to this, five technical keynote lecture sessions were offered by eminent professors, scientists and leading industry experts on both the days. Q & A sessions were provided to the delegates to update their field of expertise according to industrial and social needs.

182 quality papers were scrutinized out of 230 by the editorial team and they were presented in the conference. The quality of the technical papers is ensured by Single Blind review process carried out by the external and internal experts as a review member. Each paper is checked for its similarity by Turnitin software, and ensured to have <15%. The reviewers' suggestions are scrupulously considered as one of the criteria for accepting the 76 papers for possible publication.

The authors are solely responsible for the contents of the paper in this volume. The Publishers or Editors don't take any responsibility for the same in any manner. Errors, if any, are purely unintentional and readers are requested to communicate such errors to the editors or publishers to avoid discrepancies in future. No warranty may be created or extended by sales or promotional materials. The advice or strategies contained herein may not be suitable for every situation.

Published by

DEPARTMENT OF MECHANICAL ENGINEERING
Bannari Amman Institute of Technology, Sathyamangalam - TK
Erode District – 638401, Tamilnadu, INDIA.
Ph: 04295-226000 Fax: 04295-226666
E-mail: stayahead@bitsathy.ac.in, Web: www.bitsathy.ac.in

IOP Conf. Series: Materials Science and Engineering 1059 (2021) 011001 doi:10.1088/1757-899X/1059/1/011001

PREFACE

The Department of Mechanical Engineering, Bannari Amman Institute of Technology (BIT) is honoured to bring out the proceedings of the 2nd **International Conference on Materials, Manufacturing and Machining for Industry 4.0 (ICMMM 2020) –** ***A Virtual Conference*** in soft form. This is an extensive Mechanical Engineering associated conference to take place in India. The conference was virtually held during 09-10, October 2020 hosted by Bannari Amman Institute of Technology, Sathyamangalam.

The 2nd International Conference on Materials, Manufacturing, and Machining for Industry 4.0 (ICMMM 2.0) aims to attract like-minded people who work for the innovation in engineering applications. This conference also enhances the knowledge in the integration of advanced materials, modern manufacturing and machining processes to meet the requirements of Industry 4.0. In the present scenario, the role of Engineers/Researchers/Scientists is to enrich societal development through green engineering for the health and wealth of human life. ICMMM 2.0 extends a technical forum for academicians, industrialists, and researchers for sharing knowledge, expertise, and experience.

The conference was structured with the chief guest inaugural address by **Dr. Manoj Gupta, Professor in Mechanical Engineering, NUS, Singapore,** and a special invited talk by **Prof. Richard Yang, Western Sydney University, Australia.** Both days were slated with 5 parallel **virtual technical presentation sessions** along with **5 keynote lectures** delivered by well-known eminent personalities from various reputed *Institutions* Viz. **Prof. Brij Kumar Dindhaw Emeritus Professor, IITKH, India, Prof. Adarsh Kumar Pandey, Sunway University, Malaysia,** and **Prof. Jeyraj A/ L Selvaraj, University of Malaya, Malaysia.** Experts invited from *Government Organizations* **Dr. T. Ram Prabhu, Scientist "D" – Defence Research and Development Organisation, India,** and from *renowned industries* **Dr. Krishnan Umachandran, NELCAST, India Limited.**

More than 230 technical papers were initially received for the conference out of which 182 quality papers were selected after technical scrutiny process for the presentation in the conference and 76 accepted papers will be submitted for possible publication in Scopus and Web of Science indexed technical journal, **IOP conference series: Materials Science Engineering (ISSN: 17578981).** 75 papers were accepted for publication. The participants from various countries and states have contributed their technical papers. The conference is organized by the combined effort of a self-supporting basis.

The organizers are grateful to BIT for giving the opportunity to organize this virtual conference. The Department of Mechanical is indebted to **Dr.S.V.Balasubramaniam,** Chairman, **Dr.M.P.Vijayakumar,** Trustee, **Dr.C.Palanisamy,** Principal, **Dr.K.Sivakumar,** Dean PDS and **Dr.M.Ravikumar,** HoD, BIT for their encouragement and full support for the success of the conference.

Dr. M. Kumar | Mr. D. Dinesh
Conveners

IOP Conf. Series: Materials Science and Engineering 1059 (2021) 011001 doi:10.1088/1757-899X/1059/1/011001

Advisory Committee

- Dr C Palanisamy, Principal
- Dr K Sivakumar, Dean-Planning & Development and Student Affairs
- Dr S Jegadheeswaran, Professor-Academics and Research & Development
- Dr K L Senthilkumar, Professor & Head, Dept. of Mechatronics
- Dr M Kalil Rahiman, Professor & Head, Dept. of Aeronautical Engg.
- Dr M Bharathiraja, Associate Professor & Head, Dept. of Automobile Engg.

Organizing Secretary

- Dr M Ravikumar, Professor & Head, Mechanical Engineering

International Advisory Committee

- Prof Saidur Rahman, Sunway University, Malaysia
- Prof David Barton, University of Leeds, UK
- Prof Sang-Jae Kim, Jeju National University, Republic of Korea
- Prof Rajendra Bordia, Clemson University, South Carolina, USA
- Prof Manoj Gupta, National University of Singapore, Singapore
- Prof Senthil Sundaram, Senior Lecturer, University of Exeter, UK
- Prof Richard Yang, Western Sydney University, Australia
- Prof Toshihiko Kuwabara, Tokyo University of Agriculture and Technology, Japan
- Prof Carlo Alberto Biffi, National Research Council, CNR, ICMATE, Institute of Condensed Matter Chemistry and Technologies for Energy, Lecco, Italy
- Prof Marwan Brouche, Saint-Joseph University of Beirut, Lebanon
- Prof Kadirgama K, UniversitiMalaysia Pahang, Malaysia
- Prof Jeyraj AIL Selvaraj, University of Malaya, Malaysia
- Prof Adarsh Kumar Pandey, Sunway University, Malaysia
- Dr Kumar Kandasamy, Enabled Engineering, Virginia, USA
- Prof Balan Sundarakani, University ofWollongong in Dubai, UAE
- Prof Surojit Gupta, University of North Dakota, USA
- Prof Parviz Asadi, Imam Khomeini International University, Iran
- Prof Mahendran A/L Samykano, Universiti Malaysia Pahang, Malaysia

National Advisory Members

- Prof Brij Kumar Dhindaw, Indian Institute of Technology, Kharagpur
- Dr Eswara Prasad, N, DMSDE, Kanpur
- Dr Mrinal Pal, CSIR- Central Glass & Ceramic Research Institute, Jadavpur
- Dr SenthilKumar R, Satish Dhawan Space Centre-ISRO, Sriharikota
- Dr Venkateswaran T,Vikram Sarabhai Space Centre, Thiruvananthapuram
- Dr Gnanamoorthy R, Indian Institute of Technology, Madras
- Dr Apurbba Kumar Sharma, Indian Institute of Technology, Roorkee

- Dr Krishnan Umachandran, NELCAST Limited, Chennai
- Dr Sivapragasam C, Kalasalingam Academy of Research and Education, Sivakasi
- Dr Chandrasekhar U, Wipro Infrastructure Engineering, Bangalore
- Dr Gunasekaran V, Central University of Tamil Nadu, Tiruvarur
- Dr Harpreet Singh Arora, ShivNadar University, Uttar Pradesh
- Dr DeepakB B V L National Institute of Technology, Rourkaa
- Dr Rakesh Kumar Gautam, Indian Institute of Technology, Varanasi
- Dr Lenin Singaravau D, National Institute of Technology, Trichy
- Dr Rajamurugan G, Vellore Institute of Technology, Vellore
- Dr Gaurav Bartarya, Indian Institute of Technology, Bhubaneswar
- Dr Ponappa K IIITDM, Jabalpur
- Mr Prasad Kumar S, NSK Bearings Umited, Chennai
- Dr Anand Kumar S, Indian Institute of Technology, Jammu
- Dr Govindaraju M, Amrita Vihswa Vidyapeetham, Coimbatore
- Dr Kesavan D, Indian Institute of Technology, Palakkad
- Dr Kaushik Rajaram, Central University of Tamil Nadu, Tiruvarur
- Dr Santhy K Indus University, Ahmedabad

Conference Committee

Conveners

- Dr M Kumar, AP, Mechanical Engineering
- Mr D Dinesh, AP, Mechanical Engineering

Co-Conveners

- Dr P Sangaravadivel, AP, Mechanical Engineering
- Mr A D Pradeep, AP, Mechanical Engineering
- Mr C Sathishkumar, AP, Mechanical Engineering
- Mr A Tajdeen, AP, Mechanical Engineering
- Mr S Manojkumar, AP, Mechanical Engineering
- Mr K Kamal Basha, AP, Mechanical Engineering
- Mr B Kalidasan, AP, Mechanical Engineering
- Mr G Sundarraju, AP, Mechanical Engineering
- Mr M Sureshkumar, AP, Mechanical Engineering

Organizing Committee

- Mr C Subramaniyan, AP, Mechanical Engineering
- Mr S Velmurugan, AP, Mechanical Engineering
- Mr N Jayakumar, AP, Mechanical Engineering
- Mr A Ramakrishnan, AP, Mechanical Engineering
- Mr K B Prakash, AP, Mechanical Engineering

- Mr E Soundrapandian, AP, Mechanical Engineering
- Mr D Yuvaraj, AP, Mechanical Engineering
- Mr N Bhuvanesh, AP, Mechanical Engineering
- Mr K P Boopathiraja, AP, Mechanical Engineering

Technical Committee

- Dr K Muthukumar, Professor, Mechanical Engineering
- Dr A Amarkarthik, Professor, Mechanical Engineering
- Dr C Sasikumar, Professor, Mechanical Engineering
- Dr G Senthilkumar, Professor, Mechanical Engineering
- Dr V C Uvaraja, Professor, Mechanical Engineering
- Dr T Rameshkumar, ASP, Mechanical Engineering
- Prof M Nandagopal, ASP, Mechanical Engineering
- Dr A Anandha Moorthy, ASP, Mechanical Engineering
- Dr G Kumaresan, ASP, Mechanical Engineering
- Dr R Rajiev, AP, Mechanical Engineering
- Dr P Mathankumar, AP, Mechanical Engineering
- Mr K Ramachandra Raju, AP, Mechanical Engineering

Student Volunteers

- Mr B Sriram, 2020-Passed out, Mechanical Engineering
- Mr S Kamaraj, Computer Science and Engineering
- Mr S A Praveen, Mechatronics

Special Thanks to

Session Chairs, Co – Chairs

&

Review Members

IOP Conf. Series: Materials Science and Engineering 1059 (2021) 011001 doi:10.1088/1757-899X/1059/1/011001

Glimpses of ICMMM 2020

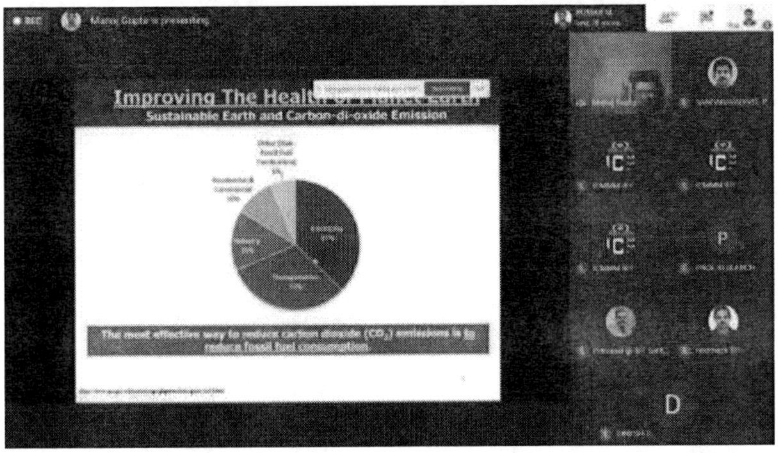

Welcome Address and Chief Guest Introduction

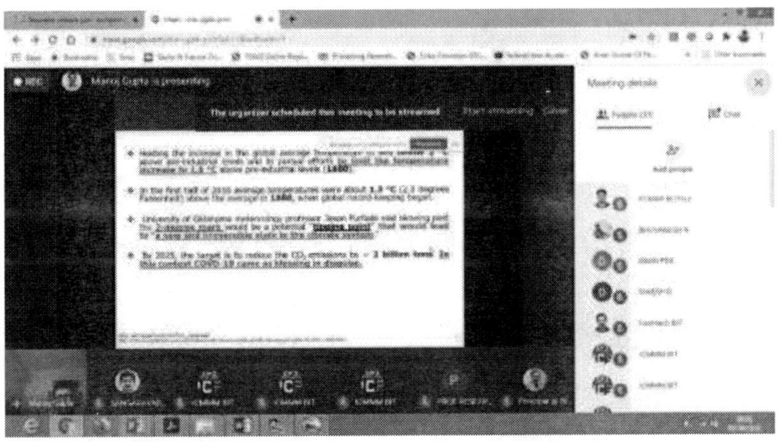

Chief Guest Address by **Dr. Manoj Gupta,** NUS Singapore

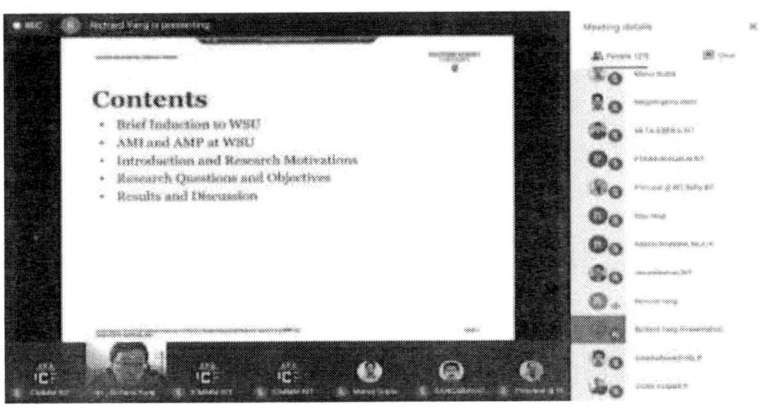

Special Invitee Talk by **Prof. Richard Yang,** Western Sydney University, Australia

IOP Conf. Series: Materials Science and Engineering 1059 (2021) 011001 doi:10.1088/1757-899X/1059/1/011001

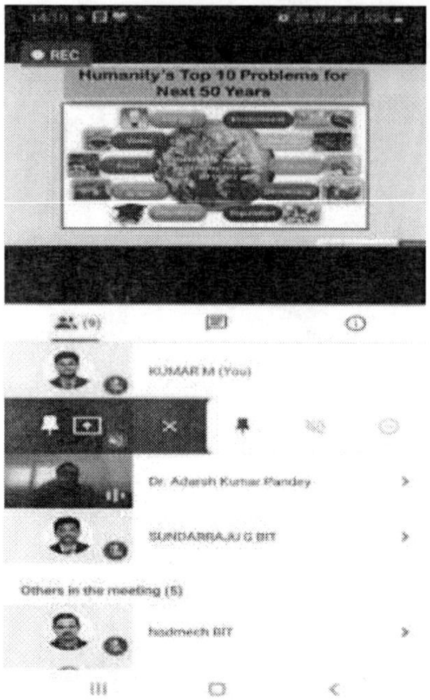

Keynote Lectures by **Dr. Adarsh Kumar Pandey**, Sunway University (left) & **Dr. T. Ramprabhu,** DRDO, Bangalore (Right)

Keynote Lectures by **Dr. Krishnan Umachandran**, CEO, NELCAST (left) & **Dr. Jeyaraj Selvaraj,** University of Malaya, Malaysia (Right)

IOP Conf. Series: Materials Science and Engineering 1059 (2021) 011001 doi:10.1088/1757-899X/1059/1/011001

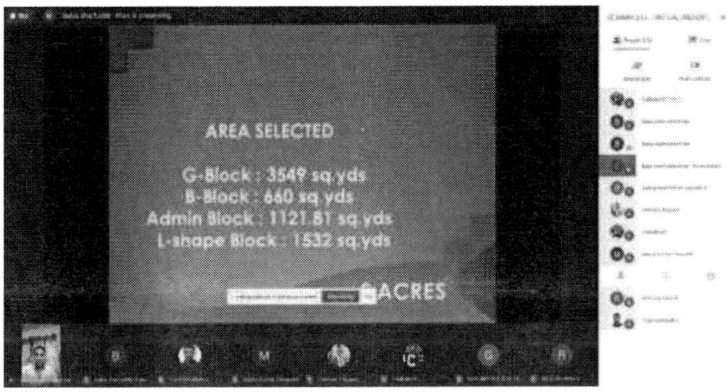

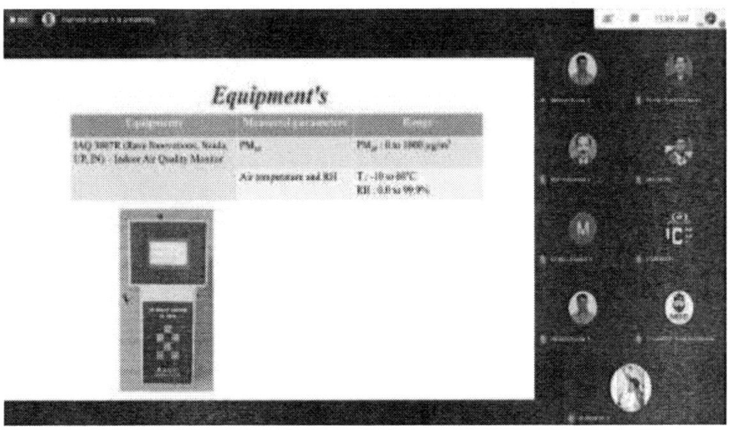

Parallel Technical Paper Presentation Sessions

IOP Conf. Series: Materials Science and Engineering 1059 (2021) 011001 doi:10.1088/1757-899X/1059/1/011001

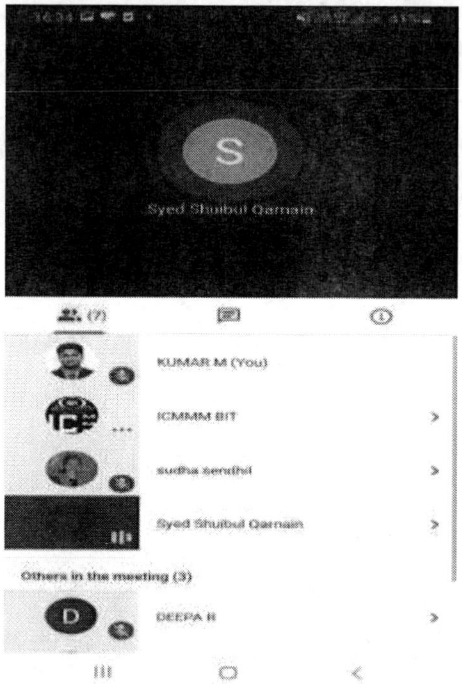

Feedback given by one of the participants

IOP Conf. Series: Materials Science and Engineering 1059 (2021) 011002 doi:10.1088/1757-899X/1059/1/011002

Peer review declaration

All papers published in this volume of *IOP Conference Series: Materials Science and Engineering* have been peer reviewed through processes administered by the Editors. Reviews were conducted by expert referees to the professional and scientific standards expected of a proceedings journal published by IOP Publishing.

- **Type of peer review: Single-blind**
- Each paper is plag checked using Turnitin software and undergone two reviews by the subject experts, and the reviewer comments are incorporated and verified to ensure the quality of the manuscript and then submitted to the **IOP Conference Series: Materials Science and Engineering** for possible publication.
- **Conference submission management system:** Easychair
- **Number of submissions received:** 110
- **Number of submissions sent for review:** 90
- **Number of submissions accepted:** 76
- **Acceptance Rate (Number of Submissions Accepted / Number of Submissions Received X 100):** 69.09 %
- **Average number of reviews per paper**: 2
- **Total number of reviewers involved:** 48
- **Any additional info on review process:** All the papers were checked for similarity before and after review process and it is maintained to have less than 15% similarity.
- **Contact person for queries:**

Dr.M.Kumar
Assistant Professor Level III,
Department of Mechanical Engineering,
Bannari Amman Institute of technology,
Sathyamangalam-638401, TamilNadu, India.
E-mail: kumarm@bitsathy.ac.in

Mr.D.Dinesh
Assistant Professor Level II,
Department of Mechanical Engineering,

Bannari Amman Institute of technology,
Sathyamangalam-638401, TamilNadu, India.
E-mail: dineshd@bitsathy.ac.in

Evaluation of mechanical and antimicrobial properties for indigenously developed Biofilm for food packaging industries

Balakrishnaraja Rengaraju[1], M Jeevanantham[2], S Shanmugaraja[2], A Selvakumar[3], and S Tamilselvi[2]

[1] Department of Food Technology, Bannari Amman Institute of Technology, Sathyamangalam, India
[2] Department of Biotechnology, Bannari Amman Institute of Technology, Sathyamangalam, India
[2] VIT Fashion Institute of Technology, Vellore Institute of Technology, Chennai, India

balakrishnaraja@gmail.com

Abstract. The preparation of the film based on aloe vera gel and polyvinyl alcohol (PVA) were carried out in this work. The film has high adhesion due to the presence of polyvinyl alcohol. The mechanical properties like tensile strength and the transparency of the film were tested. The temperature withstanding ability of the film is tested with Thermo gravimetric analysis. The functional groups of the film were identified. The films prepared were evaluated by incorporation of an antifungal and an antibacterial agent which shows marked inhibition on growth of mold, yeast and bacteria. The Shelf life of the vegetables was tested using the film.
Keywords: organic biofilm, antimicrobial film, Aloe vera

1 Introduction

Polymer and plastics produced using petrochemical inferred polymers, for example, poly (ethylene teraphthalate), poly (vinyl chloride), poly (ethylene), poly (propylene), poly (styrene), poly (amide), ABS copolymer and so on by proper embellishment strategy are assuming essential jobs in all the human exercises, for example, nourishment bundling, farming mulching, convey sacks, medical clinic bed covers, removal cups, plate, as scratch verification meagre film coatings and so forth. Polymers are generally utilized for these applications on account of their accessibility, minimal effort, great mechanical properties (tractable and tear quality) destructive opposition [1].A few biopolymers have been utilized to create eco-accommodating nourishment bundling materials. Usually, films dependent on biopolymers are profoundly touchy to natural conditions and for the most part present low mechanical opposition.

1.1 Properties of polymeric Biofilm

Therefore, a few examinations have been completed to create films dependent on blends of biopolymers and manufactured polymers. Naturally degradable plastics have pulled in developing consideration on account of their latent capacity use in the substitution of conventional non degradable plastic things getting from petroleum derivative feed stocks. Poly vinyl Alcohol (PVA) has been broadly used for the planning of mixes and composites with a few characteristic, inexhaustible polymers [2]. PVA is a material with mechanical potential which has a wide scope of utilizations because of its special physical and synthetic properties. It is a nontoxic, exceptionally crystalline issue and has great film forming and high hydrophilic properties. PVA, cellulose and Chitosan are polar polymers; in this manner a nanocomposite of PVA, Chytosin and nano crystalline cellulose (NCC) is probably going to create a material including phenomenal mechanical properties. Installing NCC into PVA framework changes the physicochemical properties of PVA and improves polymer structure in both sub-atomic and morphological levels [3]. During the time spent biodegradation the right off the bat the long polymer atoms are diminished to shorter and shorter lengths and experience oxidation (oxygen bunches append themselves to the polymer particles). This procedure is activated by heat (raised temperatures found in landfills), UV light (a part of daylight) and mechanical pressure (for example wind or compaction in a landfill). [3, 2]

1.1.2 Microbial Biofilm Degradation

Oxidation makes the particles become hydrophilic (water pulling in) and sufficiently little to be ingestible by microorganisms, making way for biodegradation to start. Biodegradation happens within the sight of dampness and microorganisms regularly found in the earth. The plastic material is totally separated into the remaining results of the biodegradation procedure. As microorganisms expend the corrupted plastic, carbon dioxide, water, and biomass are created and come back to nature by method for the bio cycle [4]. One of the primary research regions in nourishment bundling has concentrated on growing new bundling systems fit for improving nourishment conservation properties dependent on their connection with bundling. Such procedures are known as "dynamic bundling frameworks". A dynamic bundling can be characterized as a kind of material that changes its bundling conditions to broaden timeframe of realistic usability, cooperating straightforwardly with the nourishment, upgrading security and looking after quality. Specifically, the antimicrobial bundling is one of the most imaginative and promising dynamic bundling types created over the most recent decade, which incorporates frameworks fit for hindering microorganism activity and staying away from loss of nourishment quality [5]. Nonetheless, because of the couple of reports that have been made on films with high mechanical and warm properties, the planning of PVA/starch based biodegradable antibacterial films show extraordinary importance in the field of nourishment bundling. Along these lines, the reason for the current examination is to utilize starch and polyvinyl alcohol [6].

PVA material is universally applied due to its high efficiency and usability involving the alteration of physical/ chemical properties [7] Polymers are generally utilized for these applications on account of their accessibility, minimal effort, great mechanical properties (tractable and tear quality) destructive opposition [8].

2 Materials and Methods

2.1.Synthesis of Biofilm.

The films were prepared using proportions of polyvinyl alcohol i.e., 1.5%, 2.5%, and 3.5% in which *Aloe vera* is added with respect to the proportion of PVA. The film was produced using casting method. The PVA powder of about 1.5g is added in 100 ml of deionised water and kept under constant mixing in the sonicator. *Aloe vera* is added in the PVA aqueous solution 50% by mass with respect to the mass of PVA powder. It is kept under constant stirring for about 3h at 90^0C. Then the solution were poured into a dye and dried for 24h at room temperature (the same procedure is followed for preparation of 2.5% and 3.5% film).[9, 10]

2.2.Mechanical Analysis.

In this analysis, tensile strength (TS) was determined according to the standard method, works on the constant rate of transverse principle (universal testing machine is used to test the tensile strength of the films).

2.2.1 Film thickness: The thickness of the film was determined by using the Vernier calliper and measured as the average of 3 mm for each sample.

2.2.2 Opacity: The opacity of the film is determined by cut the sample into rectangular shape and the transmittance curve of various spectrums (300nm -600nm).

2.2.3 Thermal properties: The film sample were dried and cut into pieces after that Thermo gravimetric analysis (TGA), Differential scanning calorimeter (DSC), Derivative thermal analysis (DTA) analyses were performed on a NETZSCH STA 2500 and heated from 0 to 500^0C.

2.2.4 FTIR analysis: It is used to confirm the chemical nature of the film components.1.5%, 2.5%, 3.5% film were analysed by Fourier transform infrared spectroscopy at the resolution in a 4000 to 400 per cm range, using air as background.

2.3 Biocompatibility Analysis.

The biodegradation test was done by burying the film in the soil in the closed environment. Before that the mass of the film were measured and kept for five days. The weight losses of the film were then measured. For this test, tomato is taken and cut into pieces and packed with the biofilm produced and observed. The food is kept in the room temperature and photographed at different times.

3 Results and Discussion

The different concentrations of the films were produced by using the casting method that is 1.5%, 2.5%, 3.5 %. Each film shows that it can withstand high temperature and can hold more weight. It is used as to increase shelf life and to carry substance.

Film thickness: The thickness of the 1.5%, 2.5%, and 3.5% films were 0.02mm, 0.029mm, and 0.033mm respectively. It is measured by using the verniarcaliper.

Opacity: The transparency of each film is measured using UV visible spectrometer to find the opacity of the film by using transmittance curve. It shows that transparency of the film is mostly similar to each other and the PVA film is known for transparency. The transmittance curve is shown in graph 1, 2 and 3.

FTIR Analysis: It is used for the chemical characterization of the produced biofilm by FTIR Analysis. It is used to obtain spectra of the each film. The adsorption is done in the resolution of 4000-400 per cm range. The obtained results were shown in the graph 4, 5 and 6.

Mechanical properties: The tensile strength of each film was tested using the universal testing machine which shows that it has the ability to carry around one kilo gram. The results are shown in the Table 1.

Thermal properties: In the thermal decomposition test, the temperature used is ranged from 0 to 500 degree Celsius. In this analysis it is revealed that the major mass loss at 260^{0}C. These thermal analysis are shown in the graph 7, 8 and 9 respectively.

Biodegradable test: Bio degradable of the prepared film is identified by the wet soil, and each concentration of the films are weight loses every day.

Behaviour on foods: The 2.5% PVA film has increase the shelf life of the tomato by covering the pieces in the film biolayer.

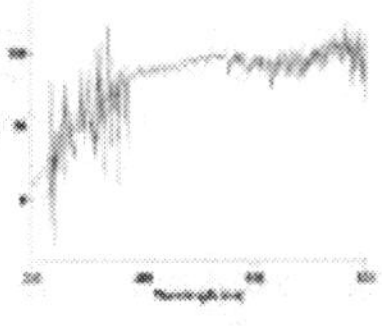
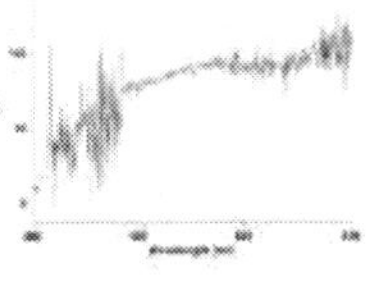
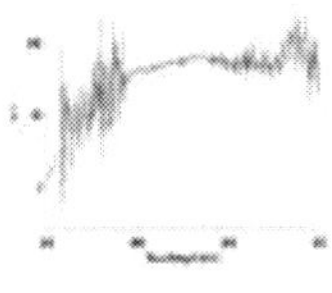

Figure 1. **Figure 2.** **Figure 3.**

Transmittance curves Figure. (1), (2) & (3)are for 1.5%, 2.5% and 3.5% respectively

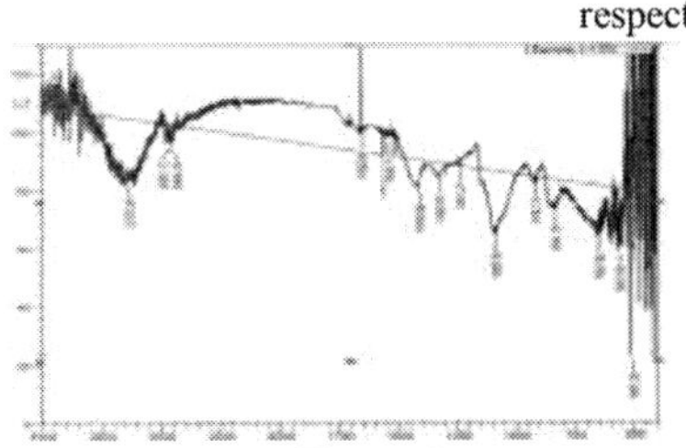
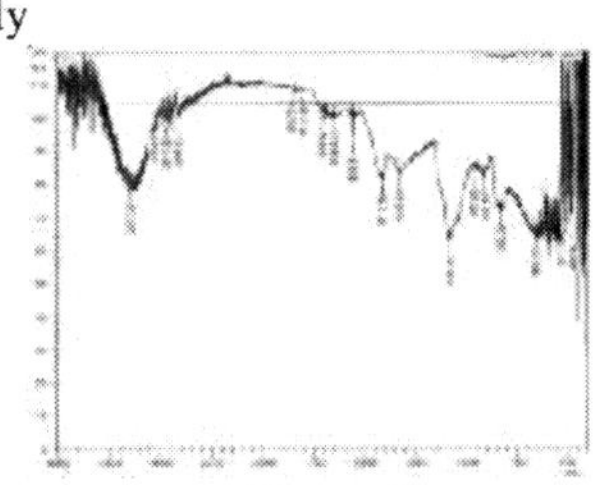

Figure 4. **Figure 5.**

FTIR Analysis Figure. (4) and (5) are for 1.5% and 2.5% respectively

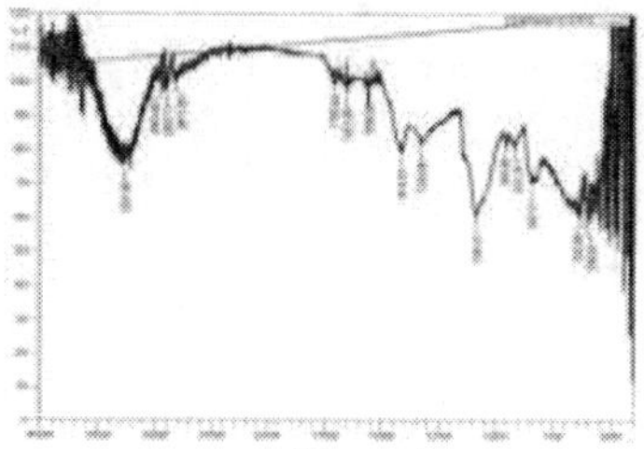

Figure6.

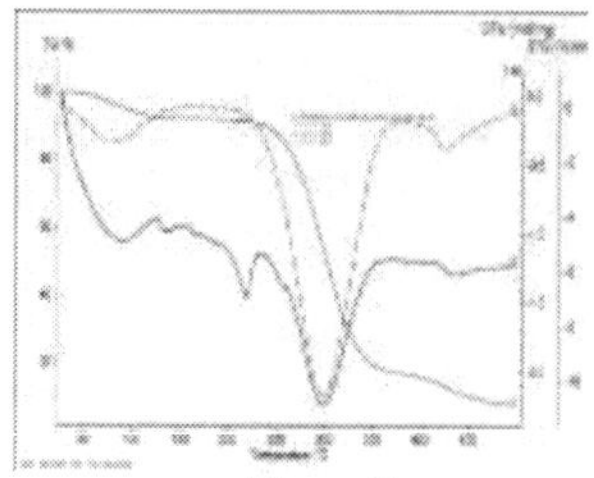

Figure7.

FTIR Analysis Fig. (6) of 3.5% and Thermal analysis (7) of 1.5% film respectively

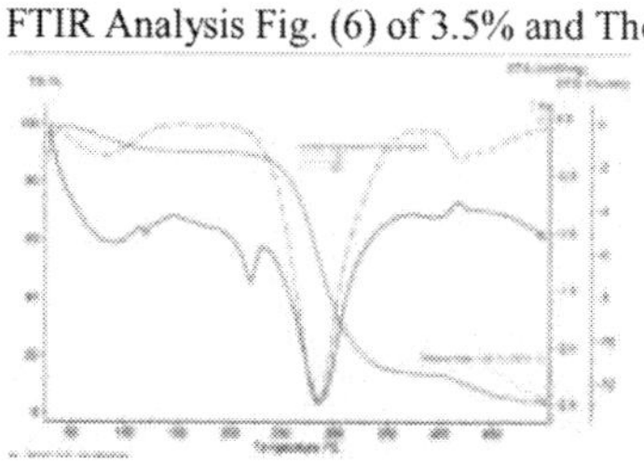

Figure8.

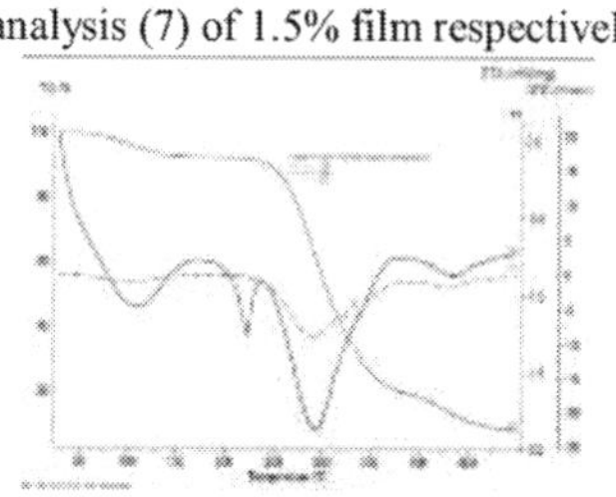

Figure9.

Thermal analysis Fig. (8) & (9) of 2.5% and 3.5% film respectively

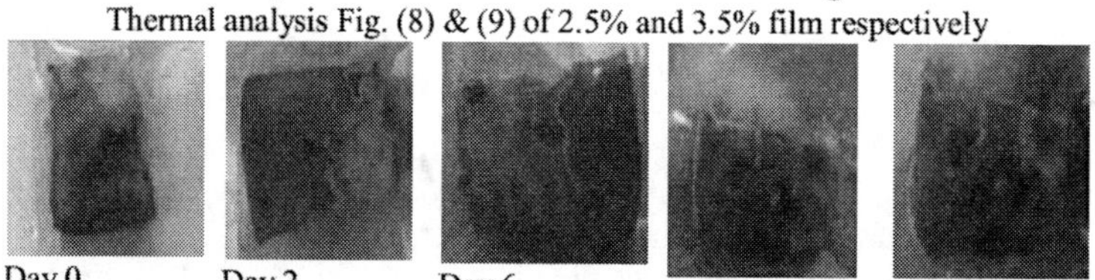

Day 0 Day 2 Day 6 Day 10 Day 15

Figure10. Shelf life test on tomato with 2.5% PVA film

Table 1. Tensile strength of each films

Film Conc. (%)	Load (KN)	Displacement (mm)
1.5%	0.0025	57.15
2.5%	0.0125	147.56
3.5%	0.0150	196.250

Table2. Biodegradation Assay of the film with different concentration

Days	Film Concentration		
	1.5%	2.5%	3.5%
1	0.0014	0.0020	0.0032
2	0.0010	0.0014	0.0027
3	0.0005	0.0009	0.0021
4	0.0002	0.0005	0.0017

4 Summary and Conclusions

Thus the produced biofilm can able to withstand high temperature and has antimicrobial activity and can hold more weight which is used for packaging and increase shelf life of vegetables. The different concentration of film shows the transparency to the light which is used for light required ailments.

Acknowledgments

The authors are thankful for the support rendered by the Principal and Management of Bannari Amman Institute of technology towards project support

References

[1] Eliassaf J 1972 Detection of small quantity of Poly (vinyl alcohol) in Poly (vinyl chloride) resins *Polymer Letters* **16** 225-235
[2] Cha D S Cookseyb K Chinnana M S and Park H J 2003 Antimicrobial films based on Na-alginate and kappa-carrageenan *Food Science and Technology Lebensmittel-Wissenschaft und-Technologies* **35**(8) 715-719
[3] Arvanitoyannis I Kolokuris I Nakayama A Yamamoto N and Aiba S 1997 Physico-chemical studies of chitosan-poly (vinyl alcohol) blends plasticized with sorbitol and sucrose *Carbohydrate Polymers* **34**(1-2) 9-19.
[4] Abugoch L E, Tapia C, Villamán M C, Yazdani-Pedram M and Díaz-Dosque M 2011 Characterization of quinoa protein-chitosan blend edible films *Food Hydrocolloids* **25**(5) 879-886
[4] Agustín González, Cecilia I and AlvarezIgarzabal 2013 Soy protein – Poly (lactic acid) bilayer films as biodegradable material for active food packaging* *Food Hydrocolloids* **33**(2) 289-296
[5] Kester J J and Fennema O R 1986 Edible Films and Coatings A Review *Food Technology* **40** 47-59.
[6] Masuhiro T, Giuliano F and John S C1994 Structure and Compatibility of Poly(vinyl alcohol)-Silk Fibroin (PVA/SA) Blend Films *Journal of Polymer Science* B: Polymer Physics **32** 243-248
[7] Sudesh K and Iwata T 2008 Sustainability of Biobased and Biodegradable Plastics *Clean Soil Air Water* **36** 433-442
[8] Yufei Han Kuang Li ID, Hui Chen and Jianzhang Li 2017 Properties of Soy Protein Isolate Biopolymer Film Modified by Graphene *Polymers* **9** 312
[9] Follain N, Joly C, Dole P and Bliard C 2005 Properties of starch based blends. Part 2. Influence of poly vinyl alcohol addition and photocrosslinking on starch based materials mechanical properties, *Carbohydrate Polymers* **60**(2) 185-192

Experimental Study and Performance Analysis of Phase Change Material Integrated Stepped Slope Solar Still

N Boopalan[1], B Kalidasan[2], K Ranjith[3], B Dhanush[3], S Anbarasu[3] and S Panner Selvam[3]

[1] Assistant Professor, Department of Automobile Engineering, Kongu Engineering College, Perundurai, Tamilnadu-638060

[2] Assistant Professor, Department of Mechanical Engineering, Bannari Amman Institute of Technology, Sathyamangalam, Tamilnadu-638401

[3] Department of Automobile Engineering, Kongu Engineering College, Perundurai, Tamilnadu-638060

E-mail: boopalan.auto@kongu.ac.in

Abstract. The major challenge in developing countries is to provide drinking water to the people, several techniques are employed but the challenge is to make cost effective and its sustainability over a period. Most of the developing countries work on saline water conversion to neat drinking water, distillation is one of the old and handy techniques in this conversion; where sunlight is used as energy drive for the process because elevating the temperature of saline water is enough for distillation and also for its free of cost availability. Solar still distilled water costs very less than the conventional method of deriving water from saline water and also people need not to worry about the purification of obtained water. Hence various analysis on increasing the performance of solar stills is required to increase the productivity to meet the rising demand for clean drinking water. The experimental work is carried out with three stepped solar stills, one without phase change material, one with paraffin wax as phase change material and other with calcium nitrate as phase change material. Solar stills are integrated with phase change materials to increase the productivity of clean drinking water by storing the solar energy in the form of latent heat to release in the dark hours for enabling solar sill distillation even during night time. There will be change in the total productivity based on the size of solar still, phase change material used, basin design and exergy efficiency, therefore these parameters have to optimized for better yield of neat drinking water from saline water. Experimentation was carried in the month of February and March of 2020 to estimate and compare the yield of three solar stills for the time period of 24 hours, certainly solar stills equipped with phase change materials resulted in higher total production than solar still without phase change material though their day time productivity was low. Solar still equipped with calcium nitrate as phase change material yielded 8.17 % more clean drinking water than the solar still equipped with paraffin as phase change material, though calcium nitrate is economical than paraffin wax but maintenance of solar still with calcium nitrate as phase change material is a hefty task.

Keywords: Phase change material, solar sill, drinking water

1. Introduction

An experimental study to estimate production of clean drinking water from conventional stepped solar still, solar still with condenser and solar still with reflector and condenser was conducted, in the above mentioned setups suction fan is connected to withdraw water vapour. Stepped solar still with condenser alone yielded about 66 % more clean drinking water than the conventional stepped solar still, whereas stepped solar still production of clean drinking water is 165 % more than the conventional one [1-3]. In the process of desalination of saline water solar stills are used because of its simple construction and economic benefits, stepped solar stills are mostly commonly used in the

IOP Conf. Series: Materials Science and Engineering 1059 (2021) 012002 doi:10.1088/1757-899X/1059/1/012002

production of clean drinking water hence stepped solar stills are investigated with humidification and dehumidification; as result of coupling this system the production of clean drinking water increased by a margin and also it is observed that increasing the air flow rate in the system significantly improved the production of clean drinking water [4-6]. Stepped solar stills are investigated with the internal and external reflectors, the reflectors are used in stepped solar still to improve the energy absorption in solar still because of proper utilization of input energy would yield more output, when the conventional solar still is compared with the stepped solar still coupled with internal and external reflectors, the later one produces 125 % more clean drinking water from saline by proper utilization of input energy [7]. When a stepped solar still is integrated with internal reflectors is compared against conventional stepped solar stills is experimented and analysed for the yield of neat drinking water from saline water in the same climatic conditions, the internal reflectors in the stepped solar still increased exergy efficiency thereby efficiency of the modified stepped solar still increased by 20 % [8-9]. A single slope stepped solar still coupled with solar air heater and aluminum as thermal storage material under the absorber plate for a still are of 0.5 m^2 is compared with the passive stepped solar still for the same area and climatic conditions, the productivity of active single slope stepped solar still is more when compared with the passive one [10]. In order to bound the limit of parameters like glass cover inclination and radiation form factor a theoretical study had been conducted in stepped solar still and it is found that clean water production from saline water is sensitive to angle of glass cover inclination and radiation form factor computed between hot saline water and glass cover [11-12]. An effluent settling tank and stepped solar still combined setup is analysed, the effluent is filtered in settling tank big and finely stable particles are settled in the tan, stepped solar still with 50 trays is analysed with two different depths; 10 mm for 25 trays and 5 mm for 25 trays, the effluent tank helped in storing thermal energy absorbed from solar radiation resulted in increased productivity when compared with conventional stepped solar still [13-14]. As the demand for clean drinking increases, need for improving the performance solar still is also an essential one in order to meet this requirement various techniques can be employed like coupling of reflectors, phase change materials for heat storage, fins, collectors, condensers and heat sinks with conventional solar still will increase the efficiency of solar still in desalination of saline water and also solar photovoltaic–thermal systems and greenhouse type stills can be employed to increase the productivity [15-16]. To transform saline water to clean drinking water is a needy one; solar desalination is one of promising technique for desalination, in this type of desalination solar radiation is used as a source of energy whether it is active or passive solar still, here the output of clean drinking water depends highly on the metrological parameters like solar radiation and wind velocity, hence these parameters are uncontrollable by humans but it would affect output of clean drinking water [17-18]. Due to cheap and convenience clean water production form saline water is carried out through solar still desalination when compared to conventional single slope solar still the output of solar still coupled with flake graphite nanoparticles, and cooling fins yielded 73 % more clean drinking water and also depth of saline water from glass cover affected single slope solar still efficiency [19]. Global warming is a serious threat to world nations hence solar thermal systems is a best choice to replace the conventional energy system in that way solar stills are used for production of neat drinking water from saline water due to their poor efficiency they are coupled with phase change materials to improve their performance and in order to reduce harmful emissions which increase global warming solar stills integrated with phase change materials would be the best choice in clean water production even after payback time economic analysis [20]. Increasing exergy efficiency of single slope solar still is an essential one to increase distilled water output quantity, hence insulation of solar still is very, and the total average benefit in the collected distilled water is 28 percent, 43 percent and 60 percent respectively for coated and uncoated metallic wiry sponges and black rocks [21 - 22]. In this work stepped solar still is chosen for experimentation, three setups conventional stepped solar still, solar still embedded with paraffin wax for energy storage and solar still embedded with calcium nitrate are investigated in same climatic conditions to optimize the parameters of stepped solar still in the production of clean drinking water from saline water.

2. Materials and Methods

As experimentation and analysis is to be carried for three solar still setups, dimensions of the solar still is to be chosen because it would affect the productivity of clean drinking water from saline water, based on the literature review three setups are made with necessary modifications, the area of the solar still is 0.81 m^2 (900 mm x 900 mm); high side wall depth is 600 mm and low side wall depth is 300 mm; the absorber plate is made of aluminum sheet with a thickness of 4mm which is divided into four steps each of 860 mm x 160 mm. The solar still basin plates are coated with black paint to increase absorptivity of solar radiation and also insulated with saw dust and thermocol to increase exergy efficiency in turn which would increase productivity of clean drinking water from saline water. The insulation layer is supported by a plywood and the basin is covered with a glass of 4 mm thick inclined at 30° horizontally. In order to embed phase change materials in the solar still copper tubes of 6mm diameter is chosen in which paraffin wax and calcium nitrate is embedded and placed in the steps under saline water which would absorb heat energy from solar radiation during daytime to release the heat energy during night time for production of clean drinking water from saline water. LM35 sensor in combination with Arduino Uno mega 2560 used to measure the basin liner, saline water, glass still inter temperature and atmosphere temperature is measured using by Arduino application of serial monitor and the average value is taken for theoretical calculation. Pyranometer and Anemometer are used to calculate solar radiation and wind velocity respectively, a flask collector of 1-liter capacity is used to measure hourly yield of clean drinking water. Figure 1 represents the experimental setup of PCM integrated Stepped Solar Still. The experimental investigation is carried at Latitude: 11°16' 12" N - Longitude: 77°36' 15.29" E - Altitude: 272 m to maximize the amount of

incident solar radiation; the whole experimental setup is kept in south direction at 185^0 to receive maximum solar radiation throughout the year.

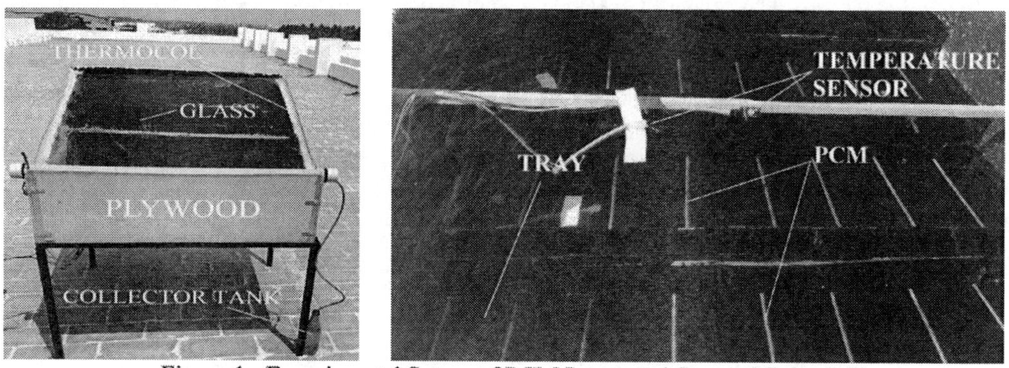

Figure 1: Experimental Set up of PCM Integrated Stepped Solar Still

3. Energy Balance Equation

3.1 Energy Balance Equation for Water Mass

Saline water in the basin can be represented by energy balance equation as shown

$$I_1 + Q_{bw} + c_{bw}\frac{dT_{bw}}{dt} = Q_{cw} + Q_{rw} + Q_{ew} + I_2$$

In the above equation Q_{bw} indicates convective heat transfer from basin to water, heat capacity of water in the basin is represented by C_{bw}, whereas basin water temperature is represented by T_{bw}, evaporative heat transferred from water to glass is indicated by Q_{ew}, heat radiated from water to glass is Q_{rw}, Q_{cw} convective heat transfer from water to glass, solar irradiation received in the still by the water after penetrating through glass cover is I_1 and solar irradiation reaching the basin liner surface

IOP Conf. Series: Materials Science and Engineering 1059 (2021) 012002 doi:10.1088/1757-899X/1059/1/012002

after passing through glass and water I_2 in the still, can be determined as follows:

$I_1 = (1-\alpha_g)I$

$$I_2 = (1-\alpha_g)(1-\alpha_w)I$$

In two modes heat is transferred from basin saline water to the glass through radiation and convection. The heat transferred through convection from basin saline water surface to glass cover happens through humid air, it is expressed by the following equation

$Q_{Cw} = h_{Cw}A_w(T_{bw} - T_g)$

Coefficient for convective heat transfer from basin saline water to glass cover is indicated by h_{cw}, Cross sectional area of water basin liner is Aw, and glass cover temperature is indicated by T_g. The difference in temperature of saline water in basin liner and glass cover arises because of radiative heat transfer, which can be estimated using Stefan – Boltmanz's law

$$Q_{rw} = h_{rw}A_w(T_{bw} - T_g) = \varepsilon_{eff}A_w\sigma(T_w^4 - T_g^4)$$

$h_{rw} = \varepsilon_{eff}\sigma(T_w^2 - T_g^2)(T_w - T_g)$

Heat transfer coefficient of radiative heat transfer from basin saline water to glass cover is h_{rw}, Stefan-Boltzman's constant $\sigma = 5.67 \times 10\text{-}8$ W/m^{2}K^4, the effective emittance of saline water surface to the glass cover is ε_{eff}. A part of heat energy is utilized for evaporating saline water from the basin that is can be represented by the following equation

$Q_{ew} = h_{ew}A_w(T_{bw} - T_g)$

3.2 Energy Balance Equation for Glass Cover

Energy balance in the glass cover can be written by the following equation, I is the solar radiation falling on glass cover, Q_{rg} is radiative heat transfer and Q_{cg} is convective heat transfer to atmosphere from glass cover respectively

$I_1 + Q_{rg} + Q_{cg} = Q_{cw} + Q_{rw} + Q_{ew} + I$

$$Q_{rg} = h_{rg}A_g(T_g - T_a) = \varepsilon_g A_g \sigma(T_g^4 - T_s^4)$$

A_g is the aperture area of glass cover, radiative heat transfer coefficient between atmosphere and glass is represented by h_{rg} is the radiation heat transfer coefficient between glass and atmosphere, atmosphere temperature and sky temperature is represented by T_a and T_s respectively and from the ambient temperature 60 C less is taken as sky temperature. From glass cover to atmosphere convective heat transfer happens it is represented by

$Q_{cg} = h_{cg}A_g(T_b - T_a)$

3.3 Energy Balance Equation for Basin

Energy balance equation for basin is represented by

$I = Q_b + \theta_{bot}$

$Q_b = h_b A_b (T_b - T_w)$

The above equation represents heat transferred from basin to saline water and also there is a loss of heat from basin to atmosphere though insulated with saw dust and thermocol which is determined by following equation, U_{bot} represents heat transfer coefficient between atmosphere and basin liner

$Q_{bot} = U_{bot}A_b(T_b - T_a)$

Clean water production from saline water hourly yield can be calculated from the following equations

$$\frac{dT_w}{dt} + T_w\left(\frac{h_{tw} + h_b}{C_{wr}}\right) = \frac{1}{C_{wr}}\alpha_w I(1-\alpha_g) + h_{tw}T_g + h_b T_b$$

$$T_w = \frac{f_1}{a_1}(1 - e^{-a_1 t}) + T_{wi}e^{-a_1 t}$$

$$a_1 = \frac{h_{tw} + h_b}{C_w}$$

$$T_g = \left(\frac{\propto_g I_1 + h_{tw}T_w + h_{tg}T_a}{h_{tw} + h_{tg}}\right)$$

$$T_b = \left(\frac{\propto_b I_1 + h_b T_w + U_{bot}T_a}{h_{tw} + U_{bot}}\right)$$

$$m_w = \frac{A_w Q_{ew}}{h_{fg}} 3600$$

For calculating theoretical hourly yield of clean drinking water; temperature of glass cover, saline water and basin liner is acquired from Arduino controller which is used to calculate hourly yield of clean drinking water.

3.4 *Energy Balance Equation for External Heat Transfer*

Solar still energy balance equation is represented by the following equation

$$I = Q_d + Q_{rg} + Q_{cg} + Q_{bw} + Q_{sw} + Q_{bot} + Q_{sww}$$

$$Q_d = m_w h_{fg}$$

From solar radiation available energy is represented by I, which is transferred to various components of still through various modes of heat transfer; portion of energy is used for distillation of saline water to produce clean drinking water but a quantity of heat is lost to atmosphere from glass cover due to convection and radiation. Quantity of clean water produced is represented by m_w and h_{fg} represents the latent heat of water. The radiation heat loss and convection heat loss is estimated by the following equations

$$Q_{rg} = \varepsilon_g \sigma A_g (T_g^4 - T_{Sky}^4)$$

$$Q_{cg} = h_{cg} A_g (T_g - T_a)$$

T_g, T_{sky} and T_a represents temperature of glass cover, sky and ambient respectively, h_{cg} is heat transfer coefficient for convection. The heat loss through walls of still is calculated by following equation

$$Q_{bw} = \frac{T_{bwi} - T_{bwo}}{R_{bw}}$$

Conductive heat resistance is represented by following equation, A_w represents front and back wall area

$$R_{bw} = \frac{1}{A_w}\left[\frac{L_1}{k_1} + \frac{L_2}{k_2}\right]$$

Heat loss through side and bottom of solar still is calculated by the following equations, apart from the mentioned losses there is also unaccounted heat loss which is represented by Q_{un}

$$Q_{sw} = \frac{T_{swi} - T_{swo}}{R_{sw}}$$

$$Q_{bot} = \frac{T_b - T_{bot}}{R_{bot}}$$

$$Q_{un} = I - (Q_d + Q_{rg} + Q_{cg} + Q_{bw} + Q_{sw} + Q_{bot} + Q_{sww})$$

The efficiency of solar still is represented by the following equation

$$\eta = \frac{Q_d}{I}$$

4. Results and Discussion

It is very important to validate the numerical model results with experimental results, after the experimentation a nice correlation was observed, though; there is 10 % deviation between theoretical and experimental model for stepped solar still step up without phase change material. The gap between theoretical model results and experimental model results is in the range of 15 % for the steeped solar still with phase change materials. Temperature of saline water is an important parameter when it is high; evaporation will be quick and also when the difference between glass cover temperature and saline water temperature is more; evaporation and condensation will be more which will increase the productivity of clean drinking water from saline water. Solar still should be placed in a region where there will not be any interruption to solar radiation incidence for better utilization of energy in producing clean drinking water. Temperature of glass cover, saline water, basin liner and atmosphere are the major parameters which involve in defining the quantity of clean water production, hence the values obtained through sensor from Arduino is plotted against time to have a clear picture. In X axis time is taken and in Y axis temperature of various components is taken, values are plotted for all the three setups made; Figure 2 represents the stepped solar still without PCM, it is observed temperature of all components rise with increase in solar radiation till noon then it tends to decrease. Basin liner temperature is more than other components as it is made of aluminum and black coated, as the saline water is on top of basin its temperature is slightly lower than the liner. The glass cover temperature and atmosphere temperature are very close to each other; atmosphere temperature influences glass cover temperature to greater extent which in turn will affect the clean drinking water production, when the difference between glass cover temperature and saline water is more then there will be possibility for increased production.

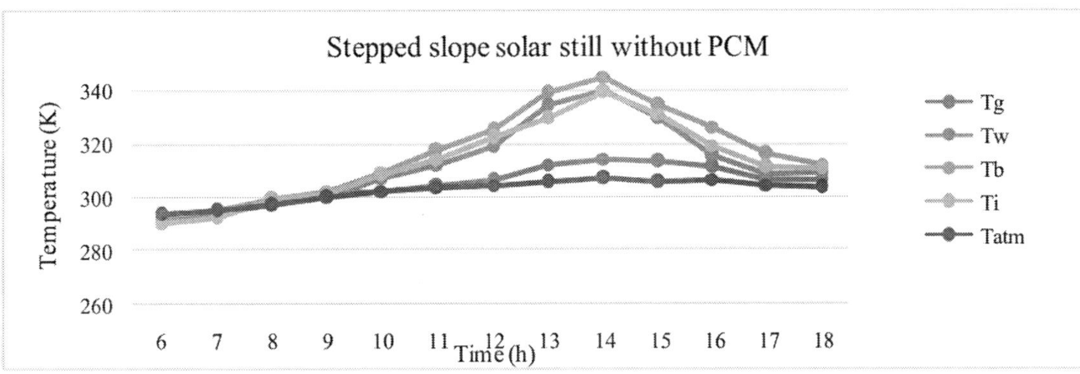

Figure 2: Temperature vs Time for Stepped Solar Still without PCM

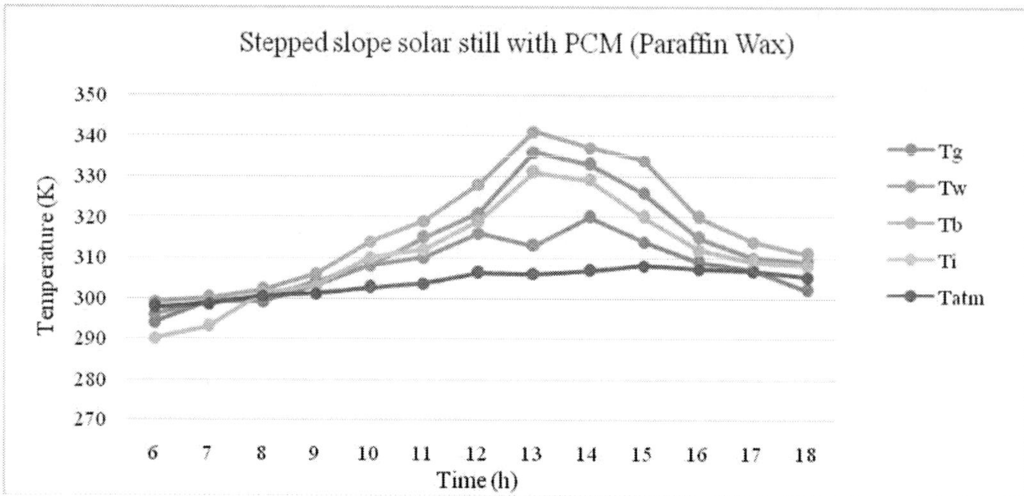

Figure 3: Temperature vs Time for Stepped Solar Still with Paraffin as PCM

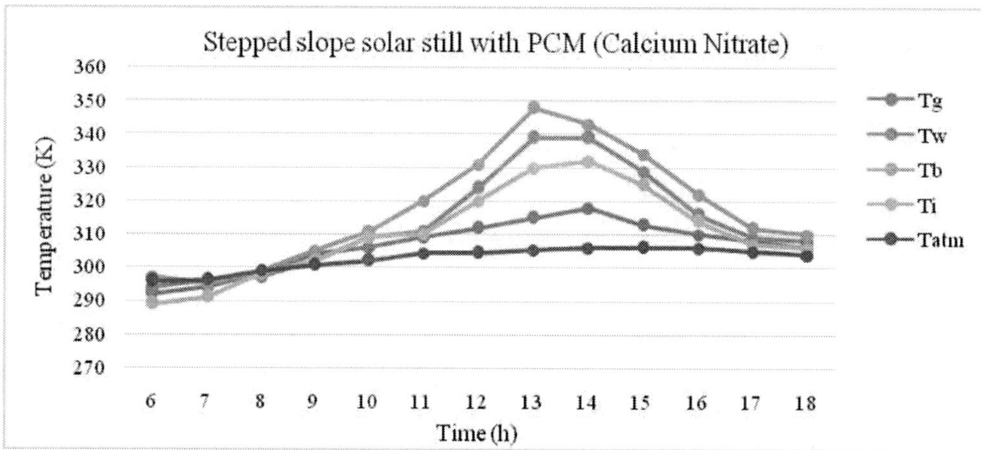

Figure 4: Temperature vs Time for Stepped Solar Still with Calcium Nitrate as PCM

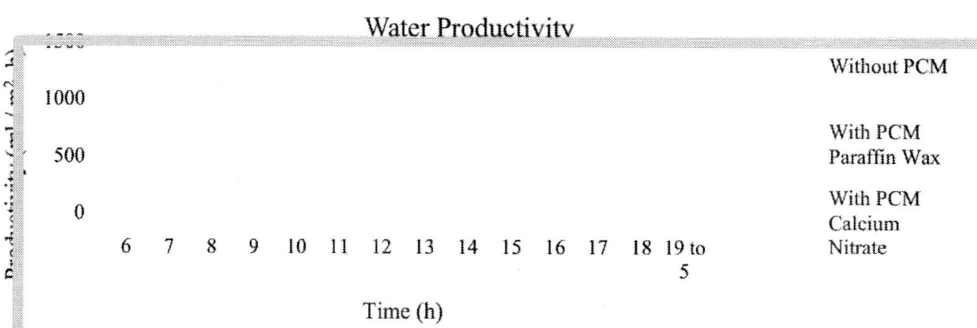

Figure 5: Water Productivity vs Time for all cases of Stepped Solar Still

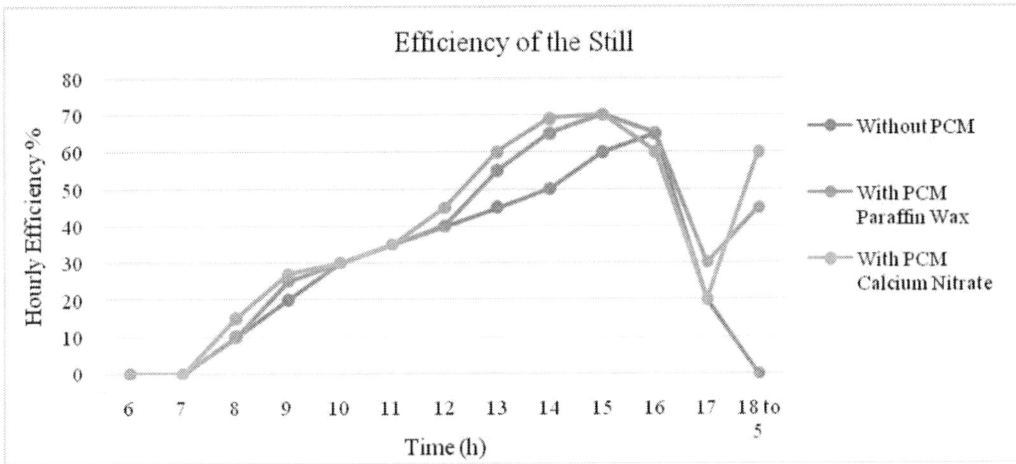

Figure 6: Efficiency vs Time for all cases of Stepped Solar Still

Temperature of stepper solar still components with paraffin wax as phase change material and calcium nitrate as phase change material is plotted against time in Figure 3 and Figure 4 respectively, as observed the temperature difference between glass cover and saline water is more in the stepped solar still integrated with calcium nitrate phase change material because of that we can expect more yield from the stepped solar still with calcium nitrate as phase change material. In order to have clear picture on production of clean drinking from saline water hourly efficiency is plotted against time; time is taken in X axis and hourly efficiency taken in Y axis, from Figure 5 the observation indicates that stepped solar still without phase change material is less efficient than other two solar stills because it cannot yield during night hours and also still with paraffin wax compared to still with calcium nitrate is less efficient. The data plotted in figure 6 has correlation with Figure 2, 3 and 4 and it is expected that stepped solar still with calcium nitrate yield will be more. Clean water productivity is plotted against time in Figure 5, it is observed that stepped solar still embedded with paraffin wax yielded 15 % more than the stepped solar still without phase change material and also still with calcium nitrate yielded 8.17 % more than the still with paraffin wax. It is also observed that still without phase change material cannot yield during dark hours but its yield during day time is more than the other two stills, still with calcium nitrate yielded more during dark hours than the still with paraffin wax.

5. Conclusion

In the present decade need for clean drinking water has been increased to meet this demand desalination of saline water would be one of the promising method, various techniques can be employed in the mentioned process but cost efficient and renewable energy source usage in desalination of saline water will be appropriate for present environmental conditions, one such method using renewable energy is solar still desalination which uses solar radiation as energy source but the disadvantage of solar still is its efficiency and inability to produce drinking water during night time. To overcome the mentioned drawbacks solar stills are integrated with thermal energy storage systems, to increase yield and to produce clean drinking water during night time as well, phase change materials embedded in solar stills to increase their efficiency, in order to optimize the performance of solar stills with thermal energy storage systems, various designs of still and heat storage techniques are employed one such technique is stepped solar still design, when it is integrated with paraffin wax in copper tubes for energy storage the output increased compared with stepped solar still without thermal energy storage systems. Stepped solar still integrated with calcium nitrate in copper tubes for thermal energy storage performed better than the stepped solar still with paraffin wax as phase change material, hence the experimental work on stepped solar stills with thermal storage systems of phase change materials concludes calcium nitrate as phase change material performed well than paraffin wax when embedded in copper tubes for integration with stepped solar still but the maintenance required for still with calcium nitrate

References

[1] Shehata, A.I., Kabeel, A.E., Dawood, M.M.K., Elharidi, A.M., Abd_Elsalam, A., Ramzy, K. and Mehanna, A., 2020. Enhancement of the productivity for single solar still with ultrasonic humidifier combined with evacuated solar collector: An experimental study. *Energy Conversion and Management*, **208**,112592.

[2] Kabeel, A.E., Omara, Z.M. and Younes, M.M., 2015. Techniques used to improve the performance of the stepped solar still—A review. *Renewable and Sustainable Energy Reviews*, **46**,178-188.

[3] Muftah, A.F., Sopian, K. and Alghoul, M.A., 2018. Performance of basin type stepped solar still enhanced with superior design concepts. *Desalination*, **435**,198-209..

[5] Essa, F.A., Omara, Z.M., Abdullah, A.S., Shanmugan, S., Panchal, H., Kabeel, A.E., Sathyamurthy, R., Alawee, W.H., Manokar, A.M. and Elsheikh, A.H., 2020. Wall-suspended trays inside stepped distiller with Al2O3/paraffin wax mixture and vapor suction: Experimental implementation. *Journal of Energy Storage*, **32**, 102008.

[6] Shmroukh, A.N. and Ookawara, S., 2020. Evaluation of Transparent Acrylic Stepped Solar Still Equipped with Internal and External Reflectors and Copper Fins. *Thermal Science and Engineering Progress*,100518.

[7] Abdullah, A.S., Essa, F.A., Bacha, H.B. and Omara, Z.M., 2020. Improving the trays solar still performance using reflectors and phase change material with nanoparticles. *Journal of Energy Storage*, **31**, 101744.

[8] Bataineh, K.M. and Abbas, M.A., 2020. Performance analysis of solar still integrated with internal reflectors and fins. *Solar Energy*, **205**, 22-36.

[9] Kabeel, A.E., Dawood, M.M.K., Nabil, T. and Alonafal, B.E., 2020. Improving the performance of stepped solar still using a graphite and PCM as hybrid store materials with internal reflectors coupled with evacuated tube solar collector. *Heat and Mass Transfer*, **56(3)**,891-899.

[10] Jobrane, M., Kopmeier, A., Kahn, A., Cauchie, H.M., Kharroubi, A. and Penny, C., 2020. Internal and external improvements of wick type solar stills in different configurations for drinking water production–A review. *Groundwater for Sustainable Development*, 100519.

[11] Bachchan, A.A., Nakshbandi, S.M.I., Nandan, G., Shukla, A.K., Dwivedi, G. and Singh, A.K., 2020. Productivity enhancement of solar still with phase change materials and water-absorbing material. *Materials Today: Proceedings*.

[12] Taheri Mousavi, S.M., Egelioglu, F. and Ilkan, M., 2020. Experimental and numerical study of the effect of various design configurations on the thermal performance of solar still desalination. Energy Sources, *Part A: Recovery, Utilization, and Environmental Effects*, 1-15.

[13] Keshtkar, M., Eslami, M. and Jafarpur, K., 2020. Effect of design parameters on performance of passive basin solar stills considering instantaneous ambient conditions: A transient CFD modeling. *Solar Energy*, **201**, 884-907.

[14] Panchal, H., Mevada, D., Sadasivuni, K.K., Essa, F.A., Shanmugan, S. and Khalid, M., 2020. Experimental and water quality analysis of solar stills with vertical and inclined fins. *Groundwater for Sustainable Development,* 100410.

.[15] Nougriaya, S.K., Chopra, M.K., Gupta, B. and Baredar, P., 2020. Stepped solar still: A review on designs analysis. *Materials Today: Proceedings*.

[16] Khalilmoghadam, P., Rajabi-Ghahnavieh, A. and Shafii, M.B., 2020. A novel energy storage system for latent heat recovery in solar still using phase change material and pulsating heat pipe. *Renewable Energy*, **163**, 2115-2127.

[17] Panchal, H., Mevada, D. and Sadasivuni, K.K., 2020. Recent advancements in condensers to enhance the performance of solar still: A review. *Heat Transfer*, **49(6)**, 3758-3778.

[18] Sohani, A., Hoseinzadeh, S. and Berenjkar, K., 2020. Experimental analysis of innovative designs for solar still desalination technologies; an in-depth technical and economic assessment. *Journal of Energy Storage,* 101862.

[19] Sharshir, S.W., Peng, G., Wu, L., Essa, F.A., Kabeel, A.E. and Yang, N., 2017. The effects of flake graphite nanoparticles, phase change material, and film cooling on the solar still performance. *Applied energy*, **191**, 358-366.

[20] Kalidasan, B., Pandey, A.K., Shahabuddin, S., Samykano, M., Thirugnanasambandam, M. and Saidur, R., 2020. Phase change materials integrated solar thermal energy systems: Global trends and current practices in experimental approaches. *Journal of Energy Storage*, **27**, 101118.

[21] Malik, M.Z., Musharavati, F., Khanmohammadi, S., Khanmohammadi, S. and Nguyen, D.D., 2020. Solar still desalination system equipped with paraffin as phase change material: exergoeconomic analysis and multi-objective optimization. *Environmental Science and Pollution Research*, 1-15.

[22] Hassan, H. and Yousef, M.S., 2020. An assessment of energy, exergy and CO2 emissions of a solar desalination system under hot climate conditions. *Process Safety and Environmental Protection*, **145**, 157-171.

Effect of Curing Methods on Strength of Self-Compacting Concrete

P V Nandhini[*1], M Nivethitha[2], A Mohanraj[3] and V Senthilkumar[4]

[1, 2] UG Student, Department of Civil Engineering, Bannari Amman Institute of Technology, Erode 638401, Tamil Nadu, India.
[3]Assistant Professor, Department of Civil Engineering, Bannari Amman Institute of Technology, Erode 638401, Tamil Nadu, India.
[4]Assistant Professor, Department of Civil Engineering, Sri Shakthi Institute of Engineering and Technology, Coimbatore 641062, Tamil Nadu, India.

*nandhinipv.ce17@bitsathy.ac.in

Abstract. In this paper three different curing techniques with variation in compressive strength of medium strength, self-compacted concrete are discussed. To achieve the mix design of medium strength self-compacting concrete initially numbers of trials were carried out. Super plasticizer of 0.7% and an internal curing agent, Superabsorbent Polymer of 0.3% together to the weight of cement were added. Once the mix design was achieved, as per ASTM standard three batches of concrete cubes were casted. Slump flow test was carried out in order to determine workability of self-compacting concrete. Concrete cube cured in a curing tank in the laboratory was taken as first batch. The second batch was cured by the application of an internal curing compound. The third batch was cured by Gunny bag Method. Three cubes were tested for compressive strength, from every batch at 7 days, 28 days and 60 days age of concrete. The split tensile strength of concrete was also considered. Graphs were drawn with analyzed results. So it was concluded that to get improved strength and sustainability of self-compacted concrete in areas with shortage of water, curing compound found to be more effective.

Keywords: self compact concrete, tensile strength, curing compound

1 Introduction

Cement hydration is one of the main reasons for concrete curing. Cement hydration is a series of chemical reaction that requires adequate water supply and proper temperature over an extended period of time [1]. It is the name given to the procedures used for promoting the hydration of cement, and consists of the control of temperature and moisture movement from and into concrete. Curing allows uninterrupted hydration of cement and leads to uninterrupted gain in strength. Once it gets cease, strength gain of concrete also stops.

The degree of hydration is affected by curing temperatures and thus it leads to distribution of pore size. For the production of high volume of concrete units, the use of elevated temperature curing techniques to accelerate early strength development has been adopted as one of the standard methods in the precast concrete industry. No much difference was found in the spatial distribution, morphology, and volume of hydrates and other micro-structural feature in the micro-structural examination of two months old concrete cured at 16°C, 42°C and 46°C. However, when it was cured at 80°C within the matrix of all the concrete a network of micro-cracks was evidenced. Concrete increases in strength with age, after

IOP Conf. Series: Materials Science and Engineering 1059 (2021) 012003 doi:10.1088/1757-899X/1059/1/012003

Setting. The strength of the concrete will be attained, if it is properly cured [2, 3]. Application of heat and/or the preservation of moisture within concrete will be affected by curing. If the concrete is casted and cured at a specific constant temperature ranges between 5°C to 46°C, the higher the temperature, the hydration of cement and the strength gain will more rapid, it is generally observed up to 28 days, [4]. Several techniques may, potentially, be used for incorporation of internal curing water in concrete [5].Several researchers have proposed the use of saturated light weight aggregates to provide "internal" curing for concrete [6, 7].

On the other hand, other researchers used poly-glycol products in concrete mixes as self-curing agent [8].In this project various curing techniques are adopted and the mechanical properties are tested. In order to improve the strength of the concrete micro silica of 15% to the cement was added. The percentage replacement was fixed based on many literatures.

2 Curing Methods

The methodology must be clearly stated and described in sufficient detail or with sufficient references. OPC 53 grade cement conforming to IS: 12267 – 1987 was used in this project. Micro silica having the chemical properties as shown in table 1 was replaced 15% to the weight of cement. River sand which is available locally with a fineness modulus of 2.8 and 12.5 mm, coarse aggregate was used. Superplasticiser (SP) used is Conplast SP 430; Sulphonated Napthalene Polymers type of 0.7% was adopted. For mixing the concrete, Potable water was used. SAP of 0.3% was used for internal curing. One set of specimen was water cured and the other set was cured by gunny bag method. Mix proportions for M40 grade of concrete is given in table.2.

Table1. Chemical Composition of Cement and Cementitious Materials

No	Chemical Analysis	Cement %	Micro-silica %
1	SiO_2	22.6	99.87
2	$Al_2 O_3$	10.77	0.045
3	$Fe_2 O_3$	5.8	0.04
4	CaO	55.6	0.001
5	TiO_2	0.55	0.024
6	MgO	1.15	-
7	Na_2O	0.37	0.004
8	K_2O	0.15	-
9	SO_2	1.6	-
10	Loss of ignition	1.5	0.016

Table2. Mix Proportions

S.No	Curing	%	Quantity in Kg/m^3
1	Cement		375
2	Micro silica	15%	66
3	Fine aggregate		975
4	Course aggregate		735
5	Water		190
6	Plastizer	0.7%	3

3 Workability Properties

As there is no standard method for SCC one has to go with several trials to achieve desired mix. The flow-ability of a fresh mix concrete is described by slump flow values in unconfined conditions. The passing ability describes the ability of the fresh mix to flow through restricted spaces and slender openings.

Table3. Workability Property

S.No	Curing	Slump Flow	L Box	V Funnel
1.	Conventional Concrete	720	0.9	10
2.	Concrete with admixtures	630	0.8	8

The finess in high level and practically spherical shape of silica fume results in good cohesion and segregation resistance (EFNARC). Table 3 shows the workability properties of SCC as per EFNARC Standard. Figure 1 shows the workability of SCC (a) Slump flow (b) V funnel test. Cubes of size 150 mm x 150 mm x 150mm and cylinder of size 150 mm x 300 mm were cast and allowed for curing.

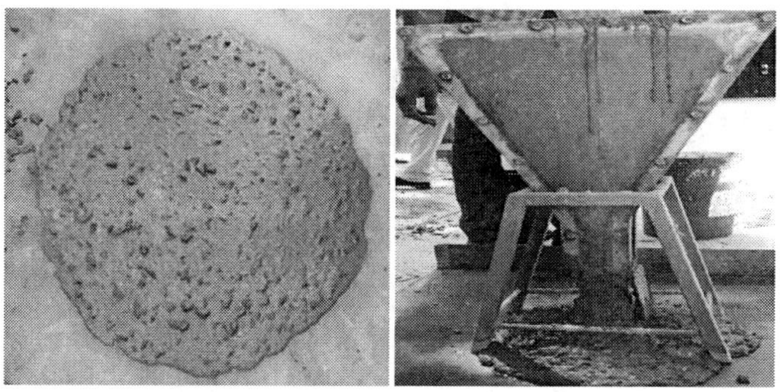

a. b.

Figure1. a. Slump flow b. V Funnel tests

Figure 2 shows the test on mechanical properties of SCC.

Figure2. a. Compressive strength b. Split tensile strength

IOP Conf. Series: Materials Science and Engineering 1059 (2021) 012003 doi:10.1088/1757-899X/1059/1/012003

Beam Details

The Beam is designed as a singly reinforced beam. The beam is designed as per Code IS 456: 2000. Thus the beam details are given below:

Beam Size	=	1000 X 150 X 220 mm,
Cover	=	25 mm
Main rod	=	3 no's of 10mm Ø rods
Hanger rods	=	2 no's of 8mm Ø rods
Stirrups	=	2 legged stirrups of8mm Ø rods @ 200mm
f_y	=	Fe 415
f_{ck}	=	40 Mpa

Figure3. Flexural strength

The beam as stated in V is tested as per the codal procedure. The two point load is applied at the distance of L/3. The clear span of beam is adopted as 800mm & the concentric loading is given during the testing of the beam. Deflecto-meter is said to measure the deflection value of the beam. The test results are given in table 3. Testing of beam is as shown in fig 3.

4 Results and discussions

The Workability properties of SCC are conducted and found to be satisfied to the EFNARC standards. The normal concrete without addition of mineral admixtures shows slump flow of 720mm. The concrete with addition of mineral admixtures shows 90 mm flow difference. The flow of concrete added admixtures is lower because of frictional effect due to more fine powder.

The concrete containing no admixtures was tested at 28 days of curing as shown in figure 4. The strength of concrete at 7 days in self curing method is decreased to about 20% and 13% respectively to water curing and gunny bag curing. At 28 days the strength of self curing decreases to about 10.5% and 8% respectively to water curing and gunny bag curing. The strength of concrete added internal curing SAP improved its strength at 60 days. This implies that sufficient amount of water [9] is supplied to concrete that confirms continuous hydration process that took place within the concrete system. Formation of hydration products makes concrete system more dense and compacted leading to improvement in strength at later days.

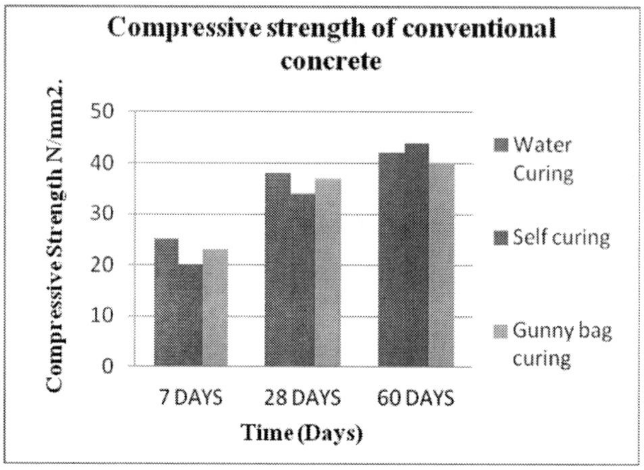

Figure.4. Compressive strength of conventional concrete in N/mm2

The strength increases by 5% and 10% respectively to water curing and gunny bag curing. This improvement in strength is attributed to formation of new materials during hydration process.

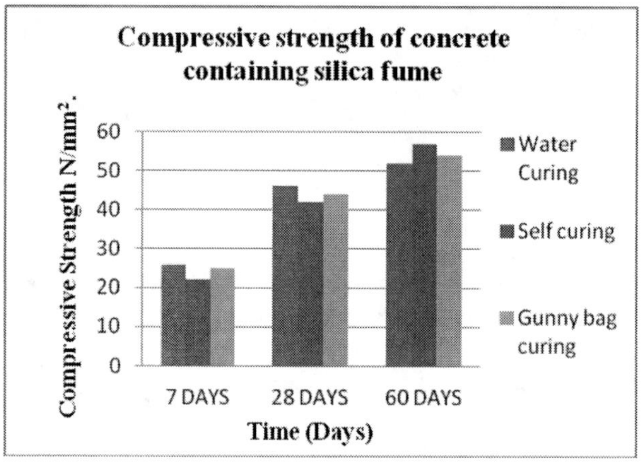

Figure5. Compressive strength of concrete containing silica fume in N/mm^2

When concrete is added with Micro-silica then the strength of concrete tends to increase drastically. The strength of concrete at 7 days in self curing method is decreased to about 15% and 12% respectively to water curing and gunny bag curing. At 28 days the strength of self curing decreases to about 9% and 5% respectively to water curing and gunny bag curing. The strength of concrete added internal curing SAP improved its strength at 60 days. The strength increases by 9% and 14% respectively to water curing and gunny bag curing. This improvement in strength is attributed to formation of new materials during hydration process that made the concrete system more dense resulting in higher strength at later ages. The other method of curing may not provide sufficient water for hydration process but continuous hydration process is possible only by internal curing. The strength of concrete containing mineral admixtures is as shown in figure 5. The split tensile strength of concrete containing admixtures is shown in figure 6.

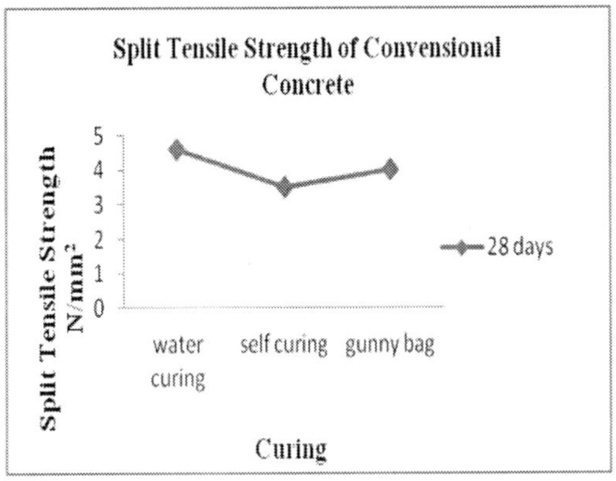

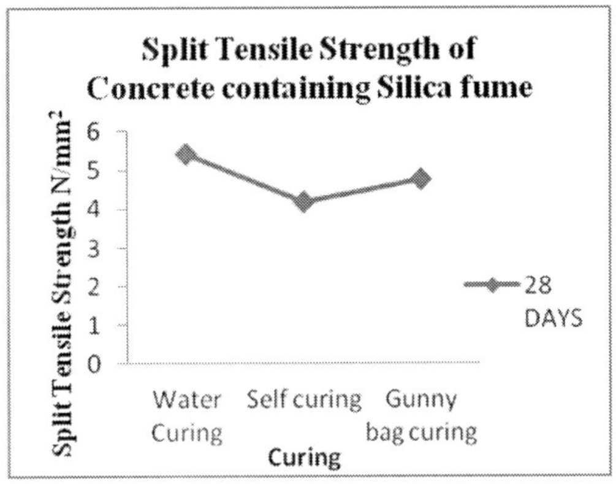

(a) (b)

Figure.6. Split tensile strength of (a) conventional (b) micro silica added concrete in N/mm^2

Table 4 gives the test result of flexural strength of concrete.

Table 4. Loads and Deflection

S.No	Loadings (KN)	Deflection (mm)	Remark
1.	0	0	
2.	10	0.02	
3.	20	0.25	
4.	30	0.46	
5.	40	0.66	
6.	50	0.87	
7.	60	1.11	First Crack
8.	70	1.34	
9.	80	1.65	
10.	90	1.98	
11.	100	2.21	
12.	110	2.53	
13.	120	2.82	
14.	130	3.15	
15.	140	3.50	
16.	150	4.00	
17.	155	4.52	Ultimate crack load & Maximum deflection

IOP Conf. Series: Materials Science and Engineering 1059 (2021) 012003 doi:10.1088/1757-899X/1059/1/012003

5. Summary and Conclusions

The concrete without silica fume showed higher flow rate than concrete containing silica fume. The strength of concrete by normal curing method can be adopted if there has availability of water as it gives higher strength than any other method of curing. However Self cuing is best sitedat the sites where there is no availability of water and the strength of concrete though decreases initially, improves after 28 days of time. The strength formation is attributed to formation of new materials during hydration process. This envisage continuity in hydration process even after 28 days of time because of the availability of internal water required for continuing the process. Also the unhrdrated residual materials that are left behind due to unavailability of water will be hydrated because of Internal curing offered by SAP. The strength of concrete achieved by gunny bag method is closer to conventional method of curing. But water can be saved by adopting this method of curing. The split tensile strength of concrete gets decresed for self curingmethod.

References

[1] Gowsika, D., Balamurugan, P. and Kamalambigai, R., 2017. Experimental study on curing methods of concrete. *Int. J. Eng. Dev. Res*, **5(1),** 2321-9939.

[2] Srinivas vasam,2018. Effect of Curing Methods on Compressive Strength of Normal, Medium and High Strength Self Compacting Concrete with and without RCA. *International Journal of Emerging Technologies and Innovative Research*, **5(2)** .7-12.

[3] Raghavendra Y.B. and Aswath M.U, 2012. Experimental Investigation on Concrete Cured with Various Curing Methods-A Comparative Study. *International Journal of Advance Scientific Research and Technology* **3(2),** 2278-0181.

[4] Nishant Yadav 2013 Internal Curing: Curing Concrete Inside Out. *International Journal of Science and Research* (IJSR) ISSN (Online): 2319-7064.

[5] Bentz, D.P., Lura, P. and Roberts, J.W., 2005. Mixture proportioning for internal curing. Concrete international, **27(2)**,35-40.

[6]Mousa, M.I., Mahdy, M.G., Abdel-Reheem, A.H. and Yehia, A.Z., 2015. Self-curing concrete types; water retention and durability. *Alexandria Engineering Journal*, **54(3)**, 565-575.

[7] Mandiwal, P. and Jamle, S., 2018. Use of Polyethylene Glycol as Self Curing Agent in Self Curing Concrete-An Experimental Approach. *International Research Journal of Engineering and Technology*, 2395-0072.

[8] Mather, B. and Hime, W.G., 2002. Amount of water required for complete hydration of Portland cement. Concrete international, **24(6)**,56-58.

IOP Conf. Series: Materials Science and Engineering 1059 (2021) 012004 doi:10.1088/1757-899X/1059/1/012004

An Experimental Performance Analysis on Robotics Process Automation (RPA) With Open Source OCR Engines: Microsoft Ocr And Google Tesseract OCR

Malathi T[1], Selvamuthukumaran D[2], Diwaan Chandar C S[3], Niranjan V[4], Swashthika A K[5]

[1]Assistant Professor, Department of Computer Science & Engineering
[2]Assistant Professor, Department of Mechanical Engineering
[3]Second Year, Department of Information Science & Engineering
[4,5]Second Year, Department of Computer Technology

Bannari Amman Institute of Technology, Sathyamangalam, Erode, Tamilnadu - 638401

Email ID:tmalathime@gmail.com

Abstract.Robotic Process Automation (RPA) plays an important role in the solution for automating human digital interaction in various software domains like IT, Business Strategy and so on. Optical Character Recognition(OCR) superimposes subtitled characters on an image.Here we use two Open source OCR engines, Google Tesseract OCR - It literally makes use of the open source Tesseract OCR Engine[1]. Microsoft OCR - This is another open source OCR engine accessible in the Robotics Process Automation tool, UiPath[1]. The technique of optical character recognition (OCR) has been used to transform printed text into editable text. In various applications, OCR is a very useful and common technique. OCR precision can be dependent on algorithms for text preprocessing and segmentation. Often because of different scale, design, orientation, complex picture context, etc., it is difficult to retrieve text from the image. We begin this paper with an introduction to the Optical Character Recognition (OCR) process, in this paper, we propose a research analysis to estimate the accuracy of two open source OCR engines with Robotics Process Automation(RPA). These RPA processes are created using the UiPath tool and our results compare the OCR engines performance during the execution with the workflow and save the results to our local storage or on cloud storage if cloud support enabled and here we used local storage for purpose, and the result analysis and error analysis was evaluated by manual process which also included excel formulas to calculate the accuracy of the extracted string with the original string.

Keywords: Robotic process automation , optical Character recognition

1 Introduction

In this 21st century, we are in the new automation era. Great advances in automation, RPA and Machine Learning (ML) algorithms are used to free the workers from mundane, boring repetitive tasks and they focus on tasks which involve creativity and human decisions. Automation is a process of replacing a manual step with one that happens automatically. It's not actually a human replacement or any cost saving play, It can save time, reduce errors, increase consistency and a way to centralise solutions and mistakes making them easier to fix to make organizations reinvent faster methods to explore their services to the society. RPA is a key way to achieve this. RPA is a tool used to imitate human digital interaction with the help of software bots.

RPA is not actually a humanoid robot which is built to resemble the human body. UiPath is helpful for beginners and marketed as "**nocoding**" solution.Learning a computer **programming** language is like learning something else. It will be easier for some people, and more **difficult** for others. But in RPA there is no such problem. **RPA neitherrequires** the development of code, nor **a** direct access to the code [4] instead we just need to drag and drop the activities in the workflow.

2 OCR engines in RPA

1 Abbyy cloud OCR
2 Abbyy OCR
3 Google Cloud vision OCR
4 Microsoft Azure Computer Vision OCR
5 Microsoft OCR
6 Microsoft Project Oxford Online OCR
7 Tesseract OCR

2.1 Microsoft OCR:
It uses built in Microsoft OCR if windows 10 is installed, else it will extract using Microsoft Office Document Imaging (MODI) OCR which is used in Microsoft Office (Up to MS office 2007)

2.2 Google Tesseract OCR:
It obtains the string information from an indicated UI element or image using Tesseract OCR Engine.

OCR Engines/Properties	Microsoft OCR	Tesseract OCR
Language	Microsoft OCR uses only "English" to take out the characters from the image file. It must be given as "english" in the Engine properties	For the Google OCR engine, this field needs to contain the language file prefix, such as "en" for English, "ita" for Italian.
Profile	To select a type for the image to recognize characters even more better. The options available in profile are: None - Recognize without a profile Screen - It is suitable for applications Scan - It is better for scanned documents. Legacy - It is the default profile in both the OCR engines	
Invert	NA	The colors of the UI element are inverted before scraping. This is more familiar with dark background and light text.
Denied Characters	NA	The OCR engine extracts the given string without taking into account the characters specified here.
Extract words info	To extract the position of the text, check the ExtractedWords in the properties window	NA
Extra Options	NA	AllowedCharacters - The OCR engine returns a given string according to the characters specified in this option
Scale	Scaling property is used to scale the image. It is used to resize the image for better performance. It is most recommended for smaller images	

Output	To assign the extracted words to a string variable.
Result	The assign the recognized words with their position on the screen to a < key, value> pair variable.

3 Data Scraping using RPA OCR Engines

Google Tesseract and Microsoft OCR were the two open source OCR engines available in UiPath. There were several differences and similarities between these OCRs. Actually Both OCR uses a common architecture, and a scalable workflow to scrape data from the image variable and recognition capabilities, this experimental research defines how different aspects such as velocity, accuracy, time are more comfortable in each engines, where each engines has been tested with set of different sources, and implemented to run on certain Machines with specifications below.

System Specifications
Minimal Hardware requirements
Central Processing Unit: Pentium 200 MHz
Random Access Memory: 32 MegaBytes
Hard Disk Space: 4 GigaBytes

Performed Specifications
Tesseract OCR
Operating system: Windows 10
Central processing unit: Core i5 i5-8265U
Random access memory - 8 GigaBytes
Processor:Intel core 15-8265 4 x 1.6-3.9GHz Whiskey Lake-U

Microsoft OCR
Operating system: Windows 10
Central processing unit: AMD Ryzen 3
Random access memory - 4 GigaBytes
Processor: AMD Pro A4-4350B R4, compute cores 2C+3G 2.50GH

4 Architecture & Process workflow
UiPath is one among the top tools available in the market of Robotic Process Automation[5]. It is used to automate Business processes and IT systems. Repetitive tasks may be boring for users. By the way UiPath is used to automate such boring tasks. Moreover UiPath is a code free environment. So, users with minimal or without coding background can automate the process using drag and drop functionality for all activities available in UiPath.

4.1 UiPath Architecture
The UiPath tool consists of three main components
- UiPath Studio
- UiPath Robot
- UiPath Orchestrator

4.1.1 UiPath Studio
UiPath is a designer tool using which a user can build workflows for automation tasks. It has several predefined activities and libraries. The users can drag and drop activities to model the workflow for the automation process. In simple terms, UiPath Studio is a tool which is used to model the automation workflow using predefined activities and libraries to automate repetitive processes.

4.1.2 UiPath Robots

UiPath Robots is a program that hosts the process that is built in UiPath studio, that lets us execute our projects on any machine with or without human monitoring (or) supervision.

4.1.3 UiPath Orchestrator

UiPath Orchestrator is a web platform that helps us to monitor our robots(i.e OCR workflows), It is a web application that lets us to manage the creation, monitoring, and deployment of our resources in our environment or machine.It allows us to launch and schedule our bots in a desktop or virtual machines,control status of the bot and analyse the results of their work.

4.2 Workflow Design

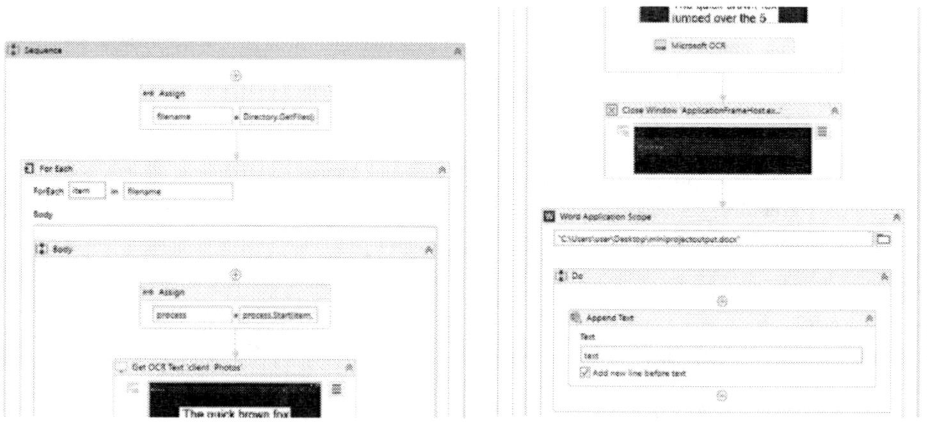

Figure 4.2

Workflows can be created in UiPath Studio and hence monitored in UiPath Orchestrator to save results to cloud storage or we can run the workflow in our local machine to save results in our Local Machine such as excel files and other word files.

Images used and Output:
https://github.com/Niranjan849/AN-EXPERIMENTAL-PERFORMANCE-ANALYSIS-ON-ROBOTICS-PROCESS-AUTOMATION-RPA-WITH-OPEN-SOURCE-OCR-ENGIN

Proposed approach:

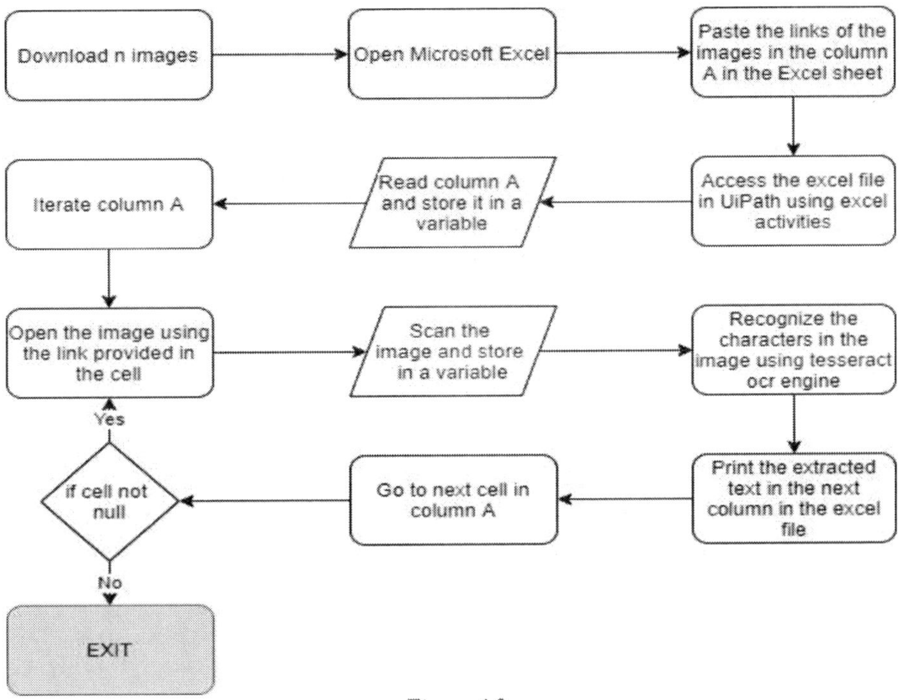

Figure 4.3

5Result analysis
Tabulation:
a.) Accuracy - The similarity between the original text and the extracted text is calculated and the average value is calculated among them
b.) Time taken - The time taken to extract 100 images is calculated and tabulated
c.) Average time taken - The average time taken to extract each image is calculated from the total time taken and is calculated and tabulated.

Characteristics/ Engine	Microsoft OCR Engine	Tesseract OCR Engine
Average accuracy(%)	76.18835979	43.05
Time taken(hrs)	00:25:00.00	00:20:32.00
Average time taken(s)	00:00:15.00	00:00:12.32

6 Error analysis
Image properties
All the images were in **.jpg** or **.JPEG** format

When images with background were scraped, many additional text and special characters which were not in the images were extracted

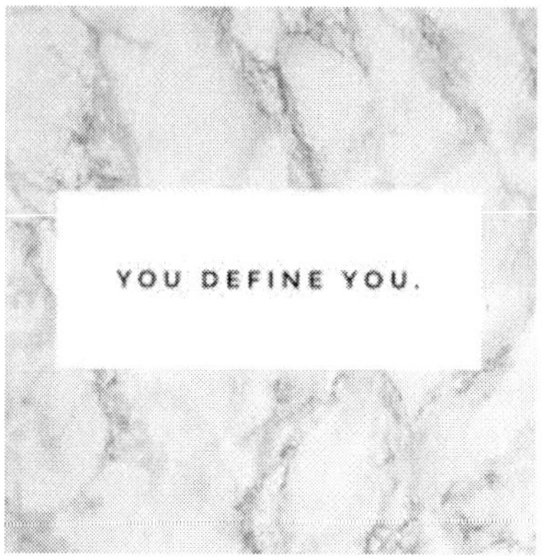

Figure 6.1

Original Text
YOU DEFINE YOU.

Output Obtained

"- 1
fl ' ,
. , r - - .w I.
~ YOU DEFINE YOU. 3
J
w , a;
a ' 7' _ ' 1 ' .4'
'~ ' ' - a"? , 1!." '

Data with more Accuracy:(one data for sample included)

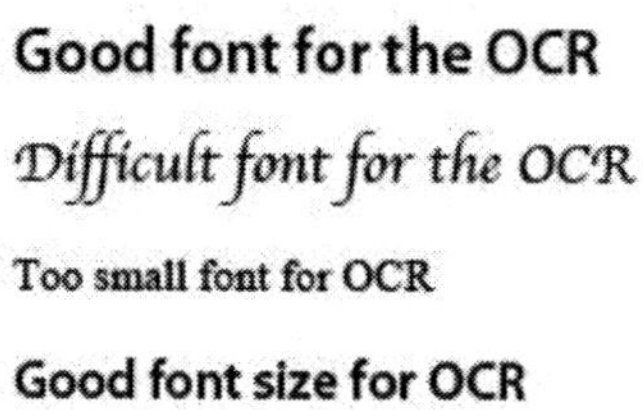

Figure 6.2

Original Text
Good font for the OCR
Difficult font for the OCR
Too small font for the OCR
Good font size for the OCR

Output obtained
Good font for the OCR
Difficult font for the OCR
Too small font for OCR
Good font size for OCR

IOP Conf. Series: Materials Science and Engineering 1059 (2021) 012004 doi:10.1088/1757-899X/1059/1/012004

Images with Dark background and light text are not recognized

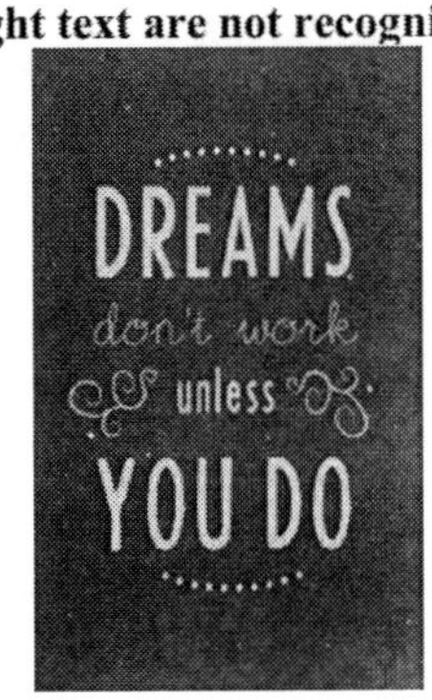

Figure 6.3

Original Text
Dreams
Don't work
unless
you do

Extracted Text
a: 1'... {-1
£9 unless'ij
/

Vertical images are not recognized

Figure 6.4

Original Text
Vertical Label

Extracted Text
oO
2
©
ad
S
2
~
dn
ov
>

Similarity is calculated manually based on the count of recognized words which may contain many errors

Different set of images are used for Google tesseract and Microsoft OCR
Tesseract OCR:

DATA/ACCURACY	Original	Extracted Text
< 20%	Family WHERE LIFE BEGINS & LOVE NEVER ENDS	: WHERE. LIl'li BUSINS 5L l.()\T. NEVER [X
< 40%	BE YOURSELF but be your best self	BE YOURS ELF M49101" MM ./'
< 60%	find your own way	{IND OWN 0 WAY $4 /
< 80%	I attend film school ny, background is acting and directing. But the acting does best, I like to work on the stage	I attend film school. my background is ucmlg and dirccling. Bul lllc acting docs best. I like to work
100%	I'm such a hypocrite I tell people to stay strong when I am the weakest person in the world.	I'm such a hypocrite I tell people to stay strong when I am the weakest person in the world

Microsoft OCR:

DATA/ACCURACY	original	Extracted Text
< 25%	life was is will be great just belive	LIFTGREAT.
< 50%	The road to success is always under construction	TO success is uwaus umer consrucTion.
< 75%	Don't let yesterday take up too much of today	DON'T LET YESTER my TIAKE UP - roo MUCU90f
100%	Hello World!	Hello xWorld!

7 Future works

There are several ways to extend this experiment. There were a lot of errors made during the experiment. 100 different images were used for Tesseract OCR and Microsoft OCR, That mistake would be rectified in the next experiment. Systems with different system requirements were used for the experiment and will use the same machine for extracting the images in the upcoming experiments. [7] The biggest problem made is that the similarity is checked manually which might not be much efficient and contains errors, so that could be rectified by using Levenshtein Distance Algorithm-String Comparison which is better in calculating the similarity between two strings and returns the similarity in percentage.

8 Conclusion

Bringing to an end of the experiment by having a squint at variousframeworks such as accuracy, time taken. The results of microsoft OCR are more accurate when compared with the results of Tesseract OCR in many instances. But also in some cases, Tesseract OCR provides better results. We have used different types of images such as images with white background, dark background, different fonts, colored fonts, grey scaled, etc.[9] This provides different values of precision in Tesseract OCR and Microsoft OCR. But while considering the time taken to recognize the characters in the images, Tesseract OCR has provided a better result compared to Microsoft OCR. Also we had made some errors in calculating values for these parameters which we would rectify in our future research.

9 Acknowledgement

The authors thank Ms. Malathi T, Assistant Professor, Department of Computer Science & Engineering, Bannari Amman Institute of Technology for all the help she made to accomplish this research.

10References

[1] https://support.bp-3.com/hc/en-us/articles/360017969373-Comparing-the-OCR-Engines-Available-in-UiPath

[2] https://www.guru99.com/robotic-process-automation-tutorial.html

[3] R.Smith,"An Overview of the Tesseract OCR Engine." In Proceedings of Document analysis and Recognition, ICDAR 2007, IEEE Ninth International Conference3.

[4] The Tesseract open source OCR engine, http://code.google.com/p/tesseract-ocr. Tesseract Vs Gocr A Comparative Study 83Published By: Blue Eyes Intelligence Engineering & Sciences Publication,Retrieval Number: D0788092413 /2013©BEIESP

[5] https://www.edureka.co/blog/uipath-rpa-architecture/

[6] R.D Lins and N.F Alves, "A New Technique for Accessing the Performance of OCRs." IADIS International Conference on Applied Computing, 2005.

[7] C.A.B Mello and R.D Lins,"A Comparative Study on OCR Tools." Vision Interface '99, Trois-Rivières, Canada, 19-21 May.

[8] R. Mithe, S. Indalkar and N. Divekar, 2013,"Optical Character Recognition." International Journal of Recent Technology and Engineering (IJRTE), ISSN: 2277-3878, **2(1)**.

[9] C. Patel, A. Patel and D. Patel, 2012,"Optical Character Recognition by Open Source OCR Tool Tesseract: A Case Study." *International Journal of Computer Applications* (0975 – 8887) 55–100.

IOP Conf. Series: Materials Science and Engineering 1059 (2021) 012005 doi:10.1088/1757-899X/1059/1/012005

Design and Development of an Automatic Unmanned Railway Level Crossing Gate

T. A. Selvan[1*], A. Viswanathan[2], S. Madhankumar[1], R. Dhanush Kumar[1]

[1]Department of Mechatronics Engineering, Sri Krishna College of Engineering and Technology, Coimbatore, India - 641008.
[2]Department of Mathematics, SNS College of Technology, Coimbatore, India – 641035.

[*]Corresponding author, Email Id: selvan@skcet.ac.in

Abstract. In rapidly developing countries like India, Railways are an important mode of transportation among all the people. In the railway sector, the level crossing gate plays a vital role, which should be taken into consideration. But in the level crossing gates, due to poor monitoring by the humans, many accidents are happening day-by-day. So, there is a need to develop technology to solve this issue. This paper deals with the automatic railway gate at the level crossing, which replaces the gates operated by the gatekeepers. By making it in an automatic way, can reduce the time consumption for closing the gate and also reduces human power. Streamlined the working of level crossing gate through a mini prototype model that solves this problem. Here the sensor is placed safely near the level crossing gate, which is kept at a certain distance in the railway track. Whenever the train is fastened in the railway track, the vibrations which are produced during its motion are being sensed by a sensor and it provides the necessary signal pulses to the controller which is kept at the level crossing column setup. Then the controller activates the motor, therefore the gate closes and opens automatically. Visualized the design calculation for making this setup. Hence by implementing this idea, can able to control the accidents in unmanned areas, therefore the precious human life can be saved.

1 Introduction

Generally, people like to travel from one place to another place, and their mode of travel may be a part of entertainment or a part of their daily life. For travelling purpose people may chose different modes of transportation based on their status and comfortless [1]. Usually, railways are the best mode of transportation among all the people, because the train travel saves a lot of time than roadways and it was cheaper too [2]. Indian railways are the fourth largest network in the world. At present 49% railway routes are electrified and 33% routes are double and multi tracked. Therefore, there is a necessary to construct a very large number of level crossing gates across India. Since there is large number of leveling gates is to be present, it's difficult to keep a separate monitoring stations and gate keeper to notify the process [3]. So, the monitoring and controlling of level crossing gates is to be automated to make it more precise and accurate [4]. Automatic level crossings are installed on the railway lines to regulate and control the road traffic [5]. Different type of level crossings is available in the market, they are named as Road Over Bridge (ROB) or Road Under Bridge (RUB) [6]. The simple representation of level crossing gate is represented in Figure 1.

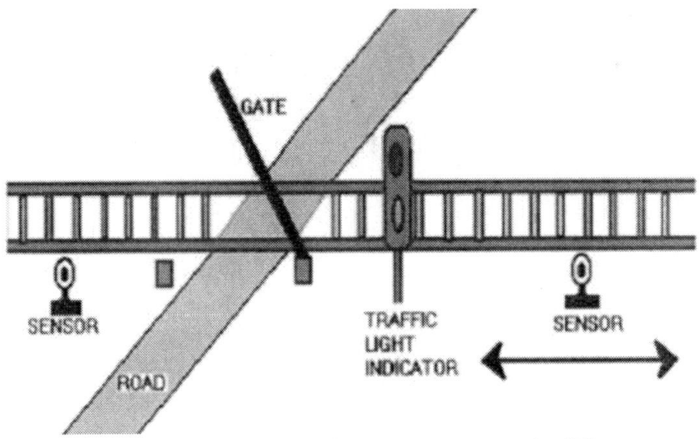

Figure 1. Level crossing gate representation[7]

In the existing level crossing gates different mechanisms like locking arrangement which includes, the hasp and staple type with padlocks, two chains with loops which is available at both the ends [8]. The gates are designed in such a fashion that it should be in the form of lifting barriers along with chains which should be in a standard design. In the existing manned level crossings, it consists of two long spare chains which is located at both the ends, and it should be placed near the tracks. The spare equipment consists of two discs painted red with the words "stop" is also fixed to the ground. The level crossings gates which is located between the outermost stop signals which should be monitored and controlled by separate Operating Department [9]. The pertaining structures and the level crossings structures shall be maintained by the Engineering Department. This prototype model deals with the automatic process of closing and opening of gates, therefore the manual operation of performing the tasks gets reduced. The main advantage in the process is that once installed this concept in the real time, there is no need of regular maintenance of this setup, hence it encourages the government to use it [10].

2 Construction

The main objective of this research work is to make the leveling gate mechanisms in an automated way, thereby reducing the manual power and to eliminate the inaccuracy in the work. Here the vibration sensor is being fitted at a certain range from the level gate. This sensor is kept near the railway track so that the vibrations produced by trains is taken into consideration by the sensor. The entire operations are being controlled by the Arduino UNO. The different level gate setup is being made with horizontal beam and vertical column. Here the DC motor is being fitted in the column, where the motor shaft gets connected to the beam. The power source for the entire operations is taken from the electric station which is used for electric trains. Therefore, there is no need to keep a separate power source for the process. The simple step-down transformer is needed to reduce the power and it provides the appropriate power suited for performing the operations. The components used in this process is briefly listed below:

2.1 Vibration sensor

Vibration sensor SW-420 is a type of sensor which includes a comparator and an adjustable potentiometer for sensitivity selection, and a LED for signal indication.This sensor produce a logic states which depends on the vibrations which being produced by the motion of trains[11].When there is no vibration it gives low output, when it senses the vibration the output will be high[12], [13].Here the sensor is fitted in such a fashion that it should not be made contact with the wheels of train, hence a precaution setup is to be there in which the sensor is being kept. The real time installation of the vibration sensor in the railway track is shown in the Figure 2.

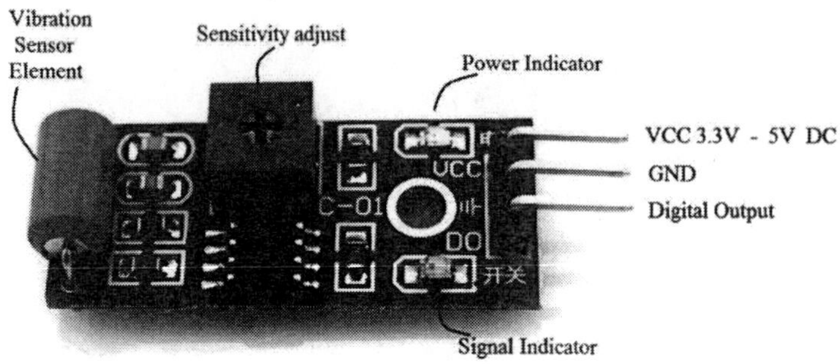

Figure 2. Vibration sensor (SW-420)[14]

2.2 Beam and Column

The column is being made up of definite length, which is used to hold the beam (lever).One of the column end is fixed to the ground and in the Centre position of it, the beam is being inserted which is used to stop the movement of the vehicles crossing the railway gate[15].Once the vibration is being sensed by the sensor, the controller sends the necessary signals to the motor and hence the beam is being actuated automatically.In the real time scenario, the actuating mechanisms is being done with the help of pulley/belt.The power source for actuating the motor is provided by the electric station from where the power is provided to run the train.

2.3 Servo Motor

A servo motor is a rotary actuator which provides the precise movement of the beam during the controller signals. The reason for choosing the servo motor is that, the direction of the beam can be controlled precisely. The motor is being controlled by providing pulse width modulation (PWM) through the control wire. It works based on the servo mechanism which basically consist of three main parts namely controlled device, output sensor and feedback system. Generally, servo motor is a kind of DC motor which is regulated and controlled by using a variable resistor called as potentiometer knob. It can be rotated from 0 to 180 degree and check the pulse in every 20 milliseconds.

2.4 Controller

The controller used in this project for controlling the entire operation is done by Arduino UNO which is the open source microcontroller. The Arduino is well equipped with digital and analog pins which is being used for different applications[16]. The programming can be easily done by using Arduino IDE (Integrated Development Environment) [17]. The power source for the controller is provided by a USB cable or by external 9-volt battery. The hardware interfacing with the board can be easily provided and user friendly. Arduino UNO differs from other board because it doesn't use FTDI USB-to-serial driver chip, instead it uses ATMEGA8U2 microcontroller chip programmed as a USB to serial converter. It has 14 digital input or output pins, 6 analog input and operate at a clock speed of 16MHz. It has a 32kb of flash memory for storing the IDE code. This board is popular among the people because it doesn't need additional hardware package it uploads the code. Arduino board generally consist of three main memories which makes the board to use it more easily. The memories are Flash memory, SRAM memory, EEPROM memory. The Arduino uses Harvard architecture in which separate memory called Program memory and Data memory [18]. In Arduino board an additional feature called reset button in which once upload the code, can able to use it again and again by pressing the reset button. The architecture of the Arduino UNO is given below in Figure 3.

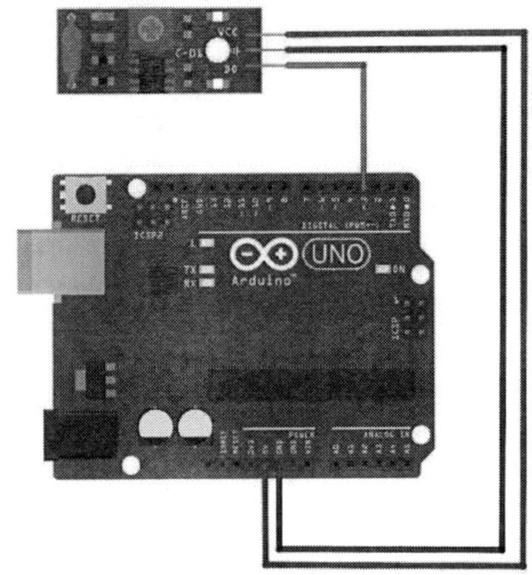

Figure 3. Arduino UNO architecture[14]

3 Working

The working prototype of an innovative project deals with the fixing of vibration sensor (SW-420), which consist of a comparator, potentiometer and a LED for indication. Here the sensor is being fitted at a certain distance from the railway closing gate. The distance between the gate and the sensor is calculated based upon the real time application scenario. Whenever the train is moving in the track at a pre-determined distance the sensor gets automatically actuated and it provides the high output signal to the controller. Here the sensor is isolated from the railway track so that it doesn't get affected by the movement of wheels of the train. Once the signal is sends to the Arduino UNO board, it suddenly actuates the servo motor which is being fitted at the middle of the column. From the column setup there is an elongation of a lever which is called as a beam, which is an external rod used to block the movement of road vehicles across the railway track during the motion of the train. Once the train crosses the level crossing gate, the closed beam automatically gets open by the instruction given by the second vibration sensor which is being fitted at the other end of railway track. In the real time scenario, wireless sensor network based on fuzzy logic is employed for more accurate way of sensing the vibrations. The beam and column setup is designed based on the calculation research which includes stress, moment of inertia, radius of gyration, slenderness ratio, shear stress, yield stress etc.., here the selection of material for beam and column plays a vital role. The motor for actuating the beam is selected by using the torque calculation[19], [20], [21], [22], [23]. The diagrammatic representation of the entire system is visualized in Figure 4 and this developed using Solidworks software [24].

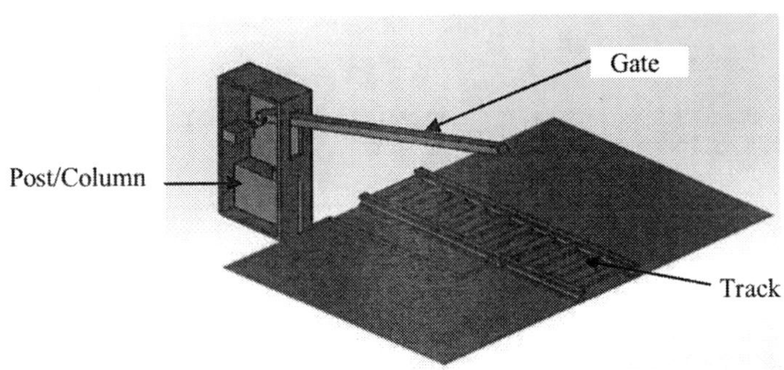

Figure 4. Assemble diagram of the system

4 Design Calculations

For checking the system model thedesign calculationsare illustrated below:

4.1 Column design

This column calculation is used for identify the column type (short column or long column).
Length of the column, $l = 837.5\,mm$
Breadth of the column, $b = 280\,mm$
Width of the column, $d = 150\,mm$
Area moment of Inertia of the column cross section can be calculated using the equation (1),

$$I = \frac{bd^3}{12} \qquad (1)$$

$$I = 7.875 \times 10^7 mm^4$$

Area of the column cross section, $A = b \times d$

$$A = 42000\,mm^2$$

The above calculated values I *and* A are used for finding radius of gyration in equation (2) and then this value is helpful for finding the slenderness ratio i.e. equation (3).

$$\text{Radius of gyration, } K = \frac{T}{A} \qquad (2)$$

$$K = 43.30\,mm$$

$$Slenderness\ ratio = \frac{l}{k} \qquad (3)$$

$$\frac{l}{k} = 19.3 \qquad (a)$$

$$\text{Yield stress, } \varsigma_{yt} = \frac{2n\pi^2 E}{\frac{L}{k}}$$

From the above equation the slenderness ratio is calculated.

$$\frac{l}{k} = 64.2 \qquad (b)$$

Comparing the values(a) and (b), (a) is less than (b). Hence it is short column. On short column the critical load is acting. This load will be calculated with the help of Johnson's formula, equation (4).

$$\text{Critical load, } P_{cr} = \varsigma_{yt} A\, 1 - \frac{\varsigma_{yt}}{4n\pi^2 E} \frac{l}{k}^2 \qquad (4)$$

$$P_{cr} = 9855300\text{ N}$$

$$\text{Allowable load, } P = \frac{P_{cr}}{FOS}$$

Where,
FOS – Factor of safety
n – Number of columns

$$P = 4927650\,N$$

Since, $P < Pcr$
Hence the column is safe.

4.2 Shear stress and Torque developed on beam

These calculations shear stress and torque value are used for selecting the motor to operate the beam.
Diameter of beam, $d = 75\,mm$
Length of beam, $L = 4000\,mm$
Density of mild steel, $\rho = 7850\,Kgm^{-3}$
Yield stress of mild steel, $\varsigma_{yt} = 247\,Nmm^{-2}$

$$\text{Shear Stress, } \tau = \varsigma_{yt}/2 \qquad (5)$$

$$\tau = 123.5\,Nmm^{-2}$$

$$\text{Torque, } T = Force \times Perpendicular\ distance \qquad (6)$$

$$T = 6 \times 10^5\,N.mm$$

5 Summary and Conclusions

Due to poor monitoring system in the railway tracks, many accidents are occurring day-by-day. So, there is need to prevent those accidents and to helps the railway department for providing a better solution in a simplest form. Hence this project, Automatic Unmanned Railway Level Crossing Gate provides a better solution by making the entire level crossing process in an automatic manner. If this project is being implemented in a real time scenario there is no need of separate person to open/close the railway gate. It also eradicates the human error completely, since there is no human involvement. Accidents can be prevented in the unmanned areas. Once installed in the real time, there is no need of regular maintenance of this setup, hence it encourages the government to use it. The developed prototype model of this system is shown in Figure 5.

Figure 5. Developed model of automatic level crossing gate

Acknowledgments
This work was supported by the Sri Krishna College of Engineering and Technology, India - 641008 as part of the motivation towards research and development of modern technologies.

References

[1] Ebrahim E A 2016 Bacteria-foraging based-control of high-performance railway level-crossing safety drives fed from photovoltaic array *J. Electr. Syst. Inf. Technol.* **3** 485–512
[2] Laapotti S 2016 Comparison of fatal motor vehicle accidents at passive and active railway level crossings in Finland *IATSS Res.* **40** 1–6
[3] Djordjević B, Krmac E and Mlinarić T J 2018 Non-radial DEA model: A new approach to evaluation of safety at railway level crossings *Saf. Sci.* **103** 234–46
[4] Kim G, Baek J, Jo H, Lee K and Lee J 2012 Design of safety equipment for railroad level crossings using laser range finder *Proceedings - 2012 9th International Conference on Fuzzy Systems and Knowledge Discovery* pp 2909–13
[5] Balamurugan K, Elangovan S, Mahalakshmi R and Pavithra R 2018 Automatic check-post and fast track toll system using RFID and gsm module with security system *Proceedings of IEEE International Conference on Advances in Electrical Technology for Green Energy 2017* vol 2018-Janua pp 83–6
[6] Borsos A, Gabor M and Koren C 2016 Safety Ranking of Railway Crossings in Hungary *Transp. Res. Procedia* **14** 2111–20
[7] Kumar V, Paul P, Sinha P and Mahato S K 2015 Automatic Railway Gate Controller with High Speed Alerting System *Int. J. Adv. Res. Electr. Electron. Instrum. Eng.* **4** 4527–31
[8] Hari Narayanan A, Brennan P, Benjamin R, Mazzino N, Bochetti G and Lancia A 2011 Railway level crossing obstruction detection using MIMO radar *8th European Radar Conference, EuRAD 2011* pp 57–60
[9] Hall S 2008 Reducing risk at automatically operated level crossings on public roads *IET* 1–5
[10] Hemanth Kumar G and Ramesh G P 2017 Intelligent gateway for real time train tracking and railway crossing including emergency path using D2D communication *2017 International Conference on Information Communication and Embedded Systems*
[11] Sharma K, Maheshwari S, Solanki R and Khanna V 2014 Railway track breakage detection method using vibration estimating sensor network: A novel approach *Proceedings of the 2014 International Conference on Advances in Computing, Communications and Informatics* pp

2355–62

[12] Dhande B S and Pacharaney U S 2017 Unmanned level crossing controller and rail track broken detection system using IR sensors and internet of things technology *Proceedings of the International Conference on Inventive Communication and Computational Technologies, ICICCT 2017* pp 206–10

[13] Madhankumar S, Rajesh S, Balamurugan R, Prawin Sankar T A, Shibin Thomas S, Nishar Ahamed B and Muhammad Nayeem M 2020 Design and modelling of self balancing electric two-wheeler using gyroscope *J. Crit. Rev.* **7** 953–7

[14] Monty Python 2018 Arduino lesson – Vibration Sensor Module *Osoyoo* 1–3

[15] Pattanaik L N and Yadav G 2015 Decision support model for automated railway level crossing system using fuzzy logic control *Procedia Computer Science* vol 48 (Elsevier Masson SAS) pp 73–6

[16] Vignesh T, Madhankumar S, Anandraj P, Anirudh Varadarajan, Arul Praveen T R S 2020 Gyro-stabilized Platform in Ambulance *Lect. Notes Electr. Eng.* **626** 281–8

[17] Syafei W A, Listyono A F and Darjat 2017 Hardware design of queuing free environmental friendly automatic toll gate using RFID *Proceedings - 2017 4th International Conference on Information Technology, Computer, and Electrical Engineering* vol 2018-Janua pp 142–6

[18] Priyanka J C, Saranya A, Shanmathi C and Baranikumar S 2015 Automatic level crossing gate with database collection *4th IEEE Sponsored International Conference on Computation of Power, Energy, Information and Communication, ICCPEIC 2015* pp 388–92

[19] S. Madhankumar, T. Vignesh, P. Anand Raj, Anirudh Varadarajan, T. Arul Praveen S R Design and Modelling of L-type Bi-directional Roller Conveyers for Glass Hauling *Lect. notes Electr. Eng.*

[20] Madhankumar S, Jishnu M, Karthick Prithiv J, Gowrishankar S, Rajesh S and Balamurugan R 2019 Design and modelling of disaster relief vehicle using rocker bogie mechanism *Int. J. Innov. Technol. Explor. Eng.* **8** 1274–6

[21] Selvan T A, Viswanathan A, Madhankumar S and Sneha S 2019 Design and fabrication of automated powder coating system *Int. J. Eng. Adv. Technol.* **8** 899–902

[22] Selvan T A, Viswanathan A, Madhankumar S and Sneha S 2019 Design and fabrication of automatic spring type chip conveyor *Int. J. Recent Technol. Eng.* **8** 917–20

[23] Gomathi P, Gowri S, Viswanathan A, Selvan T A and Madhankumar S 2019 Mathematical analysis and mechanical modeling of an inoculation system on feeder control for pouring unit in a foundry station *Int. J. Innov. Technol. Explor. Eng.* **8** 767–70

[24] Selvan T A, Viswanathan A, Sneha S and Madhankumar S 2019 Automatic Flux Coating System *Indian J. Sci. Technol.* **12** 1–7

IOP Conf. Series: Materials Science and Engineering 1059 (2021) 012006 doi:10.1088/1757-899X/1059/1/012006

A Brief Review on Effect of Nano fillers on Performance of Composites

Reshmi Olayil ,VArumugaprabu, Oisik Das and W A Lenin Anselm

Shinas College of Technology, Oman
Kalasalingam Academy of Research and Education, Krishnankoil-626126, Tamilnadu, India
Lulea University of Technology, SE-97187 Lulea, Sweden
Lecturer, Engineering Department, University of Technology and Applied Sciences-Shinas, Oman

v.arumugaprabu@klu.ac.in

Abstract. The utilization of nano materials creates a huge impact on today's scenario in development of various products. One such type of nano material developing concept is the usage of nano fillers for making composite product. This review briefs about the development of various nano fillers and its types followed by the fabrication of various composite combinations along with various nano fillers also discussed in detail. The addition of these nano fillers as a reinforcement or matrix for the composite making creates a significant impact in the mechanical properties of the composites. Special focus is given in such a way that the nano filler dispersion in which way enhances the mechanical properties. Better bonding between the matrix and reinforcement only improves the property how the nano fillers behave in that aspect also discussed in this review. The dispersion of nano fillers in various polymer matrix composites also presented in the review. Tensile, Flexural and Impact property of the composites decides its usage in various structural applications and this review presents briefly how these properties varying with respect to various nano fillers. Further the nano filler performance on the thermal properties as well as its impact in various sectors such as automobile, biomedical, sports, and aerospace applications etc., also presented in this review paper.

1 Introduction

The development of technology in modern era focuses on making materials by means of nanotechnology which attracts many researchers for finding innovative materials for various applications. One such technique derived from the nanotechnology is the usage of nano fillers. Nano fillers are dispersed into the various polymer matrixes to identify the improvement in various mechanical, thermal, electrical and moisture resistance properties. The concept of fillers in making composites evolved around the 20th century itself which acts as a binder to hold the matrix and reinforcement strongly together. The nano filler is the ones which find its place when the nano concept evolves since the main advantage offered by the nano fillers or particles is the high surface to volume ratio which in terms increases the filler and matrix bonding. The uniformly dispersed nano fillers play a vital role in deciding the mechanical properties [1]. By means of adding nano fillers the adhesion between fillers and matrixes increases which avoids the failure of the composites in early stages. This review paper addresses the fabrication of composites using nano fillers and its influence on the mechanical properties briefly based on early researches. Also to overcome the issues of dispersion proper methodology for making the nano fillers composites also discussed in this review

2 Fabrication of Nano filler based Composites

The earlier stages use nano fillers in two common forms as nanoclays and carbon nano tubes. Nano fillers used in such a way that either the matrix or reinforcement anyone has the dimension in the range of 1-100nm.Based on the shapes nano fillers used in various forms as nanoplates in1D, nanofibers

IOP Conf. Series: Materials Science and Engineering 1059 (2021) 012006 doi:10.1088/1757-899X/1059/1/012006

in 2D and nanoparticles in 3D [2]. In the beginning stages $CaCO_3$, Silica and carbon are used as the familiar types of nano fillers and researches on this is established, on the other hand few emerging materials used as nano fillers are Alumina, Magnesium, Silicon carbide,Titania and zinc oxide. All these are widely used nano filler materials for making composites. Each material has some special property in terms of increasing the mechanical properties, wear properties, thermal properties and electrical properties and depending on the applications they have chosen. The most commonly available methods used for the manufacturing of nanofiller based composites are: Insitu method, wet process-solgel method, dry process-ball milling, gas impinging etc., evaporation process-CVD, GVD and the sedimentation process [1,2].

3 Various Performance Studies on Nano filler based Composites

3.1 Mechanical and Electrical Performance

The Usage of carbon as nano fillers as reinforcement in various polymer matrixes and its effect on the mechanical, electrical, thermal, magnetic and wear properties has been analyzed in the research by Al Sheheri et al[3]. The fabrication of CNT, graphite, graphene and fullerene all are carbon based nano fillers and its influence on the polymer matrixes shows that the improvement in performance of all the properties has been evident. Also conclude that the nano fillers based composites finds a suitable place in alternating the existing ones with some excellent features. With this emphasis further research has been carried out using carbon fillers along with polyazomethine [4] as hybrid ones. The prepared composites were subjected to thermal studies and the degree of polymer crystalinity in order to know the better dispersion of nano fillers with the polymer matrixes. Also they concluded that this hybrid combination nano filler produces a better dispersion in polymer matrix that produces an enhancement in the thermal conductivity as well as low cost involved.

Further studies carried out on using polyethylene terepthalate (PET) composites reinforced on four different types of carbon fillers such a CNT, Graphite and hybrid combinations[5]. All the nano filler composites were prepared by using the high energy ball milling (HEBM) method. Mechanical, Thermal and electrical properties of the composites were compared with the pure PET in such a way that the impact of nano fillers be identified and conclude that both properties increases in all the four types of nano fillers addition. Nilagiri Balasubramanian [6] et al carried out a complete review on the usage of nano fillers and micro fillers for the electronics applications. The thermal and electrical conductivity of the prepared composites were analyzed using the filler arrangement as a key factor. The improvement in the thermal conductivity of the composites happens by the addition of nano fillers along with the mechanical properties. Also they concluded that the addition of nano fillers increased the compatibility with the matrix is the reason for the increase in all properties.

Another interesting study has been carried out using the epoxy matrix along with nano fillers cubic boron nitride and hexagonal boron nitride [7]. Hand layup method is used for the fabrication of composites and subjected to wear resistance studies. The addition of fillers increases the wear rate drastically when compared to the unfilled ones which is due to the increase in hardness. Also due to agglomeration of the filler particles varying weight percentage also leads to poor wear rate in few cases. Nano fillers based composites used in making strain sensors especially using carbon fillers. During the electrical resistance test it was noted that this hybrid nano fillers composites poses more resistance also additional few more special features such as high durability and high repeatability observed from the strain gauge sensors [8].

Polymer because of its excellent properties along with nano fillers used as a electromagnetic interference shield material since they have high electrical resistance properties. Among all the composites combinations prepared and studied using the polymer and filler the multi walled carbon nano tube based polymers and nano carbon black produces much improved electrical conductance. In addition the combination of Ag/MWCNT composites has the capability to produce a very high shielding efficiency that caters the industry needs [9].

The influence of nano fillers extended towards the biomedical sector also in such way research reported using silica nano fillers and silver nano fillers to improve the banding adhesives both in terms of physical level as well as microbial level. Among the two the experimental composites adhesives with the silver nanoparticles possess better results in terms of both physical as well as biological aspects[10]. Further studies performed using nano fillers effect in hybrid composite mechanical properties. The hybrid composites were prepared using injection molding method in which short carbon fiber used as the reinforcement and polypropylene used as the matrix. Along with this various natural fibers such as alumina, silica and CNT were added as nano fillers and the hybrid composite effect on the tensile and flexural properties has been analyzed in detail by using theoretical models. It has been noted that the adhesion properties of the composites improved by means of nano filler addition. Among all the fillers used the alumina, silica and CNT improve the inter shear strength of the composites. Based on the theoretical model inference alumina particles reflect as a best one in terms of improvement [11]. The studies using the polypropylene matrix along with nano fillers now focus towards on flame retardant capability. The fillers used are organo clay, layered double hydroxides, polyhedral oligomeric silsesquioxane (POSS), and carbon nanotubes (CNTs). All these fillers were dispersed in the polypropylene medium and studied for its flame retardancy and thermal stability. When compared to all CNTs posses better dispersion with the polypropylene, matrix in such a way there is no agglomeration observed during the TEM studies but interestingly organo clay composites possess better thermal stability and flame retardance among all the combinations [12]. The research on nano fillers using epoxy matrix and zinc oxide particles for the electrical resistance applications has been reported. It was noted that even an addition of small amount of the filler zinc oxide improves the electrical resistance drastically leads to the treep growth which extend the breakdown time and inferred that the high surface is the main reason for this behavior as well as the reduction of voids[13].

Zhifang Gao et al [14] carried out research on various nano fillers produced such as Nano-AlN particles, AlN/graphene nano-hybrids (AlN/GE) and AlN/carbon nanotubes nano-hybrids (AlN/CNTs). All the prepared fillers were subjected to the thermal conductivity studies along with the epoxy matrix. From the results it was revealed that the combination of epoxy composite with 5 wt% AlN/GE nano-hybrids and 45 wt% micro-Al2O3 particles hybrids possess enhanced thermal conductivity when compare to the pure epoxy. Edison et al [15] performed studies on the hybrid composites prepared using woven Kevlar fabric and aluminium sheets along with the epoxy matrix, in addition fillers of micro and nano size added to the combination to know its improvement in the ballistic performance. The fillers used are gamma alumina, silicon carbide, aluminium, colloidal silica and potato flour. They concluded that the addition of all fillers improves the fiber and metal bonding strength. Also it was noted that the addition of aluminium, silicon carbide and colloidal silica improves the ballistic effect when compared to the unfilled ones. The various types of nano fillers used and its classification is presented in Figure.1

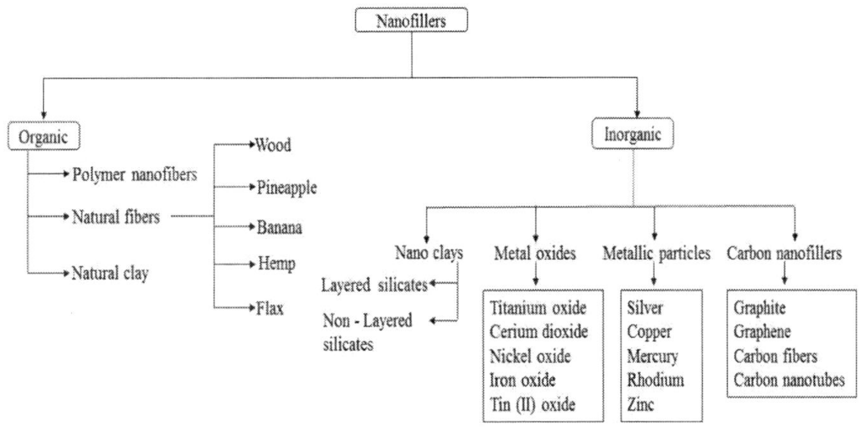

Figure 1. Classification of nano fibers

Takahiro Imai et al [16] also carried out studies on the effect of nano filler based composites for electrical resistance applications. They focused on preparing a new type composite for insulation purposes. The composites were made using LS, SiO2 or TiO2 as nano fillers and epoxy as the matrix. The epoxy based nano/micro filler composites of different weight percentages possess the better insulation capability when compared to the pure epoxy. Another type of composites developed using insitu method for the combination of epoxy, spherical silica and organo-modified montmorillonite. Mechanical and tribological studies has been carried out on the made composites showed that the addition of different shapes montmorillonite nano filler improves the toughness properties which leads to good mechanical strength and also the wear resistance gets improved a lot[17].

Ermias G Koricho et al [18] analyzed the impact resistance of the composite made using nano clay and glass bubbles along with the glass fiber reinforcement. They concluded that the composites which are tailor made using these combination posses a good impact resistance under drop weight test by means of addition of nano/micro fillers. Carbon fiber reinforced polymer composites subjected to modifications in such a way to improve its performance on toughness aspect nano fillers are incorporated. GNPs and MWCNTs were used as the fillers, the improvement in inter laminar fractures takes place due to the increase in toughness by the addition of fillers [19].

Tao Li et al [20] have done research on the preparation of composite membranes using the nano fillers. The filler used is nano-sized zeolitic imidazole framework ZIF-7.The addition of this improves both the selectivity and permeability of the prepared composite membranes. Nano composites were developed using the combination of silver nanoparticles and carboxylated cellulose nano crystals in such a improvement in mechanical properties and biological properties obtained when used with waterborne polyurethane. Bio nano composites obtained by means of these combinations which can be used as potential filler in waterborne polyurethane applications [21]. Further new composite combination prepared by using nano scopic silica fillers along with polyamide matrix. The specimens were fabricated by means of insitu polymerization. The mechanical properties gets influenced by the size of the nano fillers was reported where reduction in the filler size provides less dispersion rate which in term leads to poor bonding[22].

3.2 Thermal performance

More commonly used nano fillers now a days are clays and nano tubes. The impact of usage of nano fillers and nano particulates create a significant impact in the markets already in which the influence of polymer matrixes gets increasing day by day. With the focus towards innovation research the usage of nano crystals as fillers attracts many even though the costs involved in making this is high. The future trend will shift towards making and usage of nano fillers in order to make a green composite [23]. Sanada et al [24] used two fillers such as carbon nano tubes, alumina particles in micro filler form used in polymer matrixes for making polymer composites. The prediction of nano filler impact on the thermal conductivity of the composites has been done using the finite element analysis. From both the experimental results and the theoretical experiments it has been noticed that the addition of these micro fillers drastically improve the thermal conductivity of the prepared composites. The different types of matrixes which are used for preparing nano composites along with its applications are shown in Table.1

Table1. Various types of matrixes used with nano fillers and its applications.

Matrix	Properties studied	Applications
Polyester	Tensile Strength	Automobile
Epoxy	Flexural Strength	Biomedical
Polypropylene	Impact Strength	Packaging
Polyurethane	Thermal Conductivity	Sports goods
Polyethylene	DMA and TGA studies	Electronics

The studies on grapheme nano fillers along with the titanium composites has been carried out by song et al[25] and inferred that an improved mechanical properties such as hardness and yield strength happens. Also the nano fillers added composites has the capability to resist the various failure mechanism such as micro cutting, ploughing etc., Further Suresha et al[26] analyses the effect of micro/nano fillers on the composites made using the combination of polyamide 66 and polypropylene. The prepared composites were subjected to mechanical and tribological performance studies and observed that the filler effect creates an impact in the hardness, tensile and flexural strength in such a way all gets increased. On the other hand the wear rate also very low in the case of filler based composites. Graphene possess some excellent property that makes it as a suitable candidate for preparing the nano filler based composites in such a way that the graphene along with polymer matrixes widely used for various industrial applications. Also advanced composites were developed for electrical applications using grapheme as nano filler in polymer matrixes [27].

Another interesting study on the nano filler in such a way using the boron nitride tubes and boron nitride nano sheets combined together along with the epoxy matrix form hybrid composites. The thermal conductivity increases drastically by means of 95% and young's modulus by 57% respectively by the addition of the fillers were reported [28, 29]. Shaoxin Zhou et al [30] carried out studies on the impact on carbon based nano fillers along with the flake graphite and polymer matrix composites. A special emphasis was given to know the thermal conductivity of the prepared composites. Also by means of forming a new model the experimental results were compared theoretically in such a way that they were well fit and the addition of nano fillers size creates an improvement in the thermal conductivity behavior of the composites.

4 Summary and Conclusions

From the brief review on the importance of nano fillers following conclusions were drawn:

- The nano fillers based composites fabricated successfully by various techniques such as hand lay up, insitu method, sol gel method etc.,
- The prepared nano fillers enhance the mechanical, thermal and electrical, wear properties of the composites when compared to unfilled ones or other existing conventional materials.
- The excellent improvement in properties is due to the high surface volume ratio as possess by the nano fillers and also it increases the hardness of the material. In addition material cost also low.
- From the vast research done it has been noted that the influence of nano fillers will be high in modern days because of its interesting characteristic behavior also offering special features with low space and cost. Also the research scope will be high in the near future on this aspect.
- The nano filler based composites extend its application in major sectors such as automotive, aerospace, electronic parts and sports good.

References

[1] Monika Šupová, Gražyna Simha Martynkov, and Karla Barabaszov, 2011 Effect of Nanofillers Dispersion in Polymer Matrices: A Review, *Sci. Adv. Mate.* **3(1)** 1-25

[2] Damien M. Marquis, Éric Guillaume and Carine Chivas-Joly,2011 Properties of Nanofillers in Polymer, *Nanocomposites and Polymers with Analytical Methods*, ISBN: 978-953-307-352-1,261-284.

[3] Soad Z. Al Sheheri, Zahra M. Al-Amshany, Qana A. Al Sulami, Nada Y. Tashkandi, Mahmoud A. Hussein & Reda M. El-Shishtawy, 2019 The preparation of carbon nanofillers and their role on the performance of variable polymer nanocomposites, *Designed Monomers and Polymers*, **22(1)**, 8-53.

[4] Bronnikov S, Kostromin S, Asandulesa M, Podshivalov A, Timpu D. 2019 Morphology, structure, and segmental dynamics in polyazomethine/hybrid carbon nanofillers composites, *Polymer Composites*, **40(12)** 4638–4649.

[5] Giuliana Gorrasi, Valeria Bugatti, Candida Milone, Emanuela Mastronardo,Elpida Piperopoulos, Laura Iemmo, Antonio Di Bartolomeo, 2018 Effect of temperature and morphology on the electrical properties of PET/conductive nanofillers composites, *Composites Part B* **135** 149–154.

[6] Nilagiri Balasubramanian KB, RameshT. 2018 Role, effect, and influences of micro and nanofillers on various properties of polymer matrix composites for microelectronics:A review, *Polym Adv Technol.* **29(6)** 1568–1585.

[7] Yu J, Zhao W, Wu Y, Wang D, Feng R, 2010 Tribological properties of epoxy composite coatings reinforced with functionalized C-BN and H-BN nanofillers, *Applied Surface Science*.**434**,1311-1320.

[8] Zheng Y, Li Y, Dai K, Wang Y, Zheng G, Liu C, Shen C, 2018 A highly stretchable and stable strain sensor based on hybrid carbon nanofillers/polydimethylsiloxane conductive composites for large human motions monitoring, *Composites Science and Technology*,**156**, 276-286.

[9] Dimuthu Wanasinghe, Farhad Aslani , Guowei Ma and Daryoush Habibi 2020 Review of Polymer Composites with Diverse Nanofillers for Electromagnetic Interference Shielding,*Nanomaterials*,**10,541**,1-46.

[10]Sug-Joon Ahn, Shin-Jae Lee, Joong-Ki Kook, Bum-Soon Lim,2009 Experimental antimicrobial orthodontic adhesives using nanofillers and silver nanoparticles, *Dental materials*,**25**,206-213.

[11]Arao.Y, Yumitori.S , Suzuki.H ,Tanaka.T , Tanaka.K ,Katayama.T,2013 Mechanical properties of injection-molded carbon fiber/polypropylene composites hybridized with nanofillers, *Composites: PartA*,**55**, 19-26.

[12]Baoxian Du, Haiyun Ma and Zhengping Fang,2011 How nano-fillers affect thermal stability and

[13]flame retardancy of intumescent flame retarded polypropylene, *Polym. Adv. Technol*, **22**,1139-1146.

[14]Ding.H.Z and Varlow.B.R, 2004 Effect of Nano-Fillers on Electrical Treeing in Epoxy Resin Subjected to AC Voltage, *The 17th Annual Meeting of the IEEE Lasers and Electro-Optics Society* **doi: 10.1109/CEIDP.2004.1364255**, 332-335.

[15]Zhifang Gao, Lei Zhao 2015, Effect of nano-fillers on the thermal conductivity of epoxy composites

[16]with micro-Al2O3 particles, *Materials and Design*,**66**,176-182.

[17]Edison E. Haro , Akindele G. Odeshi, Jerzy A. Szpunar 2016, The energy absorption behavior of hybrid composite laminates containing nano-fillers under ballistic impact, *International Journal of Impact Engineering*,**96**,11-22.

[18]Takahiro Imai, Gen Komiya, Kiyoko Murayama, Tamon Ozaki, Fumio Sawa,Toshio Shimizu

[19]Miyuki Harada, Mitsukazu Ochi, Yoshimichi Ohki and Toshikatsu Tanaka,2008 Improving Epoxy-based Insulating Materials with Nano-fillers toward Practical Application, *Conference*

Record of the 2008 IEEE International Symposium on Electrical Insulation, **doi: 10.1109/ELINSL.2008.4570310**,201-204.

[20] Jia.Q.M, Zheng.M, Xu.C.Z and Chen.H.X, 2006 The mechanical properties and tribological behavior of epoxy resin composites modified by different shape nanofillers, *Polym. Adv. Technol.* **17** 168–173.

[21] Koricho, E.G., Khomenko, A., Haq, M., Drzal, L.T., Belingardi, G., Martorana, B., 2015 Effect of Hybrid (Micro- and Nano-) Fillers on Impact Response of GFRP Composite, *Composite Structures*,**134**,789-798.

[22] Kostagiannakopoulou C, Tsilimigkra X, Sotiriadis G, Kostopoulos V, 2017 Synergy effect of carbon nano-fillers on the fracture toughness of structural composites, *Composites Part B*,**129**,18-25.

[23] Tao Li, Yichang Pan, Klaus-Viktor Peinemann, Zhiping Lai, 2013 Carbon dioxide selective mixed matrix composite membrane containing ZIF-7 nano-fillers, *Journal of Membrane Science* **425-426,**235–242.

[24] He Liu,Jie Song,Shibin Shang, Zhanqian Song and Dan Wang,2012 Cellulose Nanocrystal/Silver Nanoparticle Composites as Bifunctional Nanofillers within Waterborne Polyurethane, *ACS Appl. Mater. Interfaces*,**4**,2413−2419.

[25] Reynaud.E,Jouen.T,Gauthier.C,Vigier.G,Varlet.J, 2001 Nano fillers in polymer matrix: a study on silica reinforced PA6,*Polymer* **42**,8759-8768.

[26] Rothon.R,2016 Nanofillers, *Polymers and Polymeric Composites: A Reference Series*, **ISBN: 978-3-642-37179-0,** 1-21.

[27] Sanada.K , Tada.Y, Shindo.Y, 2009, Thermal conductivity of polymer composites with close-packed structure of nano and micro fillers, *Composites: Part A* **40** 724–730.

[28] Song.Y, Chen.Y ,Liu.W.W, Li.W.L, Wang.Y.G, Zhao.D, Liu.X.B, 2016 Microscopic mechanical properties of titanium composites containing multi-layer graphene nanofillers, *Materials and Design*, **109** 256–263.

[29] Suresha.B ,Ravi Kumar.B.N, Venkataramareddy.M ,Jayaraju.T 2010,Role of micro/nano fillers on mechanical and tribological properties of polyamide66/polypropylene composites, *Materials and Design* **31**,1993–2000.

[30] Sie Chin, 2014 Tjong Polymer Composites with Graphene Nanofillers: Electrical Properties and Applications Journal of Nanoscience and Nanotechnology **14**,1154–1168.

[31] Haiyan Yan, Yanxia Tang , Juling Su, Xiaoyan Yang, 2014, Enhanced thermal–mechanical properties of polymer composites with hybrid boron nitride nanofillers, *Appl. Phys. A*, **114**,331–337.

IOP Conf. Series: Materials Science and Engineering 1059 (2021) 012007 doi:10.1088/1757-899X/1059/1/012007

Performance analysis on automating the PDF download process using Robotics Process Automation

Malathi T, Navaneethakrishnan P L,Arun E, Devakipriyadharshini V N

Assistant Professor, Department of Computer Science and Engineering, Bannari Amman Institute of Technology, Sathyamangalam,Tamil Nadu, India.

tmalathime@gmail.com

Abstract: Many people download one or two files from the internet and it does not take much time to download them. What if there is a need for more number of files. For that problem RPA could be the solution. Robotic process automation (or RPA) is a form of business process automation technology based on metaphorical software robots (bots) or on artificial intelligence (AI)/digital workers. The main objective of this paper is to download 100 random pdf files from the internet using excel, csv and datatable automations in Uipath and compare them with the manual way of downloading and finding the most efficient way of downloading the large number of files.

1 INTRODUCTION

Everyone knows how to download a pdf file manually from the internet. it does not take much time to download them. What if there is a need for more number of files. It takes a lot of time to download many files manually. For that problem RPA could be the solution. We can automate the downloading process to reduce time. Here I have used UiPath for automation. There are automation using excel, automation using csv and automation using datatable available in UiPath. Our goal is to find the most efficient way of downloading the pdfs. This paper is used to compare manual downloading , excel automation, csv automation and datatable automation and conclude which is better.

For this first you have to know what is an RPA. Robotic Process Automation(RPA) is the type of technology where everyone can configure a software robot to do any action that a human can do within the computer software system[1]. RPA robots use the computer's user interface to capture data and manipulate the applications just as humans. They interpret responses and trigger actions to perform a large variety of tasks repetitively . An RPA robot makes no mistakes and it never takes rest[4].RPA reduces expense of the work, increases the standard with greater efficiency, better value and good experience to the customers whilereducing the number of employees and their work burder to do more high-valued and meaningful work[3].

UI path is one of the widely used RPA tools and it was founded by the Romanian entrepreneur in 2015 to provide software which can eliminate the tedious timeconsuming back office work[7]. It converts boring tasks into automated ones with the help of unique tools. It enables your processes to be more effective and reliable[5].

1.1 Uipath Studio

UiPath Studio is a complete software solution that enables you to automate your back office repetitive tasks. It converts tedious tasks into complete UI automation, thus, making your work easier and faster[6].

1.2 Uipath Robot

UiPath Robot executes the processes built in Studio, as a human would. Robots can work unattended (run without human supervision in any environment, be it virtual or not) or as assistants (a human triggers the process)[8].

2 Requirements

The applications I have used in this project are

1. *UiPath Studio Pro 2020.4.3 Community license*
2. *Mozilla Firefox browser 79.0 (64-bit)*
3. *Microsoft Office 365*

The system specifications

1. Intel core i3- 7020U @ 2.30GHz CPU
2. 8 GB RAM
3. Nvidia GeForce MX110 GPU
4. 1 TB HDD
5. Wifi speed 200kbps

3 Manual method

Open the browser and paste the link in the address bar and hit the search button. After the file has opened , click the download button. Choose the destination folder and save the file.

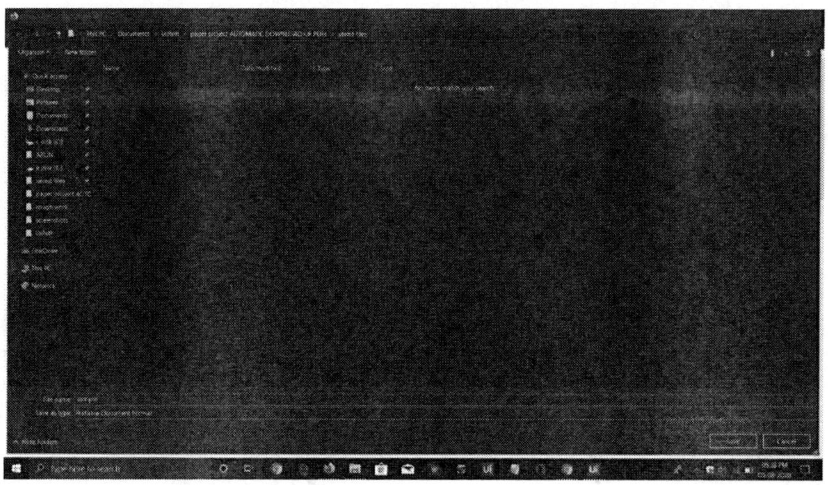

Figure 1. Image showing file manager

4 Datatable method

The activities that are used in this method are Build datatable, For Each Row, Open Browser, Click, Type into and Close tab. First,open a new project and a new sequence in UiPath Studio ,then drop 'built datatable' activity from the activities panel and create two columns in it. One containing the hyperlinks and another with the names in which the files need to be saved. After that drop 'for each row' activity and inside it drop 'open browser ' activity.Give the variable in which the datatable is stored in 'for each row' activity , also give row.item(0).ToString in the input of 'open browser' and choose the browser type.Then drop 'click' activity.Set informative screenshot as print symbol and using another 'click' activity',click print. Now another 'click' activity ,this time an informative screenshot as save symbol. Using once again 'click' activity ,choose your destination folder. Use 'type into' activity and select the box in which you are going to give the name and the value as row.item(1).ToString. Finally use 'click' activity to click save button and drop 'close tab' activity for closing tabs.

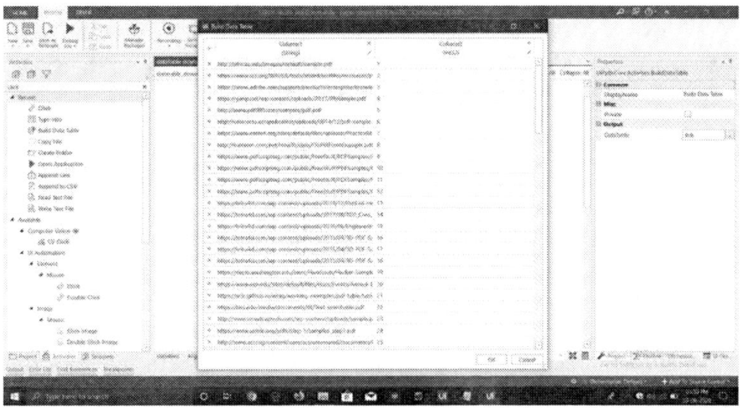

Figure 2. Image containing the links of the files and the names it has to be saved in the system in the datatable.

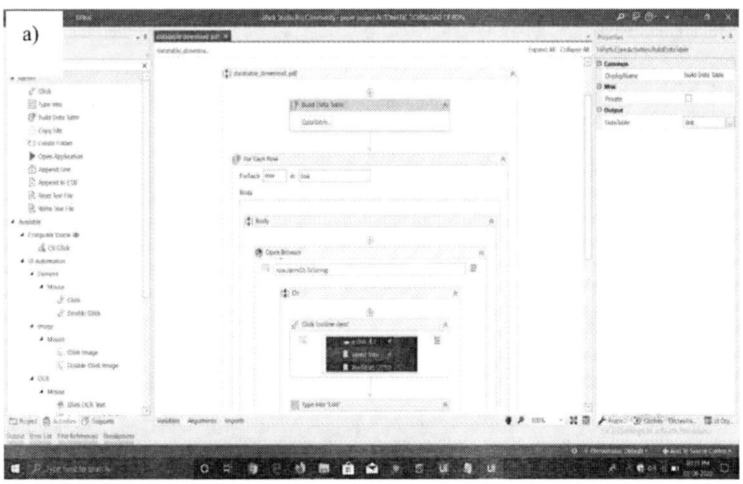

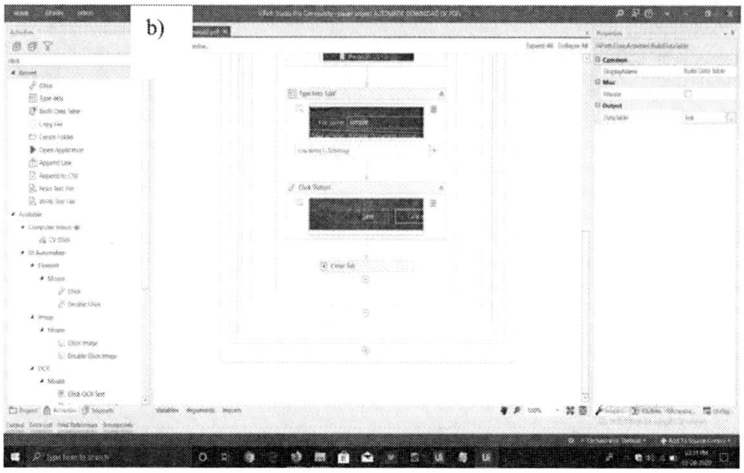

Figure 3. Images a) and b) containing the steps of the datatable automation.

5 Excel automation

The activities that are used in this method are Read range from workbook, For Each Row, Open Browser, Click, Type into and Close tab. Open a new project and a new sequence in UiPath Studio, then create an excel sheet with two columns,one containing the hyperlinks and another with the names in which the files need to be saved. Drop 'Read range' activity in the workbook from the activities panel and choose your file. After that drop a 'for each row' activity and inside it drop 'open browser ' activity. Give the variable in which the datatable is stored in 'for each row' activity also give row.item(0).ToString in the input of 'open browser' and choose the browser type.Drop 'click' activity .Set informative screenshot as print symbol and using another 'click' activity',click print. Now another 'click' activity ,this time an informative screenshot as a save symbol. Again by using 'click' activity ,choose your destination folder. Use 'type into' activity and select the box in which you are going to give the name and the value as row.item(1).ToString. Ultimately use 'click' activity to click save button and drop 'close tab' activity for closing tabs.

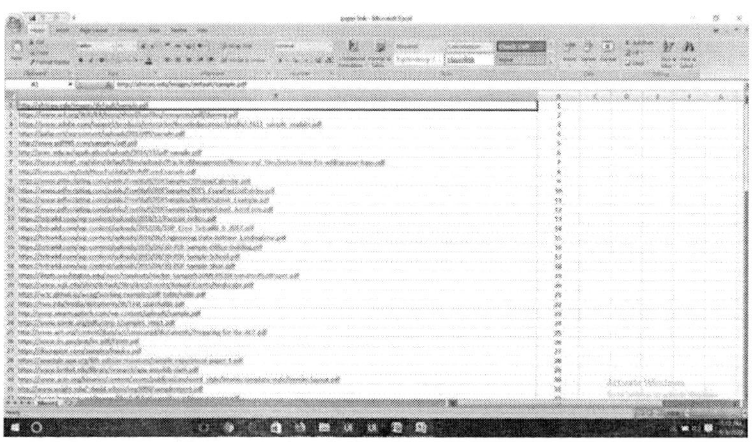

Figure 4. Image containing the links of the files and the names it has to be saved in the system in the excel sheet.

6 CSV automation

The activities that are used in this method are Read csv, For Each Row, Open Browser, Click, Type into and Close tab. Open a new project and a new sequence in UiPath Studio, then create a csv file with two columns. One containing the hyperlinks and another with the names in which the files need to be saved. Drop 'Read csv' from the activities panel and choose your file. After that drop 'for each row' activity and inside it drop 'open browser ' activity. Give the variable in which the datatable is stored in 'for each row' activity also give row.item(0).ToString in the input of 'open browser' and choose the browser type. Drop 'click' activity .Set informative screenshot as print symbol and using another 'click' activity',click print. Now another 'click' activity ,this time an informative screenshot as save symbol. Using once again 'click' activity ,choose your destination folder. Use 'type into' activity and select the box in which you are going to give the name and the value as row.item(1).ToString. Finally use'click' activity to click save button and drop 'close tab' activity for closing tabs.

Figure 5. Image containing the csv file opened in notepad.

7 Result

We have tried all the four methods and tabulated the result in the table III.E . It found that usage of excel automation for downloading pdf files is more efficient than other three methods .

Every other activity is the same , so, there is only the reading files may vary in speed. The read activity for excel workbook file is faster and effective than other read activities used here.

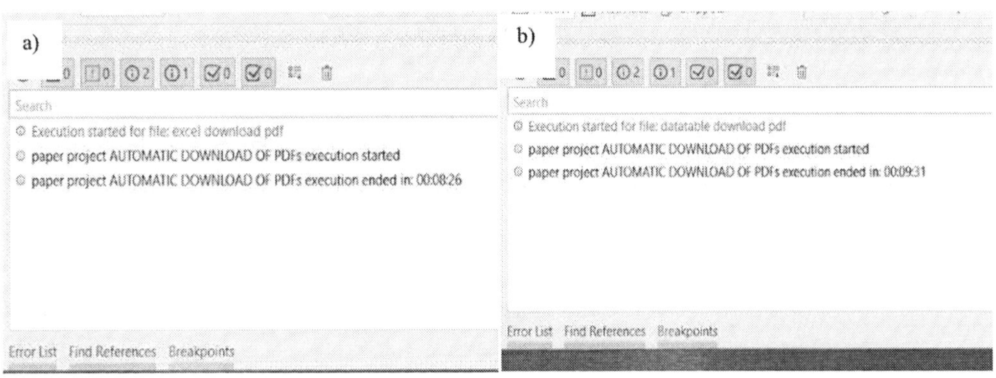

Figure 6. Time taken by each methods to download all the 100 files a) excel automation , b) datatable automation , c) csv automation and d) manual method

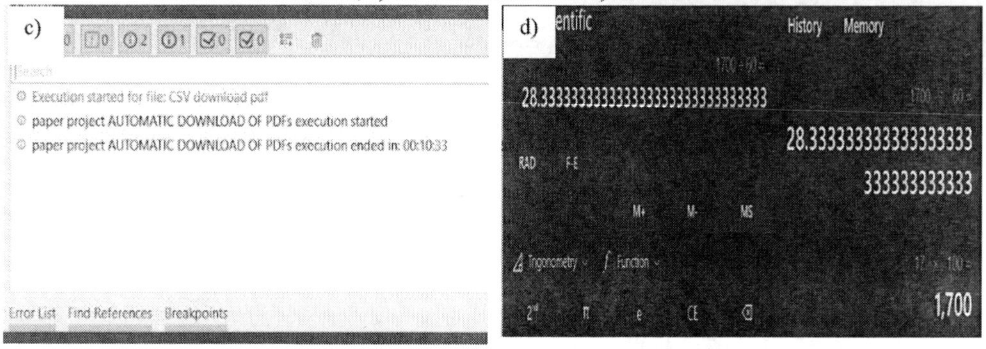

Table 1. Result of different downloading methods .

	Manual Method	Excel automation	Datatable automation	CSV automation
No. offiles	100	100	100	100
No. offilesskipped	0	0	0	0
Time taken	17 seconds x100 files 1700 seconds OR Approx 28 min	8 minutes 26 seconds	9 minutes 31 seconds	10 minutes 33 seconds

8 Conclusion

From the above result it is easy to conclude that excel automation is more efficient than the other three methods. All the four methods of downloading were tested in the same computer and connected with the same wifi network. These are some of the areas where RPA is more effective , Operational activities in sales,Procure-to-pay, Portal queries, Data extraction, Reconciliation, Price comparison,Data management and so on[2]. By the use of RPA there will be more accurate outcomes with less labour and less cost .

References

[1] Kosonen, Lotta & Lappi, Olli-Eemeli. 2020 *Impacts, benefits, and implementation of RPA and its effect on outsourced work* 10.13140/RG.2.2.26625.89440.

[2] Dr. Suryakant,Vinod L Mane and Dr. Preeti Patil 2019 *Social Innovation in the Education System by using Robotic Process Automation (Rpa)* 2278-3075

[3] Adrian-Mihai, Zaharia-Radulescu , Catalin Liviu Pricop, Darko Shuleski and Anton Cristian Ioan 2017*Rpa And The Future Of Workforce.* **11(1),** 384-392.

[4] Aguirre S., Rodriguez A., Figueroa-García J., López-Santana E., Villa-RamírezJ., Ferro-Escobar R. (eds) 2017*Automation of a Business Process Using Robotic Process Automation (RPA): A Case Study* .**742** ,62-74 https://doi.org/10.1007/978-3-319-66963-2_7

[5] Mendling J., Decker G., Hull R., Reijers H. A., and Weber I. 2018*How do Machine Learning, Robotic Process Automation, and Blockchains Affect the Human Factor in Business Process Management* **43**, DOI:10.17705/1CAIS.04319. https://doi.org/10.17705/1CAIS.04319

[6] Van der Aalst W.M.P., Bichler M. and Heinzl A.2018*Robotic Process Automation* Bus Inf Syst Eng 60, 269–272 , available online : https://doi.org/10.1007/s12599-018-0542-4

[7] Penttinen Esko, Kasslin Henje and Asatiani Aleksandre 2018 How to Choose between Robotic Process Automation and Back-end System Automation? Research Papers.66. https://aisel.aisnet.org/ecis2018_rp/66

[8] Alok Mani Tripathi 2018*Learning Robotic Process Automation: Create Software robots and automate business processes with the leading RPA tool –UiPath* ISBN 1788396510, 9781788396516

Optimization of Material Removal Rate and Surface Roughness for Micro ECM of Inconel 718 alloy utilizing Grey Relational Technique

Madhankumar S[1*], Rajesh S[2], Balamurugan R[3], Tharun Sri Ram N[1], Venuprasath S[1], Tazmeel Ahamed S[1]

[1]Department of Mechatronics Engineering, Sri Krishna College of Engineering and Technology, Coimbatore, Tamil Nadu, India - 641008.
[2]Department of Mechanical Engineering, R.M.K. Engineering College, Tiruvallur, India – 601206.
[3]Department of Automobile Engineering, Bannari Amman Institute of Technology, Erode, India - 638401.

[*]Corresponding author, Email Id: madhankumars@skcet.ac.in

Abstract.Material removal rate (MRR) and surface roughness (SR) are vital factors of micro-electrochemical machining (ECM).This machining method is focused on redox reaction. The intensity of machining relies mostly on molecular masses, its current density, the electrolytes and the duration in metal-working. In order to learn the results of MRR and SR on Inconel 718 alloy, tests were performed by considering various criteria employing stainless steel electrodes. Throughout this present context, the optimization is focused on the Taguchi functional grey method. Predictor variables are designed utilizing the Taguchi concept with 3 different process variables, namely electrolyte concentration, feed rate and voltage. In addition to the MRR and surface roughness, the overcut was used as success metrics within that report. The outcome of this study shows that the voltage appears to be the leading variable for the intended performance requirements and also that the optimum values are observed. The confirmatory examination was performed to verify the findings obtained by the grey relational method.

1 Introduction

Hard and tough to process materials like carbon steels, stainless steel, nickel-based alloys as well as other high strength temperature-resistant alloys become commonly used for the aviation and nuclear reactor sectors attributable to certain strong strength-to-weight rate, toughness and heat-resistant properties. With these materials, traditional edged machine machining becomes extremely impractical as well as the level of precision or surface quality is not upto the level. The precision and efficiency of its machining method is assessed by the processing variables. There are many processing variables available namely voltage, pulse on time, pulse off time, current, feed rate and electrolyte concentration. Therefore, it necessitates to always identify the optimal outcomes for the attainment of the finest process variables. In this respect, multiple studies have been carrying out through ECM in various materials including numerous modeling methods have been implemented in an attempt to discover an efficient level [1]. Provided multi-objective optimization for Titanium alloys in electro-chemical digging utilizing Grey relational based Taguchi process [2]. Analyzed on electrochemical machining of contemporaneous titanium, nickel-based alloys as well as reviewed for some of its application for the machining of aeromotor elements, ECM provides an effective alternative process for the preparation of derivative markets awesome post such as molds, jet engine items as well as micro dimensional parts[3]. Probed an ECM process for varying electrolyte proportions but presumed that toxic effects would lead to increased precision[4]. The ECM method for nickel-based content has been investigated and has been demonstrated that perhaps the production of a better component can be accomplished by following the same breakdown of electrical and chemical activity[5]. Explained these same generations throughout the ECM with cylinder-Steels, using different electrolytes, but also stated that even the dielectric rotational speed had been the powerful factor affecting the MRR[6]. Studied a particular feature of an ECM process, including the electrolytic rotation speed, the voltage decided to apply, the chloride output impedance in order to determine the greatest available performance assessment processing parameters, notably MRR, tool wear, excellent surface as well as machine expense [7]. The option of input variables to assess the greatest available machining efficiency is still a difficult issue. Studied numerically EDM of different nickel-based including Inconel and 718 alloys, utilizing the Taguchi concept plan to enhance efficiency features[8]. Different multi-criteria decision-making mechanism grey relational analysis used for EDM of stainless steel. The gray method focused on Taguchi was used to resolve the multi-objective optimization problem of the WEDM technique [9], [10].

Through this discussion, ECM optimization has been performed on different metals and indeed the present research investigated the different output performance variable MRR, SR and overcut as during Inconel 718 micro ECM using three input variables namely electrolyte concentration, feed rate and voltage.Afterward, certain interventions are outlined and the optimization value of its variables are eventually determined utilizing grey relational technique.

2 Materials and methods

Inconel 718 alloy has been chosen to research work material. Taguchi model (L9 orthogonal string) is designed for Inconel 718 specimen and thus the work material has a span of 50 mm, a breadth of 50 mm and a depth of 0.5 mm. The distribution of the 718-alloy material is studied from [11], [12] is mentioned in Table 1[13].

Table 1. Chemical composition of Inconel 718 alloy

Element	Ni	Cr	Nb	Cu	Ti	Si	Fe	M0	Al	Mn	C
Min	50	17	4.75	-	0.65	-	15	2.8	0.2	-	-
Max	55	21	5.5	0.3	1.15	0.35	19	3.3	0.8	0.35	0.08

Throughout micro ECM usually, the cathode tool consists of conductive material like stainless steel and copper. Within that research, stainless steel rods are used as electrocatalysts with a span of 50 mm and a diameter of 1 ± 0.02 mm. The electrolyte used in this study was sodium chloride (NaCl). The experimental setup has illustrated in Figure 1. The effectiveness of machining varies depending mostly on processing variables. As from previous research, input variables such as voltage, feed rate and electrolyte concentration have selected for the present research because they've discovered to be had a massive effect on MRR, overcutting and SR.Table 2 gives the developed Taguchi model for this experimentation. Afterward, such interventions have stated as well as the optimization process of its input variables was eventually computed utilizing grey relational grades.

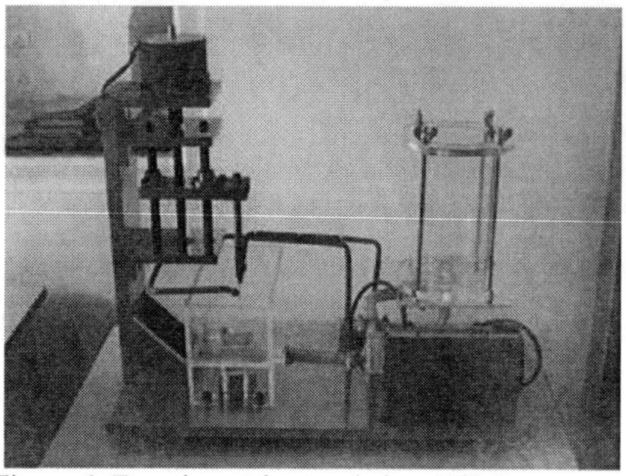

Figure 1. Experimental setup (micro ECM)

Table 2. Levels of selected input variables

Input variables	Unit	Level 1	Level 2	Level 3
Concentration	Mole/lit	0.25	0.34	0.43
Voltage	Volt	10	13	16
Feed rate	mm/min	0.03	0.05	0.07

2.1 Grey relational technique

This approach provides an expert influence of uncertainty, various inputs and inaccurate data. The grey relational technique approach is an estimation of the full significance of the context difference between residues which is often used to determine the nearly consistent correlation between residues.It becomes an efficient method that explores the association between series through virtue of fewer statistics and to analyze a variety of influences[14].At grey relational analysis, the results collected from its investigations are subsequently standardized. The grey relation coefficients are evaluated to use the datasets and the grey relation rank is measured. This protocol is explained in this section:

The source data becomes first pre-processed. The normalized response values are determined on the basis of the formula given (1) and (2), which refer to the output variables of the predictor variables[1].The desirable performance parameters for MRR is higher the better criteria; The surface roughness and overcut are indeed the lower the better criteria.Inherently, the higher the normalized values, the stronger the output qualities.

$$Higher\ the\ better, A_{pq} = \frac{Bpq - Min\,(Bpq)}{Max\,(Bpq) - Min\,(Bpq)} \quad (1)$$

$$Lower\ the\ better, A_{il} = \frac{Max\,(Bpq) - Bpq}{Max\,(Bpq) - Min\,(Bpq)} \quad (2)$$

The obvious variability in between highest possible normalized number and each normalized number is calculatedusing equation (3). Then, equations (4) and (5) used for calculation of grey relation grade and rank.

$$\emptyset pq = |Apq - \max(A_{pq})| \quad (3)$$

$$Grey\ relation\ grade, \theta pq = \frac{Min\,(\emptyset pq) + (0.5 \times Max\,(\emptyset pq))}{\emptyset pq + (0.5 \times \max\,(\emptyset pq))} \quad (4)$$

$$Grey\ relation\ grade, \gamma j = \frac{\sum_{i=1}^{n} \theta pq}{n} \quad (5)$$

3 Results and discussion

The experiment was designed according to the 3-level L9 Taguchi orthogonal model. The model was compiled and analyzed utilizing the statistical tool Minitab 19. Throughout the experiment, three variables at three different levels have been perceived. Figure 2 shows the work material after machining.

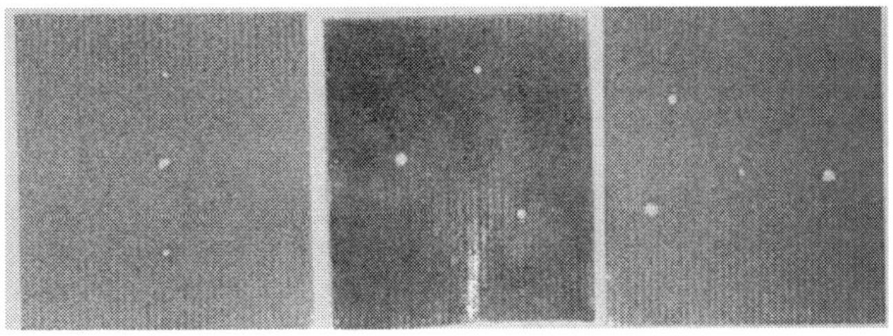

Figure 2. Work material after machining

The MRR and overcut are being determined by the equations (6) and (7) given.The original and final weights of the work material are determined by an automated measuring system with a precision of 0.001 g.The roughness of machine workpieces is reported utilizing SRG4500 hardness test machine.The optical microscope is used for measuring the diameter of specimens. The measured results of MRR, SR and overcutare included in Table 3.

$$Material\ removal\ rate = \frac{Original\ weight - Final\ weight}{Density \times Machining\ time} \quad (6)$$

$$overcut = Machined\ hole\ diameter - Initial\ diameter\ of\ hole \quad (7)$$

Table 3. Experimental results showing MRR, SR and overcut values

Run	Electrolyte Conc (mole/lit)	Voltage (volts)	Feed rate (mm/min)	MRR (mm^3/min)	SR (µm)	Overcut (mm)
1	0.25	10	0.03	0.0371	0.5592	0.13543
2	0.25	13	0.05	0.0397	0.4944	0.09862
3	0.25	16	0.07	0.0890	0.3861	0.02261
4	0.34	10	0.05	0.0403	0.6149	0.15223
5	0.34	13	0.07	0.0794	0.5313	0.07552
6	0.34	16	0.03	0.0801	0.4957	0.11263
7	0.43	10	0.07	0.0746	0.6449	0.06852
8	0.43	13	0.03	0.0574	0.6441	0.04072
9	0.43	16	0.05	0.0913	0.5442	0.03231

3.1 Impact of the operating conditions on the calculated response

Taguchi's model of the test requires lower tests to research the influence of the input variables. Taguchi is a simple, time-saving approach which could be extended closely towards any technical challenges[15], [16].While its MRR is very weak relative to traditional tooling, it is really a favored choice for heavy metal contamination-to-cut products like nickel-based Inconel 718 alloys. The efficacy of the machining can be calculated by means of the material removal rate, so it is important to know the effect of the machining parameters throughout the micro ECM of Inconel 718 alloy. Table 4 gives the material removal rate responses; the voltage is the best response output variable compared to feed rate and electrolyte concentration. Figure 3 shows the MRR main effect plot. When the electrolyte concentration and voltage increase the MRR value also increases. But there is a deviation in feed rate versus MRR.

Table 4. Response Table for MRR

Level	Electrolyte Concentration	Voltage	Feed Rate
1	-25.88	-26.35	-25.12
2	-23.94	-24.95	-25.57
3	-22.72	-21.24	-21.85
Delta	3.16	5.11	3.72
Rank	3	1	2

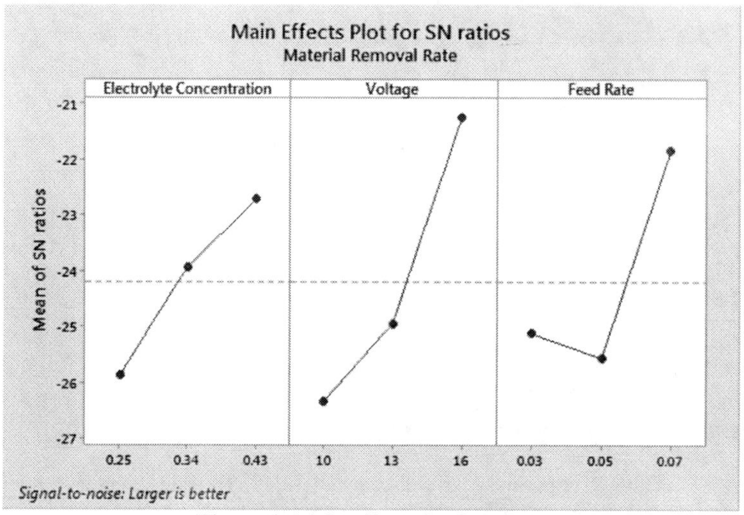

Figure 3. Main effect plot for MRR

A surface roughness including its machining process can indeed be calculated from its textured surfaces. That surface texture achieved during the machining process has been of high caliber that neither additional finishing procedure is needed again for the workpiece.From the response table 5 explains the voltage being the prominent parameter affecting the SR. Figure 4 shows the main effect plot for surface roughness. The electrolyte concentration increases from 0.25 to 0.43 mole/lit, the SR value gets reduced. But the SR increases during the voltage and feed rate increases.

Table 5. Response Table for SR

Level	Electrolyte Concentration	Voltage	Feed Rate
1	6.478	4.361	4.988
2	5.271	5.144	5.209
3	4.305	6.549	5.856
Delta	2.172	2.188	0.868
Rank	2	1	3

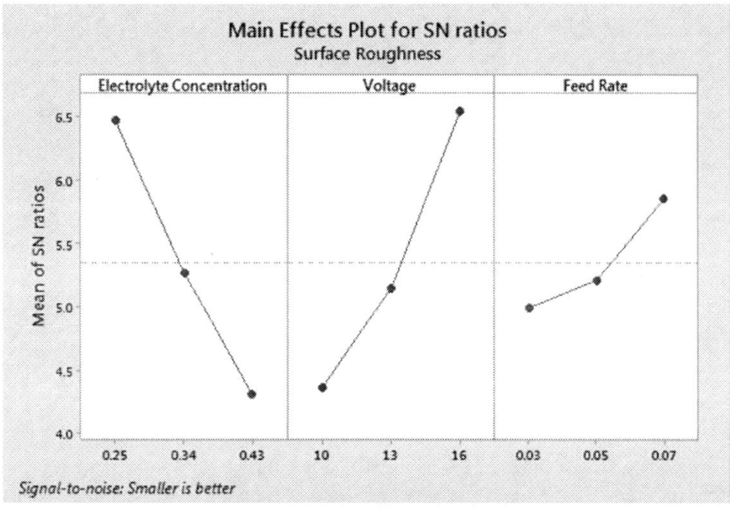

Figure 4. Main effect plot for SR

Tests are performed from an inverted optical microscope. The main impact graph for overcut as seen in Fig. 5, it is shown that while the chloride concentration rises from 0.25 to 0.34 mole/liter, the overcut rises, but perhaps the overcut reduces when the chloride content increase. A drop in the overcut with an improvement in feed rate and voltage also observed.

Table 6. Response Table for overcut

Level	Electrolyte Concentration	Voltage	Feed Rate
1	23.47	19.00	21.38
2	19.25	23.45	22.09
3	26.97	27.23	26.21
Delta	7.72	8.23	4.83
Rank	2	1	3

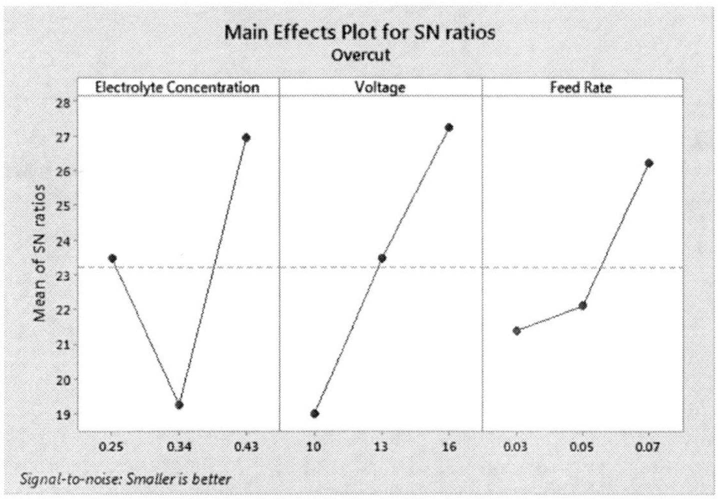

Figure 5. Main effect plot for Overcut

3.2 Multi objective optimization

The grey relation grades are calculated for each run and presented in table 7. The maximum value of grey relation grade noted as rank 1. The optimum value found in run 3 at electrolyte concentration is 0.25 mole/lit, voltage is 13 volts and feed rate is 0.07 mm/min.

Table 7. Grey relation grade and the experimental runs rank

Run	Electrolyte Conc (mole/lit)	Voltage (volts)	Feed rate (mm/min)	Grey Relational Grade	Rank
1	0.25	10	0.03	0.6245	4
2	0.25	13	0.05	0.7456	1
3	0.25	16	0.07	0.5794	6
4	0.34	10	0.05	0.5409	7
5	0.34	13	0.07	0.5957	5
6	0.34	16	0.03	0.7070	2
7	0.43	10	0.07	0.4618	8
8	0.43	13	0.03	0.4488	9
9	0.43	16	0.05	0.6227	3

3.3 confirmatory examination

The best result for a control parameter influencing the effects is being calculated by using the overall grey relation grade. Theconfirmative test has performed out once the optimum state became determined as well as the measurement was confirmed and the related MRR, SR and overcut are reported. The relation between the actual and optimal value of the input variables is seen in Table 8.From this comparison, it would be claimed that there is an increase in the grey ratio of 0.1180 between the actual run number 1 and the optimal run number 2.

Table 8. Confirmatory examination result

Responses/ Level	Actual (Run 1)	Optimum (Run 2)	% of improvement
MRR	0.0371	0.0398	7.8%
SR	0.5592	0.4939	11.7%
Overcut	0.13543	0.09858	27.2%
Grey relation grade	0.6245	0.7425	18.9%
Improvement on grey relation grade		0.1180	

4 Conclusions

This research worked towards multi-objective optimization of Inconel 718 alloy microelectrochemical machining utilizing Grey Relational Assessment dependent on Taguchi. The research outcomes were taken from its relevant study analysis and testing:

- The voltage is the primary aspect for all the necessary output parameters.
- When the electrolyte concentration and voltage increase the MRR value also increases. But there is a deviation in feed rate versus MRR.
- The electrolyte concentration increases from 0.25 to 0.43 mole/lit, the SR value gets reduced. But the SR increases during the voltage and feed rate increases.
- While the chloride concentration rises from 0.25 to 0.34 mole/liter, the overcut rises, but perhaps the overcut reduces when the chloride content increase. A drop in the overcut with an improvement in feed rate and voltage also observed.
- The optimum grey relation was identified to be 0.25 mole /liter electrolyte concentration, 13 V voltage and 0.07 mm / min feed rate.

- The confirmatory examination was performed to verify the findings obtained by the grey relational method.From this examination, it would be claimed that there is an increase in the grey ratio of 0.1180 between the actual run number 1 and the optimal run number 2.
- Its result of such a current study also highlighted a key for the manufacturers to improve productive capacity as well as quality in the handling of Inconel 718 alloy using a micro-electrochemical machining.

Acknowledgments
This work was supported by the Sri Krishna College of Engineering and Technology, India - 641008 as part of the motivation towards research and development of modern technologies.

References
[1] Madhankumar S, Manonmani K, Rajesh S, Balamurugan R and Harikrishnan M 2018 Optimization of Micro Electrochemical Machining of Inconel 625 using Taguchi based Grey Relational Analysis *Int. J. Appl. Eng. Res.* **13** 6771–9
[2] Manikandan N, Kumanan S and Sathiyanarayanan C 2015 Multi response optimization of electrochemical drilling of titanium Ti6Al4V alloy using Taguchi based grey relational analysis *Indian J. Eng. Mater. Sci.* **22** 153–60
[3] Sharma P, Chakradhar D and Narendranath S 2016 Effect of wire diameter on surface integrity of wire electrical discharge machined Inconel 706 for gas turbine application *J. Manuf. Process.* **24** 170–8
[4] Geethapriyan T, Kalaichelvan K and Muthuramalingam T 2016 Influence of Coated Tool Electrode on Drilling Inconel Alloy 718 in Electrochemical Micro Machining *Procedia CIRP* **46** 127–30
[5] Jeykrishnan J, Vijaya Ramnath B, Chenbhaga Ram N, Nikhil K, Naveen Babu R and Naveen D 2018 Investigation on the Influence of Performance Characteristics on Machining Inconel 625 alloy in Electro-Discharge Machining (EDM) *Mater. Today Proc.* **5** 20449–54
[6] Nayak A A, Gangopadhyay S and Sahoo D K 2016 Modelling, simulation and experimental investigation for generating 'I' shaped contour on Inconel 825 using electro chemical machining *J. Manuf. Process.* **23** 269–77
[7] Zeng Z, Wang Y, Wang Z, Shan D and He X 2012 A study of micro-EDM and micro-ECM combined milling for 3D metallic micro-structures *Precis. Eng.* **36** 500–9
[8] Madhankumar S, Balamurugan R and Rajesh S 2019 Investigations on austenitic nickel-chromium based super alloys - Inconel 625 and Inconel 718 from material removal rate in micro electrochemical machining *AIP conference Proceeding* vol 2128 (AIP Publishing) p 040009
[9] Rajurkar K P, Sundaram M M and Malshe A P 2013 Review of Electrochemical and Electrodischarge Machining *Procedia CIRP* **6** 13–26
[10] Kumar M, Vaishya R O, Oza A D and Suri N M 2020 Experimental Investigation of Wire-Electrochemical Discharge Machining (WECDM) Performance Characteristics for Quartz Material *Silicon* **12** 2211–20
[11] Hosseini E and Popovich V A 2019 A review of mechanical properties of additively manufactured Inconel 718 *Addit. Manuf.* **30** 100877
[12] Selvan T A, Madhankumar S and Manonmani K 2018 Optimization of Material Removal Rate and overcut in Micro Electrochemical Machining of Inconel 718 using Taguchi Analysis *Int. J. Pure Appl. Math.* **119** 869–77
[13] Qu N S, Zhang Q L, Fang X L, Ye E K and Zhu D 2015 Experimental Investigation on Electrochemical Grinding of Inconel 718 *Procedia CIRP* **35** 16–9
[14] Manikandan N, Kumanan S and Sathiyanarayanan C 2017 Multiple performance optimization of electrochemical drilling of Inconel 625 using Taguchi based Grey Relational Analysis *Eng. Sci. Technol. an Int. J.* **20** 662–71
[15] Madhankumar S, Balamurugan R, Rajesh S and Senthilkumar K M 2020 Fabrication of Al6063 alloy , silicon carbide and boron glass powder metal matrix composites in stir casting process and analysis the impact of process variables on mechanical properties *Mater. Today Proc.*

[16] S. Rajesh, S. Madhankumar, M. Harikrishnan T V 2018 Effect of Processing Parameters on Metal Matrix Composites in Stir Casting Process Using Taguchi Technique *Int. J. Res. Appl. Sci. Eng. Technol.* **6** 1610–5

IOP Conf. Series: Materials Science and Engineering 1059 (2021) 012009 doi:10.1088/1757-899X/1059/1/012009

Design and fabrication of Automated mechanism for chemical pre-treatment process in Powder coating

Ananthi K[1*], Priyadharsini S[2], Brundhavani G[2]

[1]Assistant Professor, Department of Mechatronics Engineering, Sri Krishna College of Engineering and Technology
[2]Assistant Professor, Department of Mechanical Engineering, Rajalakshmi Institute of Technology
[3]UG Scholar, Department of Mechatronics Engineering, Sri Krishna College ofEngineering and Technology

[*]Corresponding author, Email Id: ananthik@skcet.ac.in

Abstract. The paper is intended to design and fabricate an automated mechanism for chemical pre-treatment process in powder coating for avoiding the manual handling of metal substrates in chemical pre-treatment processing in order to remove dust, to avoid corrosion and to provide good adhesion for powder coating. For this, a gantry crane structure is proposed where two vertical columns and a horizontal beam setup are designed to which a movable carriage is mounted across the horizontal beam. The movable carriage is fitted with a pulley mounted by a rope, the end of the rope is connected to a hanger which has a workpiece holder to which a metal substrate is hanged. The process is controlled by a programmable controller which is programmed with appropriate delays and according to the process designated, the hanged workpiece will move up and down and get dipped into the chemical tank. Once the delay was completed, the carriage automatically moves horizontally to the next tank. The gantry crane structure with movable carriage having pulley and programmable controller are well suited for automating pre-treatment process and it can safeguard manual workers' hands from hazardous chemicals in the tank and also achieving the exact delay operations of each tank process.

1 Introduction

Pre-treatment is the method of chemical processing of different forms of metal, before adding a powder-coated surface finish. Effective pre-treatment of the surface is important in all cases of powder coating to ensure long-lasting results [1]. Pre-treatment is crucial to consistency, just as important to the finished powder-coated product's efficiency as the materials used in its production, design, and the conditions in which it is used. It is absolutely important at Powder Coating Services that high pretreatment levels are maintained to ensure a consistently high-quality end product [2]. Pre-treatment is done to prepare the surface. Hereby pre-treatment means metal pre-treatment as the powder coating is predominantly applied to metals. Preparation of surfaces includes cleaning by mechanical or chemical ways. Mechanical cleaning consists of techniques such as scratch brushing and sandblasting. It not only removes the surface impurities but also cracks and surface defects by abrasive action. Cleaning is very good, but the coating must be done immediately because the cleaned surface is in a highly reactive state and corrosion occurs very fast[3,4].Chemical cleaning includes the removal of dust, oil and grease and oxidation chemicals from the surface. The chemicals may be wiped, sprayed or dipped. Generally, chemical pre-treatment processing order to remove dust, to avoid corrosion and to provide good adhesion for powder coating. The nature of the chemicals used depends on the base metal [5]. Conversion coating application is used for three purposes: 1) Providing temporary corrosion protection in the process before Powder coating 2) Encourage the proper adhesion of the powder coating to the substrate. 3) Deliver Corrosion under the paint and thus prolong the life of the powder coating[6].

2 Literature Review

A kind of durable powder coating without any solvent is a powder coating. It is commonly used in the metal industry because of its excellent application performance and eco-friendly usage. The use of powder coatings has been introduced very rapidly in recent years, and the criteria for practical powder coatings have also been continuously improved[7]. Pre-treatment prepares a component before powder coating is added to improve adhesion and corrosion resistance. Phosphating is a method of surface treatment by which the surface of virgin steel is converted into metallic phosphate and is commonly used before painting. Phosphate coatings are produced by the chemical reaction of base metal with an aqueous solution of phosphoric acid and phosphates (Ions of Bivalent metal Zn^{+2}, Mn^{+2}, Fe^{+2}). Phosphating can be done popularly by manganese, iron and zinc phosphate[8]. In [9], authors described the coating of phosphates on stainless steel and mild steel by electrochemical method under galvanostatic conditions and the results showed that mild steel with phosphate coating had better corrosion resistance when compared to the mild steel without phosphate coating. Both traditional high zinc phosphating formulations and contemporary tricationic phosphating formulations on steel surfaces are contrasted with the characteristic features of phosphate coatings in [10]. This is suitable for the automobile finishing pretreatment process with cathodic electrodeposition primer. On analyzing, the corrosion protection of the phosphated and painted steel panels for automobiles by salt spray test and electrochemical impedance spectroscopy, the authors found that modern tricationic phosphating formulations had effective corrosion resistance. In [11], Zayed and et al. described the effect of zinc phosphating on the coating process and explained the importance of adjustment of phosphate, iron and zinc concentrations for improving the quality of phosphate coating. There are various pretreatment processes like sandblasting, grinding and polishing for applying phosphate coating on alloys. For the better formation of corrosion protective low surface roughness is needed which is not possible by sandblasting and superior in grinding and polishing [12]. Another important thing to notice about such coatings is, they do not offer complete resistance to atmosphere and are principally used as an adherent base/primer coat for paints. From a metal surface, the surface is cleaned with water for removing unreacted conversion coating chemicals and a post-treatment application. Post-treatment increases resistance to corrosion and humidity about two to tenfold when compared to conversion coatings without final rinses. Post-treatments are usually chromic acid-based. The following are the different steps involved in the process of chemical pre-treatment[13], [14].

1. Alkali degreasing
2. Cold swilling
3. De-rusting
4. Cold swilling
5. Hot phosphating
6. Cold swilling
7. Sealing/Passivation

The chemical composition and duration of each step is given Table 1.

Table 1. Standard Steps in Chemical Pre treatment Process

S.No.	Tank number	Process	Tank Capacity (in Liters)	Chemical Composition	Duration
1.	Tank 1	Alkali degreasing	7000	NaOH – 6-8% Tri sodium phosphate- 2 - 3%	15-20 minutes
2.	Tank 2	Cold Swilling	7000	Static water	2-3 times

3.	Tank 3	De-rusting	7000	H3PO4 – 25 -30%	5-7 minutes
4.	Tank 4	Cold Swilling	7000	Static water	2-3 times
5.	Tank 5	Hot Phosphating	7000	H3PO4+$Zn_1(PO4)_1$ – 30%	10-15 minutes
6.	Tank 6	Cold Swilling	7000	Static water	2-3 times
7.	Tank 7	Sealing (Passivation)	7000	Chromium tri oxide – 0.025 – 0.05%	25-30 seconds

3 Proposed method

This paper deals with the fabrication of automated material handling mechanism for chemical pre-treatment process in powder coating to avoid the manual handling of metal substrates in chemical pre-treatment processing. This intends to remove dust, to avoid corrosion and to provide good adhesion for powder coating. The gantry crane structure is proposed where the crane chassis has two vertical columns and a horizontal beam setup is designed which is a fixed structure placed on the floor to which a movable carriage is mounted across the horizontal beam. The movable carriage is attached with a pulley mounted by a rope, the end of the rope is connected to a hanger has a work piece holder to which the work piece (metal substrate) is hanged. The movable carriage is driven by a driver powered by D.C. motor using the power supply. The process is controlled by a programmable controller programmed with appropriate delays, and according to the process designated, the hanged work piece will move up and down and get dipped into the chemical tank, and once the delay was completed, the carriage automatically moves horizontally to the next respective tanks and the chemical process will take place inside the chemical tank. By doing so, it is possible to eliminate the manual pre-treatment process, and safeguard the manual workers hand from the chemicals in the tank and also achieving the exact delay operations of each tank process. The block diagram of proposed method is shown in Figure 1.

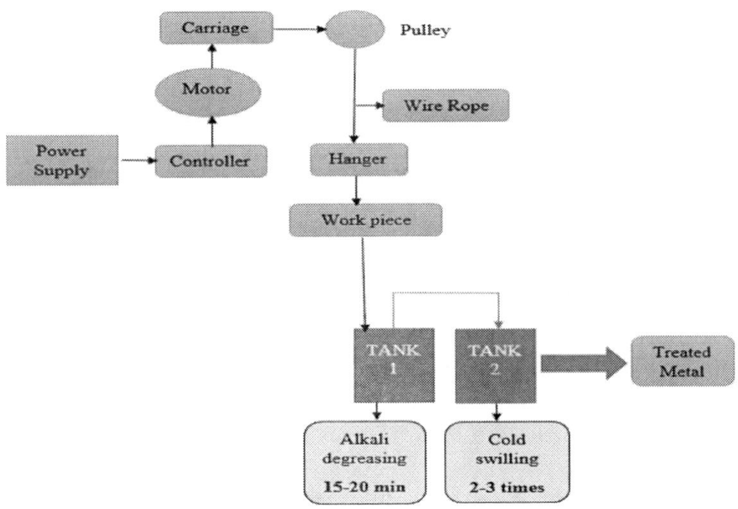

Figure 1. Block Diagram of proposed method

The various components used in the fabrication of prototype and their functions are discussed below.

3.1 Beam

While attached to a crane, the lifting beam is generally best suited for outdoor construction use, and the gantry crane is its standalone lifting device ideally suited for large warehouses or flat outdoor spaces.

3.2 Controller

The controller selected for fabricating prototype model is LilyPad Arduino.The ATmega168V (the low-power version of the ATmega168) or the ATmega328V are the basis of the LilyPad Arduino Main Board. Leah Buechley and SparkFun Electronics designed and built the LilyPad Arduino. By simply plugging the board into a USB port on the device or 5V wall charger, the built-in battery port makes it easy to choose a LiPo battery that fits the run time specifications and recharges the battery. It features include 5 Digital I/O pins,4 Analog pins, ATMega32U4, Built-in ON/OFF switch, Built-in power supply socket (JST connector) for a 3.7V LiPo battery and charging circuit (no additional battery charger needed), Simplified layout with fewer pins, giving more space for sewing or less complex projects, Micro USB connection instead of FTDI header pins

3.3 Motor

The movable carriage is driven by a driver powered by D.C. motor. Motor is one of a class of rotary electric motors that converts mechanical energy into direct current electrical energy. The forces generated by magnetic fields depend on the most common forms. Almost all types of DC motors have some internal function to regularly adjust the direction of current in a portion of the motor, either electromechanical or electrical. The first type of engine commonly used was DC motors, since they could be operated by existing direct-current lighting power distribution system.The speed of a DC motor can be regulated over a wide range, by using a variable input voltage or by adjusting the current strength of its field windings. In tools, toys, and equipment, small DC motors are used. The universal motor is a lightweight brushed motor used for compact power tools and equipment, but can run on direct current. Currently, larger DC motors are used in electric vehicle propulsion, elevator and hoist propulsion, and in steel rolling mill drives.

3.4 Crane Chassis

The crane chassis has a girder, pillars, braces and base. Some important considerations should be taken into account when designing the chassis for the gantry crane. It is important to note that the moment is usually spread evenly around the girder when the load is on the central axis of the crane and when the load is on the other side of the crane, the maximum on one side. Strong support at both legs is therefore required to keep the system in balance. In addition to braces at the top half and bottom half of the pillars, the lateral braces at the girder provide a solution to such a problem. On the other hand, the system's center of gravity should be positioned as near as possible to the bottom of the crane chassis structure, so that the center of mass allows the system to stay in balance, regardless of the size of the load, which needs to be less than the total mass of the crane. This can be achieved by designing the base of the chassis structure to cover an area that maintains an appropriate center of gravity position along the central axis of the chassis[15]. Other factors that need to be considered for the design of the crane are the resistance due to load swing, the influence of the wind and acceleration of the moving masses. Nevertheless, given the lack of wind impact due to indoor activity and the option of low speeds for load hoisting and travel, these factors can be ignored. Another important aspect is maintaining operational health, as a simple error can lead to catastrophic accidents, most of which are caused by unsecured load and exceeded load power. The various parts of the chassis are regular steel pieces that are joined to form the final chassis framework using welding and bolted joints. Using the required safety criteria, all structural components, welding and bolted joints are tested for shear stress, crushing stresses, and bending moment stresses.

3.5 Trolley Mechanism

The trolley mechanism has a wheeled carriage and hoisting mechanism. These are fitted with a ratchet wheel braking system for supporting the load placed at high places. The carriage drive mechanism consists of an electric motor, which drives an underhung wheeled carriage through a fixed coupling that moves on the rails that are attached to the girder. The mechanism of load raising and lowering consists of a drum, pulley system, speed gear reduction and electric motor. It also contains a hook that is meant to be connected to the load and steel wire ropes. All parts and mechanisms are selected using load-based formulae so that they are selected to be consistent with the requirements recommended for manufacturing gantry cranes.

3.6 Crane Mobility System

The basic function of the mobility system is to allow the user to move the crane within the area of operation. The crane's mobility system consists of two main systems; the crane driving system which is used to drive the crane forward or backward and the crane steering system which is used to guide the crane movement. That system relies on two electric motors to accomplish its operation. Using the control unit, the mobility system is controlled via the electrical wire system and the control devices. A braking mechanism is used to slow the crane or lower its speed.

4 Design Calculation

Material choice for beam and column isMild Steel.
Properties of Mild Steel to be used:
Density = $7850\ kg/m^3$
Young's Modulus = $2.08\times10^5\ N/mm^2$
Poisson's Ratio = 0.3

4.1Weight acting on the Beam

One work piece dimension = $12.5 \times 7.5 \times 0.5$ cm
One work piece Volume = $46.875\times10^{-6}\ m^3$
For three work piece, V = $140.65\times10^{-6}\ m^3$
Mass = Density × volume
= $7850\times140.625\times10^{-6}$
= 1.103kg (1)

For hanger,
Considering Golden Ratio for Rectangle, $\frac{b}{d}$ = 1.6
Here, b = 2cm
So, d = 1.24 cm (2)
L × b ×t = $11\times2\times1.24$
V = $27028\times10^{-6}\ m^3$
Mass = $7850\times27.28\times10^{-6}$
= 0.21 kg
Total mass = 1.103+0.21+2
Mass = 3.51 kg (3)
Force = mass × gravity
= $3.5\times9.81 \Rightarrow 35$ N (4)

∴ Total weight acting on beam = 35 N

4.2 To find whether the Beam Design is Safe/Not

Moment at A,
$(R_B\times0.75) = (35\times0.375)$
$\therefore R_B = 17.5$ N
$R_A + R_B = 35$
$R_A + 17.5 = 35$

$\therefore R_A = 17.5$ N (5)

To find Shear Force,

S.F at A, $F_A = 17.5$ N

(Shear Force between A and C is constant and equal to 17.5N)

S.F at C, $F_C = 17.5 - 35 = -17.5$ N

S.F at B, $F_B = -17.5$ N

Shear Forcebetween C and B is constant and equal to -17.5N

To find Bending Moment,

$M_A = 0$

$M_B = 0$

$M_C = 17.5\times0.375 \Rightarrow$ Bending Moment, $M_C = 6.5625$ Nm (6)

To find Maximum bending stress,

$\frac{\sigma}{y} = \frac{E}{R} = \frac{MB}{I}$ (DDB. 7.1)

For rectangle, from Golden Ratio,

Here, b =5cm, $\frac{b}{d} = 1.6$

b $= 1.6\times5 = 0.08$m

$\therefore \sigma_b = \frac{MB \times y}{I}$

$= \frac{6.5625\times0.025}{8.3\times10-7}$

$\therefore \sigma_b = 49.76$ N/mm^2 (7)

Allowable bending stress $= 200$ N/mm^2

$\therefore$ Calculated value < Allowable value

Therefore, the beam design is safe.

4.3 To find whether the Column Design is Safe/Not

Here, b = 5cm

For Rectangle, considering the Golden ratio, $\frac{b}{d} = 1.6$

d $= \frac{5}{1.6} = 3.125$cm

Here, L $= 70$cm

$\frac{L}{D} = \frac{70}{3.125} = 22.4$cm [$\therefore$Long column] (8)

Since it is long column, Euler's formula is used.

Consider End condition = both ends are fixed.

Crippling load, P $= \frac{4\pi^2 E\times I}{L^2}$ $[I = \frac{bd^3}{12} = \frac{5\times1.325^3}{12} = 127.1\times 10^3 mm^4]$

$= \frac{4\times\pi^2\times2.08\times10^5\times127.1}{(700)^2}$

P $= 2.12 \times 10^5$ N (9)

Crippling Stress, $\sigma = \frac{P}{A} = \frac{2.12\times10^5}{700*50}$

Crippling Stress, $\sigma = 21.0571$ N/mm^2 (10)

Allowable crippling stress, σ=40N/mm^2

The calculated crippling stress is less than the allowable crippling stress. So, the Column Design is Safe.

4.4 Calculation of Motor selection for Carriage movement

Torque, T $= 2F \times R$

F $= \mu \times R_N$

$= 0.3 \times 17.5$ [Coefficient of friction, $\mu = 0.3$]

F $= 5.25$ N (11)

T $= 2 \times 5.25 \times 0.04$ [Radius of wheel = 0.04m, Standard trolley wheel radius which withstand 35N]

$= 0.42$ Nm (12)

Power, P $= T \times \omega$ [$\omega = \frac{V}{R} \Rightarrow \frac{0.1}{0.04} \Rightarrow 2.5$ rad/s]

$= 0.42 \times 2.5$

$= 3.5\ W \Rightarrow \frac{3.5}{0.8}$

Power,P $= 5W$ [Efficiency, $\eta = 0.8$] (13)

To find Speed in rpm, consider,

$$\omega = \frac{2\pi N}{60} = N = \frac{\omega \times 60}{2\pi}$$

$$= \frac{2.5 \times 60}{2\pi}$$

$\Rightarrow$ Speed, N $= 23.87$ rpm ≈ 30 rpm (14)

From the design calculation, weight acting on the beam is determined. Based on the bending stress, the beam design and based on crippling design column design are found safe. The specification of motor required for carriage movement is also found.

5 Results and Discussion

This paper deals with designing and implementing a light duty crane, which is used to handle materials operations within small workshops. All the drawings, design calculation reports presented in this paper have been used to develop a prototype (with beam dimensions 850x50mm and column dimensions 750x50mm) that envisages the actual working of a real time gantry crane where the authors bound to simulate the real time process in the sequence using the sample tanks two in number that shows the simulation of the pre-treatment process of the powder coating mechanism. This prototype when enlarged using the real time dimensions as specified can be used for the pre-treatment of the powder coating process that involves the sequence of all the seven processes. The 3D model of the proposed system is shown in Figure 2. The fabricated model of the automated mechanism for the chemical pretreatment process is depicted in Figure 3.

Figure2. Three-Dimensional Visualization of Prototype

Figure 3. Fabricated model

6 Conclusion

The application of powder coating over the industry is very large. The effects of alkalies on the human during pre-treatment of powder coating are analyzed and came with a solution of automation. The goal of this paper is to increase the powder coating efficiency and protect the workers from harmful alkalies. The solution given here can be done in less time and it is implemented with required enhancements. Also, here the manual pre-treatment process is eliminated and it is helpful to safeguard the manual workers' hand from the chemicals in the tank and also achieving the exact delay operations of each tank process.

References

[1] Bridger R D E L 1983 Pretreatment of Galvanized Surfaces Prior To Powder Coating. *Trans. Inst. Met. Finish.***61** 13–8

[2] Reghuraj A R and Saju K K 2017 Black oxide conversion coating on metals: A review of coating techniques and adaptation for SAE 420A surgical grade stainless steel *Mater. Today Proc.***4** 9534–41

[3] Zaki M H M, Mohd Y and Isa N N C 2017 Surface pre-treatment of aluminium by cleaning, chemical ething and conversion coating *AIP Conf. Proc.***1901** 120006-1 - 12000-6

[4] Tupe P, Thorat M and Lashman K 2019 Overview of Pretreatment Process in powder coating *IRJET* **6(7)** 2118–21

[5] Bensah E C and Mensah M 2013 Chemical pretreatment methods for the production of cellulosic ethanol: Technologies and innovations *Int. J. Chem. Eng.***2013** 1-21

[6] Milošev I and Frankel G S 2018 Review—Conversion Coatings Based on Zirconium and/or Titanium *J. Electrochem. Soc.***165** 127–44

[7] Du Z, Wen S, Wang J, Yin C, Yu D and Luo J 2016 The Review of Powder Coatings *J. Mater. Sci. Chem. Eng.***04** 54–9

[8] Burduhos-Nergis D P, Vizureanu P, Sandu A V and Bejinariu C 2020 Evaluation of the corrosion resistance of phosphate coatings deposited on the surface of the carbon steel used for carabiners manufacturing *Appl. Sci.* **10(8)** 2753

[9] Sinha P K and Feser R 2002 Phosphate coating on steel surfaces by an electrochemical method *Surf. Coatings Technol.* **161** 158–68

[10] Debnath N C 2013 Importance of Surface Preparation for Corrosion Protection of Automobiles *J. Surf. Eng. Mater. Adv. Technol.* **03** 94–105

[11] Zayed M A, Nour El-Dien F A and Hassan A I 2009 Effect of chemical composition of zinc phosphating solution on phosphate coating properties *Egypt. J. Chem.* **52** 699–709

[12] Zhang C, Liu B, Yu B, Lu X, Wei Y, Zhang T, Mol J M C and Wang F 2019 Influence of surface pretreatment on phosphate conversion coating on AZ91 Mg alloy *Surf. Coatings Technol.* **359** 414–25

[13] Pal S 2017 Improvement in the Production of Powder Coating Process of Fan Blades through PDCA Review *Int. J. Res. Appl. Sci. Eng. Technol.* **5** 998–1003

[14] Li A and Liu C 2017 Lightweight design of a crane frame under stress and stiffness constraints using super-element technique *Adv. Mech. Eng.* **9** 1–15

IOP Conf. Series: Materials Science and Engineering 1059 (2021) 012010 doi:10.1088/1757-899X/1059/1/012010

Experimental study and analysis of single slope solar still integrated with Phase Change Material

N Boopalan[1], B Kalidasan[2], D Raj Kumar[3], E Ragupathi[3], M Gurumoorthy[3] and S Praveenraj[3]

[1] Assistant Professor, Department of Automobile Engineering, Kongu Engineering College, Perundurai, Tamilnadu-638060

[2] Assistant Professor, Department of Mechanical Engineering, Bannari Amman Institute of Technology, Sathyamangalam, Tamilnadu-638401

[3] Department of Automobile Engineering, Kongu Engineering College, Perundurai, Tamilnadu-638060

E-mail: boopalan.auto@kongu.ac.in

Abstract. Solar thermal systems are playing important role to encounter environmental issues like global warming, in the last two decades they replace conventional systems to create an eco-friendly environment. Lot of research have been undergone on solar thermal systems to improve their performance and efficiency, but still the area is wide open for improvement because of demand on renewable sources increase. As their is need for increasing the performance of solar thermal systems, phase change materials could serve as energy storing device during abundant available of energy. For solar thermal systems the major challenge is energy availability will not be round the clock, in order to clear this hurdle phase change materials backing served as a better support. The biggest challenge of developing countries today is to provide clean drinking water, as there are many conventional systems to convert saline water into drinking water their main disadvantage is pollution of environment due to harmful emissions. Hence, solar thermal systems with phase change materials are considered as best option in production of clean drinking water due to their operation by renewable energy, compactness and zero emissions. Phase change materials are coupled with solar system experimentally, and their behaviour and performance were characterized with results and observation, here the phase change material paraffin wax is embedded in copper tubes coupled with solar system is employed in production of clean drinking water; their performance and efficiency are compared with single slope solar still without phase change material. The thermal stability, corrosive resistance and impressive properties of paraffin wax and thermal conductivity of copper were tied together to single slope solar still for improving performance and efficiency of clean drinking water production. It is observed solar still without phase change material could generate drinking water from saline water only during day time, but the solar still coupled with phase change material embedded in copper tubes showed better results with 11 percent increase in production of drinking water even during dark hours.

1. Introduction

In most of developing countries the major challenge is their access to clean drinking water because of salinity in their water bodies, only 3 percent of water is suitable for drinking purpose from the 100 percent availability of water, the remaining 97 percent is not suitable for drinking because of higher percentage of dissolved salts in it [1-3]. In order to meet the demand of developing countries for clean drinking water, identifying new sources would be a not clearable task; whereas converting saline water to clean drinking water would be appropriate; one such method to facilitate the need is desalination of saline water. There are various methods for desalination of saline water, but the current scenario demands for a usage of clean and renewable energy sources for the process of desalination due to their emission and environment friendly characteristics, in search for the above need; refinement to one of the ancient technique solar still method to convert saline water to clean drinking water would serve the purpose [4-6]. In solar still two methods are employed namely passive and active methods to increase the distillation of saline water to obtain clean drinking water; various models of solar still like cone, pit and domo are used; however, for all these models basin design mostly would be single sloped or double sloped. Flat plate solar energy coupled with solar still of single basin yielded 24 % higher production of clean drinking water than the uncoupled one. Phase change materials are the key to increase production of drinking water in solar stills, so the ancient method is revived with new scope for efficient way because of phase change materials, solar stills embedded with PCM yielded 31 % percent more than the method without PCM and also with PCM solar stills were able to provide clean drinking water even during dark hours [7-12]. Phase change materials with high capacity of latent heat gives more productivity and there is a subsequent increase in basin temperatures after sunrise, maximum water obtained in a day by distillation in solar still embedded with phase change material with double slope design is 3.1 L/m^2 [13]. Exergy efficiency is one of the important parameter which plays dominant role in water production rate, whenever exergy efficiency is more, coupled with phase change materials and proper insulation would give better results. Paraffin as a phase change material in cascade solar still provided exergy efficiency of about 75 %, several experiments have indicated that usage of phase change materials decreased productivity in day time whereas overall productivity increased by the complement from productivity of clean drinking water even during dark hours [14-19]. Recent studies say when solar distillation systems coupled with thermal energy storage materials is one of the best solution to meet the balance between energy demand and supply, hence solar still with thermal energy storage systems would be the best option to convert saline water into clean drinking water with low maintenance and cost [20-24]. In this work paraffin wax is made as a choice for thermal energy storage because of its high latent heat capacity and non-corrosive property, in order to study its performance and role in production of clean drinking water; various modes are experimented such as paraffin wax embedded in copper tubes compared against solar still without thermal energy storage systems.

2. Materials and Methods

2.1 Design of Solar Still

The solar still design motive is to make it much exergy efficient, as we know area of solar still can influence the clean water production to greater extent with help of exergy balance so that the dimension chosen for outer area is 90*90 cm^2 and height of front column is 30 cm and rear is 80 cm, the outer cover of solar still is made of plywood which prevents heat loss and maximize exergy efficiency. The container dimension of solar still is made to be 75*75cm^2 and an aluminum plate is made as base for container with thickness of 1 cm, it is black coated to increase the heat absorbing capacity of solar still, which is a key parameter for maximizing the efficiency of clean water production from saline water. The outer cover plywood of solar still inclined at 300 is covered with a transparent glass to make it well exposable to sunlight for the better absorption of sunlight throughout the whole day. At the bottom of inclined transparent glass cover, a split PVC pipe is fitted to collect evaporated water, as this work mainly involves in study of paraffin wax embedded in copper tubes performance against ordinary solar still in the production of clean drinking water from saline water, copper tubes of diameter 6 cm are chosen to embed paraffin wax into it. The paraffin wax is melted and filled into copper tubes with help of syringe, copper tubes are placed over the black coated aluminum plate, inlet pipe and outlet pipe are fixed and sealed in the appropriate places; temperature sensors are fixed in needed places to obtain data for exergy calculations. Figure 1 shows the experimental setup of PCM integrated solar still. In order to lower heat loss the aluminum container is covered externally with thermocol sheet and further with saw dust to maximize exergy efficiency which would in turn increase the production of clean drinking water from saline water.

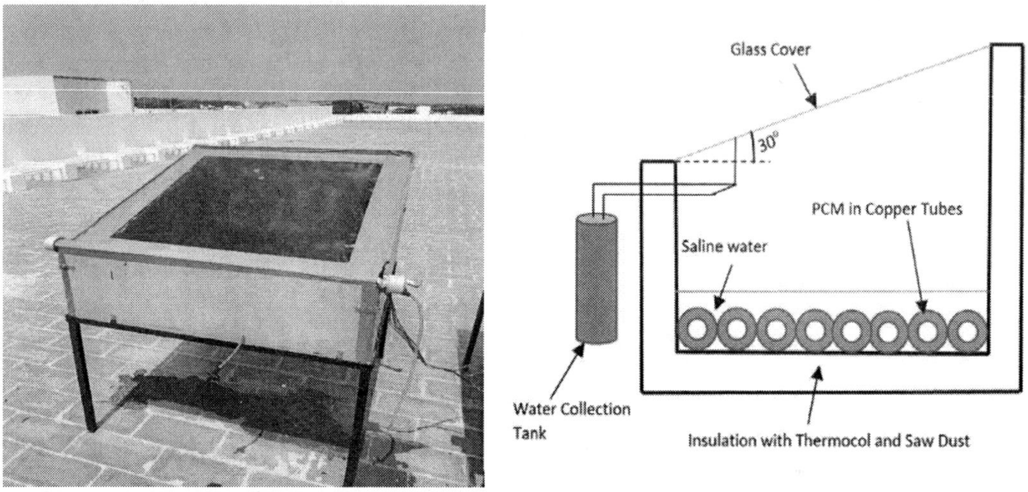

Figure 1 Experimental Set up of PCM integrated Solar Still

2.2 Methodology

The two solar still setup; one without PCM and another integrated with paraffin wax embedded in copper tubes is placed in a proper location, where there will not be any disturbance to absorption of sunlight, because from sunrise to sunset glass surface should be exposed to sunlight in order to absorb maximum of light energy as it is only source of energy. Then all the temperature sensors from both the experimental setup (with PCM and without PCM) are connected to the Arduino controller for obtaining temperature values of components like glass, saline water and copper tube. After both the solar still setup is ready for experiment, saline water is poured through the input tube to the required level, the values are not recorded immediately after commencement of experiment; readings are recorded from the next day because a day is provided for tuning of both the experimental setup. The solar radiation is absorbed by solar still through glass cover through radiation, which heats up aluminum plate, saline water and copper tubes, due to convection saline water molecules start to evaporate and pure water molecules are retained under the bottom of glass cover, as the glass cover is inclined pure water molecules roll down which is collected in a separate tank. The heat absorbed by copper tubes helps to melt stuffed paraffin wax which is encapsulated into it, there will be sufficient heat energy to keep it melt during day hours, whereas during dark hours the absorbed heat is liberated from the paraffin wax into the saline water due to this it starts to evaporate for providing clean drinking water. Saline water evaporates continuously to produce clean drinking water with help of solar energy due this salts are deposited on the surface of the aluminum plate it is removed through the drain pipe. The heat absorbed by the salt water and aluminum plate is almost the same as with PCM without PCM, but we will have clean water output with PCM setup even during night hours, temperature sensors continuously records the data in the data logger which can be used for exergy calculation.

2.3 Energy Balance Equation

Saline water in the basin can be represented by energy balance equation as shown

$$I_1 + Q_{bw} + c_{bw}\frac{dT_{bw}}{dt} = Q_{cw} + Q_{rw} + Q_{ew} + I_2 \quad ----------------(1)$$

In the equation 1 Q_{bw} indicates convective heat transfer from basin to water, heat capacity of water in the basin is represented by C_{bw}, whereas basin water temperature is represented by T_{bw}, evaporative heat transferred from water to glass is indicated by Q_{ew}, heat radiated from water to glass is Q_{rw}, Q_{cw} convective heat transfer from water to glass, solar irradiation received in the still by the water after penetrating through glass cover is I1 and solar irradiation reaching the basin liner surface after passing through glass and water I2 in the still, can be determined as follows:

$$I_1 = (1 - \alpha_g)I \text{ ---------------------------(2)}$$

$$I_2 = (1 - \alpha_g)(1 - \alpha_w)I \text{ ----------------------(3)}$$

In two modes heat is transferred from basin saline water to the glass through radiation and convection. The heat transferred through convection from basin saline water surface to glass cover happens through humid air, it is expressed by the following equation

$$Q_{Cw} = h_{Cw}A_w(T_{bw} - T_g) \text{ --------------------------- (4)}$$

Coefficient for convective heat transfer from basin saline water to glass cover is indicated by h_{cw}, Cross sectional area of water basin liner is Aw, and glass cover temperature is indicated by T_g. The difference in temperature of saline water in basin liner and glass cover arises because of radiative heat transfer, which can be estimated using Stefan – Boltmanz's law

$$Q_{rw} = h_{rw}A_w(T_{bw} - T_g) = \varepsilon_{eff}A_w\sigma(T_w^4 - T_g^4) \text{ ------------(5)}$$

$$h_{rw} = \varepsilon_{eff}\sigma(T_w^2 - T_g^2)(T_w - T_g) \text{ ---------------------(6)}$$

Heat transfer coefficient of radiative heat transfer from basin saline water to glass cover is h_{rw}, Stefan-Boltzman's constant $\sigma = 5.67 \times 10^{-8}$ W/m^2K^4, the effective emittance of saline water surface to the glass cover is ε_{eff}. A part of heat energy is utilized for evaporating saline water from the basin that is can be represented by the following equation

$$Q_{ew} = h_{ew}A_w(T_{bw} - T_g) \text{ -------------------------(7)}$$

3. Results and Discussion

3.1 Time vs Solar Radiation

As the process of distillation of saline water to produce clean drinking water using solar stills require heat energy to drive the process as a backbone; because of that focus on solar radiation intensity and their absorption increases, the experimental works were carried during month of January to March 2020, so we observe close values in the solar intensity. The required values are noted and plotted as a graph against time, in X axis time is taken and in Y axis calculated solar radiation is plotted, from the graph it is evident that in the month of March solar radiation is higher in the range of 900 W/m^2 during high intensity time of 2 to 3 pm in comparison with other two months, this effect plays a role in distillation of saline water to produce clean drinking water.

3.2 Glass Water Temperature vs Time

Temperature difference of glass and saline water in the basin is an important parameter in distillation of saline water, hence temperature of glass and saline water are recorded to have a clear view in the production of clean drinking water from saline water. Figure 2 is plotted between temperature of glass and water with time interval of 2 hrs. In X axis time interval is take with 2 hours difference and in Y axis temperature of glass cover and saline water is taken, from the plots it is observed that temperature of glass cover is in the range of 310 K in the experimental setup without PCM and the range 315 K in the experimental setup with PCM, temperature of saline water in the basin is of the range 350 K in the setup without PCM and the range 360 K is observed in the setup with PCM during 2 pm to 3 pm. When the temperature difference between saline water and glass cover is more, the distillation of saline water will be more, when the setup is integrated with PCM this difference tends to increase leading to more clean water production from saline water and also from the graph it is observed total water production is higher when PCM is integrated with solar still this is because it supplies heat energy even in the absence of solar radiation.

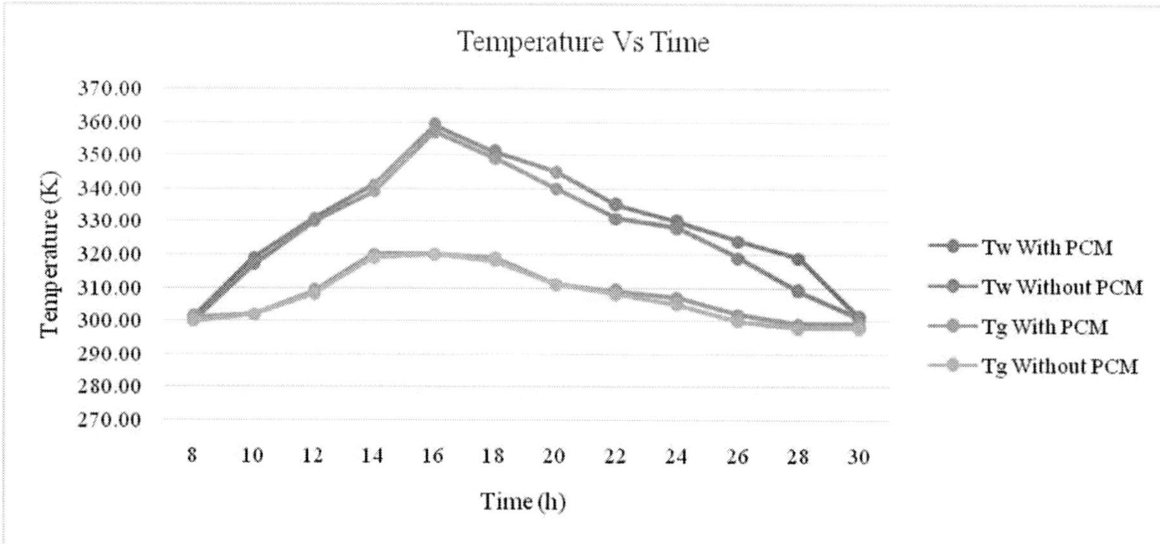

Figure 2: Glass Temperature Vs Time

3.3 Water Yield vs Time

Clean water production from saline water by using solar still depends on various parameters, the major contributor for the production of clean drinking water is solar radiation and the exergy efficiency in order to increase its various techniques are employed in solar still. In the view of implementation of this setup in distillation solar radiation is very important, hence the setup should be placed in the area where solar radiation is uninterrupted for better yield. To have the yield even during dark hours solar still is integrated with phase change materials which act as energy storing device, in this work phase change material is encapsulated in copper tubes in order increase the exergy efficiency so that the yield of clean drinking water will be more. From the Figure 3 it is observed that yield of clean drinking water in day time during intense radiation timing from solar still without phase change material because copper tubes absorb small amount of heat to melt the wax, so yield from solar still with phase change material is slightly lower during peak time of radiation. It is observed that in month of March, higher intensity of radiation is recorded and it resulted in 2264 ml of clean water production from the setup with PCM when accounted for 24 hours, whereas from the setup without PCM it yielded 2080 ml; 11 % hike in the yield from setup with PCM. In the month of January, radiation is slightly which influenced the yield of clean drinking water, 2196 ml was yielded from the setup with PCM and 2004 ml from the setup without PCM. Hence from the plots it's very clear phase change materials increased yield of clean drinking water from saline water in the solar still by a margin of 11 %.

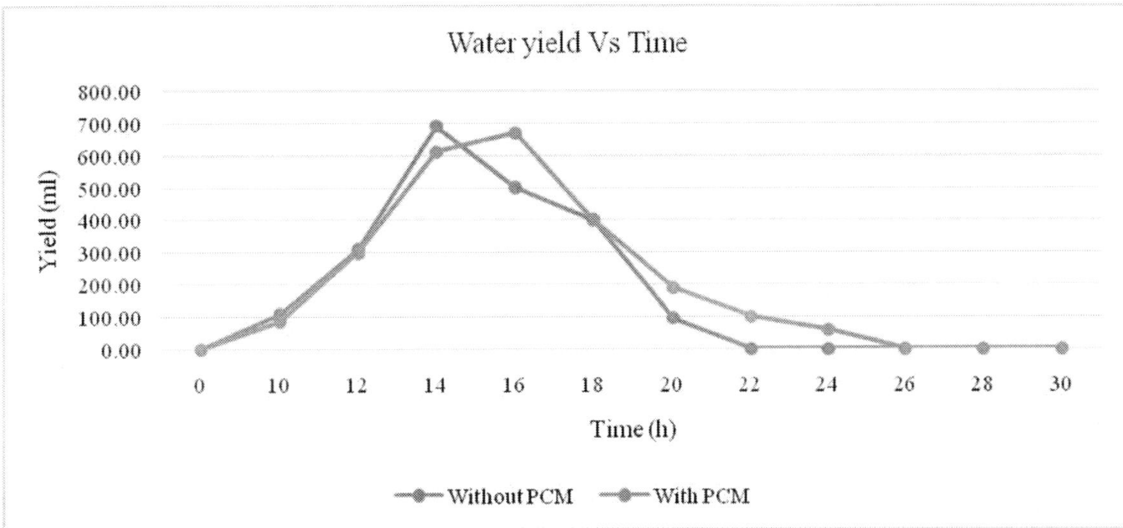

Figure 3: Water yield Vs Time for Solar still with PCM and without PCM

4. Conclusion

The need for clean drinking water arises in several developing countries, to meet that demand is an essential one; viable solution to this need is employing solar still in distillation of saline water as it depends on renewable energy source. The only problem of solar still is its efficiency and inability to produce clean drinking water during night, to overcome these problems solar still is integrated with thermal storage devices. Hence the experimental works on solar still in production of clean water with thermal storage devices is essential to have a clear view in the process, various techniques are employed to increase the yield of water through distillation of saline water. When phase change materials are used certainly there was increase in the yield of clean drinking water but on the other several efficient heat storage and transfer techniques would yield more yield, several techniques are employed and tuned to optimize the parameters those play important role in production of clean drinking water. In this work solar still insulated with thermocol and saw dust; integrated with paraffin wax encapsulated in copper tubes increased the production by 11 percent in comparison with solar still without phase change material. Hence to improve the production various techniques can be employed in future to increase exergy efficiency so that production of clean drinking water would increase from saline water to meet the demand of developing countries.

References

[1] Yousef, M.S., Hassan, H., Ahmed, M. and Ookawara, S., 2017. Energy and exergy analysis of single slope passive solar still under Egyptian climate conditions. *Energy Procedia*, ***141***, 18-23.

[2] Fathy, M., Hassan, H. and Ahmed, M.S., 2018. Experimental study on the effect of coupling parabolic trough collector with double slope solar still on its performance. *Solar Energy*, ***163***, 54-61.

[3] Al-harahsheh, M., Abu-Arabi, M., Mousa, H. and Alzghoul, Z., 2018. Solar desalination using solar still enhanced by external solar collector and PCM. *Applied Thermal Engineering*, ***128***, 1030-1040.

[4] Mettawee, E.B.S. and Assassa, G.M., 2006. Experimental study of a compact PCM solar collector. *Energy*, ***31*****(14)**, 2958-2968.

[5] Dashtban, M. and Tabrizi, F.F., 2011. Thermal analysis of a weir-type cascade solar still integrated with PCM storage. *Desalination*, ***279*****(1-3)**, 415-422.

[6] Ansari, O., Asbik, M., Bah, A., Arbaoui, A. and Khmou, A., 2013. Desalination of the brackish water using a passive solar still with a heat energy storage system. *Desalination*, ***324***,10-20.

[7] Arunkumar, T., Denkenberger, D., Ahsan, A. and Jayaprakash, R., 2013. The augmentation of distillate yield by using concentrator coupled solar still with phase change material. *Desalination*, ***314***, 189-192.

[8] El-Sebaii, A.A., Al-Ghamdi, A.A., Al-Hazmi, F.S. and Faidah, A.S., 2009. Thermal performance of a single basin solar still with PCM as a storage medium. *Applied Energy*, ***86*****(7-8)**, 1187-1195.

[9] Kabeel, A.E., Abdelgaied, M. and Mahgoub, M., 2016. The performance of a modified solar still using hot air injection and PCM. *Desalination*, ***379***, 102-107.

[10] Wolff, M.S., Teitelbaum, S.L., McGovern, K., Pinney, S.M., Windham, G.C., Galvez, M., Pajak, A., Rybak, M., Calafat, A.M., Kushi, L.H. and Biro, F.M., 2015. Environmental phenols and pubertal development in girls. *Environment international*, ***84***, 174-180.

[11] Yousef, M.S., Hassan, H., Kodama, S. and Sekiguchi, H., 2019. An experimental study on the performance of single slope solar still integrated with a PCM-based pin-finned heat sink. *Energy Procedia*, ***156***,100-104.

[12] Sonker, V.K., Singh, R.K., Chakraborty, J.P. and Sarkar, A., 2020. Performance assessment of a passive solar still integrated with thermal energy storage and nanoparticle stored in copper cylinders. *International Journal of Energy Research*.

[13] Omara, A.A., Abuelnuor, A.A., Mohammed, H.A. and Khiadani, M., 2020. Phase change materials (PCMs) for improving solar still productivity: a review. *Journal of Thermal Analysis and Calorimetry*, ***139*****(3)**, 1585-1617.

[14] Hedayati-Mehdiabadi, E., Sarhaddi, F. and Sobhnamayan, F., 2020. Exergy performance evaluation of a basin-type double-slope solar still equipped with phase-change material and PV/T collector. *Renewable Energy*, ***145***, 2409-2425.

[15] Malik, M.Z., Musharavati, F., Khanmohammadi, S., Khanmohammadi, S. and Nguyen, D.D., 2020. Solar still desalination system equipped with paraffin as phase change material: exergoeconomic analysis and multi-objective optimization. *Environmental Science and Pollution Research*, 1-15.

[16] Yousef, M.S. and Hassan, H., 2020. Energy payback time, exergoeconomic and enviroeconomic analyses of using thermal energy storage system with a solar desalination system: An experimental study. *Journal of Cleaner Production*, p.122082.

[17] Das, D., Bordoloi, U., Kalita, P., Boehm, R.F. and Kamble, A.D., 2020. Solar still distillate enhancement techniques and recent developments. *Groundwater for Sustainable Development*, 100360.

[18] Sharma, A. and Sachan, V., 2020. Effect of Phase Change Material on the Performance of Solar Still in Winter Season. *International Journal of Advanced Research in Engineering and Technology (IJARET)*, ***11*****(5).**

[19] Sharshir, S.W., Elsheikh, A.H., Ellakany, Y.M., Kandeal, A.W., Edreis, E.M.A., Sathyamurthy, R., Thakur, A.K., Eltawil, M.A. and Hamed, M.H., 2020. Improving the performance of solar still using different heat localization materials. *Environmental Science and Pollution Research*,1-13.

[20] Attia, M.E.H., Driss, Z., Abdelgaied, M., Manokar, A.M., Sathyamurthy, R. and Hussein, A.K., 2020. Performance Evaluation of Modified Solar Still Using Aluminum Foil Sheet as Absorber Cover–A Comparative Study. *Journal of Testing and Evaluation*, ***49*****(5).**

[21] Panchal, H., Mevada, D., Sadasivuni, K.K., Essa, F.A., Shanmugan, S. and Khalid, M., 2020. Experimental and water quality analysis of solar stills with vertical and inclined fins. *Groundwater for Sustainable Development*,100410.

[22] Sanserwal, M., Singh, A.K. and Singh, P., 2020. Impact of materials and economic analysis of single slope single basin passive solar still: A review. *Materials Today: Proceedings*, ***21***, 1643-1652.

[23] Bachchan, A.A., Nakshbandi, S.M.I., Nandan, G., Shukla, A.K., Dwivedi, G. and Singh, A.K., 2020. Productivity enhancement of solar still with phase change materials and water-absorbing material. *Materials Today: Proceedings*.

[24] Raj, G., Prabhansu, D., Kumar, R., Chandra, P. and Saurabh, S., 2020. Experimental study of solar still augmented with low-cost energy absorbing and releasing materials. *Energy Sources, Part A: Recovery, Utilization, and Environmental Effects*, ***42*(1)**,56-65.

IOP Conf. Series: Materials Science and Engineering 1059 (2021) 012011 doi:10.1088/1757-899X/1059/1/012011

Characterization of strength and toughness properties of metalized plastic waste fibre reinforced concrete

C ChellaGifta[1] ,R Christal Sophia[2] and R Ramya[3]

[1] Assistant Professor, Department of Civil Engineering, National Engineering College, KR Nagar, Kovilpatti ,Tamilnadu.

[2] Undergraduate Student, Department of Civil Engineering, National Engineering College, KR Nagar, Kovilpatti ,Tamilnadu.

[3] Undergraduate Student, Department of Civil Engineering, National Engineering College, KR Nagar, Kovilpatti ,Tamilnadu

E-mail: ccgcivil@nec.edu.in

Abstract. Recycling of plastic waste in concrete has drawn much attention nowadays. The influence of metalized plastic waste fibres on M30 grade concrete was investigated in this study. Strength and toughness properties were evaluated and compared with the control concrete made without addition of fibres. Fibre dosages 0%, 0.25%, 0.5% and 0.75% are taken as a parameter of study. The mechanical properties such as compressive strength, split tensile strength and flexural strength were analyzed to understand the performance of the metalized plastic waster fiber reinforced concrete. Load–deflection curves are plotted to estimate the toughness (energy absorption capacity) characteristics of the composite at the age of 28days.Increase in MPWfibre content in concrete improved the split tensile strength upto 54% and flexural strength upto 58%, but marginal improvement in compressive strength is observed. MPWfibre reinforced concrete outperformed in flexural toughness and it exhibited 2.92 folds superior toughness with 0.75% fibre addition when compared to 0.25%.

1 Introduction

Concrete is the most widely used construction material and conventional concrete exhibits high brittle behaviour because of its weak resistance to crack occurrence. The inherent nature of these cementitious materials shows less tensile strength and poor strain capacity [1]. Incorporation of short and randomly distributed fibres in composite improves the tensile strength, toughness and ductility performance and this Fibre Reinforced Concrete (FRC) has got extensive applications in Civil engineering. Fibres enhances the mechanical properties of concrete and they are available in various types such as metallic, polymeric and natural etc [2].Fibre plays a significant role in enhancing the post-cracking response and toughness of the concrete and it is proven through many theoretical and experimental stand points [3].Addition of plastic waste in concrete is a recent research and it is added in various forms namely polyethylene terephthalate bottles (PET), polypropylene and polyethylene bags etc. The development of fibre reinforced concrete with plastic waste is an effective measure in overcoming the serious shortfalls in concrete and it ensures a safe disposal of solid waste collected in huge amount. Plastic waste in concrete improves post - cracking tensile strength, brittleness and ductile behaviour of concrete.

Due to the rapid industrialization and mechanical inventions lot of waste materials are generated and among this plastic waste is non-bio degradable and it is unfit for reprocessing or reuse. Our modern food habit system generates large amount of Metalized Plastic Waste (MPW) and it is one of the recent garbage produced by the food packaging industry. These metalized plastic sheets are made from polymers coated with thin sheets of aluminium as shown in Figure 1[4]. It occupies 40% of the land fill and its rate of degradation is extremely low which persist for a long period causes serious environmental problems.

Hence utilization of this waste product in concrete will provide a better solution for the waste management strategy worldwide. In last few decades several investigations were carried out to explore the properties of various kinds of plastic waste in concrete. Considering the availability of huge amount of metalized plastic waste, an effort has been attempted to study the influence of metalized plastic waste in concrete.

Figure 1. Metalized Plastic waste

Therefore the utilization of post-consumer plastic waste such as PET bottles, polyethylene and polypropylene bags in concrete and the influence of different types of plastic waste on various properties of concrete are clearly understood through experimental studies. The improvement in impact strength and mechanical properties are studied with the incorporation of 20mm long carpet fibres at 0%-1.25% volume fractions [5]. Zhang et al. verified the addition of polypropylene fibres decreases the durability and increases the freeze –thaw resistance with the increase in fibre volume fraction [6]. Thorneycroft et al. studied the mechanical properties of concrete with recycled plastic waste material as a sand replacement [7]. Foti et al. results show the possibility of PET as fibres to improve the ductility of concrete [8]. Saikia et al. evaluated the fresh and hardened properties of plastic aggregate concrete [9]. Yin et al. reported the excellent post–cracking performance of recycled PP fibre in M40 grade concrete [10]. Ghermouti et al. examined the fresh and hardened properties of plastic bag waste fibres in self-compacting concrete and concluded the presence of these fibres delayed the location of the micro cracks. Improvement in flexural and split tensile strength of cement composite is varied between 4% to 74% [11]. Paul Borg et al. investigated the Shredded PET fibres in different lengths mitigated the plastic and drying shrinkage cracks in concrete and it becomes a potential alternate material [12]. Mohammadhosseini et al. combined the MPWfibres and palm oil fuel ash to examine the durability properties of concrete. Research results show the positive interaction between them and significant improvement in durability performance [4]. Fraternali et al investigated the fresh and hardened properties of recycled polyethylene tetraphthalatefibre reinforced concrete (RPETRC) in aggressive sea water environment and air-cured RPETRC specimens highlighted the influence of fibres in concrete [13]. Torgal et al discussed the effect of PET wastes and tyre rubber on fresh and hardened concrete properties [14]. However limited studies are available on the utilization of metalized plastic waste fibers produced from the package of food products. Therefore, this paper focused on the effect of metalized plastic waste on important mechanical properties such as strength, toughness and load- deflection behaviour of concrete.

2 Materials and Methods

2.1 Raw Materials and Mixture proportions

Ordinary Portland Cement 43 grade with specific gravity 3.13 was used in this study. Manufactured Sand (M Sand) passing through 4.75 mm sieve was used as fine aggregate and the specific gravity and water absorption are measured as 2.53 and 2.66 percent respectively. Locally available angular aggregates passing through 20 mm and retained on 12mm sieve is used as coarse aggregate and it confirms to IS 383:1970 specifications [15]. Fineness modulus of coarse and fine aggregate is 4.74 and 3.24. Based on the raw material characteristic the mixture proportion of M30 grade concrete is arrived with reference to IS 10262:2000 guidelines [16]. Mixture proportion used in this investigation is 1:1.74:2.84 with a water-cement ratio 0.4. In order to facilitate the dispersion of fibres along with the cement particles a water reducing admixture FOSROCCONSPLAST WL is used in this experimental study.

2.2 Metallized Plastic waste fiber (MPW)

MPW is a waste of packaged food products like biscuit packets, chips packets, and sweet wrappers and they are coated with metallic film inside. These plastic wastes are produced by polypropylene elements and it is not suitable for any recycling process. Hence it could be utilized as a fibre in this experimental study. They are prepared by cutting the MPW collected from chips packet, biscuit packet, and sweet wrappers and it is cut into a length of 35mm using scissors and rulers is shown in Figure 2(a). The shredded metalized plastic waste fibres 35mm long used in the casting stage are shown in Figure. 2(b). The main properties of metalized plastic waste fibre are collected from earlier literature and listed in Table 1[4].

Figure 2(a). MPW Fiber 35mm long

Figure 2(b). Shredded MPWfibres

Table1. Properties of Metalized Plastic Waste

Properties	Values
Resin category	Polypropylene
Plastic Type	LDPE
Density range	0.925 -0.94 g/cm3
Thickness	0.08mm
Water vapour resistance	Good
Oxygen permeability	High
Tensile strength (Tested)	800 N/mm2
Elongation	8-10%
Co -efficient of friction	0.45– 0.55

2.3 Preparation of test specimens

Metalized plastic waste concrete specimens are prepared as per the following procedure. All the constituents are weighed and kept ready before mixing and firstly the dry aggregates are poured and mixed in the pan mixer of 40 litres capacity. Secondly, the cement is added to the mixer and thoroughly mixed for two minutes. Soon afterward water and super plasticizer are mixed together and added to the mixture. Finally, metalized plastic waste fibres are dispersed and mixing is continued for further 2minutes to achieve the well dispersion of fibres in the concrete. Fibres are added at 0.25%, 0.5%, and 0.75% volume fractions and for each type of fibre fraction the standard size test specimens were prepared. Control concrete specimens are made without fibres. After 24 hours of casting, test specimens were demoulded and kept under immersion curing for 28 days. Specimens are labelled according to the volume fraction of fibres and they tested for various strength and toughness properties.

2.3 Test methods

In this experimental program, the hardened concrete properties are evaluated as per the relevant procedures laid in Indian standards.

2.4.1Testing of Compressive Strength

Compressive strength test is conducted on cube specimens of size 150 x150x150mm in a compression testing machine of 2000kN capacity at the age of 28days. Three cubes are tested for each group and the average of three is reported as the final results. Compression test on cubes is conducted as per the guidelines given in IS 516 -1959[17]. Test set up is shown in Figure. 3(a).

Figure 3(a). Compressive strength testing

Figure 3(b). Split tensile strength testing

2.4.2 Testing of splitting tensile Strength

Split tensile strength test is conducted on 150mm x 300mm long cylinders to determine the indirect split tensile strength as per IS 5816:1999 procedure [18]. For each series, three specimens are tested at the age of 28 days and the testing arrangement is shown in Figure. 3(b). The ultimate load at failure is noted and the split tensile strength is calculated.

2.4.3 Testing of Flexural Strength

The flexural strength test is conducted as per IS 516 -1959[17] guideline on 500 x100x100 mm (length x width x height) prism specimens and three-point bending testing is shown in Figure 4. Three specimens are tested for each group and flexural strength is calculated and reported in Table 2.

2.4.4 Flexural toughness measurement

This test is carried out to evaluate the flexural toughness on same size prism specimens at the age of 28days in a 600kN universal testing machine and the test setup is given in Figure 5. Specimens are tested for a three- point loading to obtain a load –deflection curve as per ASTMC 1018[19]. Mid span deflections are measured using a suitable data acquisition system provided with the universal testing machine. Load Vs deflection plots are made for all the series of tested specimens.

These curves are further analysed to obtain the energy absorption capacity (toughness) of these composites.

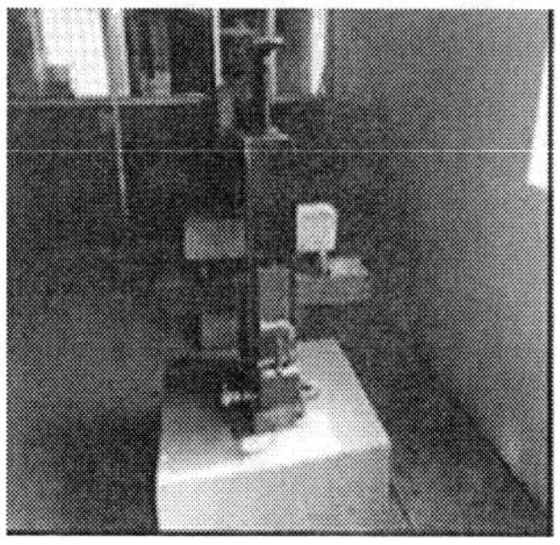

Figure 4. Three point bending testing

Figure 5. Flexural toughness testing

3 Results and Discussion

The results of compressive, split tensile and flexural strength of concrete tested at 28 days are given in Table 2.

Table 2 Strength Results of MPW fibre concrete

Series	Compressive strength MPa	Split tensile strength MPa	Flexural Strength MPa
CC	36.15	4.35	4.33
MPWC 0.25	37.30	4.75	4.83
MPWC 0.50	38.29	5.57	5.83
MPWC 0.75	39.21	6.77	6.83

3.1 Effect of MPW fibre on compressive strength

From Table 2, it is understood that the compressive strength was influenced by the addition of MPWfibres and strength improvement is fully dependent on the dosage of fibre. The compressive strength was in the range of 36.15 MPa to 39.21MPa and its variation is shown in Figure 6. When compared to Conventional Concrete series (CC) the compressive strength of the cube specimens reinforced with volume fraction 0.25%, 0.5% and 0.75% MPWfibres increases by 3.18%, 5.91% and 8.46% respectively. Since the MPWfibres are not as strong as aggregates to carry the compressive load applied, their improvement seems to be very marginal. No spalling is observed on the surface of the cube specimen because of the restrain offered by the MPWfibres in expanding the cracks.

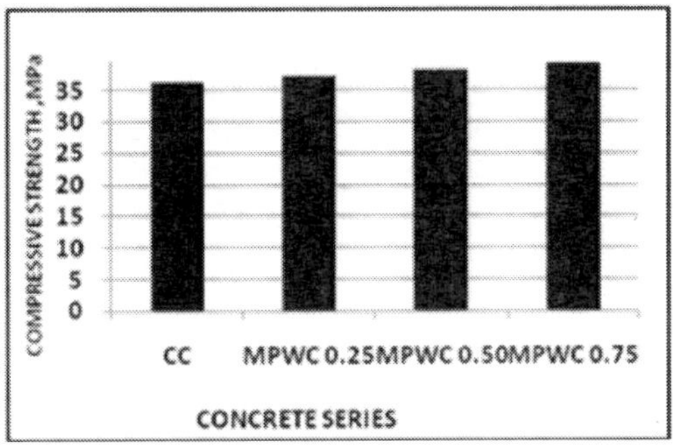

Figure 6. Compressive strength results

3.2 Effect of MPW fibre on split tensile strength

Figure.7 shows the split tensile strength of MPWfibre concrete at 28 days. Split tensile strength was highly influenced by the fraction of MPWfibres because of the crack bridging effect and presence of these MPWfibres increases the post cracking tensile properties of the cylindrical specimens. Split tensile strength was in the range of 4.35 MPa to 6.77 MPa. When compared to plain concrete mixture the addition of 0.25%, 0.5% and 0.75% MPWfibres improves the split tensile strength by 9.19%, 28.04% and 55.63% respectively. This improvement in split tensile strength is due to the increase in crack resistance offered by the MPWfibres. After the formation of the initial crack, the tensile stresses across the failure cross section reduced the tensile displacements and improved their resistance to failure. This crack bridging mechanism under diagonal tension increases the split tensile strength to greater extent.

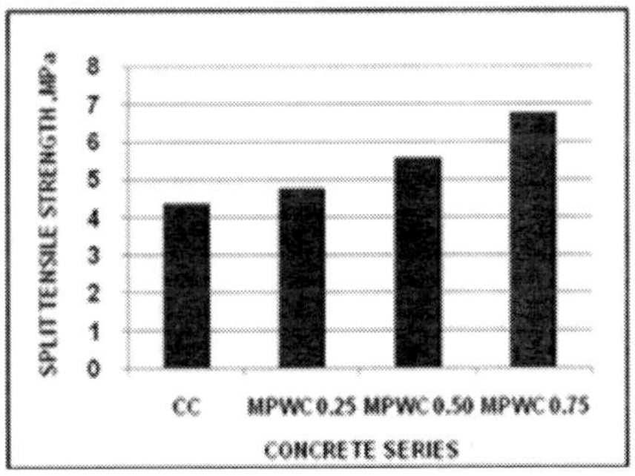

Figure 7. Split tensile strength results

3.3 Effect of MPW fibre on flexural strength

Figure. 8 shows the results of flexural performance of prismatic prisms tested under bending tests. Experimental results confirm the addition MPWfibre content increases flexural strength by 11.54% to 58% when compared to control concrete and changes the failure of the tested specimens from brittle to ductile. Plain concrete breaks into two pieces once the cracks occur , but the addition MPWfibres in the composite increases the tensile capacity even after the flexural cracking. This is because of the MPWfibres bridging across the cracks and reduced the width of the cracks thus proven its good reinforcing properties to the concrete.

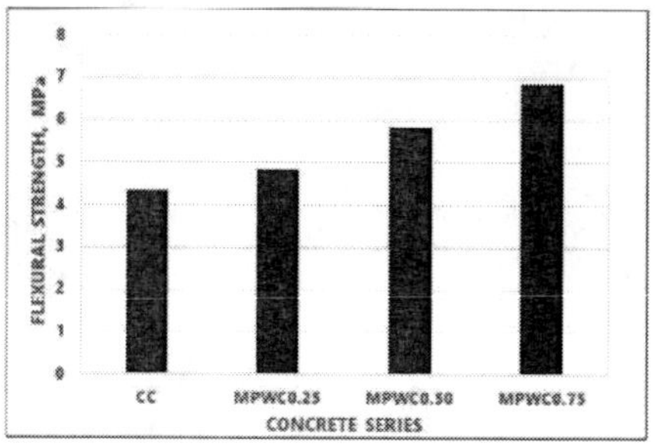

Figure 8 Flexural strength results

3.4 Load -Defection response

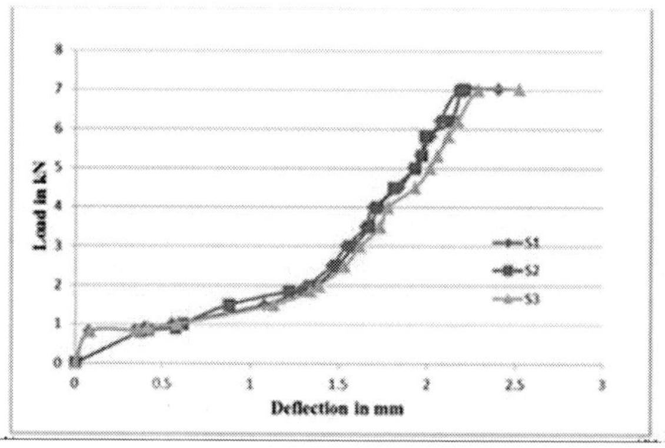

Figure 9. Load - Deflection response

Table 3 summarizes the flexural toughness capacity of standard size prisms and Figure.9 represents the load deflection plot of MPWC0.75 series. For each series three specimens are tested and the area below the load –deformation curve is measured as toughness or energy absorption capacity of concrete composites. From Table 3 it is evident that the toughness improvement in Metalized Plastic Waste fibre concrete is significant when compared to control concrete. MPWC0.5 series have given as much as 1.92 folds higher than MPWC 0.25 series. Similarly, MPWC0.75 series has given 2.92 folds greater toughness than MPWC 0.25 series. So it is inferred that the tensile behaviour of the metalized plastic fibre allows the concrete composite to undergo a large amount of deformation without any sudden failure. The bridging capacity of fibres across the crack controls the crack growth and withstands a large amount of flexural load leads to the improved toughness.

Table 3 Toughness values of MPW fibre concrete

Series	Toughness in N. mm
CC	Brittle failure
MPWC 0.25	2285.98
MPWC 0.50	4495.80

MPWC 0.75	6693.73

4 Conclusions

From the experimental investigation, the following points can be concluded:

- Fresh properties of the concrete are not affected by the inclusion of MPWfibres up to 0.75% and the mix is workable. It provides the proper dispersion of MPW fibers because of good workability.
- Addition of MPW fibres from 0.25% - 0.75% increases the compressive strength from 3.45% to 8.46% which is marginal compared to other properties. Split tensile strength increases with the addition of MPW fibres up to 55% and exhibits good resistance to tensile cracking. Flexural resistance enhanced even better than split tensile strength and 58% more than the conventional concrete.
- Toughness of MPW concrete has been improved up to 1.96 to 2.92 folds than the conventional concrete because of its ductile behaviour and large deformation.
- Since these MPWfibres are produced from the waste product they are cost- effective, when compared to other artificial fibres

References

[1] Pakravan H.R, Latifi M and Jamshidi M 2017 *Construction and Building Materials* **142** 280294
[2] Jang J and Yun HD 2018 *Composite Structures* **85** 203211
*[3]*Caggiano A, Folino P. Lima C, MartinelliE and Pepe M 2017 *Construction and Building Materials* **147** 286295
[4] Mohammadhosseini H and Mahmood Md. Tahir 2018 *Construction and Building Materials* 180 92102
[5] Awal ASMA and Mohammadhosseini H 2016 *Journal of Cleaner Production***137** 157166
[6] Zhang P and Li,Q 2013 *Composites: Part B* **45**15871594
[7] Thorneycroft,J Orr,J, SavoikarP and Ball PJ 2018 *Construction and Building Materials* **16**16369
[8] Foti,D 2011 *Construction and Building Materials* **25**1906–1915
[9] Saikia N and BritoJ 2012 *Construction and Building Materials* **34** 385–401
[10] Yin S,Tuladhar R,RiellaJ,Chung D,CollisterT,Combe,M and Sivakugan N *Construction and Building Materials* **114** 134141
[11] GhernoutiY,Rabehi B,BouzianiT, Ghezraoui H and Makhlouf A *Construction and Building Materials* **82** 89–100
[12] Paul Borg R ,Baldacchino O and Ferrara L 2016 *Construction and Building Materials***108** 29–47
[13] Fraternali F, Spadea S and Berardi VP 2014 *Construction and Building Materials* **61**293–302
[14] Torgal F DingYand Jalali S 2012 *Construction and Building Materials* **30** 714–724
[15] IS383 -1970, Indian standard Specification for Coarse and Fine Aggregates from Natural Sources For Concrete, BIS New Delhi
[16] IS10262 – 2009, Indian standard Concrete mix proportioning guidelines, BIS New Delhi.
[17] IS 516 -1959, Indian standard methods of tests for strength of concrete, BIS New Delhi.
[18] IS 5816 -1999, Indian standard Split tensile strength of concrete -Method of Test, BIS New Delhi
[19] ASTMC 1018 ASTM C 1018 1997 Flexural Toughness and First-Crack Strength of Fiber-Reinforced Concrete (Using Beam With Third-Point Loading) American *Society for Testing and Materials* West Conshohocken

Intelligent Expeditious Accident Detection and Prevention System

K P Sampoornam[1], S Saranya[2], S Vigneshwaran[2], P Sofiarani[3], S Sarmitha[3], N Sarumathi[3]

[1] Professor, Dept. of ECE, Bannari Amman Institute of Technology, Tamil Nadu, India

[2] PG Scholar, Dept. of ECE, Bannari Amman Institute of Technology, Tamil Nadu, India

[3] UG Scholar, Dept of EEE, Bannari Amman Institute of Technology, Tamil Nadu, India

E-mail: sampoornam@bitsathy.ac.in, saranya.co19@bitsathy.ac.in

Abstract. Nowadays, many of the vehicle accident occur due to over speed, drowsy driving and drunken drivers. According to statistics, for every four minutes, a person dies because of road accidents and also thousands of accidents are recorded in 2017 for which people are died in India, which is more in number compared to all other accidents. But their lives could have been saved in the presence of emergency indicators. So we are going to implement our ideas, regarding the life saving measures and precautionary measure to prevent the accident. In this proposal, by using GPS, the exact accident location is identified automatically and GSM can send the data to nearby ambulance vehicle, the police station and family members of that person. The occurrence of accident can be detected using accelerometer sensor and severity of the accident is detected using tilt sensor. It helps to monitor the tilt angle of the chassis. The drowsiness of the driver is detected using drowsiness detector. By measuring the percentage of eyelid closure over the time, the driver's drowsy state is detected and warning is given to the driver through the alarm. By this method we can measure the speed of a vehicle which gives reason for the accident and vehicle GPS location can be view through mobile application for theft detection. All the data can be processed in microcontroller with help of sensor value and send location to cloud. The major advantages of this system are cost effective, assured safety, victim life can be save quickly, low power consumption, better accuracy, efficient time consumption, reduce the chance of human error.

Keywords: GPS, GSM, Accelerometer Sensor, Mobile Application, Microcontroller, Drowsiness detector, Tilt sensor.

1 Introduction

In India the usage of the vehicles is increased every day, because of that reason accidents, stolen of a vehicle and air pollution are also increased. In India, comparing to all other cities, Chennai and Delhi are prone to many road accidents. According to the data, in a year 4,80,652 accidents leads to 1,50,785 deaths and in a month 1,817 accidents leads to 413 deaths and in a day 55 accidents leads to 17 deaths. One of the reports says, highway is the biggest killer. In the national highway, 34.5% accident death occurs and in the state highway, 29.9% accident death occurs. A vehicle was stolen every 13 minutes in the NEW DELHI. In the first three months of the year 2017, marking a sharp 44% rise over the same period in year 2016. Only around 4% of these cars are recovered. But it could be reduced by implementing our proposal.[1-7]

This system solves all above problem present in the real world and this solution is can implement in the real world. This system solve to real world problem one is to identify the accident and it can send message to the ambulance drive automatically, due to this response time is reduced from this we can save the human life. The second real time problem is to identify the theft of vehicle by this same system. This system points the exact latitude and longitude of that vehicle. By this information we can easily to find-out the vehicle place by interfacing hardware with cloud, in cloud data can be saved for future use and that data can be retrieve from the database it can be viewed in android app.

2 Literature Survey

This paper describes the control of vehicle speed under the exhaustion of the driver with the help of online safety prototype. The aim of this model is to proceed in a system to identify the signs of exhaustion in drivers and in order to prevent the accidents; the speed of the vehicle is controlled. In this system real time sensors like eye blink sensor, alcohol detector, impact sensor, fuel detector is used. GPS and Google Maps APIs for location are the software interface with the sensor networks. [8]

In this paper, vibration sensor is used to detect the accident of the vehicle and the signal will be detected using the Micro Electro Mechanical System (MEMS) sensor and it will be send to the ARM controller. The information regarding the accident will be send to the police control room or a rescue team and it will be easy for them to find the correct location of the accident. If the accident is not severe then it is waste of time for the rescue team and police. So this is one of the major drawbacks in this system. [9]

This proposal has introduced a system for detecting car theft. This system is based on computer concept technology, offering a high level of car theft prevention, and is not disruptive to drivers. These systems use the facial detection and recognition technologies. When the unauthorized driver drives the vehicle, he will be identified and alerted by sending his picture to the police and the owner of the car [10-15].

3 Proposed System

The need of introducing this system is to shorten the time required by the ambulance to reach the hospital which in turn can reduce the death rates. The main core of this proposal is to encounter the accidents using speed sensor and accelerometer sensor. GPS & GSM modules are used to find the vehicle's correct place and it is shared to ambulance and two family members of that particular person.

The same GPS locations are used to find the vehicle location and send to the cloud receive location with help of mobile application for theft detection method. This can be implemented using the embedded system and it will overcome all other existing systems since it is the emerging technology all over the world. Due to this embedded technology, everything around us can be automated which in turn can reduce the intervention of human without the loss in quality. Accident is detected using accelerometer sensor and the vehicle's speed is measured using the speed sensor and the status can be displayed in LCD.

The chassis tilt angle is monitored using the tilt sensor to ensure that the vehicle operates in the safe angle. If the angle is unsafe then the severity of the accident is measured and informs regarding the accident seriousness is sent to the nearby ambulance vehicle, police station and family members of that person. To prevent the accident, we had used eyelid sensor as a drowsiness detector. This sensor detects the drowsy state of the driver and gives the warning to the driver through the beeping alarm. By this, the drive can get conscious while driving and avoid the accident.

4 System Overview

To automate the theft and accident detection, the automation system which is based on sensor network is developed. The proposed system use one accelerometer sensor to detect the accidence of vehicle and

GPS is to find the exact place of vehicle. The system also implements the theft detection of vehicle using GPS. Sensors, GSM, GPS are programmed using Arduino and Node MCU micro-controller.

4.1 Arduino

Arduino is software which is open source. The Arduino Uno is a microcontroller containing ATmega328P in it. It has 14 digital I / O pins, a 16 MHz quartz crystal, a USB connection, a power jack and a reset button, and 6 analogue input pins. Just connect with a USB cable to a computer, or power it with the adapter or battery used to get started. The Embedded C programme is used for any application coding.

4.2 Node MCU

Node MCU has microcontroller esp8266 and inbuilt Wi-Fi module present in it. It has send data to the cloud database

4.3 Accelerometer Sensor

Accelerometer sensors detect the vehicle accident using three co-ordinates. It is an electromechanical device used to calculate acceleration forces, either static or dynamic forces of acceleration. From these change in vehicle capacitance, the acceleration can be determined.

4.4 Tilt Sensor

The seriousness of accident is measured using tilt sensor. The tilt sensor measures the chassis angle of the vehicle and calculates whether it is the safest angle or not. If it is not in the safe angle, then the signal will be given to the microcontroller.

4.5 Drowsiness Detector

The eyelid sensor is used as a drowsiness detector. Perclos is the measure of drowsiness, which is nothing but the measure of percentage of eyelid closure over time or droops rather than blinking.

4.6 GPS

Global Positioning System is a device used to track the vehicles, it is uses to determine and track its exact location. Satellite GPS helps to find out the exact location of that specific area. Accident place of latitude and longitude value can be finding with GPS.

4.7 GSM

In this system, the Global System for Mobile Communication was used in the event that an accident was detected. In this system, the purpose of GSM is to alert ambulance and nearby police station via SMS that some accident has been detected and sends accurate location.

5 Block Diagram

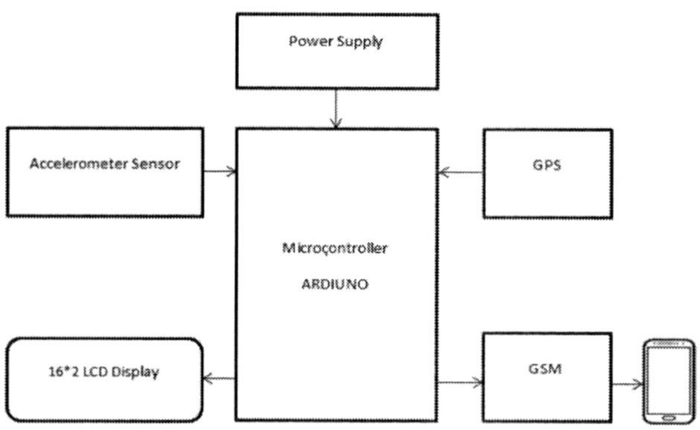

Figure1. Block Diagram for Detection of Accident

LCD – Liquid Crystal Display

GPS – Global Positioning System

GSM – Global System for Mobile communication

Above block diagram Figure.1 represent the accident detection of vehicle is to find out the accident automatically with help of accelerometer sensor and that data is feed to microcontroller. If data is reach based on user condition microcontroller get the GPS location and send the location to ambulance driver and also send to family members.

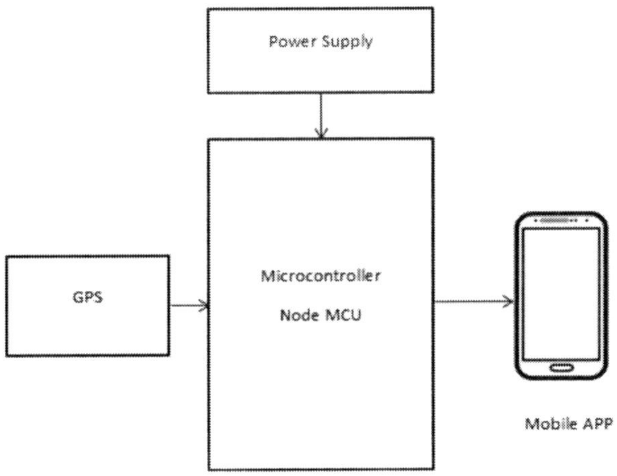

Figure 2. Block Diagram for Theft Detection

GSM – Global System for Mobile communication

Above block diagram Figure.2 represent the theft detection of vehicle using mobile application. GPS get the data from the satellite and send to node MCU, Node MCU is Wi-Fi device to transmit data to the cloud. By android app, we can see the correct location of the vehicle. By this we can find out our vehicle using this system.

6 System Architecture

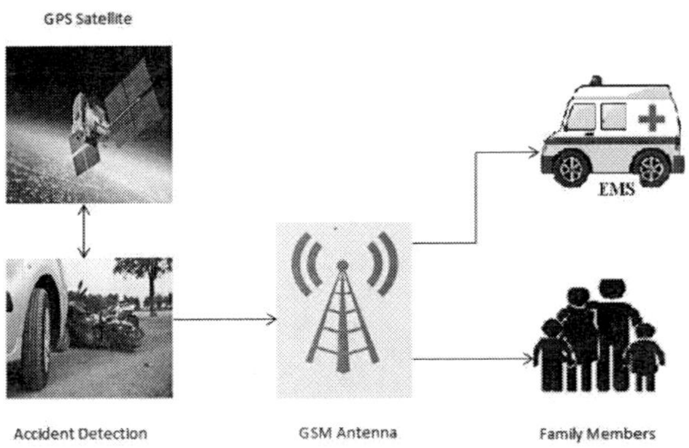

Figure 3. Architecture of the System

This system architecture is helpful to minimize the time needed for the ambulance to reach the accident area and also send user detail to that ambulance it is helpful for finding the address and also the blood group of that person. This information can send to corresponding person's family member automatically with help of this system. The accident is find out with the help of accelerometer at the same time GPS get the exact location of the accident place and that data is send to ambulance drive.

7 Flow Chart

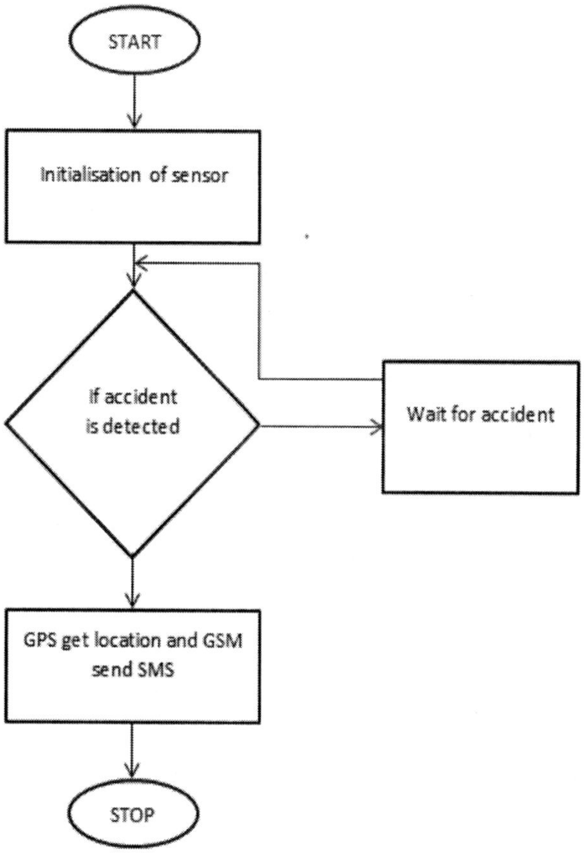

(a)

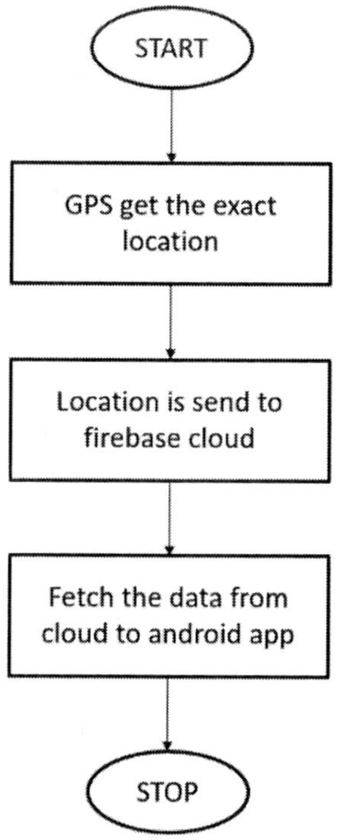

(b)

Figure 4. System Flowchart

Figure 4 (a) shows the working flow of the accident detection systems. Initially the sensors get activate, it will detect the accident occurred or not. If the accident is detected then the GPS will find the exact location. Figure 4 (b) shows, how the message will send. GSM will send the exact accident location is identified automatically and GSM can send the data to nearby ambulance vehicle, the police station and family members of that person.

8 Experimental Results

The proposed system is being practically experimented as a working model for the detection of real world accidents.

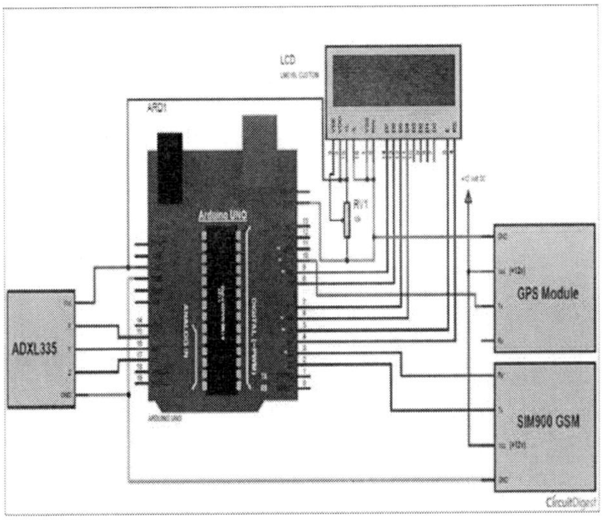

Figure 5. System Circuit diagram

Figure 5 shows the circuit diagram of the accident detection system, which includes controller, GSM, GPS circuits.

Figure 6. Hardware implementation

In Figure 6, it is shown that the practical hardware implementation of the entire system. The LCD used in this system to display the range of accident along with the vehicle speed. In the controller the sensors are connected. The controller will send the processed data to the cloud.

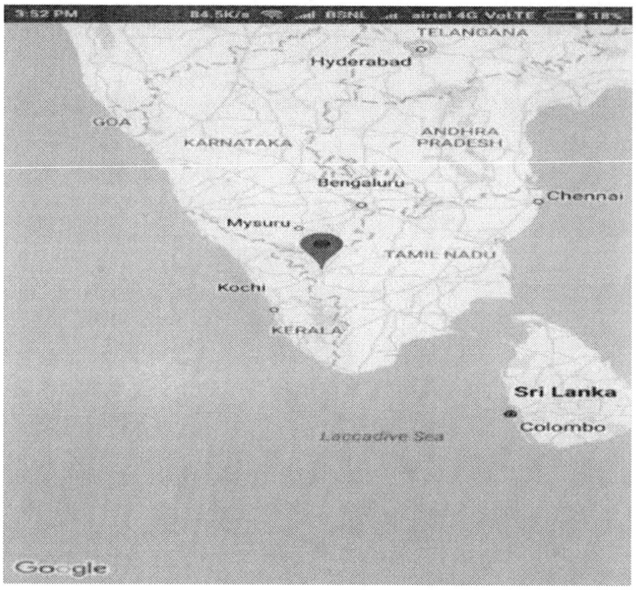

(a)

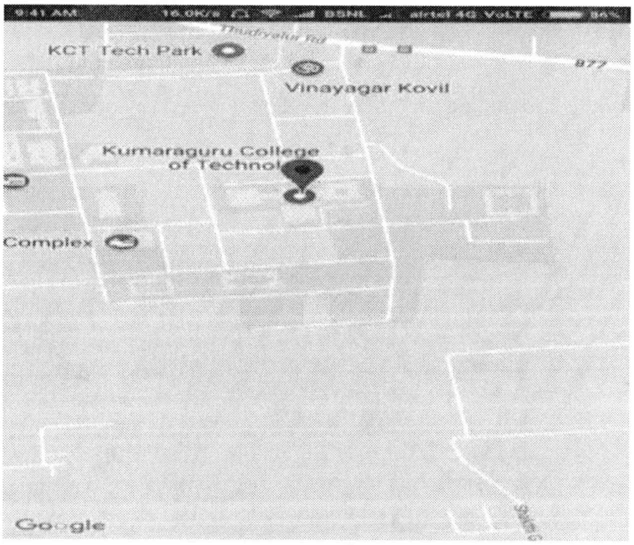

(b)

Figure 7. Android map screenshot 2

Figure 7 (a) and (b) shows the android map screenshots of the exact accident location. These data will be collected from the cloud and GSM will send this location to nearby ambulance vehicle, the police station and family members of that person. This process will takes place immediately when the accident occurs. So, the rate of survival of the accident victim will be more when compared with other techniques.

9 Conclusion

Using automatic detection of accident and theft detection system, we reduce the death rate due to accident and also find out the stolen vehicle exact location by the user. Also this technique has fast operation than older system; it saves a lot of time as it is automated whereas manual systems take some time to call the ambulance. We can also reduce the number of accident using drowsiness detector. Nowadays, most of the accident is due to the drowsy state of the driver, so by using these techniques we can reduce the accident rate. Since the design is completely automated it can be used in highway areas where less man is present. Thus this design finds its applications in many cases.

References

[1] Prabha C, Sunitha R, Anitha R 2014 Automatic Vehicle Accident Detection and Messaging System Using GSM and GPS Modem *Int. J. Adv. Res. Electri.*, **3** (7).

[2] Belhassen Akrout Walid Mahdi 2013 Blinking Measurement Method for Driver Drowsiness Detection *Proce. of the 8th Int. Conf. On Comp. Recog. Syst. CORES,* pp. 651-660.

[3] Zhixiong Liu 2005 A New Embedded Car Theft Detection System", International Conference on Embedded Software and Systems *Int. Conf. on Embd. Soft. and Syst.*

[4] Aishwarya S.R, AshishRai, Prasanth M.A, Savitha S.C An IoT Based Accident Prevention &Tracking System For Night Drivers **3** (4) pp. 2320-9801.

[5] Ji Hyun Yang, Zhi-Hong Mao et al., 2009 Detection of Driver Fatigue Caused by Sleep Deprivation *IEEE transact.on syst., man, and cyber.—part a: syst. and hum.* **39**(4).

[6] Selvathi D, Pavithra P and Preethi T 2017 Intelligent transportation system for accident prevention and detection *Int. Conf. on Intell. Comput. and Cont. Syst. (ICICCS).*

[7] Bhumkar S P, Deotare V V, Babar R V 2012 Accident avoidance and detection on highways *Int. J. Engg. Trends Techn.* **3** (2).

[8] Sri Krishna Chaitanya Varma, Poornesh, TarunVarma, Harsha 2013 Automatic Vehicle Accident Detection And Messaging System Using GPS and GSM Modems *Int. J. Scient. & Engg. Res.* **4** (8).

[9] Ishan Karande, Gandhar Deshpande et al., 2016 Intelligent Anti-Theft Tracking and Accident Detection System for Automobiles Based on Internet of Things *Int. J. Inno. Res.Comp.Comm. Engg.* **4** (3).

[10] Boto-Giralda D, Díez-Higuera J F et al., 2012 Early detection traffic flow incidents model on road networks *6th Euro Amer. Conf. on Telem. and Infor. Syst. (EATIS).*

[11] Rahul Gautam, Shubham Choudhary et al., 2015 Cloud based automatic accident detection and vehicle management system *Int. Conf. of Sci., Technol. and Manag.*

[12] Elie Nasr, Elie Kfoury, David Khoury 2016 An IoT Approach to Vehicle Accident Detection, Reporting, and Navigation *IEEE Int. Multi. Conf. on Engg. Techn.*

[13] Vijay Savania, Hardik Agravata and Dhrumil Patela 2015 Alcohol Detection and Accident Prevention of Vehicle *Int. J. Inno. Emg. Res. Engg.* **2** (3).

[14] Martinez F, Manzoni P, Garrido P, Fogue M 2012 Automatic Accident Detection: Assistance Through Communication Technologies and Vehicles *IEEE Veh. Technol. Mag.* **7** pp. 90–100.

[15] Rajkiran A, Anusha M 2014 Automatic Vehicle Accident Detection System Using Wireless Communication *Int. J. Res. Stud. Sci. Eng. Technol.* **1** pp. 98–101.

Design and implementation of carry Look ahead generator in reversible logic using nano QCA

Kannaki S[1], Karthigai Lakshmi S[2], Jaganathan S[3], Sujatha C[4], Bhoopesh D[5], Balaji S[6]

[1]Assistant Professor, Department of Mechatronics Engineering, Sri Krishna College of Engineering and Technology, Coimbatore, Tamilnadu, India - 641008.

[2,4]Professor, Department of Electronics and Communication Engineering, SSM Institute of Engineering and Technology, Dindigul, Tamilnadu, India - 624002.

[3]Professor,Department of Electrical &Electronics Engineering, Dr.NGP Institute of Technology, Coimbatore, Tamilnadu, India – 641048.

[5,6]UG Scholar, Department of Mechatronics Engineering, Sri Krishna College of Engineering and Technology, Coimbatore, Tamilnadu, India - 641008.

[1]E-mail: kannakis@skcet.ac.in
[5]E-mail: 17eumt022@skcet.ac.in

Abstract

The idea of reversible logic is another emerging architecture that has developed its ground in research area. This argument implied zero heat dissipation at the device. According to the physical laws of quantum mechanical effect and the law of coulomb the function and heat relations are satisfied. In this paper, a carry-looking generator is built using reversible logic and its QCA have been reported. To achieve the proposed design use was made of Toffoli and BJN gate. Comparing simulation outcomes to theoretical values verifies the design. The proposed model is tested and simulated using version 2.0.3 of the QCA Designer method.

Keywords: Quantum dot Cellular Automata (QCA); Reversible logic; Toffoli and BJN gate; QCADesigner

1. Introduction

A power consumption is the most difficult field in nano scale logic design. There is a growing need for a new technology that can provide less power dissipation nano size circuits. QCA provides high density applications, low power consumption and high switching speed [1]. Because of these properties quantum gates were targeted for their enabling computational reversibility functions. The computing systems' weakness of heat dissipation is the primary driving force that draws attention to reversibility. Reversible logic preserves the knowledge that is similar to energy and momentum conservation in physics.

The reversibility inherent leads to a new frame function, which results in zero heat dissipation.In QCA reversibility is one to one mapping of output inputs. Landau indicated in 1961 that the loss of one bit of information cost would be greater than the amount of KTln2 energy joules [2]. This indicates the irreversible processes do not maintain information and are loss-making. In addition to the strength of the theoretical definition, some essential interpretation was obtained by Bennet. The bennet clock mechanism achieves less power dissipation than the KTln2 switch Reversibility is a notion that requires bijective action. It indicates both backward and forward tracks for a machine are possible at any time [9]. Care should be taken when developing reversible logic circuit mapping, as fanout is not permitted. The Contribution list is as follows.

- The building blocks of a reversible system should be individually reversible.
- The number of inputs and outputs has to be always equal.

The emerging reversibility concept attracts the VLSI designer because CMOS suffers the nano-scale power dissipation problem. If used successfully to design reversible circuits, the new design will lead to ultra low-power systems.

2. Background

2.1 QCA REVIEW

Craig Lent and Al have developed a new paradigm of the calculation architecture called Quantum dot Cellular Automata to address the problem of size reduction in CMOS technology. QCA cells were realised through several physical implementations of these semi-conductor methods and metal islands are the most promising because of their high density of operation and integration [3].

There are four quantum dots to this QCA technology consisting of an array of cells in each. The electrons occupy two Quantum Points diametrically opposite. There are two polarisation levels, which correspond to the electrons. The polarisation is P=+1 when the electrons are in the lower left and higher right and the polarisation P=-1 when it is in the lower right and higher left as shown in Figure 1. The planar crossings may also be achieved by using a by using a turned version 450 of a cell [8].

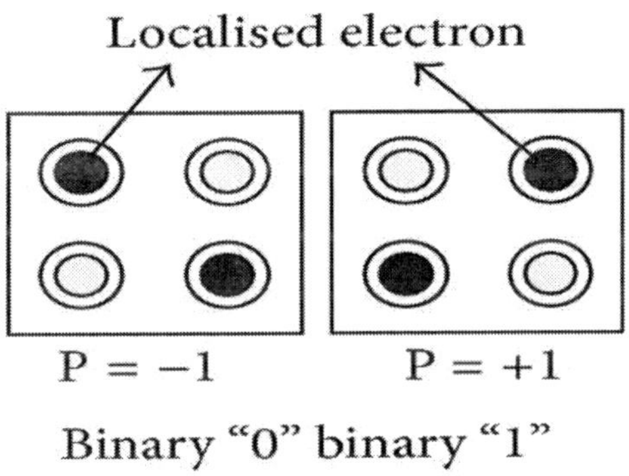

Figure 1 QCA cell with four quantum dots

The logical information can propagate from the input to the output of the cell by the force of repulsion.The fundamental QCA logic primitives are QCA wire,QCA inverter and QCA majority gate [7].The binary signal propagates from input to output because of electrostatic interactions.90° and 45° QCA wires can be used as shown in the Figure 3. The QCA inverter is created by orienting the cells at 45° to each other to take opposite polarizations as shown in below Figure 2.

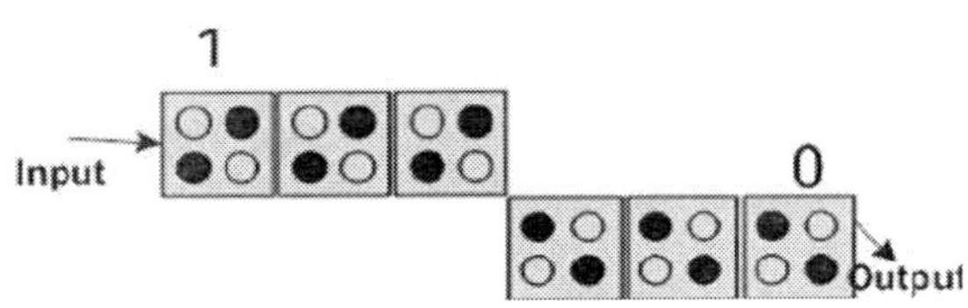

Figure 2 QCA inverter

IOP Conf. Series: Materials Science and Engineering 1059 (2021) 012013 doi:10.1088/1757-899X/1059/1/012013

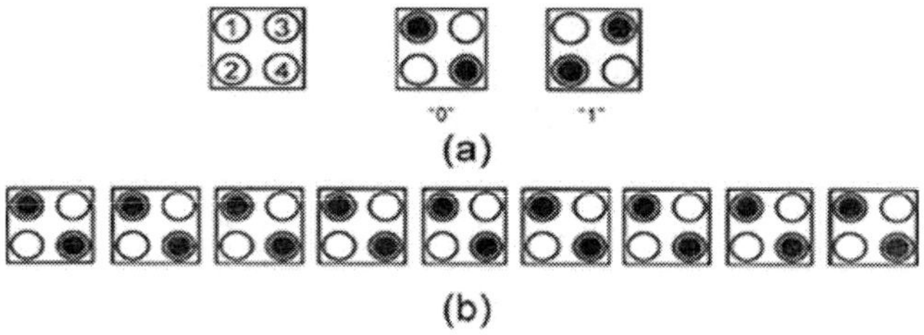

Figure 3 QCA wire

2.2 QCA CLOCKING

Synchronization and energy restorement are important to proper QCA signal propagation [4]. This can be obtained by implementing unique clocking scheme in four phases as shown in Figure 4. The Launder and Bennet are suggesting two forms of clocking. Typically, the Launder clocking scheme is used in circuits.

It has four distinct zones flipping, keeping, releasing and relaxing where each zone is moved by 90 ° relative to the previous zone as shown in Figure 5. The clocking strategies in combination with design strategies will contribute to the idea of physical reversibility. More over each QCA clock zone consists of at least two cells to maintain the clock zone effect as mentioned in Table 1.

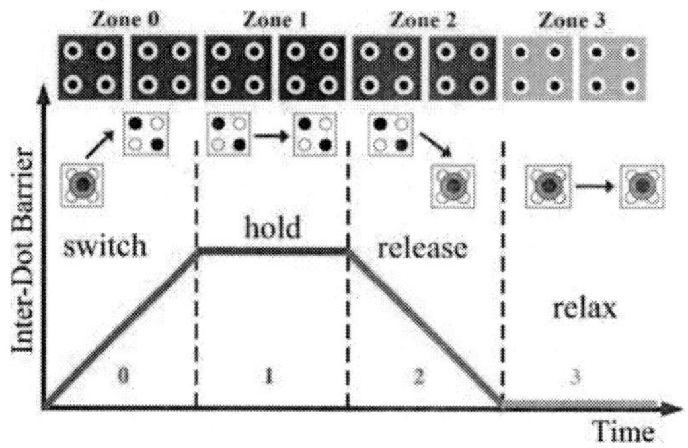

Figure 4 Clocking in QCA

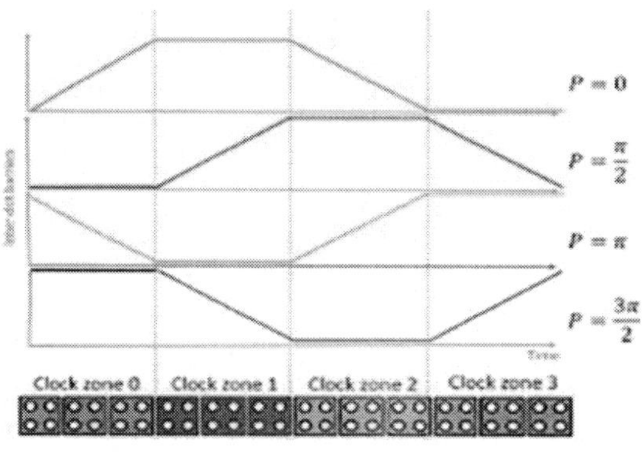

Figure 5 Clock phase shift in QCA

Clock Pulse	**Potential Barrier**	**Polarization state of the cells**
Hold	Held High	Polarized
Switch	Low to High	Polarized
Relax	Low	Polarized
Release	Lowered	Polarized

Table 1 Operation of QCA clock pulses

2.3 REVERSIBLE COMPUTING

When the input bits are lost some amount of energy is wasted. Reversible logic circuits, by recycling energy into the device, prevent energy loss [5]. When coupled with logic gates, Bennet suggested that very low power consumption is possible in logic circuits.

A gate is reversible when each separate input has a distinct output. The reversible gates are one of the benefits of being balanced. Combining the reversible gate with no constant inputs a circuit satisfies the functions of the balance.

We can use garbage outputs to satisfy unbalanced functions [6]. Reversible computing analyses the ratio of the resources used to logical computation. It can be calculated at logical level by one to one configuration between the circuit.

To prevent the destruction of data bits during logical operation, they are designed to reduce power consumption. The data bits in reversible computation are preserved that lead to the development of reversible gate. The reversible gate should have the following characteristics,

- Minimum garbage outputs.
- Minimum input constants.
- Minimum area.
- Minimum number of gates.

Several reversible gates have been developed as shown in the Figure 6. Two important types are Toffoli and BJN gate.

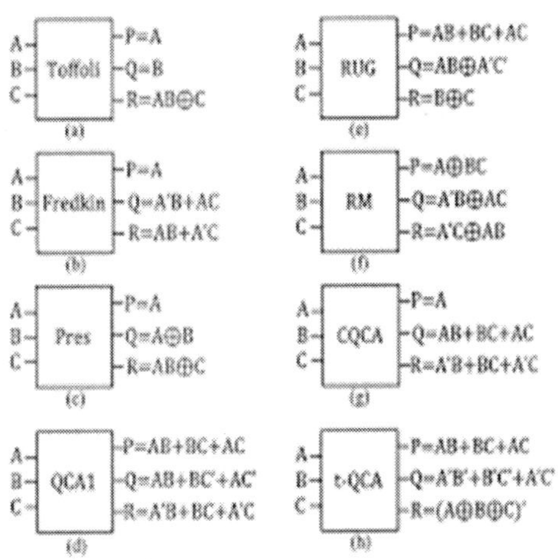

Figure 6 Reversible gates

3. Proposed Work

3.1 BJN Gate

The BJN gate is a three input reversible gate that maps input to corresponding output. Here, (A,B,C) are the inputs and (P,Q,R) are outputs . The BJN gate is shown in the Figure 7.

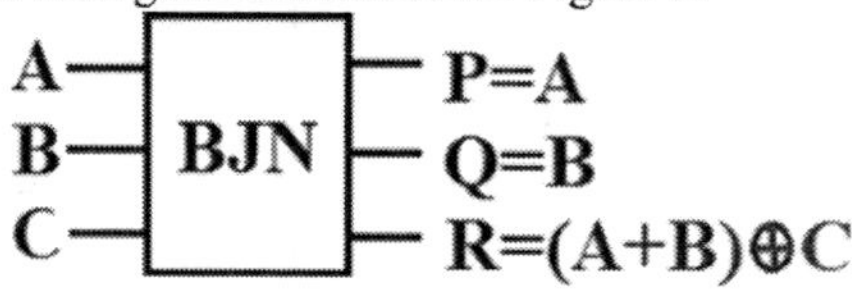

Figure 7 BJN gate – QCA implementation diagram

The Reversible carry lookahead generator operates baesd on AND and OR gates. By setting the third input to zero, the BJN gates can be mode to perform OR operation between the first two inputs. Figure 8 indicates an OR gate centred on the BJN port. The QCA representation Diagram of BJN gate is shown in the Figure 9.

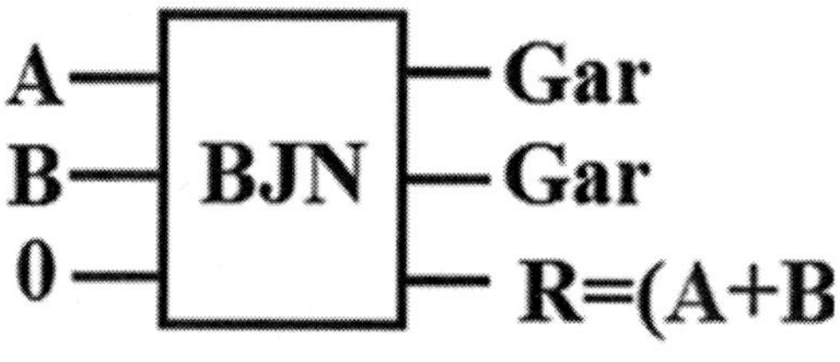

Figure 8 BJN based OR gate

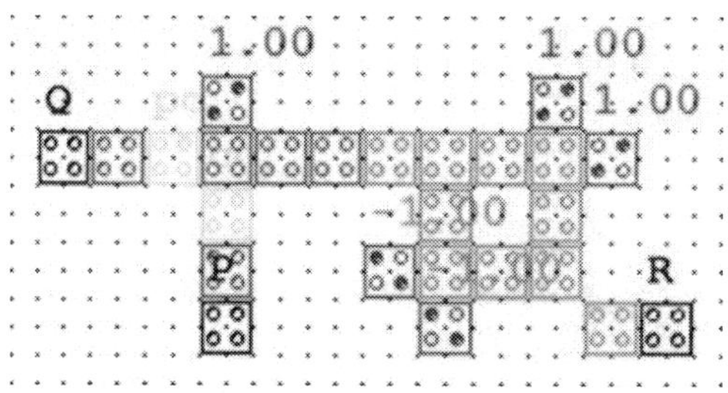

Figure 9 QCA representation of BJN gate

3.2 TOFOLLI Gate

The Tofolli gate is a three input reversible gate that maps input to corresponding output .Here (A,B,C) are the inputs and (P,Q,R) are outputs . The Tofolli gate is shown in the Figure 10.

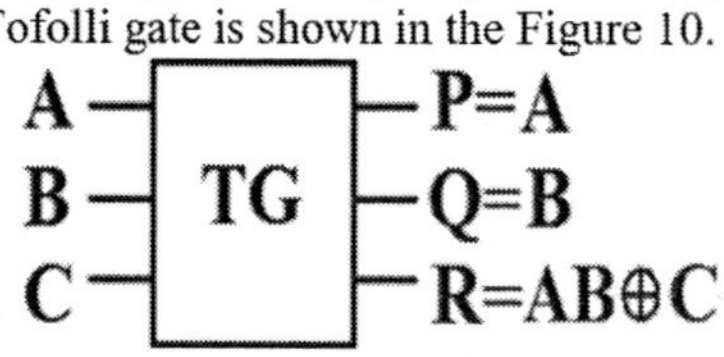

Figure 10 TG gate – QCA implementation diagram

The lookahead generator Reversible Carry operates baesd on AND and OR gates.By setting the third input to zero the Tofolli gates can be mode to perform AND operation between the first two inputs. Here the garbage values are the outputs P and Q, and R is equal to A.B. Figure 11 displays AND gate dependent on Tofolli. The QCA representation Diagram of TG gate is shown in the Figure 12.

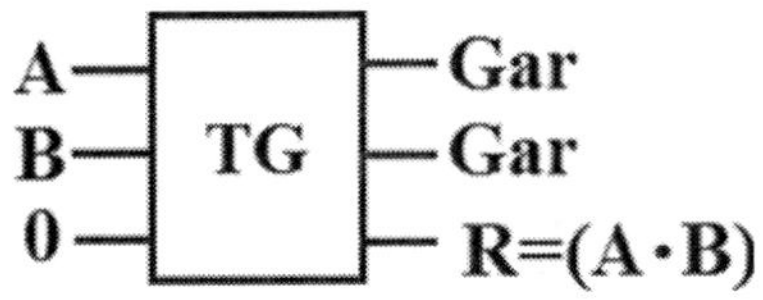

Figure 11 TG based AND gate

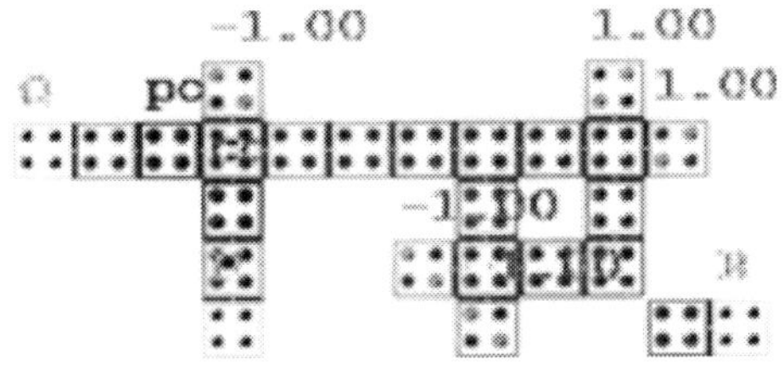

Figure 12 QCA representation of TG gate

3.3 Carry Lookahead Generator

In successive binary addition the speed of adding is restricted by propagation time for carry information. If the signal has no enough time to propagate, the outcome will not be accurate. The most commonly used technique to reduce carry propagation time is the carry lookahead logic.

The concept of generating and propagating the carries is used in this. The general logic expression carry lookahead generator is

Ci+1 = Gi + PiCi (1)

Here Ci is the input carry at stage i, Gi represents carry generate G(A,B) is used to represent the binary predicate.

G (A,B) =A.B

PiCi = A□B, it represents the carry propagate, as it shows the carry of ith signal broadcast into the (i+1)th stage (i.e) whenever Pi=1,the input carry will be propagated to the corresponding output carry. In case if the Pi=1 the carry generate is 0, then eqn(1) can be written as Ci+1=Ci. Thus if Ci=1, the input of 1 origins an output carry of 1.Similarly of Ci=0 the output carry of 0 is originated.The value of Pi depends on input of Ai and Bi,

P(A,B)=A+B

where A+B is logical disjunction. By substituting the value of 'i', the output for three bit logic circuit can be written as,

C1 = G0 + P0C0 (2)
C2 = G1 + P1C1
C2 = G1 + P1G0 + P1P0C0 (3)
C3 = G2 + P2C2
C3 = G2 +P2P1P0C0 (4)

The logic circuit is shown in figure 13.The three bit carry lookahead generator circuit requires six AND and three OR operations.

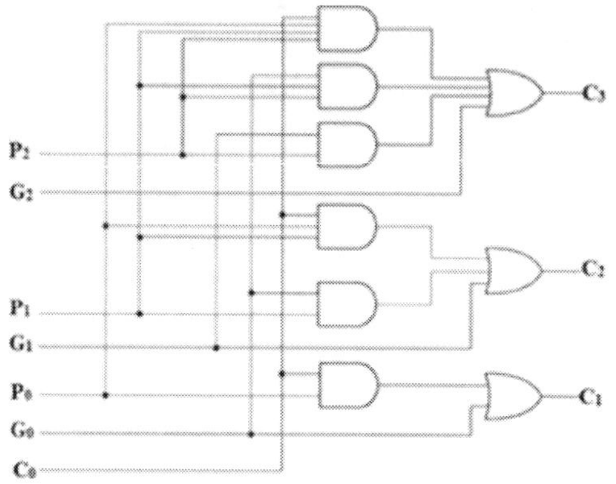

Figure 13 Logic circuit of carry lookahead generator

The reversible gates are used for these operations. The reversible carry three toffol gates and three BJN gates. QCA-based logic expression of the proposed three-bit reversible carry lookahead generator.

c1=r2(r1(p0,c0,0),g0,0)

c2=r2(r1(p1,c1,0),g1,0)

c3=r2(r1(p2,c2,0),g2,0)

The corresponding circuit in QCA in given in the Figure 14 and the outputs are compared with truth table for circuit of carry lookahead generator as mentioned in Table 2.

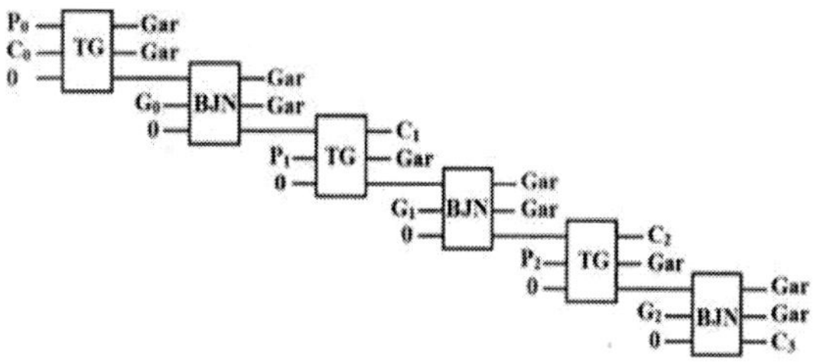

Figure 14 Structure of reversible carry lookahead generator

Input							*Output*		
C_0	P_0	P_1	P_2	G_0	G_1	G_2	C_1	C_2	C_3
0	0	0	0	0	0	0	0	0	0
1	0	0	1	0	0	1	0	0	1
0	0	1	0	0	1	0	0	1	0
1	0	1	1	0	1	1	0	1	1
0	1	0	0	1	0	0	1	0	0
1	1	0	1	1	0	1	1	0	1
0	1	1	0	1	1	0	1	1	0
1	1	1	1	1	1	1	1	1	1

Table 2 Truth table for carry lookahead generator

The carry lookahead generator circuit in QCA structure is shown in Figure 15

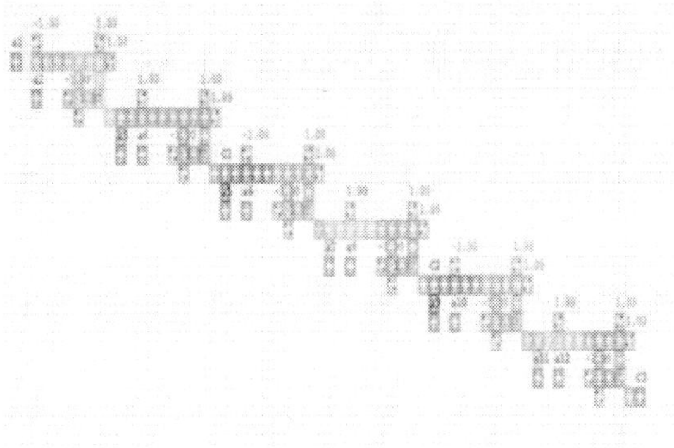

Figure 15 QCA representation of carry lookahead generator.

4. Result Analysis

4.1 Simulation Parameters

The design implementation and simulation is achieved through QCADesigner 2.0.3. The bistable simulation is carried out with 18X18 nm QCA cell.

4.2 Simulation output of Toffoli gate

The simulation result of the proposed Toffoli Reversible Gate based AND gate is shown in the Figure 16 and the result is compared with truth table as shown in the Table 2.

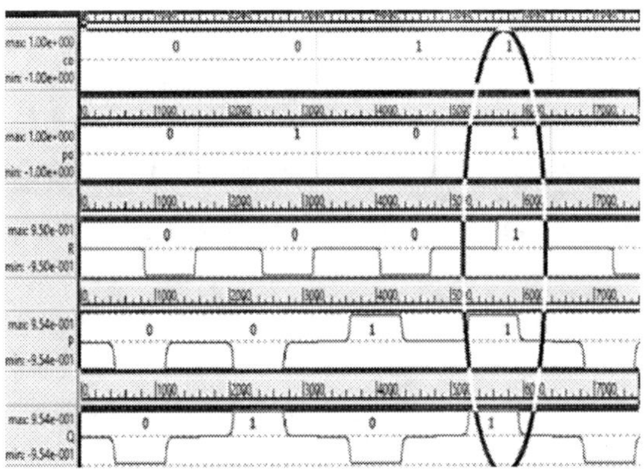

Figure 16 Simulation output for tofolli gate

4.3 Simulation output of BJN gate

The simulation result of the proposed BJN Reversible Gate based OR gate is shown in the Figure 17 and the result is compared with the truth table as shown in the Table 2.

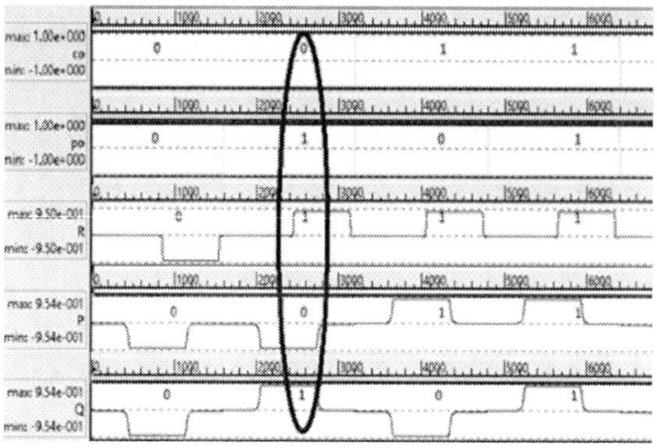

Figure 17 Simulation output for BJN gate

4.4 Simulation output of carry look ahead adder

The simulation result of the proposed carry lookahead generator using toffoli and BJN reversible gate is shown in the Figure 18 and the result is compared with the truth table as shown in the Table 2. The performance analysis of BJN, Tofolli and CLAG circuit is been shown in the Table 3 with its cell count and cell area.

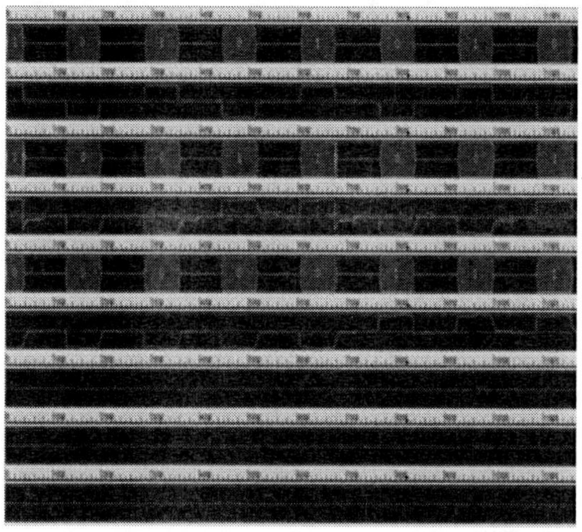

Figure 18 Simaultion output for carry lookahead generator

	Circuit	**Cell count**	**Cell area(μm^2)**
CLAG[9]	BJN	66	21384
	Tofolli	66	21384
	CLAG	478	154872
Proposed CLAG	BJN	25	8100
	Tofolli	25	8100
	CLAG	143	46332

TABLE 3 PERFORMANCE ANALYSIS

5. Conclusion

This paper has been planned and implemented using the Toffoli and BJN gates in QCA nano technology, reversible full adder and carry look forward generator.The circuit operation has been checkedaccording to the table of truths.This carry lookahead generator circuit will serve as a significant building block for the design of faster reversible parallel adder with low energy dissipation.

Acknowledgements

- Kannaki S, Assistant Professor, Department of Mechatronics Engineering, Sri Krishna College of Engineering and Technology. Specialized in Power Electronics and Drives.
- Bhoopesh D, UG Scholar Department of Mechatronics Engineering, Sri Krishna College of Engineering and Technology. Specialized in Power Electronics and circuits.

References

[1] Roohi, A., Zand, R., Angizi, S. and Demara, R.F., 2016. A parity-preserving reversible QCA gate with self-checking cascadable resiliency. IEEE Transactions on emerging topics in computing, **6(4)**,450-459.

[2] Askari, M. and Taghizadeh, M., 2011. Logic circuit design in nano-scale using quantum-dot cellular automata. European Journal of Scientific Research, **48(3)**, 516-526.

[3] Bhattacharya,S.;Ghatak,K.P.:QuantumCellularautomata:RecentTrendsandApplications.NovaSciencePublishers,Inc.,USA(2013).

[4] Bilal, B., Ahmed, S. and Kakkar, V., 2017. QCA based efficient toffoli gate design and implementation for nanotechnology applications. Int. J. Eng. Technol, **9**, 84-92.

[5] Dallaki, H. and Mehran, M., 2015. Novel subtractor design based on quantum-dot cellular automata (QCA) nanotechnology. International Journal of Nanoscience and Nanotechnology, **11(4)**, 257-262.

[6] Cocorullo, G., Corsonello, P., Frustaci, F. and Perri, S., 2016. Design of efficient BCD adders in quantum-dot cellular automata. IEEE Transactions on Circuits and Systems II: Express Briefs, **64(5)**, 575-579..

[7] Naseri, H. and Timarchi, S., 2018. Low-power and fast full adder by exploring new XOR and XNOR gates. IEEE Transactions on Very Large Scale Integration (VLSI) Systems, **26(8)**, 1481-1493.

[8] Hayati, M. and Rezaei, A., 2013. Design of novel efficient XOR gates for quantum-dot cellular automata. Journal of Computational and Theoretical Nanoscience, **10(3)**, 643-647.

[9] Das, J.C. and De, D., 2017, March. QCA based design of reversible carry lookahead generator. In 2017 Devices for Integrated Circuit (DevIC) (180-184). IEEE.

IOP Conf. Series: Materials Science and Engineering 1059 (2021) 012014 doi:10.1088/1757-899X/1059/1/012014

Exposure to particulate matter in different private commuting modes in Salem, India

Ramesh Kumar A[1], Jayabal S[2], Ramkumar P[3], Prabhakaran J[4]

[1]Assistant Professor, Department of Mechanical Engineering, Sona College of Technology, Salem, 636005, Tamil Nadu, India
[2]Professor, Department of Mechanical Engineering, Government College of Engineering, Thanjavur, 613402, Tamil Nadu, India
[3]Assistant Professor, Department of Mechanical Engineering,Francis Xavier Engineering College, Tirunelveli, 627003, Tamil Nadu, India
[4]Research Scholar, Department of Mechanical Engineering, Government College of Engineering, Thanjavur, 613402, Tamil Nadu, India

rameshkumara@sonatech.ac.in

Abstract. In our modern world, peoples are spending more time inside the vehicles in urban areas. This study assessed one of the critical pollutant of particulate matter PM_{10} (aerodynamic size $\leq$ 10 μm) in six private vehicles: hatchback car, sedan car, sports utility vehicle (SUV) car, multi utility vehicle (MUV) van, auto rickshaw, and motorcycle. PM pollutant in car and van transport modes evaluated under four different ventilation settings (air-conditioning (AC) ON/OFF; windows closed/open). In addition, air temperature (T) and relative humidity (RH) also assessed in six vehicles. The maximum PM exposure concentration was recorded in an auto-rickshaw, which is 4.28 times higher than the national pollution criteria. PM concentration lowered to 3.73, 1.94, and 1.53 timesin hatchback, sedan and SUV transport modes, respectively while altering ventilation condition from windows open to windows closed (AC ON) settings. There is no statistical significant difference ($p > 0.05$) between the transport modes of auto rickshaw-SUV, motorcycle-MUV, and motorcycle-sedan.All PM, T and RH parameters were correlated in car (AC ON) and auto-rickshaw transports.

1 Introduction

Indoor air quality (IAQ) generally describes the quality of air being in indoor buildings. However, in our today world peoples also spending more time in vehicles for a commute. Private vehicles have small spaces compared with buildings, so commuters can easily exposed to pollutant while travelling. IAQ is one of the comprised factors of indoor environmental quality (IEQ). IEQ includes thermal comfort (temperature and relative humidity (RH), lighting comfort, acoustic comfort, and IAQ. In this study, IAQ critical pollutant of particulate matter PM_{10}were evaluated in six private transport modes. The aerodynamic size for PM_{10} lies between 2.5 to 10 μm. Besides, thermal comfort parameters of temperature and RH also measured inside the vehicles.

The previous studies have shown that exposure to PM pollutants can cause cardiovascular diseases and respiratory problems [1, 2]. Only less number of studies are available related toparticulate matter exposure concentration in vehicles [3-6]. $PM_{2.5}$, ultrafine PM (particle number) and CO were measured in Jakarta, Indonesia, world largest pollutant traffic exposure, detected that mean concentration of pollutant was higher in public transport than the cars [7].A study conducted in Seoul, Republic of Korea, the restaurant's microenvironment had the highest particulate matters exposure than the offices, schools, buses and walking, as a result of cooking food in restaurants [8].D.U Park et al. reported that $PM_{2.5}$ and PM_{10} are higher in trains than on platforms, owing to the absence of mechanical ventilation systems [9]. Table 1 summarizes the mean concentration of PM_{10} in various location within the transports.

Most of the earlier studies of IAQ pollutants are carried out only in indoor buildings, and very least studies have been carried out in low volume space of indoor vehicles especially in India, which is third largest pollution country in the world. The present study was measured the particulate matter PM_{10} in various private vehicles. Also, evaluated the PM concentration under four different ventilation settings in four-wheeler transports (car and van). Independent 2 sample t-test was used to analyze the measured data.

Table1. Past studies mean PM_{10} concentration in differenttransport modes with location.

Past studies	Transports	PM_{10} ($\mu g/m^3 \pm SD$)	Location
Park and Ha [9]	Train	144 ± 13.1	Seoul, Korea
Chan et al. [10]	Tram	175	Hong Kong, China
	Bus (NAC)	112	
	Bus (AC)	74	
	Taxi	58	
	Ferry	81	
Strasser et al. [11]	Bus	41.4	Vienna,Austria
	Tram	73.2	
	Bicycle	16.7	
Zagury et al. [12]	Car	168 ± 53	Paris, France
Geiss et al. [13]	Car	48.6	Ispra, Italy

2 Experiments

2.1 Description of study location and routes

This study was carried out in the Salem urban, which is situated in the southern Indian state of Tamil Nadu. The city is located in latitude 11° 39' 51.5700" North and longitude 78° 8' 45.6396" East with an altitude of 278 m above sea level. Salem city is the 6th largest population (9, 19,150) among the 32 urban agglomerations population of one lakh and above peoplein the state of Tamil Nadu, India [14].

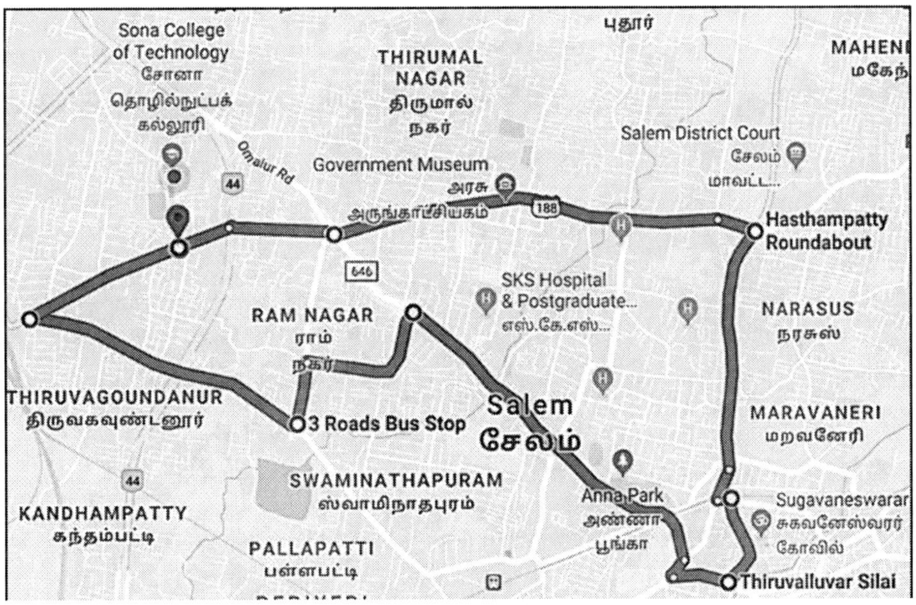

Figure1. Map showing the measured routes of Salem city.

All vehicle runs were start and end at the Sona College of Technology bus stop. Six traffic signals have on predefined routes: Hasthampatty Roundabout, ThiruvalluvarSillai, Four Roads, New Bus Stand, Three Roads and Railway Junction is shown in Figure 1. The routes covered a total distance of ~14 km length. The selected route experiences a high flow of vehicles (~6000 vehicles per hour) during rush hours in Salem city on weekdays and background site includes both anthropogenic (malls, shops, schools, colleges, banks) and non-anthropogenic (parks, no land use) activities.All vehicles were measured during the winter month of November 2019 at evening peak hour. Because, PM concentrations are mostly high during evening period compared to morning [15].A bridge construction work was undergone in predefined routes.

2.2 Measurements

A portable handheld instrument of IAQ 3007R (Rave Innovations, Noida, UP, IN) was used to measure PM, air temperature, and RH is shown in Figure 2. The maximum range of PM_{10} can measure up to 1000 $\mu g/m^3$. IAQ 3007R device is working under the principle of light scattering method. Before the experiments, the instrument was calibrated by the manufactures. The device was set to record at a one-minute time interval, and soon after, the values were dragged to the data acquisition system (laptop).

This is the first study in Salem city regarding commuter exposure assessment in transport modes. The vehicles were chosen based upon easily available private transports: hatchback car, sedan car, sports utility vehicle (SUV) car, multi utility vehicle (MUV) van, auto-rickshaw and motorcycle. The typical specification of measured transport modes is given in Table 2. The selected vehicles were often using by inhabitants in predefined routes. Before the runs, ensured all vehicles usage were within two years from the delivery date. All vehicles are manufactured in the same local company except auto rickshaw. Because that renowned company not manufactured auto rickshaw. Car and van transports are right-side driving. All private transports tested with two passengers, including the driver. The portable PM sensing device was fixed at the backseat of car and van transport adjacent to a commuter. However, the device was engaged by hand itself in motorcycle and auto rickshaw.

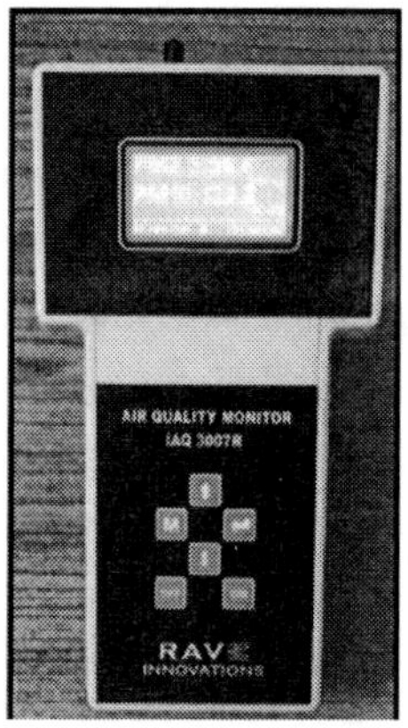

Figure2. Image of indoor air quality monitor (IAQ 3007R).

Table2. Measured transport modes.

Transport	Fuel type	Speedometer (km)	Air conditioned
Hatchback car	Petrol	12805	Yes
Sedan car	Petrol	8272	Yes
SUV car	Diesel	22273	Yes
MUV van	LPG	29320	No
Auto-rickshaw	Diesel	11096	No
Motorcycle	Petrol	3796	No

2.3 Methods

Before testing the car and van transports, doors are opened for 10 min to equalise the outdoor environment concentrations to indoor vehicles. The portable device was started as early as 15 min ahead of commencing each test. In order to avoid the interruption during measuring, smoke-free drivers are supported throughout the experiments. Therefore the drivers are fully aware of the target of the investigations. The following four feasible ventilation conditions are considered in air-conditioned cars:

- Windows open and fresh air supply mode (WO-FAS) with air-conditioning OFF
- Windows open and recirculation mode (WO-RC) with air-conditioning OFF
- Windows closed and fresh air supply mode (WC-FAS) with air-conditioning ON and
- Windows closed and recirculation mode (WC-RC) with air-conditioning ON

However, in non-air-conditioned MUV van, the above ventilation conditions are repeated without air conditioning mode. The fan speed was set to average for all the ventilation conditions. The ventilation condition are successively changed to every 3.5 km in above order to track the sudden variations of PM concentrations for different settings. For the motorcycle and auto rickshaw transports, the ventilation conditions are inappropriate. Three test runs are carried out on each vehicles in different weekdays.

3 Results and discussion

3.1 Particulate Matter

For good health conditions, National Ambient Air Quality Standards (NAAQS) recommended the acceptable limit for coarse particle PM_{10} is 100 $\mu g/m^3$ [16]. The descriptive statistical data of PM_{10} for the different transport modes is given in Table 3. The maximum PM exposure was obtained in an auto-rickshaw (428.24 $\mu g/m^3$), which is 4.28 times higher than the NAAQS. The auto-rickshaw have a vast opening in an enclosed compartment to ingress exhaust and non-exhaust particles compared to other transports.

Table3. Descriptive statistical data of particulate matter PM_{10} for the different transport modes.

Transport	Ventilation condition	Min* ($\mu g/m^3$)	Max* ($\mu g/m^3$)	Mean ($\mu g/m^3$)	SD* ($\mu g/m^3$)	CV* (%)
Motorcycle	-	52.68	334.35	126.07	67.51	53.55
Auto-rickshaw	-	178.63	428.24	266.07	54.28	20.4
Hatchback	WO-FAS	43.52	155.73	99.31	35.02	35.26
	WO-RC	50.39	148.86	101.69	33.5	32.95
	WC-FAS	38.94	160.31	69.09	33.15	47.99
	WC-RC	16.04	45.81	27.24	10.65	39.09
Sedan	WO-FAS	183.21	281.68	238.17	36.02	15.12
	WO-RC	167.18	293.13	212.98	43.48	20.41
	WC-FAS	158.02	258.78	198.41	30.11	15.17
	WC-RC	61.84	178.63	109.24	37.64	34.46
SUV	WO-FAS	194.66	380.15	275.64	53.58	19.44
	WO-RC	231.3	370.99	297.25	51.76	17.41
	WC-FAS	215.27	350.38	266.22	36.1	13.56
	WC-RC	146.57	270.23	193.52	45.12	23.32
MUV	WO-FAS	77.87	139.7	97.97	20.6	21.03
	WO-RC	96.19	183.21	118.4	26.92	22.74
	WC-FAS	98.48	139.7	112.98	13.84	12.25
	WC-RC	91.61	153.44	114.51	20.61	18

*Min - minimum; Max - maximum; SD - standard deviation; CV - coefficient of variation.

The lowest mean PM concentration of hatchback, sedan and SUV were 27.24 ± 10.65 $\mu g/m^3$, 109.24 ± 37.64 $\mu g/m^3$ and 193.52 ± 45.12 $\mu g/m^3$, respectively under WC-RC ventilation mode. Except hatchback, other two car transport (sedan and SUV) mean PM concentration exceeds the NAAQS limit. Figure 3 depicted the boxplot of PM_{10} concentration for the six private vehicles. The middle line of the boxes represents mean PM exposure. From figure 3, it is noticed that hatchback only the transport meets the PM standards (<100 μg/m^3) of NAAQS. The mean PM levels of motorcycle and auto rickshaw were 1.26 and 2.66 times higher than the national standards, respectively.

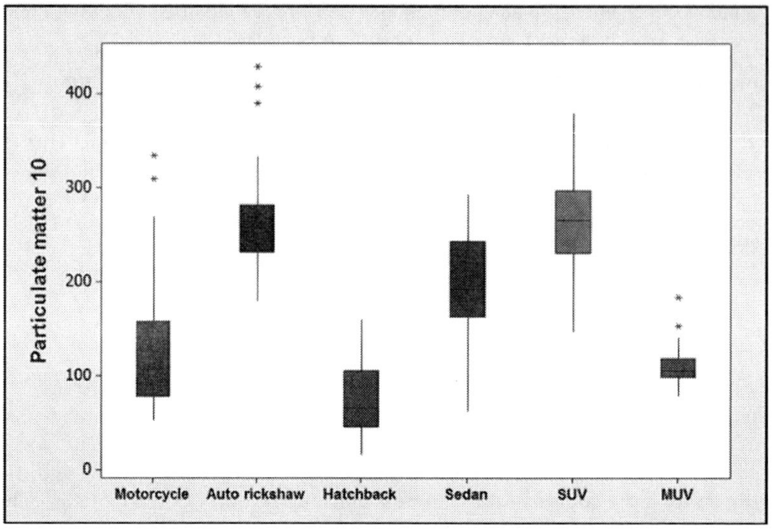

Figure3. Box plot of particulate matter PM_{10} for the six different transport modes.

World health organization (WHO) recommended 24-h exposure limit is 50 μg/m^3 [17]. According to WHO, hatchback transport meets the requirements under windows closed recirculation mode ventilation settings. When comparing the cabins of hatchback, sedan, SUV, and MUV, hatchback car have a low volume interior space. It means occupants PM exposure concentration get higher when the volume of vehicle compartments higher. SUV, auto rickshaw, and sedan mean PM concentration were 2.35, 2.11, and 1.88 times higher than the motorcycle concentrations, respectively. However, hatchback and MUV mean PM concentration were 1.23 and 1.01 times lower than the motorcycle.

In this study, independent 2 sample t-test was used to find PM exposure difference between the vehicles with a 95% confidence level. Completely fifteen t-tests were carried out among the six vehicles. The independent 2 sample t-test between the transport modes is given in Table 4. For our climatic conditions, usually, car occupants are approach windows closed (AC ON) ventilation settings and Non AC van occupants approach windows open ventilation settings. The above two settings of samples are taken for t-test to the corresponding vehicles. There is no statistically significant difference between the auto rickshaw-SUV, motorcycle-MUV, and motorcycle-sedan ($p > 0.05$) at 95% confidence level. It shows that auto-rickshaw and SUV commuters are exposed to the same level of PM_{10} concentrations. Similarly, between motorcycle/MUV and motorcycle/sedan transports.

While there is a significant difference between AC mode and non AC mode ventilation settings in car transports for PM exposure is shown in Figure 4. It is clearly shown that PM exposure was lower during AC ON setting (windows closed) than the AC OFF (windows open). Even PM concentration are lowered from fresh air supply mode (FAS) to recirculation mode (RC) ventilation settings in car transports. PM levels were 3.73, 1.94, and 1.53 times lowered in hatchback, sedan and SUV, respectively, when car settings altered from windows open to windows closed. In MUV van transport mode, the mean PM_{10} was 108.72 ± 25.71 μg/m^3 and 113.75 ± 17.05 μg/m^3 for windows open and windows closed, respectively. Therefore, it is concluded that PM exposure in Non AC MUV vehicles acts consistently on both windows opened and windows closed settings.

Table4. Independent 2 sample t - test for particulate matter PM_{10} between the transport modes.

t-test	Transports	Mean (μg/m^3)	SD (μg/m^3)	Mean difference (μg/m^3)	t	p-value
1	Motorcycle	128.4	69.5			
	Auto rickshaw	266.1	54.3	-137.7	-9.63	< 0.001
2	Motorcycle	126	65.4			
	Hatchback	51.2	33.2	74.8	4.67	< 0.001
3	Motorcycle	126	65.4			
	Sedan	155.9	56.3	-29.9	-1.59	0.119*
4	Motorcycle	129.6	64.9			
	SUV	237.1	53.3	-107.5	-5.73	< 0.001
5	Motorcycle	133.3	64.4			
	MUV	108.7	25.7	24.6	1.54	0.136*
6	Auto rickshaw	261.9	45.9			
	Hatchback	51.2	33.2	210.7	17.06	< 0.001
7	Auto rickshaw	261.9	45.9			
	Sedan	155.9	56.3	106	6.69	< 0.001
8	Auto rickshaw	262.4	47			
	SUV	237.1	53.3	25.3	1.59	0.12*
9	Auto rickshaw	262	48.3			
	MUV	108.7	25.7	153.3	12.22	< 0.001
10	Hatchback	51.2	33.2			
	Sedan	155.9	56.3	-104.7	-7.35	< 0.001
11	Hatchback	45.7	22.4			
	SUV	237.1	53.3	-191.44	-14.81	< 0.001
12	Hatchback	43.04	19.5			
	MUV	108.7	25.7	-65.66	-8.88	< 0.001
13	Sedan	155.8	57.8			
	SUV	237.1	53.3	-81.3	-4.62	< 0.001
14	Sedan	154.8	59.2			
	MUV	108.7	25.7	46.1	3.11	0.005
15	SUV	231.2	47.4			
	MUV	108.7	25.7	122.5	9.9	< 0.001

*no significant difference ($p > 0.05$)

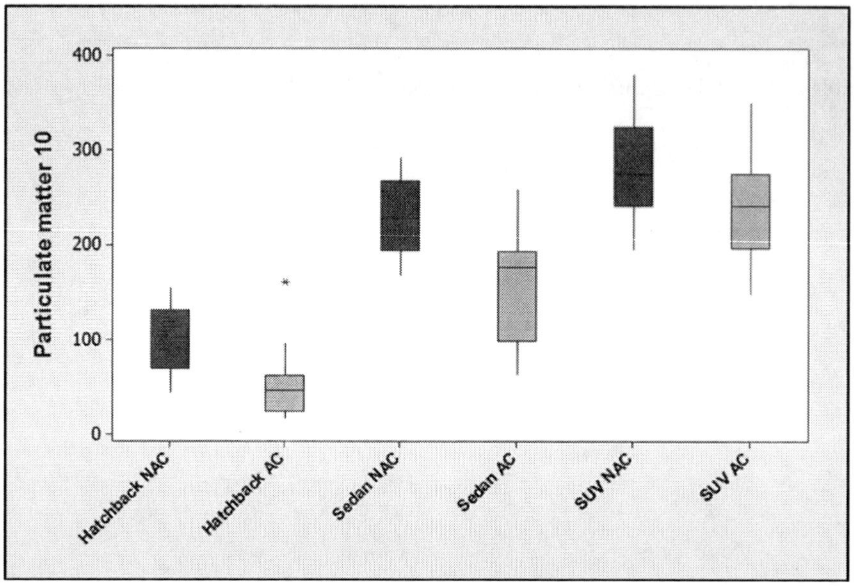

Figure4. Boxplot of particulate matter PM_{10} exposure between AC and Non AC ventilation mode in car transports.

3.2 Air temperature

The maximum air temperature of 39.9°C has obtained in the Non AC (windows closed) MUV transport mode. Temperature difference for the different transport modes is shown in Figure 5. The minimum temperature of 24.5°C was obtained in AC sedan transport under WC-RC ventilation condition. According to the ASHRAE, the indoor thermal comfort air temperature could range from 19.6°C to 27.9°C [18]. The sedan and hatchback vehicles were achieved the ASHRAE limit under WC-RC condition. Compared the sedan minimum temperature to other vehicles minimum temperature, the differences were 3.3°C, 4.4°C, 8.7°C, 8.1°C, and 10°C for the hatchback, SUV, motorcycle, auto-rickshaw and MUV, respectively.

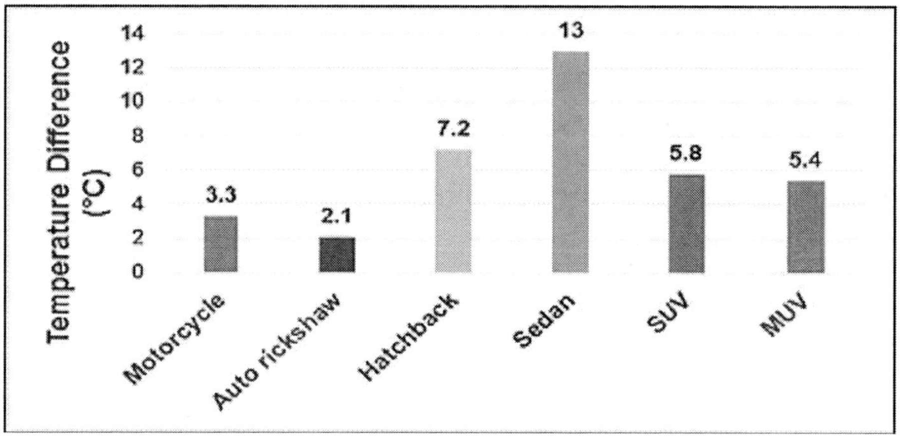

Figure5. Temperature difference obtained in six transport modes.

3.3 Relative humidity (RH)

Generally, acceptable RH is < 65% recommended by ASHRAE, because more than that can lead to microbial growth [19]. The RH may well change influenced by the air temperature. However, RH was below the ASHRAE standards in all the transport modes is shown in Figure 6. RH gradually increased when the ventilation condition changed from WC-FAS to WC-RC. The RH difference was 3%, 8% and 13% attained for the motorcycle, auto rickshaw and MUV transport modes, respectively.

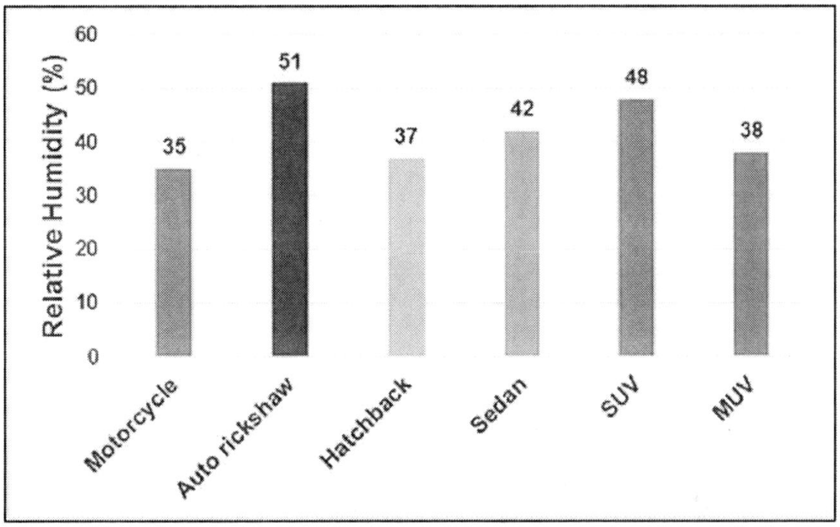

Figure6. Maximum relative humidity obtained in six transport modes.

3.4 Correlation between PM_{10}, air temperature (T), and relative humidity (RH)

The correlation analysis showsany relation among the parameters of PM, T and RH in different modes of transport. Pearson correlation between PM, T and RH of six vehicles at 95% confidence level is given in Table 5. As like before, for correlation analysis also windows closed (AC ON) sample were taken for car transport and windows open sample were considered for van. All parameters are correlated in AC cars and auto-rickshaw vehicles. As we know that, temperature and RH parameters are always correlated. According to that in all transport modes, T/RH was correlated with p-value ≤ 0.001. However, for MUV van PM_{10}/T and PM_{10}/RH parameters are not correlated due to windows open and Non AC vehicle. Similarly, PM_{10}/T and PM_{10}/RH parameters are not correlated for motorcycle due to directly exposure to outdoor concentrations.

Table5. Pearson correlation between the*PM_{10}, air temperature (T), and relative humidity (RH) in six vehicles.*

Transport	PM_{10}/T		PM_{10}/RH		T/RH	
	Pearson Correlation (r)	P-value	Pearson Correlation (r)	P-value	Pearson Correlation (r)	P-value
Hatchback car	0.918	<0.001	-0.745	<0.001	-0.931	<0.001
Sedan car	0.755	<0.001	-0.719	<0.001	-0.896	<0.001
SUV car	0.881	<0.001	-0.690	<0.001	-0.722	<0.001
MUV van	0.409	0.092*	-0.448	0.062*	-0.862	<0.001
Auto rickshaw	0.529	0.001	-0.350	0.031	-0.684	<0.001
Motorcycle	-0.222	0.164*	-0.309	0.052*	-0.486	0.001

* no significant correlation (p>0.05)

4 Conclusions

The mean PM exposure in hatchback, sedan, SUV, MUV, auto rickshaw, and motorcycle transport were 27.24 (WC-RC), 109.24 (WC-RC), 193.52 (WC-RC), 97.97 (WO-FAS), 266.07, and 126.07μg/m^3, respectively.The least PM concentration obtained in hatchback (16.04 μg/m^3) under windows closed recirculation mode (AC ON) condition. PM exposure concentration was increased according to the indoor volume of car vehicles increased. Occupants are exposed to high PM concentration during windows open than the windows closed (AC ON). The independent 2 sample t-test showed no statistically significant difference ($p > 0.05$) between the auto rickshaw-SUV, motorcycle-MUV, and motorcycle-sedan. All measured parameters of PM, air temperature (T), RH were significantly correlated ($P \leq 0.001$) in the car and auto-rickshaw vehicles. Finally, suggesting the commuters operate vehicles under air-conditioning mode to expose below acceptable limit of pollutant.

References

[1] M. Riediker, R. Williams, R. Devlin, T. Griggs, P. Bromberg 2003 Exposure to particulate matter, volatile organic compounds, and other air pollutants inside patrol cars, *Environ. Sci. Technol.* **37 (10)** 2084-93.

[2] M. Mohsen, M.B. Ahmed, J.L. Zhou 2018 Particulate matter concentrations and heavy metal contamination levels in the railway transport system of Sydney, Australia, *Transport Res D-Tr.* **62** 112-24.

[3] X. Han, M. Aguilar-Villalobos, J. Allen, C.S. Carlton, R. Robinson, C. Bayer, L.P. Naeher 2005 Traffic-related occupational exposures to $PM_{2.5}$, CO, and VOCs in Trujillo, Peru, *Int. J. Occup. Environ. Health.* **11 (3)** 276-88.

[4] Y. Cha, S. Abbasi, U. Olofsson, U 2016 Indoor and outdoor measurement of airborne particulates on a commuter train running partly in tunnels, *P I MechEng F-J Rai.* **232(1)** 3-13.

[5] B. Xu, Y. Wu, Y. Gong, S. Wu, X. Wu, S. Zhu, T. Liu 2016 Investigation of volatile organic compounds exposure inside vehicle cabins in china, *Atmos. Pollut. Res.* **7(2)** 215-20.

[6] N. Hudda, S.A. Fruin 2018 Carbon dioxide accumulation inside vehicles: The effect of ventilation and driving conditions, *Sci. Total Environ.* **610-611** 1448-1456.

[7] A.F Both, D. Westerdahl, S. Fruin, B. Haryanto, J.D Marshall 2013 Exposure to carbon monoxide, fine particle mass, and ultrafine particle number in Jakarta, Indonesia: Effect of commute mode, *Sci. Total Environ.* **443** 965-72.

[8] Y. Hwang, K. Lee 2018 Contribution of microenvironments to personal exposures to PM_{10} and $PM_{2.5}$ in summer and winter, *Atmos. Environ.* **175** 192-8.

[9] D.U Park, K.C. Ha 2008 Characteristics of PM_{10}, $PM_{2.5}$, CO_2 and CO monitored in interiors and platforms of subway train in Seoul, Korea, *Environ. Int.* **34(5)** 629-34.

[10] L.Y. Chan, W.L. Lau, S.C. Lee, C.Y. Chan 2002 Commuter exposure to particulate matter in public transportation modes in Hong Kong, *Atmos. Environ.* **36(21)** 3363-73.

[11] G. Strasser, S. Hiebaum, M. Neuberger 2018 Commuter exposure to fine and ultrafine particulate matter in Vienna, *Wien. Klin. Wochenschr.* **130(1-2)** 62-9.

[12] E. Zagury, Y. Le Moullec, I. Momas 2000 Exposure of Paris taxi drivers to automobile air pollutants within their vehicles, *Occup Environ Med.* **57(6)** 406-10.

[13] O. Geiss, J. Barrero-Moreno, S. Tirendi, D. Kotzias 2010 Exposure to particulate matter in vehicle cabins of private cars, *Aerosol Air Qual Res.* **10(6)** 581-8.

[14] Census of India 2011 Urban Agglomerations/Cities having population 1 lakh and above. Office of the Registrar General & Census Commissioner, Government of India.

[15] R.A. Abbass, P. Kumar, A. El-Gendy 2020 Car users exposure to particulate matter and gaseous air pollutants in megacity Cairo, *Sustain Cities Soc.* **56** 102090.

[16] CPCB (Central Pollution Control Board) 2009 National Ambient Air Quality Standards. India. http://cpcbenvis.nic.in/air_pollution_main.html#

[17] World health organization (WHO), Air quality guidelines – global update 2005. https://www.who.int/airpollution/publications/aqg2005/en/

[18] ASHRAE (American Society of Heating, Refrigerating and Air-Conditioning Engineers) standards (2017) 55-2017, Thermal Environmental Conditions for Human Occupancy, USA.

[19] ASHRAE (American Society of Heating, Refrigerating and Air-Conditioning Engineers) standards (2016) 62.1-2016, Ventilation for Acceptable Indoor Air Quality, USA.

Leakage Detection in High Pressure Vessels using Non Invasive Method

V Prabhu[1*], K.S Vairavel[2], R Velnath[3]
[1] Assistant Professor, Bannari Amman Institute of Technology, Sathyamangalam
[2] Assistant Professor, Bannari Amman Institute of Technology, Sathyamangalam
[3]Professor, Bannari Amman Institute of Technology, Sathyamangalam
Email ID: prabhuv@bitsathy.ac.in

Abstract: Leakages from industrial vessels can damage equipments and leads to unexpected shutdown. Production and Manufacturing operations can be affected by even small leakages. Above all the occupational safety is compromised if the leaked material is poisonous or combustible. Leaks in industrial facilities are not something to be taken lightly, as exposure to harmful gases and liquids can cause severe health effects. It is the responsibility of the engineers to identify the risks associated with unexpected leakages and ensure that the leak is found at an early stage. There are different methods available to detect leakages like acoustic, gas injection etc. This paper presents a novel method to detect leakage in high pressure vessels with combustible products using Infrared Radiation (IR) image scanning.

Keywords: leakage; detection; industry; pressure; vessels

1 Introduction

Leakage detectors forms the critical part of a plant safety system. When there is risk associated with leakages, a fixed leakage detecting system must be placed at points. The term "leakage" appears in the field of engineers working with high pressure vessels. A leak explicitly means an unintended hole or crack in an enclosed surface which allows the escape of closed medium. The critical function of leak detection system is to locate the spot of flaw. This leak test assures the integrity of a structure. Pressurized containers need to be properly sealed between external atmospheric conditions. Most of the leaks occur in joints where gaskets are used.

Infrared (IR) thermography is a well established technique used in both commercial and industrial applications. For example in commercial applications an IR thermography can be used to detect the worn out brakes in automobiles. It is a popular technique owing to its accuracy in measurement. This paper presents a method for detecting leaks in a pressurized vessel using infrared imaging. Using image processing algorithms, the region of interests where the leak is to be detected is focused and the image is segmented to determine the information about the leakage point. The output of the experimental work obtained confirms that this method is effective in detecting the presence of leaks in industrial vessels.

2 Materials and Methods

IR thermography uses infrared cameras to detect radiation in the range of 900-1400 nanometer. The image obtained through this process is called as a thermogram. Any object which is above a temperature of absolute zero emits radiation. The IR camera can detect this radiation from a surface at different angles and calculates the temperature difference using the color bar. The thermal cameras are applied in night visions scopes which are used military. The maximum acceptable leak rate for a given application depends on its nature of reaction. The image obtained from the thermal camera is used to determine the level of image using its color which in turn gives the details about an object like its shape. IR camera performs in two different modes

1) Active Thermography
2) Passive Thermography

Active Thermography: - It is a type of thermography where the area of interest is focused with the background contrast. This method requires the inspected region to be in equilibrium with the environmental surroundings.
Passive Thermography: - It is a type of thermography where the area of interest is identified with respect to the temperature difference compared to the background.

IR Thermography camera is used based on the number of pixels used to construct an image which varies upon the application. The major factor to be considered when using IR camera is the distance of target. The target to be focused should be within a limited distance such that the infrared rays emitted by the object can be detected without loss. The thermal camera distinguishes the cooler area and warmer area of an object with the differences in color.

3 Leak Identification Techniques

The first method of identifying leak is by using acoustic methods. This method detects leakage sounds using sensitive microphone and correlating techniques. Leakage in high pressure vessels creates s waves created by localized loss of energy. These waves are then recorded using an acoustic sensor installed on the target. This method of leakage detection is primarily used in targets with smaller surface area.

A gas injection technique uses a trace gas to detect leaks. Gases like helium, hydrogen or sulfur hexafluoride (SF6) is injected into the system and is traced if it comes out of the target area by reading its drop in pressure. Gas injection technique is preferably used in low-pressure conditions and is not so accurate as other methods. This is due to the number of additional equipments required to provide an output. The trace gases are chosen such that they are safe, nontoxic, and nonflammable. The major application of this method is to find leaks in a straight section of a pipe. The tracer gas is injected into the system through upstream of the suspected leakage point. A hand held sensor is used for tracing the presence of trace gas through leakage. Several other innovative techniques are prevalent today as an alternative to the conventional techniques described above. Some of these techniques include radar imaging, thermal imaging, and in-pipe acoustic technology.

Due to the critical nature of leakages there is a need to test the system integrity even for small leakages which leads to rise in production costs [1]. Infrared imaging has found its application in process industries and refineries to detect hazardous leakages in the carrying pipelines [2]. Thermal imaging is the used for detecting gas as well as liquid leaks and is one of the most accurate method of gas leakage detection [3]. This method is profoundly used in detecting gas leakages because it detects the radiation emitted from the pipeline and displays the image on the camera. The image processing algorithm sequence is, categorized into three sections which are pre-processing, feature extraction and classification [4]. Thermal imaging measures the radiation form a human skin, providing an insight into the thermal dysfunction associated with breast cancer [5]. The image obtained from the thermal camera is used to determine the level of image using its color which in turn gives the details about an object like its shape [6], [7]. A different method was proposed to detect water leakage in pipes using a low and high resolution camera which generated a thermal profile [8]. Another method was proposed where using active IR imaging was successful applied to detect blockages in pipes [9]. A work with how environmental variables allow isolating a periodic phenomenon to assist on zone determination using IR images was done which proved that it was difficult to apply due to the thermal equilibrium of the materials [10]. In the image shown in figure 1the temperature difference in cooler area has a leak. The color bar provides the details of the minimum and maximum temperature. The obtained image is then processed in MATLAB and which gives the output as difference in temperature. These existing methodologies provide us with challenges to detect the actual temperature at the leakage point and for analyzing the pattern of crack development.

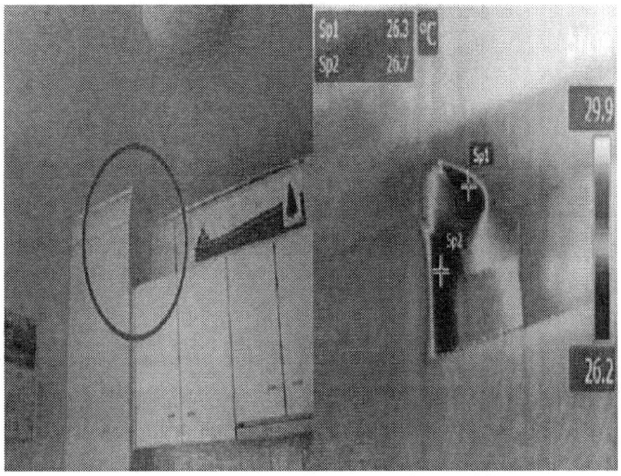

Figure 1. Thermal image of an object

4 Proposed Methodology

It is proposed to use an active infrared camera to capture the thermo graphic image of a highly pressurized vessel. The image captured is processed by image processing algorithms to detect the region where temperature exceeds a certain limit. The pseudo-colored image from the camera has colors ranging from brighter ones like orange which represents higher temperatures and cooler ones like blue which represents low temperatures. Earlier methods used feature extraction by using segmentation to partition the image or by using binary thresholding to separate the foreground pixels from background pixels and identify the difference in temperature in low temperature area which has a defect. The obtained image is then processed to get the exact point of leakage and its temperature value. In our method we have utilized the colr map feature in Matlab to detect the crack as well as the temperature.

For testing the process an IR camera was used along with a raspberry pi to capture the image of a pressure gauge connection port in a water line. A color map was used for finding image and color map regions of the image. The dimensions of the image were obtained and the number of color channels was fixed as 3. The surrounding clutter was cropped off to get the color bar. The image was divided by 255, since color map values must be between 0 and 1. The image was then flipped since the low rows are the high temperatures. Then the image was converted from an RGB image to a grayscale indexed thermal image. The highest temperature point was defined and also the temperature at the dark end of the scale was defined which would be the low temperature. The image is scaled to display its approximate actual temperatures as shown in figure 2 & 3.

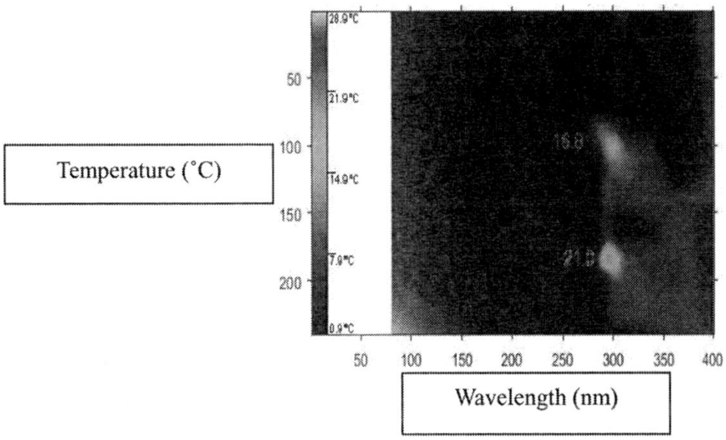

Figure 2. Original colored image displaying the temperature at leakage points

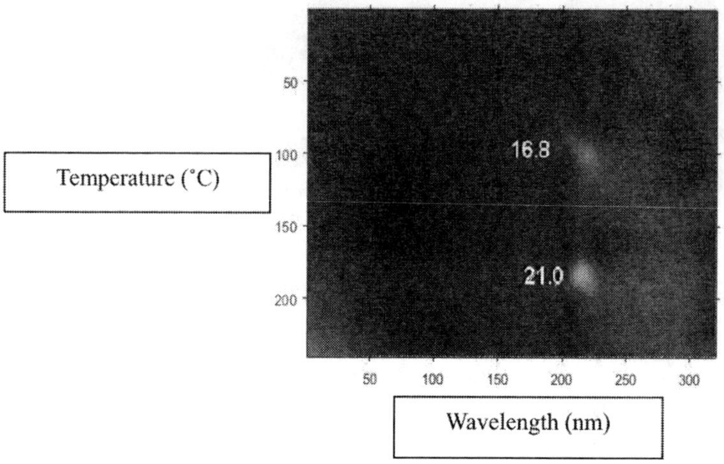

Figure 3. Grey scale image displaying the temperature at leakage points

5 Results and Conclusion

Detection of leakage in industrial vessels is an extremely important task to reduce downtime as well as avoid occupational hazard. This paper has suggested a non invasive method to identify leak without compromising the integrity of the structure. The leakage area has been detected using IR imaging and the temperature value at the leakage point was also determined. Leakage spot and its corresponding temperatures have been identified using raw pseudo coloured image obtained from IR camera. Our future scope of work is to detect the crack features such as the width, length and depth on the surface using image processing work and include spatial filters to identify and analyse the size of crack in the structure.. We have also planned to identify the direction of propagation of the crack in order to analyze the pattern of crack development and to implement corrective measures. This approach is suitable for other applications like detecting leakages in commercial pipelines and other industrial applications. This active method of using an infrared imaging can also be used for a broad range of applications such as blockage detection in pipelines, detecting erosion due to cavitation, corrosion on the inner walls of a tank and detecting cracks in industrial vessels. We have also planned to study and relate the temperature changes associated with the pattern of crack development so that appropriate actions can be taken by the engineers to seal the leakage spot.

References

[1] Manekiya, M.H. and Arulmozhivarman, P., 2016, April. Leakage detection and estimation using IR thermography. In 2016 International Conference on Communication and Signal Processing (ICCSP) (1516-1519). IEEE.

[2] Carreño-Alvarado, E.P., Ayala-Cabrera, D., Pérez-García, R. and Izquierdo, J., 2014. Identification of buried pipes using thermal images and data mining. Procedia Engineering, **89**,1445-1451.

[3] Murvay, P.S. and Silea, I., 2012. A survey on gas leak detection and localization techniques. Journal of Loss Prevention in the Process Industries, **25(6)**,966-973..

[4] Colombo, A.F., Lee, P. and Karney, B.W., 2009. A selective literature review of transient-based leak detection methods. Journal of hydro-environment research, **2(4)**,212-227.

[5] Pérez, R., Puig, V., Pascual, J., Quevedo, J., Landeros, E. and Peralta, A., 2011. Methodology for leakage isolation using pressure sensitivity analysis in water distribution networks. Control Engineering Practice, **19(10)**,1157-1167.

[6] Liu, Z. and Kleiner, Y., 2013. State of the art review of inspection technologies for condition assessment of water pipes. Measurement, **46(1)**,1-15.

[7] Yadav, G., Gupta, V., Gaur, V. and Bhattacharya, M., 2012. Optimized flame detection using image processing based techniques. Indian Journal of Computer Science and Engineering, **3(2)**, 202-211.
[8] Carreño-Alvarado, E.P., Ayala-Cabrera, D., Pérez-García, R. and Izquierdo, J., 2014. Identification of buried pipes using thermal images and data mining. Procedia Engineering, **89**, 1445-1451.
[9] Inagaki, T. and Okamoto, Y., 1997. Diagnosis of the leakage point on a structure surface using infrared thermography in near ambient conditions. Ndt & E International, **30(3)**, 135-142.
[10] Britton, T., Cole, G., Stewart, R. and Wiskar, D., 2008. Remote diagnosis of leakage in residential households. Journal of Australian Water Association, **35(6)**, 89-93.

Development of Mobile Application Incorporating Augmented Reality for E-education and Training Systems

Madhankumar S[1], Prawin Sankar T A[2*], Shibin Thomas S[2], Brundhavani G[2], Prince Paul[2], Vishnu Girishkumar[2]

[1]Assistant Professor, Department of Mechatronics Engineering, Sri Krishna College of Engineering and Technology, Coimbatore, Tamil Nadu, India - 641008.
[2]UG Scholar, Department of Mechatronics Engineering, Sri Krishna College of Engineering and Technology, Coimbatore, Tamil Nadu, India - 641008.

*Corresponding author, Email Id: 18eumt071@skcet.ac.in

Abstract.A visual platform to help the learners to enhance their knowledge of understanding. The learners can use the application as an aiding source to increase their visual understanding of knowledge through immersive 3D objects (images) in the real-world. The 3D objects are displayed in a rendered perspective aided by multimedia to be shown to the user in the real-world environment. The end-user uses an application to view the Augmented images in the real-world. The 3D visuals give the users apparently an immersive experience through user interactive environments. Complex concepts in STEM education can be apparently be displayed in 3D visuals with ease, which can improve the understanding perspective of the individual concept displayed in Augmented Reality. In the education system, this style of learning can be drastically improvised using this source of modern technology. The product is a mobile application which can be accessed offline or online according to the need and requirements of the user. The existing educational system can be complemented with this solution to enhance the knowledge and understanding of the users with technical expertise.

1 Introduction

Augmented Reality is an innovative development built over the last three decades that serves a key part in diverse use cases and implementations in all science classifications. Objects that are viewed throughout the physical world become persistent computer-generated data sources. In an AR scenario, the physical reality is not directly affected: it is an interface among the physical environment and simulated entities [1].This information can be displayed and controlled as visual, olfactory, auditory, or haptic. AR requires the integration of virtual and physical knowledge in real time across a multitude of electronic platforms, like mobile phones or tablets, to build this virtual world, which has an effect on the amount of innovations produced for AR[2].Qualitative approach assessment is employed to analyse the basic features of AR technologies in Educational leadership, the teaching methods and procedures used in the research studied, as well as the assessment approaches adopted in the strategies[3].Multiple reports have revealed patterns, benefits, prospects, threats and impacts of this innovation on learning. Consequently, almost all of the earlier research have neglected to examine pedagogies, in any way missing the fact that the effectiveness of the solution relies not just on the technological features of the systems, but rather on the instructional methods to incorporate them[4].Virtual objects are displayed in real world with user interactive positions and control of the object. The object can be controlled by general 3D object control. The primary goal of using Augmented Reality is to display digital objects which are virtual in the real-world perspective. This immersive experience can be interlinked with actual universe in such a manner that provides an interactive element to the natural setting. As AR emerges, more learners can participate interactively and share knowledge in an improved manner. Augmented reality is being used in several fields of research and conceptual practices. This has been utilized, within and between many, through: navigating (in vehicles, aircraft), defence (image monitoring, making plans), geophysical and geographical research (isometrical charts, landscape assessment), architectural (rehabilitation including its presence of destroyed structures, modelling of ventures), healthcare (visualisation of treatments), Amusement (animated exhibits, meetings) as well as cultural programmes. The thought to be asked now is whether AR technology could be utilized effectively in the teaching of library and knowledge science professionals [5]. Computer Generated Simulations of science concepts and historical content can help learners with better understanding. The visual images can be self-explanatory or user interactive.AR are being defined like an enhancement of actuality by virtual environment like Three dimensional and details. In the manufacturing setting, such material also contains development measures involved, job worksheets and repair orders.The basic elements of the AR environment entail: simulation emerging technologies, sensor system, monitoring unit and host controller[6].The best part of this AR technology is that it kindles curiosity in the minds of the leaners. With the evolution of AR, learners may engage collaboratively as well as construct quite genuinely through information. Instead of being inactive users, learners will become active participants who can engage through the educational environments. Throughout order to enrich students ' learning contexts, it is important to enhance ones sensory and intelligent participation via the use of technology, in particular until trying to understand esoteric and complicated techniques. Augmented reality modules can provide contextualised, merely-in-time guidance; self directed compilation of physical environment visuals and illustrations; as well as responses on learner behaviours [7].Knowing education processes, learning goals, performance requirements and the expertise, abilities and competencies needed in the area of AR can assist to identify patterns, insights and possibilities for more advancement of creative AR initiatives[8].Through AR innovation, the creation of successful and fascinating technology based instructional resources has become feasible. In another side, retaining overlaid details and recognising AR as an invasive technique are few of the key drawbacks of AR applications [9]. Experts and scholars often tried to extend AR to classroom-based learning in topics such as physics, arithmetic, engineering, materials science, astrophysics, and some K-12 or higher education, as well as to incorporate AR through expanded textbooks and learning manuals.In AR based interactive modelling framework which allows users to envision and change simulated premium products via elegant environments.

As previous research about the use of AR in teaching science have always concentrated on studying/educational outcomes, encouragement, and attitude, it's also assumed that perhaps the analysis of such factors may still approach the stage of exhaustion. Focusing on memory problems, experiences and social tasks outside these factors could be more useful in the long run[1]. In today's College, students are open to a range of educational stimuli, and often it is hard to identify and return the favor several principles at the same moment. This framework would both fulfill the intent and allow each student to develop in a specific teaching ability.An AR-based CAD framework allows participants to walk about in a specific 3D space to simulate digital items via HMDs and communicate with others and change digital items utilizing system that alerts. In addition, users may experience the physical country and make maximum usage of actual objects with interaction / methodological approaches[10, 11].Mobile Application-based technology through e-learning can be much more centered throughout the coming years on scientific literacy funded by AR, as this is one of the cutting-edge innovations that are trend-setting. Instituting virtual reality throughout the curriculum will turn a daily learning activity into an interactive virtual activity.AR method gives simulated models in the real-life context and incorporates game features to reinforce textbook content. Mostly as result, students are becoming more engaged with learning interest. AR lets learners to comprehend the knowledge or ideas they have mastered. As AR can indeed be implemented and accompanied by the existing school system, an effective instructional framework can be created.The core ideas behind its study, its guiding criteria, operational processes and the operating method can be clarified step by step together with the safety procedures that really need to be taken can be clearly illustrated utilizing AR. Teaching approach could be further improved by AR quizzes and exams[12].

AR has a tremendous scope for many areas of technology. While it have been widely seen in clinical or defense environments for several years, industrial technologies are mostly viewed as discrete technologies which are only available in a given and fixed workplace[13].New businesses are built with modern technology and ecosystems. In the high-tech industrial climate, authorized people are designed to attend challenging service and repair operational activities. Various learning strategies assisted by e-learning networks were intended to enhance the abilities of staff.To this end, new innovations such as AR, are changing the way in which learning materials are delivered to qualified professionals[14].

2 Materials and methods

The potential of AR systems can be divided into 3 categories: entity, model and person. Entity are all nonhuman physical environment artefacts (workpiece, equipment, climate, etc.). The model comprises certain digital resources (framework, documents, concept renderings). The person being the learner to really be assisted by its AR device. The AR improves the overall functionality among virtual and personal data leading to enhanced visual representation and regulate. The objects will be actual 3D models viewed by the student made of model with lighting and animation Its simply the model with Animation, Lighting and User Interface. The model will be a CAD model of a specific BASE TARGET, documentation, design drawing. Human will be eliminated in this context as the entire process can be automated. The AR model can be further used to train human workers by creating virtual models of different objects and adding the technical specifications and various functionals. SOLIDWORKS is used to design a 3D model and then the model is exported in FBX format. FBX format of the model is then imported in MAYA and the required animation in given the model and again the model is exported in FBX format with the animation and the model is kept imported in UNITY 3D. By using VUFORIA, an AR Tool Kit the base target (using image) or Ground Target (using ground) or User Defined Target (using any base target) for devices that does not support Ground detection. The programming for the application is done using VISUAL STUDIO the program used is C# developed by Microsoft. All of these is imported separately in same project in Unity 3D is developed using Android SDK and Android Studio.

As AR is being used as a platform for gamification, it would have the ability to serve as a motivating weapon, vast usage in studies so it can be implemented in school and college students text -books and on ground plane. Augmented Reality can be customized according to the need of the learning requirement from elementary schooling level and can be extended up to Engineering domain level. By using Vuforia the target is set as the text book or even as Ground Plain and the 3D models are uploaded for each target and when the student views the base target in the textbook using his AR camera the 3D model of the image can be viewed. And by giving animation to 3D model the exact working of the 3D model can be added and the student can view the working of it in all directing unlike a video all directions can be viewed in every frame.

By implementing Multimedia based animation, it seems like a video which can be viewed in multi directional positions and in high quality. Languages and other content can be kept constant so that the size of the application is reduced to great extent. Learners might well interpret by the AR System book through displaying visuals over the actual sections augmented with sound as well as an expressed storytelling mostly by book's author.Figure 1 shoes the overall Workflow diagram.

Implementations involving object monitoring are mainly focused on the lesser method of the label, that is focused on the identification of pixel values throughout the context of attributes and the designation of adjectives. Such obtained features are being used to apply computer graphics to the physical world[15].Google AR Core is an open source APK provided by Google. AR Core APK (Application Package) can be used to develop Baseless target Detection using which the Augmented Image can be visualized in any surface without any base or reference objects .AR Core can be used to develop cross platform Applications. Using AR Core Baseless Target Detection, the application can be made more interactive by visualizing AR in any plane surface (Plane surface can be complemented with a spatial plane which has features for detection). Before Augmented Reality image (3D visuals) is projected and can be viewed, the surface is intensively scanned for distinct features which acts as a partial reference for the AR image in the Real-World view. These features are captured are

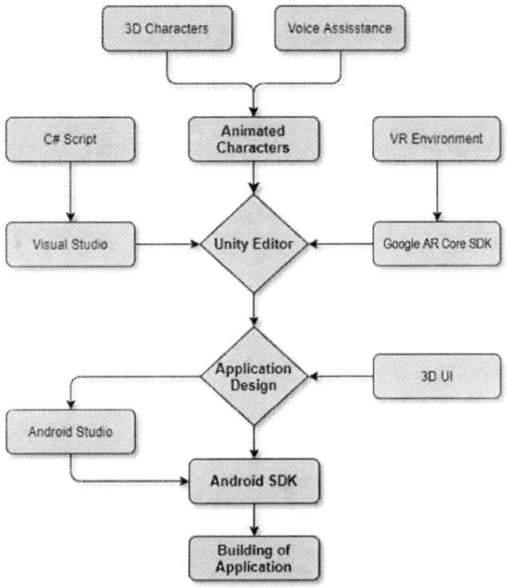

Figure 1. Overall Workflow diagram

2.1 Training Systems with implementation of Virtual Laboratory, Maintenance and Inspection

Over recent times, production firms increasingly encountered a variety of obstacles connected to the still-changing demands of buyers and producers. The goal of the interactive media plans as well as the planned initiatives is to leverage the employment opportunities arising from ongoing effect of fast-moving ICT[16].Sector is fast shifting towards digitalised 'modern manufacturing' using the IoT to develop sophisticated core technology such as condition monitoring or robust computer interaction. AR is the central platform that helps user incorporation into such a structure , providing users via an experience to communicate with the artificial world[17].Augmented Reality Laboratories can be used to train new beginners who have no previous experience in the field. When beginners are recruited, they have less knowledge about the advancements and technicality of unique(special) machines. In few cases they wouldn't have experienced this technology aided platform which can be used to learn about a particular concept i.e. a machine in detail and its various parts of the machine in an augmented environment and self-train with less dependency on external resources. In the Augmented environment they can disassemble every part of the machine, assemble them, and learn each part in detail.By way of AR tools, enables people to discover stimuli and to discover classroom opportunities that which, in certain instances, surpass that provided by conventional laboratory courses[18]. Figure 2 shows the augmented model of some concepts (Crocodile, dinosaur, ant and motor) in STEM Education. Item detection and simulation should be done in real-time so as to enable for further conversational exchange with the programme and lower the impact of failure or blurry vision.Real time technology is established to include constant guidance throughout trainings[13]. Figure 3 and 4 shows the Augmented model assisted by multi-media visuals and Multidimensional simulation of lathe machine in spatial plane

Figure 2. Augmented model of some concepts in STEM Education

Figure 3. Augmented model assisted by multi-media visuals

IOP Conf. Series: Materials Science and Engineering 1059 (2021) 012016 doi:10.1088/1757-899X/1059/1/012016

Figure 4. Multidimensional Simulation of Lathe Machine in Spatial Plane (Marker-less detection)

The Augmented environment offers the user an interactive mode of control of the machine or component by performing replica functions of the machine and simulations. They can train themselves even at comfort of their home by just using a mobile device (i.e. smartphone). This can be done with just a mobile but when the hardware is further developed the following use cases can be achieved. This type of application can equip the beginners with required amount of pre-knowledge before stepping into the real environment. AR seems to be able to achieve these goals of crumbling technical equipment, as well as IOT is used to perform studies in such a comfortable world. AR thus splits the boundaries between time and space as well as enables the topic of analysis to still be reimagined.AR adapted to various teaching situations throughout the present study offering adequate framework for evaluating knowledge and skills from components shown in this field, however in addition to cross-cutting responsibilities like: integrative skill sets (assessment and formulation skills , planning and organizational skills, problem-solving, metadata management and decision-making) are established[19]. Figure 5 shows the 3D model of Centrifugal pump with Assembly simulation. Viewing interior components in detail using virtual buttons, Depicting the solution for the ideal functioning of an Electric drive and Simulation Analysis of a system Assembly are shown in figures 6, 7 and 8 respectively.

Figure 5.3D model of Centrifugal pump with Assembly simulation

Figure 6. Viewing interior components in detail using virtual buttons

Figure 7. Depicting the solution for the ideal functioning of an Electric drive

Figure 8. Simulation Analysis of a system Assembly

2.2Assembly process guidance

The requirement to incorporate improved realism in industrial learning methods has resulted from the desire to improve the standard of comprehension in order to eliminate mistakes and misunderstandings relevant to modern protection teaching [14]. Inside the sector, production line workers face significant complicated packaging systems and complicated control processes. The ambiguity, combined with demand to finish the assembly inside a small-time frame but with the highest efficiency, makes it impossible for technicians to behave optimal way. Amazing learning instruction is used to help and enable technicians to perform assembly activities in the most effective manner[20].Maintenance and assembly line operations are distinguished by a large number of skilled labor stages.These can time consuming when done manually. Using the effective development of advancement of Industrial environment with IoT (Internet of Things) systems Infrastructure with required Sensor Instrumentation this can be achieved. Details like optimum processing methods, or controlling devices or system variables, may be shown around the operator by AR throughout these operations.

Throughout this situation, there is a significant two - way volume of contact between its single workforce as well as the hardware server on the Company production line, leading to a high degree of interaction or the digital entity, whereby technicians physically attempt to assemble the physical machine and get relevant information via controlling it and getting input via the controlled system.In order to solve the problems of complicated tasks and to create personalized statistical validity, the findings argue the use of intelligent machines. The aim is to get a professional method through understanding of basic intelligence to systematically assess the qualitative characteristics of a AR approach[21].Inspection and renewal activities are a fascinating and ability-filled trouble spot for the implementation of AR. Many operations in this area are undertaken by qualified maintenance team adapting defined techniques to recorded models in largely constant and repetitive settings[22]. Such protocols are usually structured into series of provable activities that address a specific entity at a particular place. In the recent past, attempts to achieve quality safety training in the power industry required extensive and expensive equipment for the recreation of such an environment, which in reality varied widely, given the Technical area of industries and job sites to which technicians or workers are deployed. Marked or Marker-less based systems. On either hand, in the area of automotive framework management, in specific maintenance, inspection or caregiving responsibilities, real-time process information is quite often needed in order to monitor the nation of a physical model or even of specific equipment, to identify random errors, or to protect the rights of employees. Consequently, AR devices need to be able to incorporate data collected[23].

The virtual interactive tool is composed of virtual buttons, which have meaningful assembly information of the machine in an Assembly-line where a worker is deployed. These virtual buttons along with few interactive features would be the primary control methodology to operate the machine virtually.

3 Results and discussion

Dynamically produced shortcomings of observed defects are compared to points data in the cloud defining the present position of a real-time item. The outcome of the measurement is recorded for each flaw in the virtual representation. The whole method is model-based and there is no living thing-object communication.installation procedures are distinguished by a large proportion of physical work phases and limited lead times. Details like optimal cutting strategy or control variables may be shown to the employee by AR mostly during the procedure. In this scenario, intense bi-directional contact takes place between the human worker and the physical device resulting in a high degree of communication between the person and the entity: the worker physically composes the physical model and obtains knowledge about its mechanism via controlling it and obtaining input by process variables. In the past, attempts to achieve quality safety training in the power industry required extensive and expensive equipment to recreate a work environment, which in reality varied widely, given the breadth of industries and job sites to which technicians are deployed.

AR is perceived to be one among the best technological solutions with the highest opportunity to reshape and complement the existing industrial environment.Of course, to recreate these simulated sites effectively, the same dangers and risks of injury must exist in the virtual depiction and events created.

Augmented reality training solves many of those safety issues inherent in traditional training because the simulated training environment is 100 percent safe. The exercise allows people to practice the routine hazardous activities they are expected to perform safely without any real-life consequences should they make a mistake. And service center managers have a clearer understanding of crew capabilities before dispatching a team to a customer's workplace.

4 Conclusions

Augmented Reality brings Fun (by enabling the learner to enjoy the content), Challenges (setting a task/objective allowing learners to self-understand their level of understanding), curiosity (stimulating learners to explore new concepts). Augmented reality is facilitated through mobile phone. Not just very cutting- edge but Augmented Reality can also be made extremely affordable and accessible as an open-source learning resource. It works on just mobile phone and an app, both of which are available with common people and learners. The goal is to impact as many learners through this powerful learning experience and reach out to every learner. The current education system is a passive and non-interactive learning environment where many of the learners are disengaged. Training by Virtual Reality tends to be elevated rates of interaction, due to increased educational experience and, as a result, strengthened effective training. Technology based knowledge impartment and providing Skill-based training to learners to enable their product towards a longer life. Trying to emulate learners to gain further attention by practicing game elements (able to integrate game features in educational systems. This can be asserted that AR innovations are an innovative educational tool that facilitates agile education on systematic review mostly on work place. Developers can sometimes claim that AR blends reality with insightful elements put in technical devices to construct a modern reality.Digital laboratories are immersive facilities for designing and performing type of approach based on real-world phenomena. It solves the issue of availability and the absence of personal technologies for laboratory-based learning, particularly in science and mathematics. AR, on the opposite side, allows teachers and coaches to accomplish things that they will have never or could not traditionally accomplish in a stable simulated world. Together, all of these engage learners with a curiosity driven and thus are expected to become even quite prevalent in their use and influence throughout the long term.

References

[1] F. Arici, P. Yildirim, Ş. Caliklar, and R. M. Yilmaz 2019 Research trends in the use of augmented reality in science education: Content and bibliometric mapping analysis *Comput.*

Educ. **142** 103647.
[2] J. Cabero-Almenara, J. M. Fernández-Batanero, and J. Barroso-Osuna 2019 Adoption of augmented reality technology by university students *Heliyon*, **5(5)** 10.1016/j.heliyon.2019.e01597.
[3] M. B. Ibáñez and C. Delgado-Kloos 2018 Augmented reality for STEM learning: A systematic review *Comput. Educ.* **123** 109–23.
[4] J. Garzón, Kinshuk, S. Baldiris, J. Gutiérrez, and J. Pavón 2020 How do pedagogical approaches affect the impact of augmented reality on education? A meta-analysis and research synthesis *Educ. Res. Rev.* **31** 100334.
[5] M. Wójcik 2016 Potential use of Augmented Reality in LIS education *Educ. Inf. Technol.* **21** 1555–69.
[6] M. Eder, M. Hulla, F. Mast, and C. Ramsauer 2020 On the application of augmented reality in a learning factory working environment *Procedia Manuf.* **45** 7–12.
[7] A. M. Kamarainen 2013 EcoMOBILE: Integrating augmented reality and probeware with environmental education field trips *Comput. Educ.* **68** 545–56.
[8] A. Klimova, A. Bilyatdinova, and A. Karsakov 2018 Existing Teaching Practices in Augmented Reality *Procedia Comput. Sci.* **136** 5–15.
[9] J. Garzón and J. Acevedo 2019 Meta-analysis of the impact of Augmented Reality on students' learning gains *Educ. Res. Rev.* **27** 244–60.
[10] Y. Shen, S. K. Ong, and A. Y. C. Nee 2010 Augmented reality for collaborative product design and development *Des. Stud.* **31(2)** 118–45.
[11] D. Mourtzis, V. Siatras, J. Angelopoulos, and N. Panopoulos 2020 An augmented reality collaborative product design cloud-based platform in the context of learning factory *Procedia Manuf.* **45** 546–51.
[12] N. Arulanand, A. RameshBabu, and P. K. Rajesh 2020 Enriched learning experience using augmented reality framework in engineering education *Procedia Comput. Sci.* **172** 937–42.
[13] M. Quandt, B. Knoke, C. Gorldt, M. Freitag, and K. D. Thoben 2018 General Requirements for Industrial Augmented Reality Applications *Procedia CIRP* **72** 1130–5.
[14] A. Perdikakis, A. Araya, and D. Kiritsis 2015 Introducing Augmented Reality in Next Generation Industrial Learning Tools: A Case Study on Electric and Hybrid Vehicles *Procedia Eng.* **132** 251–8.
[15] P. Khandelwal, P. Swarnalatha, N. Bisht, and S. Prabu 2015 Detection of Features to Track Objects and Segmentation Using GrabCut for Application in Marker-less Augmented Reality *Procedia Comput. Sci.* **58** 698–705.
[16] L. Damiani, M. Demartini, G. Guizzi, R. Revetria, and F. Tonelli 2018 Augmented and virtual reality applications in industrial systems: A qualitative review towards the industry 4.0 era *IFAC-PapersOnLine* **51(11)** 624–30.
[17] J. Egger and T. Masood 2020 Augmented reality in support of intelligent manufacturing – A systematic literature review *Comput. Ind. Eng.* **140** 106195.
[18] J. M. . M. Andújar A.; Márquez, M. A 2011 Augmented Reality for the Improvement of Remote Laboratories: An Augmented Remote Laboratory IEEE Transactions on Education," *IEEE Trans. Educ.* **54(3)** 492–500.
[19] J. Martín-Gutiérrez, P. Fabiani, W. Benesova, M. D. Meneses, and C. E. Mora 2015 Augmented reality to promote collaborative and autonomous learning in higher education *Comput. Human Behav.* **51**752–61.
[20] A. Syberfeldt, O. Danielsson, M. Holm, and L. Wang 2015 Visual Assembling Guidance Using Augmented Reality *Procedia Manuf.* **1** 98–109.
[21] A. Syberfeldt, O. Danielsson, M. Holm, and L. Wang 2016 Dynamic Operator Instructions Based on Augmented Reality and Rule-based Expert Systems *Procedia CIRP* **41** 346–51.
[22] S. Henderson and S. Feiner 2011 Exploring the benefits of augmented reality documentation for maintenance and repair," *IEEE Trans. Vis. Comput. Graph.* **17(10)**1355-68.
[23] L. Baron and A. Braune 2016 Case study on applying augmented reality for process supervision in industrial use cases in *IEEE International Conference on Emerging Technologies and Factory Automation, ETFA*.
[24] S. R. Sorko and M. Brunnhofer 2019 Potentials of Augmented Reality in Training *Procedia Manuf.* **31**85–90.

IOP Conf. Series: Materials Science and Engineering 1059 (2021) 012017 doi:10.1088/1757-899X/1059/1/012017

Experimental Studies on Water Absorption Behaviour of Treated and Untreated Hybrid Bio-Composites

R.Girimurugan[1*], K.G.Saravanan[2], P.Manickavasagam[3], G.Gurunathan[4], M.Vairavel[5]

[1*]Assistant Professor, Department of Mechanical Engineering, Nandha College of Technology, Perundurai, Tamilnadu, India
[2]Associate Professor, Department of Mechanical Engineering, Sona College of Technology, Salem, Tamilnadu, India
[3]Professor, Department of Mechanical Engineering, Nandha College of Technology, Perundurai, Tamilnadu, India
[4]Assistant Professor, Department of Mechanical Engineering, Mailam Engineering College, Villupuram, Tamilnadu, India
[5]Assistant Professor, Department of Mechanical Engineering, Annapoorana Engineering College, Salem, Tamilnadu, India

E-mail: dr.r.girimurugan@gmail.com[1*]

Abstract

Various experimental works were successfully accomplished over the past years on the natural resources reinforced composites. This research work is focused on the water absorption behaviour of the treated and untreated banana fiber reinforced, used camellia sinensis particles filled along with the epoxy resin matrix bio-composites. Different weight percentages of reinforcement and filler materials were taken to prepare the hybrid composite specimens, whereas the epoxy resin matrix were keeping constant. Traditional hand layup technique was adopted and the boards of the hybrid composites were effectively fabricated by using the same technique. Hybrid bio-composite specimens for water absorption tests were prepared by using the portable wood cutting machine as per the ASTM standards for the efficient justification on experimental results. Water absorption tests on the hybrid bio-composite specimens were carried out as per the ASTM standards. Water absorption behaviour of the treated and untreated banana fiber reinforced, used camellia sinensis particles filled, epoxy resin matrix hybrid composite specimens were absorbed and compared with each other. Experimental results shows that the chemically treated banana fiber reinforced with narrow amount of used camellia sinensis particle filled hybrid composite specimens exhibits the better water absorption behaviour than that of untreated banana fiber reinforced with elevated amount of used camellia sinensis particle filled hybrid composite specimens.

1 Introduction

Use of bio-composites in engineering and other fields were significantly increased due to its availability. Most of the material science research focus is now turned on the bio-materials due to the scarcity of the synthetic and traditional materials. Several researchers are established their research outputs in the field of bio-composites. Superior hardness were obtained on the camellia sinensis particles filled with epoxy resin matrix/banana fiber composites [1]. Better mechanical properties were exposed on the used camellia sinensis powder and azadirachta indica seed powder filled bio-composites [2]. 50 wt% of resin matrix and 50 wt% of banana fiber reinforced bio-composites discovered the enhanced mechanical properties than other compositions [3]. Greater moisture absorption property is revealed by the 50 wt% of banana fiber addition into the jute fiber/resin matrix composites [4]. 50 wt% of sisal fiber into the banana fiber/resin matrix composites exposed the better moisture absorption property than other combinations [5]. Amount of fiber content is the key factor which is used to determine the water absorption capacity of the composites [6]. Compared to the hybrid banana composites, ability of the composites to absorb the water content is very lesser in woven banana fiber composites [7]. Increasing glass fiber content into the matrix material causes the lesser water droplet penetration on the composite specimen's surface [8]. Enhanced water absorption properties of the composite specimens are found in ultrasonic waves treated fiber reinforcement [9]. Addition of banana fiber and glass fiber into the matrix causes the superior water absorption resistance [10]. Water absorption capability of the banana fiber reinforced composites were found good when the matrix is reinforced with limited quantity of reinforcement material [11]. Many experimental investigations were successfully conceded on the banana fiber reinforced resin matrix composites. In this experimental investigations, water absorption behaviour of epoxy resin matrix-banana fiber/camellia sinensis particles reinforced composites were established.

2 Materials and Methods

In this research work, epoxy resin was used as matrix material and it was purchased from Covai Seenu and Company limited and raw banana fibers were procured from Golden Fibres. Used camellia sinensis particles were congregated from local tea shops and bakeries in Erode city. Procured banana fibers length were in the range of 15 to 20 cm. The long raw banana fibers were then sliced into 1 to 3 cm in length using cutter. The sliced raw banana fiber was washed in well concentrated distilled water solution around 30 minutes. The washed raw banana fibers were then dried in atmospheric temperature for 48 hours. After the drying process, well cleaned banana fibers were immersed for one hour time period in 5.0 L plastic container which is filled with high concentrated Sodium Hydroxide (NaOH) and water solution. After the one hour time period the chemically treated banana fibers were dried in an oven at 110oC for 6 hours. Filler material which is used in this experimental work i.e. waste or used camellia sinensis particles were washed using the lower concentration distilled water solution around 45 minutes to remove the dust particles. After washing, the well washed camellia sinensis particles were dried at room temperature for 72 hours. Initially, epoxy resin is mixed with hardener in 2:1 ratio to enhance the epoxy resin strength using stirrer machine for 30 minutes. A rectangular moulding box (250 × 250 × 10 mm) is used to prepare the hybrid composite plates from the matrix, reinforcement and filler material. A normal hand layup technique is implemented to prepare the hybrid composite boards. Epoxy resin/hardener mixture (65 wt %) was formed as first layer in the moulding box for 2 mm thickness after that the mixture of chemically treated banana fibers (35 wt % was dispersed on the first layer i.e. Epoxy resin/hardener mixture up to 2 mm thickness. The same process is repeated up to 10 mm thickness accomplishment of the hybrid composite plate thickness. The same process is repeated again for the remaining three treated hybrid composite specimen boards preparation by keeping the weight percentage of epoxy resin/hardener mixture as constant (65 wt%) and changing the weight percentage of treated banana fibers (33, 31 and 29 wt%) and camellia sinensis particles (2, 4 and 6 wt%) correspondingly. Specimen description for both untreated and treated banana fiber reinforced hybrid composites were given in table.1 and table.2 correspondingly. The different characteristics of the water, which is used in this experimental study is given in table.3

Table1. Specimen description for untreated banana fiber reinforced hybrid composites

Epoxy resin (wt%)	Untreated banana fiber (wt%)	Camellia sinensis particles (wt%)	Specimen description
65	35	0	UTHC-A
65	33	2	UTHC-B
65	31	4	UTHC-C
65	29	6	UTHC-D

Table2. Specimen description for treated banana fiber reinforced hybrid composites

Epoxy resin (wt%)	Treated banana fiber (wt%)	Camellia sinensis particles (wt%)	Specimen description
65	35	0	THC-A
65	33	2	THC-B
65	31	4	THC-C
65	29	6	THC-D

Table3. Characteristics of the water used for Investigation

Property	Value	Remarks
Color	Na	Colourless
Turbidity	Na	Nil
Taste	Na	Tasteless
Odor	Na	Nil
Temperature	36°C	Room
Solids	Na	Nil
pH	7.5	Slightly Alkaline

Untreated banana fiber reinforced composite plates for all four samples were shown in figure.1 respectively. The above process is again followed to produce the untreated hybrid composite specimen boards using epoxy resin/hardener matrix, untreated banana fiber reinforced and camellia sinensis particles respectively. Treated banana fiber reinforced composite plates for all four samples were shown in figure.2 respectively. ASTM D-5229 standard was used to prepare the test specimens for water absorption test. The dimensions of the water absorption test was taken as 20 × 20 × 6 mm based on the ASTM D-5229 standard. Water absorption test specimens for all samples of both treated and untreated banana fiber reinforced composite specimens were illustrated in figure.3 (a) and 3 (b) correspondingly. Pure water is used to conduct the typical water absorption test under the normal atmospheric conditions. Separate water containers were used to evaluate the water absorption behaviour of both untreated and treated banana fiber reinforced hybrid composite specimens. The five different time periods were taken into account like 24, 48, 72, 96 and 120 hours respectively. Effect of chemical treatment on water absorption behaviour of both chemically treated and untreated banana fiber reinforced with waste camellia sinensis particles filled hybrid composite specimens were discussed in results and discussions chapter.

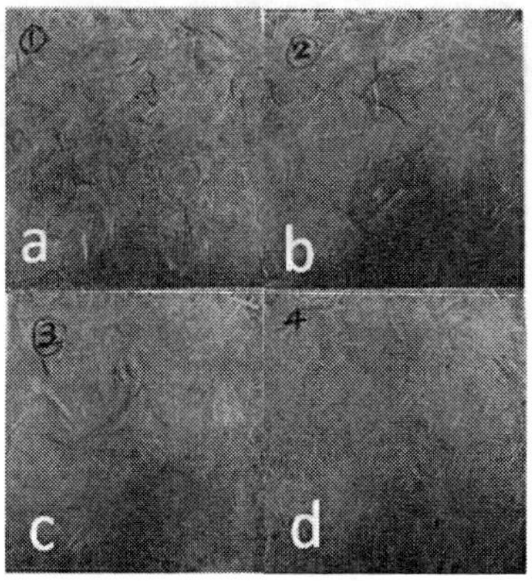

Figure 1 Untreated banana fiber reinforced composite plates for (a) UTHC-A (b) UTHC-B (c) UTHC-C (d) UTHC-D samples

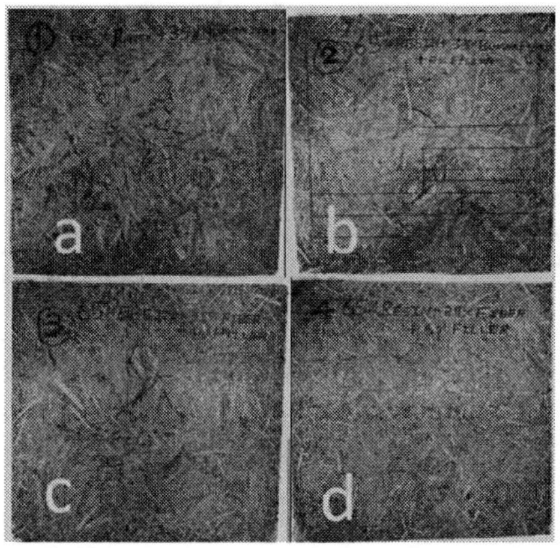

Figure 2 Treated banana fiber reinforced composite plates for (a) THC-A (b) THC-B (c) THC-C (d) THC-D samples

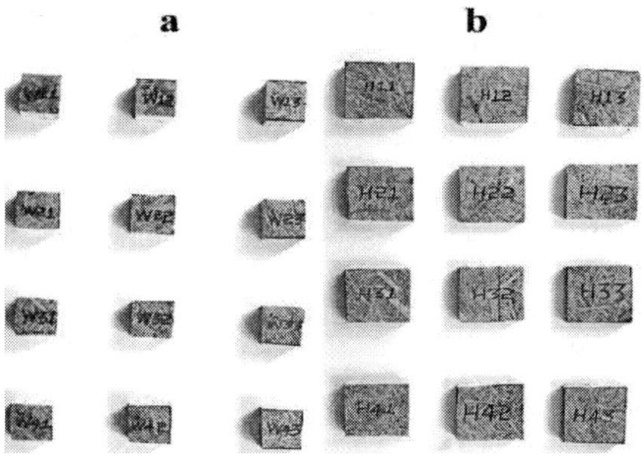

Figure 3 Water absorption test specimens (a) Treated banana fiber reinforced composite specimen (b) Untreated banana fiber reinforced composite specimen

3 Results and Discussions

The following results has been obtained through the experimental studies, which was carried out on the epoxy resin matrix, waste camellia sinensis particles filled, chemically treated and untreated banana fiber reinforced hybrid composites by varying the weight percentage of the both filler and reinforcement materials with epoxy resin matrix. Water absorption tests were performed on the treated and untreated banana fiber reinforced composite specimens to explore its water absorption behaviour. Percentage of water absorbed by the untreated banana fiber reinforced hybrid composite (UTHC) specimens after the time interval of 24, 48, 72, 96 and 120 hours were represented in figure.4. It was absorbed from the experimental results, all untreated banana fiber reinforced hybrid composite specimen's water intake percentages are in the range of 21.8, 24, 17.9 and 24.4 for UTHC-A, UTHC-B, UTHC-C and UTHC-D specimens after 24 hours respectively. After 48 hours, water absorption percentage of the all untreated banana fiber reinforced hybrid composite specimens were noticed in the range of 21.8, 24, 17.9 and 24.4 for UTHC-A, UTHC-B, UTHC-C and UTHC-D specimens accordingly. Water absorption percentage of the all untreated banana fiber reinforced hybrid composite specimens after 72 hours are found in the range of 25.12, 27.55, 20.61 and 28.06 for UTHC-A, UTHC-B, UTHC-C and UTHC-D specimens correspondingly. Percentage of water absorption by the all untreated banana fiber reinforced hybrid composite specimens after 96 hours are observed in the range of 25.78, 28.27, 21.14 and 28.79 for UTHC-A, UTHC-B, UTHC-C and UTHC-D specimens accordingly. Water absorption percentage of the all untreated banana fiber reinforced hybrid composite specimens after 120 hours were noticed in the range of 26.43, 28.98, 21.68 and 29.52 for UTHC-A, UTHC-B, UTHC-C and UTHC-D specimens correspondingly. It was noticed that, hybrid composite specimen consists of 65 wt% of epoxy resin, 31 wt% of untreated banana fiber and 4 wt% of waste camellia sinensis particle (UTHC-C) absorbs very minimal water during the water absorption test than that of other three untreated banana fiber reinforced hybrid composite specimens in all time intervals. The two important kinetic parameters like n and k was considered to understand the water absorption kinetics between the fiber and epoxr resin. The following equation shows the relationship between the two essential kinetic parameters.

$$\mathrm{Log}\,(Q_t / Q_e) = \log k + n \log t \qquad (1)$$

In the above equation (1), mole percentage of water absorbed by the fiber at a particular time t is noted as Q_t. Mole percentage of water absorbed by the fiber at equilibrium condition at time t is noted as Q_e. The constant for epoxy resin water absorption characteristics is denoted as k. If the diffusion values between the water and fiber is less than 0.5, it obeys the Fick's law and it is said to be Fickian. If the diffusion value is less than 1, the diffusion is said to be anomalous. If the diffusion value is lies between 0.5 and 1 then the diffusion is said to be nonfickian. Water kinetics between the fiber and water is indicated by the k. The mode of diffusion between the fiber and water is indicated by n. In this experimental study, the value of n is observed as less than 0.5. Due to the penetration of water molecules

through the interfacial regions and presence of void contents in the composites, the n value is very less. Diffusion coefficient exemplifies the capability of water molecule to diffuse through the fiber. Percentage of water absorbed by the treated banana fiber reinforced hybrid composite (THC) specimens after the time interval of 24, 48, 72, 96 and 120 hours were represented in figure.5. It was absorbed from the experimental results, all treated banana fiber reinforced hybrid composite specimen's water intake percentages are in the range of 19.41, 21.47, 15.56 and 21.91 for THC-A, THC-B, THC-C and THC-D specimens after 24 hours respectively. After 48 hours, water absorption percentage of the all treated banana fiber reinforced hybrid composite specimens were noticed in the range of 21.35, 23.62, 17.12 and 24.10 for THC-A, THC-B, THC-C and THC-D specimens accordingly. Water absorption percentage of the all treated banana fiber reinforced hybrid composite specimens after 72 hours are found in the range of 22.32, 24.70, 17.89 and 25.02 for THC-A, THC-B, THC-C and THC-D specimens correspondingly. Percentage of water absorption by the all treated banana fiber reinforced hybrid composite specimens after 96 hours are observed in the range of 22.90, 25.34, 18.36 and 25.85 for THC-A, THC-B, THC-C and THC-D specimens accordingly. Water absorption percentage of the all treated banana fiber reinforced hybrid composite specimens after 120 hours were noticed in the range of 23.48, 25.98, 18.83 and 26.51 for THC-A, THC-B, THC-C and

THC-D specimens correspondingly. It was noticed that, hybrid composite specimen consists of 65 wt% of epoxy resin, 31 wt% of treated banana fiber and 4 wt% of waste camellia sinensis particle (THC-C) absorbs very minimal water during the water absorption test than that of other three untreated and treated banana fiber reinforced hybrid composite specimens in all time intervals.

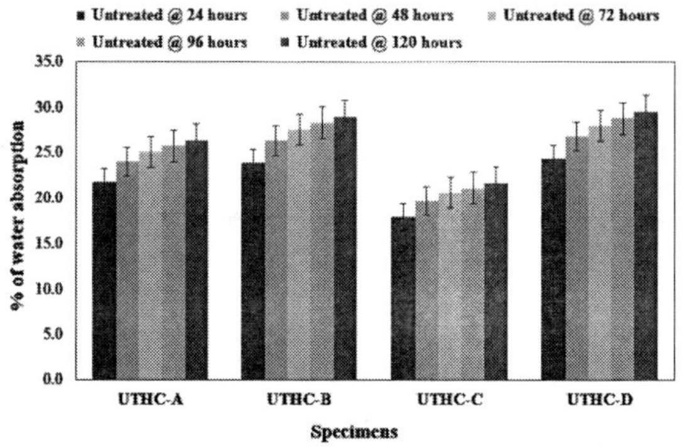

Figure 4 Percentage of water absorbed by the untreatred banana fiber reinforced hyrid composite (UTHC) specimens

Weight of all untreated banana fiber reinforced hybrid composite (UTHC) specimens after the time interval of 24, 48, 72, 96 and 120 hours were represented in figure.6. It was perceived from the experimental results, all untreated banana fiber reinforced hybrid composite specimens weight are in the range of 2.86, 2.86, 2.72 and 2.92 grams for UTHC-A, UTHC-B, UTHC-C and UTHC-D specimens after 24 hours respectively. After 48 hours, weight of the all untreated banana fiber reinforced hybrid composite specimens were noticed in the range of 3.15, 3.15, 2.99 and 3.22 grams for UTHC-A, UTHC-B, UTHC-C and UTHC-D specimens accordingly. Weight of the all untreated banana fiber reinforced hybrid composite specimens after 72 hours are found in the range of 3.29, 3.29, 3.13 and 3.36 grams for UTHC-A, UTHC-B, UTHC-C and UTHC-D specimens correspondingly.

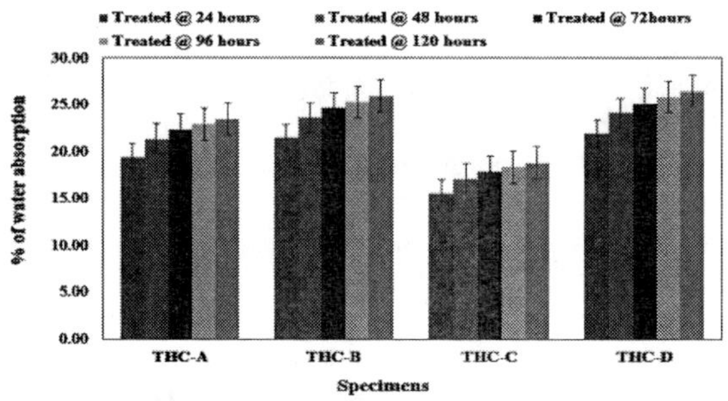

Figure 5 Percentage of water absorbed by the treatred banana fiber reinforced hyrid composite (THC) specimens

Weight of the all untreated banana fiber reinforced hybrid composite specimens after 96 hours water test are observed in the range of 3.36, 3.36, 3.20 and 3.43 grams for UTHC-A, UTHC-B, UTHC-C and UTHC-D specimens accordingly. Weight of the all untreated banana fiber reinforced hybrid composite specimens after 120 hours were noticed in the range of 3.44, 3.44, 3.26 and 3.51 for UTHC-A, UTHC-B, UTHC-C and UTHC-D specimens correspondingly. It was noticed that, weight of the hybrid composite specimen consists of 65 wt% of epoxy resin, 31 wt% of untreated banana fiber and 4 wt% of waste camellia sinensis particle (UTHC-C) possess very minimum than that of other three untreated banana fiber reinforced hybrid composite specimens in all time intervals.

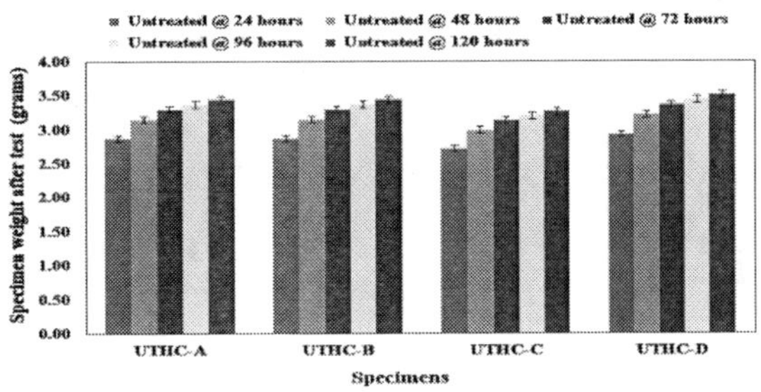

Figure 6 Weight of the untreatred banana fiber reinforced hyrid composite (UTHC) specimens after test

Weight of all treated banana fiber reinforced hybrid composite (THC) specimens after the time interval of 24, 48, 72, 96 and 120 hours were represented in figure.7. It was perceived from the experimental results, all untreated banana fiber reinforced hybrid composite specimens weight are in the range of 2.36, 2.47, 2.21 and 2.39 grams for THC-A, THC-B, THC-C and THC-D specimens after 24 hours respectively. After 48 hours, weight of the all untreated banana fiber reinforced hybrid composite specimens were noticed in the range of 2.60, 2.72, 243 and 2.63 grams for THC-A, THC-B, THC-C and THC-D specimens accordingly. Weight of the all untreated banana fiber reinforced hybrid composite specimens after 72 hours are found in the range of 2.71, 2.84, 2.54 and 2.75 grams for THC-A, THC-B, THC-C and THC-D specimens correspondingly. Weight of the all untreated banana fiber reinforced hybrid composite specimens after 96 hours water test are observed in the range of 2.77, 2.90, 2.60 and 2.81 grams for THC-A, THC-B, THC-C and THC-D specimens accordingly. Weight of the all untreated banana fiber reinforced hybrid composite specimens after 120 hours were noticed in the range of 2.83, 2.96, 2.65 and 2.87 for THC-A, THC-B, THC-C and THC-D specimens correspondingly. It was noticed that, hybrid composite specimen consists of 65 wt% of epoxy resin, 31 wt% of treated banana fiber and 4 wt% of waste camellia sinensis particle (THC-C) possess very minimal weight after the water absorption test than that of other three untreated and treated banana fiber reinforced hybrid composite specimens in all time intervals. Superior water absorption behaviour was found in Sodium Hydroxide (NaOH) treated banana fiber reinforced composite specimens, due to the elimination of lignin, cellulose and hemicelluloses contents on the alkali-treated banana fiber and it might made spaces in the banana fiber molecules. Addition of limited quantity (4 wt %) of filler contents i.e. waste camellia sinensis particle were filled up in the voids in the banana fiber and it enriches the good amount of interfacial consociate between banana fiber and epoxy resin matrix.

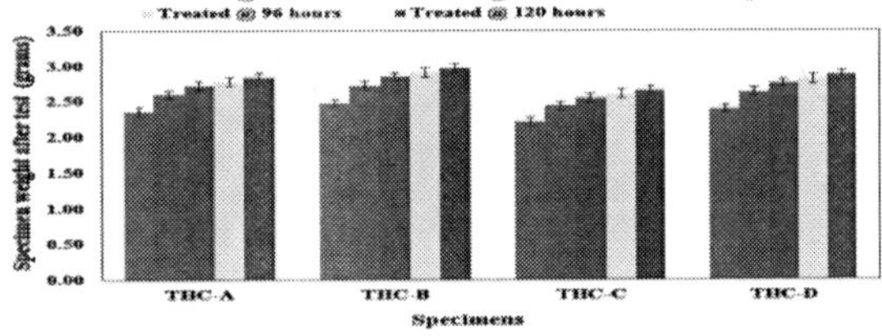

Figure 7 Weight of the treatred banana fiber reinforced hyrid composite (THC) specimens after test

4. Conclusion

The following conclusions have been made based of the experimental results that are obtained from the water absorption test on the different weight percentage of untreated and treated banana fiber reinforced, different weight percentage of used camellia sinensis powder filled, constant weight percentage of epoxy resin matrix bio-composites. Chemically treated banana fiber reinforced composite specimens were exhibited the better water absorption behaviour than that of untreated banana fiber reinforced composite specimens. The percentage of water absorption was found very minimum in 4 wt% of used camellia sinensis powder and 31 wt% of treated banana fiber composite specimens due to the enhanced compatibility of alkali-treated banana fiber with the epoxy resin matrix.

References

[1] Girimurugan R, Pugazhenthi R, Maheskumar P, Suresh T and Vairavel M 2020 Impact and hardness behaviour of epoxy resin matrix composites reinforced with banana fiber/camellia sinensis particles *Materials Today: Proceedings.* doi.org/10.1016/j.matpr.2020.07.597

[2] Vijay R Vinod A Kathiravan R Siengchin S Singaravelu D L 2020 Evaluation of Azadirachta indica seed/spent Camellia sinensis bio-filler based jute fabrics–epoxy composites: experimental and numerical studies *Journal of Industrial Textiles.* **49** 1252-1277.

[3] Ramesh M, Atreya T S A, Aswin U S, Eashwar H and Deepa C 2014 Processing and mechanical property evaluation of banana fiber reinforced polymer composites *Procedia Engineering.* **97** 563-572.

[4] Boopalan M, Niranjanaa M and Umapathy M J 2013 Study on the mechanical properties and thermal properties of jute and banana fiber reinforced epoxy hybrid composites *Composites Part B: Engineering.* **51** 54-57.

[5] Venkateshwaran N, Elaya Perumal A, Alavudeen A and Thiruchitrambalam M 2011 Mechanical and water absorption behaviour of banana/sisal reinforced hybrid composites *Materials & Design.* **32** 4017-4021.

[6] Sreekumar P A, Albert P, Unnikrishnan G Joseph K and Thomas S 2208 Mechanical and water sorption studies of ecofriendly banana fiber-reinforced polyester composites fabricated by RTM *Journal of Applied Polymer Science.* **109** 1547-1555.

[7] Venkateshwaran N and Elaya Perumal A 2012 Mechanical and water absorption properties of woven jute/banana hybrid composites *Fibers and Polymers.* **13** 907-914.

[8] Pothan L A and Thomas S 2004 Effect of hybridization and chemical modification on the water absorption behavior of banana fiber–reinforced polyester composites *Journal of applied polymer science.* **91** 3856-3865.

[9] Ghosh R, Ramakrishna A, Reena G, Ravindra A and Verma A 2014 Water absorption kinetics and mechanical properties of ultrasonic treated banana fiber reinforced-vinyl ester composites *Procedia Materials Science.* **5** 311-315.

[10] Joseph S, Sreekala M S, Koshy P and Thomas S 2008 *Mechanical properties and water sorption behavior of phenol–formaldehyde hybrid composites reinforced with banana fiber and glass fiber* Journal of applied polymer science **109** 1439-1446.

[11] Srinivasan T, Suresh G, Ramu P, Ram V G, Giresh M and Arjun K 2020 *Effect of water absorption of the mechanical behavior of banana fiber reinforced IPN natural composites* Materials Today: Proceedings.

IOP Conf. Series: Materials Science and Engineering 1059 (2021) 012018 doi:10.1088/1757-899X/1059/1/012018

Investigation of Corrosion Characteristic of Al6063 Reinforced with Al_2O_3-CBA by Using Stir Casting

Nithyanandhan T[1*], Sivaraman P[2], Prabhu M K[3], Nedumaran C[4], Narendran J[5], Naveen P S[6]
[1,2,3] Asst.Professor, Sri Krishna College of Technology, Coimbatore
[4,5,6] UG Students, Sri Krishna College of Technology, Coimbatore

Email ID: anandhan8.mech@gmail.com

Abstract: This work critiques an examination method to investigate the corrosion characteristic of composite materials with guide of Aluminium oxide and chicken bone ash as reinforcements. The weight percentage of aluminium oxide are 6%, 5%, 4% ,3% and 2% and Chicken Bone Ash are2%, 3%, 4%, 5% and 6% have been covered for this investigation. This composite casting method became made with the help of the stir casting method. The "Aluminium chloride" solution is used to examine the detail corrosion behaviour of the metal matrix composite by using of the immersion test. The result of this research work produce the corrosion rate would be decreased with increase of the weight percentage of Reinforcements. This work concluded that the estimation of rate of corrosion free with increment in maximum weight percent of hybrid composites.

Keyword: Al (6063), Aluminium oxide, chicken bone ash, Aluminium Chloride, Immersion test

1. Introduction

Aluminum and aluminum-primarily based metal matrix composites (MMCs) were exceedingly studied with the aid of diverse professionals lately, parallel to the growing needs to substances with decrease weight in combination with higher mechanical properties along with wear resistance, hardness, impact toughness and Corrosion resistance. Corresponding to the expanding requests to substances with decrease weight in mixture with higher mechanical homes, for example, put on obstruction, hardness, sway sturdiness and ballistic competition. Due to those prevalent houses, MMCs are becoming fascinating in designing packages, for instance, protect, car industry, aviation, athletic equipment and digital bundling. The hybrid composite overall performance is a considered sum of the separate additives in which there may be a greater favorable stability between reinforcement introduced and the advantages and downsides of its feature. Materials which are consolidated of in any occasion special substances, which enhance their properties. The article deals with the Al-4.5% Cu blend were used because the Matrix was bonded with ash and silicon carbide (SiC) as support. The combination metallic community composite became introduced utilizing everyday foundry methods. The ash and SiC were incorporated 5%, 10%, and 15% by using weight to the fluid metal. (K. V. MAHENDRA, 2010). The effects of welding boundaries had been reviewed with the guide of mechanical and microstructural levels. The microstructure of the composite and dispersing of particle fortresses had been tried by means of optical amplifying focal point for example Optical magnifying instrument and also the properties of yield quality, extreme elasticity and extension had been completely investigated the utilization of all-inclusive testing machine. The upgraded grinding mix weld boundaries have been perceived for 20% weight division fortified cross breed composites (B.SURESHBABU, 2018). In this modern-day research, Al/Chicken Bone Ash (CBA) metal community composites (MMCs) have been created via vortex technique. Distinctive weight portions of fortification have been utilized to produce the composites. Checking electron microscope furnished with energy dispersive X-ray analyzer is applied for the microstructural characterizations. The residences like thickness, hardness, and severe elasticity have been researched (D. SIVA PRASAD, 2011). The paper deals with the fabrication of aluminium based hybrid metal matrix composite and then characterized their chemical properties which are recently developed modification of stir casting has been used in the present investigation to produce aluminium (6063) alloy reinforced with Silicon carbide (Sic).In this paper, Aluminium (6063) alloy is used as a base metal and Silicon carbide (SIC) is been used as reinforced material with the hybrid reinforcement as chicken bone ash (HARIDASS.R, 2018).This study focuses on the Mechanical properties of Al6063 with reinforcement of coconut shell ash and Aluminium oxide particulates are prepared by stir casting techniques. In this experimental, reinforcement Aluminium oxide has various altitude of 5,8&3 wt% and other one coconut shell ash has1,1 & 2 wt% for all specimens. The result concludes that the tensile strength and fracture toughness would be improved in reinforcement material compared to pure aluminum alloy. (T. NITHYANANDHAN, 2017). Metal matrix composites represent a new generation of engineering materials which combine metallic properties of matrix alloys with the properties of reinforcements, which is desired for the industries today. Among various particulate reinforcements like SiC, Al2O3, Al2O3, SiC, TiC, AlN, fly ash etc, fly ash presents as a natural source, low cost, toxic industrial waste particulate reinforcement. Reinforcement of fly ash particles in the aluminium melt diminishes the cost of the composites, incorporation of hybrid particle reinforcement shows couple the properties of primary and secondary reinforcements (K. KAVIYARASAN, 2018).

2. Material Used

1. Aluminium 6063
2. Aluminum Oxide (AL_2O_3)
3. Chicken Bone Ash (CBA)
4. Aluminium Chloride ($ALCL_3$)

2.1 Aluminium 6063

Al 6063 is an aluminum compound, with magnesium and silicon as the alloying components is shown in Fig 1. 6063 is the most widely recognized combination utilized for aluminum expulsion. It permits complex shapes to be shaped with smooth surfaces fit for anodizing as is famous for obvious engineering applications, for example, window outlines, entryway casings, rooftops, and sign edges.

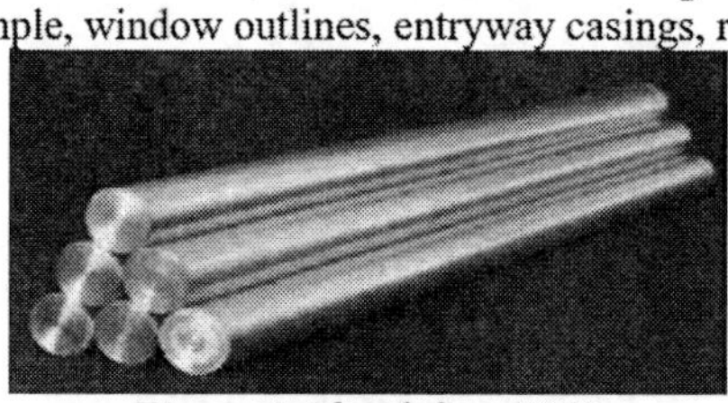

Figure.1 Aluminium 6063

2.2 Aluminium Oxide Al_2O_3

The vast majority of the aluminum oxide delivered is utilized to shape aluminum metal. Oxygen regularly catalyzes consumption in response with the metal aluminum. Nonetheless, when reinforced with oxygen to shape aluminum oxide, a defensive covering structures and forestalls further oxidation. This includes quality and makes the material less defenseless against decay regularly delivered by the Bayer cycle, which means refining bauxite to create alumina. Bayer measure this cycle begins by drying squashed and washed bauxite, ordinarily containing 30–55% Al2O3 is shown in Fig 2. The bauxite is disintegrated in scathing soft drink to frame a slurry, warmed to temperatures of around 230–520 °F (110–270 °C). The aluminium oxide which would increase the hardness of the materials, for that the different weight percentage of aluminium oxide was added to the aluminium alloy 6063.

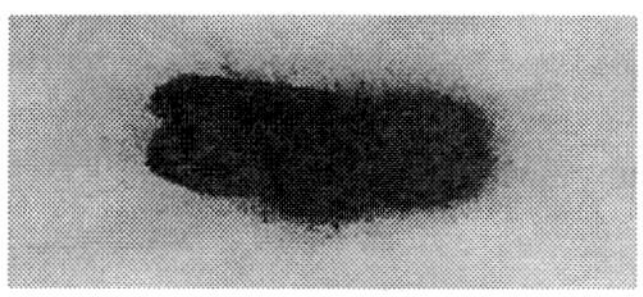

Figure.2 Aluminium Oxide

2.3 Preparation of CBA

Chicken bone Ash (CBA) is the burning buildups of chicken bone and the impact of different elements as contact time, measure of CBA, arrangement pH and beginning focus on adsorption cycle such of metal particles was learned at room temperature to improve the conditions for most extreme adsorption. Bone debris is a white material created by the calcination of bones. Commonplace bone debris comprises of about 55.82% calcium oxide, 42.39% phosphorus pentoxide, and 1.79% water. The specific structure of these mixes fluctuates relying on the kind of bones being utilized, however by and large the recipe for bone debris is: Ca5 (OH) (PO4)3. Bone debris typically has a thickness around 3.10 g/mL and a dissolving purpose of 1670 °C (3038 °F) is shown in Fig 3. Most bones hold their cell structure through calcination. Chicken bone ash is the natural waste agriculture product which would increase the mechanical properties of the materials, in this paper the different weight percentage of the CBA was added to improve the corrosion resistance of the Aluminium composites.

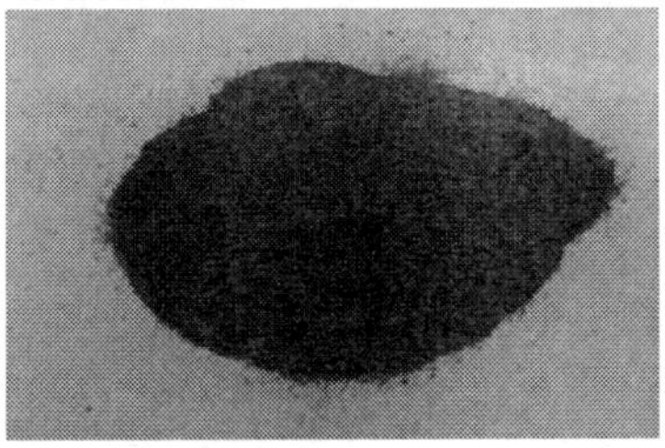

Fig.3 Aluminium CBA

2.4 Aluminium Chloride ($AlCl_3$)

Aluminum chloride (AlCl3) is the arrangement which is utilized to analyze the consumption wear rate for the aluminum composite. It is white, however tests are regularly sullied with iron chloride, giving it a yellow shading. It is chiefly delivered and devoured in the creation of aluminum metal, however huge sums are likewise utilized in different regions of concoction industry. Aluminum chloride is made for an enormous scope by the exothermic response of aluminum metal with chlorine or hydrogen chloride at temperatures between 650 to 750 °C (1,202 to 1,382 °F)

$$2 Al + 3 Cl2 \rightarrow 2 AlCl3$$
$$2 Al + 6 HCl \rightarrow 2 AlCl3 + 3 H2$$
$$2Al + 3CuCl2 \rightarrow 2AlCl3 + 3Cu$$

Aluminum chloride may be formed via a single displacement reaction between copper chloride and aluminum metal.

$$Al (H2O) 6Cl3 \rightarrow Al (OH) 3 + 3 HCl + 3 H2O$$

3. Experimental Process

During preheating degree, the fortresses are warmed freely towards number one cycle temperature of 400°C simultaneously as aluminum is disintegrated in a remarkable cauldron at a temperature of 830°C. The chicken bone ash and Aluminum oxide and their combo were preheated to 280°C for 60 Minutes to discard soddenness. Unadulterated aluminum becomes softened creation utilization of obstruction heater. The mollify temperature become raised to 750°C and later on the melted aluminum changed into joined with the assistance of a mix projecting Technique. The blending got kept up between 3 to 5 min at a mix speed of 200 rpm. The softening temperature kept up was to 700°C for the term of extension of AL_2O_3 and chicken bone ash blend particles. The mollify with fortified particulates had been filled the preheated interminable metallic shape. The pouring temperature got set aside at 700°C. The dissolving aluminum composite was then licensed to solidify inside the structure. The stir casting setup was shown in Fig 4. The Process parameters which are used in this setup were noted in table 1. Table 2 shows the different composition of composite materials which are solidified from the casting.

Figure.4 Stir Casting Unit

The testing method of this paper was done by immersion test. The consumption register turned with completed in aluminum chloride arrangement with pH estimation of 2.5 with 1N arrangement which had been readied following typical normalized approaches. The composite materials were sliced to the components of Ø 30x32mm.The examples were de-lubed with acetone after which flushed in refined water sooner than inundation inside the readied regardless arrangements of AlCl3 with 1N. The answer for example surface area proportion turned out to be around 80 ml cm-2. The erosion rates have been assessed through weight decrease and further more consumption rates are estimated by one day spans. The examples were uncovered inside the aluminum chloride in environmental factors condition for 48 hours individually. Consumption charge and erosion hindrance execution for each example become assessed from the weight decrease estimations observing Standard methodologies

Table.1 Process Parameters

Speed of stirrer	200-250 Rpm
Stirrer Time	3to 5mins
Preheat Temperature	300°c
Stirrer Temperature	750°c

Table.2 Composition of Composites

Sample No	Weight Percentage of AL6063	Weight Percentage of Al2O3	Weight Percentage of CBA
Sample 1	100	0	0
Sample 2	92	6	2
Sample 3	92	5	3
Sample 4	92	4	4
Sample 5	92	3	5
Sample 6	92	2	6

4. Experimental Procedure

The 600 Grams of pure aluminium was melted in a resistance heated muffle furnace and casted in a clay graphite crucible. For this the melting temperature was raised to 800°C and then the Aluminium oxide (5% ,7.5%) and chicken bone ash (5%,7.5%) reinforcement was preheated in the preheater for about 300°C to remove the moisture for two hours. Then reinforced materials were added in the ratio of 0:1,1:1,1:0 to the 5% weight of aluminium matrix for the production of three different composites and also reinforced materials were added in the ratio of 0:1,1:1,1:0 to the 7.5% weight of aluminium matrix for the production of another three different composites. Commercial pure aluminum was melted by raising its temperature to 800°C and then the melt was stirred using a mild steel stirrer. Reinforcement particles were added to melt at the time of vortex in the melt due to stirring. The soften temperature was kept up between 750°C-800°C during the expansion of fortified particles. At that point the soften was poured in to the predetermined kick the bucket to get cooled. The composite was expelled from the bite the dust. The consumption attributes of Al-Al2O3-CBA composites were controlled by plunging the six examples of composites in AlCl3 (Aluminum chloride) with ordinariness of 1N, for this AlCl3 were taken in six Glass holder and these six examples of composites were dunked in the AlCl3 answer for three days. Weight reduction of composite is resolved from beginning and last loads. Erosion rates and consumption inhibitor efficiency of the samples were assessed.

5. Result and Discussions

5.1 Result

Corrosion inhibition efficiency and Corrosion rate of Al6063 with Al2O3and CBA at room temperature was listed in Table 3.

Table 3. Corrosion inhibition efficiency and Corrosion rate

Sample. No	Weight Of Alloy (Mg) Before (W_B)	After (W_I)	Weight Difference (Mg)	Corrosion Inhibition Efficiency [IE =(W_B-W_I)/W_B*100]	Corrosion Rate (Mmpy)
Sample 1	55.66	54.23	1.43	0.02570	0.1372
Sample 2	56.24	55. 92	0.32	0.56899	0.0307
Sample 3	56. 73	56.26	0.47	0.82848	0.0451
Sample 4	57.21	57.98	0.23	0.40202	0.0220
Sample 5	57.53	57.48	0.05	0.08691	0.0479
Sample 6	57.78	57.75	0.03	0.05192	0.0287

Above results shows that, the sample 4(4%of $Al_2 0_3$ & 4% of CBA) has less corrosion rate compare to the other reinforced and non-reinforced aluminium alloy. In the sample 4 component have a high percentage of reinforcements. The above results show that the maximum percentage of the CBA in the aluminium alloys has less corrosion rate compare to the others.

5.2 Discussions

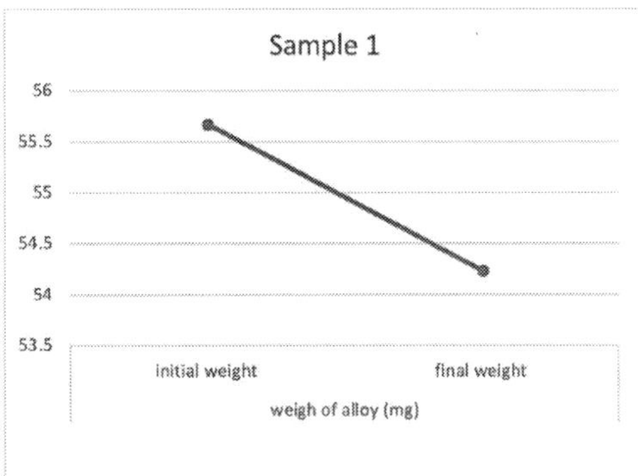

Figure.5 Weight loss of sample 1

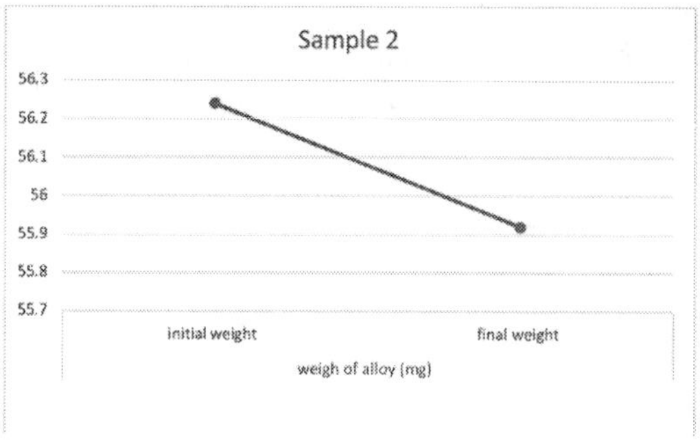

Figure.6 Weight loss of sample 2

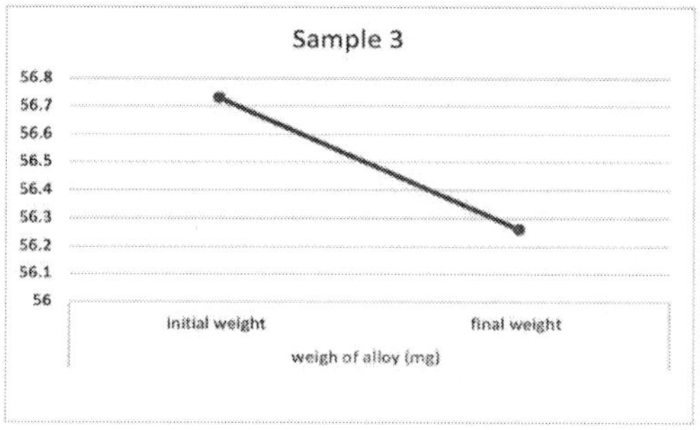

Figure.7 Weight loss of sample 3

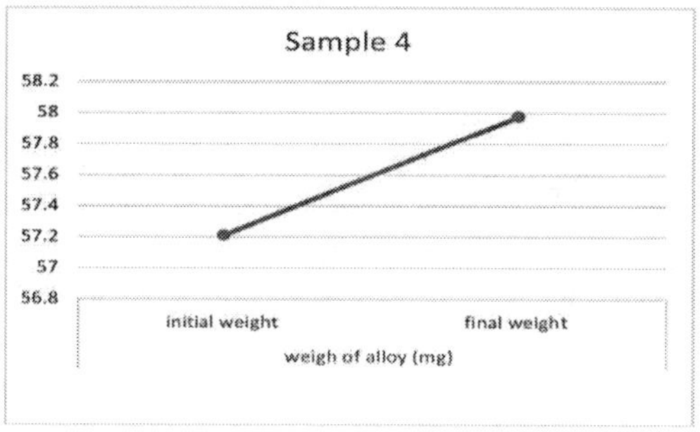

Figure.8 Weight loss of sample 4

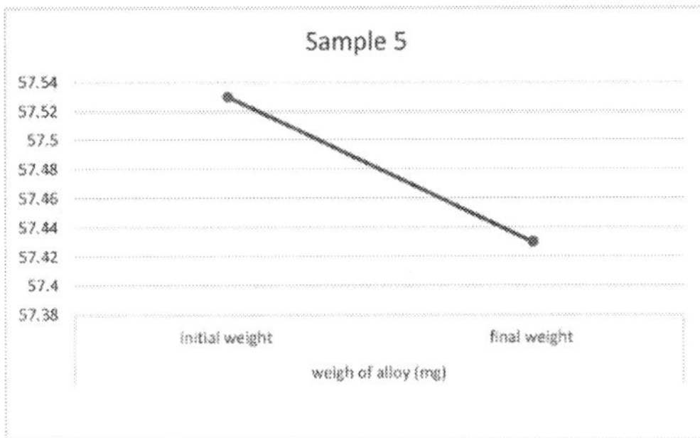

Figure.9 Weight loss of sample 5

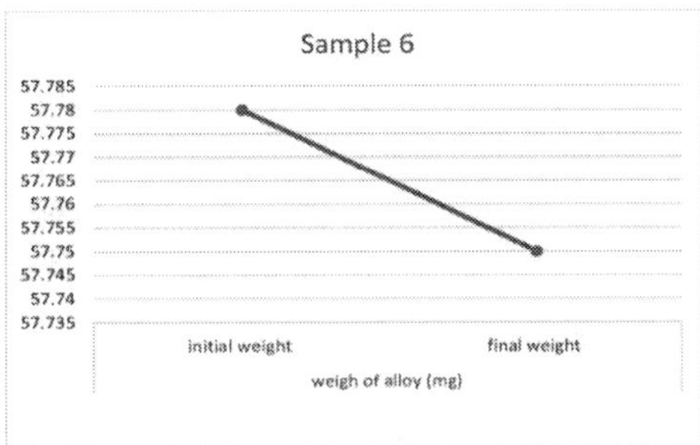

Figure.10 Weight loss of sample 6

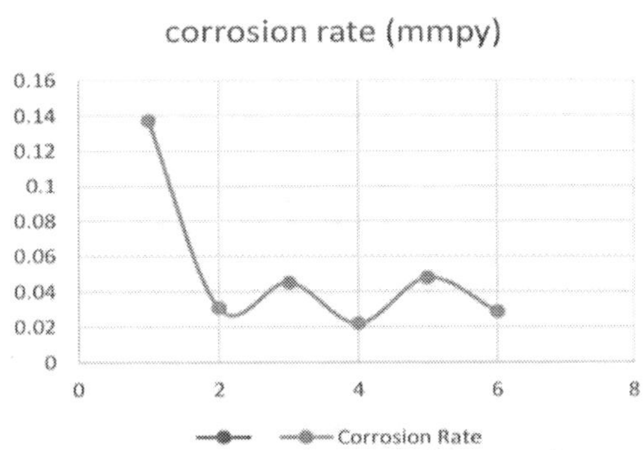

Figure.11 Corrosion rate of the samples

In the Figure 5. shows that the weight loss gradually decreases for the sample 1 starting from 55.66 to 54.23 mg ranging between initial and final weight because of weight percentage of Reinforcements was not considered. In the Fig 6. it is observed that the weight loss decreases slightly from 56.24 to 55.9 mg which is from initial weight to final weight for sample 2 because of weight percentage of Aluminium oxide (6%) higher than the CBA (2%) but the difference between the weight percentage of two reinforcements is higher than sample 1. In the Fig 7. it is observed that the weight loss decreases from 56.73 to 56.26 mg which is from initial weight to final weight for sample 3 because of weight percentage of Aluminium oxide (5%) less than the CBA (3%), the reason for more weight loss due to more difference percentage of reinforcement content in this sample. In the Fig 8. it is observed that the weight loss decreases from 57.21 to 57.98 mg which is from initial weight to final weight for sample 4 because of weight percentage of Aluminium oxide (4%) and the CBA (4%), due to the same percentage of

Reinforcement compare to the other samples.In the Fig 9. it is observed that the weight loss decreases from 57.53 to 57.48 mg which is from initial weight to final weight for sample 5, because of weight percentage of Aluminium oxide (3%) higher than the CBA (5%) but the difference between the weight percentage of two reinforcements is less. In the Fig 10. it is clear that the weight loss decreases very minute from 57.78 to 57.48 mg which is from initial weight to final weight for sample 6, because of weight percentage of Aluminium oxide (2%) less than the CBA (6%), the reason for more weight loss due to more percentage of CBA content in this sample. In the Fig 11. it is clear that the corrosion rate increased to 0.1372mmpy at sample 1 (0% Aluminium oxide &0 % CBA), the corrosion rate decreased to 0.03mmpy at sample 2(6% Aluminium oxide &2% CBA), the corrosion rate increased to 0.045mmpy at sample 3(5% Aluminium oxide &3% CBA), the corrosion rate decreased to 0.02 at sample 4(4% Aluminium oxide &4% CBA), the corrosion rate increased to 0.047 mmpy at the sample 5(3% Aluminium oxide &5% CBA), the corrosion rate decreased to 0.028mmpy for the sample 6(2% Aluminium oxide &6% CBA).

6. Conclusion

The erosion characteristics of composites materials have been finished by distinctive weight percent like 6%, 5%, 4%, 3% and 2% wt of Aluminium oxide and reinforcement with 2%, 3%, 4%, 5% and 6% of chicken bone ash. The rate of corrosion and corrosion resistance was carried out by Aluminium chloride solution through the usage of immersion test. The end result as follows: -

- About the paper it is miles found that we can utilize farming waste like chicken bone debris, rice husk debris for the composites assembling and ready to change over industrial squanders into business riches.
- The Erosion rate was decreased in the reinforcement with 4% of Aluminium oxide and 4% of CBA with the correlation of unadulterated Aluminum 6063.
- The Erosion resistance of Reinforced composites have been improved with increased in of 5% Aluminium oxide and 3% chicken bone ash.
- The Maximum Percentage of reinforcements is extended the corrosion resistance and decreased the corrosion rate.

REFERENCE

[1] Kandpal B C, Kumar J and Singh H 2014 Production technologies of metal matrix composite: a review *International Journal of Research in Mechanical Engineering & Technology,* **4,** 27–32

[2] Mr.Kaviyarasan .K, 2018 "Extensive Review on Properties of Metal Matrix Composites Reinforced with Fly Ash", *International Journal of Mechanical Engineering and Technology (IJMET)*, **9**, 0976-6340.

[3] Kainer, K.U., 2006. *Basics of metal matrix composites*. Wiley-VCH GmbH & Co. KGaA, Weinheim, Germany.

[4.] Deuis, R.L., Subramanian, C. and Yellup, J.M., 1997. Dry sliding wear of aluminium composites—a review. *Composites science and technology*, **57(4)**, 415-435.

[5] T.Pridhar, C.Boopathi 2016,"The effect of particle hybridization on micro structure analysis and mechanical behavior of metal matrix composites: an experimental approach" *Digest Journal of Nanomaterials and Biostructures,* **3**, 845 – 852.

[6] Christy TV, Murugan N, Kumar S, 2010, "A comparative study on the microstructures and mechanical properties of Al 6063 alloy and the MMC Al 6063/TiB2/12p". *JMMCE,* **9**, 57–65.

IOP Conf. Series: Materials Science and Engineering 1059 (2021) 012018 doi:10.1088/1757-899X/1059/1/012018

[7] Macke A, Schultz BF, Rohatgi P, 2012 "Metal matrix compositesoffer the automotive industry anopportunity to reducevehicle weight, improve performance" *Adv Mater Processes*, **170**, 19–23.

[8] Pridhar T, 2019, "Evaluation on properties and characterization of asbestos free palm kernel shell fibre (PKSF)/polymer composites for brake pads" *Materials Research Express*, **6**, 11-19.

[9] Siva Prasad.D, Rama Krishna.A, 2011," Production and mechanical properties of A356.2/ CBA composites", *International Journal of Advanced Technology*, **33**, 51–57.

[10] Nithyanandhan T. 2017, "Investigation of Mechanical Properties on Aluminium Based Hybrid Composites". *International Journal of Innovative Research in Science, Engineering and Technology*, 7, 118-126.

[11] Haridass.R ,2018, "Determination of corrosion behaviour of Al6063-Sic-CBA metal matrix composites", *International Journal of Pure and Applied Mathematics*,**118**, 907-915.

[12] Pridhar T, 2018, "Mechanical properties and characterization of zirconium oxide (ZrO_2) and coconut shell ash(CSA) reinforced aluminium (Al 6082) matrix hybrid composite"*Journal of Alloys and Compounds*,**765**,171-179.

[13] Murthy K S, Girish D P, Keshavamurthy R, Varol T and Koppad P G 2017 ''Mechanical and thermal properties of Al6063/bone ash hybrid composites obtained by hot forging'' *Progress in Natural Science: Materials International*, **27**, 474–81

[14] Nithyanandhan, T., Sivaraman, P., Ramamoorthi, R., Kumar, P.S., Kannakumar, R. and Kumar, A.N., 2020. Enhancement of corrosion behaviour of AL6061-B4C-RHA reinforced hybrid composite. *Materials Today: Proceedings*.

[15] Nithyanandhan.T,2020," Mechanical Properties and Tribological Behaviour of Reinforced Aluminium Metal Matrix Composites", *International Journal of Recent Technology and Engineering*,**8**, 1191-1195

Experimental Investigation and Weld Characterization Analysis on Carbon Steel by using Nd:Yag Laser Welding

Palanivendhan M[1*], Chandradass J[1], Dhruv Dhingra[1], Madhuritha R[1]

[1]Centre for Automobile Materials, Dept. of Automobile Engineering, SRM Institute of Science and Technology, Kattankulathur-603203, Tamil Nadu, India.

*Email: palanivendhan@gmail.com

Abstract. Laser welding is a compulsion in attaining certain structures and in situations demanding high-temperature resistance in one area and corrosion resistance in another. The objective of this study is to increase the efficiency of the weld joints and to study their properties. This study is aimed at optimizing the fastening method parameters to improve the life and efficiency of C45 carbon steel joint. Laser welding of C45 carbon steel plates is done without any prior heat treatment with a Nd:YAG laser pumped by a flash lamp that produces a pulsed output. Laser welding is carried out by varying parameters like Laser Power, Pulse Width, and Frequency thereby varying the laser penetration power and the bead formation that occurs in the weld area. Optimum laser welding parameters that yield maximum weld strength and minimum heat affect zone are found experimentally. The focal diameter was fixed at 1.8mm and the other three parameters: Power, Pulse Width, and Frequency were varied to evaluate the best combination. Experimental parameters were selected through Taguchi design using the L9 orthogonal array. The welded joints have also been subjected to microstructure test, tensile strength, and hardness to evaluate its metallurgical and mechanical properties.

Keywords: Mini Tab; Optimization; Taguchi Design; Welding.

1. Introduction

Laser fastening is a non-conventional and non-traditional way to affix materials. The heat-affected zones are very minimal in ray fastening which may be attributed to factors such as the detailed density of laser DIA, rapid and high heating, and cooling rates. Lasers have a wide range of industrial applications and are used for a variety of machining and surface treatment operations. Joining of components by laser welding is commonly used in industries as different groups or compositions of materials can be fastened or joined together using laser welding [1]. Amongst the different materials, the demand for joining of dissimilar materials is more as the requirement of having varying properties in different parts of the same weldment exist. Of all the different welding techniques available, laser welding is preferred because of factors that include rapid heating and cooling rates, minimal fastening defects, minimal HAZ's, minimized post- processing stress(residual), and no requirement for fillers. Due to these advantages, laser welding of materials favors the formation of a finely detailed microstructure resulting in increased strength. For this study, the Nd: YAG solid-state optical device is employed. The optical device welding parameters namely: Power, Pulse Width, Frequency, and focus position on the weld joint strength have been investigated. [2] Optical devices are widely used for thermal supply to industrial applications; owing to their native treatment, precise operation, and short interval. A crucial advantage in the industrial application of the optical device process is that, the optical device attachment offers substantial benefits over the standard attachment strategies. The optical device also provides flexibility in pulse shaping at various pulse repetition rates as well as in length, in the range starting from 0.5ms up to a maximum of 20ms that enables the thermal input to be managed efficiently. The demand for manufacturing joints of dissimilar materials is ever-growing, as it is possible to obtain acceptable mechanical properties for an economical price. Fastening of materials by the optical device are usually characterized by a parallel fusion zone, a very slim bead, and high level of penetration [3]. This study includes Style of the experiment (DOE) and applied mathematics techniques for the optimization of method parameters to maximize the strength of the weldment and also reduce experiments. The Taguchi technique victimization Associate in Nursing L-9 orthogonal array has been employed for the design of experiments. This style supports the specified three fastening parameters with four levels. The three parameters selected for optimization were: power, pulse width, and frequency. The S/N magnitude relation for every level of the parameters is evaluated using the S/N analysis and the three classes of quality characteristics involved in the analysis are the smaller-the-better, the larger-the-better and nominal-the-best. [6]. The target is set to obtain the maximum possible durability. Maximum durability is achieved at a particular level of the method parameters, i.e., the optimal S/N magnitude relation [5, 6, 9]. Also, to point out the statistical significance of the square measure of the method parameters to evaluate the best combination of the method parameters statistical analysis of the variance has also been performed.

2. Literature Review

M.J.Torkamany et al investigated optical device fixing of low steel and aluminium alloy with the steel overlapping on the aluminium, using hole attachment. The alloy of aluminium used was 5754 and the intermetallic compounds that were formed throughout optical device attachment were reduced, a result of optical device SPOT United States intelligence agency, pulse period, and the overlapping issue was investigated. [1] Tensile test, spectroscopic analysis, Vickers hardness test were performed, along with the microstructure test by optical and scanning electron microscopes. It was found that with the increase in the peak with constant pulse energy and increase in pulse period at constant peak SPOT DIA and overlapping issue with the other two kept constant produces an increase in the formation of the intermetallic compounds. Contradicting to this, they found out that any decrease in the mentioned parameters would cause effects like inadequate penetration depth- where the laser doesn't penetrate to the entire joint, spattering which causes the laser beam to spatter all around the weldment. It also leads to formation of cavities which usually have negative and damaging effects. It is critical to have reduced or lower intermetallic compounds or parts to have increased strength of the weld metal. Finally, an acceptable level of optimized peak SPOT DIA, pulse period, and overlapping issue was reported. [9] Xiu-Bo Liu, et.al administered solid cast Ni-based superalloy K418 turbo disk laser welded with alloy steel 42CrMo shaft. The weld was subjected to microstructure analysis using scanning electron microscopy, X-ray diffraction technique, and a energy dispersive spectroscope. [2] The weld seam was also subjected to microhardness test and test to check life of the weldment, to find out the mechanical properties. It was found out that the weld seam showed the presence of solution dendrites having solid microstructures of $FeCr_{0.29}Ni_{0.16}C_{0.06}$, a less dense fine spread Ni_3Al section, some quantity Laves particles and also as very little quantity of carbides. The lastingness of the bottom metal was achieved through partial penetration fastening which also attributes to the mix of plasticity and brittleness during fracture. [1] The initiation and propagation of micro voids and cracks within the weld seam were found to be hastened by the presence of Laves particles, which ultimately affect the long-lasting nature of the welded joint. S.A.A. Akbari Mousavi, et al were involved in a study of Nd: YAG laser butt welding of AISI 321 and AISI 630 stainless steel having a rounded geometry. [3] Results discussed the impacts of laser SPOT DIA, pillar measurement and heartbeat length on width. The outcomes demonstrate the increment of weld profundity and weld width with voltage [12] and also showed that the weld dot profundity and width effectively depended on the beat length. A weld zone composing of fine cellular dendritic structures was obtained while in the metal neighboring the AISI 630 a martensitic microstructure was seen, in the weld metal neighboring to the AISI321 a ferritic microstructure was established. It was also found out that a part of the weldment on the AISI630 side had the most extreme hardness and the part of weld on the 321 stainless steel side had base hardness through the microhardness tests. [10] Han Guo Ming et al involved the study of temperature field in optical laser welding of stainless steel and discussed the field's formation and distribution by dynamic simulation in ANSYS using finite element analysis [5]. The transition modelling was employed for the components due to the high non-linearity in the fastening using optical lasers. A residue management technique was adopted to facilitate precise node selection. It was proved through calculations that coherent results were obtained in the simulation and experimentation. A.P. Tadamallea, et al investigated interaction between optical devices, the dominance of method parameters and their impact on melting, process potency, and its relation with the behaviour of the weld joints. This paper aims to check the impact of fastening speed on different method efficiencies in 304L stainless-steel. [6] The bead on 304L solid solution stainless-steel plate is formed by variable fastening speed. A novel semi-empirical technique supporting weld pool volume measured from experimental results is employed to predict the melting potency of Nd: YAG optical device fastening method. The dimensionless parameter models are accustomed to estimate numerous forms of measurable dominant parameters. These parameters have a smart agreement with the varied accessible models in the literature. Shuhai subgenus Chen, et al investigated the microstructure and mechanical properties of laser butt welding of Ti alloy with stainless steel. The effect on fracture behaviour of the welded joint and the changes in structure due to the laser beam have been discussed. The weldment showed two layers of uniformly thick inter-metallic compounds. The durable association is achieved on the condition that the beam of sunshine is offset by 0.6 metric linear units towards the Ti alloy. [7] Jose Roberto Berretta, et al investigated laser welding of AISI 304 stainless-steel with AISI 420 stainless-steel using Nd: YAG optical device. The effect of the position of the ray of a beam on the welded joint was studied. Welding was done with the laser, incident on the joint and then varied in position by 0.1mm and 0.2mm on both sides of the joint to measure the changes due to change in position of the ray. [12] The joints were examined in an optical magnifier for cracks and pores. A scanning electron microscope coupled with an energy dispersive spectrometer was used to perform the microstructure analysis of the weld and identify the heat-affected zones. The variations were indicated in weight percentages which were also verified. A. Arun Mani, et al involved a microstructural and characteristic study of welded joint produced by CO2 laser beam fastening of AISI 430 ferritic stainless steel with AISI 304 austenitic stainless steel. Experimentation involved laser beam welding with heat inputs of 1500W and welding speeds in the range of 50-100-150cm/min in an argon-helium gas atmosphere. The welded joint was subjected to microstructural analysis using optical microscopy and SEM analysis and the heat-affected zones were studied. A tensile strength test was also performed to measure the mechanical properties of the weldment. It was found that with an increase in welding speed the width of the welding zone and the heat-affected zones reduced. On the contrary, results showed that when the heat input was increased there was a consequent increase in the width of the weld joint which was also backed by the results of the tensile strength test [13].

3.Experimental Procedure

3.1 Nd: YAG Laser

The solid-state Nd: YAG optical maser is used due to factors like high power density and easy repair and due to its ability to generate pulses in short periods. The high repetition rate of pulses offered by the laser has applications in the material processing industry. Also, the Nd: YAG optical maser provides the highest possible SPOT DIA. The beam quality and output SPOT DIA depend on the resonator length. The quality of the beam profile is improved by scaling down the effective radius of producing some degree of aperture inside the resonator. As the absorbing capability of metals increases with decrease in wavelength, the Nd: YAG optical maser is best suited for conductivity fastening. Since conductivity fastening is often used for smaller parts, a small range of optics is used to feed or focus the beam on the workpiece.

3.2 Process parameter

The process is always done at a temperature with a clean setting. Acceptable fixturing is required to ensure that the components don't move out of their position relative to each other which might cause a gap. Sometimes an inert Shielding gas like Argon is required to provide protection to the weld metal against contamination from surroundings. Gases are employed in the process of fastening for several reasons such as; 1) Shielding gas which protects against contaminants like water and oxygen, 2) Plasma suppression which is used to minimize the laser beam attenuation and scattering, and 3) Protection of optics. Each one of the weld bead face and root face needs shielding. Plasma formation is especially noticeable where fastening is done with a high spot dia and with a speed lesser than 1m/min. Helium, Argon, Nitrogen, and Carbon dioxide were used as shielding gases. Argon gas is employed as a plasma suppression gas as it provides smooth arc starting characteristics owing to its lower ionization potential and is also cost-effective. For the experiment, a square butt-joint is preferred due to its good optical maser fastening. Accurate fixturing is critical in the optical maser fastening gap. Fixturing is a long and costly process, however, it is required to achieve good product quality and reduced post-welding remodelling.

3.3 Material properties (C45)

C45 is a pure carbon tool steel, used for the development of plastic mould assemblies. DIMO C45 is intended for holder block applications requiring intensive machining and high dimensional stability as mould assemblies, ejector plates, supports, clamping plates and housings for die casting dies.

3.4 Composition

The chemical composition of C45 carbon steel is shown in the below table 1.

Table. 1. Composition of C45 Steel

Material	% of Carbon	% of Silicon	% of Manganese	P% (Max)	S% (Max)	% of Chromium
C 45	0.42 – 0.50	0.15 – 0.35	0.50 – 0.80	0.025	0.025	0.20 – 0.40

3.5 Mechanical Properties

The mechanical properties of C45 steel in different conditions are listed below in table 2.

Table. 2. Mechanical Properties

Grade	Condition	Yield Strength MPa	Tensile Strength MPa	Elongation AS%	Hardness HRC	Quenching Temperature ^{0}C	Bendability	Nominal Thickness(t) 1.95<t<10mm
C45	Rolled Annealed Water - Quenched Oil Quenched	460 330	750 540 2270 1980	18 30	58 55	820 860	<90deg min required	Rolled-2t Annealed 1t

4. Methodology

4.1 Taguchi Design

The Taguchi design employs the use of orthogonal arrays(L9) and checks the parameter with very minimal experimental variations. The results are remapped into a signal to noise (S/N) magnitude relation which is then used to point out the deviations from the required values. The classes of quality characteristics within the analysis are 1) the lower the better, 2) the higher the better, and 3) nominal the best. Irrespective of the class, higher quality characteristics are obtained in correspondence with higher S/N magnitude relation. Therefore, the level with the best S/N magnitude relation is the level having the best optimal process parameters. The results obtained from the Taguchi style indicating the optimum process parameters have been verified by a final experiment. The formulae for signal to noise ratio is formulated such that the person performing can perpetually choose the biggest factor level setting to optimize the standard characteristics of an experiment.

4.2 Design *of L9 Orthogonal Array*

First, Taguchi Orthogonal array is designed in minitab-17 to calculate S/N magnitude relation and suggested steps are given below. Created Taguchi style has been chosen and is shown in table 3.

Table. 3. Design of L9 Orthogonal Array

Levels	Process Parameters		
	Power	**Pulse Width**	**Frequency**
1	180	4	10
2	185	4.5	11
3	190	5	12

4.3 Process parameters for L9 Array

The process parameters for the design of L9 Array are described in the below table 4.

Table. 4. Process Parameters for L9 Array

Sl. no.	Power	Pulse Width	Frequency
1	180	4	10
2	180	4.5	11
3	180	5	12
4	185	4	11
5	185	4.5	12
6	185	5	10
7	190	4	12
8	190	4.5	10
9	190	5	11

4.4 Rockwell Hardness Test

Rockwell hardness test was conducted to find the hardness of the specimen. The specimens were tested at three different spots using B-Scale and the results of which are shown below in the table 5.

Table. 5. Rockwell Hardness Test Results

S.No	Power	Pulse Width	Frequency	Hardness
1	180	4	10	69
2	180	4.5	11	68
3	180	5	12	71
4	185	4	11	72
5	185	4.5	12	58
6	185	5	10	82
7	190	4	12	80
8	190	4.5	10	79
9	190	5	11	67

4.5 Tensile Test

A tensile check helps in deciding tensile properties like tensile strength, yield strength, proportion of elongation and reduction in the area, and modulus of elasticity. The fastening parameters were randomly chosen within the machines limits. The joints were created with random parameters and valuate tensile strength and burn off. Then the joints were created and measured. The friction welded specimens were ready as per the ASTM standards. The check was administered in a universal testing machine (UTM) 40 tones FIE make. The results of Tensile Test are depicted in the below table 6.

Table. 6. Tensile Test Results

S.no	Power	Pulse width	Frequency	Designation	Tensile Strength
1	180	4	10	$A_1B_1C_1$	255.20
2	180	4.5	11	$A_1B_2C_2$	275.60
3	180	5	12	$A_1B_3C_3$	285.60
4	185	4	11	$A_2B_1C_2$	276.80
5	185	4.5	12	$A_2B_2C_3$	270.80

6	185	5	10	$A_2B_3C_1$	275.60
7	190	4	12	$A_3B_1C_3$	276.80
8	190	4.5	10	$A_3B_2C_1$	280.80
9	190	5	11	$A_3B_3C_2$	293.60

4.6 Analysis of Variance
Analysis of variance and the percentage contribution of the welding method parameters towards the optimization are shown in Table7.

Table. 7. Analysis of Variance

Source	DF	Seq SS	Adj SS	F	P	% of contribution
Power	2	226.81	113.40	2.38	0.296	26
Pulse Width	2	357.37	178.68	3.75	0.211	41
Frequency	2	201.53	100.76	2.11	0.321	22
Error	2	95.40	47.70			11
Total	8	881.10				100

4.7 Microstructure Test
The microstructure test was conducted. The material was subjected to polishing with 1/0, 2/0, 3/0, 4/0 grades of emery paper. Then it was polished on the velvet cloth. 5% Nital solution was dropped on the weld. The specimen was kept under the Upright Metallurgical Microscope Image Analyzer. The eyepiece was 10X and the magnification range 50X to 500X. The software used for the image analyzing was Metaplus. Optical Microscope is used for face identification and failure analysis of any material. Information on unit cell dimensions is obtained through this rapid analytical technique. In every category of specimens two samples were analyzed by using optical microscope.

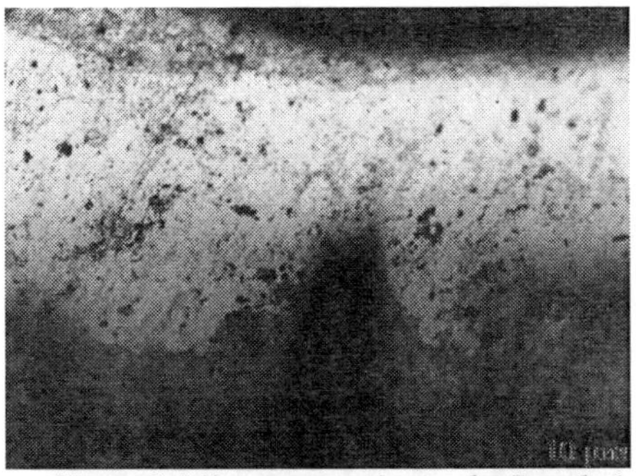
Figure 1. Microstructure image for Sample 1

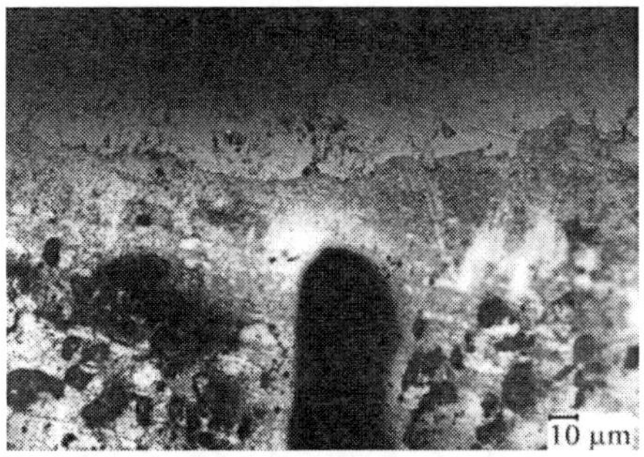

Figure 2. Microstructure image for Sample 2

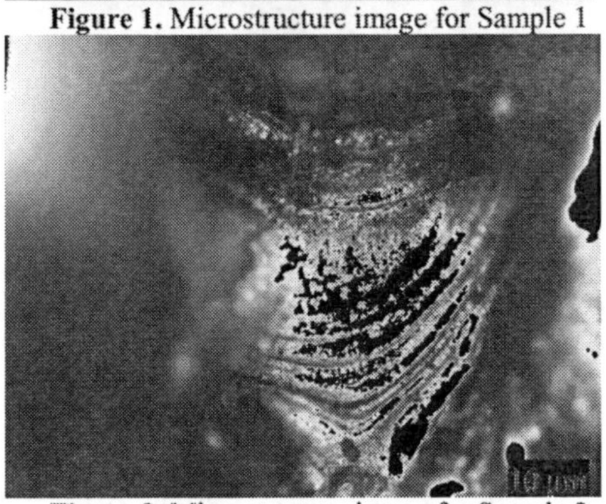
Figure 3. Microstructure image for Sample 3

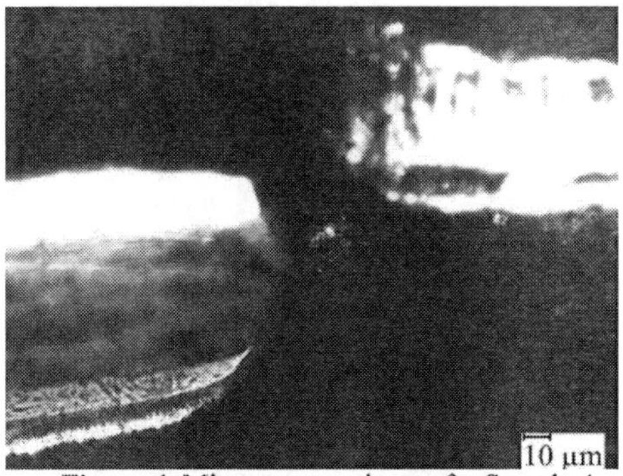

Figure 4. Microstructure image for Sample 4

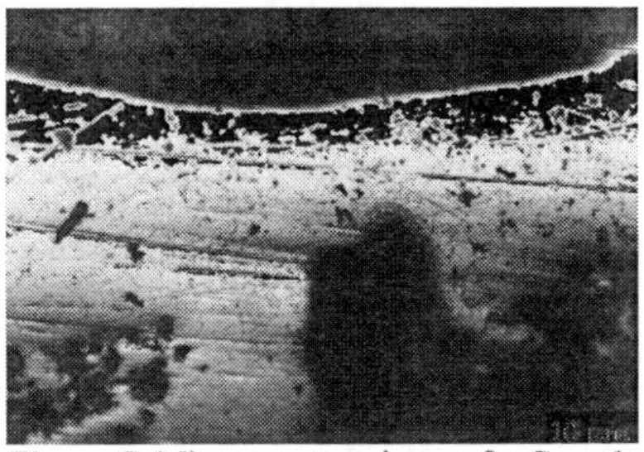
Figure 5. Microstructure image for Sample 5

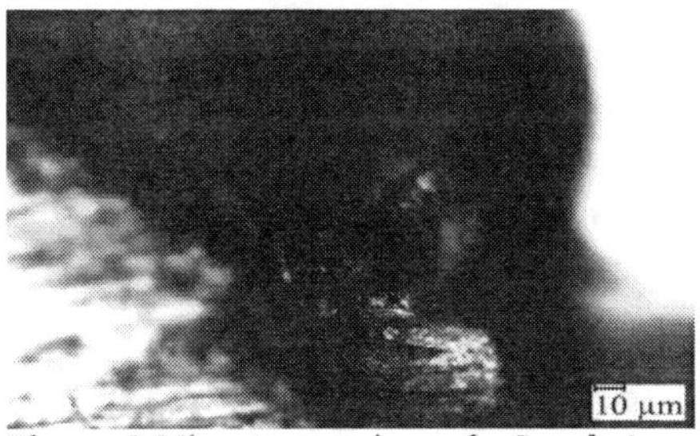

Figure 6. Microstructure image for Sample 6

Figures 1 to 6 shows the stage wise development of the weld joint between the two C45 sheets. The HAZ remained small and did not show any deformity.

5. Conclusion

From the results of the tests performed on the laser butt welded joint of C45 steel, the following conclusions were observed. The three method parameters – Power, Pulse Width and Frequency were varied while the focal diameter was kept at a constant of 1.8mm. Design of Experiment was done using Minitab Software to select the parameters. Microstructural images represent the penetration depth of the weldment. Finally concluding the experimental investigation, the optimum input parameters for Nd-YAG LASER welding process of C45 steel were – power: 190 watts; Frequency: 11Hz and Pulse width: 5µs. From the results of Taguchi design, optimized parameters for maximum Tensile strength are as follows:

5.1. Optimal Control Factor

Tensile Strength –A2 (Power 185W) B1(Pulse width 4µs) C3 (Frequency 12Hz).

5.2. Percentage Contribution of Process Parameter

Tensile Strength- Pulse Width- 41%.

References

[1] Torkamany, M.J., Tahamtan, S. and Sabbaghzadeh, J., 2010. Dissimilar welding of carbon steel to 5754 aluminum alloy by Nd: YAG pulsed laser. *Materials & Design*, ***31*(1)**, 458-465.
[2] Liu, X.B., Yu, G., Guo, J., Gu, Y.J., Pang, M., Zheng, C.Y. and Wang, H.H., 2008. Research on laser welding of cast Ni-based superalloy K418 turbo disk and alloy steel 42CrMo shaft. *Journal of Alloys and Compounds*, ***453*(1-2)**, 371-378.
[3] Mousavi, S.A. and Sufizadeh, A.R., 2009. Metallurgical investigations of pulsed Nd: YAG laser welding of AISI 321 and AISI 630 stainless steels. *Materials & Design*, ***30*(8)**, 3150-3157.
[4] Palanivendhan, M., Chandradass, J., Rajendran, A., Govindarajan, N.M. and Krishnaa, M.A., 2019. Experimental Investigation of Alumina Reinforced Copper Metal Matrix Composite by Stir Casting Method. *International Journal of Vehicle Structures & Systems (IJVSS)*, ***11***(1).
[5] GuoMing, H., Jian, Z. and JianQang, L., 2007. Dynamic simulation of the temperature field of stainless steel laser welding. *Materials & design*, ***28*(1)**, 240-245.
[6] Tadamalle, A.P., Reddy, Y.P., Ramjee, E. and Reddy, V., 2014. Evaluation of Nd: YAG laser welding efficiencies for 304L stainless steel. *Procedia materials science*, ***6***, pp.1731-1739.
[7] Arivarasu, M., Vishnu, G., Hari, P.R., Vipin, V.P., Katiki, K., Gokulkumar, K., Manikandan, M., Ramkumar, K.D. and Arivazhagan, N., 2014. Micro-segregation studies on the continuous Nd: YAG laser beam welded AISI 316L. *Procedia Engineering*, ***97***, pp.892-901.
[8] Khan, M.M.A., Romoli, L. and Dini, G., 2013. Laser beam welding of dissimilar ferritic/martensitic stainless steels in a butt joint configuration. *Optics & Laser Technology*, ***49***,125-136.
[9] Palanivendhan, M., Yash Nigam, and Kshitij Trivedi 2015 Experimental evaluation of aluminium alloy reinforced with silicon carbide *International Journal of Applied Engineering Research* **10**.91.
[10] Chen, S., Zhang, M., Huang, J., Cui, C., Zhang, H. and Zhao, X., 2014. Microstructures and mechanical property of laser butt welding of titanium alloy to stainless steel. *Materials & Design*, **53**, 504-511.
[11] Vigneshwaran, K., Murugadoss, P. and Gokul, K., 2017. *Extraction and characterization of microfibers obtained from banana waste* (No. 2017-28-1987). SAE Technical Paper. Vigneshwaran, K., Murugadoss, P. and Gokul, K., 2017. *Extraction and characterization of microfibers obtained from banana waste* (No. 2017-28-1987). SAE Technical Paper.
[12] Berretta, J.R., de Rossi, W., das Neves, M.D.M., de Almeida, I.A. and Junior, N.D.V., 2007. Pulsed Nd: YAG laser welding of AISI 304 to AISI 420 stainless steels. *Optics and Lasers in Engineering*, ***45***(9), 960-966.
[13] Mani, A.A., Kumar, T.S. and Chandrasekar, M., 2015. Mechanical and Metallurgical Properties of Dissimilar Welded Components (AISI 430 Ferritic-AISI 304 Austenitic Stainless Steels) by CO2 Laser Beam Welding (LBW). *J. Chem. Pharm. Sci*, ***974***, 2115.

IOP Conf. Series: Materials Science and Engineering 1059 (2021) 012020 doi:10.1088/1757-899X/1059/1/012020

Numerical Analysis of Wear Characteristics of Zirconia Coated Aluminum 6061 Alloy

Renjish Vijay[1*],V N Aju Kumar[1], A Sadiq[1] andR Rakesh Pillai[1]

[1]DepartmentofMechanicalEngineering, TKM College of Engineering, Kollam 691005,APJ Abdul Kalam Technological University, Kerala, India

E-mail id*: renjishvijay@gmail.com

Abstract.Any mechanical systems with moving parts encounter the problem of wear and industriesacross the world are trying to reduce its detrimental effects. In this paper, Finite Element Method (FEM) is used for analysis of wear characteristics of Yttrium stabilized Zirconia coating on Al6061 Aluminum alloy. Aluminum alloy 6061, having high strength to weight ratio, is used as the substrate material upon which Thermal Barrier Coating (TBC) is applied. Friction and wear coefficient obtained from Continuous Sliding Motion (CSM) of Yttrium stabilized Zirconia coating over Aluminum oxide disc material is used for the numerical analysis inputs. Tribological characteristics such as contact pressure, frictional stress and wear thickness are evaluated numerically using ANSYS Mechanical based on Archard wear model.

1Introducion

Aluminum alloy 6061, with its high strength to weight ratio and comparatively lower manufacturing costs, finds application in combustion system components. In order to improve the thermal resistance of these components, Thermal Barrier Coatings (TBC) are normally employed. Enhancement of tribological performance is also an indispensable requirement where sliding contact is predominant as in piston and cylinder arrangement [1&2].Sliding results in wear, which is one of the most common engineeringproblems. The material loss due to wear along with thermal behavior can be controlled to an extend by surface modifications [3-5].Conventionally there are different types of surface modification (coating) processes available such as Physical Vapor Deposition (PVD) which includes sputtering [6] and evaporation [7], Chemical Vapor Deposition (CVD) [8-10], electrodeposition [11], sol-gel process [12], etc. High temperature plasma spray coatingtechniques used for the deposition of TBC Yttrium stabilized Zirconia, offers the possibility of operating with metals and alloys having high melting points.[13&14].
Researchers have analyzed the improvement in tribological characteristics of plasma coated zirconium using a tribometer, where pin on disc (ASTM G99) test is the most commonlyused [15]. There are several factors which affect the tribological performance of materials likecoefficient of friction, contact between surfaces, contact pressure, contact temperature, frictional force, etc. and several works [16-18] were conducted to identify the influencing parameters and its effect on overall tribological performance with and without coating.

This paper focuses on analysis of tribological characteristics such as frictional stress, contact pressure, penetration, wear thickness and sliding status of Yttrium stabilized Zirconia coating over Al6061 using nonlinear Finite Element Analysis (FEA).

2 Model Setup

Aluminum alloy 6061 is used for the present study and the tribological characteristics of this alloy coated with Yttrium stabilized Zirconia, are evaluated numerically using FEM (Finite Element Method). Steps involved for the numerical analysis of wear sample are (a) Finite Element Modelling (b) Meshing - ANSYS Mechanical (c) Applying Boundary conditions – Multi Physics Problem (d) Solving and Interpretation.

The Table 1 shows the experimental parameters considered for the evaluation of tribological properties of Yttrium stabilized zirconia coating on aluminum 6061 alloys using Pin on disc tribometer.

Table 1. Experimental Parameters for Pin on disc tribometer [19]

Sliding Velocity (m/sec)	1.04
Sliding Distance (m)	315
Load (N)	15
Coating thickness (μm)	450
Speed of disc rotation (rpm)	250
Wear track diameter (mm)	80

Using the experimental parameters listed in Table 1, the tribological properties wear rate, friction coefficient and wear coefficients are estimated from Krishnamurthy et al. [19] and the results are tabulated as in Table 2.

Table 2. Experimental results used for numerical analysis

Load (N)	Wear rate (μm)	Friction Coefficient	Wear Coefficient*10-2
5	8	0.18	2.872
10	17	0.22	3.051
15	20	0.25	2.392

Archard wear theory [20&21] is used for the determination of wear coefficient andit states that the wear rateis a function of material hardness, contact pressure and sliding distance as shown in equation 1;

$$K=\frac{\dot{w}H}{PD} \qquad (1)$$

where H, K, P, D and $\dot{w}$ are Hardness (MPa), Wear coefficient, Pressure (MPa), Sliding distance (mm) and wear rate (mm) respectively. The experimental results, shown in Table 2, is used as an input in the current work for the numerical simulation to determine contact pressure, penetration, frictional stress and wear thickness for coated sample. It is found thatthe calculated wear coefficient is significantly low for 15N load because of increased contact pressure at the surface. In thiswork FEA is used to determine the tribological characteristics like contact pressure, penetration, frictional stress andwear thickness for a load corresponds to 15N. The following sections describes theprocedures adapted in this work for the determination of the tribological characteristics.

2.1 Finite Element Modelling

The Three-Dimensional model of Pin on disc tribometer was modelled using Design Modeler of ANSYS workbench.For the modelling of Pin on Disc Tribometer, Aluminum oxide disc is modelled as rotating disc of 135mm diameter and 8mm thickness,whereas Yttrium stabilized zirconia coated aluminum alloy 6061 is modelled as per ASTM G99 (30mm height and 12mm diameter pin) which is held stationary as shown in Figure 1 (a). The coating thickness of 450μmis provided at one end of the pin surface, where it is in sliding contact with the disc material.

The stiffness behavior of disc and pin material is another important consideration to be considered and rigid and flexible stiffness behavior is assigned for respective model. The mechanical properties such as Elastic modulus, Poisson's ratio also physical property like Density of materials are assigned to the geometry under consideration.

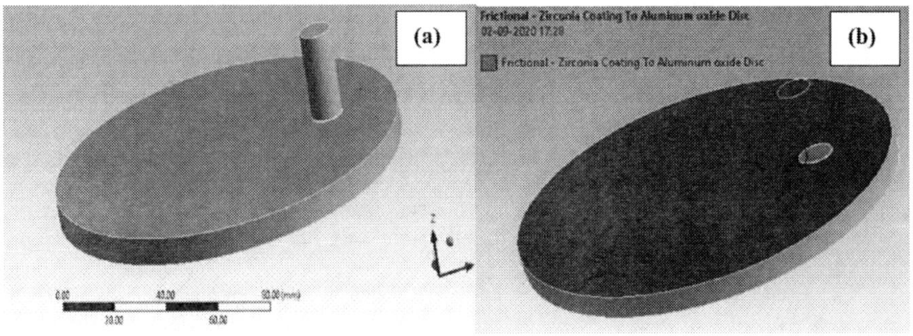

Figure 1 3D FE Model (a) Pin on Disc geometry (b) Frictional -Zirconia coating to Aluminum oxide disc

(a)

Scope	
Scoping Method	Geometry Selection
Contact	1 Face
Target	1 Face
Contact Bodies	Zirconia Coating
Target Bodies	Aluminum oxide Disc
Definition	
Type	Frictional
Friction Coefficient	0.25
Scope Mode	Manual
Behavior	Asymmetric
Trim Contact	Program Controlled
Suppressed	No

(b)

Scope	
Scoping Method	Geometry Selection
Contact	1 Face
Target	1 Face
Contact Bodies	Zirconia Coating
Target Bodies	Aluminum Alloy 6061
Definition	
Type	Bonded
Scope Mode	Manual
Behavior	Asymmetric
Trim Contact	Program Controlled
Suppressed	No

(c)

Advanced	
Formulation	Augmented Lagrange
Detection Method	Nodal-Normal To Target
Penetration Tolerance	Program Controlled
Elastic Slip Tolerance	Program Controlled
Normal Stiffness	Manual
Normal Stiffness Factor	0.1
Update Stiffness	Each Iteration
Stabilization Damping Factor	0.
Pinball Region	Program Controlled
Time Step Controls	None

Figure 2 Contact connections (a) Frictional (b) Bonded (c) Advanced Formulation

The wear model consists of two contacts namely frictional contact and bonded contact. The frictional contact, with a frictional coefficient of 0.25as shown in Figure 1 (b) & 2 (a) exists between the Yttrium stabilized zirconia coating and disc whereas bonded contact in Figure 2 (b) exists between coating and pin. For contact analysis,Augmented Lagrange formulation with nodal normal to target detection method is used for the convergence of the solution and isshown in Figure 2 (c)[22&23].

2.2 Meshing - ANSYS Mechanical

Meshing is the most important step in FEA simulationas it determines thetime for convergence and accuracy of solution. Here, the model is optimized for the convergence of solution by controlling the total number of elements and nodes.Figure 3(a)shows that the meshing is done only at the upper face of the disc which is in contact with pin and the rest of the portions are free from meshing in order to reduce the computational time. Figure 3 (c) provides the details of meshing with

2350 number of nodes and 396 number of elements altogether is provided for both Pin and disc model.

Figure 3 (b) shows the meshing done for the surface of pin which is in sliding contact with the aluminum oxide disc material. Increment in number of elements increases the computational time for solving the problem.

2.3 Applying Boundary Condition – MULTIPHYSICS

This section deals with the application of boundary condition for the modelled pin and disc tribometer. Force, joint load and cylindrical supports are the boundary conditions in this mathematical model. The force of 15N as mentioned in Table 1 is applied at the top face of the pin facing downward direction, with a cylindrical support such that the axial motion of pin is set to be free to move along z-direction.

The simulated data of magnitude and direction of force applied and the cylindrical supports to the pin are shown in Figure 4 (a) and (b). Thealuminum oxide disc material is groundedand is set to rotate 360°anticlockwise direction along z- axis with an angular velocity of 1500 degree/sec (250 rpm) (Figure 4 (c)). The analysis is done for single revolution of the disc. The time required for the single revolution was calculated as 0.24 seconds.Pin on disc tribometer is now subjected to force, support and rotation and the combined representation of boundary condition is shown in Figure 4 (d).

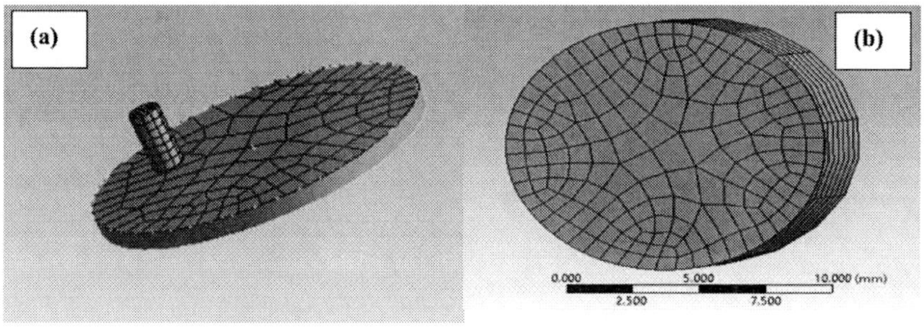

(c)

Sizing	
Use Advanced Size Function	Off
Relevance Center	Coarse
Element Size	Default
Initial Size Seed	Active Assembly
Smoothing	Medium
Transition	Fast
Span Angle Center	Coarse
Minimum Edge Length	37.6990 mm

Figure 3 Meshing (a) Geometry default mesh (b) Contact surface of Pin (c) Default meshing

(a)

Definition	
Type	Force
Define By	Components
Coordinate System	Global Coordinate System
X Component	0. N (ramped)
Y Component	0. N (ramped)
Z Component	-15. N (ramped)

(b)

Definition	
Type	Cylindrical Support
Radial	Fixed
Axial	Free
Tangential	Fixed
Suppressed	No

(c)

Scope	
Joint	Revolute - Ground To Aluminum oxide Disc
Definition	
DOF	Rotation Z
Type	Rotation
Magnitude	360. ° (ramped)
Lock at Load Step	Never
Suppressed	No

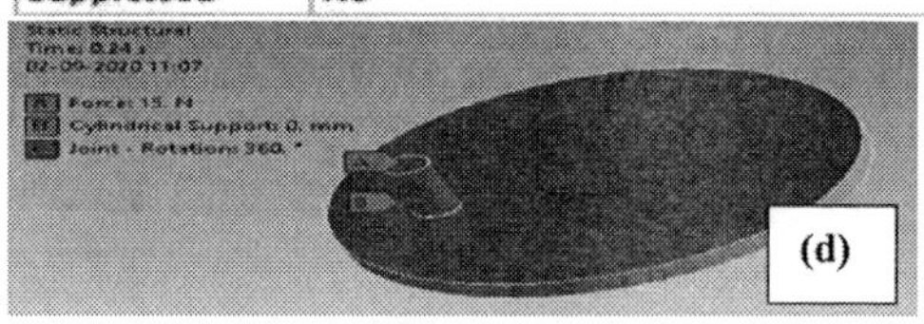

Figure 4 Boundary condition (a) Force application (b) Cylindrical Support(c) Revolution – Joint Load (d) Static Structural boundary condition applied altogether

3 Results and Discussion

This section deals with solution and interpretation of tribological properties such as status of sliding, contact pressure, penetration, frictional stress and wear characteristics under Continuous Sliding Motion (CSM) of Yttrium stabilized Zirconia coating over aluminum alloy 6061 with the help of ANSYS Multiphysics analysis software.

3.1 Status of Sliding

This sub-section deals with the analysis of pure sliding condition occurring between Yttrium stabilized zirconia coating and Aluminum oxide disc material. Throughout the experimentation, the pin is maintained to be in contact with disc by providing cylindrical support with free axial movement as shown in Figure 4 (b). As mentioned in section 2.3, the disc is maintained to be revolving along z-direction, where pure sliding is observed between coating and rigid disc material. This aspect is shown in Figure 5 (a). The average value of sliding distance was found to be 0.243mm which has 3.29% deviation from the calculated value of 0.251mm. This shows that the disc material undergone one complete revolution. The Figure 5 (b) shows the sliding distance contour for the simulation.

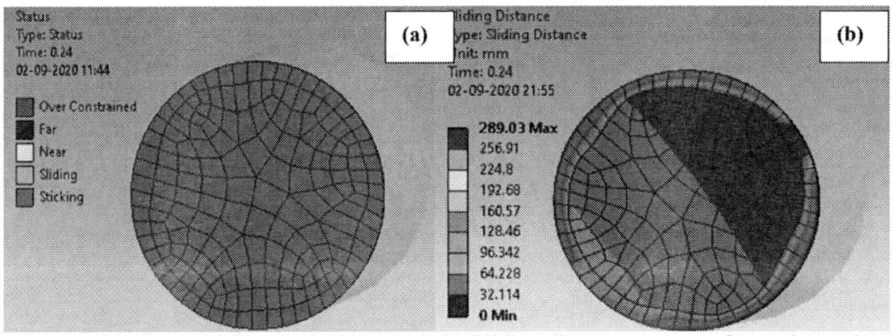

Figure 5 (a) Status of frictional sliding and (b) Sliding distance contour of coated sample

3.2 Contact Pressure and Penetration

Archard wear theory deals with the direct proportionality relationship between wear rate and contact pressure. The major assumption for this theory lies in the concept of uniform pressure throughout the cross section.

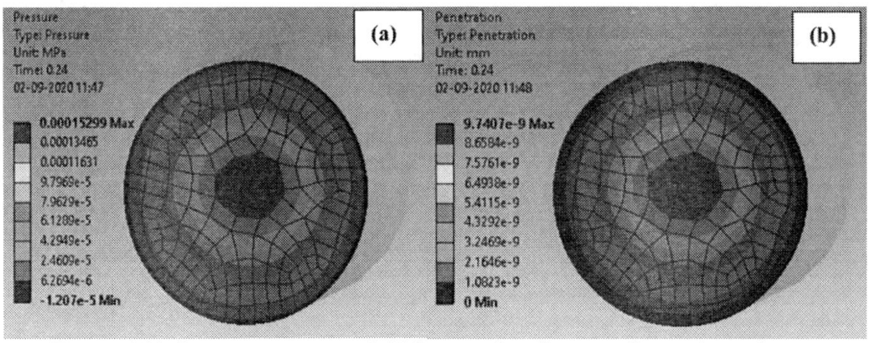

Figure 6 (a) Contact pressure and (b) Penetration Contour of coated sample

Figure 6 (a) and (b) shows contact pressure and penetration contour for Yttrium stabilized zirconia coating subjected to 15N loading condition. The maximum value of contact pressure is found to be 1.52×10^{-4} MPa at the center portion of the pin material, which results in penetration of 9.74×10^{-9} mm at the center.The contact pressure and penetration are correlated to each other as shown in Figure 6. The depth of penetration is found to be in the order of 10^{-9}, which is significantly less, even though penetration is found to exist in between contact surface is due to continuous sliding of pin with harder aluminum oxide disc material.

3.3 Frictional Stress

The stress due to friction is an important factor which affects the wear rate and wear volume. The frictional stress concentration due to continuous sliding of Yttrium stabilized zirconia coating subjected to 15N loading against aluminum oxide disc is shown in Figure 7 (a) and (b). Asmentioned in the section 3.2, the contact pressure is accumulated heavily at the center of pin surface. This hints at increased frictional stress acting on the surface under consideration, since contact pressure is directly related to frictional stress acting. The graphical representation of frictional stress is shown in Figure 7 (a) and contour represented in Figure 7 (b), where it increases up to 4.41×10^{-5}MPa as time lapses from 0 to 0.24 seconds. Maximum and minimum value of frictional stress is shown with the aid of green and red lines in the frictional stress vs time plot.

3.4 Wear

This section deals with wear characteristics of Yttrium stabilized zirconia coating subjected to 15N loading against aluminum oxide disc. ANSYS workbench does not provide wear directly, so it is recommended to use user defined results for its analysis and interpretation. For arriving at the solution "CONTNMISC183" is used, as it the expression for wear analysis along z-direction. Thegraphical representation of wear is shown in Figure 8 (a), which increases up to $8.17\times10^{-8}$for time varying from 0 to 0.24 seconds. Maximum and minimum value of wear is shown with the aid of green and red lines in the wear vs time plot. Wear at the center surface of pin is found to be maximum from the contour shown in Figure 8 (b) and it is due to increased frictional stress at the center as mentioned in the section 3.3. It is evident from the analysis that, when contact pressure at the contour is increased, there will becorresponding increase in penetration, frictional stress and wear.

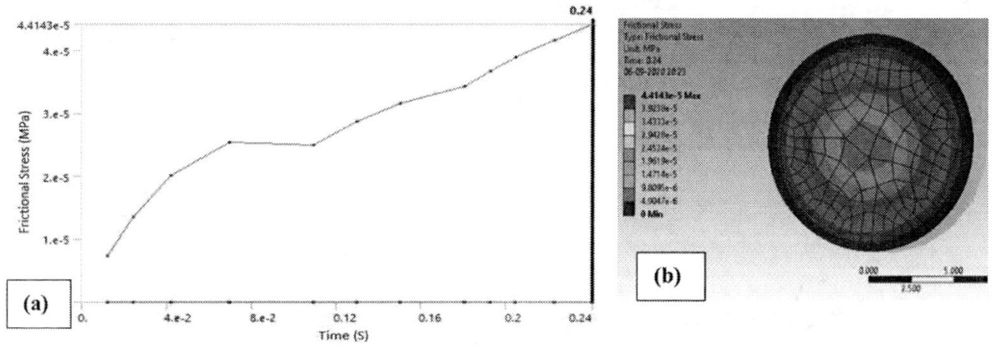

Figure 7 Frictional Stress Characteristics (a) Graphical representation (b) Contour of coated sample

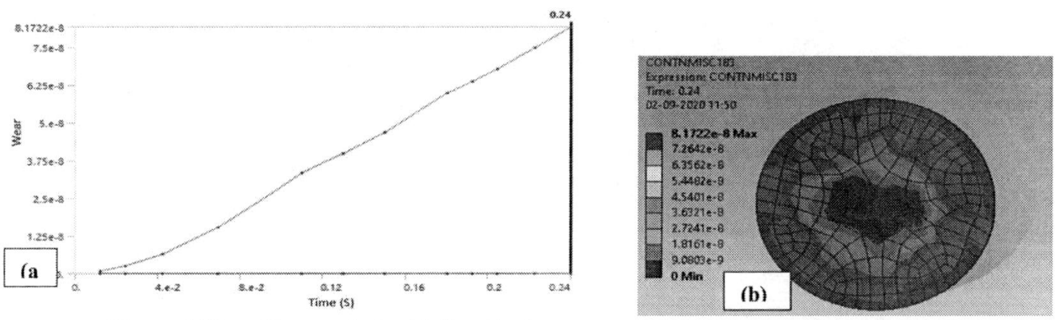

Figure 8 Wear Characteristics (a) Graphical representation (b) Contour of coated sample

4 Summary and Conclusions

The wear characteristics of Yttrium stabilized Zirconia coating is analyzed based on Archard wear theoryand using Finite Elemental Analysis. The study was able to explain the phenomenon of sliding, contact pressure, penetration, frictional stress and wear thickness observed during pin on disc experiments. It was observed that sticking was absent in any loads since the loads that were applied are not high enough to cause high contact stresses leading to sticking phenomenon. The contact pressure and penetration are found to be $1.52x10^{-4}$ MPa and $9.74x10^{-9}$ mm which is high at the central region of the contact area, which can lead to local wear. Maximum wear of $8.17x10^{-8}$was observed in the region of maximum contact pressure and vice versa. Due to increased contact pressure, the frictional stress is also found to be maximum at the central region of the pin contact surface say$4.41x10^{-5}$ MPa. The complete reproduction of experimental results was not able to observe in numerical simulation, as adhesive effects, surface roughness and debris formation are not included in the model. The accuracy of the results can be improved by finer meshing, but compromising with time required for obtaining the solutions.

Acknowledgments

The present research was conducted at TKM College of Engineering, Kollam under APJ Abdul Kalam Technological University, India. Theauthorsthank Dr. Mohammed Sajid N. K, Professor &Head of Department of Mechanical Engineering for providing facilities for computation. Authors would also like to acknowledge,Dr. K A Shafi, Research Dean andDr. Krishnakumar T Sand all the faculty members and research scholars for the technical support renderedin the completion of work.

References

[1] Mehran Q M, Fazal M A, Bushroaand Saeed Rubaiee A R 2018 A critical review on physical vapor deposition coatings applied on different engine components, *Critical Reviews in Solid State and Materials Sciences*, **43,** 158-75

[2] ABUtusov and N.Yu. Dudareva 2016 Improving the reliability of aluminum-alloy pistons in internal-combustion engines by microarc oxidation, *Russian Engineering Research.* **36** 116–8

[3] J. Vetter, G. Barbezat, J. Crummenauer, and J. Avissar 2005 Surface treatment selections for automotive applications *Surface and Coating Technology*,**200**,1962–8

[4] W. Paatsch 2003 Recent trends in surface finishing for automobile industry in Germany, *Surface and Coating Technology,***169**,753–7

[5] Jianqin Fu, Qijun Tang, Jinping Liu, Banglin Deng, Jing Yang, and Renhua Feng 2014 Acombined air cycle used for IC engine supercharging based on waste heat recovery, *Energy Conversion and Management.* **87** 86–95

[6] H. Schafer, H.R. Stock 2005 Improving the corrosion protection of aluminum alloys using reactive magnetron sputtering. *Corrosion Science* **47** 953–64

[7] Wiklund U, Wanstrand O, Larsson M and Hogmark S 1999 Evaluation of new multilayered physical vapour deposition coatings insliding contact. *Wear* **236** 88–95

[8] Hintermann H E 1981 Tribological and protective coatings by chemical vapour deposition, *Thin Solid Films.* **84** 215-43.

[9] Yiguang Wang, Qiaomu Liu, Jinling Liu, Litong Zhang, and Laifei Cheng 2008 Deposition mechanism for chemical vapor deposition of zirconium carbide coatings *Journal of the American Ceramic Society.* **91** 1249–52

[10] Kohzaki M, Higuchi K, Noda S, Uchida K 1992 Tribological characteristics of polycrystalline diamond films produced by chemical vapor deposition *Journal of Material Research.* **7** 1769-77

[11] Arghavanian R, Bostani B, Parvini-Ahmadi N, Yazdani S 2014 Field-enhanced co-electrodeposition of zirconia particles with a magneticshell during ni electrodeposition, *Surface and Coating Technology.* **258** 1171–5

[12] Tlili B, Barkaoui A, Walock M 2016 Tribology and wear resistance of stainless steel. The Sol-Gel coating impact on the friction and damage *Tribology International.* **102** 348-54

IOP Conf. Series: Materials Science and Engineering 1059 (2021) 012020 doi:10.1088/1757-899X/1059/1/012020

[13] Pantelis D I, Psyllaki P, Alexopoulus N 2000 Tribological behavior of plasma sprayed Al_2O_3coating under severe wear condition *Wear.* **237** 197-204

[14] Stanley M Jerome, Senthilkumar Sundararaj 2017 Experimental study on the effect of thermal barrier coating on cylinder head of a semi- adiabatic diesel engine, *SAE International*, https://doi.org/10.4271/2017-28-1978

[15] Wenguang Zhang, Weimin Liu, ChengtaoWang 2002 Characterization and tribological investigation of sol-gel Al_2O_3 and doped Al_2O_3 films *Journal of the European Ceramic Society.* **22** 2869-76

[16] Molinari J F, Ortiz M, Radovitzky R, Repetto E A 2001 Finite element modelling of dry sliding wear in metals *Engineering Computation.* **18**, 592-610

[17] Priit Podra and Soren Andersson1999 Simulating sliding wear with finite element method *Tribology International.* **32** 71-81

[18] Bortoleto E M, Rovani A C, Seriacopi V, Profito F J, Zachariadis D C, Machado I F, Sinatora A, Souza R M 2013 Experimental and numerical analysis of dry contact in the pin on disc test *Wear.***301** 19-26

[19] Krishnamurthy N, Prashanthareddy M S, Raju H P and Manohar H S 2012 A study of parameters affecting wear resistance of alumina and yttria stabilized zirconia composite coatings on al-6061 substrate *International Scholarly Research Network ISRN Ceramics.* https://doi.org/10.5402/2012/585892

[20] Archard J F 1953 Contact and rubbing of flat surfaces *Journal of Applied Physics.* **24** 981-8

[21] Mahesh B Davanageri, Narendranath S, Ravikiran Kadoli 2018 Finite element wear behavior modelling of super duplex stainless steel AISI 2507 using Ansys *IOP Conference Series: Material Science and Engineering.* **376** 012131

[22] Suresh R, Prasanna Kumar M, Basavarajappa S, Kiran T S, Mahesh Yeole and Naresh Katare 2017 Numerical simulation & experimental study of wear depth and contact pressure distribution of aluminum mmc pin on disc tribometer *Materials Today: Proceedings.* **4** 11218-28

[23] Sumit Khot, Utpal Borah 2013 Finite element analysis of pin on disc tribometer test, *International Journal of Science and Research.* ISSN (Online): 2319-7064

Tribological Behaviour of Mos2 and Graphite Reinforced Aluminium Matrix Composites

A.Saravanakumar[1], D.Ravikanth[1], L.Rajeshkumar[2*], D.Balaji[2], M.Ramesh[3]

[1]Department of Mechanical Engineering, KSRM College of Engineering, Kadapa, AP 516003
[2]Department of Mechanical Engineering, KPR Institute of Engineering and Technology, Coimbatore, Tamilnadu 641407
[3]Department of Mechanical Engineering, Kalaignarkarunanidhi Institute of Technology, Coimbatore, Tamilnadu 641402

*Corresponding Author e-mail: lrkln27@gmail.com

Abstract. Tribological behaviour of MoS_2 and graphite reinforced aluminium matrix composites were investigated in this research. In the current scenario, applications in which either metal based alloys or composites are used, requires lubrication to avoid friction. In such cases, self-lubricating materials based aluminium composites can be fabricated to exhibit better tribological properties. These materials can be fabricated by reinforcing self-lubricating materials like graphite (Gr), molybdenum disulphide (MoS_2) and so on within the aluminium matrix. In this tribological study, two different solid lubricants (Gr and MoS_2) were taken at equal proportions to fabricate aluminium matrix composites individually through stir casting method. Pin on Disc apparatus was utilized to record the rate of wear of the composite specimen by considering the process parameters like sliding distance, load and counter-face disc hardness. Experiments were designed by Taguchi's approach and analyzed by means of response values and analysis of variance (ANOVA). Results depicted that graphite based aluminium alloy composites possessed better wear resistance than MoS_2 reinforced aluminium alloy composites.

Keywords: *Tribology, MoS2, Graphite, Aluminium matrix composites, Taguchi Method*

1. Introduction

Aluminium metal matrix composites (AMMC) are the class of materials that indicate enormous uniqueness that enumerate the requirements of applications of AMMC in some notable areas like transportation, sports and aviation [1, 2]. Utilization of AMMCs in nuclear, aerospace and automobile fields steadily have increased owing to their admirable physical, mechanical and thermal characteristics [3]. On account of several better properties like less density, high modulus value, high corrosion resistance, good thermal properties and high ductility of aluminium based alloys possibly uphold their capability as the materials resisting friction [4, 5]. Aluminium alloys has to work under severe conditions like intense temperature, vacuum environment and in many cases these materials has to undergo sliding contact with another metallic surface. During extreme conditions of sliding and Metal to metal contacts, application of liquid lubrication is ineffective or difficult [6]. Further liquid lubricants are non-biodegradable which may release harmful materials into the atmosphere upon disintegration. Under those conditions, aluminium alloy materials can be made to resist friction by means of inherently present materials which are characterized by low coefficient of friction and rate of wear and these materials are termed as solid self-lubricating materials [7]. Incorporation of such solid lubricants into aluminium alloys may open avenues to many other areas such as green tribology and pave way in developing sustainable and reliable materials.

Few authors stated that self-lubricating materials arise either by reinforcing solid lubricants into the matrix of a base metal or by coating the substrate material with the self-lubricating material. Many-a-times, reinforcing the solid lubricants into aluminium alloys to obtain AMMCs render greater resistance towards fatigue cycles, excellent resistance towards corrosion and incredible tribological characteristics when compared with the materials prepared through coating method. Increasing percentage of self-lubricating materials like Gr and MoS_2 in matrix material increases the wear resistance because of the development of a solid lubricant tribo-layer amidst the contact sliding surfaces [8-10]. Several studies reported the influence of MoS_2 and Gr particles on the wear behavior of AMMCs as hybrid reinforcements. Tribological properties of Al-Al_2O_3-MoS_2 hybrid composite were examined and it was noted that the inclusion of MoS_2 particles reduced the rate of wear and friction coefficient [11]. Few authors observed the tribological behaviour of AMMCs reinforced with graphite and results showed that addition of Gr particles was a significant parameter that influenced the decrease in wear rate [12, 13].

Researchers investigated the wear behaviour of AMMCs reinforced with silicon carbide and Gr and observed that the addition of Gr, applied load, sliding speed and distance influenced the wear rate. Further statistically analyzed the friction characteristics of Al-SiC-Gr hybrid composites and observed that the friction coefficient is influenced by solid lubricants [14, 15]. Experimenters carried out tribological study on Al-Al_2O_3 -Gr composites with the help of Taguchi method and concluded that graphite particles influencing the wear properties and reducing the wear loss considerably [16-18]. Tribological studies were conducted on AA2219/Gr matrix composites by varying percentage of Gr particles, rotational speed and applied load. Results revealed that tribological properties influenced by solid particles percentage in matrix material [19]. AA2219/MoS_2 matrix composites were fabricated by reinforcing of MoS_2 with different percentage with stir casting as the processing route. Tribological behaviour of the AA2219/MoS_2 AMMCs was examined both in dry and lubricated wear conditions. Taguchi DOE technique was utilized to conduct and analyse the experiment results and noted that the wear behaviour of the AA2219/MoS_2 composites under lubricated condition was superior with the inclusion of MoS_2 [20-22]. Literature review shows that the tribological test parameters and the consideration of solid lubricant particles greatly influence the tribological performance of the hybrid Aluminium matrix composites. The tribological characteristic studies of composites reinforced with solid particles as a first reinforcement are limited. Mechanical properties of Al 2024-MoS_2 composites were evaluated by varying weight percentage of MoS_2 and found that 4 wt. % MoS_2 reinforced aluminium composites shows better mechanical properties compared to other weight percent of MoS_2 particles [23-26]. For this reason, in the current study, the tribological characteristics of AA2029-4 wt. % Gr composite and the AA2029- 4wt.% MoS_2 composite is analyzed through Taguchi method to understand the significance of MoS_2 and Gr particles adding together on the wear resistance.

2. EXPERIMENTAL DATA

AA2219 material was procured as billets and was further processed to make composite. Solid lubricating particles Gr and MoS_2 utilized in the form of powder with average particle size of 40 μm. MoS2 and Gr were reinforced with AA2219 at constant weight percentage of 4 for getting better tribological properties [18].

2.1. *Composite fabrication*

AA2219/Gr and MoS_2 fabricated by means of conventional stir casting technique in two steps: AA2219 was melted at a temperature of 780 °C in a crucible and Gr & MoS_2 particles were preheated to 200 °C by using an electrical furnace for about 20 minutes. The preheated solid lubricants were added as reinforcements into the molten matrix material [19]. A mechanical stirring action at a rate of 525 rpm was imposed to the molten matrix and reinforcement mixture in order to obtain a homogenously dispersed solid solution for a time period of 8 minutes. The molten metal was then poured into mould which is preheated around 100 °C. Both samples AA2219 / 4 wt. % Gr and AA2219 / 4 wt. % MoS_2 were independently fabricated and cooled at atmospheric temperature.

2.2. *Conduct of Experiment*

Dry wear test was performed in a CONMAT pin-on-disc tribometer at room temperature with the specimens of dimensions φ 10 × 30 mm. Experiments were conducted following ASTM G99 standards. Contact face of the composite specimen was polished with a 10 grit size emery paper to obtain smooth surface finish. Sliding distance, applied load, counter-facing disc hardness and two different solid lubricants are considered as the process parameters to obtain the wear results. Table 1 shows the levels and factors of Taguchi's L18 (2^1 x 3^7) mixed type orthogonal array (OA). L18 OA was chosen above other arrays in order to have an accurate optimization of factors. The main purpose this study is to observe the less wear and hence the objective function for the response was chosen as 'smaller the better' and the values of signal to noise ratio (S/N) values was inferred corresponding to the objective function. Table 2 shows the combination of levels as per Taguchi DOE and results obtained from the experiments of wear.

Table 1. Wear test Factors and levels

Level	Solid lubricant (A)	Load (B) in N	Sliding distance (C) in m	Disc hardness (D) in HRC
I	Gr	19.62	1000	56
II	MoS_2	29.43	2000	62
III		39.24	3000	68

Table 2. Response values for L18 orthogonal array

Exp.No	Solid Lubricant (A)	Load (B)	Sliding distance (C)	Hardness (D)	Wear (µm)	S/N ratio
1	Gr	2	1000	56	212	-46.53
2	Gr	2	2000	62	246	-47.82
3	Gr	2	3000	68	274	-48.76
4	Gr	3	1000	56	237	-47.49
5	Gr	3	2000	62	271	-48.66
6	Gr	3	3000	68	302	-49.60
7	Gr	4	1000	62	285	-49.10
8	Gr	4	2000	68	326	-50.26
9	Gr	4	3000	56	294	-49.37
10	MoS_2	2	1000	68	251	-47.99
11	MoS_2	2	2000	56	283	-49.04
12	MoS_2	2	3000	62	316	-49.99
13	MoS_2	3	1000	62	292	-49.31
14	MoS_2	3	2000	68	323	-50.18
15	MoS_2	3	3000	56	355	-51.00
16	MoS_2	4	1000	68	377	-51.53
17	MoS_2	4	2000	56	353	-50.96
18	MoS_2	4	3000	62	366	-51.27

IOP Conf. Series: Materials Science and Engineering 1059 (2021) 012021 doi:10.1088/1757-899X/1059/1/012021

3. RESULTS AND DISCUSSION

Results obtained in pin on disc are tabulated and shown in table2. These observed wear results are analysed with MINITAB software through S/N ratio values, Main effect plot, Response table and ANOVA.

3.1. Main effects plot and response table

Previous studies revealed that, Taguchi technique is a well-known method to obtain optimal the process parameters. L18 orthogonal array is meant for understanding the effect of four independent variables having mixed factor (2^1 X 3^3) level values. Objective function from which the SN ratio is inferred is classified into three categories: lower the better, nominal the better, and higher the better. In Wear test, the less wear rate could be attributed to the good tribological behaviour of the material. Hence the objective function, for attaining optimum tribological characteristics, was chosen as "Lower the Better" and based on this S/N ratio calculated for each test [19]. Table 2 lists the calculated S/N ratios for each experiment based corresponding response value of wear which indicates the maximum and minimum values of variance.

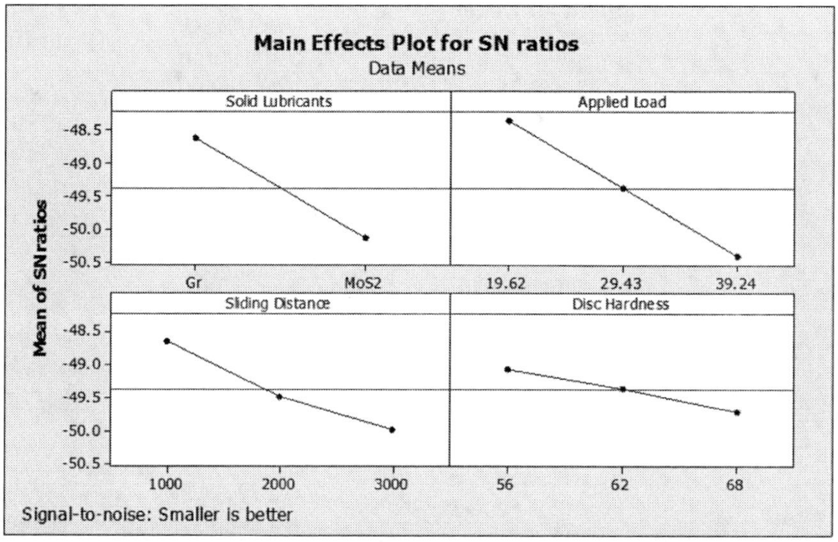

Figure 1 Main effect plot for wear

Figure 1 illustrates the main effects plot for S/N ratios and the effect of load, distance, hardness and solid lubricants on the rate of wear. This plot results revealed that wear reduced with decreasing in sliding distance, applied load and low disc hardness. Further it is observed that less wear achieved in graphite based material compared to MoS_2 particles. From this plot, it was clearly found that maximum value of S/N ratio gives optimal value of the better wear characteristics. The observed optimal values are 19.62 N load, 1000 m distance and 56 HRC hardness of disc with graphite solid lubricate to achieve the better wear characteristics.

Table 3. Response table

Level	Solid Lubricants (A)	Load (B)	Sliding Distance (C)	Hardness (D)
1	-48.62	-48.35	-48.66	-49.06
2	-50.14	-49.38	-49.49	-49.36
3		-50.41	-50.00	-49.72
Delta	1.52	2.06	1.34	0.66
Rank	2	1	3	4

The effect of wear parameters on tribological characteristic has been presented using response table. Response tables are utilized to analyze the observed data in a simple manner. The complete response table for the experimental design is shown in Table 3 with significance of the factor by appropriate ranking. The values in table 3 depicts the significant responses when the level of process parameters move to higher levels from lower level. Ranking of parameters indicates that effect of parameter on the response of wear loss which reveals that applied load is the most influencing factor when it changes form one level to other level followed by solid lubricant, sliding distance and disc hardness.

3.2. Analysis of variance for wear

Analysis of variance (ANOVA) was utilized for determining the process parameter's significance which were given as input to measure the wear rate [27]. ANOVA was performed to find out the optimum values for obtaining less wear rate and relative significance of wear parameters affecting wear of all the compositions.

Table 4. ANOVA table for wear rate

Source	Degree of Freedom	Sequential Sum of Square	Adjusted Mean Square	F-value	P value	Percentage contribution
Solid Lubricants (A)	1	10.409	10.409	43.37	0.000	32.2%
Load (B)	2	12.724	6.362	26.51	0.000	39.4%
Sliding Distance (C)	2	5.492	2.746	11.44	0.003	17.0%
Disc hardness (D)	2	1.298	0.649	2.7	0.115	4.0%
Residual Error	10	2.4	0.24			7.4%
Total	17	32.323				

It was also carried out to verify the factors which are statistically significant at 95% confidence level with the help of the P-value. The values of P which are less than or equal to 0.05 indicates higher level of significance. According P-value, except disc hardness all other remaining factors are less than 0.05 which shows that those are significant parameters on wear. Disc hardness shown not much significant in this wear test, it may be due to utilization of fixed weight percentage of reinforcement. The last column of the ANOVA table shown in table 4 indicates the percentage of contribution (Pc) in which Applied Load has the highest level of contribution (39.4%) followed by solid lubricants (32.2%), sliding distance (17%) and Disc hardness (4.0%).

3.3. Influence of Solid lubricants on wear

In wear analysis AA2219 / 4 wt. % Gr and AA2219 / 4 wt. % MoS_2 samples were used and results are tabulated in Table 2. It revealed that AA2219 / 4 wt. % Gr specimens exhibited better tribological characteristics compared to AA2219 / 4 wt. % MoS_2 samples. Further it was illustrated from the figure 2 and 3 that Gr based specimens had undergone less wear when compared with MoS_2 specimens almost for all levels the applied load (Figure 3 a) and sliding distance (Figure 3 b).

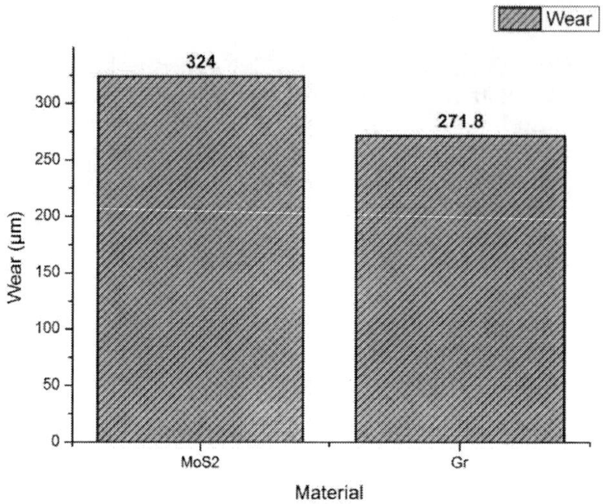

Figure 2 Mean wear loss of MoS_2 and Gr specimens

Many literatures revealed that graphite acts as an efficient lubrication when ample amount of vapour pressure of water was available while MoS_2 suits better for the vacuum atmospheres [23, 24]. Many experimental results also revealed that the wear rate and friction of the materials with MoS_2 as a reinforcement increased with the humidity in the working environment while the phenomenon was vice-versa when the materials was reinforced with Gr. MoS_2 samples showcased excessive wear than graphite samples which could be due to accelerated oxidation at normal environment by the small particles of MoS_2 [20].

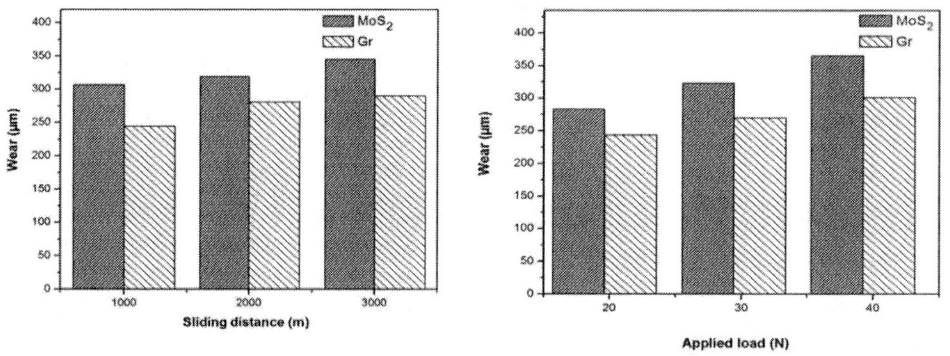

Figure 3 Mean wear loss of MoS_2 and Gr specimens for a) sliding distance and b) applied load

Further, Graphite is structurally hexagonal in orientations that are arranged together with planes of polycyclic carbon atoms. The distance between planes and carbon atoms are lengthy and due to this bonding between atom and plane becomes weaker. The energy available with the water vapor atmosphere would be enough to reduce the available energy of graphite within the hexagonal plane bonding to a level that is lesser than that of Gr adhesion energy with the matrix material [21]. These reasons are attributed to better tribological performance of graphite samples than MoS_2 samples at normal atmospheric conditions.

3.4. Confirmation tests

A confirmation tests were conducted at the optimal level of wear parameters [22]. The S/N ratio γ' for the optimal level of parameters can be attained from the equation 1,

$$\gamma' = \gamma_m + \Sigma(\overline{\gamma_i} - \gamma_m) \qquad \text{-------------------------- (1)}$$

Where, γ_m - total mean S/N ratio, γ_i - mean S/N ratio of the optimum value.
In order to confirm the calculated values from the optimal levels, a confirmation test was carried out. Table 5 shows the error difference between the predicted and experimental responses. The optimal level of wear parameters obtained for minimum wear rate was at lowest applied load (19.62N) load, sliding distance (1000 m) and 56 HRC hardness of disc with graphite reinforced AA2219 sample.

Table .5 Confirmation tests

	Optimum parameters	
	Experimental	Predicted
Level	A1B1C1D1	A1B1C1D1
Wear (μm)	212	207
S/N raio	-46.53	-46.55

4. CONCLUSION

AA2219 reinforced with 4 wt. % of Gr and MoS_2 were fabricated by stir casting technique and were subjected to wear tests. Optimization of wear rate was carried out by Taguchi's method and the following conclusions were arrived.

- Addition of Gr solid lubricants shown better wear characteristics than MoS_2 particles based composite for the selected parameters at normal environment.
- Applied load and solid lubricants are the main factors, which is affecting the wear properties in all samples for all experimental conditions.
- Wear rate increases with the increase in applied load and sliding distance in all experimental conditions. It was observed that disc hardness less significant on wear compared to applied load and sliding distance.
- The lowest wear are recorded at the lowest applied load (19.62N) load, sliding distance (1000 m) and 56 HRC hardness of disc with graphite reinforced AA2219 sample.

REFERENCES

[1] Preetkanwal Singh Bains, Sarabjeet Singh Sidhu, H.S. Payal, Fabrication and machining of metal matrix composites: a review, *J. Mater. Manuf. Proc.* 31 (2016) 553–573,

[2] L.Rajesh kumar, K.S.Amirthagadeswaran, "Corrosion and wear behaviour of nano Al2O3 reinforced copper metal matrix composites synthesized by high energy ball milling." *Particulate Science and Technology,* Vol.38 (2) (2020), pp. 228-235, DOI: 10.1080/02726351.2018.1526834.

[3] B.Ya. Sachek, A.M. Mezrin, O.O. Shcherbakova, T.I. Muravyeva, D. Zagorski, Studies on the tribological properties and structure of antifrictional ironcontaining aluminum alloys, *J. Frict. Wear* 39 (2018) 206–214, https://doi.org/10.3103/S1068366618030108.

[4] Kamyar Shirvanimoghaddam, Salah U. Hamim, Mohammad Karbalaei Akbari, Seyed Mousa Fakhrhoseini, Hamid Khayyam, Amir Hossein Pakseresht, Ehsan Ghasali, Mahla Zabet, Khurram Shahzad Munir, J. Shian Jia, Paulo Davim, Minoo Naebe, Carbon fiber reinforced metal matrix composites: fabrication processes and properties, *J. Compos.*: Part A 92 (2016) 70–96, https://doi.org/ 10.1016/j.compositesa.2016.10.032.

[5] L.Rajesh kumar, K S Amirthagadeswaran, (2019), "Variations in the properties of copper-alumina nanocomposites synthesized by mechanical alloying." *Materiali in Technologije*, 53(1), pp. 57-63.

[6] S.K. Khatkar, N.M. Suri, S. Kant and Pankaj, "A review on mechanical and tribological Properties of graphite reinforced self Lubricating hybrid metal matrix composites", *Rev. Adv. Mater. Sci.* 56 (2018) 1-20.

[7] P.L.Menezes, In: Tribology for Scientists and Engineers: From Basics to Advanced Concepts, ed. by C.J. Reeves, M. R. Lovell,and P.K. Rohatgi ,*Springer Science*, London (2013), p. 341.
[8] P. Ravindran, K. Manisekar, S. Vinoth Kumar and P. Rathika // *Mater. Des.* **39** (2013) 42
[9] A. Baradeswaran and A. Elaya Perumal //*Tribol. Trans.* **58** (2015) 1.
[10] D. S. R. Smart, N. P. Johns and J. J. Smart // *Mater. Sci. Forum* **889** (2016) 56
[11] Dharmalingam, S., Subramanian, R., & Kök, M. (2013). Optimization of abrasive wear performance in aluminium hybrid metal matrix composites using Taguchi–grey relational analysis. Proceedings of the Institution of Mechanical Engineers, *Part J: Journal of Engineering Tribology*, **227**(7), 749-760
[12] Ravindran, P., Manisekar, K., Narayanasamy, P., Selvakumar, N., & Narayanasamy, R. (2012). Application of factorial techniques to study the wear of Al hybrid composites with graphite addition. *Materials & Design*, **39**, 42-54.
[13] L.Rajesh kumar, A.Saravanakumar, R.Saravanakumar, S.Sivalingam, V.Bhuvaneswari. (2018) "Prediction capabilities of mathematical models for wear behaviour of AA2219/MgO/Gr hybrid metal matrix composites." *International Journal of Mechanical and Production Engineering Research and Development*, **8**(8), 393-399
[14] Monikandan, V. V., Jacob, J. C., Joseph, M. A., & Rajendrakumar, P. K. (2015b). Statistical analysis of tribological properties of aluminum matrix composites using full factorial design. *Transactions of the Indian Institute of Metals*, **68**(1), 53-57.
[15] Suresha, S., & Sridhara, B. K. (2012). Friction characteristics of aluminium silicon carbide graphite hybrid composites. *Materials & Design*, **34**, 576-583.
[16] A.Saravanakumar & Sasikumar, P. (2016). Dry sliding wear behaviour of Al6063/Al2O3P/grp hybrid metal matrix composites. *J. Balk. Tribol. Assoc*, **22**(2-I), 1253-1264.
[17] A.Saravanakumar,S. Sivalingam and L.Rajeshkumar, "Dry Sliding wear of AA2219/Gr Metal matrix composites." *Materials Today: Proceedings*, **2** (2018): 8321-8327.
[18] L.Rajeshkumar, A.Saravanakumar, V.Bhuvaneswari, M.P.Jithin karunan, N.Karthick Raja, and P. Karthi, Tribological Behaviour of AA2219/MoS_2 Metal Matrix Composites under Lubrication, *AIP Conference Proceedings* 2207, 020005(2020);doi.org/10.1063/5.0000042.
[19] L.Rajeshkumar A.Saravanakumar, V. Bhuvaneswari, and G. Gokul. "Optimization of wear behaviour for AA2219-MoS2 metal matrix composites in dry and lubricated condition." *Materials Today: Proceedings* **27** (2020): 2645-2649.
[20] Bhargavi Rebba, N.Ramanaiah, "Evaluation of Mechanical Properties of Aluminium Alloy (Al-2024) Reinforced with Molybdenum Disulphide (MoS_2) Metal Matrix Composites", *Procedia Materials Science* **6** (2014) 1161 – 1169.
[21] A.Saravanakumar, P. Sasikumar, and S. Sivasankaran. "Synthesis and mechanical behavior of AA 6063-x wt.% Al2O3-1% Gr (x= 3, 6, 9 and 12 wt.%) hybrid composites." *Procedia Engineering* **97** (2014): 951-960.
[22] A. Saravanakumar, A. Avinash & R. Saravanakumar (2016) Optimization of biodiesel production from Pungamia oil by Taguchi's technique, Energy Sources, Part A: Recovery, Utilization, and Environmental Effects, **38:17**, 2524-2529.
[23] Chen, Z.; He, X.; Xiao, C.; Kim, S.H. E_ect of Humidity on Friction and Wear—*A Critical Review. Lubricants*,2018, **6**, 74
[24] Saravanakumar, A., Rajeshkumar, L., Balaji, D. et al. (2020), Prediction of Wear Characteristics of AA2219-Gr Matrix Composites Using GRNN and Taguchi-Based Approach, *Arabian Journal for science and engineering*, https://doi.org/10.1007/s13369-020-04817-8.
[25] Fahlman, B.D. Materials Chemistry – *Solid State Chemistry, Springer, Berlin*, 2011:13 – 156.
[26] A,Saravanakumar, P. Sasikumar, and N. Nilavusri. "Optimization of machining parameters using taguchi method for surface roughness." *J. Mater. Environ. Sci* 7, no. **5** (2016): 1556-1561.
[27] Saravanakumar A, Rajesh kumar L, Sanjaykumar NS, Sivakumar R, Sriramthilak N.S, Vignesh R (2020), "Multi response Optimization of Aluminium alloy Machining using Full Factorial Design of Experiments", *AIP Conference Proceedings*, **2207**, 020002. https://doi.org/10.1063/5.0000221

Tribological Properties of Coated and Uncoated Materials for Piston Ring Applications

Boopathi D[1]*, M Palanivendhan[1], Jennifer Philip[1], Dhulipalla Srikar Hari Vishnu[1]

Department of Automobile Engineering, College of Engineering and Technology, SRM Institute of Science and Technology, Kattankulathur, Kanchipuram – 603203, Tamil Nadu, India.

*E-mail: boopathmech@gmail.com

Abstract. Coating is an essential method to improve surface properties such as corrosion resistance, lubrication of material used. The simulated laboratory tests is a more feasible alternative to experimental or real-time testing. A study conducted under controlled parameters of load, surface coating, and temperature showed the changes in the tribological characteristics with and without functionally graded cermet coatings and their influence on stainless-steel piston rings. Altogether simulations for six cycles were taken into consideration for each of the test cases – with and without coating. The obtained data from the microstructure imaging in optical microscope following the investigation conducted employing tribometer showed that the resistance of the coated rings was comparatively more than the uncoated rings, hence proving the coating to be beneficial for real-time use.

Keywords: Frictional losses; Functionally-graded cermet coating; Microstructure Analysis; Plasma Spray Coating; Wear Resistance.

1. Introduction

The tribological characteristics of piston rings have been under investigation for decades concerning internal combustion engines. It is a well-known fact that there are frictional losses inside the cylinder of the engine over a while. Piston ring materials play an essential role and are in continuous contact with the cylinder liner. Piston rings act as a seal inside the cylinder to prevent the passage of combustion gases towards the crankshaft that would otherwise reduce the efficiency of the engine. For this very reason, piston rings must be of materials that show high corrosion resistance, wear, and friction resistance. With the advancement in science and technology, research, as well as the development of various ways to improve the tribological behavior of the piston rings and cylinder liners, is in progress such that they provide a hard-outer structure along with a soft inner-structure for withstanding the wear and absorbing the energy respectively. Coating is a useful technique to prevent these factors even though there have been previous attempts to replace stainless-steel pistons with aluminum pistons. Researchers have been conducting investigations and experimental studies to bring about changes in the material of the pistons such that they have favorable tribological characteristics for real-time uses. For this purpose, Kenny M, Hoppe, et al. developed new piston ring coatings [1]. Shanhong Wan et al. [2] studied the effects on the scuffing resistant capacity due to the integration of traditional ceramic coatings and their load-carrying capacity. Zheng Ma et al. [3] are well known for studying the tribological irregularities not only in piston rings but also in cylinder liners. Ronghua Wei et al. [4] are well-known for the assessment of nanocomposite coatings characterized by improved mechanical strength, increased life of performance as well as decreased rates of friction for piston rings. M. Palanivendhan et al. [5] conducted an experimental evaluation of composite coatings of an aluminum alloy whose properties showed improvement by using silicon carbide for use in the automobile sector. Hyo-Syo et al. [6] investigated the tribological properties of composite coatings that had the presence of Al_2O_3, TiO_2, ZrO_2 employing plasma spray technique. Huadong et al. [7] proposed that cast iron having low phosphorous content when closely compared with a matching Al_2O_3 – 40% ZrO_2 coating applied to employ plasma spray coating in the presence of lubricated conditions showed better resistance with high and low phosphorous content than with the medium amount of phosphorous.

Tribological characteristics, specifically ceramic coatings applied using plasma spray technique was investigated by Giovanni Bolelli et al. [8] and compared with the microstructure properties of more commonly used ceramic coatings. Experimental studies were carried out by Sun Yong [9] to understand the effect of SMA treatment on the wear in stainless-steel of grade AISI 304 both with and without lubrication. The theory that composite coatings are known to have increased the resistance to wear due to high temperatures were analyzed by Subba Rao Medabalimi et al. [10] during the investigation conducted on the effects of NiCr coating containing Al in a partially oxidized form. The characteristics of piston ring functioning and the challenges faced during their operations due to irregularities in tribological properties and manufacturing defects were studied and summarized by Rita Ferreria et al. [11].Wenping Li et al. [12] compared the influence of four different types of coatings for piston rings namely - Physical Vapor Deposition Coating, Chromium, Goetze Diamond Coating, and Plasma Chemical Vapor Deposition Coating. They concluded that GDC or Goetze Diamond Coating showed the least wear rate while Chromium showed the highest. Marcus Kennedy et al. [13] developed a special coating for piston rings used in high-charged gasoline engines. This coating was named Carbodlidge and was known to reduce losses incurred due to friction and wear. Swain et al. [14] studied the mechanical properties of NiTi smart alloy coatings when applied to mild-steel by atmospheric plasma-spray technique. Hwang et al. [15] subjected piston ring specimens to high temperatures and studied the wear characteristics of these specimens when coated using the plasma spray technique. The wear properties of pistons ring. The crux of these experiments was to understand the wear and friction of piston rings under various tribological conditions. Based on these works, an investigation was carried out to understand the influence of functionally-graded cermet coating on stainless-steel piston rings.

2. Materials and methods

2.1 Materials used

The piston uses stainless-steel as the substrate, which provides corrosion resistance, cryogenic toughness, and higher ductility, strength, and hardness. But it wears off due to continuous usage. A ceramic coating of 60% Al2O3 + 40% TiO2 applied through the plasma spray coating method ensures less wear. The ceramic compounds used as heat-resisting barriers because of low conductivity along with other factors like the high coefficient of thermal expansion that brings about less wear in the substrate. The properties of this coating and its easy availability make it is a suitable choice for conducting this particular experimental study.

2.2 Plasma Spray Coating Procedure

Plasma spray coating is a technique of applying heated and ionized material on the surface of a substrate by injecting the powder form of the coating material into a high-temperature plasma plume. The material heated to a high temperature of nearly 16000o K makes this a feasible option for applying a coating of uniform thickness. The material is accelerated through the nozzle at a high velocity onto the substrate while maintaining a fixed temperature. When the hot coating material comes within proximity of the substrate, heat transferred from the hot particle to the cold surface causes the particle to shrink and solidify. The plasma spray gun is composed of an anode and a cathode. The presence of gases like He, H2, or N2 is necessary between the anode and the cathode. A high voltage discharge initiates the plasma spray causing ionization. That created a path for a DC arc between the anode and the cathode. The heat generated from the arc forms a plasma, and the gas dissociates and ionizes due to extreme temperatures. The powder sprayed can reach a 25 to 150mm distance. Plasma spray coating provides the benefit of high corrosion and wears resistance and conductivity. Based on these advantages, the composition of the functionally-graded cermet coating was 60% Al2O3 + 40% TiO2. The thickness of the ceramic coating obtained employing this procedure was 350 μm. The thickness of the bonding material zirconium oxide (ZrO2) was 150 μm.

Figure 1 shows the virtual representation of the plasma spray coating technique.

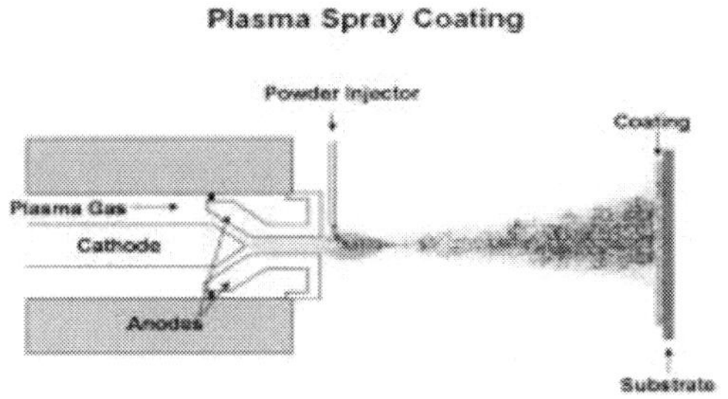

Figure.1 Plasma Spray Coating

2.3 Ceramic Coating Characteristics

The ceramic coating applied on the stainless-steel piston rings consists of 60% Al2O3-Alumina and 40% TiO2-Titania in powder form. Due to the presence of high hardness and strength determined by strong ionic bonds in Al2O3, alumina has a considerable amount of resistance to different types of wear and is one of the components for the functionally graded cermet coating. It is well-known that high wear resistance and hardness brings about substrate corrosion and thermal protection. The addition of Al2O3 increases the bonding strength without bringing about any changes to the ductility. Properties like wear resistance, coefficient of friction, and fracture toughness in AL2O3 show improvement with the use of TiO2. The addition of TiO2 leads to the formation of a dense and hard ceramic coating as it is chemically inert and provides protection against a significant decrease in ductility or embrittlement. TiO2 is known to enhance protection against wear by hard surfaces when combined with Al2O3. The combination of Alumina and Titania has a melting point of 3340 F that served the purpose of high wear-resistant and high-quality finishing. The properties of Alumina-Titania ceramic coating such as high dielectric strength, resistance to heat up to 1000 F make it a viable option for reducing wear and friction in stainless-steel piston rings. Resistance to wetting and corrosion by aqueous and dilute alkali solutions bring about a very smooth finish with high bond strength. Based on the previously mentioned properties, the coating is composed of 60% Al2O3 + 40% TiO2. For the bond coating, 150 microns of ZrO2 was used, which provides a protective layer owing to the presence of high thermal, mechanical, and electrical characteristics. Zirconium, being lighter than steel, has high thermal stability. This property makes it fit for being used in ceramic coatings at high temperatures. The quoted and unquoted specimen are shown in the figure 2(a) and 2(b).

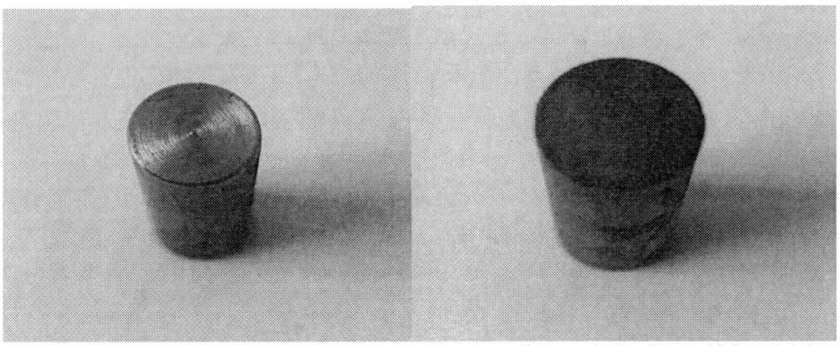

Figure.2 (a)-Uncoated Material Surface **Figure.2 (b)**-Coated Material Surface

2.4 Pin-on-disc test

Tribometers are instruments or devices used for the tribological investigation of the materials. One of the most common tribometers used is the pin-on-disc setup, a high-frequency reciprocating machine. A dry-testing method was employed to ensure easy installation and cleaning purposes. The pin installed in the fixture holder resembled the material of the piston ring, and the rotating disc resembles a section of the cylinder liner. For this particular experimental investigation, the material used for the disc is EN31, high carbon steel known for its hardenability. The pin of 25mm length, and diameter, 9mm, was used. Under a constant load of 40N, the pin is pressed against the rotating disc of diameter 60 mm. The disc rotates at the speed of 400 rpm for 1 hour. The test was for six cycles, each cycle having a duration of 1 hour. The heater block present in the setup consists of a piezoelectric transducer that measures the frictional force occurring between the pin and the disc.

Different discs were used for each cycle, while the same pin was used by cleaning it ultrasonically before and after each cycle. The graphs from the previously mentioned process gave the wear rate, and the frictional force encountered. Table 1 shows the details of the equipment used. The original setup of the apparatus is in Figure 3 and the schematic diagram in Figure 4.

Figure .3 Pin-on-disc setup

Table 1. Equipment details

Wear and Friction Monitor (Pin on Disc Monitor)	
ASTM Standard	G99
Wear Disc diameter	Diameter - 100mm & Thick - 6 to 8mm
Pin diameter	4 - 8mm diameter in steps of 2mm
Length	20 - 30mm length
Frictional Force	Max 100 N
Wear	0 - 2000 microns

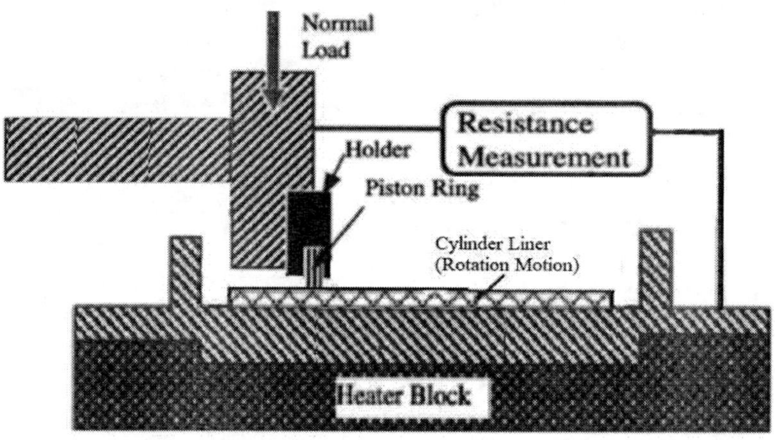

Figure.4 Schematic representation of pin-on-disc setup

2.5 Microstructure Analysis

The microstructural analysis process evaluates the microstructure in a metal alloy. The observation made shows the crystal analysis and possible metal deformations in the specimen after the pin-on-disc test. The wear encountered was observed at 100x magnification after every cycle. The microstructure analysis paves a path for a better understanding of the irregularities observed in the tribological characteristics during the pin-on-disc test.

3. Results and discussion

3.1 Pin-on-disc setup readings and observation

The data obtained were analyzed for changes in the wear rate and the frictional force. The pin was ultrasonically-cleaned before and after performing the test. The graphs plotted for the frictional force and the wear rate against the data points showing the results obtained.

3.2 Observation of frictional force

From the graphs given below, it is visible that after using the same specimen for six cycles of lab testing each of 1 hour, the amount of frictional force on both the uncoated and coated specimen increased from Cycle 1 to Cycle 6. When the data obtained were compared, it was evident that, although the amount of frictional force increased from Cycle 1 to Cycle 6 individually for both the coated and uncoated specimen, the frictional force in the case of the coated specimen was comparatively less than that of the uncoated specimen. Even that the highest mean value of frictional force, the uncoated specimen had a value of 27.1, as seen in **Figure 5.11,** while the coated specimen had a value of 16.7 in **Figure 5.12**. The same was the case at the lowest values. The uncoated specimen had a mean frictional value of 13.6, shown in **Figure 5.1,** whereas the coated specimen had a value of 12.9, evident from **Figure 5.2**.

Cycle 1

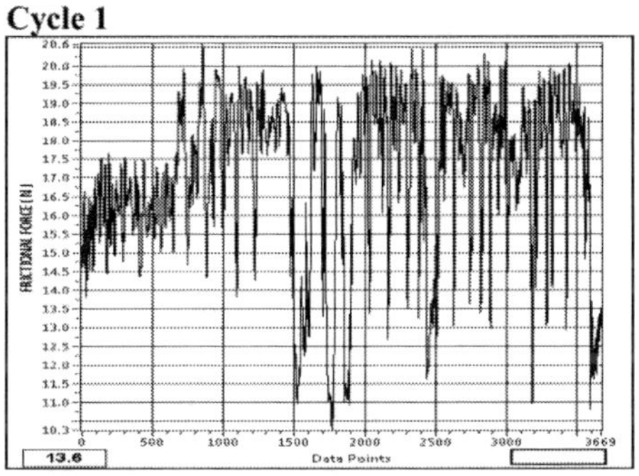

Figure 5.1 Uncoated – Friction

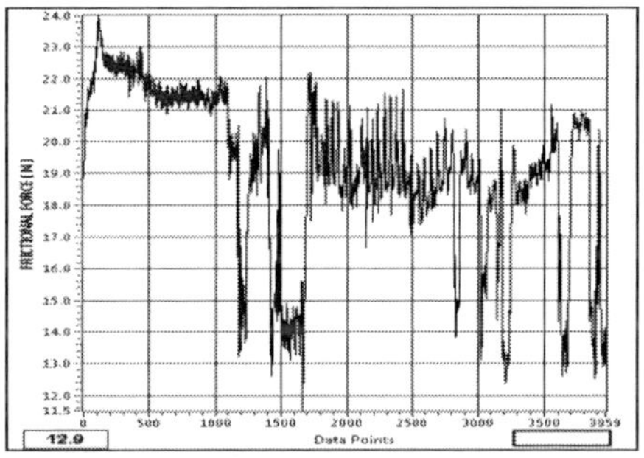

Figure 5.2 Coated – Friction

Cycle 2

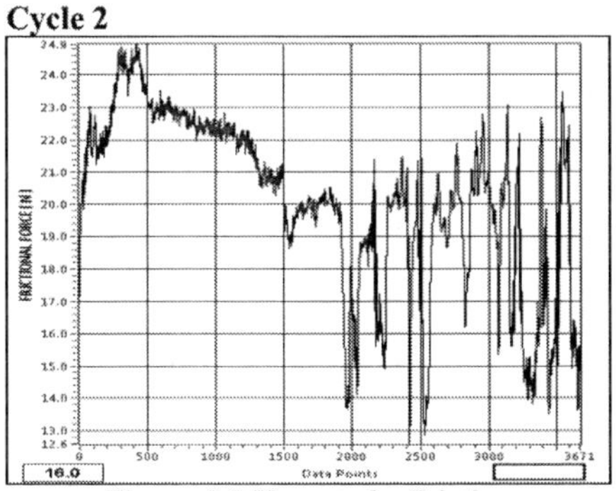

Figure 5.3 Uncoated – Friction

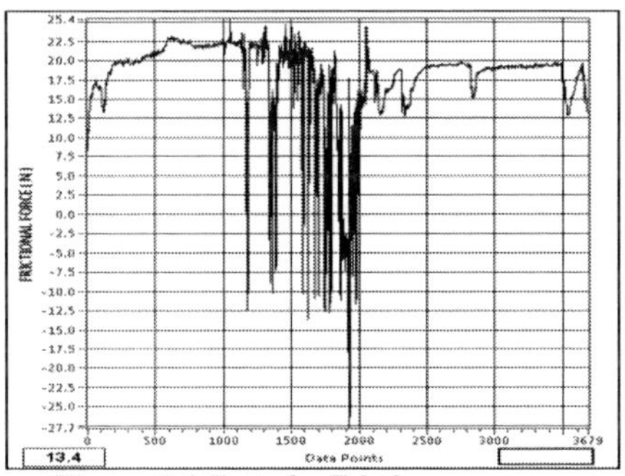

Figure 5.4 Coated – Friction

Cycle 3

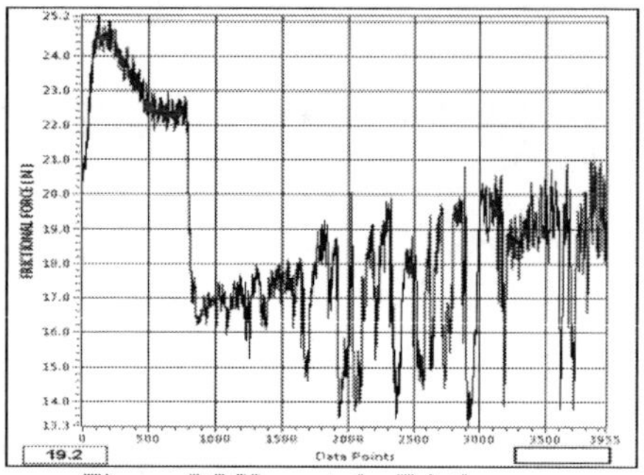

Figure 5.5 Uncoated – Friction

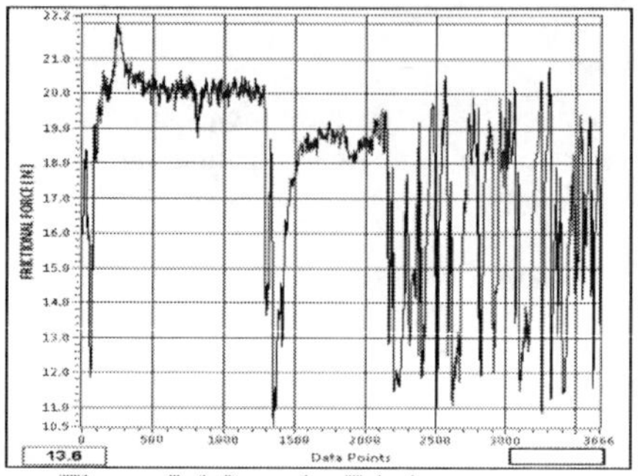

Figure 5.6 Coated – Friction

Cycle 4

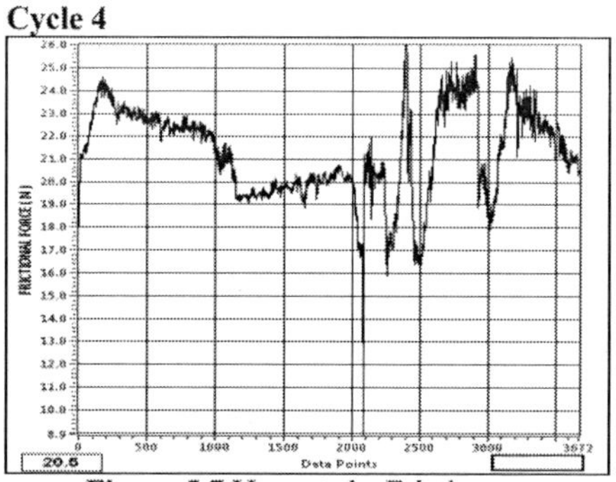

Figure 5.7 Uncoated – Friction

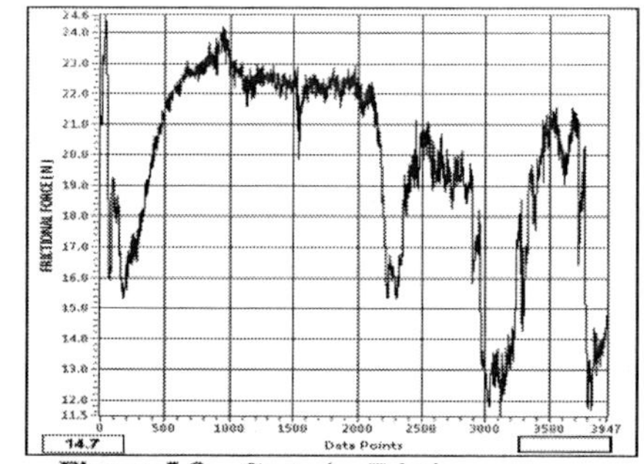

Figure 5.8 – Coated – Friction

Cycle 5

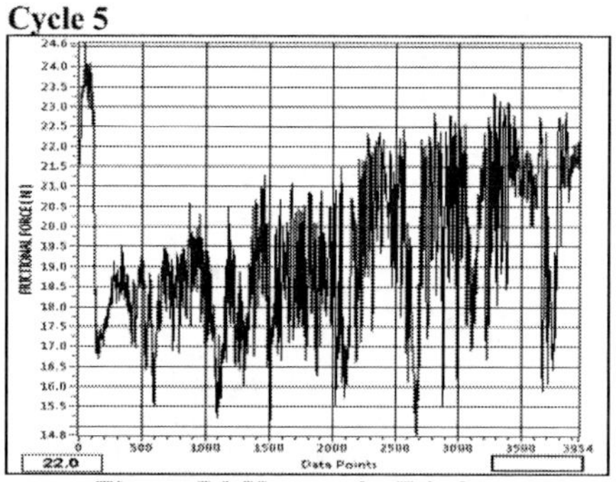

Figure 5.9 Uncoated – Friction

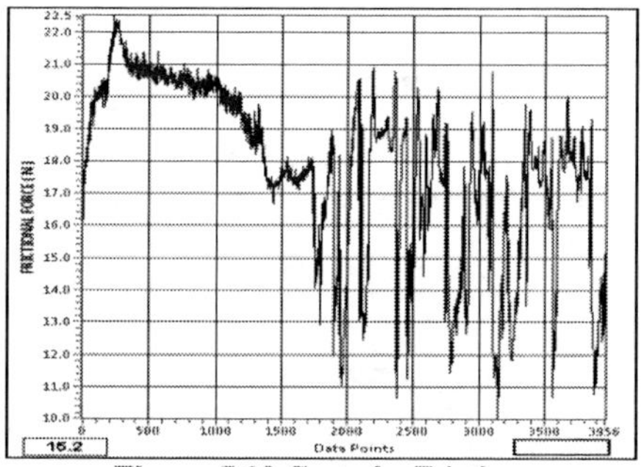

Figure 5.10 Coated – Friction

Cycle 6

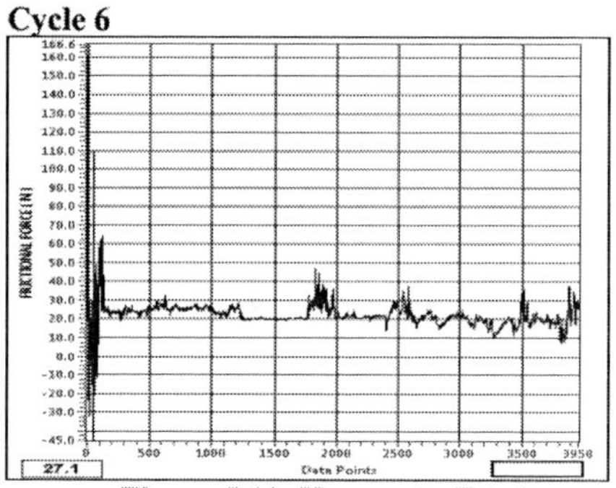

Figure 5.11- Uncoated – Friction

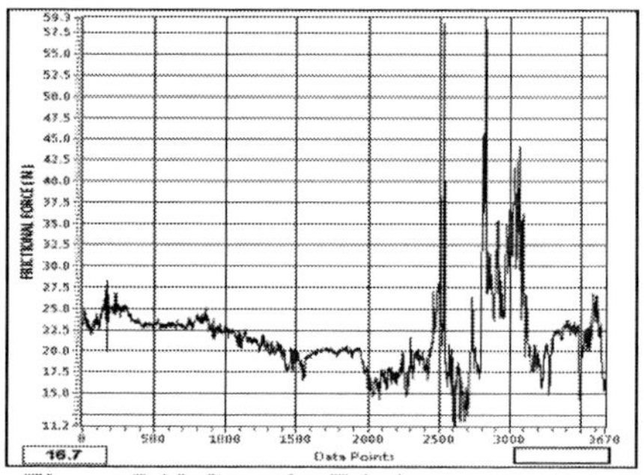

Figure 5.12 Coated – Friction

3.3 Observation of wear encountered

Corresponding to the amount of friction between the pin and the disc, both for the uncoated and coated specimen, the amount of wear observed from the graphs plotted between wear encountered against the data points shows that the value of wear shows a continuous increase and after one stage, it tends to become constant. The only exception to the previously stated point is the graphs obtained from Cycle 1. There is an uneven increase in the wear rate, which is visible through the microstructure imaging. This happens due to sudden application of load on the specimen without any previous contact with any other surfaces. The mean value of wear in the uncoated specimen, after cycle 6, is found to be 1585 from **Figure 5.23,** and in Cycle 1 is 374 from **Figure 5.13,** respectively. For the coated specimen, the mean value of wear in the coated specimen in Cycle 6 is 1341, evident from **Figure 5.24,** whereas in Cycle 1 is 111, as seen in **Figure 5.14**. As the testing process progresses from Cycle 1 to Cycle 6, the value of both the uncoated specimen and the coated specimen increase individually due to continuous application of pressure, but at the highest and the lowest values of the wear, the amount of wear in the coated specimen is comparatively less than that of the uncoated specimen.

Cycle 1

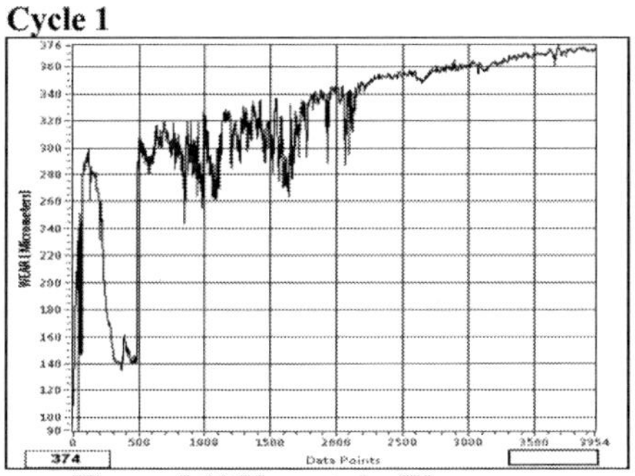

Figure 5.13 Uncoated – Wear

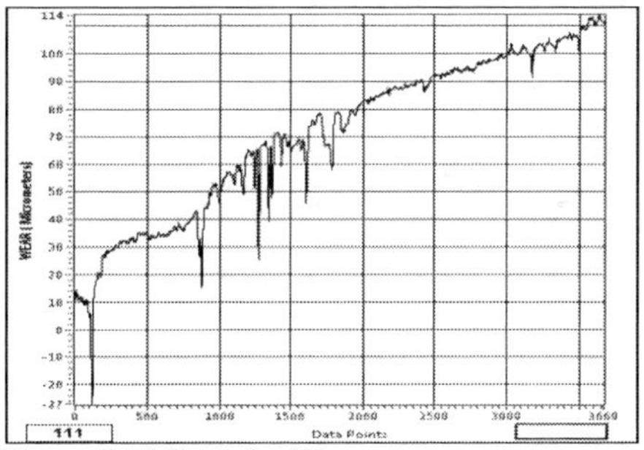

Figure 5.14 Coated – Wear

Cycle 2

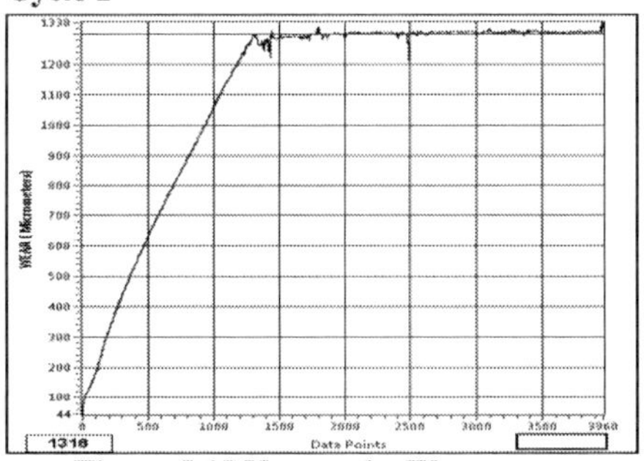

Figure 5.15 Uncoated – Wear

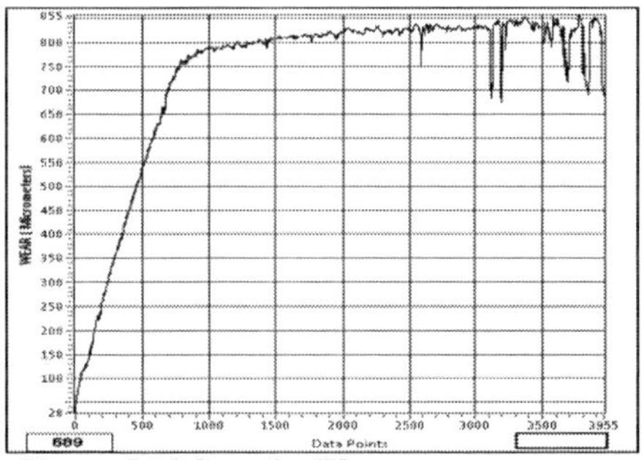

Figure 5.16 Coated – Wear

Cycle 3

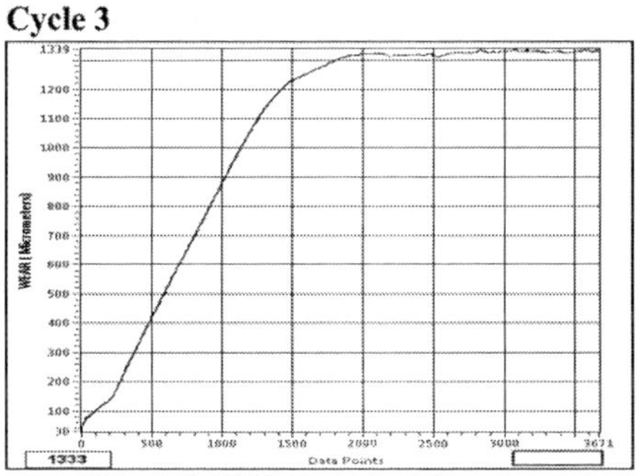

Figure 5.17 Uncoated – Wear

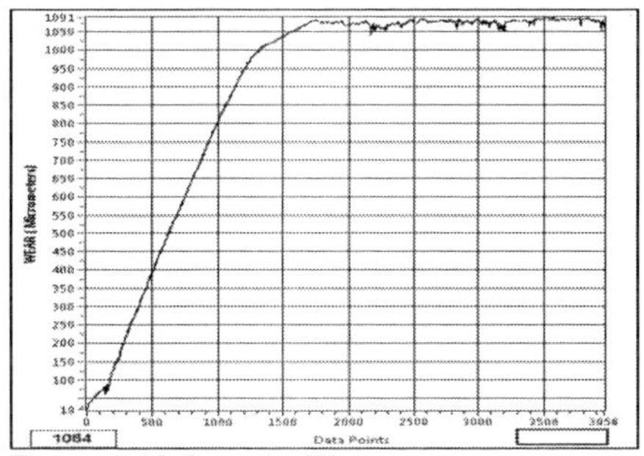

Figure 5.18 Coated – Wear

Cycle 4

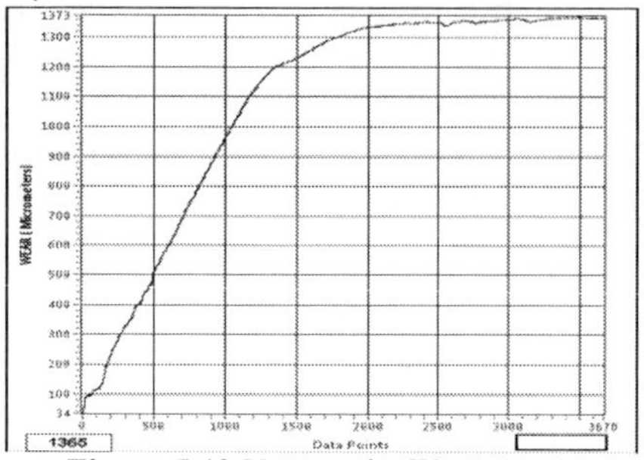

Figure 5.19 Uncoated – Wear

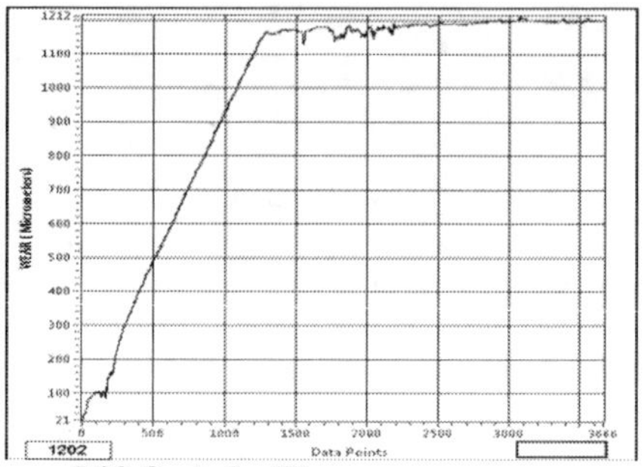

Figure 5.20 Coated – Wear

Cycle 5

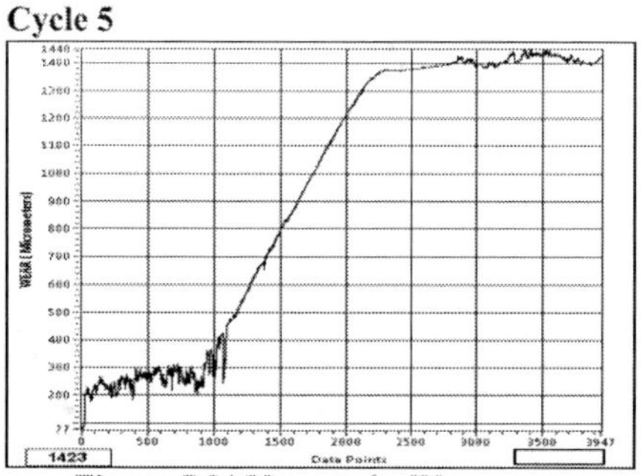

Figure 5.21 Uncoated – Wear

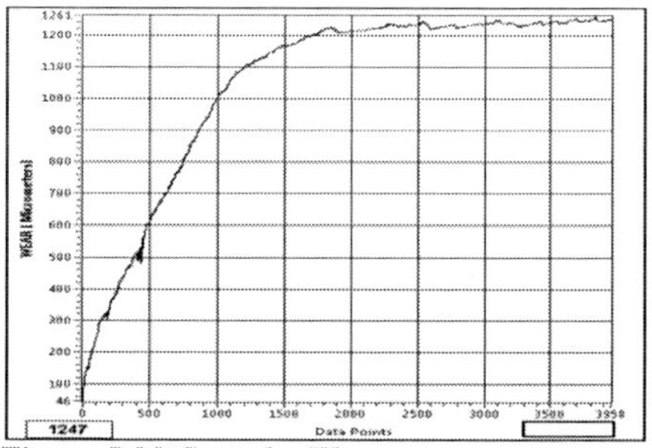

Figure 5.22 Coated – Wear

Cycle 6

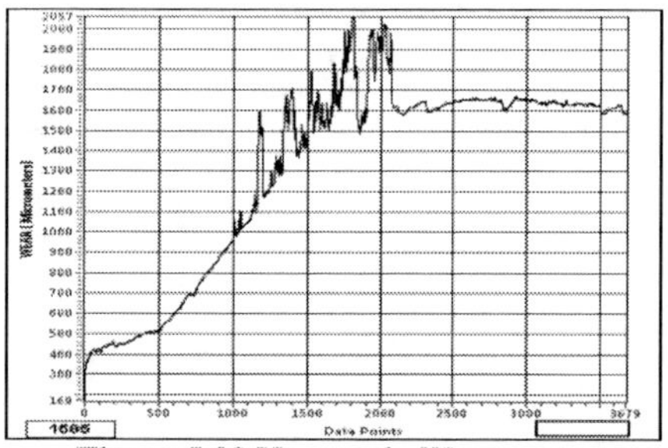

Figure 5.23 Uncoated - Wear

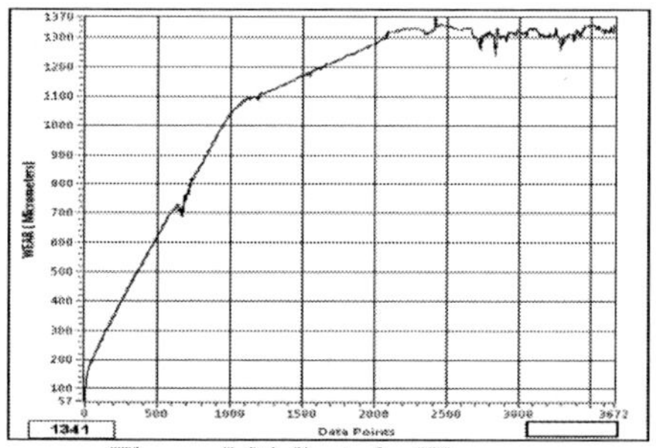

Figure 5.24 Coated – Wear

3.4 Microstructure Imagining and Analysis

A microstructure analysis performed using an optical microscope just before and after each cycle for both- the coated and uncoated specimen. The lab test showed the presence of voids and visible cracks in the uncoated material as well as the coated material. However, the irregularities observed in the uncoated specimen were higher as compared to the coated specimen due to the defects resulting from the fabrication processes. In the uncoated specimen, due to setbacks like stainless steel, it is observed to be susceptible to easy scratching off of material due to wear and friction or easy scuffing as seen in **Figure 6.1,6.3,** and **6.5** during cycles 1,3 and 6, respectively. The irregularities found during the 1st cycle causing large portions of uneven wear larger in number as compared to be more than the aberrations noticed during the 6th cycle. In the case of the coated specimen, cracks can also be the result of the high range impact on the specimen during the plasma spray coating or the high temperatures used for applying the powdered-coating. This results in excessive thermal stresses, which are not visible to the naked eye, but with the help of an optical microscope. Another reason for these cracks can be the sudden uneven solidification of the bonding material zirconium oxide (ZrO2) and the ceramic coating (60% Al2O3 + 40% TiO2). It leads to the formation of uneven layers of the coating that haven't blended well into each other, thereby causing uneven wear at some places but, when compared to uncoated specimens, they are less. Uneven layers are visible in the microstructure imaging of the 1st cycle for the coated specimen, which reduces further in the 6th cycle owning to the value of wear becoming constant after the continuous increase. However, the appearance of these irregularities seems to be less in the case of the coated specimen when compared to the uncoated specimen, indicating that the zirconium-based functionally graded Alumina-Titania cermet coating provides better resistance to both wears as well as friction. Therefore, the reduction of the amount of wear that would take place in an ordinary stainless-steel piston ring specimen in real-time cases as seen in Figures **6.2**, **6.4**, and **6.6** during cycles 1,3 and 6 respectively.

Cycle 1

Figure 6.1 Uncoated specimen

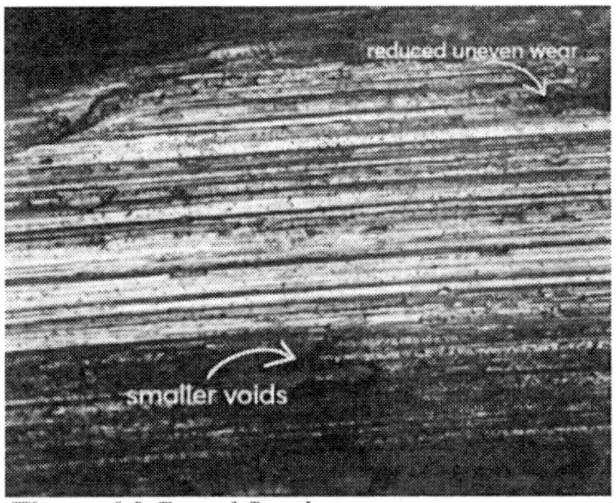

Figure 6.2 Coated Specimen

Cycle 3

Figure 6.3 Uncoated Specimen

Figure 6.4 Coated Specimen

Cycle 6

Figure 6.5 Uncoated Specimen

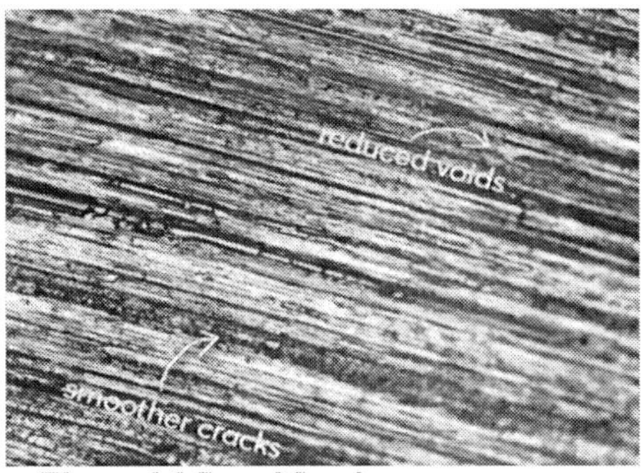

Figure 6.6 Coated Specimen

4. Conclusion

From the results of the tests performed on coated and uncoated piston rings, it can be concluded that the resistance of the coated piston ring was found to more than the uncoated aluminum piston by 14.5%. The resistance of the piston was found to more than the uncoated steel piston by 11.3%. The frictional force as well as the wear increased starting from Cycle 1 to Cycle 6 each of a duration of 1 hour. This was evident both in the case of uncoated and coated specimen. Although the values of frictional force and the wear increased over a period of time, the amount of wear in the coated specimen was less than the uncoated specimen even during the 6th Cycle. The microstructure images obtained, showed aberrations like, voids and visible cracks both in uncoated and coated specimens. However, these aberrations were less in the coated material as compared to the uncoated material. From the results obtained, the graphs plotted and the microstructure imaging, it is evident that the coated piston rings yield comparatively less wear with the uncoated piston as a result of a reduced amount of friction during the movement of the piston rings against the cylinder lines in real-time usage.

References

[1] Kennedy, M., Hoppe, S. and Esser, J., 2014. Lower friction losses with new piston ring coating. *MTZ worldwide*, *75*(**4**), 24-29.
[2] Wan Shanhong, Dongshan Li Anh Kiet Tieu and Bo Zhang 2017 Comparison of the scuffing behaviour and wear resistance of candidate engineered coatings for automotive piston rings. *Tribology International,* **106,** 10-22.
[3] Ma, Z., Henein, N.A. and Bryzik, W., 2006. A model for wear and friction in cylinder liners and piston rings. *Tribology transactions*, ***49*(3)**, 315-327.
[4] Lin, J., Wei, R., Bitsis, D.C. and Lee, P.M., 2016. Development and evaluation of low friction TiSiCN nanocomposite coatings for piston ring applications. *Surface and Coatings Technology*, ***298***, pp.121-131.
[5] Palanivendhan, M., Nigam, Y. and Trivedi, K., 2015. Experimental evaluation of aluminium alloy reinforced with silicon carbide. *International Journal of Applied Engineering Research*, ***10*(91)**, p.2015. [6] Ahn, Hyo-Sok, Jang-Yun Kim, and Dae-Soon Lim 1997 Tribological behaviour of plasma-sprayed zirconia *coatings Wear,* **203,** 77-87.
[7] Huadong, W., Yuansheng, J., Nicoll, A.R. and Barbezat, G., 1994. Friction and wear of a plasma sprayed Al2O3-40% ZrO2-cast iron system. *Wear*, ***176*(1)**, 49-60.
[8] Bolelli, G., Cannillo, V., Lusvarghi, L. and Manfredini, T., 2006. Wear behaviour of thermally sprayed ceramic oxide coatings. *Wear*, ***261*(11-12)**, 1298-1315.
[9] Sun, Y., 2013. Sliding wear behaviour of surface mechanical attrition treated AISI 304 stainless steel. *Tribology International*, ***57***, 67-75.
[10] Medabalimi, S.R., Ramesh, M.R. and Kadoli, R., 2019. High-temperature wear and frictional behavior of partially oxidized Al with NiCr composite coating. *Materials Research Express*, ***6*(12)**, 126599.
[11] Ferreira, R., Martins, J., Carvalho, Ó., Sobral, L., Carvalho, S. and Silva, F., 2020. Tribological solutions for engine piston ring surfaces: an overview on the materials and manufacturing. *Materials and Manufacturing Processes*, ***35*(5)**, 498-520.
[12] Li, W. and Lu, Y., 2016, October. The Study of Friction and Wear Capability of Piston Ring Coatings. In *Society of Automotive Engineers (SAE)-China Congress* 241-247. Springer, Singapore.
[14] Swain, B., Mallick, P., Bhuyan, S.K., Mohapatra, S.S., Mishra, S.C. and Behera, A., 2020. Mechanical Properties of NiTi Plasma Spray Coating. *Journal of Thermal Spray Technology*, **1**-15.
[15] Hwang, J.H., Han, M.S., Kim, D.Y. and Youn, J.G., 2006. Tribological behavior of plasma spray coatings for marine diesel engine piston ring and cylinder liner. *Journal of Materials Engineering and Performance*, ***15*(3)**, 328-335.

Development of Sound Absorbing Recycled Nonwoven Composite Materials

S. Sakthivel [1], S. Senthil Kumar [2*], N. Mekala [3] and G. Dhanapriya [4]
[1]Department of Textile and Garment Technology, Federal Tvet Institute, Addis Ababa, Ethiopia
[*2]Department of Handloom and Textile Technology, Indian Institute of Handloom Technology, Salem, Tamilnadu, India
[3,4]Department of fashion Technology, Bannariamman Institute of Technology, Tamilnadu, India
[*] Corresponding Author: Email ID : thil.tex@gmail.com

Abstract: The Sound absorbent textile materials, especially nonwoven composite structure of reclaimed materials which have low manufacturing costs, low relative density and are more attractive. In this research the use of reclaimed cotton and polyester fiber for production of sound absorption nonwoven composite materials has been investigated. Three different blend ratios of reclaimed cotton and polyester fibers that is 25:75, 40:60 and 50:50 have been used. The reclaimed cotton and polyester nonwoven composites are specified for their physical properties, such as thickness, areal density, bulk density, porosity and sound absorption characteristics in the frequency range of 250HZ-2KHZ. The values of the sound absorption coefficient and noise reduction coefficient acquired signifies that the reclaimed polyester fiber nonwoven composite holds excellent sound absorption behavior in the entire frequency range. Before compressed reclaimed cotton/polyester nonwoven composite of 25:75 blend ratios with greater bulk density and lower porosity is seen to give excellent performance while using the provided air gap behind the reclaimed cotton/polyester nonwoven composites.

Keywords: Reclaimed fiber, nonwoven composite, Sound absorption behavior, Impedance tube method

1 Introduction

Reclaimed fibers have proven to be suitable reinforcement materials for composites recognition to a combination of good mechanical properties and environmental advantages such as renewability and biodegradability expressed that Arbelaiz (2005). Xiaoqun Zhang (2004) acknowledged that a composite material is consists of more than one component, in which at least one of the components remains in solid state when it is manufactured. Several applications of reclaimewdl fiber composites can be found in construction, packaging, furniture and automotive fields. (Govardhan Goud 2013). Says that more environment friendly composite materials for automotive manufacturing have been made by substituting coir fibers for the widely used polyester fibers to make nonwoven fabric composites of coir fibers and recycled polypropylene fibers. (Stanten Greer et al 2011). Kozłowski et al (2008) stated that the nonwoven composite from reclaimed fibers that can be applied in buildings as a filling and facing insulation materials. Structural and insulation materials is made of natural raw materials (e.g. flax, hemp, wool) ensure full and active "breathing" of the whole structure element giving users a real feeling of building comfort. (Stanciu 2012). Indicate that today much importance is given to the acoustical environment. For creating an acoustically pleasing environment noise control and its principles plays an important role. This can be achieved practically when the intensity of the sound is brought down to a level which is not harmful to human ears.

(Seddeq 2009). Mention about Noise is a major cause of industrial fatigue, irritation, reduced productivity and occupational accidents. The continuous exposure of 90dB or above is considered as the most dangerous for hearing.

IOP Conf. Series: Materials Science and Engineering 1059 (2021) 012023 doi:10.1088/1757-899X/1059/1/012023

One of the main methods of noise control is to install the noise absorbent barriers (made from wood and textiles) between the source and the subjects. Measurement techniques used to characterize the sound absorption properties of a material are reverberant field method, impedance tube method and steady state method. Acoustic insulation and absorption properties of nonwoven composite depend on fibre was found to increase with the decreasing fiber diameter and length and increasing surface area i.e. large surface area and shorter length increase the absorption of sound. It was also observed that when sound reduction increases with the distance between the nonwoven fabric and the sound source, and this effect also increases with the increase in the areal density of fabric geometry and fiber arrangement inside the fabric structure which was expressed by Vinay Kumar Midha et al (2012). Carvalho et al (2012) verified that the sound absorption coefficient (Kannan Allampalayam 2005) The sound absorption is one of the most important acoustical properties of the porous materials which was used for sound insulation products commonly have the role of sound barriers, walls, road surfaces.

2 Materials andMethods

2.1 Reclaimed fibers

The raw material for the nonwoven is 'Trimmed waste' of garment industries. Garment industries are producing the garments in a high volume, which in turn gives high volume of trimmed waste. This waste fabric converted in to reclaimed fiber using recycled fabric waste opener machine. These waste fibers are classified into cotton and polyester and used as raw material to develop adhesive bonded nonwoven fabrics.

Figure 1. (a) Re-Used Fabric Opener Machine

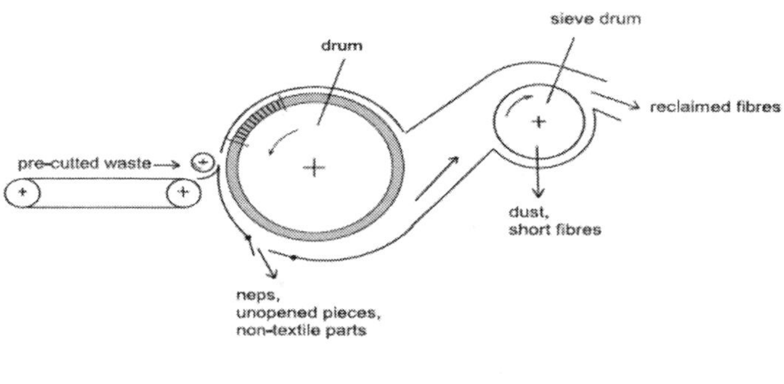

Figure 1. (b) Line Diagram of Re-Used Fabric Opener

2.2 Web formation

The web which is fed in the machine is opened further to achieve the very thin layer of fiber, which will be deposited over the circumference of the condensing cages (by the aerodynamic principle of web formation) and thus, the thin fibrous layer is delivered. This web former will produce about 138 meters of fibrous layer per hour. The fibrous layer from the web former is sprayed with polyvinyl acetate (PVA) at a constant pressure and flow as shown in the Figure 1. The adhesive add on percentage is taken care to maintain at 20%. Precaution is taken to avoid excessive or lesser flow of adhesive through the sprayer. By the calendar roller pressure, the fibrous layer is converted into nonwoven fabric. The spray adhesive bodings are an exact measure of the amount of binders applied, uniform binder distribution and a soft fabric handle. The preferred samples of proportions with 6-8 mm thick, 80 mm wide and 200 mm long were developed to nonwoven composite materials.

Table 1. Reclaimed Cotton Polyester blend ratio

Sample	Recycled cotton	Recycled Polyester
S1BC	25	75
S2BC	40	60
S3BC	50	50
S1AC	25	75
S2AC	40	60
S3AC	50	50

2.3 Preparation of nonwoven composites

The table 1 shows that the blend proportion of reclaimed nonwoven composite with fabric loading was fabricated using chemical bonded nonwoven samples with size mould of 28cm(length, L)x 28cm

(width, W)x 0.7cm (Thickness, T) initially, epoxy and hardener were mixed in ratio of 4:1 to form a matrix. Then, nonwoven fabric was spread into mould and covered with the matrix. The curing time of 6 hours was maintained uniformly for the composite materials at (60°). Two different types of nonwoven composite samples, namely before compressed form (BC) under zero load condition and after compressed form (AC) with the application of 22.27/kg/m^{2}loaded were produced. This load was selected with the view to reduce the thickness of around 20% compared to before compressed form.

2.4 Testing Methodology

The nonwoven composite fabric thickness was determined in accordance with ASTM D5729 standard method. The average of 20 observations was taken for each sample. Areal density which was a study by Koizumi *et al.* (2002) showed that in the middle of sound absorption get increased and the density of the sample got increased when the frequency is higher. The number of fibers present in the chemical bonded nonwoven samples from reclaimed cotton and polyester fibers increases per unit area when the apparent density is large. The areal density was determined in accordance with ASTM D 6242. The developed chemical bonded nonwoven composite samples using reclaimed fibers are subjected to bulk density analysis for the influence of sound absorption behavior.
Shoshani *et al.* (2003) stated that designing a nonwoven web to have a high sound absorption coefficient, porosity should increase along with the propagation of the sound wave. The porosity was found with ASTM B809.The sound absorption coefficients (SAC) of the nonwovens composite were tested by the impedance tube method on ASTME 1050 at Indian Institute of Technology, Madras. A sound source (loud speaker) is mounted at one end of the impedance tube and at the other end the

nonwoven and composite materials will be placed. The loud speaker generates broadband and stationary random sound. This sound propagates as planner waves in the tube, hits the sample and gets absorbed. Thus a standing wave interference pattern gave a result due to the superimposition of forward and backward travelling waves inside the tube. The sound pressure at two fixed location is measured and by using the two-channel digital frequency analyzer. The sound reduction coefficient NRC by the specimen can be calculated by the following derivation by Teli et al (2007) and surajait and Senguptha et al (2010).

$$SR\% = \frac{dB_{wos} - dB_{ws}}{dB_{wos}} X100$$

where,

SR	– Sound reduction
dBwos	– Sound level without sample and
dBws	- The sound level with sample

The samples were placed in three different positions: 25 cm, 50 cm, and 75 cm from the sound source and measured for sound source and measured for sound reduction with three decibel value of 60 dB, 70dB, and 80dB. The average sound reduction percentage of these three decibel values are calculated and compared for all the samples.

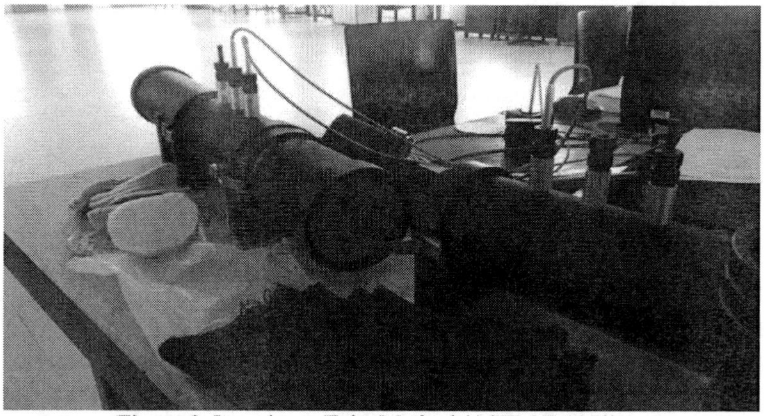

Figure 2. Impedance Tube Method (ASTM E 1050)

3 Results and Discussion

3.1 Analysis on physical characteristics of reclaimed nonwoven composites

Table 2. Physical properties of reclaimed nonwoven composite materials

Sample	Thickness (mm)	Areal Density (g/m^2)	Bulk Density (g/m^3)	Porosity (H)	SAC
S1BC	7.72	330.5	110.29	0.62	0.72
S2BC	8.21	653	92.61	0.73	0.76
S3BC	8.42	980.77	64.71	0.81	0.79
S1AC	4.34	323.11	142.31	0.56	0.81
S2AC	4.64	648.1	110.72	0.59	0.84
S3AC	4.71	960.21	70.22	0.61	0.87

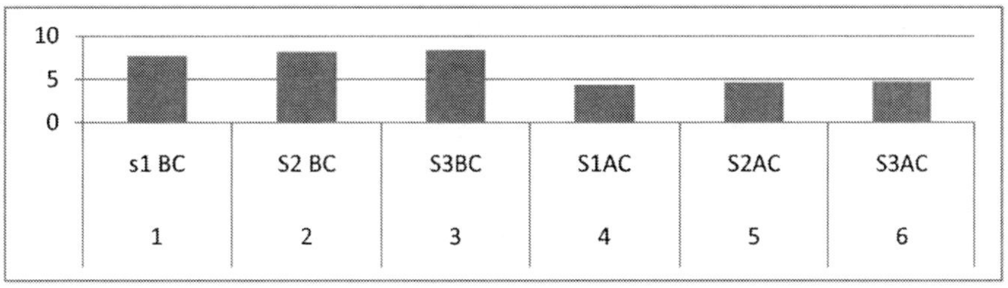

Figure 3. Thickness of Reclaimed Cotton /polyester nonwoven composites

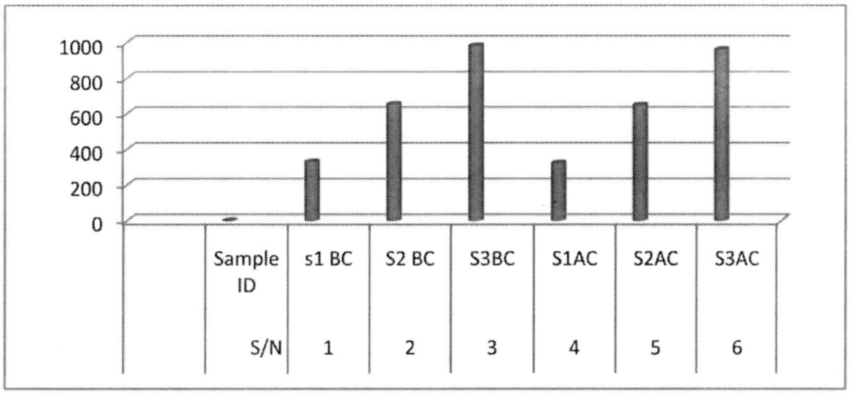

Figure 4. Areal density of Reclaimed Cotton /polyester nonwoven composites

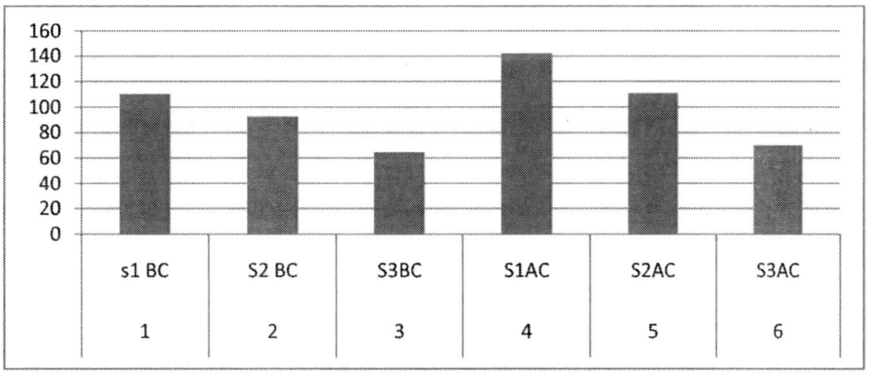

Figure 5. Bulk density of Reclaimed Cotton /polyester nonwoven composites

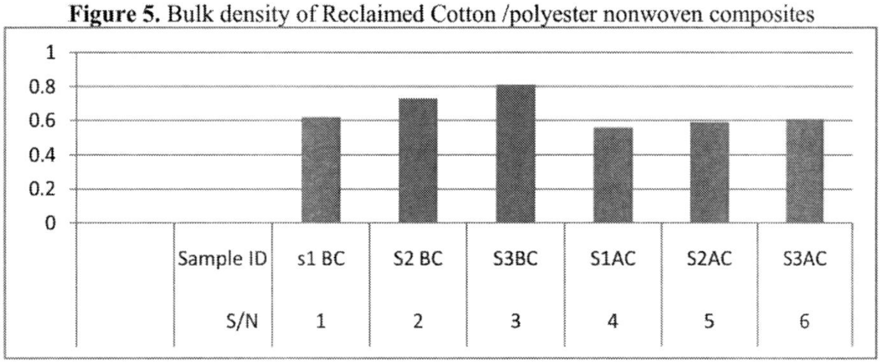

Figure 6. Porosity of Reclaimed Cotton /polyester nonwoven composites

Table 2 shows that as the content of polyester fiber in the nonwoven composite increases from 20% to 40% (from s1to s3) The figures 3, 4, 5and 6 shows that the result reveals increase in thickness of the nonwoven composites. When there is an increase in areal density there is an increase in sound absorption coefficient for reclaimed cotton and polyester nonwoven composites. The bulk density of after compressed and before compressed samples decreases. Further, it can be observed that the bulk density of after compressed composites is higher than that of before compressed composites. Increase in polyester content in the blend increases the porosity but it decreases on compression. Moreover, they showed that the following effects of density on sound absorption behavior in nonwoven fibrous materials gave the results of less dense and more open structure absorbs sound of lower frequencies (500Hz). Denser structure performs better in the frequencies above 2000 Hz. The above quoted results are in line with the bindings of Koizumi et al (2002)

3.2Performance in the without air gap

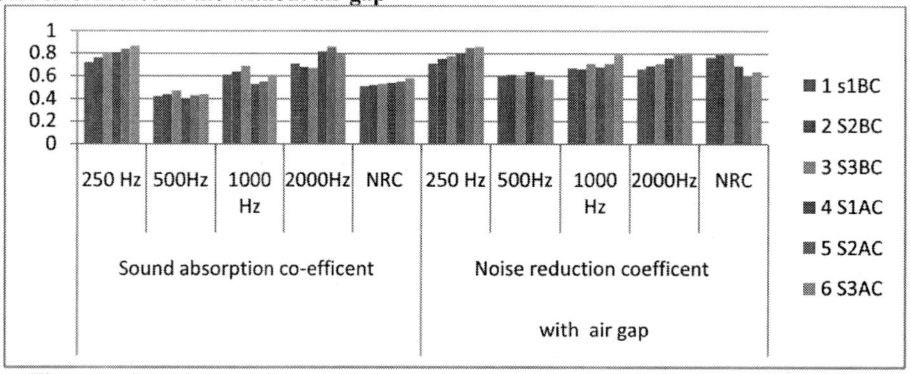

Figure 7. Sound absorption coefficient of Reclaimed Cotton /polyester nonwoven composites

3.2.1 Before compressed reclaimed nonwoven composites

The figure 7 shows the increase in the reclaimed polyester content in the reclaimed nonwoven composites does not show significant changes in SAC, at all the frequencies. These composites will be superior absorption behavior towards 250Hz and 2000Hz which is reasonable to low behavior at average frequency levels. However, the diversity in behavior of the nonwoven composites with respect to frequency levels is not reflected in its NRC. This same trend was observed by Andrea Zent et al (2007).

3.2.2 After compressed reclaimed nonwoven composites

With respect to reclaimed polyester content, theses composites are found to perform like before compressed composites. The figure 7 shows that their behavior on sound absorption at dissimilar frequency levels is also found to be the same as in before compressed composites. But absorption at lower and average frequency levels than that of high frequency level is lower and higher than before compressed composites respectively. The NRC values obtained are found to be approximately the same as in before compressed composites. Kosuge (2005) in his findings of sound absorption with combination of nonwoven fabric of reclaimed fiberother natural fibers.

3.3 Performance with air gap

3.3.1 Before compressed reclaimed nonwoven composites

The figure 7 shows the increase in reclaimed polyester content in composites does not have effect at low and upper middle, and high frequencies, whereas it has a distinctive effect at lower middle frequency (500Hz).These composites have greater sound absorption behavior at higher middle and high frequency levels compared to that at low average frequency. However, these effects are not significantly reflected in the NRC values. The overall performance of these nonwoven composites with respect to SAC and NRC is enhanced with before compressed and after compressed samples without air. Mingzhang Ren and Finn Jacobsen, (1993) in their findings fibers which got interlocking in nonwovens are the frictional elements that provide resistance to acoustic wave motion.

3.3.2 After compressed reclaimed nonwoven composites

The figure 7 shows the Reclaimed polyester content in these composites does not show any effect at low and high frequency levels, whereas it decrease at low average frequency and increase at high average frequency levels. Sound absorption is originated to be high at low frequency levels compared to that at all other frequency levels. It is all so found that the sound absorption of these composites is lower than that of before compressed samples at all frequencies. In this case also, the difference with respect to sound absorption at dissimilar frequency levels is not establish to reflect significantly in NRC but the values obtained are lower than that of before compressed reclaimed nonwoven composites. Shoshani *et al.* (2000). In his research findings in designing a reclaimed nonwoven web to have a high sound absorption coefficient, porosity should increase along the propagation of the sound wave.

4 Summary and Conclusions

Considering reclaimed polyester content in nonwoven composites and the role of compression on the behavior of composites in the with and without of air with respect to SAC at dissimilar frequency levels and NRC, it can be understood that the highest sound absorption at any particular frequency between 250Hz and 2000Hz and highest absorption of sound waves with varying frequency in higher than frequency range can be obtained from the before compressed reclaimed cotton /polyester nonwoven composite of 25:75 blend ratio having high bulk density and low porosity. This strength has resulted in high frequency of the nonwoven composite a preferred property for sound absorption, in with of air gap behind the composite. As the prospective in sound absorption which appears bright, in the further examination the direction of understanding the behavior of reclaimed polyester fiber in the composite would lead positively to a superior solution for decrease in noise. The same trend was observed by Yang et al (2001) in their research work developed a porous laminated composite material by molding of premix, pre heating and lamination exhibited a very high acoustic coefficient property on the frequency range of 250Hz and 2KHz.

References

[1] Andrea Zent, John T Long 2007, Automative sound absorbing material survey results. In *International Conference on Noise and Vibration*, United States

[2] Arbelaiz A, Fernandez B, Cantero G, Llano-Ponte R, Valea A & Mondragon I, 2005, 'Mechanical properties of flax fibre/polypropylene composites. Influence of fibre/matrix modification and glass fibre hybridization', *Applied Science and Manufacturing,* vol. 36, no 12, pp. 1637-164

[3] Carvalho R, Rana S, Fangueiro R & Soutinho F, 2012, 'Noise reduction performance of thermobonded nonwovens' in 12th *World Textile Conference AUTEX*, Carotia, pp. 597-600.

[4] Govardhan goud & Rao RN, 2011, 'Effect of fibre content and alkali treatment on mechanical properties of *Roystonea regia*-reinforced epoxy partially Biodegradable composites', *Journal of Engineering Fiber and Fabrics*, vol. 6, pp. 4

[5] Ioan Curtu, Mariana, Domnica, Stanciu, Camelia Cosereanu & Vasile Ovidiu, 2012, 'Assessment of acoustic properties of biodegradable composite material with textile inserts', *Material Plastic,* vol. 49, pp. 68-72. 31-436.

[6] Koisszumi ,2002,"The development of sound Absorbing materials using natural bamboo fibers", *High performance structure and composites book, WIT Press.*
[7] Kozłowski R, Mieleniak B, Muzyczek M & Mańkowski J, 2008, 'Development of Insulation Composite Based on FR Bast Fibers and Wool'. *Austria*, pp. 553-556.
[8] Stanciu M, Curtu D, Cosereanu I, Lica C & Nastac D, 2012, 'Research regarding Acoustical properties of recycled composites', *New York.*
[9] Stanton Greer D, Walter L, Bradley, Danny Natividad & Ryan J Vano, 'More sustainable Non-woven fabric composites for automotive using coir (coconut) fibers'. *In international conference of Sydney*. Pp.78-85.
[10] Subhash Chand, Gajanan, Bhat D, Joseph E. Spruiell & Sanjiv Malkan, 2000, 'Structure and properties of polypropylene fibers during thermal bonding'. *Indian Textile Journal*, vol.9, issue 3, pp.52-59.
[11] Surajit sengupta 2009-10, 'Sound reduction by needle-punched nonwoven fabrics' *Indian journal of fiber & Textile Research*, vol. 35, pp 237-242.
[12] Teli,M.D.,2007,"Efficacy of nonwoven materials as sound insulator", *Indian Journal of Fiber and Textile Research*.Vol.32,June.
[13] Yang "Development of a novel porous laminated composite material for high sound absorption"*J. Vib Contr.*Vol.7(5),pp.675-8,2001.
[14 Youneung Lee & Changwhan Joo 2003, 'Sound absorption properties of recycled polyester fibrous assembly absorbers'.
[15] Wang "Experimental Study Of The Absorption Charecteristics Of Some Porous Fibrous Materials"*Appl Acoust*,Vol.62,pp.447-59,2001.
[16] Verma BB 2009, 'Continuous jute fiber reinforced laminated paper composite and reinforcement-fiber free paper laminate', Bull. *Mater. Sci.*, vol. 32, no. 6, pp. 589-595
[17] Utkarsh Sata 2013 'Development of Cotton Nonwoven Composite Fabric for Toxic Chemical Decontamination and Characterization of its Adsorption Capabilities', *Journal of Engineered Fibers and Fabrics,*Volume 8, Issue 1, pp 94-106.
[18] Vasanta v cholachagudda 1, udayakumar p a 2, ramalingaiah 2013, 'Mechanical characterisation of Coir and rice husk reinforced Hybrid polymer composite' *international journal of innovative research in science, Engineering and technology*, vol. 2, issue 8, pp 3779-3786.
[19] Seyed Abbas Mirjalili & Mohammad-Shahi M, 2012, 'Investigation on the Acoustic Characteristics of Multi-Layer Nonwoven Structures', *Part 1 - Multi-Layer Nonwoven Structures with the Simple Configuration.*
[20] Sagartzazua, Hervellab L & Pagaldaya JM, 2001, Review in Sound Absorbing Materials, vol. 4, pp. 25-32.

[21] Olynyk D & Northwood TD, 1964, 'Comparison of Reverberation-room and Impedance-tube Absorption Measurements', *The Journal of Acoustical Society of America*, vol. 36, no. 11, pp. 2171-2174.
[22] Murugesan "recycled polyolefin based plastic wastes for sound absorption..*Polym Plast Technol Eng.*,Vol.45(7),pp.885-8,2006.
[23] Mingzhang Ren "A method of measuring the dynamic flow resistance and reactance of porous materials", *Applied Acoustics,* Vol.39 (4),pp.265-276,1993.
[24] Merve Kucukali Ozturk, Banu Uygun Nerg & Cevza Candan 2010, 'A study on the influence of fabric structure on sound absorption behavior of spacer knitted structures", *7th International conference - Texsci.*
[25] Maya jacob john1 and rajesh d. Anandjiwala 2009, 'Chemical modification of flax reinforced Polypropylene composites', *textile research journal,* vol. 4, pp 56- 63.

[26] Lyndon B Johnson, 2001, 'Acoustic Noise Control and Analysis Plan for Human Research Facility Payloads and Racks', *National Aeronautics and Space Administration*, pp 43-49.

[27] Lundahl A, Fangueiro R, Soutinho F & Duarte F, 2010, 'Waste fiber reinforced ecocomposites', *Material science forum*, vol. 636-637, pp. 1415-1420.

[28] Kozłowski R, Mieleniak B, Muzyczek M & Mańkowski J, 2008, 'Development of Insulation Composite Based on FR Bast Fibers and Wool'. *Austria*, pp. 553-556.

[29] Kenneth D Langley, Yong K Kim & Armand F Lewis, 2000, 'Recycling and reuse of mixed – fibre fabric remnants' (spandex, cotton, polyester), *Chelsea Center for Recycling and Economic Development, University of Massachusetts Lowell*, pp. 28.

[30] Kath U, 1983, 'A reverberation room round robin on the determination of absorption coefficients', *Acustica*, vol. 5, pp. 211-215.

[31] Jing Li et al"A study on the relationship between the thickness of nonwoven and sound absorption capability", *Modern Applied Science*, Vol.1 No.4, pp.74-79, 2007.

[32] Jia-Horng Lin1, Chen-Hung Huang, Kuo-Cheng Tai, Ching-Wen Lin & Jin-Mao Chen Ching-Wen Lou, 2006, 'Recycling Nonwoven Fabric and Polypropylene Selvage to Create Composite Fabric', *Journal of Engineered Fibers and Fabrics*, vol. 8, issue 1, pp 65-67

[33] Janusz Szosland, Anna Koziorowska & Adam Kabziński, 2005, 'Utilisation of Linear Textile Wastes with Use of a Rotary Loom', *Fibres & Textiles in Eastern Europe*, vol. 13, no. 4, pp. 34-37.

[34] Hoda S. Seddeq, 2009, 'Factors Influencing Acoustic Performance of Sound Absorptive Materials', *Australian Journal of Basic and Applied Sciences*, vol. 44, no. 4, pp. 145-150.

IOP Conf. Series: Materials Science and Engineering 1059 (2021) 012024 doi:10.1088/1757-899X/1059/1/012024

Design and analysis of High-Speed Low-Power Vedic Multiplier with 3-1-1-2 compressor Using Reversible Logic gates

Ms.Dharani S[1], Mr.Abin Satheesan[2], Asuvanti M A[3], Ranjith Kumar[4], Vidhya S[5]

[2] Assistant Professor, Dept. of ECE, Bannari Amman Institute of Technology, Tamil Nadu, India

[1345] PG Scholar, Dept. of ECE, Bannari Amman Institute of Technology, Tamil Nadu, India

E-mail: Dharnisubramaniam273@gmail.com

Abstract. The FFT Function in digital signal processing is one of the most important function in several applications such as Image Processing, Wireless Communications and Multimedia. FFT Processors consisting of butterfly structure operations involving necessary operations such as Addition, Subtraction, and Multiplication of complex values. The FFT Butterfly Structure work is designed with a "Vedic Multipliers" for applications at high speed. In this Vedic Multiplier, an algorithm called "Urdhva Triyabhyam" was used to improve its efficiency by optimizing the number of logic gates, constant inputs and garbage outputs. The Data Computation time is reduced by an 3-1-1-2 compressor using reversible logic gates. Hence reducing the surplus power consumption of 11.24% and summation of the partial products is done with less delay factor of about 5.28%. The area, power, delay, area delay product and power delay product are calculated using cadence virtuoso and is implemented in Spartan-6 device family using Xilinx ISE.

Keywords: Area; power; delay; Vedic multiplier; reversible logic gates.

1. Introduction

In forth coming generation, Wireless technology and technology (IOT) will play a crucial role in digital era. New access structures for next generation digital systems are the FDMA multi-access frequency division, OFDMA, "single carrier FDMA", and "interleaved FDMA".

These programs are based on effective, Fast-Fourier transforming algorithms consisting of fewer arithmetic operations. The present-day circuits dissipating power of order of nano watts which is comparatively 3-4 magnitude larger than the power lost due to entropy change. The high-speed data processing system for several multimedia applications with lower power consumption and area is the most critical limit. The adder and multiplier are basic arithmetic unit most commonly used essential operation. The digital world of the next generation expects high quality digital information with high accuracy. For example, the system requires high-quality images. To satisfy the quality of image criteria pixel value is increased. Meanwhile increasing the computation time in the digital system.

2. Literature Survey

The configurable power multiplier with booth algorithm, dynamic range, adjuster and partial product generator corrections. This powerful design detects redundant calculations and reduces power consumption, more areas and the partly producing product.

The architecture of the multiplier with a fixed width consists of modified booth encoders, an error reward circuit and partial generation of products, which meet the needs for area and power consumption with a high degree of precision for lossless application with limited design complexity. Approximate power efficiency which includes a bit pre computation reselect and 2-1 mux. The presets are used to choose anyone from MSB or LSB. The device that achieves 99% precision in the application of audio and recognition.

The construction of approximately 15-4 compressors using 5-3 circuit is designed to reduce power and delay and increases the throughput. The construction of approximately 15-4 compressors using 5-3 circuit is designed to reduce power and delay. This concept is most commonly used in the image processing framework and the result shows improved performance.

In order to reduce the combinational time in a compressor 3-2,4-2,5-2, the XNOR-XOR logic is implemented.4-2 compressors remain as high-speed multiplexers to decrease the time-consuming delay in the summation part of the product and are constructed using XNOR-XOR gates and 2:1 mux.3-1-1-2 compressor consisting of a reduced area and less delay by summing up the partial products.

3. Existing System

3.1 Vedic Mathematics

Vedic maths is an assortment of procedures to illuminate scientific mathematics in simple and quicker manner which can be utilized for issues involved in arithmetic, algebra calculus, polynomial math, geometric analytics, conics and so on. Vedic mathematics is a method that Indian mathematicians have developed to solve problems with standard mathematical measures. Often complex and time consuming, but using general Vedic mathematical principles and particular techniques, numerical calculations can be carried out very quickly. Vedic science is an arithmetic framework which was found by Indian mathematician utilizing standard numerical advances, considering various issues.

Advantages

- Vedic arithmetic can definitely solve scientific, numerical figuring in quicker manner
- More than 1700 per cent faster than ordinary math
- The procedure Speeds up and performance
- It improves scholastic execution and instant moment results

3.2 Nikhilam Navatashcaramam dashatah

For the concept of "computing quickly." If you say the number of bit-wise multiplications (and additions) is reduced, the answer is that they don't. Both methods take multiplications and additions around n2 bit-wise, where n2 is the number of bits in the multiplicand and the multiplier each. Unless you mean the need for memory then the sutra would be significantly stronger. You first write down n2 intermediate rows in the general method, and move them column by column, from right to left. This method is one of the 16 sutras from Vedic mathematics. This sutra can be used to simplify multiplication of large digit numbers into smaller numbers with only a few subtracts, adds, and shifts, the steps are followed below:

- We pick the power of 10 as our foundation depending on the numbers closer to the power of 10
- Apply the nikhilam theory, eliminate the digit from the base-all the digits 9 and 10 last. Such numbers are referred to as excess if positive and if negative, as deficiencies
- Apply cross addition to multipliers deficiencies or excess results

- The second stage must also have the same number of digits as the number of the chosen base's zeroes If zeroes are less than pre-appended values. If more, pass the initial numbers to the 1st stage

Advantages

- Converting large numbers to corresponding small digit multiplication.
- Efficient if multiplier and multiplicand are similar to the same base power.

3.3 Urdhva Triyabhyam

Urdhva Triyabhyam is the popular and well-established equation suitable for all occurrences of multiplication and moreover in the division of an enormous value by another enormous value. It signifies "vertically and crosswise".

Steps:

For non-equidigit numbers, order the numbers vertically, prefix the same number of zeroes in front.

- Consider arrangements of vertical segments, preferably from left to right, increasing in order and raising in order afterwards.
- Recognize the sum of the edge-equidistant digit cross-results to show each sequence.
- Reformat the number in base 10, whenever possible. Sets are acquired by using each vertical section in turn from left to right (ascending) and then again with the exception of each vertical section in turn from right to left (descending).

For calculating n-digit set, the cross product is:

- For even number of digits: the entire number of digit outcomes, equidistant to the end.
- For the odd number of digits: the total number of digit tests, equidistant from the finishes applied with the primary digits multiplied by each other.

For a single-digit set,

For multiplying two-digit number, we can simply multiply two values, such as m1.m2, which brings cross-product of m1 x m2.

Advantages

- Easily realized in hardware
- Time taken to perform multiplication operation and design complexity of circuit is very less.
- Main advantage of using this type of multipliers reduces the number of the bits, area and speed gradually compared to the other fast multipliers.

3.4 Implementation of Vedic multiplier using urdhva Triyabhyam

Vedic multipliers are one of the fastest and efficient methods in the arithmetic function processing that Indian mathematicians developed and used. In this Vedic multiplication is logically expressed with "Boolean equation". Bitwise summation equation is made a comparative list for 4x4 and 8x8 multiplier. In the summation equation '+' symbol which represents the binary addition operation and it is executed by implementing and making use of half and full adders. The number of unwanted switching that can be reduced by using a detection unit. The detection unit for 4x4 and 8x8 input data are calculated by "2*(N/2-1) 2input logic OR gate", "four 2 input AND gate", and "2 NOT gate".

Table 1. Detection unit logical gate count for N-BIT multipliers

Multiplier NxN	**No.of.input AND gates**	**No.of.input OR gates**	**No. of. NOT gates**	**Total gates Required**
4X4	04	02	02	08
8X8	04	06	02	12

3.5 Multiplier design with 3-1-1-2 compressor

The first stage compute is based on the bitwise logical AND operation of the 4x4 multiplier based on Vedic math's. 16 logical AND gates from LSB to MSB are arranged to generate the single output in the first step. For the multiplication result S0 to S7 with carry propagation, the half adder's and the full adder's computation are made used. The first stage 16 "AND" gates are arranged based on input data type.

If the input date be X=00xx and Y=00xx data, only four logical AND gate operation is necessary during the 1st stage. Therefore, LSB positions these four logical AND operations from LSB. Also, when the X=00xx input data and Y = xxxx inputs 8 logical AND operation is required to replace the 16 logical AND operation for the following bit at the 1st stage of computation. Consequently, in 2nd phase and propagation phases half adder's and full adder's have been used. The 4x4 multiplier, based on 3-1-1-2 compressor, has a partial product propagated by the logical AND process. The 4x4 multiplier generates sixteen logical AND operations when the data is full-range, and it is ordered from 0 to 15.

Table 2. Logical equation for 4x4 multiplier with Detection unit

Type of Multiplier	Input data type	With Detection unit
4x4	X = 00xx ; Y=00xx	$Z_0 = x_0y_0$; $Z_1 = x_0y_1 + x_1y_0$; $Z_3z_2 = x_1y_1 + c_1$;
	X = 00xx ; Y = xxxx	$Z_0 = x_0y_0$; $Z_1 = x_0y_1 + x_1y_0$; $Z_2 = x_0y_2 + x_1y_1 + c_1$; $Z_3 = x_0y_3 + x_1y_3 + c_2$;

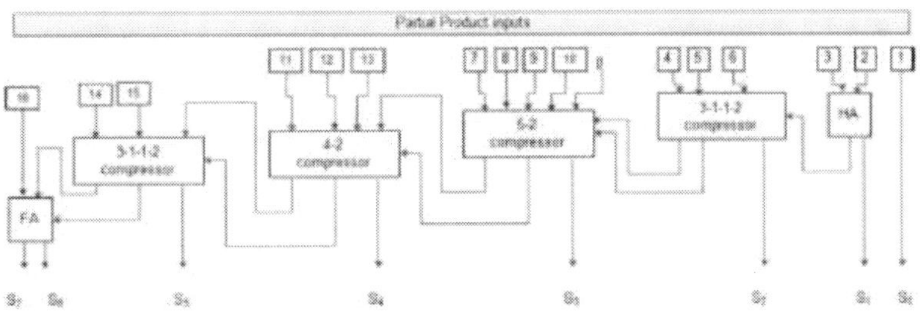

Figure 1. 4x4 Multiplier design with 3-1-1-2 compressor

4.Proposed Design

According to Landuer theory, Energy is dissipated if we have a logical operation on the unit. The cause for this operation is due to thermodynamic laws, in which entropy of system tends to increase whenever there is an operation that doesn't have identical and different solution. This forms the basis of reversible logic concept where there will be correspondence between only input and output of the system. The one of the important weighted function in ALU is multiplication. The multiplier is the circuit which is tremendously used factor in many DSP applications. This multiplier block this involves two major process namely partial product generation and addition of partial products. Vedic Multiplier that offers two popularly used algorithm named "Urdhva Triyabhyam" and "Nikhilam" algorithm for multiplication. Here we use "Urdhva Triyabhyam" which is popularly used technique for fast multiplication by multiplying "vertically and crosswise".

The algorithm is derived for 4x4 multiplication of binary numbers is given below:

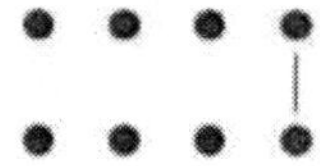

A0.B0 = P0.C0 ⟶ q0 [1]

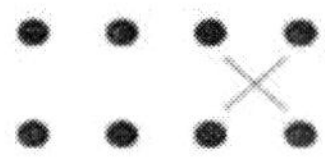

(A1.B0) + (A0.B1) = P1.C1 ⟶ q1 [2]

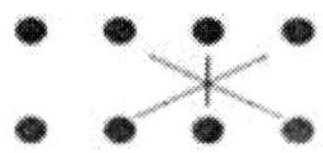

(A2.B0) + (A0.B2) + (A1.B1) = P2.C2 ⟶ q2 [3]

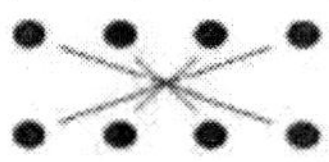

(A3.B0) + (A0.B3) + (A2.B1) + (A1.B2) = P3.C3 ⟶ q3 [4]

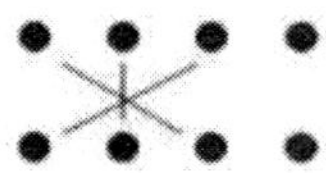

(A3.B1) + (A1.B3) + (A2.B2) = P4.C4 ⟶ q5 [5]

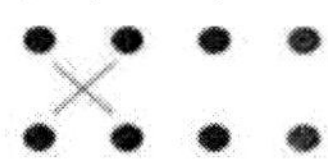

(A3.B2) + (A2.B3) = P5.C5 ⟶ q5 [6]

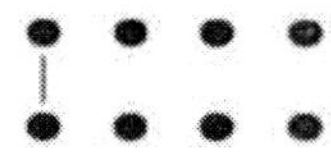

A3.B3 = P6.C6 ⟶ q6 [7]

This may generate combinational delay in the circuit, which may decrease the speed of the system. In order to avoid delay in the circuit we introduce a block called compressor 3-1-1-2 using reversible logic. The reversible logic circuit is said to be reversible which is most popularly used to predict the input vector by output value this is because of previous value of information bit is not lost. The reversible logic is a new and promising field which reports the problem of power dissipation. Also proven to consume zero power theoretically.

The simple multiplication operation may consume many cycles to complete the operation. The main goal of embedded system is low power dissipation, high speed, less Area. There are various reversible gates available but we manipulate the fredkin gate which holds of quantum cost value of 5 and garbage value of only 3.

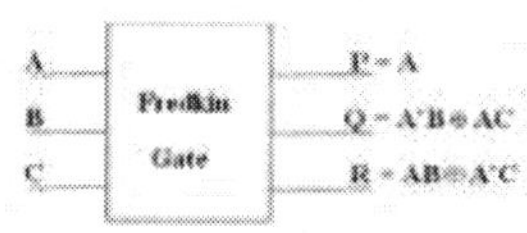

Figure 2. Fredkin gate (Reversible logic gate)

Computational circuit suitable for reversible computing consists of equal no.of. input and output that permits the first bit to remain unchanged and reciprocate the last two bits only if, the first bit is 1 and quantum cost of gate is 5.

The proposed circuit is mainly designed to compute the lower part of summation with three inputs and one carry input compressor are required to compute summation result with smaller combinational delay. It is a 3*3 "Fredkin Gate". If P=A, Q=C if A=1 else B, R=B if A=1 else C. Hence this gate is known as Controlled SWAP Gate (CSWAP). Quantum cost is 5 and its delay is 5. This gate is popularly known as Universal Reversible gate i.e., Any logical or Arithmetic operation can be performed entirely using Fredkin Gate. We can observe the drastic change, when every change of state two bits goes in and one bit comes out. Therefore, the above resistive losses of material, the gate must dissipate:

T_E = K log 2T/ Joules per transition. [8]

Where,

K -Boltzmann constant

T - Temperature

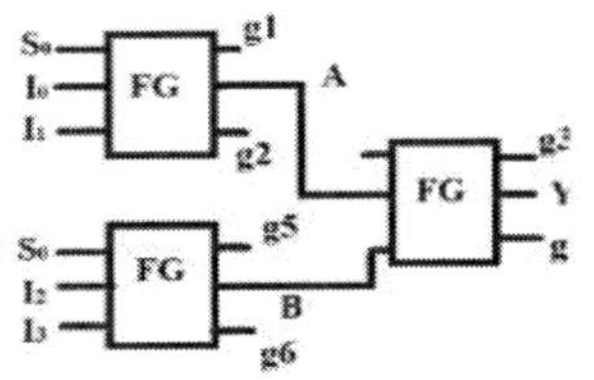

Figure 3. Mux design using reversible logic fredkin gate

In the above figure,

$$A = \overline{So}I_0 + SOI_1 \quad [9]$$

$$B = \overline{s_0}I_2 + S_0I_3 \quad [10]$$

$$y = \overline{S_1}\,\overline{S_0}I_0 + \overline{S_1}S_0I_1 + S_1\overline{s_0}I_2 + \overline{s_1}S_0I \quad [11]$$

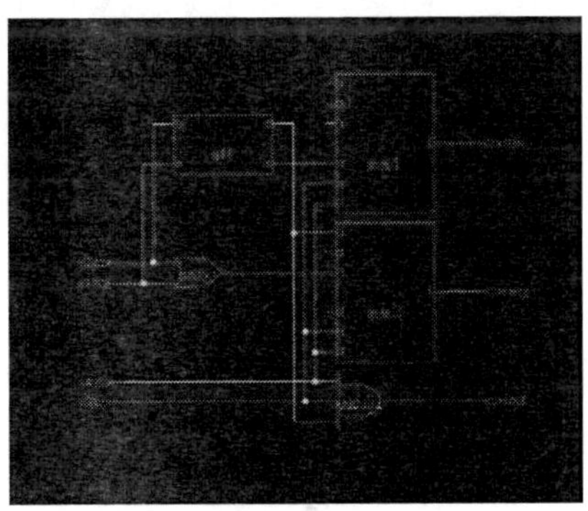

Figure 4. Design of compressor 3-1-1-2 using Reversible logic Gate

The above designed fredkin gate among other reversible logic gates to design mux which is efficiency made use to design 3-1-1-2 compressor using reversible logic gate is given in below

Sum=$((n_1 \oplus n_2)\overline{n_3}\,\overline{n_4}) + ((n_1 \oplus n_2)\overline{n_3}n_4) + ((n_1 \oplus n_2).n_3.\overline{n_4}) + ((\overline{n_1 \oplus n_2}).n_3.n_4$ [12]

Carry = $(n_1.n_2.\overline{n_3}.\overline{n_4}) + ((n_1 + n_2).\overline{n_3}.n_4) + ((n_1 + n_2).n_3.\overline{n_4}) + (\overline{n_1.n_2}).n_3.n_4)$ [13]

$C_{out} = n_1.n_2.n_3.n_4$ [14]

The above equation has been proposed using k map and Boolean expression is derived.

The proposed 3-1-1-2 compressor achieved reduction in delay upto 11.98% compared to the previous existing compressor and also achieved reduced power dissipation, during the summation of partial products.4x4 Bit multiplier is simulated using proposed “3-1-1-2 compressor” using reversible logic gates.

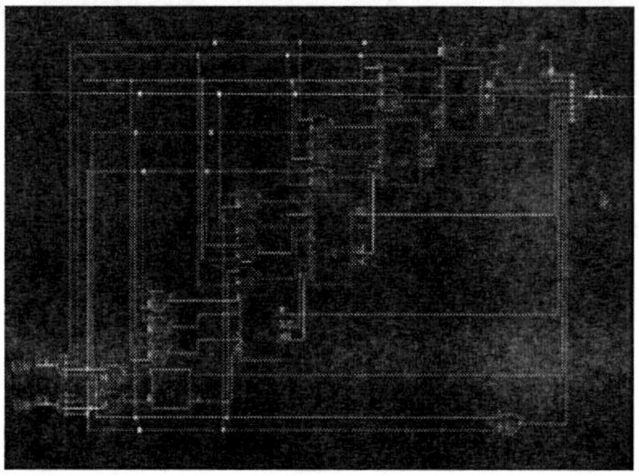

Figure 5. 4x4 multiplier using compressor 3-1-1-2 with reversible logic gates

OUTPUT WAVEFORM

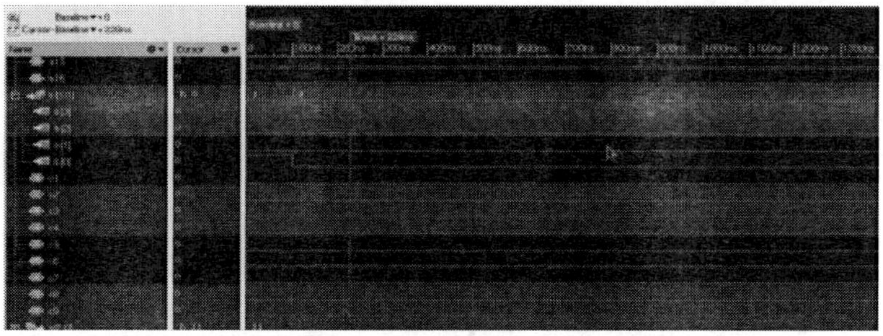

Figure 6. Output waveform of 4x4 multiplier using compressor 3-1-1-2 with reversible logic gates

4. Results and Discussion

The below chart that makes a perspicuous view about the researches that were made on various multipliers with different algorithm and achieved performance of the multipliers. The proposed system that achieved a better performance results comparing to the existing design. While it achieved improved area of 31%, power of 89.35% and delay of 11.98% for 4x4 multiplier and also achieved better results for high order multipliers and achieved improved area of 33.6%, power of 69.86% and reduced computational time of 10.93%.hence the proposed multipliers can be used in image processing applications where the signal-to-noise ratio(PSNR) of the image quality measured is compared with the accurate multipliers and the results obtained illustrates that our proposed multiplier operates better than the previous approximate multipliers.

Table 3. Comparison table and performance analysis of various multipliers with proposed multiplier

		AREA	POWER	DELAY	ADP	PDP
4x4	Array	118.33	2835.93	12.37	146267	35053
	Baugh wooley	131.52	2991.97	14.48	190296	43294
	Wallace	134.83	3055.23	12.26	165177	37425
	Dadda	115.74	2782.35	12.44	143853	34585
	Existing Vedic multiplier	330	12514.94	12.63	108405	26034
	Proposed Vedic multiplier	310	11183.08	11.98	1857.8	66986
8x8	Array	544.63	13056.48	36.29	19.758	473688
	Baugh wooley	605.95	13863.21	37.83	22917	524307
	Wallace	619.27	14154.46	35.62	22052	504041
	Dadda	534.18	12708.37	36.88	19706	468812
	Existing Vedic multiplier	598.53	14373.29	20.93	12521	300688
	Proposed Vedic multiplier	1259	40683.82	18.62	7657.6	367431

4.1 Comparison chart of various multipliers

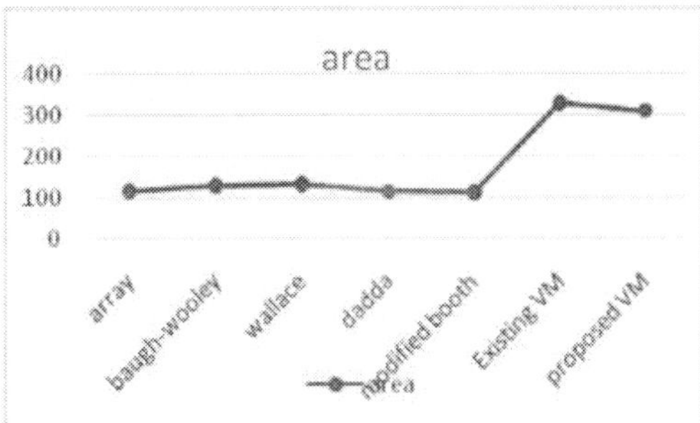

Figure 7. Area comparison graph of various multipliers

From the above chart we can observe the performance analysis of various multipliers and the proposed multipliers can be achieved the optimized performance parameters comparing to other existing multipliers such as area.

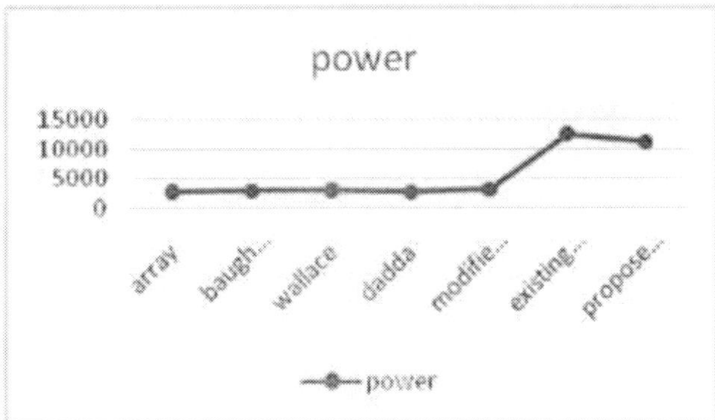

Figure 8. Power comparison graph of various multipliers

The above chart gives a clear conclusion about the optimized power performance of proposed Vedic multiplier using reversible logic gates than all other existing Vedic multipliers.

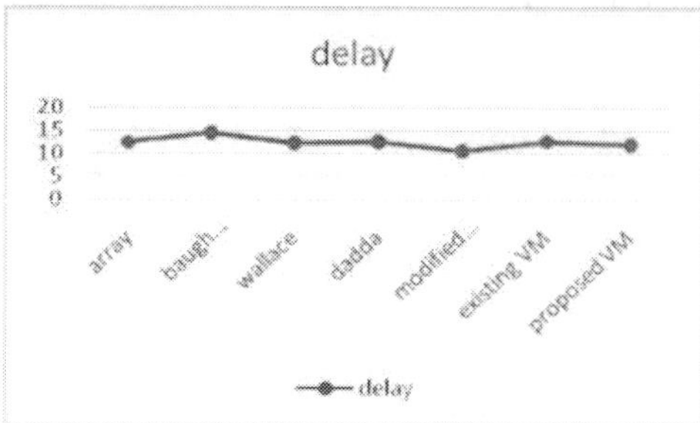

Figure 9. Delay comparison chart of various multipliers

From the above chart gives a clear conclusion about the optimized delay performance of proposed Vedic multiplier using reversible logic gates than all other existing Vedic multiplier this is due to decrease in combination delay of 2nd stage of multipliers algorithm. In contrast to this increases the area delay product and power delay product of the design.

4.2 FPGA implementation of Vedic multiplier

The proposed improved design and existing available multipliers are synthesized and simulated using VHDL Coding style and it is implemented using cadence virtuoso EDA tool. The proposed multipliers are implemented in Spartan-6 FPGA family devices. The performance of the FPGA Device family Spartan-6 is analyzed by combinational Path delay of the multipliers. The proposed 4x4 Vedic multiplier achieved reduced combinational path delay of about 9.11% while implementing in FPGA device family. The below table shows the improved delay performance of various multiplier.

Table 4. FPGA Delay Performance comparison with various multipliers

Multipliers	Spartan-6
Array-Multiplier	11.388
Baugh-Wooley Multiplier	12.3
Wallace-tree Multiplier	12.11
Dadda Multiplier	12.15
Vedic Multiplier	9.53
Existing Vedic Multiplier	9.503
Proposed Vedic Multiplier	9.112

4.3 Image Processing Application

To analyze the performance analysis of multiplier which is designed using 3-1-1-2 compressor Vedic multiplier using reversible logic gate. Image contrasting is implemented with proposed Vedic multiplier. MATLAB tool is used to simulate the image. For this we have choose a color images of size 512x512 having RGB pixels. Therefore, the available count of pixels is 786432 and pixel holding value of 8bits. Each pixel holds the range of (0-255). The size of our multiplier is also 8x8 bit therefore it is not necessary to convert pixels. All input pixels are mapped from the value (0-255) to (0-65535).

Contrasting of image is done using equation given below

$$A = (\text{Input Image} \times \pi) \div 65535$$

$$\text{Contrast Image} = (1 - \cos(A) \div 2)$$

Quality of image is measured based on peak signal noise ratio (PSNR) of the output image is computed based on mean squared error (MSE). The PSNR value above 30 is considered as good quality of image. Input Image simulated also holding the PSNR value above 30 considered to be good quality image which is outcome of 8x8 proposed Vedic multiplier.

$$\text{MSE} = (1 \div mp) \times (\sum_{x=0}^{x=1} \sum_{y=0}^{y=1} [x(i,j) - y(i,j)]2) \quad [15]$$

$$\text{PSNR} = 10 \log_{10} \times (MAX_I^2 \times MSE) \quad [16]$$

(a) Original Image

(b) Contrast Image

(c) Proposed Multiplier

Table 5. PSNR and MSE calculation of proposed multiplier

Multiplier	PSNR (dB)	MSE
Proposed Vedic Multiplier	34.6	12.6

5.0 Conclusion

In this article, we proposed a compressor 3-1-1-2 using reversible logic which is a field that mainly concentrating to reduce the power, area and the combinational delay in summation part of partial products. This proposed design achieved reduced value of surplus power consumption upto 11.98% and limited combination delay upto 5.28%. The power, delay, and area of any device is always being an uncompromising parameter while designing any circuit in VLSI field. This approach showed a contrasting difference between the existing and proposed modified design which reduced the functioning logic gate count of 31%, during data computation for 4x4 multiplier and 33.6% for 8x8 multiplier. Therefore 3-1-1-2 compressor-based result can be most widely applied in real time applications such as Filters, cryptography, DSP processors, multimedia applications etc.

References

[1] Chen O, Sheen R and Wang S 2000 A low-power adder operating on effective dynamic data ranges *IEEE Trans. Very Large Scale Integr.* (VLSI) Syst. **10** 435–453

[2] Kung S R and Wang J P 2010 Design of power-ecient configurable booth multiplier *IEEETrans.* Circuits. Syst.–I: RegularPapers **57** 568–580

[3] Wang J P, Kuang S R and Liang S C 2011 High-accuracy fixed-width modified booth multipliers for lossy applications *IEEE Trans. Very Large Scale Integr.* (VLSI) Syst. **19** 52–60. 583

[4] Cilardo A, Caro D D, Petra N, Caserta F, Mazzocca N, Napoli E and Strollo A G M 2014 High

speed speculative multipliers based on speculative carry- save tree *IEEE Trans.CircuitsSyst.–I:RegularPap.* **61** 3426–3435
[5] Sivanandam K and Kumar P 2015 Run time reconfigurable modified Vedic multiplier for high speed multimedia applications *IEEE Int. Conf. Intell. Syst. Control.* 2109–2113
[6] Wen M C, Wang S J and Lin Y N 2005 Low-power parallel multiplier with column by passing Electron *Lett.* **41** 581–583
[7] Chang C H, Gu J and Zhang M 2004 Ultra low-voltage low-power CMOS 4-2 and 5-2 compressors for fast arithmetic circuits *IEEE Trans*. Circuits Syst.-I **51** 1985–1997.
[8] Ramakrishnan M and Mallick P S 2017 Design and analysis of multiplier using approximate 15-4 compressor *IEEE Access* **5** 1027–1036
[9] Veeramachaneni S, Avinash L, Puppala S R and Srinivas M B 2007 Novel architecture for high-speed and low- power 3-2, 4-2 and 5-2 compressors *Inter- national Conference on VLSI design (VLSID'07)* 324-329
[10] Kuang S R and Wang J P, 2010 Design of power-ecient configurable booth multiplier *IEEE Trans. Circuits Syst.-I*:Regular Papers **57**(3) 568–580
[11] Kuang R S, Wang J P and Cang-Yuan 2009 Modified booth multiplier with a regular partial product array *IEEE Trans. Circuits Syst.-II* **56** 404–408
[12] Qian Z and Margala M 2016 Low-power split-Radix FFT processors using Radix-2 Butterfly units *IEEE Trans. Very Large Scale Integr.* (VLSI) Syst. **24** 3008–3012
[13] Lamberti F, Andrikos N, Antelo E and Montuschi P 2011 Reducing the computation time in (short bit-width) two's complement multipliers *IEEE Trans.* Comput. **60** 148–156
[14] Lai S C, Juang W H, Lee Y S, Chen S H, Chen K H, Tsai C C and Lee C H 2016 Hybrid architecture design for calculating variable- length Fourier transform *IEEE Trans.CircuitsSyst.II* **63** 279–283

IOP Conf. Series: Materials Science and Engineering 1059 (2021) 012025 doi:10.1088/1757-899X/1059/1/012025

Design of High Speed and Low Power Approximate Multiplier for Image and Digital Signal Processing Applications

Ranjith Kumar T, Saravana Kumar R, Dharani S and Asuvanti M A

Department of Electronics and Communication Engineering , Bannari Amman Institute of Technology, India

ranjithkumar.vd18@bitsathy.ac.in

Abstract. Truncation based approximate multiplier is a scalable multiplier in which truncation is performed based on the leading one bit positions in each and every input operands, hence the number of partial products are reduced. In this multiplier design the multiplication process is done by addition, shifting and multiplication processes which has a very good improvement in space occupation and power consumption when compared to the exact multiplier. To enhance the multipliers speed each and every input operand in the multiplication side are truncated using the method given below. Based upon the leading 1's the input operands, the inputs are truncated show that the accuracy does not depend on input operand's width. Hence this multiplier is scalable. Very large improvement in multiplier's design parameters like power and area consumption are obtained which is independent of the width of the input operand. To find the working efficiency of this multiplier we have compared the design parameters with some recently proposed multipliers and the exact multiplier. They obtained result states that the proposed multiplier enhances power consumption, delay and area up to 90%, 80% and 95% when compared to be exact multiplier. Also our design outsmarts other multipliers in parameters like power consumption, speed and area. Proposed approximate multiplier has zero mean value Gaussian distribution. We have figured it out it by MATLAB applications and a good PSNR and MSE is obtained. The output shows that the degradation in quality is almost negligible. Also we recommend Truncation based approximate multiplier where power consumption has to be adjusted in multiplication operation.

1 Introduction

Power consumption is very much important parameter in the design of digital circuit systems. To reduce power consumption and to increase the speed one of the methods called approximate computing is used. It's result may not be correct so it is used in error Resilient applications. Here are some of the examples of those applications which includes audio processing, data mining, image processing and machine learning. In case of signal processing applications, major portion in power consumption is because of the arithmetic operations. This multiplication operations are repeatedly used and has high latency and power consumption. Hence this approximate multipliers for very good suitable for reducing time delay and faster processing.

There are three stages in binary multiplication process. The first stage is where the partial products are produced based upon the input operands. The partial product terms are stored in the next stage until only two rows exist which are then added by using fast adders in the final third stage. The approximation technique is introduced in the first stage in order to reduce the number of partial products. It may also be applied during the second step in order to reduce the latency for energy consumption. These two parameters in multiplication operation is affected by the adders architecture which are used in final step of multiplication.

Important parts in this paper are summarised as follows

1) A scalable approximate multiplier which finds leading 1's in the bit terms to perform truncation in order to improve multiplication operation's accuracy
2) Truncation(t) and rounding(h) parameters to decide power consumption, delay and accuracy
3) Circuit level design of the Approximate multiplier
4) Image processing application in MATLAB software

2 Related Work

We discuss some research efforts to develop estimated multipliers in this section.

A Booth multiplier is studied. Thus, approximate encoders can be used to generate partial products to reduce the complexity [6]. In [7] the partial products were generated and accumulated about by an approximate radix-4 booth multiplier. In addition, in the [8] radix-8 booth multiplier, which produced the least significant triple-multiplicand bits using approximate adjusters, was proposed earlier.

Then an exact Wallace tree multiplier is analysed as shown in Figure 1. The delay for a N X N Wallace tree multiplier is reduced to log2N [9]. Hence it is faster than an exact multiplier but making the multiplier much complex thereby taking a huge computational time reducing its efficiency. Here during reduction process, each row were grouped into three set numbers. (3,2) counters are reduced to (2,2) sets [10]. The RTL schematic diagram is obtained as in Figure 2.

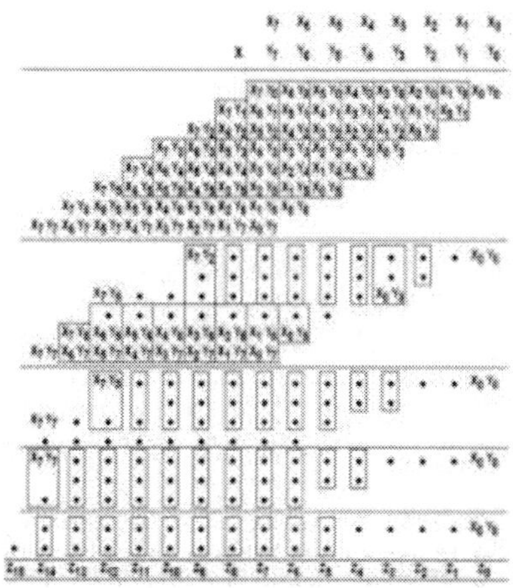

Figure 1. Multiplication products by Wallace multiplier

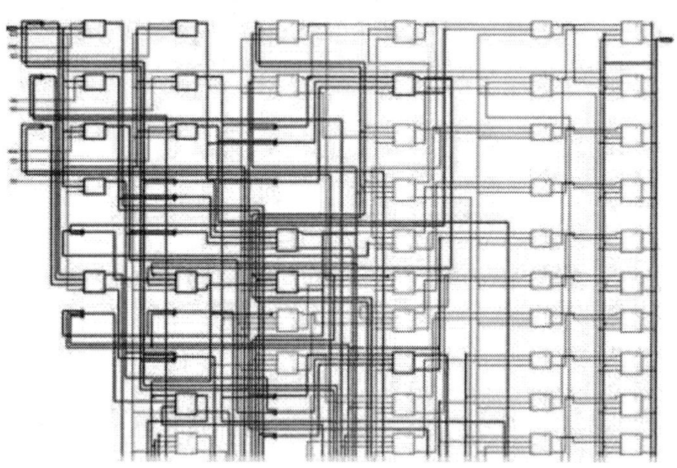

Figure 2. RTL schematic of existing Wallace multiplier

3 Proposed Approximate Multiplier

3.1 TAM

The multiplying result of A and B is calculated as

$$A\times B = 2^{kA+kB} \times D_A \times D_B . \quad (1)$$

k is the position of the leading 1 bit .D_A and D_B widths are equal to that of A and B for the exact time and power of $D_A\times D_B$. The remaining part shows the fractional portion of D in

$$R= D- 1. \quad (2)$$

The range - 0,0–1.0 is divided into T equal halves where T is given by

$$T = 2^h \quad (3)$$

in which h indicates the positive integer(an arbitrary value) which is there in our design parameters. The duration of every section is naturally equivalent to 1/T.

In order to find Y_{APX}, only h most important Y bits must be considered. If, for instance [11], S 4(h-2) is zero in two of the most significant bits of Y, this means $0 < Y < 1/4$. So, $1/8= (0.001)_2$ is chosen as Y_{APX}. If $2/4< Y < 3/4$, YAPX is thus roughly approximate as $5/8= (0.101)_2$, then Y_{APX} is the highest bits of Y.

$$A \times B = 2^{kA+kB} \times(1 + Y_A + Y_B + Y_A \times Y_B) \quad (4)$$

Its approximate is written as

$$A \times B \approx 2^{kA+kB} \times (1 + Y_A + Y_B + (Y_A)_{APX} \times (Y_B)_{APX}) \quad (5)$$

"1" in the term $1+(Y_A)_t+(Y_B)_t+(Y_A)_{APX}X\ (Y_B)_{APX}$ is added as shown in Figure 3. The number of partial product terms in a 16-bit multiplier is 256, which must be totalled in order to produce the final result while only 31 of the partial products are maintained in the proposed process.

Finally, for non-signed operands, the proposed multiplication approach is feasible.

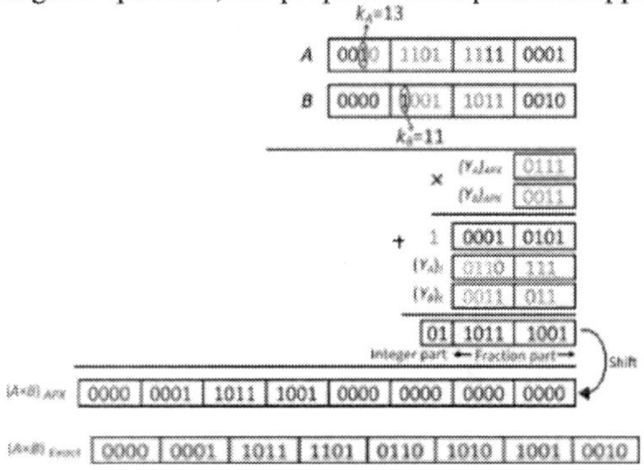

Figure 3. Example of Truncated 16 bit approximate multiplier (3, 7).

A = 11761 and B =2482.

Approximate result $[(A\times B)_{APX}]$ =28 901 376 .

Exact result $[(A\times B)_{Exact}]$ = 29 190 802.

Absolute error = 289 426 which is about 0.99% of the exact output (the error is less than 1% in this case).

The precision of the approach proposed is dependent on the validity of t and h and the RTL schematic is shown in Figure 4.

Figure 4. RTL schematic of truncated approximate multiplier

4 Implementation

The approximate multiplier block consists of two inputs A and B which are fed to the absolute approximate unit [12]. This block inverts the input value if it is high. So many LUT's and gates are formed up for the circuit design. Then these two outputs are given as input to leading 1's detector and truncation unit. Leading 1's detector output is given to truncation and shifter unit through arithmetic unit and then the multiplied approximate result is obtained. The RTL schematic is given in Figure 5.

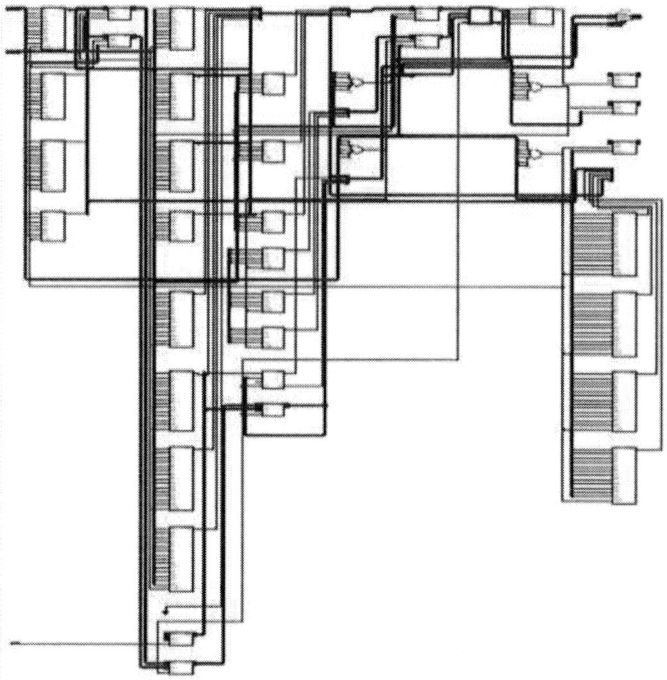

Figure 5. RTL schematic of truncated mux based approximate multiplier

5 Results and Accuracy Analysis

5.1 Design Parameters Extraction

5.1.1 8 Bit Multipliers

The use of this approximate multiplier can increase by up to 32%, 77%, 84%, and 90% in terms of area, power, speed and delay.

5.1.2 16 Bit Multipliers

The area is 72% smaller than that of Booth multiplier. With an efficiency value of 90%, this structure improves the multiplier speed by approximately 18% while consuming 86% less energy, while it occupies 70% less of the area than the exact Wallace multiplier.

5.1.3 32 Bit Multipliers

The use of the truncated muxed approximate multiplier structure results in an improvement in the latency, power, space, energy, EDP, and PDA respectively of approximately 94%, 85%,95% and 96%.

The Figure 6, Figure 7 and Figure 8 shows the outputs of proposed multiplier and other multipliers.

The multiplier's parameters like area, power, speed are compared with some other existing approximate multipliers in this section as shown in table 1.

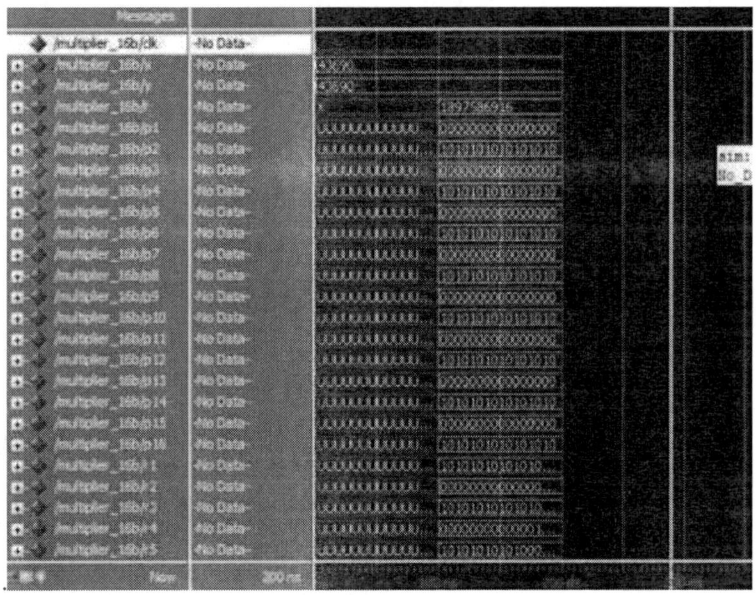

Figure 6. Output result for Wallace multiplier

IOP Conf. Series: Materials Science and Engineering 1059 (2021) 012025 doi:10.1088/1757-899X/1059/1/012025

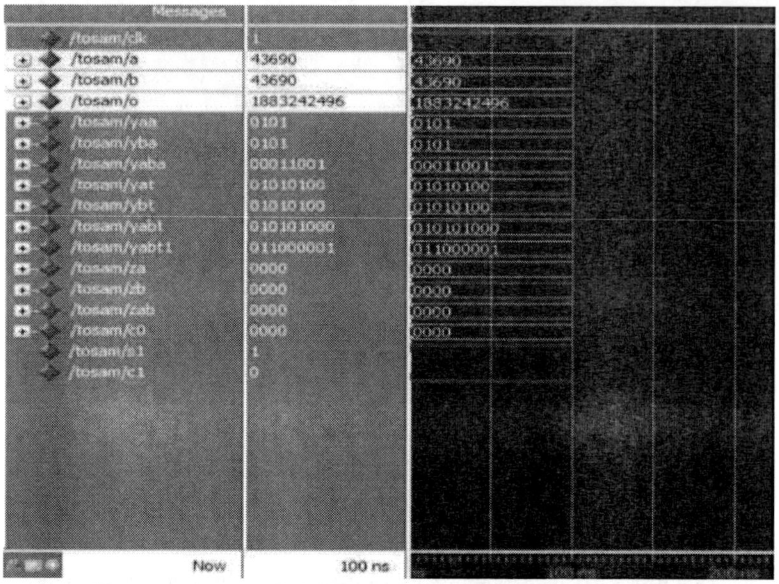

Figure 7. Output result for Truncated approximate multiplier

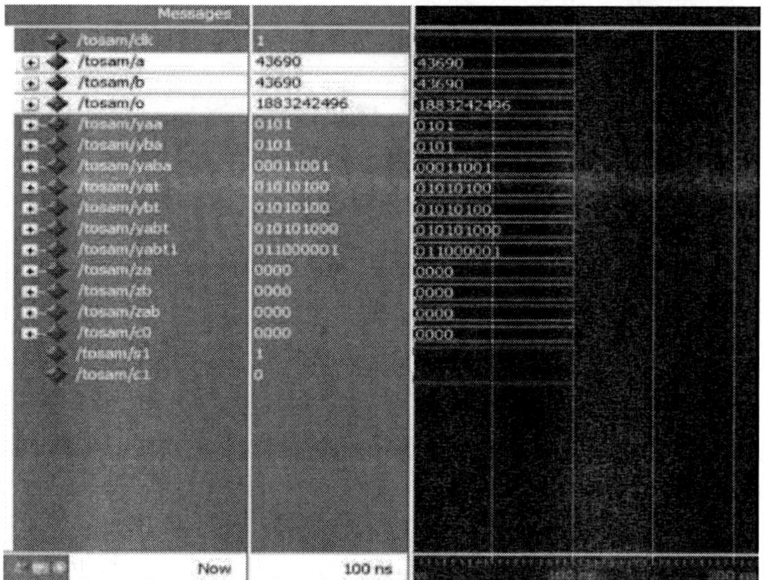

Figure 8. Output result for truncated mux based approximate multiplier

Table 1. Comparison of area and power consumption for the proposed multiplier with other multipliers

MULTIPLIER	AREA	POWER(mW)
Wallace	20361	184
Tuncated Apprx. Multiplier	3656	83
Modified Truncated Apprx. Multiplier	2735	76

IOP Conf. Series: Materials Science and Engineering 1059 (2021) 012025 doi:10.1088/1757-899X/1059/1/012025

6 Applications

In this section the results are analysed for the proposed algorithm with the help of MATLAB. The scalability of the multiplier is examined by modifying the length of the bits and measuring the relative mistake of the multiplier. The efficiency of the proposed truncated muxed approximate multiplier is compared with other existing exact multipliers. With 8bit multipliers, 65 536 numbers have been used to extract accuracy [13].

6.1 Image Processing Application

AC is a solution to enhance computation efficiency in image processing applications, as mentioned previously. We have used this in different applications of image processing, such as sharpening and JPEG compression, to evaluate the efficiency of the proposed multipliers. There have been a number of benchmarks in this section, including Baboon and Lena .The PSNR ratio and the Structural Similarity (SSIM) of approximate results were extracted using MATLAB simulations in the sharpening application compared with precise output images.

Furthermore, the Multiplier and Accumulator (MAC) module was installed to implement a sharper unit and the MAC energy consumption was measured for various non-signed 8-bit multipliers.

The output in the below Fig 9 shows that the difference between the MATLAB output and the VLSI output is almost negligible thus increasing the efficiency of the proposed multiplier structure. A very good signal to noise ratio and mean square error is obtained is compared in table2.

Figure 9. Application in Image multiplication process using MATLAB.

Table 2. Comparison table of PSNR and MSE for the multipliers

MULTIPLIER	PSNR	MSE
Wallace	48.1309	1.0000
Tuncated Apprx. Multiplier	9.2893	7.6587e+03
Modified Truncated Apprx. Multiplier	9.2102	7.7994e+03

7 Conclusion

Thus in this paper the input operands were truncated to 2 different lengths: t and h. So in order to maintain the image quality even though compressing the image this multiplier can be used. In terms of speed, area and energy the proposed multiplier could be employed. When comparing the Wallace multiplier with the proposed multiplier, the 32-bit variant of the proposed has an improvement in power consumption of about 92%. The multiplication latency and power usage are increased and decreased by 4% –41% and 89% –97% respectively, in comparison with that of the other multipliers. The area, delay, speed and the power consumption of the proposed multiplier is increased as the multiplier width increases in comparison to that of the other multiplier. So in image processing and in JPEG, MPEG compression applications this multiplier with high accuracy is highly advantageous and is always recommended.

References

[1] Narayanamoorthy S, Moghaddam H A, Liu Z, Park T, and Kim N S 2015 Energy-efficient approximate multiplication for digital signal processing and classification applications *IEEE Trans. Very Large Scale Integr. (VLSI) Syst.* vol. **23** no. 6 pp. 1180–1184.

[2] Vahdat S, Kamal M, Afzali-Kusha A and Pedram M 2017 LETAM: A low energy truncation-based approximate multiplier *Comput. Elect. Eng.* vol. **63** pp. 1–17.

[3] Pedram M 2017 RoBa multiplier: A rounding-based approximate multiplier for high-speed yet energy- efficient digital signal processing, *IEEE Trans. Very Large Scale Integr. (VLSI) Syst.*, vol. **25** no. 2 pp. 393–401.

[4] Leon V, Zervakis G, Soudris D and Pekmestzi K 2018 Approximate hybrid high radix encoding for energy-efficient inexact multipliers, *IEEE Trans. Very Large Scale Integr. (VLSI) Syst.*, vol. **26** no. 3 pp. 421–430.

[5] Ko H J and Hsiao S F 2011 Design and application of faithfully rounded and truncated multipliers with combined deletion, reduction, truncation, and rounding, *IEEE Trans. Circuits Syst. II, Exp. Briefs* vol. **58** no. 5, pp. 304–308.

[6] Liu W, Qian L, Wang C, Jiang H, Han J and Lombardi F 2017 Design of approximate radix-4 booth multipliers for error-tolerant computing, *IEEE Trans. Comput.*, vol. 66 no. **8** pp. 1435– 1441.

[7] Venkatachalam S, Lee H J and Ko S B 2018 Power efficient approximate booth multiplier in *Proc.IEEE Int. Symp. Circuits Syst. (ISCAS)* Florence Italy pp. 1- 4.

[8] Jiang H, Han J, Qiao F and Lombardi F 2016 Approximate radix-8 booth multipliers for low power and high-performance operation, *IEEE Trans. Comput.*, vol. **65** no. 8 pp. 2638–2644.

[9] Combet M, Van Zonneveld H and Verbeek L 1965 Computation of the base two logarithm of binary numbers *IEEE Trans. Electron. Comput.* vol. EC-**14** no. 6 pp. 863–867.

[10] Mitchell J N 1962 Computer multiplication and division using binary logarithms *IRE Trans. Electron. Comput.* vol. EC-**11** no. 4 pp. 512–517.

[11] The USC-SIPI Image *Database*. Accessed: Jun. 2017 [Online]. Avail-able http://sipi.usc.edu/database

[12] Wang Z, Bovik A C, Sheikh H R and Simoncelli E P 2004 Image quality assessment: From error visibility to structural similarity *IEEE Trans. Image Process.* vol. **13** no. 4 pp. 600–612.

[13] Myler H R and Weeks A R 2009 *The Pocket Handbook of Image Process- ing Algorithms in C.* Englewood Cliffs NJ USA: Prentice-Hall.

Low Power Kogge Stone Adder Applied in FIR Filter using Cadance Tool

Asuvanti M A[1], Nirmal Kumar R[2], Dharani S[3], Ranjith Kumar T[4]
[1,3,4] PG Scholar,[2]Assistant Professor Level III Bannari Amman Institute of Technology, Sathyamangalam.

Corresponding Author: asuvanti.vd18@bitsathy.ac.in

Abstract: A maximum width configuration AT is accomplished by post truncation and direct truncation by using a maximum width adder-tree (AT) configuration. The full width Output AT will be truncated by one lower order bit each adder and the lower order bits of the final stage Adder output, when p = dlog2 Ne & N is the size of the Input-Vector, will be truncate in the case of a post-truncation. These two methods are not efficient in design. Carry look forward adder is used for the current process. In this article we propose a new method to obtain truncated feedback of the fixed-width tree type. A probabilistic approximation formula to account for the truncation error is provided. In the AT architecture suggested we have used Kogge stone adder to approximate the output with almost exactly the same precision as the AT fixed-width post-truncation, with the best accuracy in the current AT fixed-width. And in the FIR filter we enforce the current process. Walsh-Hadamard also noted that, based on its proposed fixed width architecture AT, transforming higher-textured images of higher PSNR and moderate textures with approximately the same PSNR that were obtained using current PSNR models, reconstructing images with higher high noise ratio (PSNR). The proposed design provides a further gain, much as with information, to optimise additional blocks that occur in an upstream adder tree (AT) complex architecture.
Keywords: Truncation; Kogge stone Adder; Filter; Peak Signal Noise Ratio; Dynamic architecture

1.0 Introduction

The architecture of the fixed-width tree (AT), using post truncation and direct truncation, is accomplished by full-width AT architecture. The output of maximum width of AT in direct-truncation will be truncated by one bit of each adder in a lower order and in case of pos-truncation, fpg will be truncated with less order bits of the final stage adder output where p = dlog2 Ne and N have the input vector dimension. The effective implementation of all these approaches is not feasible. Carry look ahead is used for the current system. A new proposal is introduced in this document to receive truncated feedback for the fixed-width adder tree architecture. In order to compensate for truncation error, a probabilistic estimation formulation is presented. We used the Kogge stone adder in our proposed fixed-width AT design to measure the output in approximately the same precision as the AT post-truncated, with the maximum accuracy of the current fixed-width AT.And we use the current FIR filtre system. Furthermore, the Walsh-Hadamard transform supported a suggested re-constitution of high texture images with higher PSNR and moderate textures with almost equivalent PSNR (PSNR) compared with those obtained using the current AT designs. In this situation, it was also noticed that the Walsh-Hadamard transform was supported by a fixed width AT architecture. The proposed architecture also gives the optimization of other blocks in a dynamic design upstream of the adder tree.

In recent years, several schemes for estimated addition measurement using ripple carry adder (RCA) have been proposed to save crucial delay in travel (CPD) and area. Approximate logic suggests the organically influenced lower section OR adder. It proposes four different types of approximate adder designs. For the rough computation of triple multiplicand without transportation, a rough 2-bit adder is proposed. These preliminary structures may be used in order to incorporate RCA with less time and a certain level of accuracy. These preliminary structures may be used in order to incorporate RCA with less time and a certain level of accuracy. The estimated RCA architecture can be used to achieve AT with set width after truncation. Therefore, an area delay efficient configuration does not have an estimated set AT width (APX-FX-AT).

The use of shifting operation is recommended for bit-level FL-AT optimisation for multi-constant multiplication (MCM). The use of the estimated adder of Gaussian filter for imprecise implementation for image processing applications is proposed to an optimal FL-AT system. We consider that the designed AT of is MCM-specific, and no current architecture deals with problems relevant to the fixed-width implementation of AT. Direct cutting and truncation approaches do not have an optimal FX-AT interface. This is observed. A specific approach is needed to improve effective, currently lacking FX-AT architecture in the literature. A successful FX-AT architecture definitely helps increase the performance of dedicated DSP algorithm-based VLSI systems.

More precious than ever is the advantage of using low-energy and low-energy design techniques. Now power is a major concern because personal computer equipment and wireless communication systems are remarkably growing and succeeding, requiring high-speed computing and complex functions with low power consumption. Implementation of incentives to reduce energy use varied. The goal is to provide battery life, proper weight and low packaging costs for small battery-powered applications such as cell phones. The goal is to reduce the power delivery of electronic device components to just a half the total energy distribution for compact high-performance computers such as laptops. Finally, the ultimate aim of energy efficiency is to minimize battery costs while maintaining long-term computer stability in devices that do not require high-performance batteries such as working stations. Room, performance, price and reliability were the key concerns of VLSI designers in the past. The most important elements of electricity are secondary. However, this has begun to improve in recent years and is seen as weighing rather than space and distance. This pattern is backed by other influences. Perhaps the main factors are the incredible performance and development of electronic hardware and wireless networking networks involving high-speed processing and sophisticated low-powered functions. The average power consumption is crucial to the design in these applications. The transfer of energy from the environmental point of view of smaller electronic devices, heat pumps into the house, power usage will be decreased and the global climate will suffer.

In most optical signal processors for real time use additives are the most critical components. Additives are also found in the geometric framework of other logical functions, such as the subtractors, multipliers, shifters, macs etc. Even today, Adder 's architectural research is looking to use energy-efficient devices for something interesting. The general survey [3, 5-9] on adder architecture has proved to be better suited.

2.0 Existing Technology

The computational model is unique to the range of DSP block hardware implementations. It is shown that 70% of energy is consumed by an inbuilt, reduced computer processor in the supply of data and instruction, and only 6% is consumed by arithmetic. If the whole system is taken with an estimated calculation in this case, there are no energy gains. In general purposes, programmable processors with no specific application are built. If the entire system is viewed with an estimated arithmetic in this case, there are no energy gains. Programmable processors with no application specific specialization are designed for general purpose applications. Therefore, owing to provisional programming, there may not be many programs that may withstand errors. It also renders processors for general purposes not appropriate for the use of the building blocks. This matter has been discussed already. In this article, we find implementation of error tolerance techniques such as image and video compression by application-specific integrated circuit systems. We are designed to achieve major power consumption gain, with low loss in performance efficiency, using estimated hardware for such applications. Some works focused on low-energy programming, algorithms and architecture level computer-based computation include algorithmic noise (ANT), meaning drived computation (SDC), and non-uniform scaling voltage (VSS). The basic VOS principle, along with additional circuits to fix or limit the resulting errors, are used in all these techniques. In, it proposes a fast but "inaccurate" adder. This uses 2 x 2, incorrect multiplier blocks arising from Karnaugh map simplification, which is proposed to create an efficient multiplier architecture. This paper examines the reduction of logical complexity with Karnaugh maps. Also proposed logical complicated reductions through simplification of the Karnaugh map by Shin and Gupta and Phillips et al. Other works focusing on the reduction of logical complexity in the door are. Many methods use algorithm-level complexity reduction to fulfill real-time energy requirements.

In LSBs, we only use approximate FA cells, while the MSBs are using precise FA cells. Consequently, the errors made by VOS are much greater in isofrequency as compared with the hypothetical supplements proposed. Since truncations are a renowned method for easing voltage scaling, the efficiency of the estimated suppliers proposed has been compared to truncated suppliers. Our medium error expressions indicate the supremacy of the suggested truncation approximations. In addition, higher order compressors are used for aggregation of partial products in a variety of tree multipliers (use hold save trees).

3.0 PROPOSED SYSTEM

Two single binary digits A and B are added to the half adder. The outputs are two; Sum (S) and C. This transmitting signal is the result of a multi-digit leakage into the next digit. The cumulative value is 2C + S. The simplest half additional design, shown on the right side, includes an XOR door for S and an AND door for C. The logic for the boolean number (in this case S) is A'B+AB', and the logic for the boolean number (C) is AB. Two half additionals may be combined to create a complete adder with an OR gate for combining their output. The half integrates two input bits and produces a carrying value and a sum of the two outputs. Elements and add-bits are considered the input variables of a half adder. Amount and hold are the output variables. Figure 1 is a half-adder block diagram and figure 2 is a half-adder chain diagram.A full adder incorporates binary numbers and accounts for both the in and out values. A 1-bit complete extension introduces three single-bit digits, sometimes written as A, B and Cin; A and B are the operands of the previous less important level, and Cin is a little more detailed. In a cascade of add-ons that incorporate 8, 16, 32, etc. bit binary numbers, the complete addition is typically a part. A two-bit output is given by the circuit. The output and the balance of the Cout and S signals, where the sum of 2Cout + S is the same.

TABLE 1 Full Adder Truth Table

TRUTH TABLE				
A	**B**	**Cin**	**SUM**	**Cout**
0	0	0	0	0
0	0	1	1	0
0	1	0	1	0
0	1	1	0	1
1	0	0	1	0
1	0	1	0	1
1	1	0	0	1
1	1	1	1	1

By modifying the correct logic in this specification, the final OR gate before output may be superseded by an XOR gate. Having two gates are simpler to use where a circuit has just one form gate for a pin using simple built in circuit circuits.

A full Adder can also be generated by connecting the input of a half Adder with an A and B, then connecting the outputs from each halved Adder and S as an input to the second half Adder and Cin as their other entry. The last digit of the second half adder is the second half adder. The completed product (Cout) is the complete adder and product of the OR Window. The vital way of finishing a complete adder goes all by XOR doors and finishes in total.At the transistor level we propose reducing the logical complexity. It is due when inserting the mirror (MA) circuit on the bit level. We have developed unreliable yet simpler arithmetic units that save extra power than traditional low-power architecture techniques. The reduced logical complexity of the new numerical units is due to this. Please note that the average arithmetic units include not only a minimal number of transistors,

but also that the efficiency of the internal nodes is decreased substantially. Reduction of uncertainty leads to two separate methods of rising control. The effect of smaller equipment was a significant drop in shifted capacitance and leakage. Second, the reduction in complexity often leads to shorter critical travel, making it easier to reduce voltage without timing errors. In short, our research is substantially different from other experiments (SDC, ANT and non-single VOS), because the implementation of error tolerance takes a particular approach. Our target is a simple, approachable logic for low-performance architecture. As the DSP blocks mainly consist of addressees and multipliers, we propose some additional adders, which can be used easily within these blocks.
By offering two simpler MA copies, we expand our article. In addition, we propose to use an approximately adder for measuring the "price" of a DSP unit. We also suggest a technique for using provisional adders to achieve optimum power savings according to a strict cap on efficiency.

3.1 Truncation and Bias Estimation Formula
To describe the trunking system, N-point input vector bits are arranged into a matrix (N v) size A, so that the I row of bits is the I input fxig x and w bit width. In Figure we show bit-matrix A for N = 8 and w = 8. 4. 4. The bits of A and carrying bits are inserted in a column to measure one bit of AT output, and bits of FL-AT output fyg from w columns of A are measured in lower order. The remaining p top bits of fyg are based on the carrying bits produced from the w-th column of A. Figure FL-AT structure for N = 8 and w = 8. 5(a) where the RCA is used to execute each adder.
FX-AT-PT and FX-AT-DT structures are shown in Figure. 4. 5(b) Fig. 5(b). 4. 5(c), each. 5(c). FX-AT-PT and FX-AT-DT both do not have an effective FXAT interface for area delay. A different type of truncation method and error compensation scheme are discussed here in order to address this issue.

					MSP				LSP	
10	9	8	7	6	5	4	3	2	1	0
			x_{17}	x_{16}	x_{15}	x_{14}	x_{13}	x_{12}	x_{11}	x_{10}
			x_{27}	x_{26}	x_{25}	x_{24}	x_{23}	x_{22}	x_{21}	x_{20}
			x_{37}	x_{36}	x_{35}	x_{34}	x_{33}	x_{32}	x_{31}	x_{30}
			x_{47}	x_{46}	x_{45}	x_{44}	x_{43}	x_{42}	x_{41}	x_{40}
			x_{57}	x_{56}	x_{55}	x_{54}	x_{53}	x_{52}	x_{51}	x_{50}
			x_{67}	x_{66}	x_{65}	x_{64}	x_{63}	x_{62}	x_{61}	x_{60}
			x_{77}	x_{76}	x_{75}	x_{74}	x_{73}	x_{72}	x_{71}	x_{70}
			x_{87}	x_{86}	x_{85}	x_{84}	x_{83}	x_{82}	x_{81}	x_{80}
y_0={y_{010}	y_{09}	y_{08}	y_{07}	y_{06}	y_{05}	y_{04}	y_{03}	y_{02}	y_{01}	y_{00} }
y_1={y_{17}	y_{16}	y_{15}	y_{14}	y_{13}	y_{12}	y_{11}	y_{10}	0	0	0 }

Figure .1. Input bit-matrix A of adder tree for N = 8 and w = 8. y0 and y1, respectively, represents the output of FL-AT and FX-AT-PT

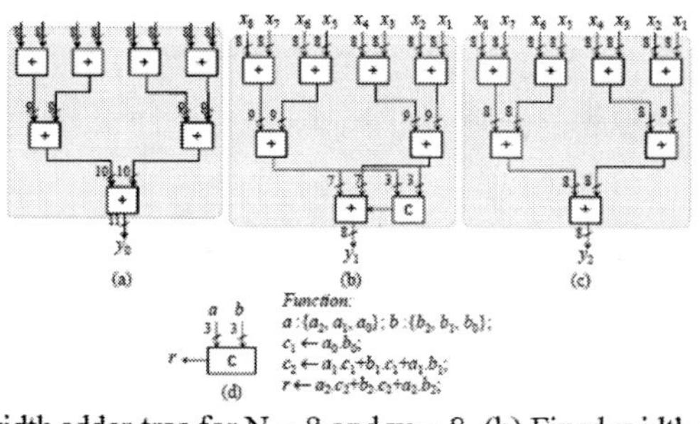

Figure .2.(a) Full-width adder-tree for N = 8 and w = 8. (b) Fixed-width post-truncated adder-tree (FX-AT-PT) for N = 8 and w = 8. (c) Fixed- width with 1-bit direct-truncated adder-tree (FX-AT-DT) for N = 8 and w= 8. (d) Function of carry-generator used in final stage of FX-AT-PT

The bit matrix A is divided into two parts, called "MSP" and "Less Significant Part" (LSP). The columns of the lower p of the A forms the LSP and the column of the upper p form the MSP. In order to design a Fixed width AT which is addressed in the next section, LSP columns are partially or entirely cut down.

3.1.1 Truncated Fixed – width AdderTree

The MSP of the vector of the input bit is only used to measure the value of the proposed TFX-AT. A significant quantity of error is entered in the AT output due to LSP truncation. Regarding error compensation, a fixed bias for the truncated LSP is applied to the MSP. The TFX-AT production is shown as:

$$y = MSP + 2^p \cdot \sigma \quad (1)$$

where, the term (2p.σ) corresponds to the LSP and (σ) represents the estimated bias. The expected value of the LSP is calculated using the formula:

$$E[LSP] = N \sum_{i=1}^{p} \left(\frac{1}{2}\right) . 2^{p-i} = \left(\frac{N}{2}\right) \cdot 2^p (1 - 2^{-p}) \quad (2)$$

The quantized value is calculated as

$$\sigma = round\left[\left(\frac{N}{2}\right)(1 - 2^{-p})\right] \approx \left(\frac{N}{2}\right) \quad (3)$$

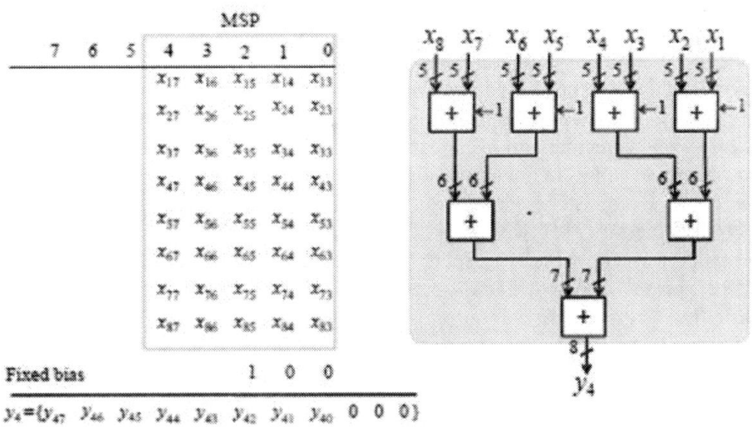

Figure. 2(a) Input bit-matrix of proposed truncated fixed-width adder-tree (TFX-AT) with fixed-bias for error-compensation. (b) Structure of proposed TFX-AT design

Using (3) the fixed-bias for N = 8 and 16 is found to be {σ=4 and 8}. The input bit-matrix ofTFX-AT with fixed bias is shown in Fig.2(a) for N = 8 and ω = 8. The binary values of σ=4 is added to the least significant column of MSP for error-compensation. Structure of proposed TFX-AT with fixed-bias is shown in Fig.2(b).

3.1.2 Proposed Improved Truncated Fixed-widthAdder- Tree

The LSP of A is further divided into two parts as major part and minor part to estimate more exactly the bias of the truncated part. The main part (MJP) is the most important column of the LSP, while the other columns (p 1) below are the minimal part (MNP). The partiality is measured in this case with LSP's MJP and MNP as:

$\sigma = \sigma_{major} + \sigma_{minor}$

where, $\sigma_{major}$and $\sigma_{minor}$are the estimated bias of MJP and MNPofLSP.$\sigma_{major}$is estimated accurately using the actual signal value of MJPwherethe$\sigma_{minor}$is estimated usingtheprobabilistic approach. The quantized value ofσ_{major} is estimated using the relation,

$$E[MJP] = \lfloor \sum_{i=1}^{N} 2^{p-1} x_{i,p-1} \rfloor = 2^p \lfloor \sum_{i=1}^{N} \left(\frac{x_{i,p-1}}{2}\right) \rfloor$$

$$\sigma_{major} = \lfloor \sum_{i=1}^{N} \left(\frac{x_{i,p-1}}{2}\right) \rfloor$$

(5)

The quantizedvalueof σ_{minor} is estimated as:

$$E[MNP] = \left(\frac{N}{4}\right) \cdot 2^p (1 - 2^{-p+1})$$

$$\sigma_{minor} = round\left[\left(\frac{N}{4}\right)(1 - 2^{-p+1})\right] \approx \left(\frac{N}{4}\right) \quad (6)$$

The suggested truncated fixed-width adder-tree input-bit matrix (ITFX-AT) "for N = 8 and = 8 is introduced. This µmajor logic block measured the bits {c0;c1;c2;c3;c4;c5;c6}. The logic block is used in this sense. These carry bits and the set bias corresponding to the smallest MSP column are applied.

3.2 Kogge Stone Adder

Adder is one of the automatic system's core elements. This is used both in microprocessing devices and in other signals like filtration and convolution. Basic blocks like image and video decoding, object recognition, modulation, etc are also present in large signal processing equipment. It allows other technologies to catch up by speeding nutrients in these systems. About the deployment of VLSI hardware, a variety of different architectural modes have been identified, each with its own strengths and weaknesses. The principal aim of this article is to make sure of its efficiency, power and ranking compared with other adder architectures, including username, look forward and the normal library faulty adder, using Kogge stone adder technology. This paper offers the fastest addition architecture. The aim is to provide the VLSI with a fast adder that can be applied and executed in larger systems. Some research performed by the KSA includes the introduction of a 32-bit KSA adaptive power ratio. However, for general supplementary operations and only partly idle loops, technology is not successful. In the analysis is carried out on FPGAs with the KSA, but in VLSI it is not as successful as the implementation is primarily gate-level. The work in [3] is very similar to that of the paper, in which the KSA is carried out using 0. Technology of 18µm method. The emphasis, however, was primarily on simulation and temperature analysis in Monte-Carlo rather than on benchmarking other additional architectures. The paper suggests that the standard KSA addder and benchmark should be implemented in VLSI against some of the common and frequently used architectures.

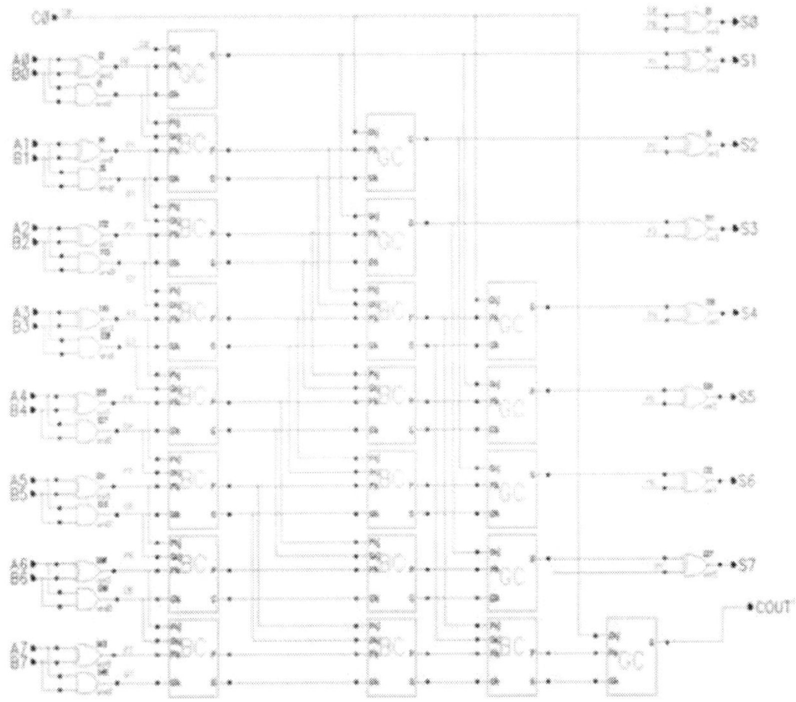

Figure.3 8 Bit Kogge stone Adder

Table 2 Parameters Comparison

	Area	Power	Delay
Existing	3656	86	37.53
Proposed	2816	74	20.75

4.0 FIR Filter

For almost every form of wireless frequency response, a FIR filter is used. Such filters are created usually by using a compressor, external structures, and a series of delays to produce the filter output. The following figure shows the simple FIR filters of N frequency. To delay the test, input samples are used. The values of hk are the components of multiplication. The o / p is then determined by the corresponding coefficients and the sum of all samples postponed. It is the process to decide the length and coefficients of the filter. The filter design is defined. The aim is to set the parameters to provide results from running the filter to the correct parameters such as a stopband and passband. Many engineers use filter architecture for MATLAB tools. Typically, their responses to different frequency components that have the I / P signal are described by filters. The responses to the filters are categorized according to frequencies such as stop band, pass band and transition band in three different forms. The passband 's reaction is the influence of the filter on the frequency components which are normally unimpacted. At a particular point of view, frequencies in a filter stop band are reduced. The transition band denotes the middle frequencies, which might be the, but which are not completely removed from the o / p signal.

In this article, the following can be defined. 1) We propose a decrease in functional complexity of transistor speeds as an alternate approach for DSP implementations. (2) They illustrate how conceptual complexity is reduced by reducing a traditional MA cell with the number of transistors and internal capacities. In view of this, we suggest the condensation of 3 separate versions of the MA to ensure minimum errors on the whole truth table for adder (FA). (3) They are suggesting multiple imprecise or provisional multibit adders to be used as building blocks for the DSP framework using a simplificated FA cell version. For retaining a decent rate of performance, we only use approximate FA cells in the lowest bits (LSB). In specific, we concentrate on more systems that use their essential building blocks FA cells. We used estimated Carry Save Adders (CSAs) in 4:2 and 8:2 compressor architectures (Section III). The mean error in the rib carrying adder (RCA) output is eliminated by the approximate LSB number from the mathematical functions. 4) VOS is a popular method for significant energy consumption improvement. However, in the most important bits (MSBs) the VOS contributes to delays. This could result in significant errors in the corresponding outputs and significantly harm the application's output efficiency.

We use only provisional FA cells in LSBs while precision FA cells are used by MSBs. Consequently, isofrequencies will result in VOS errors significantly greater than the approximated additional expected. As truncation is a well-known voltage scaling facilitator technique, the performance of the approximate supplements proposed is often contrasted with truncated suppliers. Therefore, our method is also useful for constructing estimated tree multipliers commonly used in DSP systems. Typically, we can extend our approach to every FA-built arithmetic chain. 5) We present designs for video compression algorithms, output and power dissipation efficiency check for predicted arches, with the suggested approximate arithmetic units. 6) Suggest that the DSP block performance rate can be statistically defined by rough adders. We do have simple math models to predict the amount of voltage and power usage these calculated adders will raise. On the basis of these models, we create a design problem to achieve maximum power savings under some cost constraints.7) Finally, with two examples, the discrete cosine transform (DCT) and the finite impulse response (FIR) filter we demonstrate the optimisation methodology.

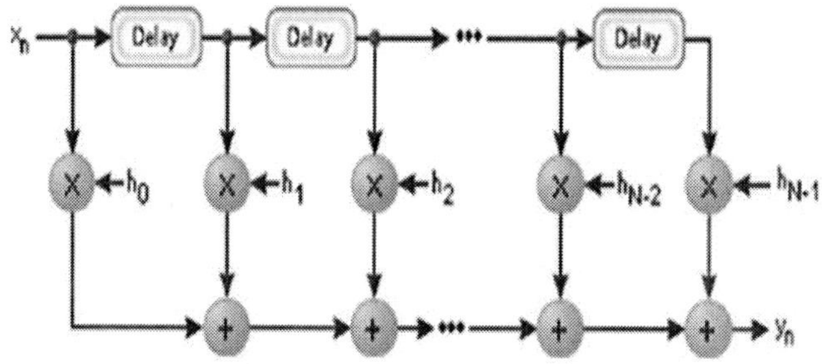

Figure 4 FIR Filter

5.0 Conclusion

In this paper, we proposed truncated fixed width adder and improved truncated fixed width adder using Kogge stone adder. This reduces area, power and delay with 13% from the existing system. Therefore, the results can be most widely utilized in real time applications such as Filters, DSP Processors, Multimedia Applications etc. The same is applied in FIR Filter in this process and the expected output is gained.

REFERENCES

[1] Efficient Design for Fixed-Width Adder-Tree Basant Kumar Mohanty, Senior Member,IEEE
[2] B. K. Mohanty and V. Tiwari, 2014 Modified probabilistic estimation bias formulation for hardware efficient fixed-width Booth multiplier, Circuits, Systems and Signal Processing, Springer, **33**, 3981– 3994.
[3] H. R. Mahdiani, A. Ahmadi, S. M. Fakhraie and C. Lucas,2010 Bioinspired imprecise computational blocks for efficient VLSI implementation of soft-computing applications", IEEE Transactions on Circuits and Systems-I, Regular Papers, **57,** 850–862.
[4] V. Gupta, D. Mahapatra, A. Raghunathan, and K. Roy,2013 Low-power digital signal processing using approximate adders, IEEE Transactions on Computer Aided Design of Integrated Circuits and Systems, **32**, 124–137.
[5] J. Liang, J. Han and F. Lombardi, 2013 New metrics for the reliability of approximation and probabilistic adders, IEEE Transactions on Computers, **62**, 1760–1771.
[6] H. Jiang, J. Han, F. Qiao and F. Lambardi, 2016 Approximate radix-8 Booth multipliers for low-power and high- performance operations", IEEE Transactions on Computers **65**, 2638–2644.
[7] Y. Pan and P. K. Meher,2014, Bit-level optimization of adder- trees for multiplie constant

multiplications for effcient FIR filter implementation", IEEE Transactions on Circuits and Systems-I, Regular Papers, **61**, 455–462

[8] R. O. Julio, L. B. Soares, E. A. C. Costa and S. Bampi, 2015 Energy-efficient Gaussian filter for image processing using approximate adder", In. Proc. International Conference on Electronics, Circuits and Systems, 2015, 450-453,.

[9] J.-P. Wang, S.-R. Kuang, and S.-C. Liang, 2011 "High-accuracy fixedwidth modified booth multipliers for lossy applications," IEEE Transactions on VLSI Systems, **19**, 52– 60.

[10] C.-Y. Li, Y.-H. Chen, T.-Y. Chang, and J.-N. Chen, 2011, A probabilistic estimation bias circuit for fixed-width booth multiplier and its dct applications," IEEE Transactions on Circuits and Systems II, **58**, 215–219.

[11] Y.-H. Chen, C.-Y. Li, and T.-Y. Chang, 2011, Area-effective and power efficient fixed-width booth multipliers using generalized probabilistic estimation bias," IEEE Journal on Emerging and Selected Topics in Circuits and Systems, **1**, 277–288.

[12] Y.-H. Chen and T.-Y. Chang, 2012 A high-accuracy adaptive conditional-probability estimator for fixed-width booth multipliers," IEEE Transactions on Circuits and Systems , **59**, 594–603.

[13] P. Kulkarni, P. Gupta, and M. Ercegovac, 2011 ,Trading accuracy for power with an underdesigned multiplier architecture,24th International Conference on VLSI Design, .346–351.

[14] C.-H. Lin and C. Lin, 2013, High accuracy approximate multiplier with error correction in ICCD, IEEE, 2013, pp.33.

[15] B. Millar, P. E. Madrid, and E. Swartzlander,1992 "A fast hybrid multiplier combining booth and wallace/dadda algorithms," in Proceedings of the 35th Midwest Symposium on Circuits and Systems, 158–165.

[16] J. Hidalgo, V. Moreno-Vergara, O. Oballe, A. Daza, M. Mart´ınVazquez, and A. Gago, 1998 "A radix-8 multiplier unit design for spe- ´ cific purpose," in XIII Conference of Design of Circuits and Integrated Systems, **10**, 1535–1546.

[17] J. Liang, J. Han, and F. Lombardi, 2013,New metrics for the reliability of approximate and probabilistic adders," IEEE Transactions on Computers, **62**, 1760–1771.

[18] H. Jiang, J. Han, and F. Lombardi, 2015 ,A comparative review and evaluation of approximate adders," in Proceedings of the 25th edition on Great Lakes Symposium on VLSI. ACM, 343–348.

Implementation of Industrial Engineering concepts in Apparel Industry to improving Productivity and it's cost reduction

N Mekala [1], S D Sanju[2], V Thamaraiselvan[2] & M Kavya[2]
[1]Assistant professor ,Department of Fashion Technology, Bannari Amman Institute of Technology, Sathyamangalam , Erode, Tamilnadu, India.
[2]Department of Fashion Technology, Bannari Amman Institute of Technology, Sathyamangalam , Erode, Tamilnadu, India.
mekalan@bitsathy.ac.in

Abstract: In current scenario the fashion industry (industry 4.O) needs to satisfy the various needs of customer in the aspect of quantity as well as quality of the products. In traditional methods the apparel industry used basic sewing machines for all operations with the help of man power so on that time to introduce different styles in the apparel is very difficult. In recent days the automation as well as improved techniques known as industrial engineering concepts is applicable in the field to improve productivity as well as quality to overcome the competitive market. Manufacturing of garment involves number of operations in different departments. In apparel sector to improve productivity as well as quality one must concentrate on machine, method, material and man. So the industrial engineering concepts will cover and manage the all above said factors and help to improve the productivity as well as quality of the products that will satisfies the consumer needs. This paper is going to discuss about the existing (manufacturing) methods in apparel units and help them to improve productivity by applying the one of recent techniques such as Capacity Study, Work study, Time study, Operator Performance, Follow-ups ,Work-in-Progress and line balancing and they are already proven techniques to improve the productivity in different industries.

Key Words: Productivity, Cost reduction, Material & Machine management.

1. Introduction

Industrial engineering is basically related to the efficient use of machines, manpower, material and energy. The IE team is engaged in cutting, sewing and finishing work to reduce manpower and to increase the machine capacity to complete the work in progress [1]. Time research of the manufacturing process is measured to out the scope of improvement. The time it takes to complete each step in garment manufacturing is measured by averaging 20 cycles per process [2]. Setting the production in circular layout can improve work efficiency, productivity. It is done by demand forecasting, develop new designs, develop layouts by using different quality tools and blend product and seasoning development. An industrial engineer's work is to manage the efficiency, wages and income calculation, CNC machine operation, scheduling, capacity analysis, production line balance, production planning [3]. Calculation of NPT deals with the social, mathematical and physical aspects of integrated production systems. Two important attributes must be considered, one is the possible standard method and the other one is the considerable method. Time research takes time to record the individual capabilities of each employee and production line using stop watch. Work metrics will record the time (start and end of the operation) and work rate in the process [4]. Then they analyze and record the results of the job measurements to implement the job performance level.

Through time study research, each worker's actual personal ability in process line is calculated. There are two processes in manufacturing a garment, one is completion of whole garment by an individual operator (tailor) but there will be not proper or standard measurements will follow. The other one is a group of people joined together to complete a whole garment (industry people) they follow standard size charts for preparation [5]. The concepts outlined here are the most an important factors in improving industrial engineering methods, such as method engineering, work research, capacity research, production line planning and other operation management systems, which will ultimately high profits and develop the industry. Working environment modular design of garment manufacturing is based on expanded elasticity and flexibility modular product are used to increase the economies of scale of end user value. Combines flexibility, modularity, design adaptability to provide greater freedom to adjust and adapt to change for improvement productivity, this research will improve various productivity according to the government's initiative, the garment industry will become part of manufacturing.

2 Materials and methods

2.1 Material

Men's formal shirt has been used as material. This is the very common costume mostly used by men in various scenarios in south India locations. So, the most of the manufactures focused on Shirt manufacturing process. So it will help the industry to implement the IE concepts in Shirt manufacturing process. [6].

2.2 Problem Identification:

In particular Apparel industry the productivity of the concern product getting delay. So that the problem has been identified through the method study and time study concepts belongs to the IE concepts. The following problems have been identified for lag or excess time required completing different operations of the product.1. Machine Breakages, 2.Improper training to the tailors (labour), 3. Improper Layout plan, 4.The poor knowledge on Machine handling and maintenance etc...

2.2.1 Solution to the problem

This project, have been analyzed that the major lag and defect occurs mainly in the production area. After cutting and numbering the fabrics are passed to sewing department, the major operations of sewing have been carried out manually. Production of the manufacturing unit always depends the skilled and trained labour, proper machine alignment with respect to the operations breakdown of the garment, Good working conditions of sewing machines; The above all factors are very important to save the time and also increase the productivity of the company. In traditional method the companies only relied on man power and machineries, they used to hard work to complete the order given by buyer not by smart work.

The Industrial engineering concepts will help the technicians to improve their work as well as productivity by doing smart work with the help of industrial engineering concepts. The Automation is the Key factor of Industry 4.0 and that can increase the productivity and can reduce the labour cost but the developing companies can't invest huge money on machineries so the IE concepts will help them to improve their production with no cost.

This work analyzed the problem behind the sewing department then finds the solution for them without affecting the production of the particular product. The following obstacles were found in the production area. The Sewing machines are not aligned in proper manner with respect to the style specifications of the particular mentioned garment due to this unwanted movement has taken place in the sewing line the production gets delay.[1]

To rectify this step by step operations has been listed out then identified the required good working conditioned machineries replaced to complete that particular operation. In these keeping machines in good condition through systematic maintenance the idle time can be consumed without any disturbance the productivity as well as quality of the garments can be achieved. [7]

To avoid machine breakages in sewing line the bottom level labours are equipped with proper maintenance like proper oiling, Needle change, proper threading & proper winding of bobbin case [2].In this case also the time study has been carried out before and after the training to the labour

The Unwanted materials from the line have been removed in the sewing area to avoid the unwanted movement. Handling of materials has been studied and improved in easy manner. New concepts like lean manufacturing, 5S management process are advised to implement inside the factory to avoid the wastages of time which affects the production. [3].So that the sewing operator cannot mesh up with other unwanted materials. They can stitch the operation easily and without delay time in searching the next material.

2.2.2 Time & Method Study In Industry

It is the technique used to record the time & method of performing certain activity to identify the time taken by the operator to complete the particular task in certain method. Time Study and Method study is the main techniques of Industrial engineering concepts which are needed to conduct to improve the productivity in the garment industry. So in this project we have been studied the methods as well as time to improve the production by reducing the time by implanting the necessary changes in sewing line.

To study the time study first the design needs to be analyzed with their specifications properly. Once the design analyzed properly then operation breakdown is needed for that particular style. After that the calculation of required machineries and operator efficiency need to be calculated. Then suitable layout needs to be planned according to the style of the garment. The time should be recorded for each and every change happened in the line to improve the productivity to identify the improvement that happened in the industry.

3 Result and discussion

After observation of method and time to complete the garment the following line balance concept has been introduced to reduce the idle time as well as improve the quality in production process. The construction process has been broken into so many steps and aligned the machinery with respect to the operation.

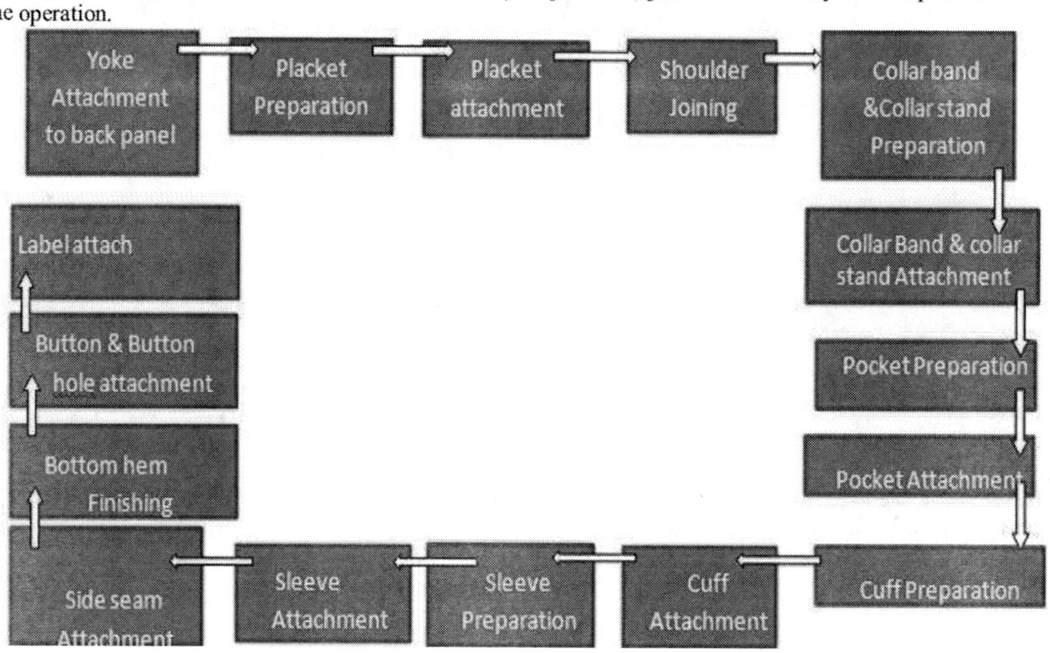

Flow process 1: Circular Line balancing System

By using circular line balancing system, the operation of the garment is able to complete soon with inspection table (or) machine located at the middle of the circle [11]. For difficult operations it is supposed to add an extra operator to that process so we can finish the garment quickly. Thus, the productivity will also increase. The Machineries has been arranged in circular manner to reduce the unwanted movement to carry the pieces one after another operation and the operators can handle the machine without delay.

The SAM value of each operation differ, So we can add one more machine in the operation which has less SAM value easily so the productivity will improve automatically. In addition to that we can introduce some work aids in the sewing machine to reduce the SAM value and it also helps to reduce the fatigue of labour so this will leads to improve work efficiency of the worker.

3.3 SAM calculation before and after implementation of time study method

SAM- Standard Allowance Minute. It is a measuring unit to calculate the efficiency. SAM= Basic time + Bundle allowance+ machine allowance

Table 1. SAM Calculation before and after Time Study

S.no	Operation list	Standard SAM Value (sec)	Industries SAM Value (sec)	After implementation of SAM value (sec)
1	Pinning to profile	0.234	0.459	0.345
2	Run stitch collar	0.219	0.382	0.285
3	Trim collar	0.285	0.340	0.312
4	Clip and turn collar	0.223	0.372	0.245
5	Crease collar	0.381	0.428	0.392
6	Top stitch collar	0.42	0.84	0.53
7	Run stitch collar band	0.291	0.465	0.301
8	Crease collar band	0.317	0.419	0.324
9	Insert collar in neck band	0.799	0.986	0.810
10	Turn and crease collar	0.452	0.580	0.462
11	Attach bias piece to cuff	0.874	0.910	0.882
12	Crease cuff	0.458	0.657	0.461
13	Hem cuff	0.501	0.744	0.522
14	R/S cuff	0.565	0.726	0.572
15	Trim cuff	0.36	0.52	0.42
16	Turn cuff	0.527	0.655	0.530
17	Crease cuff bottom	0.566	0.759	0.574
18	Attach placket to sleeve	0.795	0.993	0.813
19	Lock and make diamond	1.026	1.346	1.032
20	Sew pleats(4.no)	0.389	0.575	0.392
21	Hem right front	0.494	0.736	0.501
22	Attach front placket	0.603	0.880	0.612
23	Create pocket	0.542	0.799	0.553
24	Hem pocket mouth	0.28	0.53	0.35
25	Attach pocket to front	0.861	1.102	0.865
26	Attach brand label and tack loop	0.589	0.794	0.591

IOP Conf. Series: Materials Science and Engineering 1059 (2021) 012027 doi:10.1088/1757-899X/1059/1/012027

27	Crease patch pieces	0.236	0.657	0.243
28	Attach patch piece	0.333	0.696	0.356
29	Sew pleats	0.278	0.771	0.286
30	Attach yoke to back	0.475	0.825	0.479
31	Top stitch bake yoke	0.365	0.725	0.368
32	Join shoulder	0.64	0.91	0.72
33	Top stitch shoulder	0.656	0.840	0.676
34	Attach collar	0.535	0.819	0.542
35	Close collar with size label	1.01	1.42	1.06
36	Sleeve attach	0.862	1.109	0.867
37	Top stitch armhole	0.678	0.987	0.682
38	Top stitch side seam	1.036	1.435	1.039
39	Attach &close cuff	0.696	0.922	0.698
40	Top stitch cuff	0.524	0.792	0.527
41	Hem bottom	0.947	1.109	0.954
42	Total SAM	22.322	32.011	23.184
	Excess time consumed per garment with respect to industrial time	32.011-23.184= 8.827 seconds		

3.4 Outcome:

Operation Time (Before implementation of IE concepts (Time study) =32.011 – Operation time (After Implementing IE concepts) 23.184 = Reduction time: 8.827 seconds.
To manufacture one men's formal shirt the time has been reduced from 32.011 to 23.184 seconds, Obviously it states that production & productivity of that particular product will improve.
There are nearly 25 machines being used in the industry, for constructing a men's formal shirt it requires 18 machines.
Equation used to calculate the production:
Total machine productivity = (total no of output per day per line)/ no. of machines used .... [1]
By using equation [1], we get after implementation of time study concept
Daily machine productivity = working time per day* no. of sewing machines/total time per garment
= 480*18/32.011 =270
By using equation [1], we get after implementation of time study concept
Daily machine productivity = working time per day* no. of sewing machines/total time per garment
=480*20/23.184 = 414
Here we get 270 garments per day as a total productivity before time study implementation, 414-270=144garments were increased after using this method. Thus the productivity of constructing a garment is being increased.
The above chart shows the individual time required for each and every operation we can also able to calculate the target or lead time from this study concept 8.827 seconds has been reduced from the industrial SAM value [12].

4. Conclusion

The above research clearly states that the IE concepts are very effective tools to improve the productivity in the competitive fashion market. Current industry mostly looking forward of new machines which will reduce the man power as well improve the quality as well as production. The cost is very high to invest on machineries instead of that it is very easy to implement the IE concepts and improve the productivity. Anyhow if we invest on advanced machineries also we need to utilize very effectively to reduce the manufacturing cost of the product and productivity of the particular order. In this research it clearly states that by implementing the Industrial concepts, which can reduce the idle time of the particular sewing line as well as can improve the production of the product without cost reduction.

Reference

[1]. Khatun, M.M., 2013. Application of industrial engineering technique for better productivity in garments production. *International Journal of Science, Environment and Technology, **II***, 1361-1369.

[2]. Price, J.W., 1994. Industrial Engineering: Important for Management in Australia. In *Fourth International Conference on Engineering Management, 1994: Preprints* (59). Institution of Engineers, Australia.

[3]. Haque, A., 2009. Lead time management in the garment sector of Bangladesh: an avenues for survival and growth. *European Journal of Scientific Research*, ***33*(4)**, 617-629.

[4]. Kader, S. and Akter, M.M.K., 2014. Analysis of the factors affecting the lead time for export of readymade apparels from Bangladesh; proposals for strategic reduction of lead time. *European Scientific Journal*, ***10*(33).**

[5]. Basak, A., Seddiqe, M.I.S., Islam, M.R. and Akanda, M.O.F., 2015. Supply chain management in garments industry. *Global Journal of Management and Business Research*.

[6]. Babu, V.R., 2012. *Industrial engineering in apparel production*. Woodhead Publishing Limited.

[7]. Stevenson, W.J., 2015. *Operations management*. McGraw-hill.

[8]. Park, H. and Kincade, D.H., 2011. A historical review of environmental factors and business strategies for US apparel manufacturing industry 1973-2005. *Research Journal of Textile and Apparel*, ***15*(4)**, 102.

[9]. Patil, N.S., Rajkumar, S.S., Chandurkar, P.W. and Kolte, P.P., 2017. Minimization of Defects in Garment during Stitching. *International Journal on Textile Engineering and Processes*, ***3*(1)**, 24-29.

[10]. Chandurkar, P., Kakde, M. and Bhadane, A., 2015. Improve the productivity with help of industrial engineering techniques. *International Journal on Textile Engineering and Processes*, ***1*(4)**, 35-41.

[11]. Khatun, M.M., 2014. Effect of time and motion study on productivity in garment sector. *International Journal of Scientific & Engineering Research*, ***5*(5)**,825-83

[12]. Sheffi Yossi, Supply chain management under the threat of international terrorism, The International Journal of logistics management, Vol. **12(2)**, 2001, 1-11.

Development of Acoustic Absorption Composite with Application of Coir and Viscose

V Thamaraiselvan, N Mekala and G Dhanapriya
Faculty of Fashion Technology, Bannari Amman Institute of Technology, Sathyamangalam, Erode, Tamilnadu, India.
mekalan@bitsathy.ac.in

Abstract. In the current scenario the industry looking for good quality products at low cost as well as recyclable ones. The non-woven products, especially reclaimed nonwoven manufacturing cost is very low when compared to other fabric manufacturing cost and it can be recyclable & attractive too. The sound absorption property is very essential in recent years. So it is very advisable to manufacture the sound absorption material with the help of non-woven fabric. This study speaks about the improvement in sound absorption characteristics of the non-woven composite. To improve the acoustic property of the composite the natural fiber coir has been blended with recycled viscose fiber in six different ratios using a needle-punching technique. The six different (in %) blend ratios of coir and reclaimed viscose is 70:30; 60:40; 50:50: 40; 60: 30:70& 20:80.The composite has been introduced at a minimum of 200 grams so that adequate acoustic property is achievable. The composites are examined with GSM, Abrasion Test, Moisture test, and Acoustic coefficient. The GSM of the composite increases, when the viscose ratio increases in the sample. The abrasion-resistant decreases when the viscose increases. The moisture regains increases when the viscose content increases in the composite. The values of the sound absorption coefficient are high in sample 6 (Coir: reclaimed viscose, 20:80) when compared to other samples, it reflects that the reclaimed viscose has higher acoustic property than coir.
Keywords: Coir, Reclaimed viscose fiber, Non-woven, Acoustic property, Needle Punching machine.

1 Introduction:

Undesirable and potentially hazardous noises cause side effects, due to the wide scope of the present-day building structure and different procedures. The continued advancement of innovations has brought about the natural effect of commotion to turn into a significant concern. The expanding interest for quietness in building and assembly halls has animated the utilization and improvement of powerful methods for commotion decrease.

Sound absorption characteristics of nonwoven materials have been the subject of several works nowadays related to research. Many researchers have explained the dissipation mechanism of sound in fibrous materials. Faulkner proved that when sound passes through porous materials, acoustic energy loss takes place due to frictional losses, momentum losses, and temperature fluctuations. Experimental data indicate that the fiber mass, thickness, and porosity of the nonwoven fabrics seem to be the most influential factors in this respect. It was found that nonwovens with a higher proportion of fine fibers dissipate more sound. This causes more resistance due to frictional viscosity as a result of air vibration. It was also established that the sound absorption coefficient of the absorptive system incorporating with anisotropic nonwovens is marginally higher than the systems containing isotropic webs in their structure.

Nonwoven structures contain pores in the micron and submicron scale, making them the preferred medium for sound-insulating and sound-absorbing applications. Moreover, their sound absorbency is the principal controlling factor in environmental sound pollution. Nonwoven fabrics can serve as noise absorption elements in a wide range of applications including acoustic ceilings, noise-reducing quilts, and noise proof barriers.

Coir fiber from coconut husk is a vital rural waste in Malaysia. The acoustic ingestion co-efficient of the fiber is taken as a permeable material and it's considered in this paper for proceeding. Sound absorption characteristics of non-woven materials concerning fabric thickness have been thoroughly discussed in the literature. Two investigative models are followed to test the fiber's acoustic property. They are Delany– Bazley, and Biot– Allard is utilized for examination. Test estimations in impedance tubes are led to approve the scientific results. Results demonstrate that new coir fiber has a normal ingestion coefficient of 0.8 at f > 1360 Hz and 20 mm thickness [14]. Expanding the thickness has moved forward the sound ingestion in lower frequencies, having a similar normal ingestion coefficient at f > 578 Hz and 45 mm thickness. These results show a poor acoustic property in low frequencies.

Therefore, the fiber must be blended with another material to upgrade its qualities, like sound absorption, firmness, and combustibility. Consequently, different methodologies such as including air holes or punctured plates are to be utilized to enhance the acoustical properties of modern treated coir fiber [6]. Coir is blended here with viscose and made into a non-woven composite, and then tested for its acoustic properties.

2 Materials and Methods

2.1 Materials

2.1.1 Coir

Coconut coir fiber was utilized as the main raw material which is prepared from the coconut husk. The coir fiber is the thickest and most resistant of all commercial natural fibers. Low decomposition rate is the key advantage of making durable products. The visual analysis results and cross-section shows there are a lot of cavities inside the fiber, and roughly one-third of the bulk of fiber is filled by air. This entrapped air gives rise to the pronounced springiness (resilience) of the fiber, its buoyancy in water, and increases the time water takes to penetrate the fibers. The properties of coir are less affected by wet conditions than other hard fibers. The thickness of coir fiber limits the products made by coir being coarser and heavier.

2.2.2 Viscose

Viscose is one of the regenerated fibers in textile phenomena. Here the fibers are used from recycled viscose fiber. Viscose fibers are prepared from cellulose from wood pulp. Wood pulp cellulose is reacted with caustic soda, after a matured waiting period, the developing process will take place. During that time, depolymerization occurs and carbon disulphide will be added. It forms a yellow crumb, called Xanthate which can be easily dissolved in more caustic soda to give a viscous yellow solution.

2.2 Methodology

2.2.1Manufacturing non-woven

The fibres (Coir & Recycled viscose) need individualization before the non-woven fabric manufacturing process called carding. First, coir is processed and lay of two varying proportions are made. Then, viscose is processed and lay of two varying proportions are made. The two fiber lays are blended then applied with a needle punching process.

2.2 Sampling

The non-woven composite should be prepared about 200grams and this is the minimum thickness required to attain the acoustic properties of the particular application to attain the same minimum 30% of viscose needs to blend with coir fabric.

The samples (non-woven composites) are prepared with six different ratios of coir & viscose is 70:30; 60:40; 50:50; 40:60; 30:70; 20:80 respectively.

2.3 Material preparation

The pre-process of the fibers and the needle punching process are shown. The conversion of the fibre into the lay form using a miniature carding machine. The Carding machine operated with 0.61 feeder roller, 394rpm & Doffer 4.14 settings. The web coming out a form that machine is processed again with a cross lapper machine to form the lap. The various proportionated weight of coir and viscose fibre is blended thoroughly and given as input for the machines. Then the lap is subjected to a needle punching machine operated under 20 mm/stroke speed to get a composite layer.

Figure 1. Miniature Carding M/C (conversion of fiber into lay form)

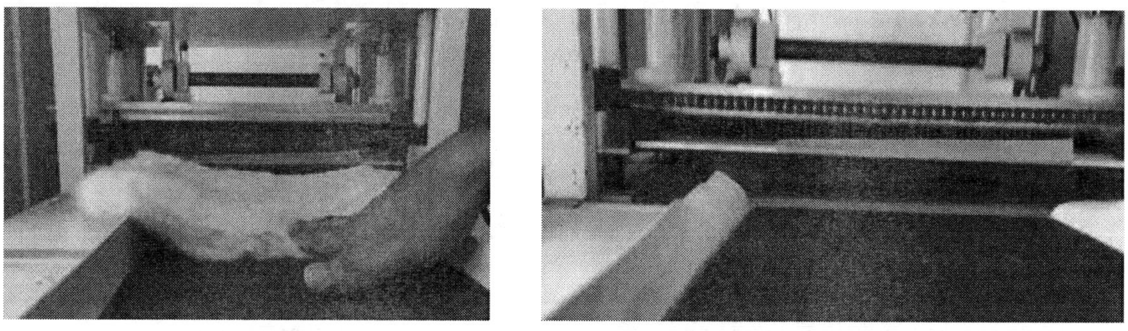

Figure 2. Feeding blended fibres fed into the needle punching machine.

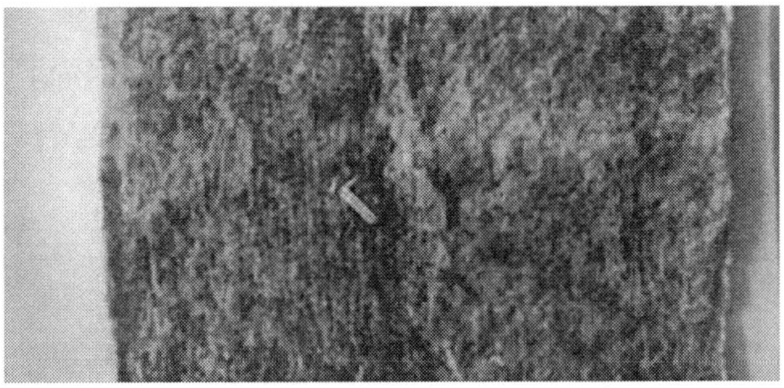

Figure 3. Non-Woven Composite

2.4 Testing

The developed non-woven composite undergoes a Sound absorption test, Abrasion test, and moisture test to identify the sound absorption (acoustic) property.

2.4.1Sound absorption co-efficient

The sound absorption coefficient of materials is correlated with frequency, and it varies with different frequencies. The sound absorption co-efficient frequency characteristic curves can be used to illustrate the sound absorption properties of different frequencies exactly. It is not convenient to compare and state, so the average sound absorption coefficient, which is the average of an acoustic material's absorption co-efficient at a specified set of frequencies, is used for simplification. The average sound absorption coefficient is represented by α. The sound absorption coefficient of the material is measured using an impedance tube. The results will be in the form of a graph.

2.4.2 Abrasion test

Abrasion is just one aspect of wear and is the rubbing away of the component fibers and yarns of the fabric. It is a series of repeated applications of stress; therefore a capacity to absorb punishment is required to the fibres. Inherent fibre properties such as work of rupture may give a high resistance to abrasion. This test is done to test the resistance of the material to other materials in the process of repeated friction between the materials.

2.4.2.1 Procedure

The samples are weighed before testing then kept in an abrasion tester. Then the material was subjected to rotation about 50 times. Once the process is done the samples are taken out from the machine and weighed again to measure the resistance of the composite using the following formula

Abrasion (%) = (loss in weight /initial weight) *100................................. [1]

Where Loss in weight = initial weight- final weight.

2.4.3. Moisture test (hot air oven)

This test is done to analyze the moisture content absorbed by the material, by comparing its weight before evaporation of water till those get dry.

Moisture % = (loss in weight / initial weight)*100................................... [2]

Where, Loss in weight = initial weight- final weight.

3. Results and Discussion

3.1. Sound Absorption test

Table 1 shows that as the content of viscose fiber in the nonwoven composite increases from 30% to 80% (from Sample 1 to sample 6). When there is an increase in areal density there is an increase in sound absorption coefficient for reclaimed viscose and coir nonwoven composites. This shows that coir has less sound absorption coefficient when compare to viscose since it is a synthetic fibre.

Table 1. Sound absorption test: (Co efficiency value of the material)

Sample. No	Frequency (in Hz)					
	1000	2000	3000	4000	5000	6000
Sample 1	0.09	0.1	0.13	0.15	0.35	0.40
Sample 1	0.10	0.19	0.35	0.38	0.42	0.45
Sample 3	0.19	0.25	0.38	0.42	0.45	0.49
Sample 4	0.22	0.29	0.40	0.46	0.53	0.58
Sample 5	0.35	0.39	0.45	0.57	0.68	0.75
Sample 6	0.45	0.53	0.64	0.75	0.81	0.89

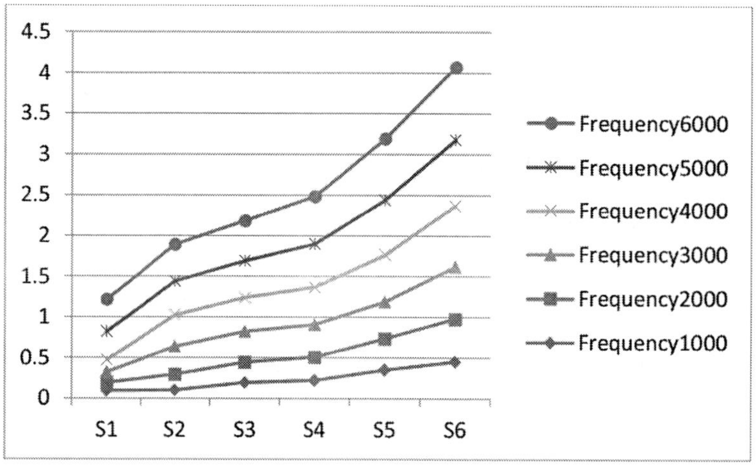

Figure 4. Graph - Sound absorption test: (Co efficiency value of the material)

3.2 Abrasion Test

Table 2 shows that as the content of viscose fiber in the nonwoven composite increases from 30% to 80% (from Sample 1 to sample 6). When there is an increase in areal density there is an increase in abrasion, which means the coir has higher abrasion resistance than viscose it leads to the better acoustic property of the product.

Table 2. Abrasion test

Sample.no	Initial weight	Weight after abrasion	Change in weight	Abrasion (%)
S1	8.730	8.38	0.35	4.0
S2	8.930	8.460	0.47	5.26
S3	8.980	8.450	0.53	5.90
S4	9.050	8.550	0.55	6.07

S5	9.140	8.570	0.57	6.23
S6	9.220	8.620	0.60	6.50

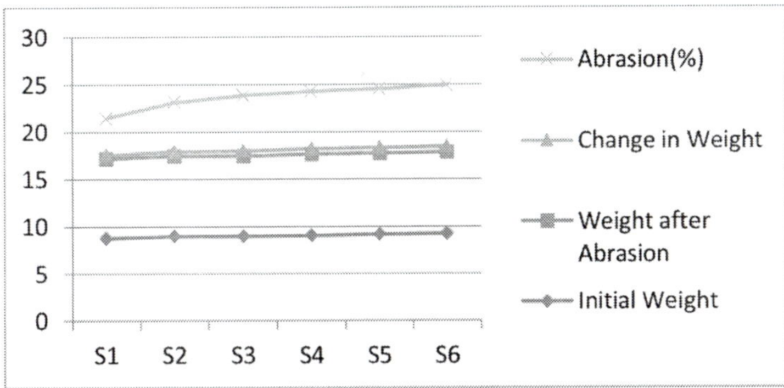

Figure 5. Graph -Abrasion test (% of abrasion)

3.3. Moisture Test

Table 3 states that when there is a higher percentage of viscose in composite material the moisture content of the product is when compared to less percentage of viscose in the non-woven composite. So it's clearly said that the viscose has higher moisture regain than coir fiber. So the same reflects in the acoustic property of that particular composite. The acoustic property increased when the viscose proportion increased in the composite.

Table 3. Moisture Test

Sample.no	Weight before moisture absorption (in gms)	Weight after moisture absorption (in gms)	Change in weight (in gms)	Moisture (%)
S1	8.610	8.42	0.19	2.20
S2	8.675	8.425	0.25	2.88
S3	8.690	8.3	0.39	4.48
S4	8.750	8.24	0.51	5.86
S5	8.790	8.21	0.58	6.59
S6	8.808	8.178	0.63	7.79

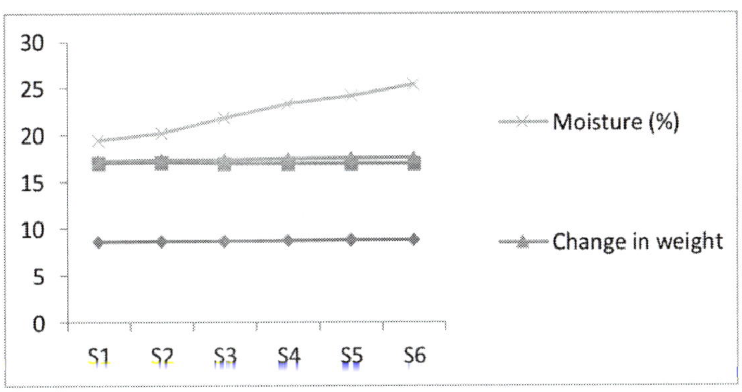

Figure 6. Graph -Moisture Test in %

IOP Conf. Series: Materials Science and Engineering 1059 (2021) 012028 doi:10.1088/1757-899X/1059/1/012028

3.4 GSM Test

Due to the compact structure of the viscose, sample 6 has high GSM when compare to other samples. Table 4 shows the nonwoven composite GSM for different samples. It clearly states that when the viscose content increases then GSM increase respectively.

Table 4. GSM Test

Sample. No	GSM (Average) in grams
S1	180
S2	205
S3	212
S4	223
S5	230
S6	234

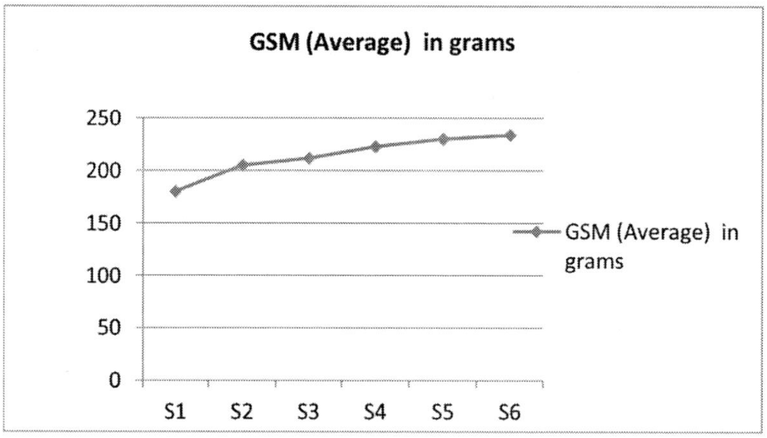

Figure 7. Graph - GSM of the Fabric in gms

4 Conclusion

This paper shows the acoustic property of the product is increasing, concerning the thickness of the end product. If the viscose percentage increases in the product, the quality of the product i.e, acoustic property (Acoustic coefficient) of the product will improve when compared to others. Since the higher in proportion viscose leads to higher moisture, weight addition, and more abrasion weight loss. This is due to viscose is a very light fiber that can make a compact object and not resistant to abrasion due to its friction property. Since the presence of coir fiber makes air porosity due to denser fiber which leads to avoiding echo in frequency application. The concept behind the thickness of the layer is to enhance the incident sound waves to lose more energy as they take a longer path through the material.

References

[1] Nor M J M, Jamaludin N and Tamiri F M 2004 A preliminary study of sound absorption using multi-layer coconut coir fibers *Electronic Journal Technical Acoustics* *3* 1-8.

[2] Shahani F, Soltani P and Zarrebini M 2014 The analysis of acoustic characteristics and sound absorption coefficient of needle punched nonwoven fabrics *Journal of Engineered Fibers and Fabrics* **9**(2)

[3] Barron R F 2003 Industrial Noise Control and Acoustics, New York: Marcel Dekker.

[4] Delany M E Bazeley E N 1970 Acoustic Properties of Fibrous Absorbent Materials Applied Acoustics Vol **3** No 2 105-116.

[5] Gopalakrishnan D Processing and characterization of nonwoven structured coir and palmyra for acoustic application.

[6] Majid Ali 2010 Coconut Fibre – A Versatile Material and its Applications in Engineering, Second International Conference on Sustainable Construction Materials and technologies Universitapolitecnicadelle Marche Ancona Italy.

[7] SurajitSengupta 2010 Sound Reduction by Needle-punched Nonwoven Fabric *Indian Journal of Fibre & Textile Research* Vol. **35** pp. 237-210.

[8] Mamtaz H, Fouladi M H, Al-Atabi M and Narayana Namasivayam S 2016 Acoustic absorption of natural fiber composites. Journal of Engineering 2016.

[9] Yang W D and Li Y 2012 Sound absorption performance of natural fibres and their composites, *Science China Technological Sciences* vol. **55** no. 8 pp. 2278–2283.

[10] Shahani F, Soltani P and Zarrebini 2014 The analysis of acoustic characteristics and sound absorption coefficient of needle punched nonwoven fabrics *Journal of Engineered Fibres& Fabrics* vol. **9** no. 2 pp. 84–92 2014.

[11] Allard F 1993 Propagation of Sound in Porous Media: Modelling Sound Absorbing Materials Elsevier Application Science. NewYork. NY. USA.

[12] Ingard K U 1994 *Notes on Sound Absorption Technology* Noise Control Foundation, Poughkeepsie, NY, USA.

[13] Coates M and Kierzkowski M 2002 Acoustic textiles-lighter, thinner and more sound-Absorbent *Technical Textiles International* vol. **11** No. 7 pp. 15–18.

[14] Soltani, P and Zarrebini M 2013 Acoustic Performance of Woven Fabrics In Relation To Structural Parameters and Air Permeability, Journal of the Textile Institute, Vol. **104** No 9 1011–1016.

Effect of Stainless Steel Powder as Reinforcement on Mechanical and Tribological Properties of Stir Cast Zn-Al Alloy

Soumyan Skariah Alex[1], S. Ilangovan[2], A.Shanmugasundaram[*3] and Subinaya Mohapatra[4]

[1 & 4] Post Graduate Student, [2] Associate Professor, [* 3] Assistant Professor.

Department of Mechanical Engineering, Amrita School of Engineering, Coimbatore,

Amrita Vishwa Vidyapeetham, India

E-mail: a_shanmugasundaram@cb.amrita.edu

Abstract. In this study, Zn-40Al alloy is reinforced with 310 stainless steel particle of size 45 μm and its effect on tensile strength, hardness and wear rate were investigated. The Zn-40Al-SS composites were fabricated with 310 tainless steel particle of 3, 6, and 9 wt. % of volume fractions using stir casting technique. Tensile strength and vickers microhardness were tested for the base alloy and for the various combinations of Zn-40Al-SS composite. Dry sliding wear test using Pin-on-Disc machine were carried out for the base alloy and for all the combinations of the composites. Characterization techniques like optical microscopy, scanning electron microscope (SEM) and elemental distribution by energy-dispersive X-ray spectrometry (EDS) were carried out to analyse the composite. The tensile strength and hardness were high for the reinforcement of 9 wt. %. Also the wear resistance was maximum for the reinforcement of 9 wt. %. The microstructure shows the fine uniform dispersion of the reinforcing particles over the entire matrix. Overall the composite shows a significant improvement in the wear resistance and mechanical properties.

Keywords: Metal matrix Composite, Zinc Aluminum, Mechanical properties, Wear, Friction, Stir casting

1. Introduction

Zinc Aluminum (Zn-Al) combinations are being used in numerous modern applications because of their better castability and good mechanical properties [1-4]. These alloys are functionally limited above the temperature of 100 ºC. Usage of these alloys are limited in high temperature applications. These alloys experience lower mechanical properties at elevated temperature [5, 6].
A feasible methodology of enhancing the properties is the incorporation of thermally stable second phase particles [7,8]. The impact of size and introduction of the reinforcement phase on mechanical and wear properties have been explored over the past. Ceramic dispersoids and glass fibers are particularly reinforced in the zinc based composites [9, 10]. Sliding mechanical components like bearings, piston rings are subject to wear. Reinforced second phase particles in Zn-Al alloy have been suggested to sliding wear applications as compared to Metal Matrix composites (MMC) [11-13]. MMC increases the wear resistance in most of the cases. Moreover, these composites exhibit excellent mechanical

properties. Preheating the reinforcement helps enhance the interface strength and facilitates the dispersion of the particles throughout the alloy.
In view of above, an attempt to fabricate improved materials with enhanced properties in comparison to those of the matrix. The present work investigates the effect of 310 stainless steel (SS) powder additions on the mechanical and Tribological behavior of Zn–40Al alloy system. The wear surfaces were examined using SEM.

2. Experimental Procedure

2.1 Material Preparation

In the present study, commercial pure zinc (Zn - 99.96 %) and LM0 Aluminium were used to fabricate the Zn-Al alloy. The composites were prepared by dispersing 3 - 9 % in steps of 3 wt. % of the 310 SS powder of size 45 μm in the melt of the base alloy. In a previous study, the similar 3 to 9 wt. % alumina is added in AlSi10Mg aluminium alloy along with 3 wt. % of graphite and the effect of these reinforcements on the mechanical and wear properties were studied [14]. In general, when the wt. % of reinforcement is increased up to some level, it will improve the mechanical and wear properties and after that properties will gradually decrease. Samples fabricated through stir casting techniques with more wt. % of reinforcement particles are difficult to machine. During machining of such more wt. % of reinforced samples, the particles will detach itself from the composites due to abrasive nature [15]. Table 1 shows the detail formulation of four different types of Zn-Al-310 SS powder. The liquid metallurgy process was used to manufacture the composites.

Table 1. Chemical Composition

Description	Chemical Composition (wt. %)		
	Zn	Al	Stainless steel powder
Zn-40Al	60	40	-
Zn-40Al-3SS	57	40	3
Zn-40Al-6SS	54	40	6
Zn-40Al-9SS	51	40	9

The chemical compositions of these alloys were weighed according to ratios and melted in a graphite crucible. The SS powder was weighed as per the ratio and preheated separately at the initial stage to a temperature of 450 °C. This helped to remove the moisture and improves the wettability of the Zn-Al alloy. Aluminium is heated to 670 °C in a crucible furnace until it melts completely. Then before the Zinc was added, the furnace temperature was lowered to 500 °C. After Zinc had melted completely, it was cooled to a semi solid state (to about 450 °C) and stirred at 200 rpm for 5 min to achieve homogenization. The preheated SS powder is then charged into the melt which is stirred for 5 min. The semi-solid composite slurry formed was superheated to a temperature of 530 °C and stirred again at 400 rpm for 10 min. The melt was released through an orifice at the base of the crucible. Mould made of cast iron is used to shape the melt in the form of 30 mm diameter and 300 mm long cylindrical casting. Figure 1 shows the base substrate and the composites.

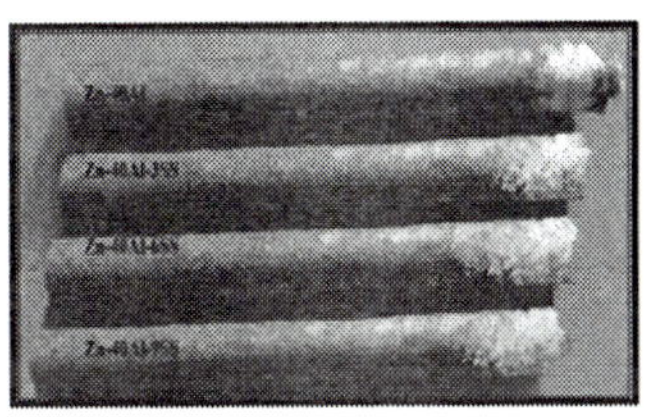

Fig. 1. Pure alloy and reinforced composites.

Hydraulic press runs for 30 s time with 6 MPa pressure applied on the composite to reduce porosity and to improve the bonding force between the Zn-Al alloy and SS particles. Finally, the mould was opened after 5 min and the fabricated composites were air-cooled to room temperature.

2.2 Mechanical and Physical Characterization

2.2.1 Density and Void

The effect of SS powder wt. % on composite density was analyzed by resolving the density of composite fabricated. The experimental density is calculated using Archimedean water displacement method. The theoretical density was calculated according to rule of mixture [16]. Porosity was determined by comparing the experimental density and theoretical density of individual weight proportion of SS reinforced composite fabricated. An accurate electronic weighing balance with an accuracy of 1 mg was used to analyze the density of the specimen. The void fraction of the composites was evaluated using the relation Eq. (1) where ρcal is the theoretical density and ρexp is the experimental density.

$$voidfraction = \frac{(\rho_{cal} - \rho_{exp})}{\rho_{cal}} \quad (1)$$

2.2.2. Tensile test

Tensile test specimens as shown in Figure 2 are prepared according to ASTM standard E8 – 04. Tests were conducted on Tinius Olsen testing machine with the capacity of maximum 25 kN and at the extension rate of 2 mm/s. Tests were repeated three times to guarantee reliability of the data. Dimensions for tensile test specimen were tabulated in Table 2.

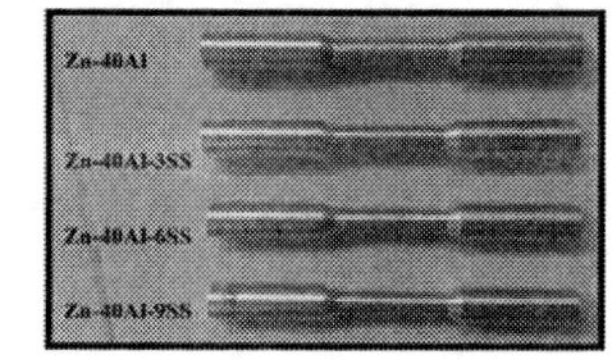

Figure 2. Tensile test specimens

Table 2. Dimensions for Tensile Test Specimen

Specimen	Dimension (mm)
G – Gauge length	30
D – Diameter	6
R – Radius of fillet	6

2.2.3. Hardness

The Mitutoyo MVH-H11 equipment was used to measure the microhardness. The specimens were prepared as per ASTM E384 standards. Before the measurement of hardness the surface of the specimen is polished by using emery paper ranging from 300 μm – 1200 μm. A load of 100 gf is then applied to the sample for a period of 15 sec.

2.2.4. Dry Sliding wear test

Pin-On-Disc machine was used to conduct the dry sliding wear test of the base material and the stainless steel powder reinforced composite. (Make: Ducom; Model:TR-20 with Wear and Friction Monitor). The contact load was varied by 10, 20, 30 and 40 N. Sliding velocity was varied by 1, 2, 3 and 4 m/s. Sliding distance was varied by 500, 1000, 1500 and 2000 m. Track diameter was kept constant at 60 mm. ASTM 99-95a standard was followed. The sample (Ø8 mm × 25 mm) is held stationary against the rotating disc made of En-31 steel having hardness of 60 HRC. Load was applied to the sample by a cantilever mechanism. The disc is cleaned to remove debris after every test. The weights of specimens were measured by an electronic weighing balance before and after the wear test. Wear rate were calculated by weight loss method.

2.2.5. Microstructure
An inverted metallurgical microscope (Zeiss Axiovert 25 CA) was used to investigate the microstructure of the composite. Specimens were initially polished using emery paper followed by a disc polishing to get a fine finish and finally etched using mixture of Ethanol – HCl solution. Microscopic examinations were done using Scanning Electron Microscope (SEM) to characterize the wear surface and wear debris by using JEOL JSM - 6490LA. The distribution of elements was carried out by EDS at 15 kV.

3. Results and Discussion

3.1. Elemental Composition and Microstructure
Figure 3(a) shows the microstructure of the matrix alloy. From Figure 3(b, c and d), it is evident that the nearly uniform dispersion of the particles in the matrix. Since the SS particles melting temperature (1510 °C) was higher than the melting point of Zn-Al (490 °C), the SS particles does not melt during stir casting and also it does not form any other inter-metallic particle. This reveals that the reinforcement is stable and non-reactive in the matrix alloy. The line EDS spectrum shown in the Figure 4 consists of elements like Iron, Chromium and Nickel.

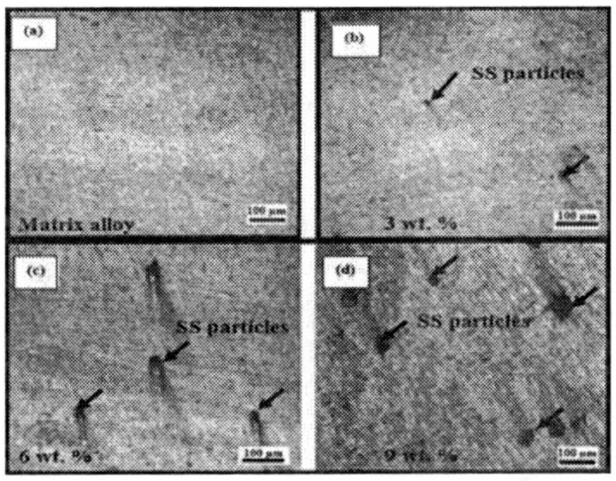

Figure 3. Optical micrograph of (a) matrix alloy, (b-d) composite showing stainless steel particles

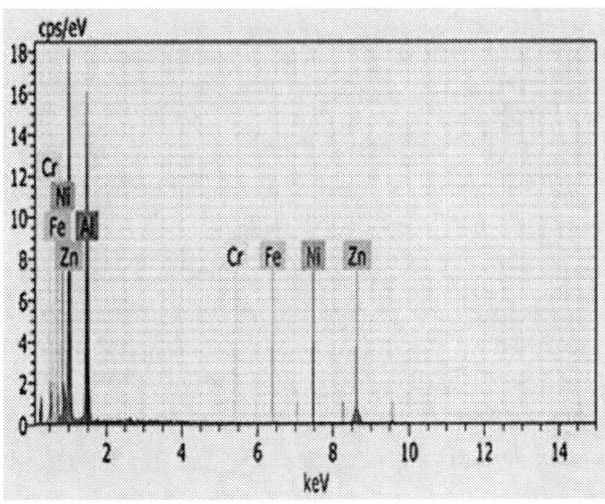

Figure 4. EDS spectrum composite.

3.2. Density

From the Table 3, it can be inferred that when the content of SS powder increases the overall density of the composite also increases linearly. Also it can be noticed that the experimental values of the density are lower than the theoretical one. This shows that these composites contained some porosity. The porosity of the composites was calculated by using the equation (1) and is between 0.6 – 3.33 %. However these values are within the acceptable range of 4 % [17]. The instant pressure applied after casting in this study has reduced the porosity.

Table 3. Density

Composition (wt. % reinforcement)	Theoretical Density ρcal (kg/m^3)	Experimental Density ρexp (kg/m^3)	Void fraction
0	4198	4172	0.006
3	4262	4205	0.013
6	4336	4242	0.021
9	4413	4267	0.033

3.3. Mechanical Properties

The tensile strength and the hardness of the Zn-Al-SS composite for various wt % were tabulated in the Table – 4. From the table it can be inferred that the tensile strength improves from the base by 47 %, 58 % and 64 % with respect to 3 wt. %, 6 wt. % and 9 wt. % of SS powder reinforcement respectively. The maximum improvement in the tensile strength was with the 9 wt. % of SS particle. The tensile strength increases with increase in wt. % of reinforcement, because of the increase in resistance to deformation. Stress generated due to thermal mismatch causes the hard reinforcement in the matrix to bring about an increase in density. Because of this there is an increase in tensile strength of the composites [18, 19]. The increase in tensile strength can be justified by the dispersion of SS particles which hinders the dislocation motion.

Table 4 Mechanical Properties

Mechanical Properties	Composition (wt. % reinforcement)			
	0	3	6	9
Tensile strength (MPa)	240	353	379	395
Micro hardness (HV)	83	95	98	106

The hardness values given in Table 4 were the average of five readings. From the table 4 and figure 5 it can be inferred that the hardness increases from the base by 14 %, 18 % and 28 % with respect to wt. % 3, wt. % 6 and wt. % 9 of SS powder reinforcement respectively. Due to the presence of SS particles, an increase is observed in the hardness of the composites from the base substrate as the wt. % of reinforcement increases. The reason behind the increase in the hardness can be inferred that when the hardness measurements were done, the material resists the penetration of indent because of the hard SS powder reinforcement on the soft Zn-Al matrix. Also the increase in the hardness of the Zn-Al-SS composite can be justified because of the higher density than the base material. The enhancement in the hardness properties of the composites can be credited to the factors such as higher hardness of SS particles as well as the active interfacial adhesion between the matrix and SS interfaces.

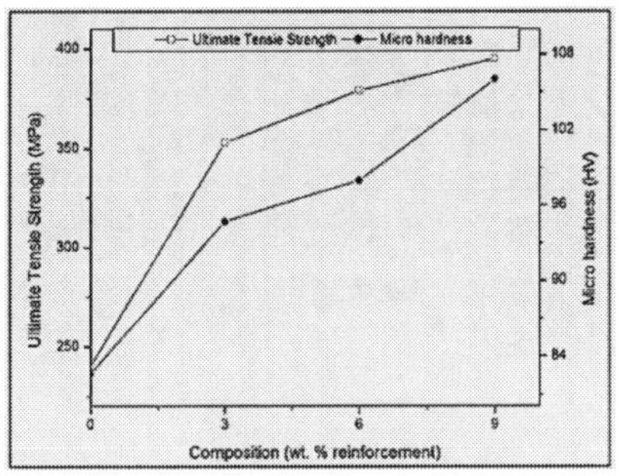

Figure 5. Effect of composition (wt. % reinforcement) on mechanical properties.

3.4. Dry Sliding Wear test

Figure 6 shows the effect of applied load on the sliding wear of the Zn-Al-SS composite as a function of reinforcement wt. %. With reference to Figure 6, it can be inferred that when load increases, the wear rate also increases [20-22]. When load applied is low, the wear loss is quite small for the composites. This is due to the load bearing capacity of reinforcement phases. On increase of applied load, the wear mechanisms changes from mild to severe wear and hence wear rate increase [23]. It is clearly noticed that the wear rate of the each composite reduced with the addition of particle content. This demonstrates that the wear behaviour of the composites was found improved significantly with the addition of SS particles as reinforcement. Because of the good interface between the SS particle reinforcement with the base substrate of Zn-Al, the Zn-Al-SS MMC responds with a lower wear rate [24,25].

It is clearly observed from the Figure 7 as the load increases, the coefficient of friction (COF) increases. An increase in the load increases the deformation of frictional layer. This tends to the increase of frictional force which in turn increases the temperature of material. Hence COF increases. Among the various combination of fabricated MMCs, 9 wt. % composite is having the lowest COF and the highest wear resistance at room temperature.

The lowest wear rate (0.0009 mm^3/m) was obtained with the composite with 9 wt. % SS particles which also had the lowest COF (0.48), whereas the base substrate shows the highest wear rate (0.0081 mm^3/m) and with the highest COF (0.565). From the above it is clear that the wear resistance of Zn-Al-SS MMC is 88 % higher than that of Zn-Al base alloy. Also it can be inferred that the COF of Zn-Al-SS MMC is 17 % lower than that of Zn-Al base alloy. This indicates the inverse relationship between wear resistance of alloys and COF i.e., as wear resistance increases, there will be considerable decrease in the COF [26-28].

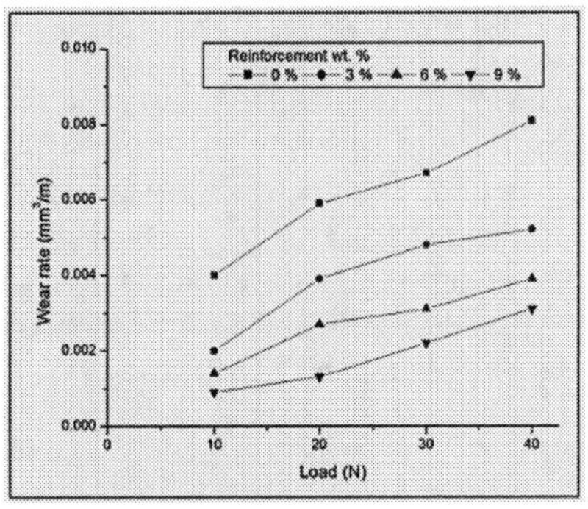

Figure 6. Effect of load on wear rate

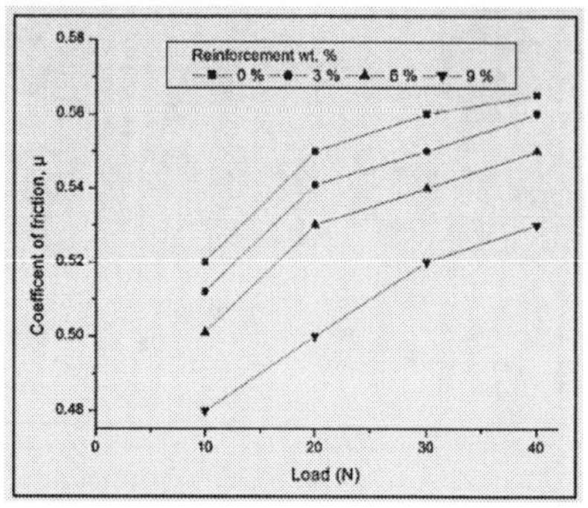

Figure 7. Effect of load on coefficient of friction

The lowest wear rate (0.0009 mm^3/m) was obtained with the composite with 9 wt. % SS particles which also had the lowest COF (0.48), whereas the base substrate shows the highest wear rate (0.0081 mm^3/m) and with the highest COF (0.565). From the above it is clear that the wear resistance of Zn-Al-SS MMC is 88 % higher than that of Zn-Al base alloy. Also it can be inferred that the COF of Zn-Al-SS MMC is 17 % lower than that of Zn-Al base alloy. This indicates the inverse relationship between wear resistance of alloys and COF i.e., as wear resistance increases, there will be considerable decrease in the COF [26-28].

3.5. Microscopic Examination using SEM on the Wear surface

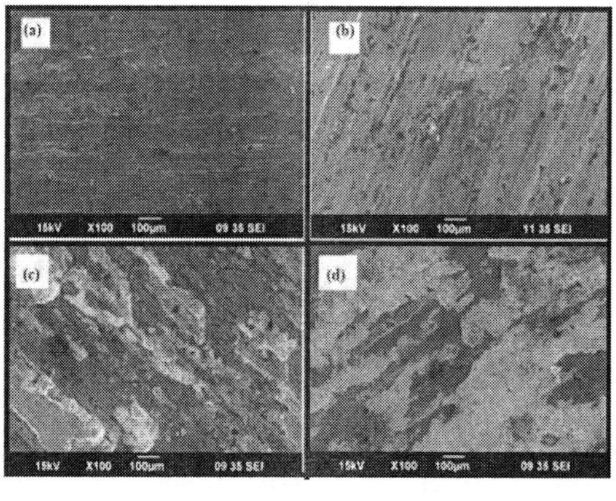

Figure 8. SEM images on the worn surface of matrix alloy at (a) 10N (b) 20N (c) 30N and (d) 40N.

Figure 8 (a-d) depicts the scanning electron micrograph of worn surface of the base material at different loading conditions. The wear surface of samples, shown in Figure 8 (a), tested at 10 N for the unreinforced alloy consists of ridges and grooves which are the typical characteristics of sliding wear. When the load increased to 20 N, Figure 8 (b), a slight growth in number of grooves were observed. On further increment of load to 30 N, Figure 8 (c), micrographs indicates the formation of continuous wear grooves. At 40 N damaged spots as shown in Figure 8 (d) were observed from the micrographs. These are indication of severe deformation, resulting high wear rate.

Figure 9. SEM images of the wear debris (a) Zn-Al alloy (b) 3 wt. % composition (c) 6 wt. % composition (d) 9 wt. % composition.

Figure 9 (a-d) represents the SEM images of Zn-Al-SS MMC worn surfaces. From the Figure 9 it is clear that some debris formed over the surface. This debris generated mostly from the pin material. The presence of material plastic flow could be noticed from these micrographs. During wear, the particles are pulled off from the surface and smeared onto the worn surface. Figure 9 (a) depicts the morphology of the Zn-Al-alloy worn surface with imprints of ploughing marks. The micrographs of composite with 3, 6 and 9 wt. % stainless steel particles were shown in Figure 9 (b), (c) and (d) respectively. Figures 9 (b, c and d) shows lack of deep marks on the surface compares to Figure 9 (a). Zn-Al-SS MMC worn surfaces shown in the Figure 9 (b), (c) and (d) depicts that less grooves and scratches were found when compare to Figure 9 (a). The size of the debris is reduced with 9 wt. % stainless steel particles as shown in Figure 9 (d) [29].

Figure 10. Morphology of the worn surface of composite at 40N.

But at high loads, protective layer is no longer remaining stable. The worn surface of the composite shown in Figure 10 loses its ability to withstand the load of 40 N. This is due to the development of strain when counter face comes in direct contact with the matrix alloy which causes delamination of wear [30, 31].

4. Conclusions

The following conclusions can be reported from the results of this study of Zn-Al-SS MMC.

• Stir casting was done successfully with fine dispersion of SS particle as reinforcement in the matrix alloy.

• The tensile strength of the Zn-Al-SS MMC increases by 64 % for 9 wt. % of SS particles.

• The hardness of the Zn-Al-SS MMC increases by 28 % for 9 wt. % of SS particles.

• When compared with the Zn-Al base alloy, the wear resistance of Zn-Al-SS MMC improves by 88 % for 9 wt % of SS particle.

• Wear rate of the composite and the base alloy possess direct relation. It increases with increase of load. Wear takes place much faster in the base substrate than in composites.

• From the studies, it can be concluded that Zn-Al alloy reinforced with 310 stainless steel powder exhibits higher hardness and higher wear resistance.

References

[1] Savaskan T, Purcek G and Murphy S 2002 *Wear* **252** 693–703.
[2] Zhu Y.H, Man H.C and Lee W.B 2003 *Journal of Materials Processing Technology* **139** 296-301.
[3] Lee P.P, Savaskan T and Laufer E 1987 *Wear* **117** 79-89.
[4] Prasad B.K, Patwardhan A.K and Yegneswaran A.H 1996 *Wear* **199** 142-151.
[5] Savaskan T and Murphy S 1987 *Wear* **116** 211-224.
[6] Purcek G 2005 *Journal of Materials Processing Technology*. **169** 242-248.
[7] Purcek G, Savaskan T, Kucukomeroglu T and Murphy S 2002 *Wear* **252** 894-901.
[8] Prasad B.K, Patwardhan A.K and Yegneswaran A.H 1997 *Materials Transactions JIM* **38 (3)** 197-204.
[9] Ranganath G, Sharma S.C and Krishna M 2001 *Wear* **251** 1408-1413.
[10] Sharma S.C, Satish B.M, Girish B.M, Kamath R and Asanuma H 1998 *Tribology International* **31** 183-188.
[11] Murphy S and Arikson R1992 *Wear* **155** 105-115.
[12] Savaskan T, Hekimoglu A.P and Purcek G 2004 *Tribology International* **37** 45-50.
[13] Lo S.H.J, Dionne S, Sahoo M and Hawthorne H.M 1992 *Journal of Material Science* **27** 5681-5691.
[14] Radhika N and Subramanian R 2013 *Tribology* **7 (1)** 36-41.
[15] Radhika N, Balaji T.V and Palaniappan S 2015 Journal of Engineering Science and Technology **10 (2)** 134-144.
[16] Sahin Y and Acilar M 2003 *Composite,* **34** 709-715.
[17] Ghosh P.K and Ray S 1988 *AFS Trans.* **88** 775-782.
[18] Chawla N and Shen Y 2001 *Advanced Engineering Materials* **3 (6)** 357-370.
[19] Ehsani R and Seyed Reihani S.M 2004 *Scientia Iranica* **11(4)** 392-397.
[20] Ilangovan S, Shanmugasundaram A and Arul S 2016 *Journal of Surface Science and Technology* **32 (3-4)** 93-98.
[21] Ritesh P and Shanmugasundaram A 2018 *Transactions of the IMF* **96 (2)** 95-102.
[22] Ilangovan S and Jiten S 2014 *International Journal of Research in Engineering and Technology* **3 (7)** 28-32.
[23] Shanmugasundaram A, Arul S and Sellamuthu R 2017 *Transactions of Indian Institute of Metals* **71 1** 117-125
[24] Sirong Yu, Zhenming He and Kai Chen 1996 *Wear* **198(1-2)** 108-114.
[25] Modi O.P, Rathoda S, Prasad B.K, Jhaa A.K and Dixitb G 2007 *Tribology International* **40** 1137-1146.
[26] Prasad B.K, Patwardhan A.K and Yegneswaran A.H 1997 *Zeitschrift fuer Metallkunde* **88 (4)** 333-338.
[27] Modi O.P, Yadav R.P, Prasad B.K, Jha A.K, Dasgupta R and Dixit G 1998 *Materials Transactions JIM* **5** 582-586.
[28] Iwai Y, Hou W, Honda T and Yoneda Y 1996 *Wear* **196** 46-53.
[29] Mandal A, Chakraborty M and Murty B.S 2007 *Wear* **262 (1–2)** 160-166.
[30] Das S, Prasad S.V and Ramachandran T.R 1998 *Wear* **133(1)** 173-187.
[31] Suh N.P 1977 *Wear* **44(1)** 1-16.

Mechanical and Tribological Properties of Al 6061 Alloy Reinforced with Gr – WC Particulates Using Powder Metallurgy Technique

Srikanth K M 1, Sagar S R 2 and Jayasimha R 3
[1, 2] Faculty, Department of Mechanical Engineering, Amruta Institute of Engineering and Management Sciences Bidadi, Bangalore, India.
[3]Technical Specialist, Altair Engineering Pvt. ltd Bangalore India.

srikanthkm88@gmail.com

Abstract. There is a tremendous demand for composites based on metal matrix, as they give low density, high wear resistance, stiffness and specific strength as compared to pure aluminium. In the present work, we investigated the mechanical and tribological properties by considering metal matrix as Aluminium 6061 and reinforcements as Gr and WC. The composites were prepared based on powder metallurgy technique by varying the weight proportions of Gr particulates for 0%, 2%, 4%, 6%, 8%, and 10% with this we kept WC particulates weight proportion as 3% constant for added reinforcement specimens. From hydraulic compaction machine, compaction process was performed and specimens were sintered in a muffle furnace at 5500C. The mechanical properties were performed for BHN, compression strength of the prepared composites were improved as compared to the pure matrix material Al6061 alloy and tribological properties like wear and SEM were examined for Al 6061 alloy results in less wear resistance as compared to Aluminium 6061 with the addition of reinforcements Gr - WC.
Keyword: Aluminium 6061, Tungsten Carbide, Graphite, Wear rate, SEM.

Nomenclature - Al 6061 = Aluminium 6061, WC = Tungsten Carbide, Gr = Graphite, SEM = Scanning Electron Microscope, L = Load, S = Speed, SD = Siliding Distance, WR = Wear Resistance.

1. Introduction

Due to its various advantages over manolithic materials are replacing conventional Engineering materials. The major developments of the new begining of composites was the creations of metal matrix composites. The MMCs is a material consisting of metal alloy reinforced with cermaic particulates, whiskers, continous fibers etc., because of their increased wear resistance and rigdity these MMCs are commonly used in automobile, marine, aerospace and transporation industries. Aluminium alloys are widely used in MMCs. The few examples of new combinations of metal matrix with reinforcements are: aluminium – silicon carbide, aluminium – graphite, aluminium – Flyash, aluminium – Bottom ash which are good in wear resistance. Powder metallurgy is a process of mixing fine powdered composite materials, pressed to form a desired shape(structure) through compacting and then prepared composite specimens are heated in a muffle furnance (sintering) at room temperature. Powder metallurgy technique basically consists of four steps: preparation of powder, powder blending, compaction and sintering.

A.R.K Swamy.et.al the weight of reinforcement particlates in the Al – matrix ranges from 0% - 4%. The composites are fabircated through a process called 'Vortex'. From this study results shows that, as the context of WC particlates significantly increases in hardness and tensile strength while decreasing in its ductility.

Z.Zulkoffli.et.al the author examined by considering reinforcement silicon carbide and fabricated with pre – alloyed powder AZ61. Compaction process carried in hydraulic machine with out binder agent under 200Mpa compaction pressure about 20 minutes by varying temperatures of 350^{0}C, 300^{0}C, 250^{0}C and 200^{0}C.

2. Experimental Work

The flow chart gives information about the fabrication and experimental procedure of the present work. In this Aluminium 6061 considered as metal matrix alloy and reinforcements as combinations of Graphite and Tungsten Carbide, also the specimens were prepared according to ASTM Standards. By using planetary ball milling and compaction machines mixing and compaction process was carried out and finally sintering process was done by Muffle Furnace. Here we investigated the mechanical and tribological properties like BHN Test, Compression Test, Wear Test and performed SEM. For all these tests results obtained are tabulated and graphs were plotted and ended with conclusions.

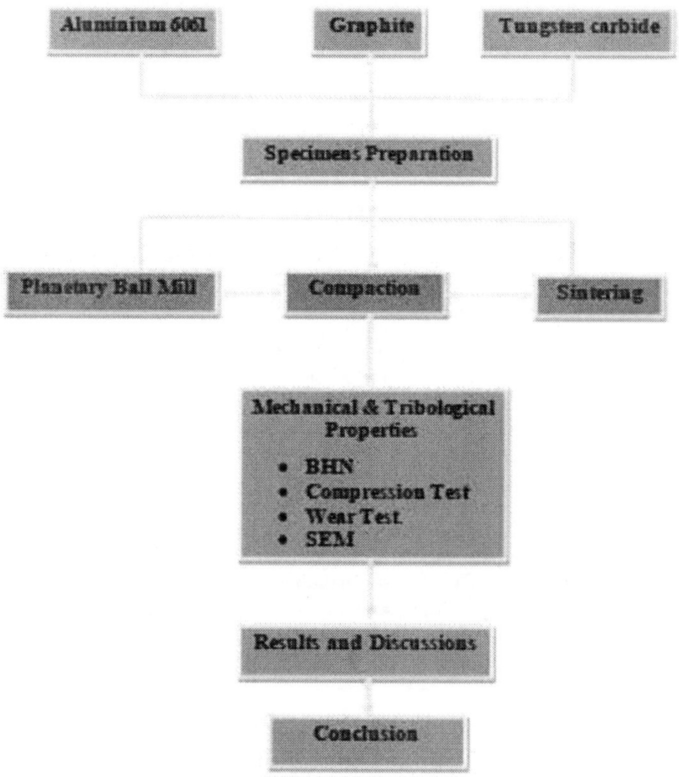

3. Materials and Methodology

3.1 Aluminium 6061

Aluminium 6061 is on the whole commonly used alloys for general purpose. It is developed in 1935 and called as "Alloy 61s". 6061 is an aluminium alloy which contains silicon and magnesium elements as its main alloy. It has excellent properties of mechanical and shows excellent weldability and is frequently extruded.

Applications: Castings, Extrusions, Forging, Welding etc.

TABLE I. Chemical Compositions of Aluminium 6061

Chemical Compositions of Aluminium 6061

Chemical Composition	Ti	Zn	Cr	Mg	Mn	Cu	Fe	Si	Al
(%)	0.2	0.25	0.04 - 0.35	0.8 - 1.2	0.15	0.15 - 0.40	0.7	0.4 - 0.8	Bal

TABLE II. Mechanical Properties of Aluminium 6061

Sl. No	**Mechanical properties**	**Values**	**Units**
1	Density (f)	2.7	(g/cm3)
2	Young's Modulus (E)	68.9	(Gpa)
3	Tensile Strength (Σt)	124 – 290	(Mpa)
4	Poisson's Ratio (V)	0.33	
5	Melting Temperature (Tm)	586	(^{0}C)
6	Thermal Conductivity (K)	151 – 203	(W(m-k))
7	Specific Warm Capacity (C)	897	(j/Kg-K)
8	Elongation Break (€)	13 – 26	(%)

3.2 Tungsten Carbide

WC is a chemical compound that contains atoms of tungsten and carbon in equal proportions. Basic form of a WC is a fine gray powder. By means of a process called sintering, where it can be pressed and shapes can be formed to used in industrial machinery's like cutting tools, Amor piercing shells, abrasives and jewellery.

Applications: Surgical Instruments, Sports equipments, Machinery cutting tools, mining tools etc.

3.3 Graphite

It is also known as "Plumbago". The crystalline shape of the carbon elements with its hexagonally arranged atoms naturally occurs in this form. Also it converts in to diamond under the conditions of high temperature and high pressure.

Applications: Batteries, Steel making, Brake linings, Foundry facings, lubricants etc.

Figure.1 Aluminium 6061

Figure.2 Tungsten Carbide

Figure.3 Graphite

3.4 Specimens Preparation

In the existing work, Hybrid composites like Aluminium 6061 + Gr + WC were fabricated through the technique of powder metallurgy. First step is the blending of particles like Al6061 + Gr + WC in a Ball mill (Fig.4) at a speed of 260 rpm for about 2 hours at 22Hz. In the Next step extracted powder is compressed in the die under optimum load through compaction machine (Fig.5). Finally, all the prepared Specimens will be kept in the muffle Furnace (Fig.6) and heated up to 550^{0}C about 2 hours. At last all the heated specimens were cooled at room temperature.

Figure.4. Planetary Ball Mill

Figure.5 Compaction Machine

Figure.6 Muffle Furnace

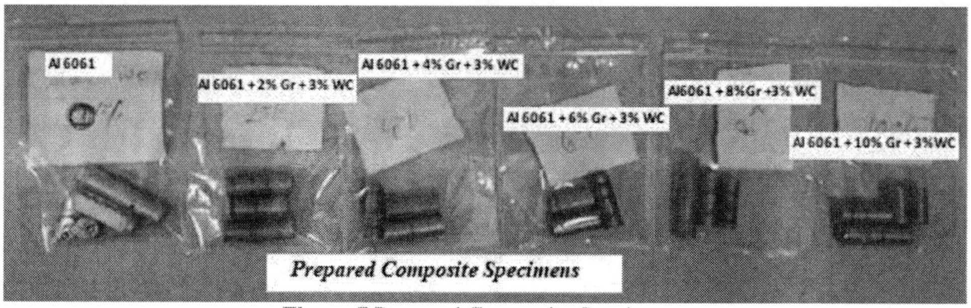

Figure.7 Prepared Composite Specimens

4. Results and Discussions

The specimens had been prepared according to standards of ASTM. The experiments were carried out to investigate the mechanical and Tribological properties such as BHN, Compression, Wear and SEM.

4.1 BHN Results

The aim of experiment is to find out the Brinell Hardness Number (BHN) of the prepared composite specimens with ASTM B-925 Standards having dimensions of 20mm depth and Diameter of Φ15mm with the help of BHN testing machine. The primary test of hardness by varying the reinforcement ratio of Gr such as 0%, 2%, 4%, 6%, 8%, 10% and WC reinforcement particulates is kept constant with a ratio of 3% with aluminium alloy 6061 for all prepared composites. Hardness is found out by considering the indentations on the specimens. BHN test is commonly used to measure hardness of metals.

Table.3 Numerical Results of BHN Test

Specimen Percentages	BHN
Al 6061	40
Al 6061 + 2% Gr + 3% WC	43
Al 6061 + 4% Gr + 3% WC	44
Al 6061 + 6% Gr + 3% WC	44
Al 6061 + 8% Gr + 3% WC	51
Al 6061 + 10% Gr + 3% WC	54

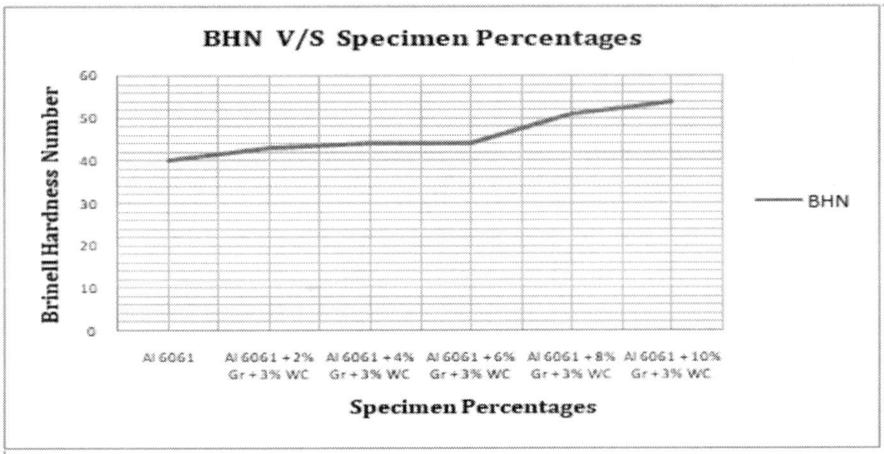

Figure.8 BHN V/S Specimen Percentages

4.2 Compression Results

The experimental aim is to investigate the compressive strength (Mpa) and observations are made to know the behaviour of the prepared specimens with ASTM E9-95 Standards having dimensions of 25mm depth and dia of Φ25mm were subjected to continually increasing uni-axial compressive force with the help of UTM machine. The tests were performed by varying the reinforcement ratio of Gr such as 0%, 2%, 4%, 6%, 8%, 10% and WC reinforcement particulates is kept constant with a ratio of 3% with base material Al - 6061 for all prepared composite specimens.

Table.4 Numerical Results of Compression Test

Specimen Percentage	Compression Strength
Al 6061	449.34
Al 6061 + 2% Gr + 3% WC	569.84
Al 6061 + 4% Gr + 3% WC	668.74
Al 6061 + 6% Gr + 3% WC	671.94
Al 6061 + 8% Gr + 3% WC	679.94
Al 6061 + 10% Gr + 3% WC	686.71

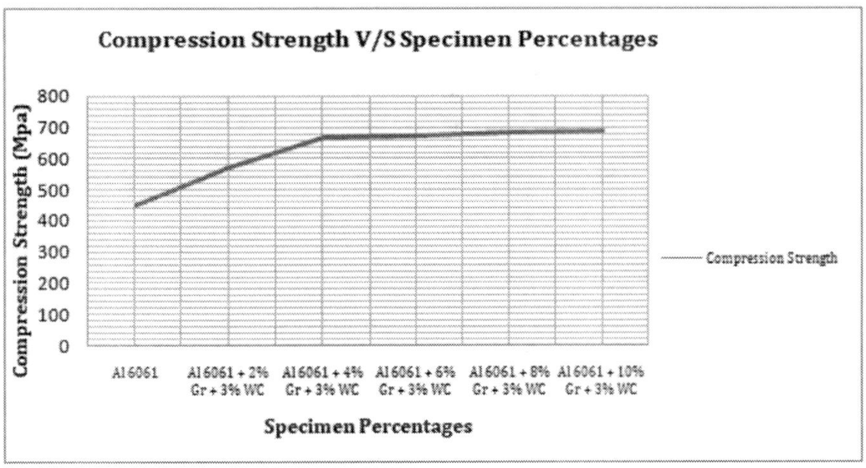

Figure.9 Compression Strength (Mpa) V/S Specimen Percentages

4.3 Wear Results

The experimental aim is to investigate the significant factors of wear mechanism, in order to achieve the minimum wear rate for prepared specimens with ASTM G99–95 Standards having dimensions of 30mm depth and dia of Φ6mm with the help of pin – on – disc apparatus. The preliminary test of wear mechanism, by varying the reinforcement ratios of Gr 0%, 2%, 4%, 6%, 8%, 10% and WC reinforcement particulates is kept constant with a ratio of 3% for all prepared specimens. Wear resistance were found out by considering parameters like load, speed, sliding distance and weight loss for composite specimens with high abrasion resistance will have less volume loss. Therefore the prepared specimens were tested by considering with different varying wear parameters.

Table.5 Numerical Results of wear

Specimens Percentages	L (N)	S (rpm)	SD (m)	WR (N - m/mm3)
	10	200	1000	6962.92
Al 6061	15	400	1500	12345.15
	20	600	2000	16701.73
	10	200	1000	6344.21
Al 6061 + 2% Gr + 3% WC	15	400	1500	7925.81
	20	600	2000	12625.75
	10	200	1000	2127.12
Al 6061 + 4% Gr + 3% WC	15	400	1500	9187.58
	20	600	2000	13982.16
	10	200	1000	2911.18
Al 6061 + 6% Gr + 3% WC	15	400	1500	9245.18
	20	600	2000	15786
	10	200	1000	7271.04
Al 6061 + 8% Gr + 3% WC	15	400	1500	12204.06
	20	600	2000	18446.09
	10	200	1000	3676.55
Al 6061 + 10% Gr + 3% WC	15	400	1500	3827.37
	20	600	2000	4235.89

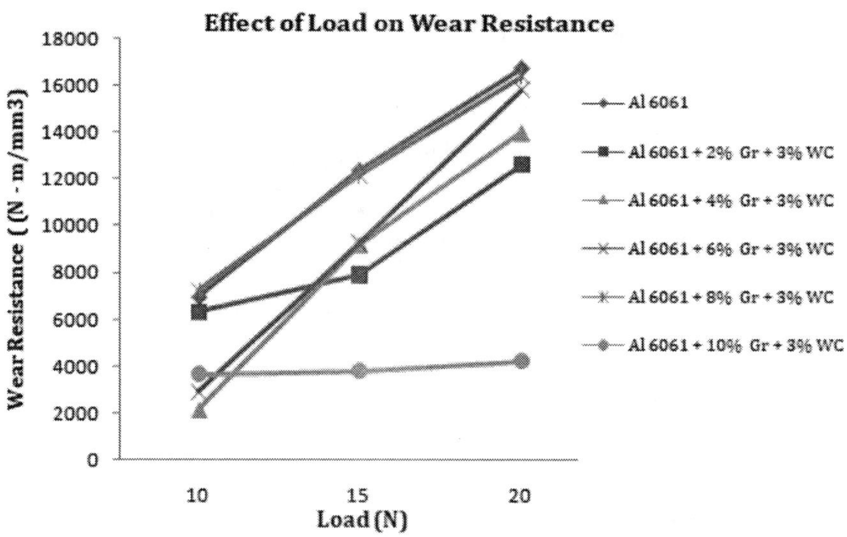

Figure.10 Wear Resistance (N –m/mm3) V/S Load (N)

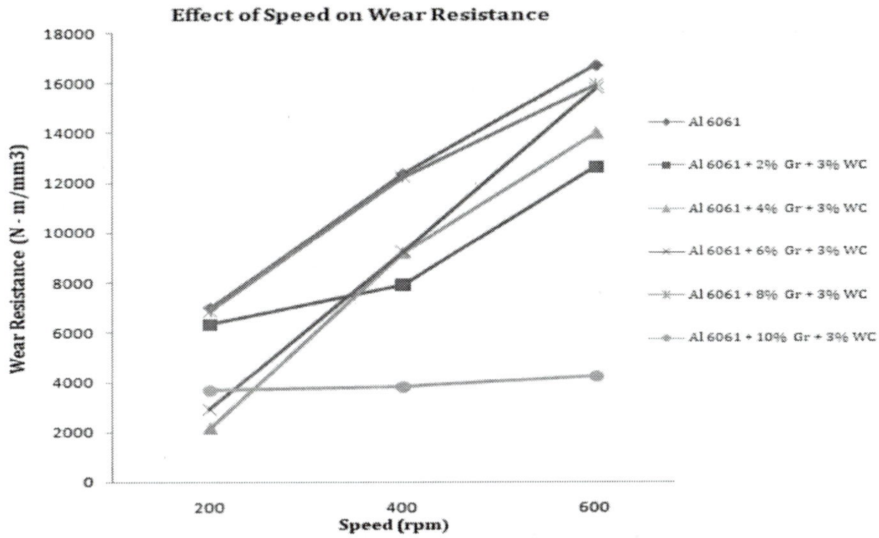

Figure.11 Wear Resistance (N –m/mm3) V/S Speed (rpm)

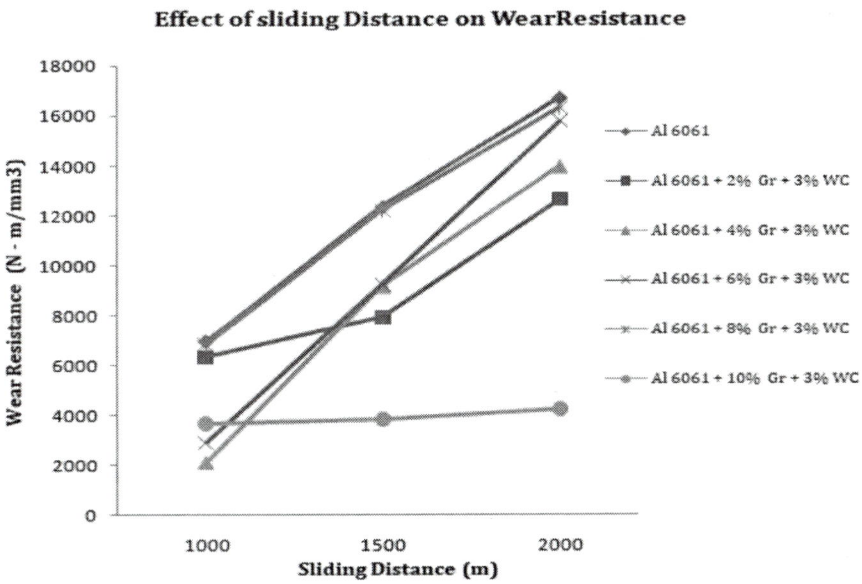

Figure.12 Wear Resistance (N –m/mm3) V/S Sliding Distance (m)

Fig. 10 shows the effect of load on wear resistance, Fig.11. shows the effect of speed on wear resistance and Fig.12 shows the effect of sliding distance on wear resistance for different weight proportions of reinforcements. The various levels of wear rate for prepared composites as shown in the above Figures. Wear rate will significantly increase by the effect of load, speed and sliding distance. It can be notice that plastic deformation is greatly increased which leads to higher wear at different levels. Usually plastic deformation leads to splits in the material surface, which in turn leads to removal of more material. As the load, speed and sliding distance increases with the sliding movement of prepared composite specimens on the top of the pin – on – disc apparatus. It develops interfaces of sliding for Al6061 + 8% Gr + 3% WC exhibits less frictional co –efficient on loads related to alloy materials.

4.4 SEM Results

The worn surfaces were examined by SEM to understand the wear mechanism of the prepared composites after performing the preliminary wear test for 6 Specimens with ASTM E407-07 Standards having dimensions of 20mm depth and dia of Φ15mm. During sliding the pin's surface area will be in contact with the steel disc's surface, and machine marks can be observed.

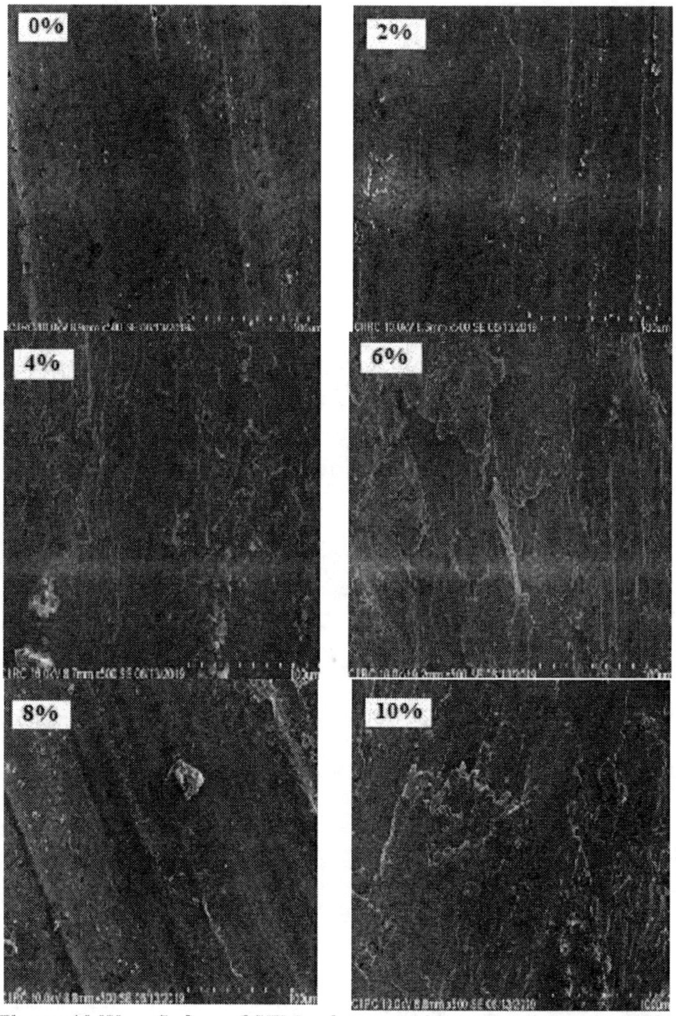

Figure.13. Wear Suface of SEM of prepared wear specimens at 500X

In this study we investigated the wear of particulate reinforced MMCs under varying applied loading conditions were identified. At lower loads (10N), the particles support the applied load in which MMCs wear resistance is better than alluminium alloy in the order of magnitude. At higher loads (20N), transition of severe wear takes place. We can notice the wear debris will be in the size of the order of micrometers, at lower loads it is of the order of a few hundred micrometers as the load increases the proportion of metallic wear debris also increases. Wear debris includes coarse metallic particles of the surfaces, WC and Gr particles, which acts as a abrasive particles between the specimen and disc. Figure.8 shows the SEM photographs of wear specimens with the varaition of reinforcement porportions of Gr with 0%, 2%, 4%, 6%, 8% and 10% by keeping WC as a constant proportion of 3% with the base metal at a load of 20N , with a speed of 600 rpm and with siliding distance of 2000m. The crack exhibits randomized longitudnal hollows, parallel towdards sliding distance. This indicates voids on the surface found to be damaged. It can be notice, that a little amount of material was removed from the prepared specimens. The surface displays different appearance with a percentages rising in reinforcements.

5. Conclusions

- The successful fabrication between base material aluminium 6061 with particulates of reinforcements including WC and Gr by using a powder metallurgy technique.
- By varying weight proportions of reinforcements with aluminium 6061 alloy, where we observed a good mechanical and tribological results.
- As increasing the context of reinforcements with in the base material has been found to results in significant improvements in its mechanical properties such as BHN and compression.
- At Al 6061 + 10% Gr + 3% WC, the properties of mechanical such as BHN and Compressive strength were found to be high.
- The properties of wear behaviour can be improved by increasing the reinforcement ratio of Graphite. The results of preliminary test shows highest wear resistance for Al 6061 + 8% Gr + 3% WC.
- Wear loss is minimum at load 10N, with a speed of 200rpm and with siliding distance of 1000m for each wear rate.
- The most affected specific wear rate were observed at applied load 20N, speed 800rpm and siliding distance 2000m for Al6061 + 8%Gr + 3%WC.
- Images of SEM clearly reflects the microscopically worn surface such as microcracking due to its applied loads, speed and siliding distances.

Acknowledgements

The authors would like to acknowledge the AMRUTA INSTITUTE OF ENGINEERING AND MANAGEMENT SCIENCES providing Lab facilities for our research work.

References

1. R. K. Swamy, A. Ramesha, G.B. Veeresh Kumar, and J. N. Prakash, 2011, "Effect of Particulate Reinforcements on the Mechanical Properties of Al6061-WC and Al6061-Gr MMCs", *Journal of Minerals & Materials Characterization & Engineering*, **10**,12, 1141-1152.
2. Z. Zulkoffli, J. Syarif and Z. Sajuri, 2009 "Fabrication of AL61/SiC composites by powder metallurgy process", *International Journal of Mechanical and Materials Engineering (IJMME)*, **4**, 2, 156 -159.
3. Hari Prasad Rao Pydi, Balamurugan Adhithan, 2013, "Microstructure Exploration of the Aluminium-Tungsten Carbide Composite with different Manufacturing circumstances", I*nternational Journal of Soft Computing and Engineering (IJSCE)*, **2**, 6, 2231-230.
4. M. Asif, K. Chandra, and P.S. Misra, 2011"Development of Aluminium Based Hybrid Metal Matrix Composites for Heavy Duty Applications", *Journal of Minerals & Materials Characterization & Engineering*, **10**,14, 1337-1344.
5. S Jerry Andrews Fabian, and B. Selvam, 2011, "Densification behaviour of Aluminium reinforced with Tungsten Carbide Particulate Metal Matrix Composite processed by P/M", *IOSR Journal of Mechanical and Civil Engineering*, e-ISSN: 2278-1684, p-ISSN: 2320-334X, 24-29.
6. Dinesh Kumar and Jasmeet Singh, 2014, "Comparative Investigation of Mechanical Properties of Aluminium Based Hybrid Metal Matrix Composites", *International Journal of Engineering Research and Applications (IJERA)* ISSN: 2248-9622.
7. M. Ravichandran, A. Naveen Sait and V. Ananda krishnan, 2014, "Synthesis and forming behaviour of aluminium-based hybrid powder metallurgic composites", *International Journal of Minerals, Metallurgy and Materials*, **21**, 2, 181.
8. Ramesh.B.T, 2014, "Characterization of Al Based Nano composites Using Powder Metallurgy Technique" *International Journal of research in Aeronautical and Mechanical engineering*, 2, 131-147.
9. H. Izadi, A. Nolting, C. Munro, D.P. Bishop, K.P. Plucknett and A.P. Gerlich, 2013, "Friction stir processing of Al/SiC composites fabricated by powder metallurgy", *Journal of Materials Processing Technology*, **213**,1900-1907.
10. P.C Angelo and R Subramanian, 2008, Text Book of Powder Metallurgy: Science, Technology and Applications, publications of *PHI Learning Pvt. Ltd*, ISBN 8120332814, 9788120332812, Length 312 pages.

IOP Conf. Series: Materials Science and Engineering 1059 (2021) 012030 doi:10.1088/1757-899X/1059/1/012030

11. Bhaskar. Raju S A, A R K Swamy and A Ramesha, 2019, "Mechanical Characterization of AL6061 – Tungsten carbide composites using powder metallurgy technique", *international journal of Engineering Applied Sciences and Technology,* **4**, 7, ISSN NO.2455 – 2143, 325 – 331.
12. K Sinha Text Book of Powder Metallurgy, Publications of Dhanpat Rai (P) Ltd, ISBN – B 97893831822145, ISBN – 10 9383182148, 292.
13. Rajesh purohit, R.S.Rana and C.S Verma, 2012, "Fabrication of Al – SiCp composites through powder metallurgy process and testing of properties", *International journal of Engineering Research and Applications*, **2**, ISSN: 2248 – 9622.

Vibration control in boring process using a constrained viscoelastic layer damper

V.Vishal Krishna[1], S.Saravanamurugan[2], P. Sanjeev Kishore[3], K.J.Yedhu[4], Goutham K Iswar[5], A.Shanmughasundaram[6*]

1,3,4 & 5 Under Graduate Student, 2 & 6 Assistant Professor
Department of Mechanical Engineering, Amrita School of Engineering, Coimbatore
Amrita Vishwa Vidyapeetham, India
E-mail: a_shanmugasundaram@cb.amrita.edu

Abstract A passive vibration control method is adopted to control vibrations of a boring bar. A viscoelastic layer sandwiched between main steel boring bar and aluminium tube, known to be constrained viscoelastic layer damper (CVLD), is designed and developed in order to reduce the vibration produced during the machining operations. The low density aluminium tube and natural rubber as viscoelastic layer are used in the boring bar to increase its natural frequency and damping property respectively. The finite element simulation of boring bar with and without CVLD are carried out using ANSYS and the results are used to develop a model to predict the influence of thickness of viscoelastic layer and aluminium tube on natural frequency of boring bar using full factorial design of experiments (DOE) and regression analysis .The DOE and regression analysis, carried using Minitab package, provides optimum thickness of viscoelastic layer and aluminium tube. These optimum thickness values are used to fabricate CVLD and its effectiveness to control boring bar vibration is tested by conducting machining experiments. The surface roughness of the machined components is also measured and the results show that the boring bar with CVLD is more efficient than conventional boring bar in controlling vibrations during machining. Moreover, stability lobes, which are plots of spindle speed vs. depth of cut of a machining process, are also constructed. Stability lobes indicate that the machining stability may be improved by 15 % by using boring bar with CVLD.

Nomenclature

δ	Deflection
K	stiffness
ω_n	Natural Frequency
ξ	Damping ratio
I	Area moment of inertia
E	Young's Modulus
L	Length of the boring bar
F	Force at the endpoint
ρ	Density
FEM	Finite Element Method
CLD	Constrained Layer Damper
DOC	Depth of Cut
FRF	Frequency response function

1 Introduction

The vibrations between the boring tool and work piece alter the magnitude of machining forces that lead to a self-excited vibration known as regenerative chatter. The regenerative chatter causes instability during boring process. This form of instability can damage the cutting tool and deteriorates the surface of the work piece [1]. Passive vibration control methods such as attachment of vibration absorbers and addition of damping using constrained layer dampers can be adopted to control regenerative chatter in boring process. The use of Constrained Viscoelastic Layer Damper (CVLD) is a passive method of controlling vibration. It involves a damping material, such as viscoelastic layer, placed between two elastic materials that normally have very low damping [2]&[3]. Tuned mass dampers, made of constrained viscoelastic dampers, were used to control regenerative chatter in machining processes [4]&[5]. Carbon fibre epoxy composite was used to fabricate a rotating boring bar and optimum geometry of the boring bar was found using experimental studies [6]. Damping coating using nano composite materials was applied to boring to improve chatter stability [7]. Constrained layer damping tool holder was developed to control chatter in end milling process by improving the dynamic stiffness of the end mill cutter [8]. Constrained layer damper was added to boring bar to improve machining stability of boring process. Beam theory and finite element simulations were used to optimize the constrained layer damper [9]. Finite element model of three layer composite bar consisting of viscoelastic layers was described to estimate the modal damping ratio [10]. Constrained layer damper was applied to thin walled member during milling process and the experimental results showed the improvement in surface finish [11]. The influence of chatter frequency on machining dynamics of boring bar with constrained layer damper was studied [12]. Chatter stability in boring process was analyzed by plotting stability lobes. Constrained layer damper was added with conventional boring bar to control regenerative chatter [13]. Chatter stability in deep hole boring process was analyzed. Constrained layer damper was used to improve the chatter resistance of a slender boring bar[14]. In this work, a ring type constrained viscoelastic damper is used to control machining vibrations in boring process.

2 Hollowing of boring bar

In order to place a ring type CVLD, a hole is drilled in boring bar. The removal of material increases the natural frequency and deflection of the boring bar whereas the stiffness is reduced. The variations of stiffness, deflection and natural frequency with inner diameter of the hole of boring bar are plotted as shown in fig-1, 2 & 3. Stiffness is calculated theoretically from the formula [12]:

$$K = \frac{3EI}{L^3} \qquad (1)$$

where K= Stiffness of bar, E= Young's modulus of the bar, I= Moment of Inertia, L= overhang length of the bar.

$$I = \frac{\pi(d1^4 - d2^4)}{64} \qquad (2)$$

where d1 = Outer diameter of boring bar, d2 = Inner diameter of boring bar.

Deflection is calculated theoretically using the formula [12]:

$$\delta = \frac{FL^3}{3EI} \qquad (3)$$

where F= Force on the boring bar, E= Young's modulus of the bar, I= Moment of Inertia, L= overhang length of the bar.

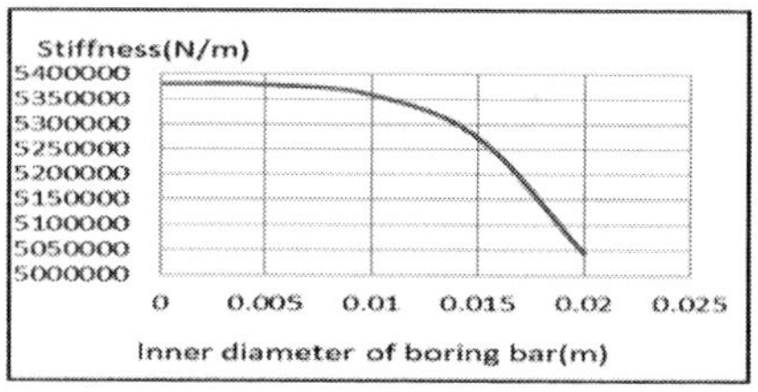

Figure 1 Inner diameter of boring bar vs deflection.

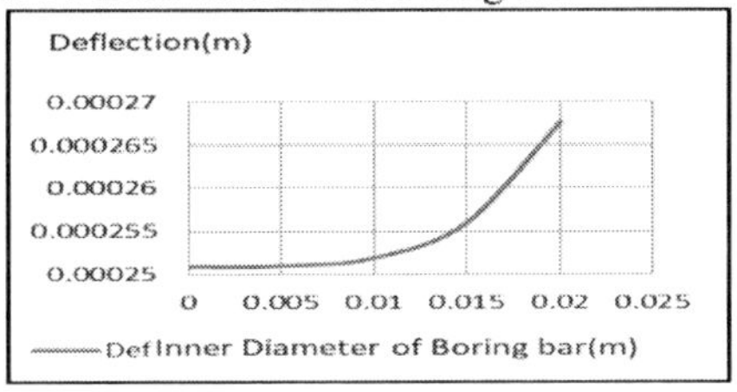

Figure 2 Inner diameter of boring bar vs Deflection.

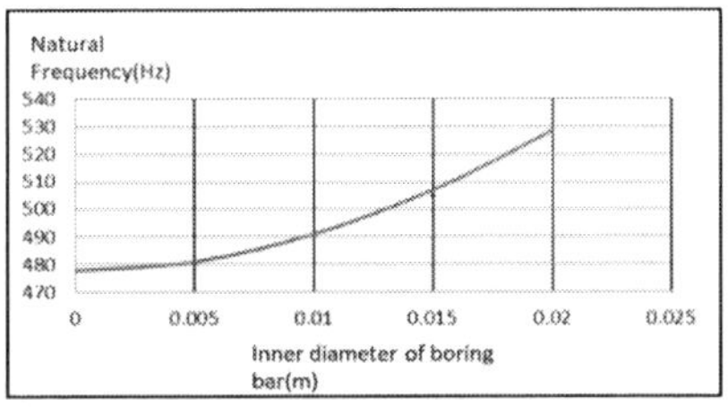

Figure 3 Internal diameter of boring bar vs Natural Frequency

From the Fig 1, 2 & 3, it is inferred that natural frequency increases steadily due to the reduction in stiffness and mass where the latter plays a major role. Also, Steady increase in deflection is found. The inner diameter and outer diameter of boring bar is fixed to be 20mm and 40mm respectively. The 10% reduction in stiffness is compensated by adding CVLD which also enhance the damping. The boring bar dimensions used in this study are given in the table-1.

Table 1 Boring bar dimensions

Attributes	Quantity
Diameter	40mm
Length	300mm
Material	Hardened steel

The bar was modelled in ANSYS APDL software with the above dimensions. The material properties of boring bar and CVLD are given in table 2:

Table 2 Material properties

Properties	Steel	Rubber	Aluminium
Young's Modulus	210 GPa	0.001 GPa	71.7 GPa
Density	7850kg/m^3	900 kg/m^3	2810 kg/m^3
Poisson's Ratio	0.3	0.47	0.33

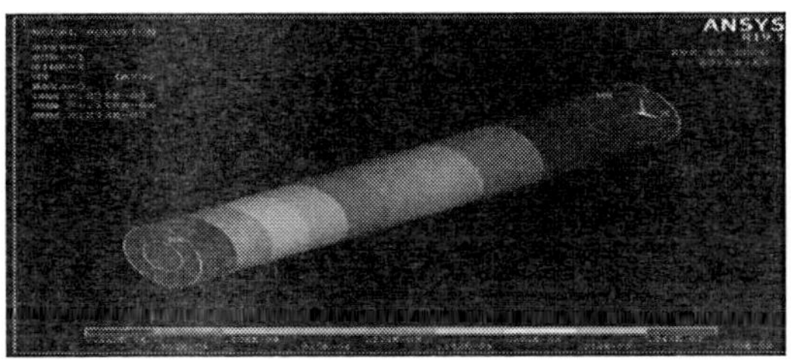

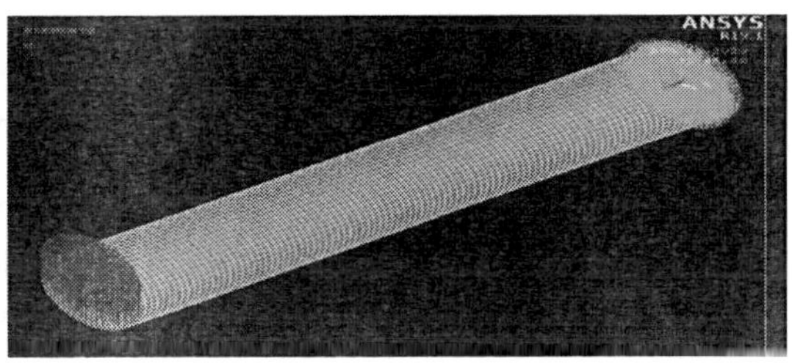

4(a) 4(b)

Figure 4 Finite element Mesh of boring bar in ANSYS with loads and results

The meshed finite element model of the boring bar is shown in Fig-4. Static and modal analyses are used to find deflection and natural frequency of boring bar with CVLD respectively. For static analysis, the cutting force of magnitude 1350 N is used [14]. Finite element simulations are carried out by increasing the inner diameter of the boring bar from 0mm to 20mm with the step of 5mm. Two other combinations of material used to make CVLD are given in the table 3 and 4.

Table 3 Materials used in finite element simulation

Viscoelastic material	Elastic Material
Rubber	Aluminium
Foamed-Aluminium- -material(FAM)	Carbide
	YG20C

Table 4 Material properties

Material	Density	Young's modulus	Poisson's ratio
FAM	650 kg/m3	12 GPa	0.33
YG20C	13400 kg/m^3	400 GPa	0.3
Carbide	12180 kg/m^3	606.7 GPa	0.3

Using ANSYS APDL, three hollow cylinders are constructed and they are glued together in the order of boring bar material as outer layer, viscoelastic layer as middle layer and elastic material as inner layer. The results of natural frequency and deflection for different thickness values of rubber and metal layer are used in Design of Experiments (DOE) to find the optimum geometry of CVLD to control regenerative chatter in boring process.

3 Design of experiments

Design of experiments is performed to determine the variation of natural frequency and static deflection with thickness of both the viscoelastic and elastic layers. Minitab software is used to perform DOE and it basically works on statistical background. Each factor has 4 levels and there are totally 16 experiments. The Minitab randomizes these experiments and provides the thickness to be varied. The combinations obtained through full factorial method are given in the table 5.ANSYS results are exported to Minitab to carry out the statistical analysis and the results are plotted as shown in fig 5. With regression analysis, the optimum thickness of viscoelastic layer and elastic layer are found.

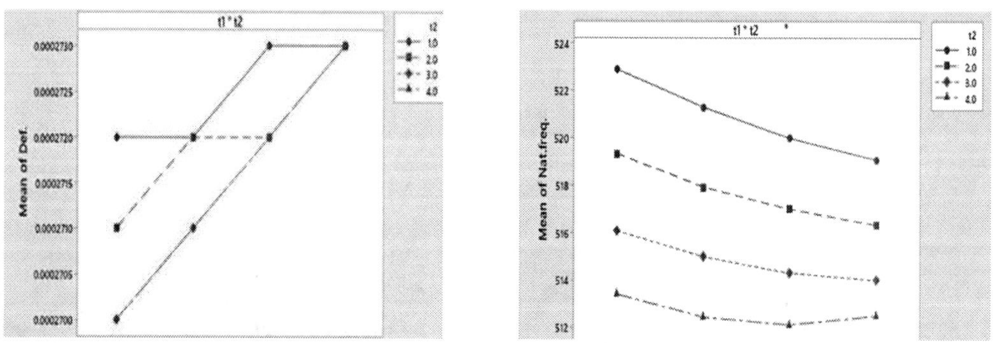

(a)Natural Frequency (b)Deflection

Figure 5 Interaction Plots of the natural frequency and deflection of MS-rubber –Aluminium combination.

The interaction plots, obtained from regression analysis, represent the variation of natural frequency and deflection trends with thickness of rubber and aluminium. It is found that there is a steady decrease in natural frequency with increase in thickness and marginal change in deflection exists with changes in thicknesses.

Table 5 Design using full factorial Method

t1 (mm)	t2 (mm)	Natural Frequency (Hz)	Deflection (m)
0.4	1.2	523.425248	0.000271
0.4	2.4	518.948288	0.000268
0.4	3.6	515.133728	0.000266
0.4	4.8	511.981568	0.000263
0.4	6	509.491808	0.000261
0.8	1.2	522.597312	0.000271
0.8	2.4	518.272992	0.000268
0.8	3.6	514.611072	0.000266
0.8	4.8	511.611552	0.000263
0.8	6	509.274432	0.000261
1.2	1.2	521.840192	0.000271
1.2	2.4	517.668512	0.000268
1.2	3.6	514.159232	0.000266
1.2	4.8	511.312352	0.000263
1.2	6	509.127872	0.000261
1.6	1.2	521.153888	0.000271
1.6	2.4	517.134848	0.000268
1.6	3.6	513.778208	0.000266
1.6	4.8	511.083968	0.000263
1.6	6	509.052128	0.000261
2	1.2	520.5384	0.000271
2	2.4	516.672	0.000268
2	3.6	513.468	0.000266
2	4.8	510.9264	0.000263
2	6	509.0472	0.000261

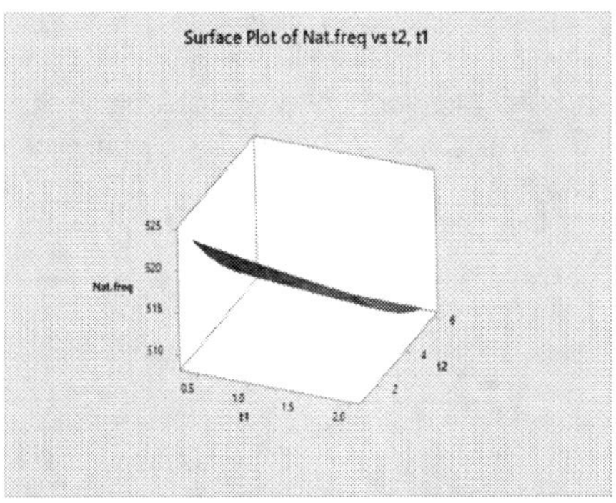

(a)Natural Frequency

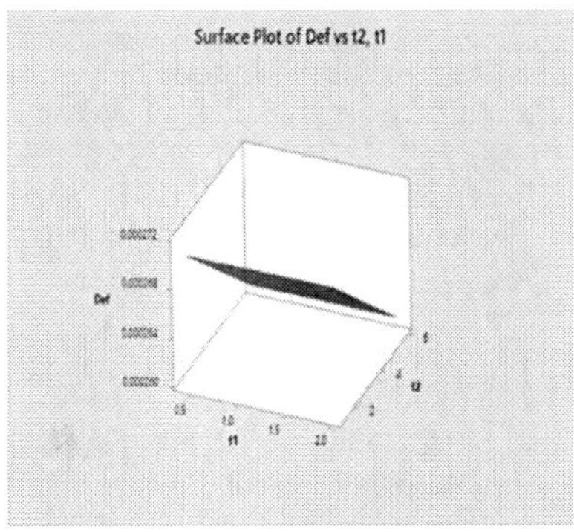

(b)Deflection

Figure 6 Surface plot of the natural frequency and deflection of MS-rubber –Aluminium combination.

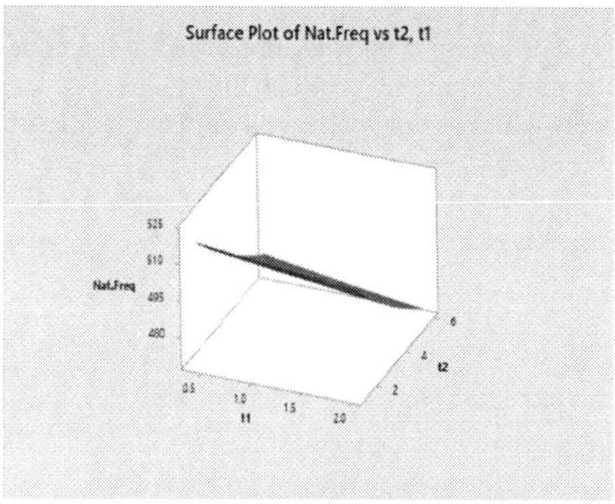

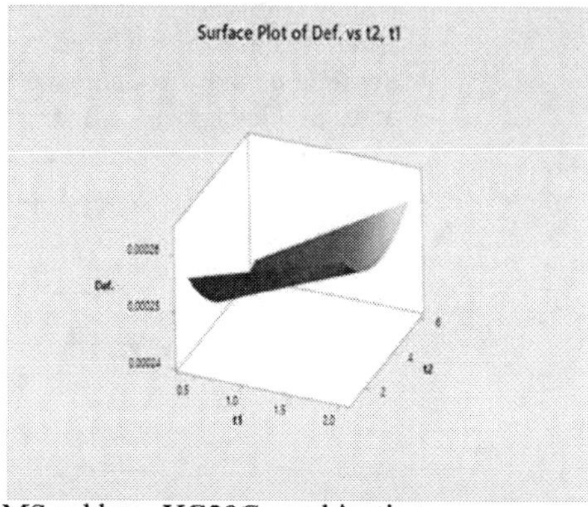

(a)Natural Frequency

Figure 7 Surface plot of the natural frequency and deflection of MS-rubber –YG20C combination.

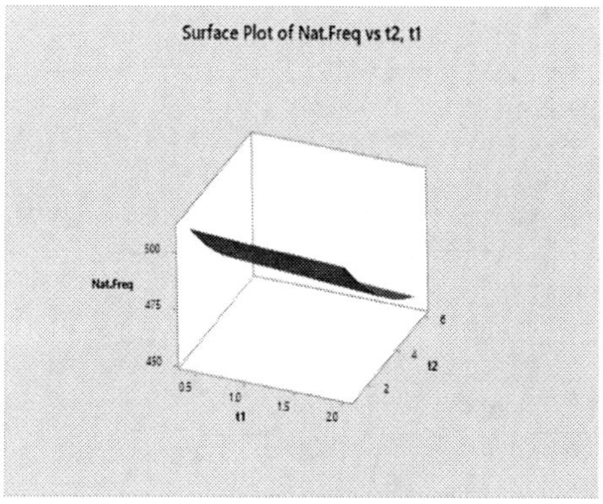

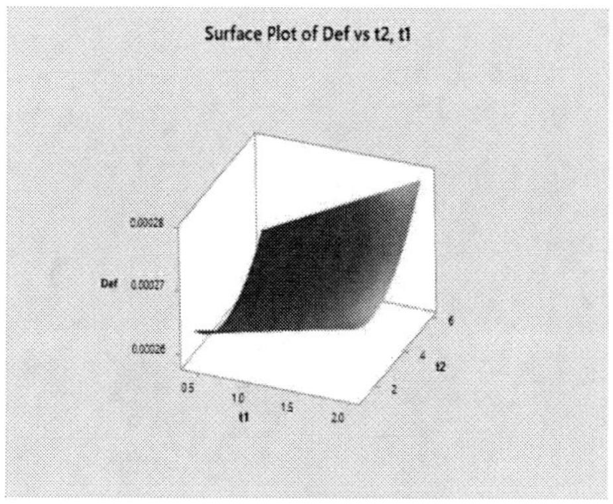

(a)Natural Frequency (b)Deflection

Figure 8 Surface plot of the natural frequency and deflection of MS-rubber –Carbide combination

Figures 6,7 &8 show the surface plots depicts the variation of natural frequency and deflection. The surface plots reveal that there is a steady decrease in natural frequency over the gradual increase in thickness of viscoelastic and elastic layers whereas the deflections of the boring bar increases with decrease in thickness of two layers. By choosing mean deflection and mean natural frequency as basis, the thicknesses of both viscoelastic and elastic layers are chosen to fabricate the CVLD.

4 Experimental Method

In order to assess the effectiveness of CVLD in improving damping, two hollow boring bars with and without CVLD were fabricated. The CVLD was press fitted into one of the hollow boring bars and machining experiments were conducted. The surface roughness of the machined surface was also measured. It was also ensured that the both the boring bars have the same specification for the comparison purpose.

4.1 Selection of boring bar

The important parameter to be considered while selecting boring bar is the diameter to length ratio. Generally, Steel boring bars have a diameter to length ratio of 4:1. Choosing an appropriate boring bar material is also needed for an efficient vibration resistant operation. Some of the common materials used to make boring bar are steel, heavy metal or carbide. The specification of the boring bar S40TDCLNR12F3 is given in the following table. The density and Young's modulus of the boring bar is taken as 7800 kg/m3 and 210 GPa respectively. To increase the natural frequency of boring bar, boring bar is hollowed inside the conventional bar so that the decrease in mass results in increase the natural frequency. The conventional boring bar is bored with inner radius of 18.7 mm. Tool description of the boring bar is elaborated below in table 6.

Table 6 Tool description

Symbol	Description
S	Shank material (steel)
40	Diameter(mm)
T	Shank Length (300 mm)
D	Double clamp
C	80 deg. Apex angle diamond type
L	Major cutting edge angle (95 deg.)
N	Insert relief angle (0 deg.)
R	Feed direction right hand

4.2 Fabrication of CVLD

Boring bar with CVLD was fabricated by die casting. The composite bar of natural rubber and aluminium was press fitting inside the hollow boring bar. Aluminium 6xxx series was used in fabrication. It contains magnesium silicate which makes it heat treatable and it has medium strength and weld ability and machinability properties. Aluminium rod of 19mm diameter is turned to 14.7mm outer diameter and drilled to 6.5mm inner diameter. Length of the rod is 240mm. Natural rubber is die casted upto 2mm thickness over the hollow aluminium rod. Then the aluminum rad with natural rubber was press fitted ress.

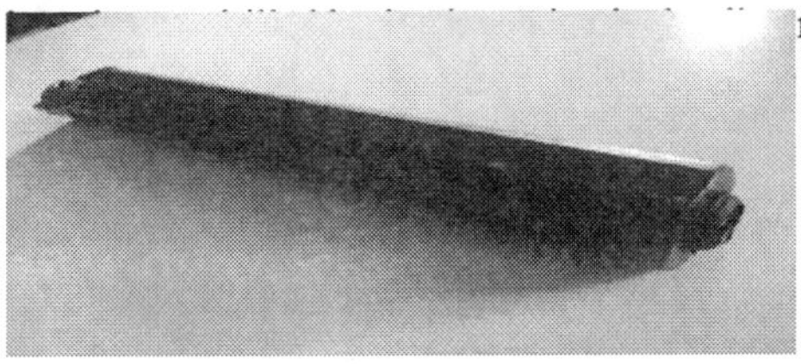

Figure 9 Boring bar fitted with CLD

The boring bar with CVLD is shown in figure 9. The second part of experiment, which was machining, was done to assess the vibration control capabilities of CVLD by determining the surface roughness. The two hollow pipes were machined by conventional boring bar and boring bar with CVLD. The spindle speed of 280 rpm, feed of 0.2 mm, depth of cut of 0.25 mm and boring bar overhang length of 200 mm were chosen for machining experiments. Experimental setup of the machining experiments is shown in figure 10.

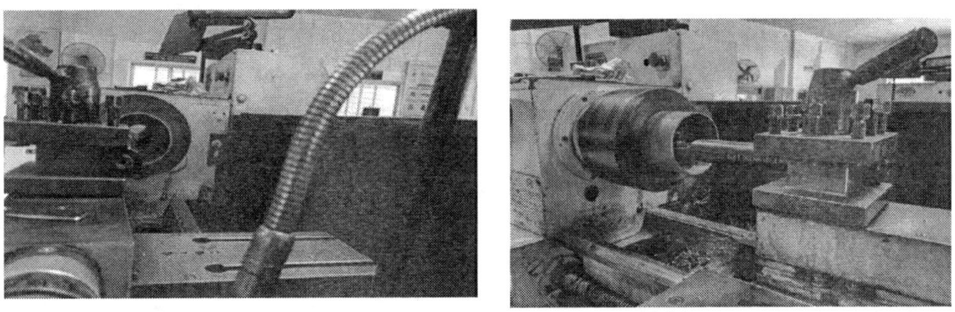

10(a) 10(b)

Figure 10 Experimental setup -Machining process using boring bar without and with CVLD

4.3 Harmonic analysis

Harmonic analysis is done by using ansys. Ansys offers different solution methods among which the full method was used to obtain the harmonic response of the boring bar with and without CVLD. The Frequency range was chosen from 200-800 Hz with 1Hz as sub step. The harmonic analysis gives the plot of frequency response function (FRF) of Y-component displacement vs frequency. The mode shapes are defined as a specific pattern of vibration occurring at a specific natural frequency of the structure. They are determined by defining the stiffness and mass and also the boundary conditions of the structure. The harmonic analysis is used to predict the steady-state response of the structure by applying external harmonic force over a specified frequency range based on results of modal analysis. In order to include damping to the system, the structural beta damping is applied. The damping factor value 0.02 is applied for conventional boring bar whereas 0.05 is used for boring bar with CVLD. Using the equation 4[13], the beta damping value is calculated and used in the harmonic analysis.

$$\beta = \frac{2\xi}{\omega} \qquad (4)$$

The FRF of the boring bar with and without CVLD is shown in Fig 11(a) and 11(b).

11(a) 11(b)

Figure 11 Harmonic response of boring bar without and with CVLD

The peak amplitude of FRF of boring bar with CVLD is lower than that of with CVLD. Hence the results of harmonics analysis establishes the effectiveness of CVLD in controlling vibration.

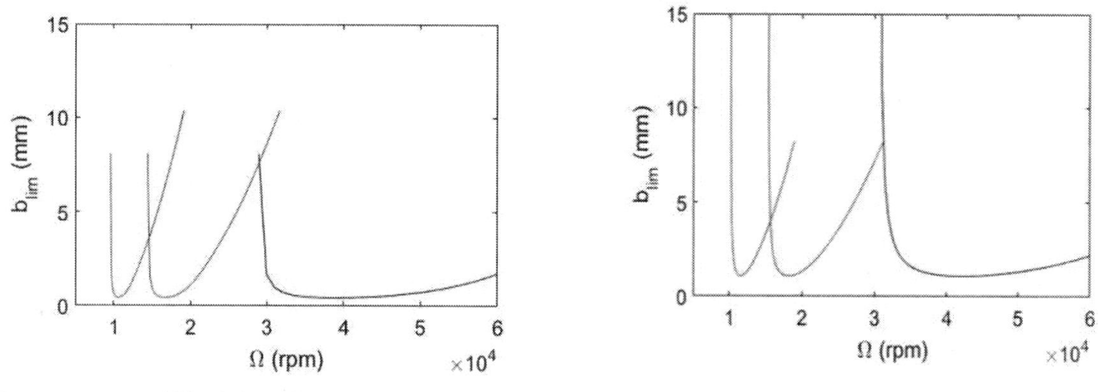

Figure 12(a) Stability lobes of boring bar without CVLD **Figure12(b)** Stability lobes of boring bar with CVLD

Figures 12a & 12 b represent the Stability lobe diagrams of boring bar without and with CVLD respectively. Stability lobes are plots of spindle speed vs depth of cut [14]. The frequency response function, which is required for construction of stability lobes, is obtained from harmonic analysis carried out in ANSYS. From the stability lobe diagrams, it may be inferred that the one with the damper has higher stability when compared to that of conventional boring bar without damper.

5 Results and discussion

From the modal analysis of the boring bars, it is found that the natural frequency of boring bar with damper is 504 Hz whereas the natural frequency of conventional boring bar is 477 Hz. Increase in the natural frequency leads to better machining stability. Furthermore, the results from harmonic analysis also prove that the CVLD minimizes the amplitude of FRF of boring bar. Machining experiments were also conducted and surface roughness values were measured. The machined components are shown in figure 13 (a) and (b). Surface roughness centreline average (R_a) was measured using CARL ZEISS (E-35A) portable surface measuring equipment. The surface profile of the component machined without and with damper are shown in the figure 14 and 15 respectively

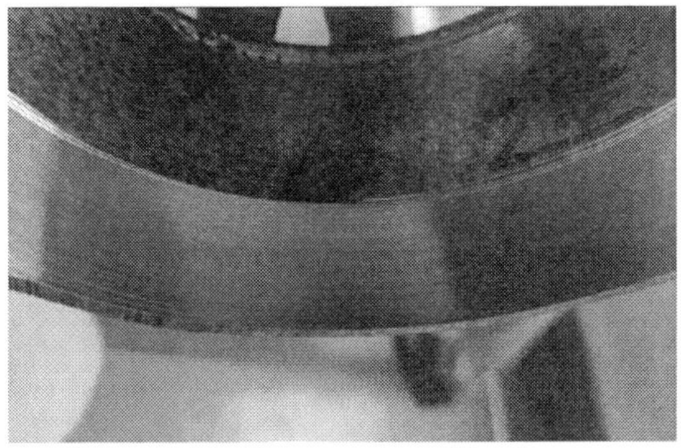

Figure13(a) Component machined without damper **Figure 13(b)** Component machined with damper

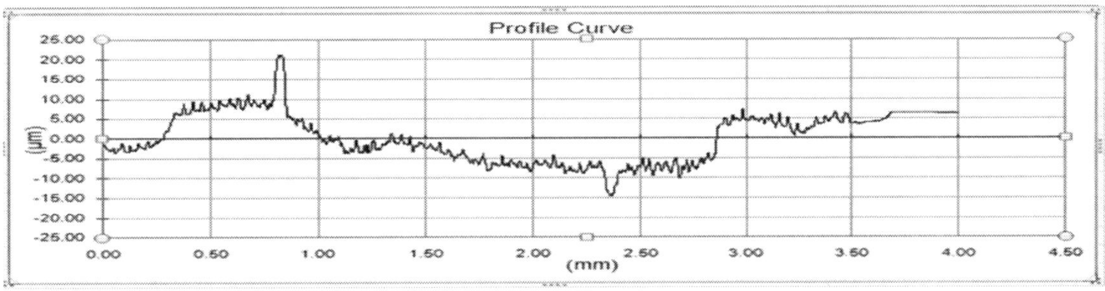

Figure 14 Surface profile of component machined without damper.

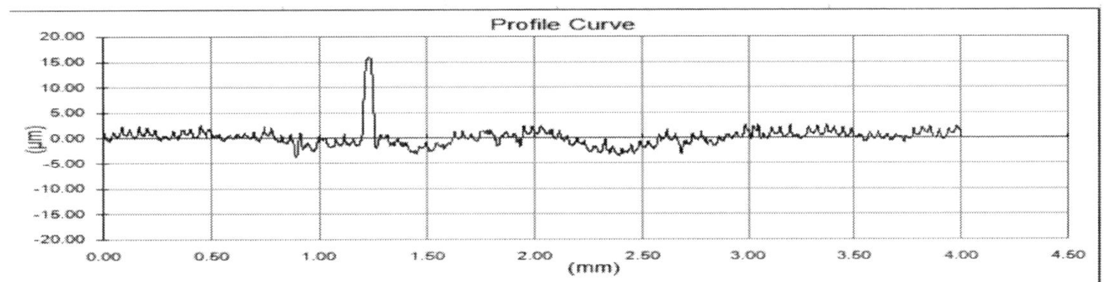

Figure 15 Surface profile of component machined with damper.

From the two profiles, it is inferred that the boring bar with CVLD provides a smooth surface finish. The component which was machined without the damper has Ra value of 5.36 µm and with damper has Ra value of 1.06µm.

6 Conclusions

In this work, a constrained layer damper was designed, fabricated and tested for a boring bar to control vibrations during boring operations. The optimum thickness of natural rubber and aluminum tube were obtained using design of experiments and regression analysis. The finite element simulation results of boring bar with CVLD were used to develop a statistical model which was analyzed for maximization of natural frequency. The CVLD was press fitted into the hollow boring bar and the hollowing of boring bar and addition of CVLD leads to increase in natural frequency and damping of the boring bar. Machining experiments were conducted to assess the effectiveness of boring bar with and without CVLD. The surface profile of the component machined without damper depicts that there was a large variation in the surface roughness (R_a) with reference to mean centre line, whereas the surface profile of the component machined with damper shows very small variations. It may be inferred that there was an improvement of 65% in the surface roughness of the machined component with damper.

References

[1] Tobias S.A 1965 *Machine-tool vibration J. Wiley.* .
[2] Jones, David IG. 2001 *Handbook of viscoelastic vibration damping. John Wiley & Sons.*
[3] Zhou X Q, DY Yu, XY Shao ,SQ Zhang, and S. Wang 2016 *Composite Structures* **136**, 460-480.
[4] Saravanamurugan S, Thirunarayanaswamy A, and Devarajan K 2015 *Journal of Vibration and Control* **21**, **5**, 949-958.
[5] Ram, Ganapathy HV, and Saravanamurugan S 2019 *In IOP Conference Series: Materials Science and Engineering* **577**, **1**, 012152.
[6]Hwang HY and Kim JK 2003 *Composite Structures*, **60(1)**, 115-124.
[7]Fu Q, Lundin D and Nicolescu CM 2014 *Journal of materials engineering and performance* **23(2)**, 506-517.
[8] Liu Yang, Liu Z, Song Q, and Wang B 2016 *International Journal of Mechanical Sciences* **117**, 299-308.
[9] Song Q, Jiahao Shi J, Liu. Z, Wan Yi and Xia. F 2016 *The International Journal of Advanced Manufacturing Technology* **83, 9-12**, 1951-1966.
[10] Johnson Conor D, and David A Kienholz 1982 *AIAA journal* **20, 9**, 1284-1290.
[11] Shi Jiahao, Song Q, Liu. Z and Wan. Yi 2017 *International Journal of Mechanical Sciences* **128**, 294-311.
[12] Liu, Yang, Liu Z, Song Q, and Wang B 2019 *Journal of Materials Processing Technology* **266**, 687-695.
[13] Yuhuan, Zhang, Yongsheng R, Jishuang T, and Jingmin Ma 2019 *Journal of Vibration and Control* **25, 16**, 2204-2214.
[14] Zhang, Jinfeng, Wang H, Yongsheng Ren, Chao Feng, and Chunjin Zhang 2020 Applied Sciences **10, 13**, 4537.
[15] Young, Clarence W, Budynas R.G and Ali M. Sadegh 2002 *Roark's formulas for stress and strain*. Vol. 7. New York: McGraw-Hill.
[16] Altintas Yusuf and A. A. Ber. 2001 *Manufacturing automation: metal cutting mechanics, machine tool vibrations, and CNC design. Appl. Mech. Rev.* 54, no. 5 : B84-B84.
[17] Manual, ANSYS User'S 1994 Swanson analysis systems. Inc., Volumes I, II, III, IV, Revision 5.

Design and Fabrication of Robotic Arm for the assembly of Phase Selector Switch

C. Sowmya Dhanalakshmi1, P. Madhu, N. Hemachandran, V.E. Muthukumar and Harish Arvinth.L.B

B.E. Mechanical Engineering, SNS College of Technology, Coimbatore

chandruntr430@gmail.com

Abstract. This paper explores the robot arm where the ability of human to do a project is restrained however now not with the aid of his mental electricity however through his bodily energy. A humanoid mechanical technology is another difficult field. To co-work with people, humanoid robots not just need to include human like structure and structure, yet more significantly, they should arrange human like conduct with respect to the movement, correspondence and intelligence. In environmental factors where human exchange can't be conceivable to do a specific errand, the robots can do. Robots are used for lean industrial processes and have diversified their contributions to meeting lines inside the production international. The main focus of this project is to design and develop the mechanism for robotic arm for lifting. It is a type of mechanical arm, usually programmable, with similar functions to a human arm. The arm might be a unit system or might be an aspect of a more unpredictable mechanical cycle. The end effectors or mechanical hand can be intended to play out any ideal undertaking, for example, welding, holding, turning etc., depending on the application. The mechanical arm is planned and designed with four degrees of freedom and modified to achieve precisely basic light material lifting undertaking to aid the creation line in any industry. 3D printing strategy is utilized in this undertaking to create the parts of the automated arm. Hence, it gave more exact measurements and gigantic time and cost expensive in creation. The automated arm is furnished with 4 servo engines to interface the parts and bring arm development. A sequential construction mechanical system can improve its efficiency through developing the assembling pace and its consistency. They additionally spare individuals from dreary and tedious mechanical production system employments. Many industries are yet to implement such automation in their meeting traces, as they do no longer have the technical recognize how of the changeover, or the concern of failure of investment. It is proposed to layout and increase a low price assembly robotic which could overcome the above drawbacks. We have also included Design Thinking process before starting to do this experiment. The stages of Design Thinking such as Empathize, Define, Ideate, Prototype & test, Evolve are performed in the complete fabrication of the robotic arm. Many industries are yet to implement such automation in their assembly lines, as they do not have the technical knowhow of the changeover, or the fear of failure of investment. To proceed in this direction, a case study was conducted at a phase selector switch making company. Presently the company does manual assembly of the switches, which is a hindrance to the growth of the company. This process required a pilot study to explore the possibility as well as the implementations issues of robots in the assembly. The present system uses a belt conveyor system which has no feedback signals and the speed of the conveyor is not synchronized with the rate of assembly. It is generally not advised to transfer small size components in a conveyor since picking, orienting and inserting the components will be cumbersome. When the size of the component is considerably small, manual assembly is not recommended as per the guidelines of DFMA (Design for manufacture and assembly). Manual error is more prone to assembly of the part in a wrong orientation, especially for those components that are symmetrical.

The parts to be assembled are both axi-symmetric as well as prismatic in shape. Constraints in loading and unloading of casing were also reported. In this work, the design and development of a low cost assembly robot is done which can overcome the above drawbacks.

1 Introduction

Robots are used for lean business strategies and have diversified their contributions to meeting lines within the production world. An assembly line robotic can improve its productivity through growing the manufacturing pace and its consistency. They also keep employees from tedious and monotonous meeting line jobs. End of arm tooling or End effectors can be custom designed for every meeting robotic to cater to the producing necessities. Additional alternatives, like robot imaginative and prescient, can also be incorporated to enhance performance and accuracy of part orientation or sorting identifiers.

Many industries are but to implement such automation of their meeting traces, as they do not have the technical knowhow of the changeover, or the fear of failure of investment. To continue on this path, a case examine changed into conducted at a segment selector switch making agency in Coimbatore. Presently the enterprise does guide meeting of the switches, which is a difficulty to the boom of the agency. This manner required a pilot take a look at to explore the possibility in addition to the implementations problems of robots in the assembly.

The following inconsistencies were discovered by means of the initial have a look at.

1. The present system uses a belt conveyor machine which has no comments indicators and the velocity of the conveyor is not synchronized with the price of assembly.
2. It is normally now not counseled to transfer small length additives in a conveyor considering the fact that picking and orienting and putting the additives might be cumbersome. When the scale of the element is considerably small manual assembly isn't recommended as in keeping with the tips of DFMA (Design for manufacture and meeting).
3. Manual errors is greater liable to assembly of the part in an incorrect orientation, particularly for those additives which might be symmetrical.
4. The parts to be assembled are each axi-symmetric as well as prismatic in shape
5. Constraints in loading and unloading of casing have been additionally suggested.

In the light of the above situation, a Project become envisaged to automate the entire meeting of PS-forty phase selector transfer at M/s Salzer Electronics, positioned at Coimbatore.

It is proposed to design and develop a low cost assembly robotic which can conquer the above drawbacks. Since the no. of components to be assembled is huge multi-stage meeting is endorsed.

This idea is submitted for investment the assignment for this meeting line. The knowhow and the experienced gained with the aid of this educational work may be shared with different such requirements of meeting in numerous small and medium scale industries.

2 Objective & hypothesis of work

2.1 *Objective*

The task is aimed at design, improvement and implementation of a totally automated the switch meeting unit for PS-forty (Phase Selector Switch) the usage of robots and allied devices. The

Predominant purpose of this proposed research work is to construct automation within the switch meeting unit. The Pick and place mechanism, Servo managed conveyor, Vibratory bowl for loading the additives is to be applied for assembling the transfer gear. The implementation of robot arm mechanism is anticipated to improve the performance inside the meeting unit of Salzer Electronics.

2.2 *Hypothesis of the work*

Automation has come to be a figuring out component in whether or not an organization will remain aggressive in the production industry. In the past twenty years, technology has modified the nature of manufacturing. In the vintage days, production and fabrication have been all accomplished by using hand by means of people. Now the computer systems and generation have penetrated the enterprise, automation has emerge as the competitive benefit in today's manufacturing world. Automation has allowed corporations to produce mass products at notable speeds and with notable repeatability and nice.

IOP Conf. Series: Materials Science and Engineering 1059 (2021) 012032 doi:10.1088/1757-899X/1059/1/012032

3 Review of literatures

Bhavesh Kaila, (2014) used a choose and area robotic for moving components in a ceramic industry and feature efficiently applied the same.

Sushanth Kanbail Srinivas, (2013) used a select and place mechanism actuated via an electromagnetic device for picking components from a conveyor. This required the synchronisation of the conveyor movement along with the collection of the robotic operations. .

Ravikumar Mourya, (2015) had proposed a four-DOF articulated pick and vicinity robot arm that does operations like gripping, lifting, placing and liberating. Articulated arm consisted of revolute joints that allowed angular movement among adjoining joints. Four servo automobiles were used on this mission to perform four diploma of freedom (four-DOF) and successfully established its usage.

Santosh C, (2016) designed and fabricated a pneumatically operated pick out and vicinity arm for moving non- cylindrical gadgets from a decrease plane to a higher plane. It consists of cylinders, a shaft with a lead screw mechanism converting the piston moves right into a rotary motion using compressed air. The pneumatic arm model can be without problems set at intermediate positions by way of regulating the pressure the use of the flow manipulate valve. It can be used in loading and unloading of goods in shipping harbors because the motion of products is achieved from decrease plane to better aircraft.

Visual Vasistha, (2014) has designed and implemented a Pick and Place operational Robot with stud mechanism for loading and unloading machines, setting components on circuit forums, moving components off conveyor belts and so forth. The disadvantage is of those sorts of robots are, they should be custom made to a particular software.

4 Scope and methodology:

The following are the scope of this take a look at which can be listed in conjunction with the method of implementation:

- Working version Automation gadget
- Prismatic joint for automation
- Stepper motor for conveyer belt
- Stack for loading of additives
- Servo motor for conveyer belts rather than stepper motor
- Vibratory bowl for loading of additives in preference to stack
- Pick and vicinity mechanisms
- Manual assembly of switchgear deliberating the assembly problems
- Minimization of Cost to make the device lean
- Timing diagram and Electronic layout to be considered
- Finalization of Bill of substances
- Prototype checking out
- Prototype testing at the industry web site
- Implementing of the device on the enterprise and analyzing the time reduction and compare with the existing guide feed of additives
- Training of enterprise employees and turning in of the challenge

5 Tools used for implementation

The Raspberry Pi is a low price, credit-card sized laptop that plugs right into a trendy display device. To add to it, GPIO guide additionally method it has to interface and coordinate among the OS and low-level hardware. Thus, to get most effective overall performance, Raspberry Pi 3 Model B+ which has excellent computational strength amongst all of the editions. Raspberry Pi runs at the Raspbian OS, a Linux kernel essentially. Thus, any external webcam will work (Using a general USB webcam), but now not with the equal performance level as the digicam module v2 (Camera Module V2 - Raspberry Pi) provided through Raspberry Pi which has an eight MP sensor. It is ideally suited for pleasant real-time processing responsibilities the use of RPi. Also, electricity and modern-day consumption, as well as latency issues, are deterrent to the usage of a webcam with a Raspberry Pi for High-best belief tasks like pick and place.

IOP Conf. Series: Materials Science and Engineering 1059 (2021) 012032 doi:10.1088/1757-899X/1059/1/012032

6 Design Calculations

Table 1. Specifications of the components

Sl. No	Items	Specifications
1	Pneumatic Cylinders	5mm Diameter
2	C – Frame	10 meters
3	Arm Gripper	4mm displacement
4	5/2 Solenoid operated DCV	
5	FRL Unit	Medium duty
6	FRL adapters	Medium duty
7	Pneumatic Link Wires	8 meters
8	Pneumatics @ 20,000	7 sets
9	Conveyor Belts (Rubber/leather)	LXW – 4 X 0.300 m
10	Pulleys	100 mm dia
11	Compressor Unit	1 HP
12	PIC-MICROCONTROLLER (interfaced with drives)	(7 no's)7
13	Servo motor for conveyor belt with feedback	(if necessary)
14	Servo motor for grip.	
15	Motor controlled drive	(for controlling DCV)
16	Raspberry Pi 3 Model B+ Complete kit with Accessories	Model B
17	Waveshare Newest Raspberry Pi IR-CUT Camera Module Kit Sensor to Support Night Vision	5MP OV56471
18	Raspberry Pi 7TOUCH Touchscreen LCD Display	1 No
19	Raspberry Pi	No

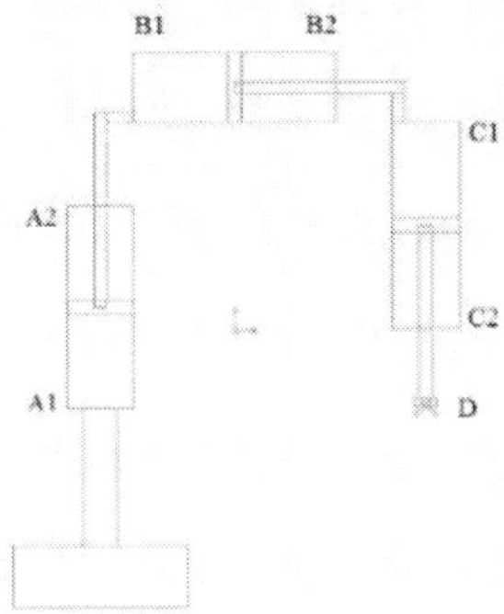

Figure 1. Block Diagram of the proposed system

Table 2. Valve sequencing

TIME (Sec)	DCV A	DCV B	DCV C	D
0	-	-	-	-
1	-	B1	-	-
2	A1	-	-	-
3	-	-	C1	-
4	-	-	-	CLOSE
5	A2	-	-	-
6	-	B2	-	-
7	A1	-	-	-
8	-	-	-	OPEN
9	-	-	C2	-
10	-	-	-	-

Table 3. Components used

Sl.No	Part Name	Quantity	Image
1.	Cam	3	
2.	Bush	2	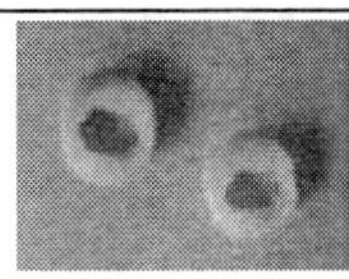

IOP Conf. Series: Materials Science and Engineering 1059 (2021) 012032 doi:10.1088/1757-899X/1059/1/012032

3.	Circlip	2	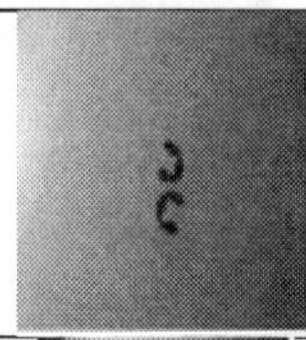
4.	Hexagonal Shaft	1	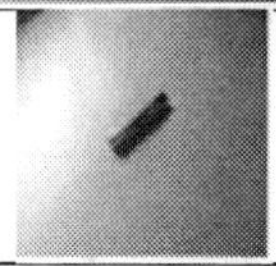
5.	Brass Plate A	1	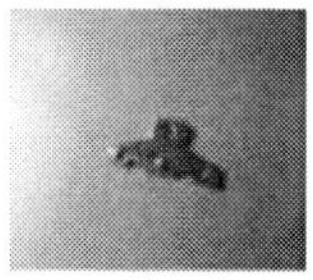
6.	Brass Plate B	3	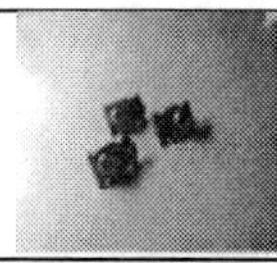
7.	Brass Plate C	3	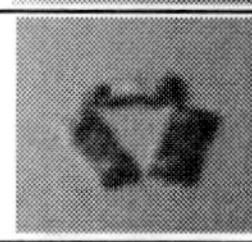
8.	Spring	3	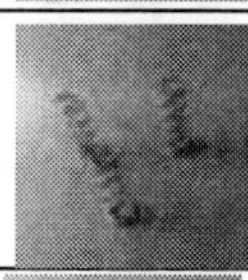
9.	Follower	3	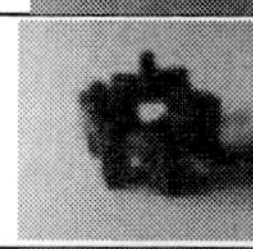
10.	Screw	2	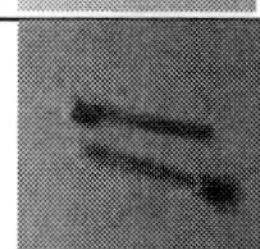
11.	Casing A & B	1	

6.1.1 *Design of Cylinders*

Horizontal, vertical, rotary cylinders & arm lifting cylinder are of same design and selected from the product catalogue (Janatics pneumatic products). Cylinders are of double acting type. So double acting cylinders are preferred. This time interval cannot be achieved by single acting cylinders. According to the applications the forward and return stroke of the piston has to be controlled with some time interval.

Force to be exerted is 40N

Pressure x area = 0.4 x 105 N/m2

Force/pressure = 40/40000 = 0.001m2

From Janatics pneumatic products catalogue we have selected 40mm bore diameter cylinder

For forward stroke

For 40mm bore diameter

Corresponding rod diameter = 16mm
Area of the piston = (πd2)/4 = (πx402)/4 =1256.8mm2
Force (modified) to be exerted = Pressure x area
=0.4x105 x 1256.8 = 50N

For return stroke

On the return stroke, when the pressure is applied to the reverse direction, the force on the piston due to the pressure is = P x (A-a)
Where,
P = Pressure in the cylinder (N/m2) A = Area of the piston (m2)
A = Cross sectional area of the piston rod (m2) Therefore
Area of the piston (A-a) =1256.6 - 2011055 mm2
Force to be converted on the reverse direction
pressure x area = 0.4 x 105 x 1055 = 42.2 N
For working pressure of 0.4 x 105 N/m2
Extending force = 50.3 N
Retracting force = 42.2 N

1. *Design of Arm Lifting Cylinder of the prototype*

Force to be exerted is 30 N
Force= pressure x area
Pressure in the cylinder = 0.4 x 105 N/m2
Area of the piston = (π x d2)/4
= 30/0.4 x 105
= 0.75 x 105 m2
=750mm2
Bore diameter = 27.38 mm
From Janatics pneumatic product catalogue we have selected 32mm as bore diameter of arm lifting cylinder

For forward stroke

For 32 mm bore diameter Corresponding Rod Diameter = 12mm
Area of the Piston = $(\pi \times d^2)/4$
= (π x 122)/4 = 804 mm^2
Force to be exerted = 0.4×10 5 x 804 = 32.2 N

For return stroke

On the return stroke, when the pressure is applied to the reverse direction, The force on the piston due to the pressure is = P (A-a)
Therefore,
Actual area of the piston (A-a)
= $\{(\pi \times d^2)/4\}-\{(\pi \times d_1^2)/4\}$
= {(π x 402)/4}-{π x (122)/4}
= 690 mm^2
Force to be exerted in the reverse direction
= Pressure x area = 0.4 x 105 x 390 = 27.6 N
For working pressure of 0.4105 N/m^2
Extending Force = 50.3 N, Retracting Force = 42.2 N

7 Conclusion

Thus, a pneumatic pick out and location robot arm is evolved which helps to recognize how to obtain low fee automation. The ability of our material coping with gadget is 35 N. The software of pneumatics produces clean operation. The paintings extent is produced within the attitude of 120°. With connection with remark made in this work, it could be concluded that employing the choose and place mechanism will carry accuracy and efficiency to the manufacturing machine. It is controlled with the aid of Arduino microcontroller. It makes use of solenoid valve and pneumatic cylinder for motion and rotation of robotic arm. Automation is the manner to bring those modifications. The gadget will lessen human attempt notably, hence lowering the price incurred by using human errors. The prototype can be used in a whole lot of industries which includes micro technology, manufacturing, packaging, canned food industry for inter transportation. The fabricated model will reduce the want for the presence of humans in industries. It will reduce the time and growth productivity.

In a comparable manner, any other 11 robot palms are deliberate to be fabricated and six arms in each transfer tools assembly line is to be carried out for assembling correctly. This will growth the Production price of the enterprise and will lessen errors which in flip will increase the income of the industry.

References

[1] Sushanth Kanbail Srinivas, 2013 "Belt Conveyor with Pick and Place Mechanism", International Journal of Engineering Research & Technology (IJERT), **2 (12)**,2278-0181.
[2]Bhavesh Kaila & Hardik Acharya, 2014 "Modeling (Mechanical) and Analysis of Robo-Arm for Pick and Place Operation in Ceramic Industry", International Journal for Scientific Research & Development, **2(9)**,344-346.
[3]Velineni, P., Suresh, J., Kumar, N. and Suresh, M., 2020. Design of Pneumatic Gripper for Pick and Place Operation (Four Jaw). International Research Journal of Multidisciplinary Technovation, **2(2)**,1-8.
[4]Mourya, R., Shelke, A., Satpute, S., Kakade, S. and Botre, M., 2015. Design and implementation of pick and place robotic arm. International Journal of Recent Research in Civil and Mechanical Engineering (IJRRCME),**2(1)**.233-234.
[5]Vasistha, V., Kulshrestha, S. and Vasistha, C.V., 2014. PICK AND PLACE OPERATION ROBOT WITH STUD MECHANISM(2014)

Authors profile

Hemachandran.N 3rd year Mechanical engineering at SNS college of Technology, Attended International conference and Presented paper on 'An overview on 3d printing technology, Materials, and its Applications at PSG institute of technology (chandruntr430@gmail.com, +91 9942978889)

Harish Arvinth.L.B, 3rd year Mechanical engineering at SNS college of Technology

Muthu Kumar.V, 3rd year Mechanical engineering at SNS college of Technology

IOP Conf. Series: Materials Science and Engineering 1059 (2021) 012033 doi:10.1088/1757-899X/1059/1/012033

Tensile Behaviour of Hybrid Polymer Composites – An Experimental Study

R.Girimurugan[1*], N.Senniangiri[2], B.Pitchia Krishnan[3], Dr.S.Kavitha[4], M.Vairavel[5]

[1*,2,3]Assistant Professor, Department of Mechanical Engineering, Nandha College of Technology, Perundurai, Tamilnadu, India
[4]Assistant Professor, Department of Mechanical Engineering, Kalasalingam University, Krishnankoil, Tamilnadu, India
[5]Assistant Professor, Department of Mechanical Engineering, Annapoorana Engineering College, Salem, Tamilnadu, India

[1]*dr.r.girimurugan@gmail.com

Abstract

In recent years, developments of hybrid polymer composites by adding the natural and synthetic resources as reinforcement materials are significantly increased to enhance the required properties of the composites. In this research work, effects of Nano alumina particles and water hyacinth particles addition into the HDPE polymer matrix for its tensile properties were studied through experimentation. Fine Nano alumina particles are used as filler and the well water hyacinth particles are used as reinforcement materials. Hybrid composite specimens were prepared by varying the weight percentage of filler and reinforcement materials and keeping the HDPE's weight percentage as constant. Hybrid composite boards were fabricated by using hydraulic type injection moulding machine. Hybrid polymer composite specimens for tension test is cut away from the hybrid composite boards as per the ASTM standards with the help of water jet machining process. Experimental results reveal that the limited addition of fine Nano alumina and water hyacinth particles into the HDPE polymer matrix improves the tensile properties of the hybrid composites significantly when compare to the pure HDPE polymer composite.

1 Introduction

In recent years the usage of polymer materials for both domestic and industrial purpose are drastically increased due its superior properties than other materials. In order to fulfil the huge demand and immediate requirements for the lightweight materials to the society, different polymeric materials were introduced day-by-day by adding the different reinforcement materials with them. High Density

Polyethylene (HDPE) is one of the most promising light weight polymeric material due to its higher bonding behaviour with natural and synthetic reinforcement materials. Different researchers are attempted to enhance the mechanical properties of the High Density Polyethylene (HDPE) by adding the different kind of reinforcement materials with it. Loading rate of mechanical strain and the amount of reinforcement materials which are the two key factors greatly affected the yield stress and ultimate compression strength of the composites [1]. Addition of water hyacinth particles in 6 wt% into the HDPE matrix exhibited the superior strength during tensile load application and reveals the better Young's modulus values than other combinations [2].
Superior mechanical properties were established by adding the sisal fiber as reinforcement materials with the High Density Polyethylene (HDPE) matrix [3]. Addition of sawdust particles in smaller sizes with HDPE matrix exhibited the superior mechanical properties than that of larger size of sawdust particles [4]. Higher interfacial bonding between the reinforcement particles and recycled HDPE matrix composites were found in XRD analysis [5]. Tensile strength and Young's modulus of the natural rubber and chicken feather particles reinforced with recycled HDPE matrix composites were found maximum [6].

Chemically modified water hyacinth particles which added with the poly (methyl methacrylate) matrix reveals the superior tensile strength and Young's modulus values than others also the bonding between water hyacinth particles and poly (methyl methacrylate) matrix were found good in SEM analysis [7]. Strength of the composites during the tensile test and modulus of tension were found maximum in low density polyethylene/acrylonitrile butadiene styrene matrix reinforced with modified water hyacinth particles than that of an unmodified water hyacinth particles reinforced composites [8].

Lower elongation, higher strength and higher modulus of elasticity were observed in high density polyethylene (HDPE) matrix/water hyacinth particles reinforced composites with NCO-polyol coupling agent also the distribution of reinforcement material i.e. water hyacinth particles in the high density polyethylene matrix were found good through SEM analysis [9]. Strength of the composite specimens during the tensile test was noticed maximum in recycled high density polyethylene/natural rubber/chicken feather fiber composites with caprolactam also lower elongation was observed during the same tensile test on these composites [10].

In this experimental work, an innovative approach is initiated to improve the tensile properties of the water hyacinth particles and nano alumina particles reinforced with HDPE matrix composites.

2 Materials and Methods

Matrix material and one of the reinforcement materials which is used in this experimental work i.e. High Density Polyethylene (HDPE) and Nano alumina powder was purchased from Covai Seenu and Company Ltd, Coimbatore, Tamilnadu, India. Another reinforcement material i.e. water hyacinth powder is directly procured from the eco green unit, Coimbatore, Tamilnadu, India.

Table1. Matrix and reinforcement compositions

Sl.No.	High density polyethylene (wt %)	Nano alumina powder (wt %)	Water hyacinth powder (wt %)	Sample description
1.	100	-	-	Pure HDPE
2.	95	01	04	Sample - A
3.	95	03	02	Sample - B
4.	95	05	-	Sample - C

Pure HDPE specimen consists of 100 wt% pure HDPE particles. Sample – A consists 95 wt% of pure HDPE particles along with 1 wt% of alumina powder and 4 wt% of water hyacinth powder. Sample – B consists 95 wt% of pure HDPE particles along with 3 wt% of alumina powder and 2 wt% of water hyacinth powder. Sample – C consists 95 wt% of pure HDPE particles along with 5 wt% of alumina powder and 0 wt% of water hyacinth powder. The different matrix and reinforcement compositions, which are used to prepare the pure and hybrid HDPE polymer composite bars are shown in table.1 correspondingly. The different basic physical and chemical properties of the nano alumina powder is shown in table.2. Initially, the procured alumina power size in not meet out the required size. In order to fulfil the requirement, the procured alumina powder was processed in ball mill. The particle size confirmation test also carried out and the measured diameter of the nano alumina particles was found in the range of 100 nm.

Table 2. Properties of Nano Alumina (Al_2O_3) Powder

Description	Unit	Values
Aluminium	%	52.42
Oxygen	%	47.04
Density	g/cm^3	3.90
Molar mass	g/mol	101.96
Particle size	nm	100

Before feeding these mixtures into the double screw extruder machine, HDPE particles were heated above its melting temperature. After heating the HDPE particles were changed into liquid stage. Required weight percentage of reinforcement materials i.e. Nano alumina powder and water hyacinth particles were added into liquid state HDPE. After the addition of both reinforcement materials into the liquid state HDPE matrix material, the entire mixture was thoroughly mixed in a blender machine for 30 minutes. A double screw extruder machine is used to prepare the composite mixture of both matrix and reinforcement materials in the form of smaller length pieces (1 to 2 cm) in length.

A hydraulic injection moulding machine is used to prepare the composite bars in the form of rectangular shape from the well prepared hybrid composite pieces as per the ASTM D638 standard for tensile test (165 mm in length, 13 mm in length and 6 mm in thickness). Hybrid composite bars which are attained from the hydraulic injection moulding machine is shown in figure.1. Hybrid composite specimens for tensile test (before test) were shown in figure.2. Tensile test specimens after the test were depicted in figure.3.

Figure 1 Hybrid composite bars

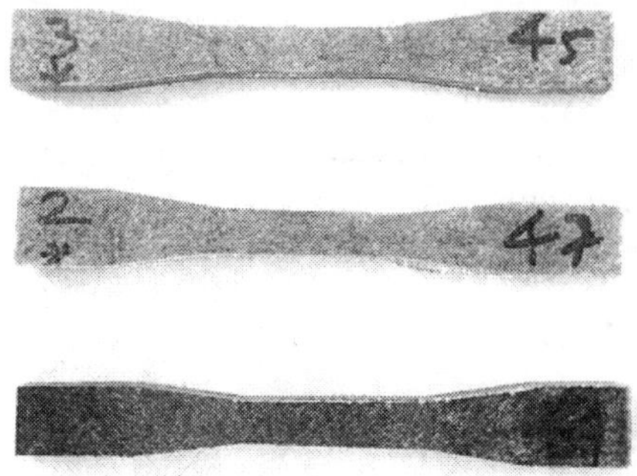

Figure 2 Hybrid composite specimens (before tensile test)

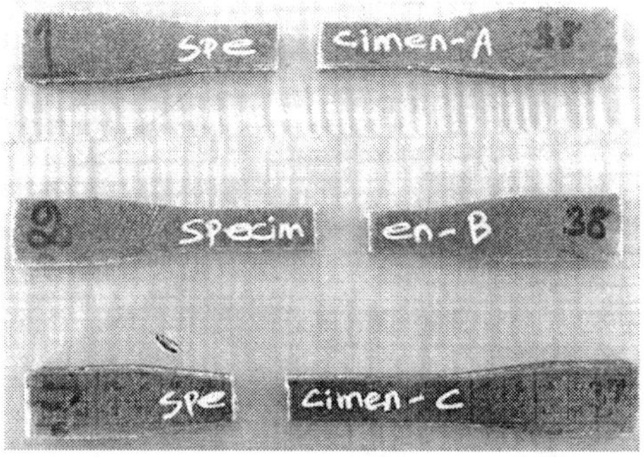

Figure 3 Hybrid composite specimens (after tensile test)

3 Results and Discussions

An experimental investigations on the HDPE polymer matrix which is reinforced with the limited amount of Nano alumina powder and fine water hyacinth particles was successfully carried out and the following results and discussions were made based on the experimental results. Peak tensile load for pure and hybrid HDPE polymer composite specimens were illustrated in figure.4.

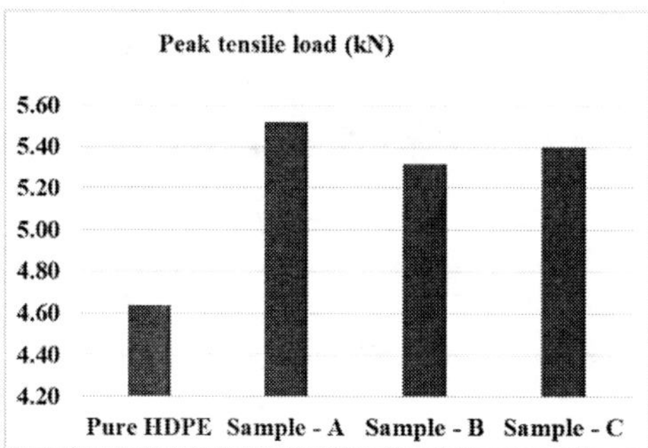

Figure 4 Peak tensile load for pure and hybrid HDPE composite specimens

It was noted that the peak tensile load for pure HDPE, sample-A, sample-B and sample-C in the range of 4.64, 5.52, 5.32 and 5.40 kN respectively. Enhanced load shifting between the matrix and reinforcement material causes the highest tensile load capacity for the hybrid HDPE composites than that of pure HDPE specimens. Maximum tensile load of 5.52 kN was observed in sample-B and minimum tensile load of 4.64 kN was observed in pure HDPE composite specimen respectively.

Maximum displacement attained by the pure and hybrid HDPE composite specimens during the tensile test were illustrated in figure.5. Displacement of 7.00, 8.40, 8.00 and 8.90 mm were observed in pure HDPE, sample-A, sample-B and sample-C correspondingly. Higher concentration of alumina powder in the HDPE matrix causes the higher displacement magnitude (sample-C) than other three composite specimens. Maximum and minimum displacement of 7.00 and 8.90 mm were noticed in pure HDPE and sample-C specimen correspondingly.

Yield strength exposed by the pure and hybrid HDPE composite specimens were shown in figure.6. Stress exhibited by the pure and hybrid HDPE composite specimens during the yield point were observed in the range of 37.27, 41.92, 40.39 and 35.39 MPa correspondingly. Maximum and minimum yield strength of 41.61 and 53.08 MPa were noticed in pure HDPE and sample-A specimen disparately. Larger quantity of water hyacinth particles addition into the HDPE polymer matrix (sample-A) is the reason for superior yield strength than other three composites. Yield load observed by the pure and hybrid HDPE composite specimens were depicted in figure.7.

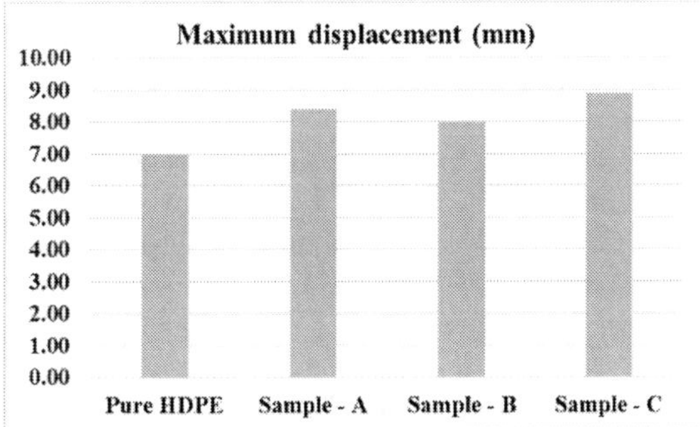

Figure 5 Maximum displacement attained by pure and hybrid HDPE composite specimens

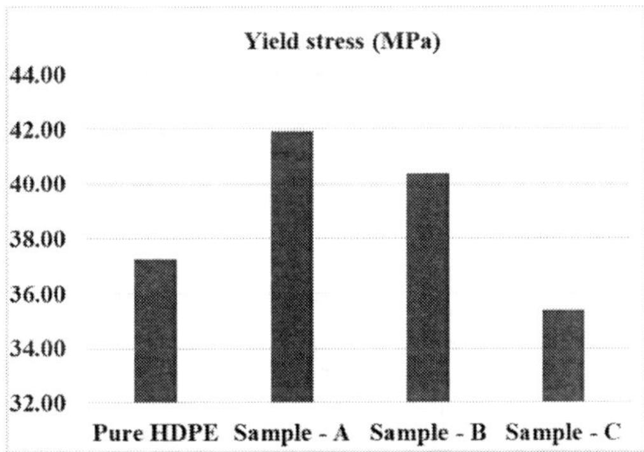

Figure 6 Yield stress for pure and hybrid HDPE composite specimens

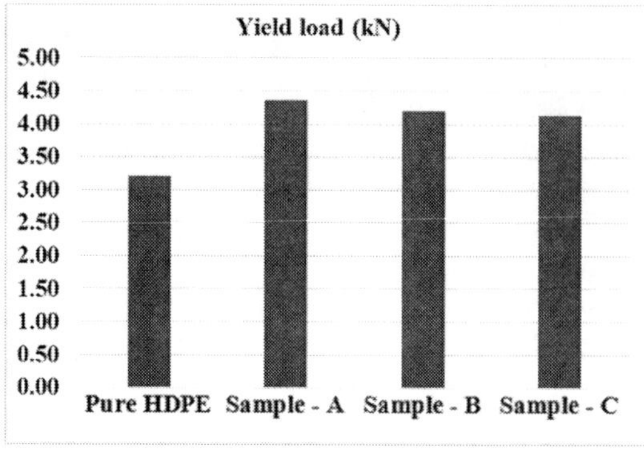

Figure 7 Yield load for pure and hybrid HDPE composite specimens

Load obtained by the pure and hybrid HDPE composite specimens during the tensile test at yield conditions were found in the range of 3.21, 4.36, 4.20 and 4.14 kN respectively. Maximum and minimum yield load of 3.21 and 4.36 kN was noticed in pure HDPE and sample-A specimen respectively. Higher concentration of water hyacinth particles into the matrix material is enhances the higher yield load carting capacity due to the superior interfacial bonding between the reinforcement and matrix. Maximum tensile strength exhibited by the pure and hybrid HDPE polymer composite specimens were represented in figure.8 correspondingly. Ultimate tensile strength for pure and hybrid HDPE polymer composites specimens were found in the range of 41.61, 53.08, 51.15 and 46.15 MPa respectively. Elongation percentage exposed by the pure and hybrid HDPE composite specimens were demonstrated in figure.9. Percentage of elongation obtained by the pure and hybrid HDPE polymer composite specimens during the tensile test were absorbed in the range of 1.44, 2.61, 2.17 and 2.17 correspondingly. Minimum (1.44) and maximum percentage of elongation (2.61) was found in sample-A, pure HDPE composite specimens respectively. Reduction percentage in pure and hybrid HDPE composite specimen's area were represented in figure.10. The reduction of area during the tensile test were found in the range of 5.76, 7.69, 10.10 and 9.19 for pure and hybrid HDPE polymer composite specimens correspondingly. Maximum and minimum area reduction percentage of 5.76 and 10.10 was noticed in pure HDPE and sample-b hybrid composite specimens accordingly.

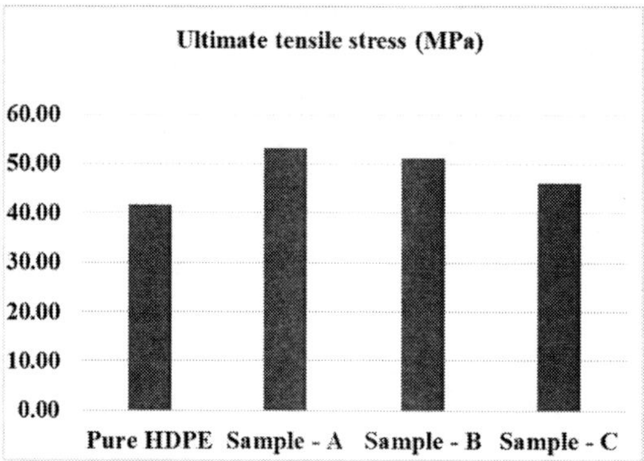

Figure 8 Ultimate tensile stress for pure and hybrid HDPE composite specimens

Ratio between the yield stress and ultimate tensile stress for Pure and hybrid HDPE polymer composite specimens were shown in figure.11 respectively. Pure and hybrid HDPE polymer composite specimen's yield stress to ultimate tensile stress ratio were noticed in the range of 0.90, 0.79, 0.79 and 0.77 respectively. Maximum and minimum ratio of yield stress to ultimate stress ratio of 0.77 and 0.90 were noticed in sample-c hybrid HDPE and pure HDPE polymer composite specimens considerably.

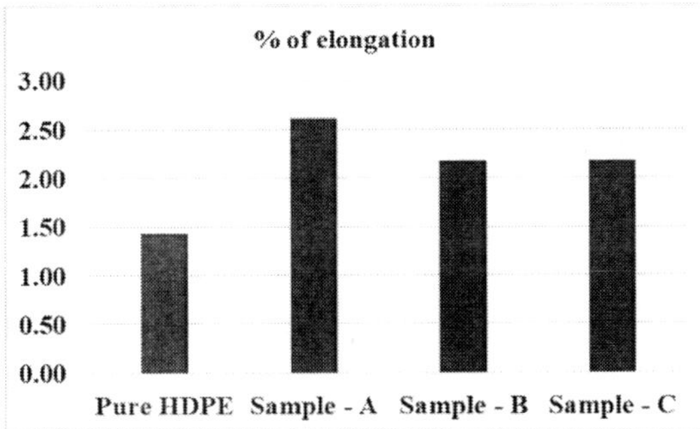

Figure 9 Elongation percentage for pure and hybrid HDPE composite specimens

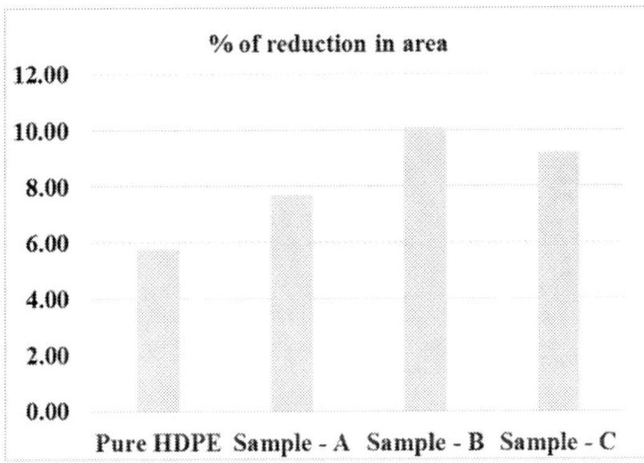

Figure 10 Area reduction percentages for all composite specimens

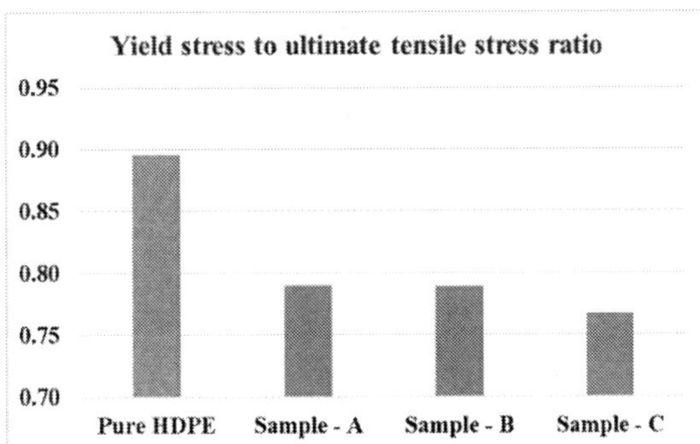

Figure 11 Yield stress to ultimate tensile stress ratio for all composite specimens

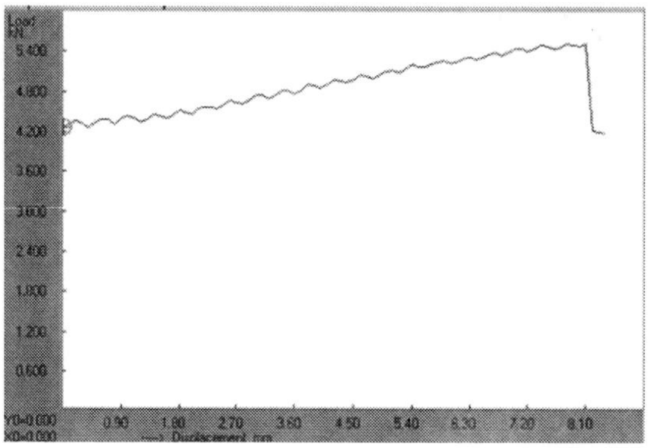

Figure 12 Load vs deflection graph for sample A

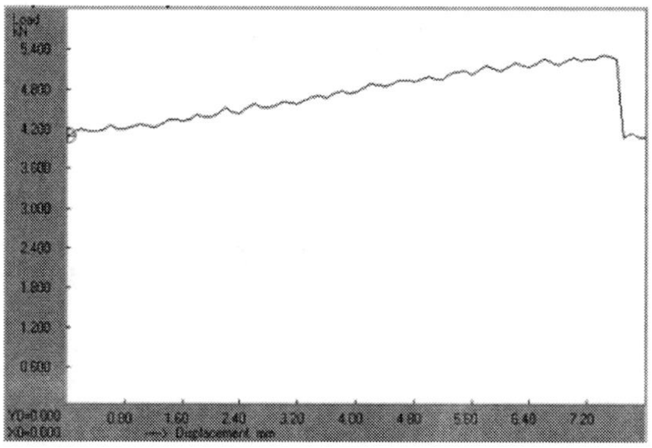

Figure 13 Load vs deflection graph for sample B

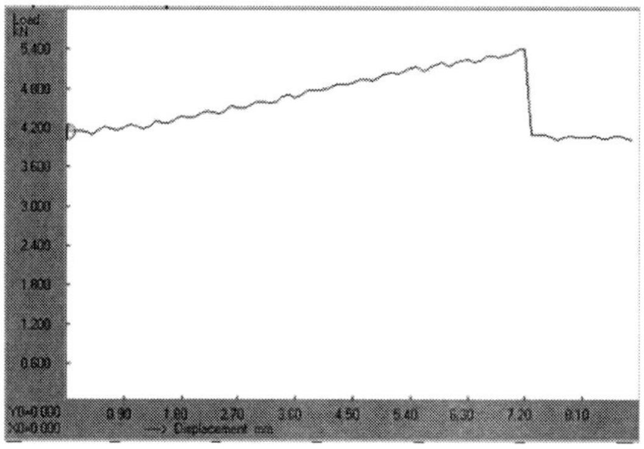

Figure 14 Load vs deflection graph for sample C

During the tensile test, the relationship between the load vs deflection is taken as a graph from the software. The load vs deflection graphs for sample A, B and C is shown in figure.12 to 14 correspondingly. Effective interfacial bonding between the matrix and reinforcement material causes the superior performance of hybrid HDPE composite specimens than that of the pure HDPE specimens during the tensile test. Narrow amount of water hyacinth particles addition into the HDPE polymer matrix leads to the enhanced tensile properties of the hybrid composite specimens significantly.

4 Conclusion

The following conclusions were made based on the experimental studies which is carried on a pure and hybrid HDPE composite specimens. Among the all four composite specimens, hybrid HDPE specimen-A exhibits the better tensile behaviour than other three composite specimens. Superior interfacial bonding between the reinforcement and matrix materials was the origins for better tensile behaviour of hybrid HDPE composite specimen. Limited inclusion of Nano alumina and water hyacinth particles quantity into the HDPE matrix is enhanced the tensile behaviour of the composites significantly due to an effective load shifting between the matrix and reinforcement materials. Matrix fiber-filler interaction is the major reason for the variation in tensile properties of the composites. Excess reinforcement quantity addition into the matrix material is reduce the tensile behaviour of the hybrid composites considerably.

References

[1] Omar M F, Jaya H, Akil H, Md Ahmad Z A and Noriman N Z 2015 Mechanical Properties of High Density Polyethylene (HDPE)/Sawdust Composites under Wide Range of Strain Rate *Applied Mechanics and Materials* **754–755**, 83–88

[2] Tan S J, Supri A G, Chong K M 2019 Properties of recycled high-density polyethylene/water hyacinth fiber composites: the effect of different concentration of compatibilizer *Polymer Bulletin* **72** 2019–31

[3] Abou-Kandil A I, Shehata A, Yousef E, El-Sabbagh I and Salem K 2017 Effect of microstructural changes on thermal, electrical, and mechanical properties of HDPE reinforced by sisal fibers *Journal of Thermoplastic Composite Materials* **30** 1373-92

[4] Jaya H, Omar M, Md Akil H, Arifin Ahmad Z, and Zulkepli N 2016 Effect of Particle Size on Mechanical Properties of Sawdust-High Density Polyethylene Composites under Various Strain Rates *Bio Resources* **11** 6489-504

[5] Aizat E A and Supri A G 2014 Tensile Properties, Swelling Behaviour and XRD Characteristic of R-HDPE/Tyre Dust and R-HDPE/Chicken Feather Fiber Composites *Advanced Materials Research* **925**, 215–8

[6] Yazid M I M, Supri A G, Firuz Z and Musa L 2013 Recycled High Density Polyethylene/Natural Rubber/Chicken Feather Fibers (RHDPE/NR/CFF) Composites: The Effects of Fiber Loading and Benzyl Urea on Tensile Properties and Morphology Analysis *Advanced Materials Research* **795** 582–6

[7] Supri A G, Tan S J, Ismail H and The P.L 2011 Effect of Poly (methyl Methacrylate) Modified Water Hyacinth Fiber on Properties of Low Density Polyethylene/Natural Rubber/Water Hyacinth Fiber Composites *Polymer-Plastics Technology and Engineering* **50** 898-906

[8] Supri A G and Ismail H 2011 The Effect of Isophorone Diisocyanate-Polyhydroxyl Groups Modified Water Hyacinth Fibers (Eichhornia Crassiper) on Properties of Low Density Polyethylene/Acrylonitrile Butadiene Styrene (LDPE/ABS) Composites *Polymer-Plastics Technology and Engineering* **50** 113-20

[9] Abdul Ghani Supri and Hanafi Ismail 2010 The Effect of NCO-Polyol on the Properties of Low-Density Polyethylene/Water Hyacinth Fiber (Eichhornia crassiper) Composites *Polymer-Plastics Technology and Engineering Volume* **49** 766-71

[10] Supri A G and Izzuddeen M 2013 Tensile Properties, Swelling Behavior, and Morphology Analysis of Recycled High Density Polyethylene / Natural Rubber / Chicken Feather Fibers (R-HDPE / NR / CFF) Composites: The Effects of Caprolactam *Advanced Materials Research* **844** 293–6

Structural Analysis of Chassis using AISI 4130 and AA 7068

Arun G V[1], Kishore Kumar K[2], Velmurugan S[3]

[1,2] Mechanical Engineering, Bannari Amman Institute Of Technology, Sathyamangalam.
[3] Assistant Professor Level 2, Mechanical Engineering, Bannari Amman Institute Of Technology, Sathyamangalam.

E-mail: arungv.me19@bitsathy.ac.in,
kishorekumar.me19@bitsathy.ac.in,
vsvmurugan@gmail.com

Abstract. The chassis frame is the backbone of a vehicle to safely carry the maximum load for all the designed operating conditions. This paper describes the design and analysis of vehicle chassis using different materials for different cross sections. The main objective of this attempt is to reduce the self-weight of the chassis. In this work a truck chassis is taken for modeling and analysis. Two different alloy materials of steel and aluminium are considered for three different cross sections namely I, C and regular box type. SOLIDWORKS and ANSYS software are used for modeling and analysis of this work respectively.

Keywords: Chassis Design, Weight Reduction, Alloy Materials, Modeling and Analysis

1 Introduction

Automotive chassis is a skeletal frame on which various mechanical parts like engine, axle assemblies, brakes, steering and various other parts are bolted. The chassis is considered to be the most significant component of an automobile. It provides strength and flexibility to the automobile. Steel alloy AISI 4130 is the most commonly used material in the production of automotive chassis. In this work AA 7068[1-6] (aluminium alloy) is used as a replacement material for the chassis, even though the young's modulus of AISI 4130 is significantly higher than that of AA 7068, still it is taken for this work because of its cost reduction and weight reduction properties, which has similar physical properties to AISI 4130[7-13]. Three different cross sections namely I, C and Box type are analysed with these two materials in order to find the best suitable replacement for AISI 4130 in the aspects of strength, weight reduction and cost. For this work Eicher E2 truck's ladder chassis[14-23] is taken to find the equivalent stress and total deformation of the chassis. The total load applied on the chassis for analysis is 112270 N[15]. For modeling of the chassis SOLIDWORKS 2020 is used and for the purpose of analysing ANSYS R2 is used.

2. Properties of Materials

The AA 7068 [1-6] is one of the strongest aluminium alloys with a yield strength of 590 MPa, poisson ratio of 0.23, young's modulus of 73.1 MPa, Ultimate strength of 641 MPa and density of 2.85 g/cm^3, used as a replacement for AISI 4130 [7-13] steel alloy with a yield strength of 460 MPa, poisson ratio of 0.29 young's modulus of 205 MPa, ultimate strength of 560 MPa and density of 7.85 g/cm^3. From the physical properties of these materials it is clearly understood that the density of AA 7068 is very low which will help in weight reduction.

Table1. Physical Properties of AA 7068 and AISI 4130

Properties	Materials	
	AA 7068	AISI 4130
Young's modulus(MPa)	73.1	205
Poisson ratio	0.23	0.29
Density (g/cm^3)	2.85	7.85
Yield strength (MPa)	590	460
Ultimate strength (MPa)	641	560

3. Modeling and Analysis

3.1 Modeling.

The ladder chassis of truck (Eicher E2) is modeled in solid works 2020 for the different cross sections namely I, C and box type are shown in the below figures.

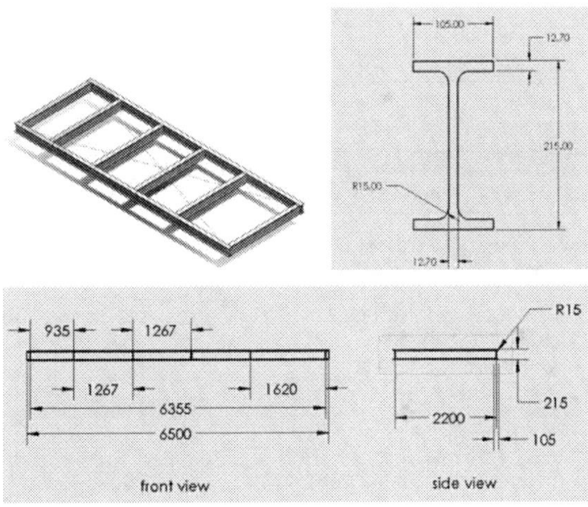

Figure 1: Modeling of I section

The above figure 1 shows the modeling of chassis using I cross section in solidworks, the dimensions of the model is given for reference.

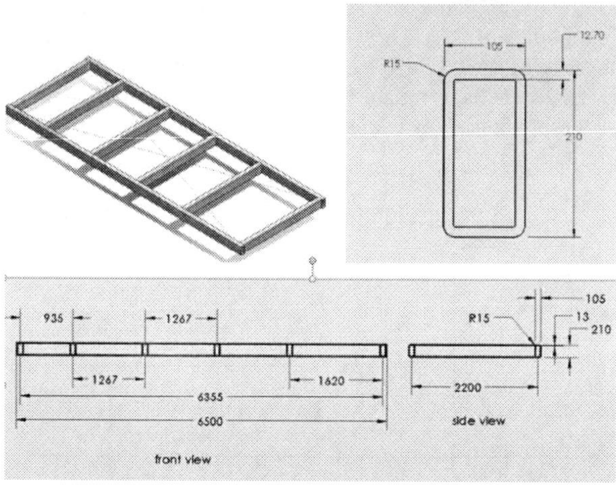

Figure 2: Modeling of Box Type Cross section

The above figure 2 shows the modeling of chassis using box type cross section in solidworks, the dimensions of the model is given for reference.

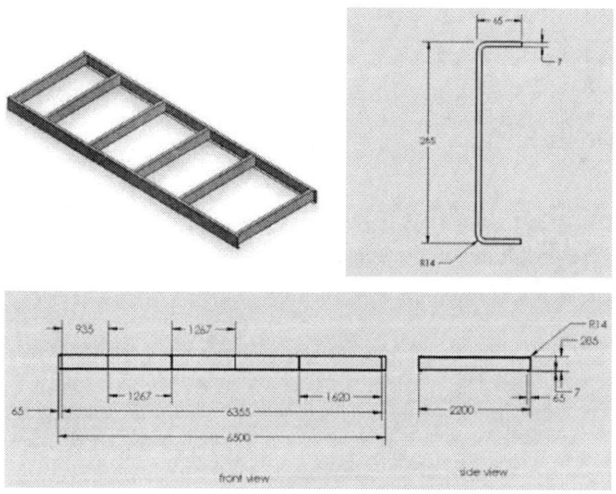

Figure 3: Modeling of C Type Cross section

The above figure 3 shows the modeling of chassis using C cross section in solidworks, the dimensions of the model is given for reference.

3.2 Analysis.

For the purpose of analysing the models,Ansys R2 is used, the solidworks file is imported into ansys in IGS format. Equivalent stress and total deformation is calculated for three different cross sections (I, C and Box type) using two different materials (AISI 4130 and AA 7068) to find the best material and cross section for the truck chassis. Second and fourth beam of the chassis is fixed and a load of 112270 N is applied on the chassis for analysis. For the purpose of analysis the load is considered to be static.

Figure 4: Fixed support

Figure 4 shows the fixed support of the chassis during analysis.

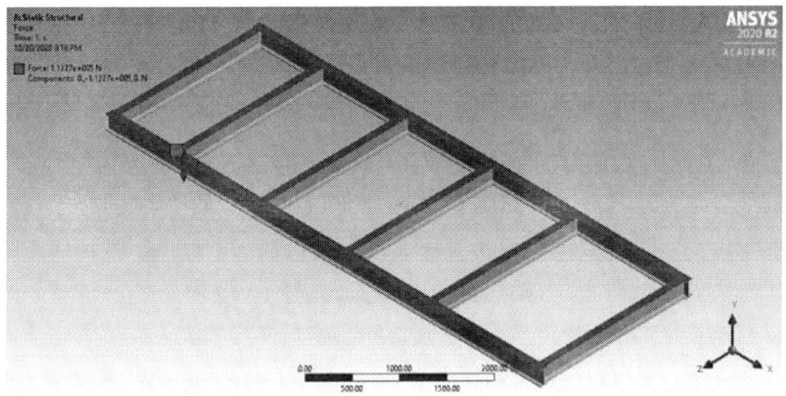

Figure 5: Load applied

Figure 5 shows where the load is applied on the chassis during analysis.

4. Result and Discussion

The result on analysing the chassis for different materials in different cross sections is given below.

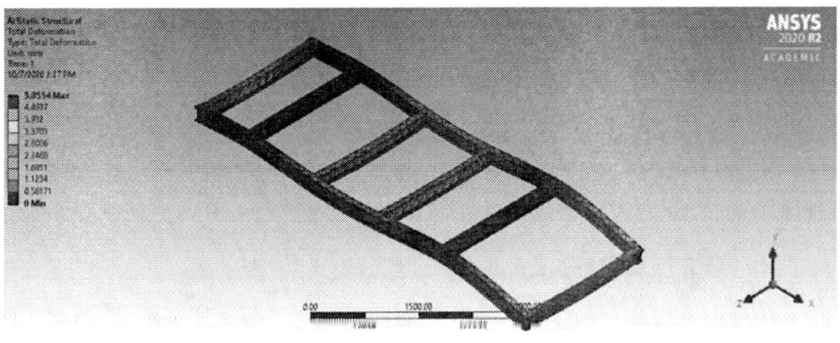

Figure 6: Deformation of AA 7068 – I section

IOP Conf. Series: Materials Science and Engineering 1059 (2021) 012034 doi:10.1088/1757-899X/1059/1/012034

The figure 6 shows the total deformation of chassis when the force is applied for AA 7068 in I cross section, the maximum deformation obtained is 5.0554 mm.

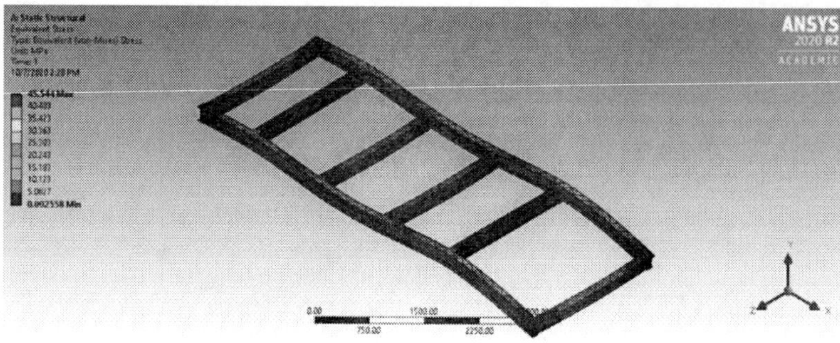

Figure 7: Equivalent stress (von-mises) of AA 7068 – I section

The figure 7 shows the stress deformation of chassis when the force is applied for AA 7068 in I cross section, the maximum stress value obtaines is 45.544 MPa.

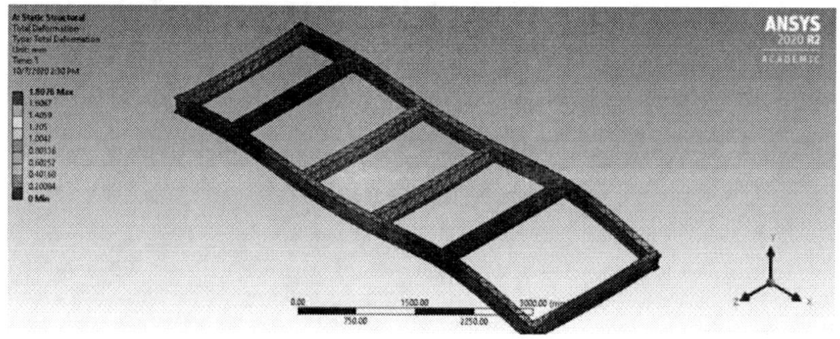

Figure 8: Deformation of AISI 4130 – I section

The figure 8 shows the total deformation of chassis when the force is applied for AISI 4130 in I cross section, the maximum deformation obtained is 1.8076 mm.

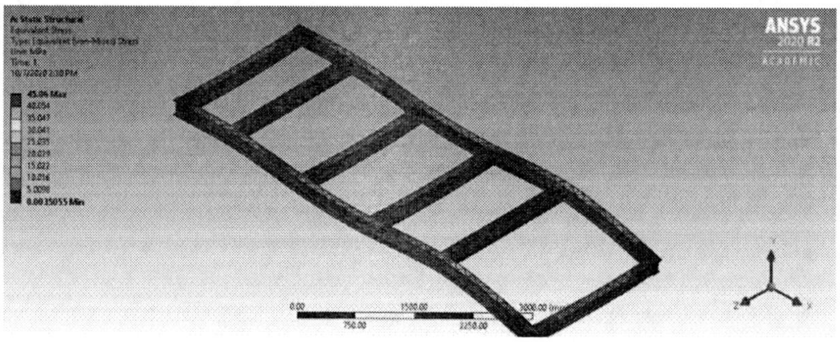

Figure 9: Equivalent stress (von-mises) of AISI 4130 – I section

The figure 9 shows the stress distribution of chassis when the force is applied for AISI 4130 in I cross section, the maximum stress value obtained is 45.06 MPa.

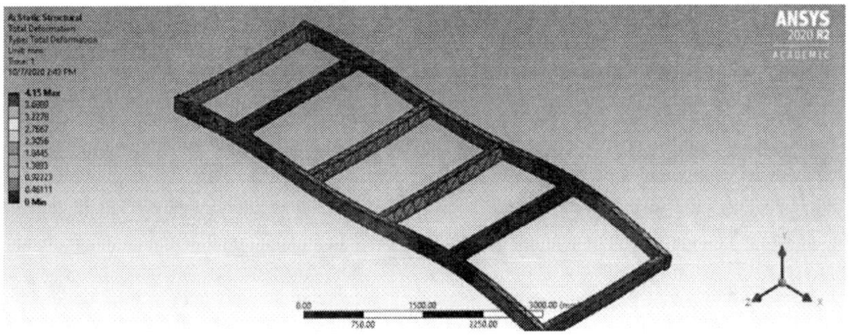

Figure 10: Deformation of AA 7068– Box type cross section

The figure 10 shows the total deformation of chassis when the force is applied for AA 7068 in box type cross section, the maximum deformation obtained is 4.15 mm.

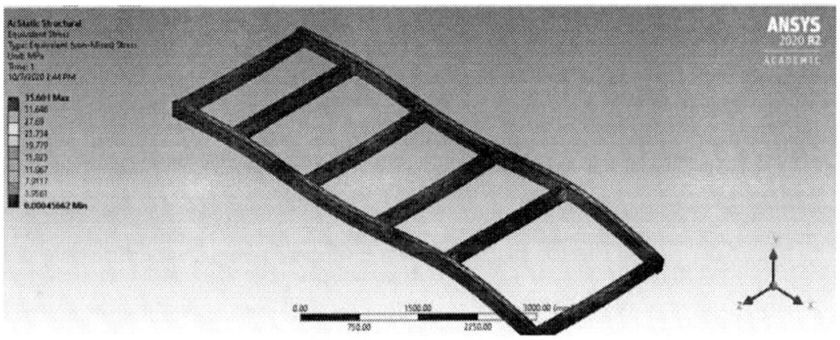

Figure 11: Equivalent stress (von-mises) of AA 7068 – Box type cross section

The figure 11 shows the stress distribution of chassis when the force is applied for AA 7068 in box type cross section, the maximum stress value obtained is 35.601 MPa.

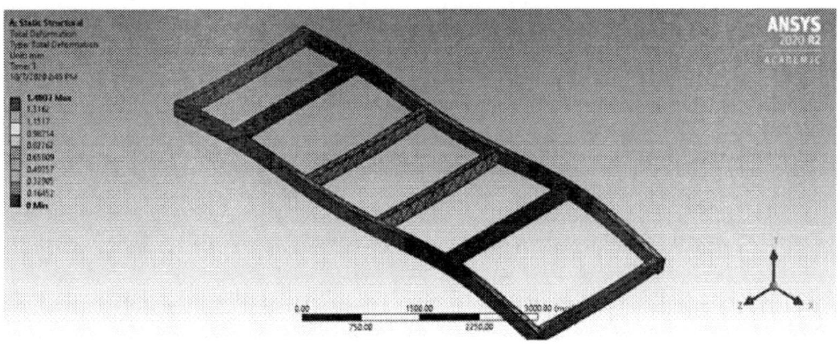

Figure 12: Deformation of AISI 4130– Box type cross section

IOP Conf. Series: Materials Science and Engineering 1059 (2021) 012034 doi:10.1088/1757-899X/1059/1/012034

The figure 12 shows the total deformation of chassis when the force is applied for AISI 4130 in Box type cross section, the maximum deformation obtained is 1.4807 mm.

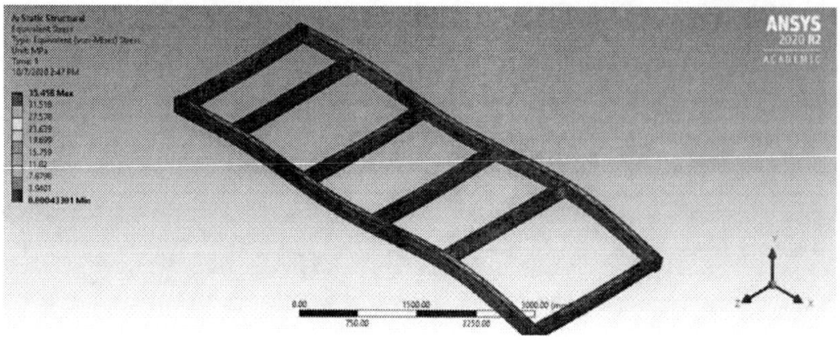

Figure 13: Equivalent stress (von-mises) of AISI 4130 – Box type cross section

The figure 13 shows the stress distribution of chassis when the force is applied for AISI 4130 in Box type cross section, the maximum stress value obtained is 35.458 MPa

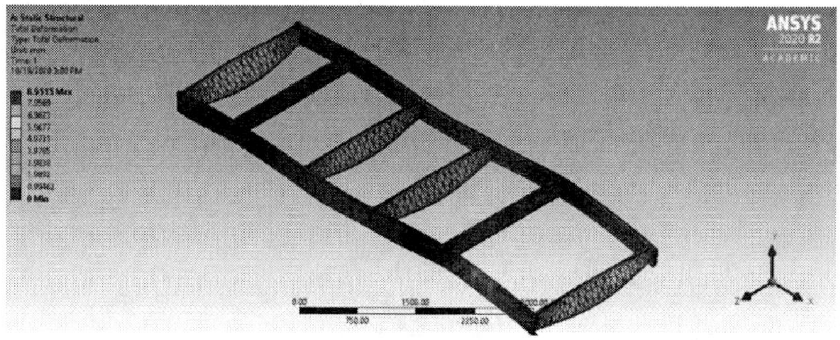

Figure 14: Deformation of AA 7068– C section

The figure 14 shows the total deformation of chassis when the force is applied for AA 7068 in C cross section, the maximum deformation obtained is 8.9515 mm.

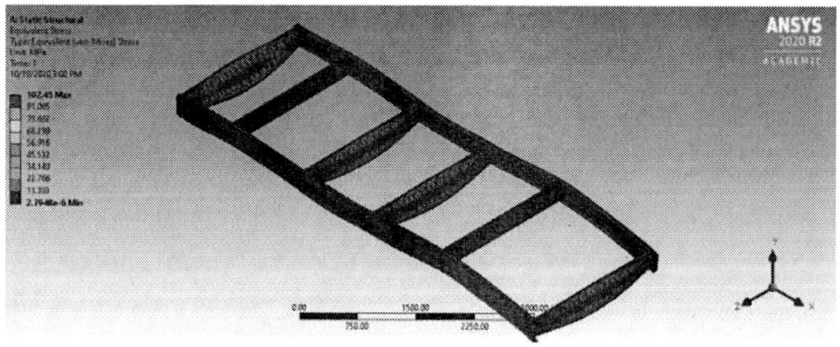

Figure 15: Equivalent stress (von-mises) of AA 7068 – C section

The figure 15 shows the stress distribution of chassis when the force is applied for AA 7068 in C cross section, the maximum stress value obtained is 102.45 MPa.

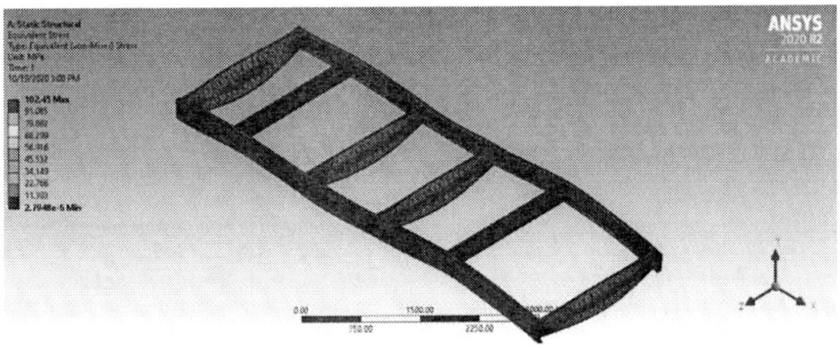

Figure 16: Deformation of AISI 4130– C section

The figure 16 shows the total deformation of chassis when the force is applied for AISI 4130 in C cross section, the maximum deformation obtained is 3.1999 mm.

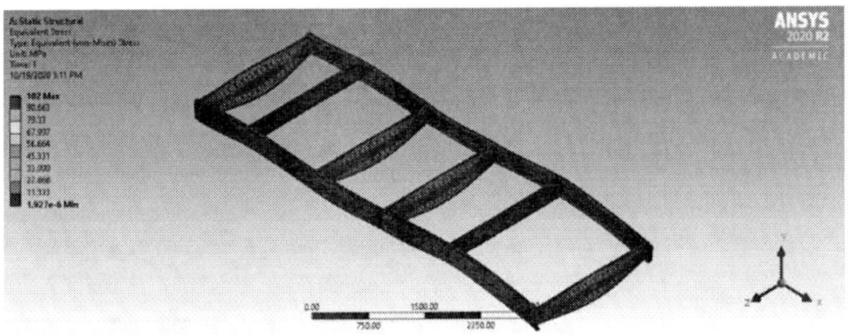

Figure 17: Equivalent stress (von-mises) of AISI 4130 – C section

The figure 17 shows the stress distribution of chassis when the force is applied for AISI 4130 in C cross section, the stress value obtained is 102 MPa.

By doing analysis in ansys the following results of total deformation, equivalent stress and weight of the chassis is obtained.

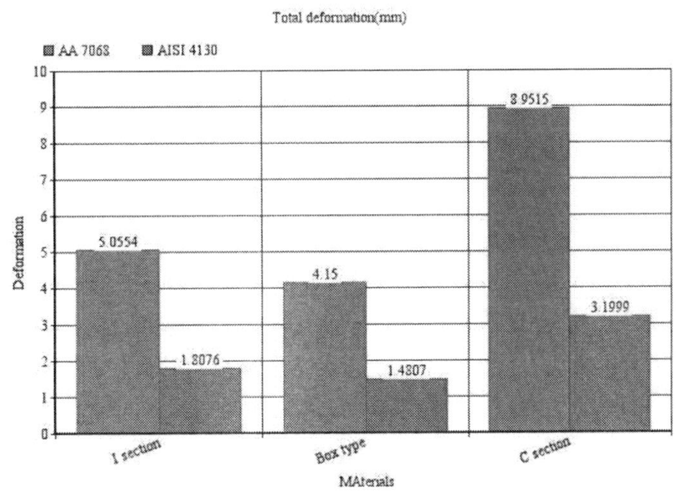

Figure 18: Total deformation of chassis

Table 2: Total deformation of chassis(mm)

Cross sections	**Materials**	
	AISI 4130	**AA 7068**
I section	1.8076	5.0554
Box type	1.4807	4.15
C section	3.1999	8.9515

The table 2 and figure18 shows the total deformation of the chassis in three different cross sections in both materials. From the graph it is clear that the steel alloy (AISI 4130) has lower deformation rate when compared to aluminium alloy. The box type cross section in aluminium alloy has lower deformation when compared to other cross sections in aluminium alloy.

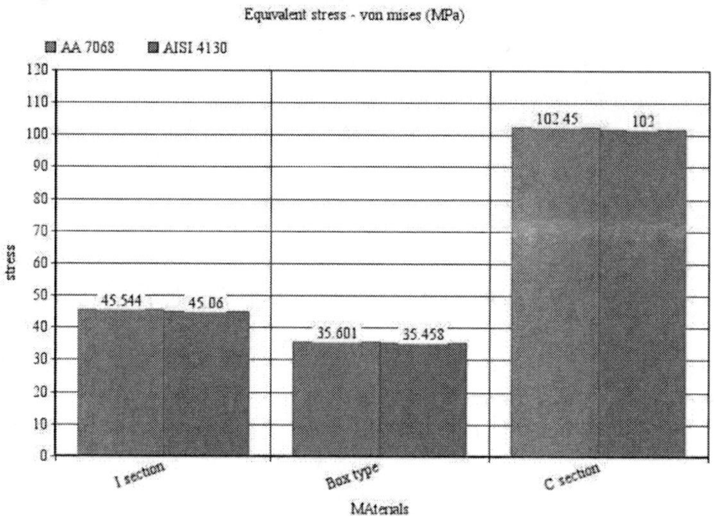

Figure 19: Equivalent stress – von mises

Table 3: Equivalent stress – von mises (MPa)

Cross sections	**Materials**	
	AISI 4130	**AA 7068**
I section	45.06	45.544
Box type	35.458	35.601
C section	102	102.45

IOP Conf. Series: Materials Science and Engineering 1059 (2021) 012034 doi:10.1088/1757-899X/1059/1/012034

The table 3 and figure 19 shows the equivalent stress – von mises on the chassis for different cross section in both materials, it is visible that the steel alloy (AISI 4130) has lower stress distribution when compared to AA 7068, but there is no significant difference.

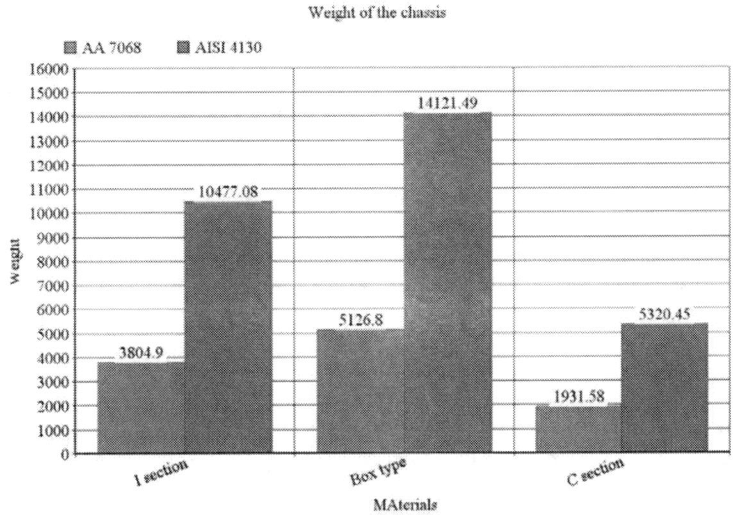

Figure 20: Weight of the chassis(kg)

Table 4: Weight of the chassis (kg)

Cross sections	**Materials**	
	AISI 4130	**AA 7068**
I section	10477.08	3804.9
Box type	14121.49	5126.8
C section	5320.45	1931.58

The figure 20 and table 4 shows the weight of the chassis, it is noticed that there is more than 50% weight reduction in AA 7068 (aluminium alloy) when compared to AISI 4130.

Conclusion

This paper explains about the design and analysis of chassis using different materials in different cross sections. It is proven that the AA 7068 has higher deformation when compared to AISI 4130, but still AA 7068 can be considered as a suitable replacement for steel alloy (AISI 4130) because it provides us with more than 50% of weight reduction, which will increase the efficiency of the vehicle considerably. Aluminum alloy (AA 7068) with box type cross section can be used as a replacement because it has less deformation (4.15mm) when compared to other cross sections in AA 7068, greater moment of inertia and it will be helpful in wire harness to organise the cables.

Future Scope

In this work, only the static load is taken into consideration and the load is considered to be uniformly distributed. For future works dynamic load can be taken into consideration and varying load distribution must be taken for analysis.

References

[1] Amin Azimi, Ali Shokuhfar Omid Nejadseyfi, Hamid Fallahdoost and Seid Salehi 2015, Optimizing consolidation behavior of Al 7068–TiC Nano composites using Taguchi statistical analysis. *Transaction of Nonferrous Metals Society of China* **25 :** 2499–2508.
[2] Mohammed Javad Ghasemi, Mohammed Silani, Ali Maleki and Mostafa Jamshidian 2020, Micromechanical simulation and experimental investigation ofAluminum - based Nano composites. *Defence Technology.* https://doi.org/10.1016/j.dt.2020.03.010.
[3] Madhusudhan M, G J Naveen and K Mahesha 2016, Mechanical Characterization of AA7068-ZrO2 reinforced MetalMatrix Composites. *Materials Today Proceedings* **4 :** 3122–3130.
[4] LushaTian, YongchunGua, Jianping Li, Jianli Wang, HongboDuan, Feng Xia, Minxian Liang 2018, Elevated re-aging of a piston aluminium alloy and effect on the microstructure and mechanical properties. *Materials Science & Engineering A* **738** : 375-379.
[5] F. Hughes, A. Prudom and G. Swallowe 2013, The high strain-rate behaviour of three molecular weights of polyethylene examined with a magnesium alloy split-Hopkinson pressure bar. *Polymer Testing* **32(5) :** 827-834.
[6] Julia Osten, Benjamin Milkereit, Michael Reich, Bin Yang, Armin Springer, Karina Nowak and Olaf Kessler 2020, Development of Precipitation Hardening Parameters for High Strength Alloy AA 7068. *Materials 2020* **13(4) :** 918.
[7] ShwetankAvikal, AtulBisht, Devender Sharma, HimanshuHindwan, SiddharthYadav, K.C. Nithin Kumar and Padmanabh Thakur 2020, Design and fatigue analysis of front axle beam of a heavy duty truck using ansys. *Materials Today : Proceedings* **26(2)** : 3211-3215.
[8] Y Reda, A M El-Shamy and Ashraf K Eessaa 2018, Effect of hydrogen embrittlement on the microstructures of electroplated steel alloy 4130. *Ain Shams Engineering Journal* **9(4)** : 2973-2982.
[9] Matthew Connolly, May Martin, Peter Bradley, Damian Lauria, Andrew Slifka, Robert Amaro, Christopher Looney and Jun-Sang Park 2019, In situ high energy X-ray diffraction measurement of strain anddislocation density ahead of crack tips grown in hydrogen. *Acta Materialia* **180 :** 272-286
[10] D. B. Kececioglu and L B Chester 1976, Combined Axial Stress Fatigue reliability for AISI 4130 and 4340 steel. Journal of Manufacturing Science and Engineering **98(1)** : 153-160
[11] Li Fan, Hai-yan Chen, Yao-hua Dong, Li-hua Dong and Yan-sheng Yin 2018, Wear and corrosion resistance of laser-cladded Fe-based composite coatings on AISI 4130 steel. *International Journal of Minerals, Metallurgy and Materials* **25** : 716-728
[12] A. Kumar and P.P. Patil 2017, Free vibration and connecting boltconstraint-based FEA analysis ofheavy vehicle medium dutyTransmission gearbox housingmade from AISI 4130 alloy material. *Mathematics Applied to Engineering* 133-146.
[13] Zhipeng Pan, YixuanFeng, Xia ji and Steven Y. Liang 2017, Turning Force Prediction of AISI 4130 Considering Dynamic Recrystallization. *International Manufacturing Science and Engineering Conference*
[14] Mohamed Sithik M, Vallurupalli R., Lin B and Sudalaimuthu S. 2014, Simplified Approach of Chassis Frame Optimization forDurability Performance. *Technical Papers*
[15] Hidaitulla Mangole 2017, Cross-Section and Material optimization of an automotive chassis using FEA
[16] B. Stalin, M. Ravichandran, C. Ramesh Kannan and K. SathiyaMoorthi 2019, Design and Analysis of Stringer on the Chassis Frame in Load Carrying vehicle. *Advances in Manufacturing Technologies*
[17] Jessica GissellaMaradeyLázaro,HelioSneyder Esteban Villegas,Braulio José Blanco Caballero 2018, Finite Element Analysis (Fea) For Optimization The Design Of A Baja Sae Chassis. *ASME International Mechanical Engineering Congress and Exposition*
[18] Barry (Baizhong) Lin 2013, A New Method of d'Alembert's Principle Finite Element Based Fatigue Calculation with Input ofLoads and Accelerations. *Technical papers*
[19] Fang LIU, Jinwen DONG and Yongzhi JING 2012, Structure analysis of levitation chassis of medium-to-low speed maglev vehicles based on left-right decoupling. *Journal of Modern Transportation* **20** : 82-87.

[20] MohdAzizi Muhammad Nora, HelmiRashida, Wan MohdFaizul Wan Mahyuddinb, MohdAzuanMohdAzlanc and Jamaluddin Mahmud 2012, Stress Analysis of a Low Loader Chassis. *Procedia Engineering* **41** : 995 – 1001.
[21] R. V. Patil, J. U. Marathe, A.P.Tadamalle, Dr. Y. P. Reddy 2016, Optimization of Tractor Trolley Chassis Using Orthogonal Array Method. *Material Today : Proceedings* **4(8)** : 8796–8805.
[22] Yuan Ren, Yongchang Yu, Binbin Zhao, Chuanhui Fan, He Li2016 Finite Element Analysis and Optimal Design for the Frame of SX360 Dump Trucks. *Procedia Engineering* **174** : 638 – 647.
[23] Antonio Flavio Aires Rodrigues, André Luis Rosa Poncio, André Cervieri, Luiz Carlos Gertz, Moisés Da SiluePeraira, Goessling Silva Pereira and Anderson Bonsanto Oliveira 2015, Comparative Analysis of Torsional Rigidity of a Chassis of an Automotive Prototype Obtained by Numerical Analysis and Experimenta. *Technical Papers*

IOP Conf. Series: Materials Science and Engineering 1059 (2021) 012035 doi:10.1088/1757-899X/1059/1/012035

Cloud based smart wireless and health care monitoring system

Nithya Priya S, Ranjith Kumar M and Nibin Sabari Anand N

Assistant Professor, Sri Krishna college of Engineering and Technology, Coimbatore
Student, Sri Krishna College of Engineering and Technology, Coimbatore-641008

nithyapriya@skcet.ac.in

Abstract People are facing several health issues due to the stress and unhealthy food habits. But the greater issue being the ignorance towards these two. On this regard we proposed a new idea that is a smart blood pressure and glucose level monitoring system through sensor monitoring system anybody can measure the blood pressure and glucose level, this sensor was interfaced with IOT through raspberry-pi. At first the data of the person's average blood pressure and glucose level is stored as a data in cloud, and as per the measurements of daily or monthly basis the data will be compared with the cloud through HTTP and MQTT Protocol. If there any changes found in the data, it will automatically forward the message to the hospitals through that the doctor can suggest the person regarding daily routine food through this we can avoid major health problems.

1. Introduction

In this Generation the technology is developing gradually, but the awareness of health among the people were not created enough. The unhealthy food habits and also the stress causes one of the reason of increasing blood pressure and the glucose level, ultimately the upcoming generation also followed by these problems. The government also made many initiatives to create awareness the only awareness not going to help the people, further the precautions should also have been given to the individuals. The paper proposes an idea that the data of the following patients is been retrieved and the following data is handled by the medical consultancy. The required precautions will be given by the doctor to that person, before that the retrieved data will be compared with the average blood pressure and glucose level of that person then if it found any abnormalities it will intimate the consultancy through IOT this process going to take place.

The remaining contents of the paper framed as follows: 2. Related works of the proposed system. 3. It consists of implementation and it is also covering the different components that is going to be installed along with the module. 4. we conclude.

2. Related Works

In recent days IOT has attracted the researchers and it also creates the potential to investigate, as well as it acts as a mediator of healthcare system. In medical field there is more demand for IOT it is used to share the information and also acts as a tool to communicate with others

The review about the future scope is now a day the precautions will be given by the medical consultancy or doctors handling the particular person data. In future this won't be done by the doctor or the consultancy. The precautions will be predicted and it will be analyzed through machine learning there will be no person behind the communication to instruct. Artificial intelligence will take care the person data as well as his health care problems

Glucose level can be measured according to three types which is invasive technique, minimally invasive and non-invasive. In this invasive technique will be more accurate the fluids from the body is taken out and it is measured using glucometer. The finger prick technique is the inexpensive technique it is most familiar and followed by most of us. The fluids might also be extracted from forearm, upper arm, base of thumb and thigh, but the accuracy of glucose level won't be clear as compared to the fluids taken from the fingertip.

In this paper we proposed an idea of using non- invasive glucose monitoring sensor and blood pressure monitoring sensor. In which the non-invasive glucose level monitoring technique is to reduce the pain in blood extraction and insulin injection. The methods used to detect the glucose level are ultrasound, infrared, fluorescence these are the various methods to detect the glucose level using non-invasive method. The blood pressure monitoring sensor is used to analyze the pressure level by and it is easily accessed by the individual, it is also used to change the medications and the changes in daily routine food can be followed in order to reduce the blood pressure level.

Advantages: There won't be any physical interaction between the doctor and the patient so visiting hospitals for minor problems can be reduced.

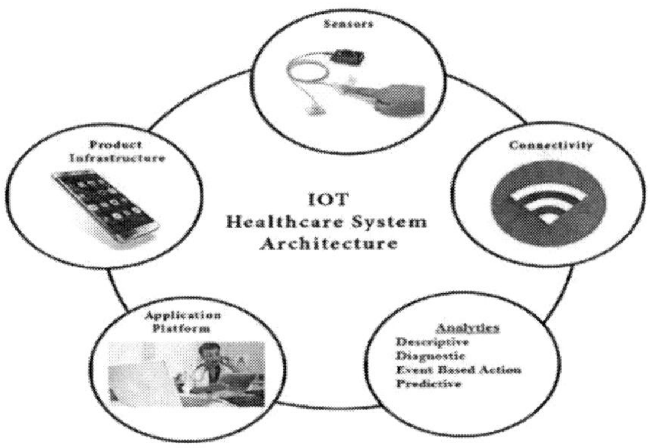

Figure 1 diagram of health care monitoring system

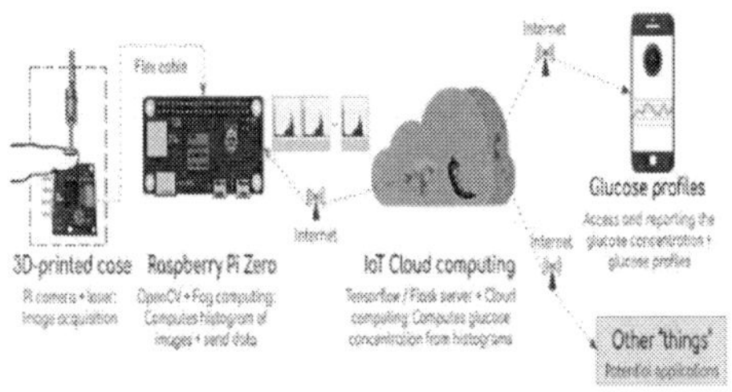

Figure 2 Diagram of cloud data transmission

3. Methodology

The ultimate aim of the proposal is to

- To get the precautions easier so that the major health problems can be avoided in the upcoming years
- To provide faster and easier way of communication to the person who's accessing the device
- Collection of accurate data from the user
- It is an user friendly system and it provides simple way of accessing the device.

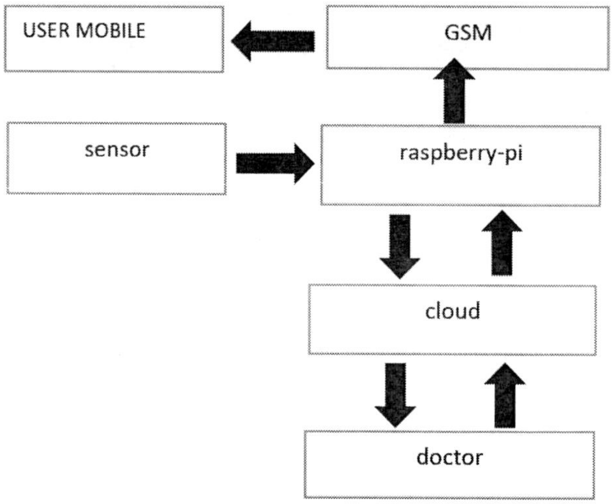

Figure.3 Represents block diagram

3.1 System workflow

At first the sensors get the data and the next the data is been processed and finally the data transmited

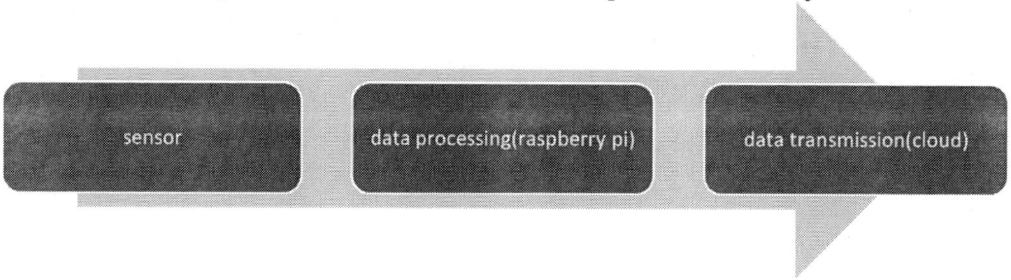

Figure.4 Represents the workflow process

As per the work flow chart, the sensors get the input and it gets processed after which the following data gets compared with raspberry pi. If the programmed average level of blood pressure and glucose level of the person is exceeded, it will display the alert message in the display as well as the processed data gets stored in the cloud as per monthly or daily basis the data will get repeatedly updated in the cloud. The doctor can access the data and the following precautions will be given by the one who's accessing the particular person's data. The precaution in the sense is not only giving the appropriate and necessary medicine, but also taking the medical history of the particular patient, which can be monitored continuously over periods of time and the best possible report along with proper diagnosis can be submitted to the concerned doctor or medical consultancy.

The protocols used for data transfer and communication are HTTP [hypertext transmission protocol] and MQTT [message queuing telemetry transport] where used in connectivity. The reason of the selection of this protocol is because this protocol is mainly used for sensor based application, based on the application requirement.

3.2 Device used

The system architecture consists of blood pressure sensor, glucose level monitoring sensor and raspberry-pi.

Raspberry pi

The raspberry pi is the smallest and the cheapest computer that can be easily interfaced with the pc. It can be accessed by the people at any of ages provides the easy way of computing and it provides the platform to learn languages like python. It can be also controlled through IOT.

The data will get received to the raspberry-pi through the sensor and the data manipulation takes place and the manipulated data get transmitted to the cloud database from that the doctor or medical consultancy can be able to read and write the data.

Fig 5: Represents the picture of raspberry-pi

3.3 Blood pressure sensor

The blood pressure is the pressure of blood from arteries as it pumped around the body by the heart. The heart contract and pushes during this process the blood is get forced through the arteries. The pressure is been recorded as two type- systolic pressure during heart beat over the diastolic pressure (the heart relaxes between the beats). This part is measured using sphygmomanometer.

The SUNROM-1437 is used to measure the blood pressure. The pressure won't remain constant at all times it will change in day to day life. Because of the food habits we following in our day to day life.

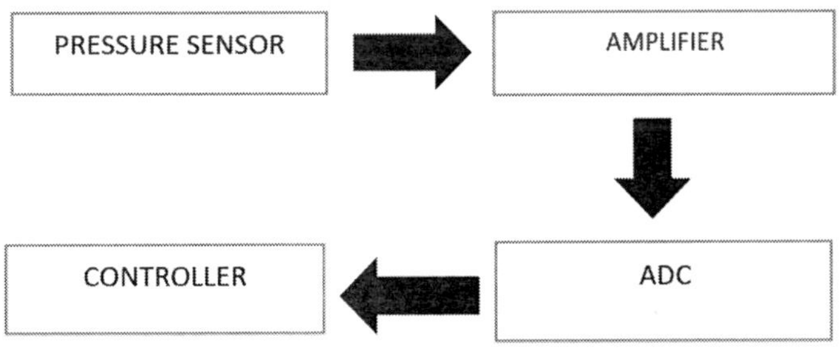

Figure.6 block diagram of blood pressure sensor

3.4 Glucose level sensor

The non-invasive glucose level sensor is used to measure the glucose level of the patient without pricking the blood. It works on the principle of beer lambert's law if an optical transmitter and receiver is placed opposite to each other on the alongside of the finger the receive light intensity at RX will be depending on the following parameters reflection by skin and absorption scattering by platelets and obviously the amount of glucose present in the blood sample. The above states properties are utilized to develop the prototype model the glucometer has a sensor which captures the reflected light intensity after it passes through the skin and affects the value in terms of voltage. The following code in raspberry-pi will convert the voltage to equivalent glucose level. The technology is in developing stage in future the non-invasive sensor going to play a important role in medical world.

Global system for mobile

GSM was developed as time division multiple access for communication process. A GSM digitizes and send the data to the different streams of client data. If we want to access the raspberry-pi at any part of the world without any wi-fi access point a digital cellular network maybe used

On hardware side GSM module is used it is also called as GSM modem. The modem is get interfaced with raspberry-p and similarly moreover all the modems use3 a serial RS-232 communication. The GSM is not only accessed through WLAN network it can also be accessed through internet through TCP server. The SIM800 chip had lot of variance and it is available in several module and also it is accessed through 4.5v power supply. The SIM800 chip is used because it consumes only less power and it is used for faster way of communication

Figure.7 Represents GSM module

4. Conclusion

The review about the future scope is now a days the precautions will be given by the medical consultancy or doctors handling the particular person data. In future this won't be done by the doctor or the consultancy. The precautions will be predicted and it will be analyzed through machine learning there will be no person behind the communication to instruct. Artificial intelligence will take care the person data as well as his health care problems

References

[1] Boukerche, A. and Ren, Y., 2009. A secure mobile healthcare system using trust-based multicast scheme. *IEEE Journal on Selected Areas in Communications*, *27*(4), pp.387-399.

[2] Milenković, A., Otto, C. and Jovanov, E., 2006. Wireless sensor networks for personal health monitoring: Issues and an implementation. *Computer communications*, *29*(13-14), pp.2521-2533.

[3] Ashwini gutte, Ramakrishna vadali IOT based health monitoring system using Raspberry-pi.

[4] Varshney, U., 2007. Pervasive healthcare and wireless health monitoring. *Mobile Networks and Applications*, *12*(2-3), pp.113-127.

[5] boukerche, "performance Evaluation of routing protocols for Ad hoc wireless networks", ACM/springer mobile networks and application.

[6] Hutchinson, C., Ward, J. and Castilon, K., 2009. Navigating the next-generation application architecture. *IT professional*, *11*(2), pp.18-22.

[7] Lin, Y.H., Jan, I.C., Ko, P.I., Chen, Y.Y., Wong, J.M. and Jan, G.J., 2004. A wireless PDA-based physiological monitoring system for patient transport. *IEEE Transactions on information technology in biomedicine*, *8*(4), pp.439-447.
[8] A.V. Halteren, R. Bults, K. Wac, D. Konstantas, I. Widya, N. Dokovsky, G. Koprinkov, V. J ones, R. Herzog, (2004), Mobile patient monitoring: The MobiHealth system. *Journal on Information Technology in Healthcare*, **2** (5), 365-373
[9] Ali, A. and Khan, F.A., 2013. Energy-efficient cluster-based security mechanism for intra-WBAN and inter-WBAN communications for healthcare applications. *EURASIP Journal on Wireless Communications and Networking*, **1**,.216. E.E.
[10] Marinelli, E.E., 2009. *Hyrax: cloud computing on mobile devices using MapReduce* (No. CMU-CS-09-164). Carnegie-mellon univ Pittsburgh PA school of computer science.
[11] Mayrhofer, R. and Gellersen, H., 2007, May. Shake well before use: Authentication based on accelerometer data. In *International Conference on Pervasive Computing* (144-161). Springer, Berlin, Heidelberg.

An Experimental Investigation on Performance and Emission Characteristics of PCCI Engine Using Biodiesel-Ethanol Blends in Dual Fuel Mode

Gowtham R[1], Akash V[2], Ganesh Kumar P[2], Bhaskar K[1]

1 Faculty, Rajalakshmi Engineering College, Chennai-602105, India.

2 UG student, Rajalakshmi Engineering College, Chennai-

602105, India. E-mail: gowtham.r@rajalakshmi.edu.in

Abstract. The global automotive industry is faced with the task of reducing global greenhouse gas emissions generated by vehicle exhaust. Heavy machinery like trucks, trailers, off-road vehicles, etc. powered by the diesel engines. These vehicles emit large quantities of NOx and smoke emissions. The simultaneous reduction of both emissions is the key challenge for the automotive industry. Burning a homogeneous charge at relatively low temperature seems to be the means to control both the emissions simultaneously. In this experimental study, the combustion, performance and emissions characteristics of a premixed charge compression Ignition Engine (PCCI) with ethanol injected in the intake manifold of the engine along with biodiesel (Palm oil methyl ester) injected directly into the combustion chamber. The experiments were carried out in a four-stroke, single cylinder vertical water cooled, constant speed diesel engine with the range of 10- 30% premixed ethanol fuel from no load to full load condition is studied. The experiments were conducted with an engine operating in PCCI mode with biodiesel – ethanol blends in dual fuel mode. The experimental results showed that there was a slight increase in Hydrocarbon (HC) and Carbon monoxide (CO) but there was a reduction in Nitrous-oxide (NOX) and Smoke emissions for the blend containing 70% biodiesel and 30% ethanol.

Keywords: Biodiesel;Dual blend;ethanol;PCCI engine;Hydrocarbon;Smoke emission;Ethanol.

1. Introduction

The constant challenges facing day to day are increased emissions and fuel rate consumption which are the environmental as well as energy challenges respectively. To supply the demand for fuel economy, environmentally less dangerous emissions, conservation of energy and, partial Premixed Charge Compression Ignition combustion system idea is ideal. Engine output releases include Unburnt Hydrocarbons, CO, Oxides of Nitrogen, Smoke and matter of particulates, etc; Emissions from the engines remain a condemnatory situation damaging the layout of design and working of Internal Combustion engines. One favourable way to resolve that kind of problems is to understand Homogeneous Charge Combustion (HCC) at maximum low temperature and with a compression ratio which is higher at that time, which is the last word goal [1]. In HCCI the homogenous mixture and control of combustion are difficult. Hence we have moved to partial premixing of charge which is called Partially Premixed Charge Compression Ignition (PCCI) engine. The aftermath of petroleum fuel which has premixed and correct timid direct injection on half the amount of HCCI were noted and inspected. The answers showed a constant lowering of emission and oxides of nitrogen with a modest CO increase and unburned carbon. To get out of the tough situation, the PCCI idea is taken. Despite the fact that HCCI can give out smoke and benefits of fuel consumption compared to DI-CI Engine it is now still a concept due to its operating difficulties [2]. To come out of these hardships, the PCCI idea is taken for decreasing the emissions of the NOx and soot. The operation of PCCI involves the pre-mixed charge preparation outside of the cylinder. A little fuel is sent inside with the incoming air at the intake manifold and the mixture then goes inside the combustion chamber and the remaining fuel is ejected inside as usual. The diesel injection using port injection is not easy for the surroundings if it is at extreme low temperature for the fuel to dry up. In the CI, the emission and combustion are affected by the atomisation happened and particularly, by the mixture inside the chamber[3]. Several processes are done to achieve correct vaporization which is taking place in the intake manifold. During EGR, air is preheated and premixing chamber was used. Every processing has both success and failures. Employment of hot GCR vaporization of large scale is noted, but it would increase the chamber temperature also, therefore NOx production is increased, and hence the EGR needs a non-manual control mechanism, In the case of EGR is used under various loads conditions[4]. The estimated density, kinematic viscosity, and flashpoint for Palm based oil and biodiesel to mix with the diesel. Results tell the quality of fuel that blends are very high to that of diesel until 30% less than other similarities inside certain restrictions. The experimented data were related as a quantity fraction of oil within the mix. Different comparisons are experimented to note the bio- oil-diesel mix properties supported by our experimental result analysis. The notes relations were calculated by comparing the predictions with experimental data. The equations which were analysed can be used as a frontier for giving diesels engines the best test blending mix. They analysed the total qualities of the C.I. engine charged with the blended diesel. Biodiesel and palm oil methyl ester mix were made in volume percentages of 20 and 100%. The biodiesel mixtures qualities were the same as diesel oil. The experimentation is set on a CI engine and it equals with diesel[5]. UBHC and CO emissions for biodiesel blends have decreased but vice-versa for oil blends comparing it to diesel. NOX emissions have been increased for biodiesel. Blends of diesel to biodiesel up to percentage of biodiesel by volume are recommended [3-6]. The inspected ratios of emulsified fuels on all basic properties of single cylinder 4 stroke Kirloskar engine under various loads[6]. For this particular test Benzalkonium chloride is mixed to the diesel blend as an emulsifier to resist formation of layers and to be made as a blend of homogenous mixture. At top most BP, the contrast between the diesels and emulsified fuels shows improved properties in brake power efficiency in results with depletion in smoke and SFC. The CO_2, HC, cylinder force, and warmth rate of release for D50 E40 emulsions are greater than diesel [7-8]. The partial Premixed Charge Compression Ignition-Direct Injection (PCCI-DI) Engine with premixed fuels such as ethanol and gasohol is injected with mixed diesel and all the basic properties such as performance, emission etc has been analysed and investigated. The experiments were done on all load conditions in a naturally aspirated air-cooled 4 stroke engine. The premixed fuel additive strengthens the strength of fuel- air mixture and the combustion duration is made lesser in bi-fuel injection .The above study, it had been noted that 70% reduction in smoke from mixed gasoline and 67% from ethanol fuel in comparison to pure diesel. The emission of nitrogen oxides were also decreased to 30% and 24% for ethanol fuel and premixed gasoline. In Particular smoke, NOx emissions are reduced in premixed gasoline slightly equal to the amount of ethanol, brake thermal efficiency was increased gradually and noted in 20% and ethanol and premixed gasohol in bi-fuel mode, in comparison to pure diesel experiment [9]. They checked the power output and emitted pollutant properties of a CI engine filled with ethanol and palm-oil methyl ester. The performance and exhaust emission characteristics of Palm oil methyl ester biodiesel (PB10, PB20, PB30) adding ethanol (E5% and E10%) in an sudden Injection – Compression Ignition engine over the changing loading conditions on the engine and found lesser CO emissions. The corrections can be done through a smaller amount of Carbon Monoxide emissions and Exhaust Gas centigrade reduced inside the raises of doping of ethyl alcohol content in mixture of Methyl Ester and its mixture [10]. They investigated the causes of double bio-fuel (Jatropha Methyl esters and purified Turpentine oil) on a cylinder normally aspirated diesel (Kirloskar) unaccompanied exhaust gas recirculation.

IOP Conf. Series: Materials Science and Engineering 1059 (2021) 012036 doi:10.1088/1757-899X/1059/1/012036

Jatropha Methyl esters and purified turpentine oil have a more and less viscosity fuel addition with comparative boiling values to that of diesel; so this will offer use in a compression-ignition engine. The further conclusion of BT50 in all load conditions, depletion for 2.9%, 4.56%, 4.72%, 29.16%, 42.5% in brake energy efficiency, smoke, CO, NO_x and HC even though the CO_2 emissions raises 10.7% [11]. This review suggests that fuel injection is a suitable way for HCCI mode, so its dividend to the compression-ignition engine requires be analysing and studying. Inside a diesel engine compression process takes place based fuel injection into the combustion chamber. The necessary properties are timing and period of the ejection of fuel working, the giving out of the fuel and atomization in the combustor, the ignition timing, the injected fuel are proportional to the rotation of the crankshaft, and the complete fuel injected in connection to powerhouse(engine) capacity. The arrangement idea has developed and introduced to the matter more emission levels of conventional direct injection compression-ignition engine systems. During early stages the premixed charge compression ignition technique is used in injections for the burning process when a higher amount of fuel is burned, which disturbs combustion noise. The combustion noise reduced by using a tool of pilot injection usage .The most important motive of this article is to analyse by experimenting the combustion noise, emissions, and diesel performance in operation below premixed charge compression ignition in this operation of pilot injection. Besides, a special process supported the atomization of the in- cylinder force signal, combustion noise analysis was experimented.

2. Experimental setup

Internal combustion liquid cooled which has vertically fixed engine type which rotates at 1500rpm. The addition of loads is done through an electrical dynamometer. In an unit time the fuel quantity can be measured by a burette which is linked to the tank. The exhaust gas was measured by AVL437C smoke meter. A test rig is connected with a data acquisition system for obtaining different curves and results during experimental investigation. A five gas analyzer aligned with a stationary engine to provide composition details of emission in exhaust gas are CO, CO_2, UBHC, NO_x, and other non-useful oxygen. In this method, the cable at one end is connected to the exhaust gas outlet while the other is at the end of the inlet of the analyzer. Figure 1 and 2 shows the experiment setup. Tabulation 1 and 2 shows Engine Specification and properties of fuel.

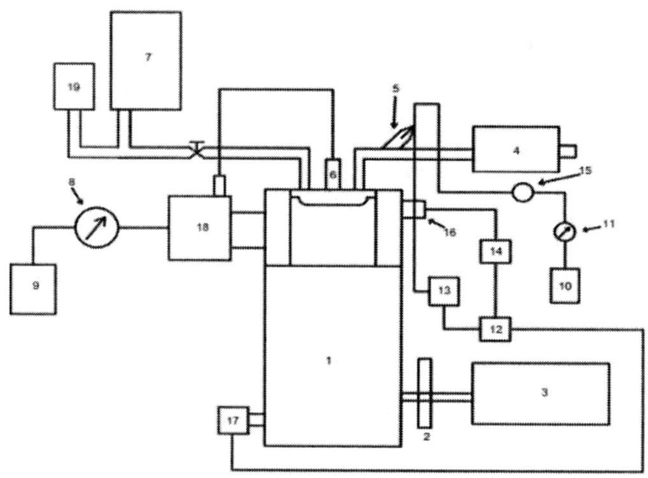

Fig. 1 Experimental Setup

IOP Conf. Series: Materials Science and Engineering 1059 (2021) 012036 doi:10.1088/1757-899X/1059/1/012036

1- Engine 10- Port Fuel Tank
2- Flywheel 11- Flow Meter
3- Dynamometer 12- Data Acquisition
4- Air Surge Tank 13- Relay Controller 5-
Auxiliary Fuel Injector 14- Charge Amplifier 6-
Main Fuel Injector 15- Fuel Control Valve
7- Exhaust Gas Analyzer 16- Pressure Transducer 8-
Flow Meter 17- Crank Angle Encoder
9- Main Fuel Tank 18- Fuel Injection Pump

Fig.2 Experimental Setup Table

1 Technical specification of the test engine

Engine type	Kirloskar Oil Engine TV1
Bore	87.5 mm
Stroke	110 mm
Rated power output	5.20 kW
Rated speed	1500 rpm
No. of cylinder	1
Cooling system	Water-cooled
Injection pressure	200 bar
Injection timing	23 deg bTDC
Compression ratio	17.5 : 1
Connecting rod length	234.00(mm)

Table 2 Comparison of Fuel properties

Properties	Units	Diesel	Palm Oil Methyl Ester	Ethanol
Auto Ignition Temperature	$_o$C	210	343	363
Octane Number	Nil	-	-	102
Cetane number	Nil	51	47	-
Density at 15oC	kg/m^3	873	904	786
Flash Point	$_o$C	65	196	26
Fire Point	$_o$C	75	210	34
Calorific Value	kJ/ kg K	45000	37500	29700

The below equation is used to calculate the premixed fuel ratio (m_r);

$$r_p = \frac{m_p h_p}{m_p h_p + m_d h_d}$$

Where m_p be the premixed fuel mass consumed and m_d be the directly injected conventional fuel correspondingly, where h_p , h_d are calorific values of premixed fuel ratio and diesel. The experimental tests were taken out for a premixed fuel ratio of 0.1, 0.2 & 0.3 respectively. The duration of fuel injection was calculated within milliseconds and position in an electronic control unit. The premixed fuel ratio is decreased by using electronic control unit(ECU) .The ratio for the experiment is the diesel – 100%, Biodiesel (Palm oil methyl ester) 100%, Biodiesel 90%, and 10% Ethanol, Biodiesel 80%, Ethyl alcohol 20%, Methyl esters 70% and Ethanol 30%.

3. Results and Discussion

In the paperwork converse the burning characteristics, amount of heat release for the crank angle manifold injected premixed ethanol, and diesel oil was compared with palm oil. In the same way, the performance emission characteristics of curb energy efficiency and curb particular fuel expending and discharge attribute such as curb particular CO, NOx, curb particular HC and pollution emission was compared with diesel oil. In this paper the result shows the full experimentation of alternative-fuel to advanced want of HCCI ignition.

Combustion Analysis

Load condition at 50%

The crankshaft angle for pure diesel is compared with cylinder pressure figure 3 is given below, biodiesel, biodiesel-ethanol mixture at 50% load condition. It is observed that the neat diesel produced more cylinder pressure compared to B100 and biodiesel-ethanol blends. The lowest cylinder pressure was produced by the B70+E30 blend. As mentioned earlier, the cooling effect of alcohol combined with higher octane number has increased the delay period which led to the attainment of peak pressure after TDC.

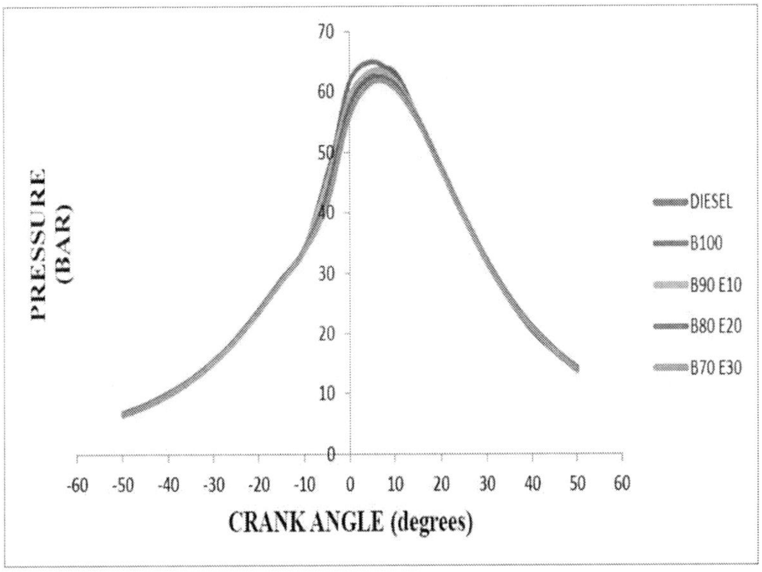

Fig. 3 CrankAngle vs. Cylinder Pressure (50% Load Condition)

Load condition at 100%

In the above graph 4 shows that the crankshaft angle for pure diesel is compared with cylinder pressure, Methyl esters, biodiesel-ethanol mixture at filled load state. This result showing that more cylinder pressure is produced by B70+E30 than pure diesel and other Methyl ester-ethanol mixture. The B100 has lower cylinder pressure. Under full load, higher combustion temperature overtakes the cooling effect of ethanol, and B70+E30 will produce higher combustion pressure thereby leading to better combustion.

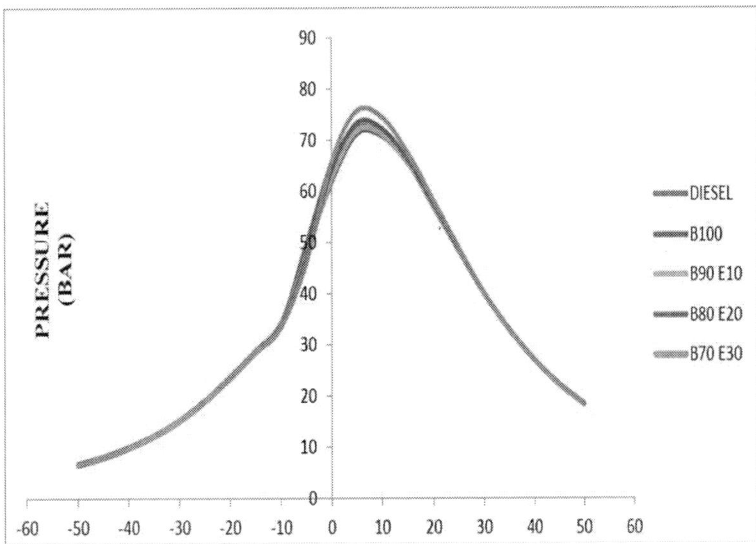

Fig. 4 CrankAngle vs. Cylinder Pressure (100% Load Condition)

Performance Characteristics
Brake Thermal Efficiency

Figure 5 shows differences in the Brake thermal efficiency with various loads for pure diesel, methyl ester, and methyl ester-ethanol mixture. The capability of combustion system is indicated by curb energy efficiency and how proficient the fuel energy is transformed to mechanical energy. From the above graph fig 5, it is seen that the maximum curb energy efficiency was produced by the pure diesel compared to biodiesel and other biodiesel-ethanol blends. However, the B70+E30 blend produced brake thermal efficiency which is closer to neat biodiesel at full load. Also, this blend performed better than other biodiesel-ethanol mixture for the whole operating load of the engine.

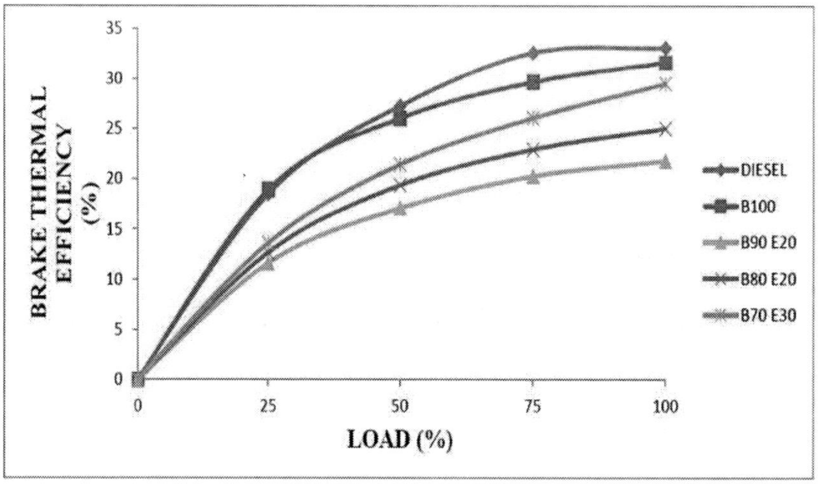

Fig. 5 Variation of Brake thermal efficiency with the load

Brake Specific Energy Consumption

Fig 6 indicates difference in curb energy efficiency consumption to various loads for neat diesel, biodiesel, biodiesel-ethanol blends.

$$\mathbf{BSEC = BSFC \times CV}$$

From the above graph figure 6, it is seen that the particular energy consumption is found to be more when the engine was operating with B90+E10 blend and lowest with neat diesel. The B70+E30 blend produced lower specific energy consumption compared to other biodiesel-ethanol blends. This may be due to improvement in fuel vaporization characteristics leading to better combustion efficiency. Since the calorific value of methyl ester , methyl ester-ethanol mixture are found to be lower than diesel, the engine will consume more fuel to produce the same power output, leading to the increased specific fuel consumption for biodiesel and methyl ester-ethanol mixture. Also, the fact that during dual fuel mode, more energy is required for injecting both the fuels separately. This is the reason for higher energy consumptions by biodiesel-ethanol blends. The SEC is highly dependent on SFC and Calorific value.

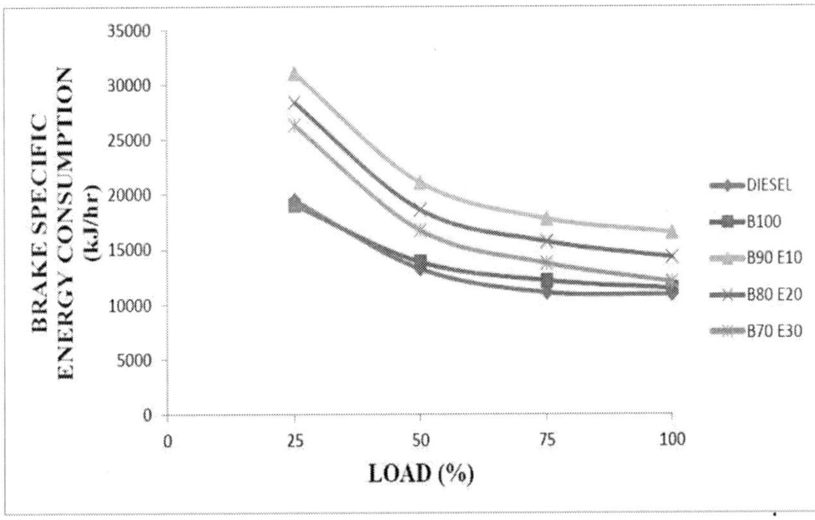

Figure 6 Variation of Brake specific energy consumption with the load.

Emissions Characteristics
Brake Specific Unburnt Hydrocarbon (BSUBHC)

Figure 7 indicates the difference in the Hydrocarbon emissions to different loads for neat diesel, biodiesel, biodiesel-ethanol blends. The neat diesel and biodiesel have lower HC emissions than the Biodiesel-Ethanol blend. The B70+E30 have higher HC emissions than neat diesel, biodiesel, and other methyl ester-ethanol mixture. The increase in HC emissions along with premixed ratio increment because a reduction in the temperature cylinder. The cooling effect of ethanol can lead to flame quenching which is the most significant reason for higher HC emissions. However, B70+E30 produced slightly higher HC emission at full load compared to other blends, because of the increased cooling of induced alcohol. The escape of premixed fuel through the crevice volume is the reason for increased hydrocarbon emissions at the low and part loads. The lower latent heat of ethanol reduces the combustion temperature and affects the burning of biodiesel by forming a quenching layer. At full load, the temperature inside the combustion chamber is necessary to burn all the fuels. This may be since some hydrocarbon components are composed on the surface of the carbon particles which contribute of the engine.

IOP Conf. Series: Materials Science and Engineering 1059 (2021) 012036 doi:10.1088/1757-899X/1059/1/012036

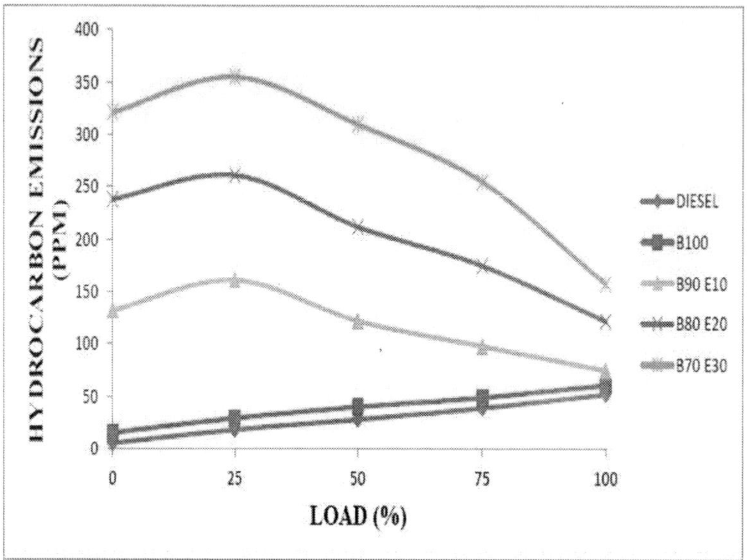

Figure 7 variation of HC emissions with respect to load 3.2.3LOAD

VS CARBON MONOXIDE

The conventional diesel fuel produces low amount of CO emission for biodiesel and its blends. There is a higher amount of CO emission in light load situation for B70 +E30, conversely there promising of decreases in Co emission at maximum engine load at same blends. This is due to rich air fuel production in cylinder. The usage of ethanol blends along with this fuel may reduce the peak cycle temperature leads to decrease carbon dioxide conversion that may left of larger amount of carbon monoxide in the exhaust at minimum load. This inference that HCCI the conversion of Carbon dioxide from carbon monoxide depends upon the peak to peak cycle combustion temperature increase above leads to 1500 K for conversion.

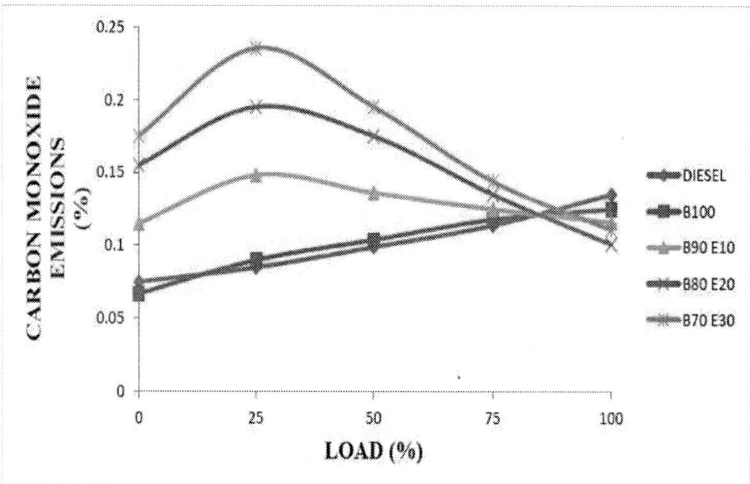

Figure 8 variation of CO emissions with respect to load

Oxide of Nitrogen

Figure 9 indicates the variation of Nitrous Oxide emissions to various loads for neat diesel, biodiesel, and biodiesel-ethanol blends. The neat diesel has higher NOX than neat biodiesel and other blends. From the experimental observation that NOx is reduced when there is an increase in the amount of premixed ratio when related to neat diesel and biodiesel. The concept of increasing ethanol in the intake manifold has reduced temperature and this pre-cooling might have reduced the in-cylinder combustion temperature. The values of NOx at no load to full load for B70+E30% are lower than neat diesel and biodiesel. The cooling effect of higher ethanol content in the B70+E30 blend combined with a reduction in cetane number led to increase in ignition delay which caused the peak pressure to occur slightly after top dead center, this is the main reason for a decrease in NOx emission with the higher premixed ratio of ethanol. But, there is an increase in NOx emission at the full load. This is due to increased cylinder pressure which causes a high peak combustion temperature. The increase in cylinder temperature causes increased oxides of nitrogen emission.

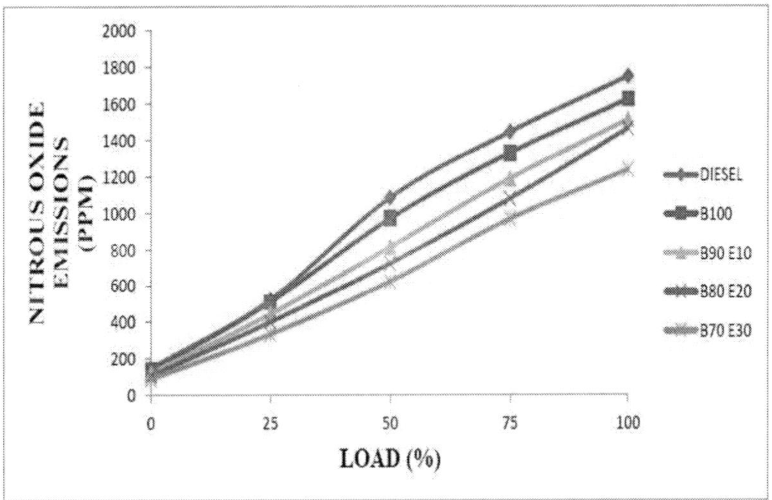

Fig. 9 variation for oxides of nitrogen emissions with respect to load.

Smoke Emission

Figure 10 shows the difference in smoke emissions of various loads for neat diesel, biodiesel and biodiesel ethanol blends. The neat biodiesel has more smoke than pure diesel and methyl ester ethanol mixture. A smoke emission for B70+E30 is reduced due to more amount of oxygen and a lower amount of carbon content in the mixture. The higher viscosity and density of methyl ester is the reason for increasing smoke in pure biodiesel. The indication of maximum energy release and peak pressure for the B70+E30 mixture enhanced the oxidation of carbonaceous soot produced during the mixing controlled combustion which led to lower smoke emission than others mixtures at higher engine power.

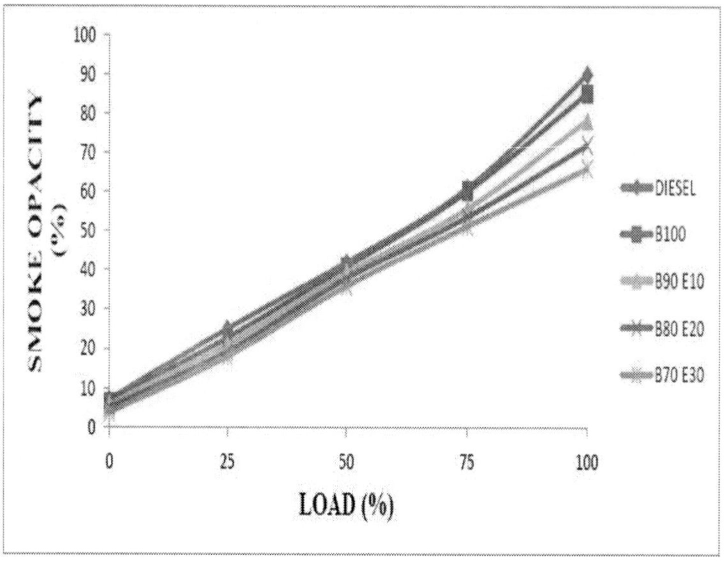

Fig, 10 variations of Smoke emissions with respect to load.

4. Conclusion

From the experimental study the following conclusions can be determined:

- The cylinder pressure was found to be higher in full load for B70+E30 (75.8 bar at 6o aTDC). The B100 has lower cylinder pressure (71.25 bar at 9°aTDC). Whereas in the case of part load, the cylinder pressure was found to be higher for neat diesel (64.95 bar at 5° aTDC) and lower for B70+E30 (61.46 bar at 8° aTDC).
- The brake thermal efficiency was found to be higher for neat diesel (33.75%) at full load. And lower brake thermal efficiency was found to be B90+E10 (21.95%) at full load. However, the B70+E30 (30.22%) blend produced brake thermal efficiency which is closer to neat biodiesel at full load.
- The specific energy consumption was found to be higher for B90+E10 (16513.75 kJ/hr) at full load. And lower specific fuel consumptions were found to be neat diesel (10893 kJ/hr) at full load. The B70+E30 blend (12028.22 kJ/hr) produced lower specific fuel consumption compared to other biodiesel-ethanol blends. This may be due to improvement in fuel vaporization characteristics leading to better combustion efficiency.
- The HC emissions were found to be higher for B70+E30 (158 PPM) at full load. And lower was to be found for neat biodiesel (52 PPM) at full load. The cooling effect of ethanol can lead to flame quenching which is the most significant reason for higher HC emissions.
- The CO emissions were found to be higher for neat diesel (0.156 %) at full load. And lower was found to be for B80+E20 (0.099 %) at full load. The main reason for higher CO emission is because of fuel-rich zones and insufficient oxygen availability inside the combustion chamber.
- The NOx emissions were found to be higher for neat diesel (1745 PPM) at full load. And lower emissions were found to be B70+E30 (1225 PPM) at full load. The cooling effect of higher ethanol content in the B70+E30 blend combined with a reduction in cetane number led to an increase in ignition delay and reduction in in-cylinder temperature.

- The smoke emissions were found to be higher for neat diesel (90 %) at full load. And lower emissions was found to be B70+E30 (65 %) at full load. Due to maximum heat release and peak pressure for the B70+E30 blend enhanced the oxidation of carbonaceous soot produced during the mixing controlled combustion which led to lower smoke emission compared to other blends at higher engine load

5. References

[1] Ganesh, D., Nagarajan, G. and Ibrahim, M.M., 2008. Study of performance, combustion and emission characteristics of diesel homogeneous charge compression ignition (HCCI) combustion with external mixture formation. *Fuel*, *87*(17-18), pp.3497-3503..

[2] Channapattana, S.V., Pawar, A.A. and Kamble, P.G., 2017. Optimisation of operating parameters of DI-CI engine fueled with second generation Bio-fuel and development of ANN based prediction model. *Applied energy*, ***187***, pp.84-95.

[3] Zhang, C. and Wu, H., 2016. Combustion characteristics and performance of a methanol fueled homogenous charge compression ignition (HCCI) engine. *Journal of the Energy Institute*, ***89*(3)**, pp.346-353.

[4] d'Ambrosio, S. and Ferrari, A., 2015. Effects of exhaust gas recirculation in diesel engines featuring late PCCI type combustion strategies. *Energy Conversion and Management*, ***105***, pp.1269-1280.

[5] Gad, M.S., El-Araby, R., Abed, K.A., El-Ibiari, N.N., El Morsi, A.K. and El-Diwani, G.I., 2018. Performance and emissions characteristics of CI engine fueled with palm oil/palm oil methyl ester blended with diesel fuel. *Egyptian Journal of Petroleum*, ***27*(2)**, pp.215-219.

[6] El-Araby, R., Amin, A., El Morsi, A.K., El-Ibiari, N.N. and El-Diwani, G.I., 2018. Study on the characteristics of palm oil–biodiesel–diesel fuel blend. *Egyptian Journal of Petroleum*, ***27*(2)**, pp.187-194.

[7] Rao, P.V., 2011. Experimental investigations on the influence of properties of jatropha biodiesel on performance, combustion, and emission characteristics of a DI-CI engine. *World academy of science, engineering and technology*, *75*, pp.855-868.

[8] Parthasarathi, R., Gowri, S. and Saravanan, C.G., 2014. Effects of ethanol-diesel emulsions on the performance, combustion and emission characteristics of DI diesel engine. *American Journal of Applied Sciences*, ***11*(4)**, p.592.

[9] Saravanan, S., Pitchandi, K. and Suresh, G., 2015. An experimental study on premixed charge compression ignition-direct ignition engine fueled with ethanol and gasohol. *Alexandria Engineering Journal*, ***54*(4)**, pp.897-904.

[10] Srinidhi, C. and Madhusudhan, A., 2017. A diesel engine performance investigation fuelled with nickel oxide nano fuel-methyl ester. *International Journal Of Renewable Energy Research*, *7*, pp.677-686..

[11] Dubey, P. and Gupta, R., 2018. Influences of dual bio-fuel (Jatropha biodiesel and turpentine oil) on single cylinder variable compression ratio diesel engine. *Renewable Energy*, ***115***, pp.1294-1302.

[12] Himabindu, M., Sathyanarayanan, G., Thomson, S. and Mahalakshmi, N.V., 2008. *Effect of premixed charge of ethanol and exhaust gas recirculation on the performance and emission characteristics in an ethanol-diesel fueled HCDC engine-An experimental investigation* (No. 2008-01-2609). SAE Technical Paper.

[13] Shi, L., Cui, Y., Deng, K., Peng, H. and Chen, Y., 2006. Study of low emission homogeneous charge compression ignition (HCCI) engine using combined internal and external exhaust gas recirculation (EGR). *Energy*, ***31*(14)**, pp.2665-2676..

[14] Torregrosa, A.J., Broatch, A., García, A. and Mónico, L.F., 2013. Sensitivity of combustion noise and NOx and soot emissions to pilot injection in PCCI Diesel engines. *Applied Energy*, ***104***, pp.149-157..

[15] R.H. Thring, Homogeneous charge compression ignition (HCCI) engines, SAE Paper 8902068 (1989).

IOP Conf. Series: Materials Science and Engineering 1059 (2021) 012037 doi:10.1088/1757-899X/1059/1/012037

Influence of Solution Treatment and Artificial Aging onFracture Load of Friction Stir Welded Lap Joints of AA2014-T6

C Rajendran[1], Parthiban. G[2], Pranav.K[3] and Nithi Balaji. S.P[4]
[1] Associate Professor, Sri Krishna College of Engineering and Technology, Department of Mechanical Engineering, Coimbatore, Tamil Nadu, India.
[2-4.] UG Scholar, Sri Krishna College of Engineering and Technology, Department of Mechanical Engineering, Coimbatore, Tamil Nadu, India.

rajendranc@skcet.ac.in

Abstract.High strength aluminium alloys (2xxx and 7xxx) are widely used for many applications. Particularly, aircraft and automobile industries. The fabrication of this alloy is little bit difficult using traditional welding process. It produces porosity, alloy segregation, partially melted zone etc. In last few decades, these alloys are welded using a solid-state welding process, friction stir welding. However, the strength of the alloy is more than 80% of the base material strength. Although, the FSW joints have lowest hardness in the weld region due to heterogeneous properties in the microstructure. The lowest hardness region is the soft region that is the fracture-originating region during tensile test. Several techniques were used to enhance the lowest hardness region properties such as the aging process, the solution treatment process, cryogenic treatment in joints. In this research, the solution treatment followed by artificial aging on lap shear strength of friction stir lap-welded joints were performed. The heat-treated joints were compared with welded joints. From the experimental results, the heat-treated joints yield higher lap shear strength than the as welded joints. The reason for the higher lap shear strength was reprecipitation of precipitates in the welded joints during an artificial aging process. However, the aluminium matrix retains fine precipitates by the solution treatment process (under rapid cooling).

1 Introduction

High strength to weight ratio and corrosion prevention [1], high-strength Al alloys (7xxx, 6xxx and 2xxx series) are employed in many industries, including cars, aircraft and military vehicles. The welding of Al alloy-containing Cu (AA2014) is, however, a difficult task using a convention welding method, as it results in hot shots, alloy segregation, improper melt in zones and cluster of pores [2]. In addition, there is a substantial decrease in strength [3] due to the growth of the dendritic structure in the welding area caused by traditional welding. FSW is a solid-state process of which hot shear joins metals. Non-consumable pin profiles of the tool are revolved and gently penetrate into the region around the overlapping sheet material's joint line. FSW is used to weld aluminium alloys, including the 2xxx and 7xxx series, which were previously considered to have low weldability characteristics. Aluminium alloy AA2014 is an age-hardened alloy and has characteristics such as high strength as a result of heat treatment processes such as quenching, tempering and artificial aging treatments due to the development of Al2CuMg and CuAl2 (Copper Aluminide) stages [4]. All the phases or precipitates previously formed are dissolved during FSW. However, a fraction of the precipitates was coarsened because of the applied thermal cycle in the thermo-mechanically affected zone (TMAZ). Although the FSW joints yields greater strength than fusion welded joints, as a result of precipitate distribution in the FSW joints. It was found that there was a significant reduction in the distribution of precipitates in the HAZ and TMAZ[5]by different welding speed.

IOP Conf. Series: Materials Science and Engineering 1059 (2021) 012037 doi:10.1088/1757-899X/1059/1/012037

Water cooling was investigated and its influence on the basic properties of AA2014, and the lowest hardness zone (LHZ) was observed on both side of work piece under different cooling conditions.However, the LHZ was located on different zones for air cooling and water cooling. The LHZ was, however, located in separate air and water-cooling zones. The LHZ was found in the HAZ for joint welding in air and in the TMAZ for joint welding in underwater FSW [6]. Rajendran et al. [7] stated that solutionization followed by aging was effective in restoring the joint's tensile properties compared to treatments with AA and AW. Rajendran et al. [8] conducted on heat treatment experiments after welding on strength properties of AA2014 and reported solutionizing followed by aging improves strength of AA2014 alloy. From the observations [1-8], it is inferred that lot of research works have been focused toward heat treatment after weld. In this work, analyse the influence of solution treatment followed by ageing on fracture load in lap welded joints.

2 Materials and Methods

The Parent metal (PM) considered for this study was rolled AA2014 alloy sheets of 2 mm thickness. The sheet was cut in to 5 mm x 75 mm sections and locked in place tightly to establish lap joints structure. PM properties were confirmed as per the ASTM standard (Table 1 and Table 2).

Table 1 Chemical composition (%wt.) of PM

Cu	Mn	Si	Fe	Cr	Zn	Mg	Ti	Al
4.2	0.65	0.81	0.21	0.02	0.02	0.51	0.02	96

Table.2 Mechanical properties of PM

0.2 % YS (MPa)	UTS (MPa)	% Elong. in 50 mm gauge length	Hardness (0.5 N, 15 s)
432	460	7.5	152

Using FSW the joints were placed right angles to the sheets load direction. The shoulder diameter of 12 mm, tilt angle of 1.5 °, tool movement of 90 mm / min, and tool speed of 900 rpm are the optimal parameters used for manufacture of welds (Table 3). The joints were employed to solutioning (496 C for 1h.) followed by artificial ageing (160 C for 12 hr.) to enhance the strength of joints (Figue.1a, b).
Table.3 Nomenclature FSW tool

Table.3 Nomenclature FSW tool

shoulder		Pin		
D (mm)	Type	D_1 (mm)	D_2 (mm)	L (mm)
16	Concave	3	2	3

The lap shear specimens prepared in compliance with the specification ANSI / AWS / SAE / D8.97 [7] and the lap shear testing were performed at a slide movement of 1.5 kN / mm using UTM. High magnification microscopes (scanning electron and transmission electron microscopes) were employed to analyze the feature of various regions in welded joint. The elemental presences in the weld region were carried out using energy dispersive spectroscopy.

Figure.1 a) Fabricated joints b) Tensile specimens (before testing)

3 Results and Discussion

3.1 Lap shear strength

The shear fracture load of FSW joint and heat-treated joints is offered in Table 4. Three specimens were tested in each condition. PM exhibits a maximum shear fracture load of 18.34 kN. The heat-treated and as-welded joints yielded a maximum shear strength of 15.27kN and 12.76 kN, respectively. It is observed that the heat-treated joints conceived higher lap shear strength than as welded joints. Figure.2a-c shows the optical micrograph of PM in as received condition and SZ of as weld and heat-treated joints. PM microstructure constituted rough and enlarged grains with a mean diameter of 30 μm.

Table 4. Lap shear strength of welded joints under different condition

Sl.No	Condition	Microhardness "HV"			Fracture load "kN"
		SZ	AS-TMAZ	RS-TMAZ	
1	As weld joint	123	106	115	12.76
2	Solution treatment + artificial aged joint	135	115	117	15.27
3.	As received		155		18.34

3.2 Microstructure

FSW joint is divided into three regions, such as thermo-mechanically affected zone (TMAZ), stir zone (SZ) and PM. Figure.2b shows the SZ micrograph of as welded joint. Similarly, SZ micrograph of heat treated joint as revealed in Figure2c. SZ of both joints framed fine and recrystallized grains. The SZ of the heat-treated joint (Figure. 2c) indicates a substantial increase in the size of the grain owing to a combination solution treatment process followed by an artificial aging process.

Figure.2 Optical micrograph of a) As received condition, stir zone of b) As weld, c) Heat treated joint

The microstructure of as welded joints in advancing and retreating sides TMAZ is exposed in Figure. 3a, b. Although, the microstructure of TMAZ in advancing side and retreating sides of heat-treated joint is shown in Figure. 3c, d. Both the microstructure under different conditions showed different grain sizes by material flow in retreating and advancing sides. Moreover, the grains in the advancing side of both conditions were revealed elongated and slanted to the upper side due to the pushing of plasticized materials in SZ by high straining action. The grains in the retreating side showed less slanted and coarsened in both conditions. The slanting and coarsening of grains happened by the velocity differences of plasticized material in the SZ by tool geometry and finer precipitates precipitated during heat treatment process [9].

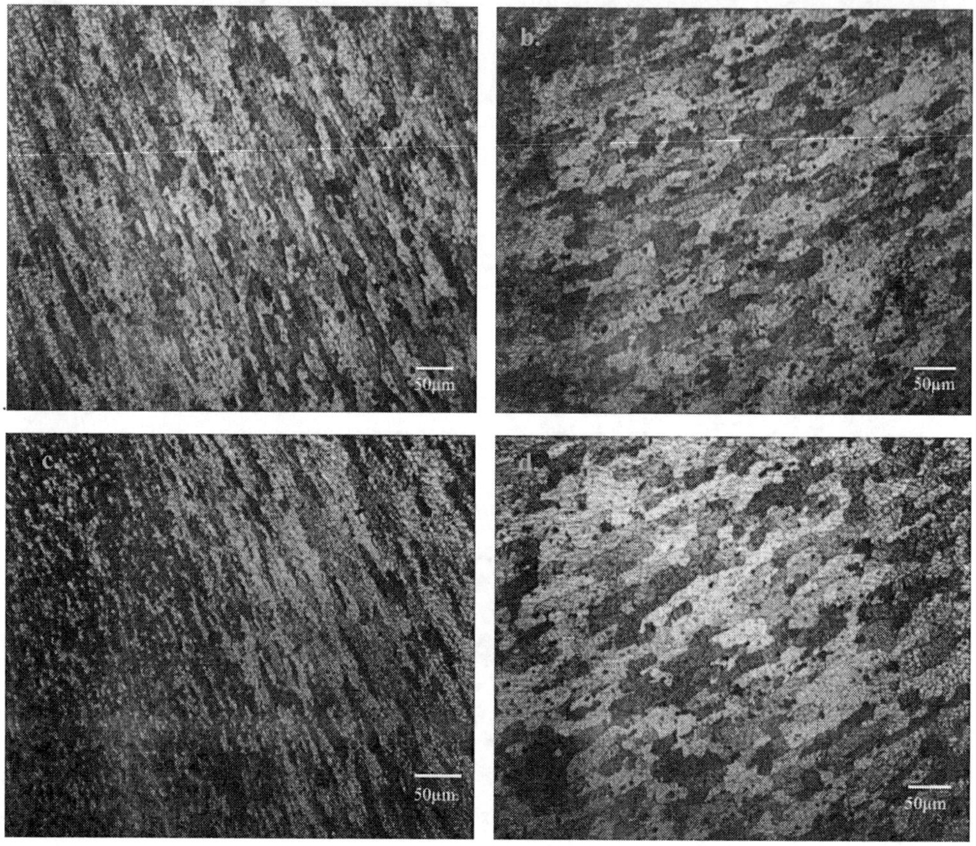

Figure.3 Optical micrograph of as weld joint a) AS-TMAZ b) RS-TMAZ, heat treated joint c) RS-TMAZ, d) AS-TMAZ

3.3 Hardness distribution

The hardness of high strength aluminum alloy (2014) in as received condition showed a maximum hardness of 155 HV. Figure.4 show the hardness distribution of as welded and heat-treated joints and SZ of as welded joints and heat-treated joint exhibited a peak hardness of 123 HV and 135 HV, respectively. The hardness of the FSW joint is based on the size of the grains and spreading of precipitates. In the as welded joint, lowest hardness region was observed on both sides of advancing and retreating TMAZ. Similarly, heat-treated joints formed two lowest hardness regions on both advancing and retreating sides. The lowest hardness of 106 HV was recorded on the advancing side of as welded joint and 115 HV was observed on the retreating side TMAZ. In the heat-treated joint, the minimum hardness of 115 HV on the advancing side TMAZ and 117 HV on retreating side TMAZ were observed. The pattern of hardness distribution in the FSW weld is as symmetrical due to variation in the microstructure in the welded joint.

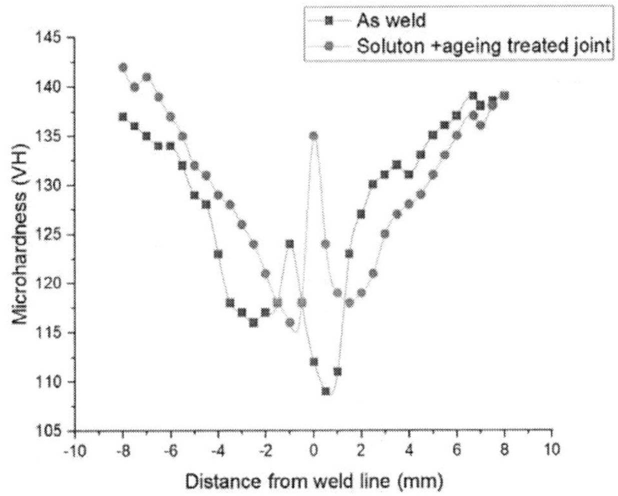

Figure. 4 Hardness distribution of FSW joints under different condition

3.4 Fractograph

The fracture region of welded joints under different condition is presented in Table 5. The fracture region of as welded joint was occurred at HAZ and the fracture point of heat-treated joint was found at SZ/TMAZ interface. A close inspection of the failure surface of the specimen provides valuable data on the function and influence of the inherent micro-structural characteristics to the strength and ductility of the joint [10,11]. The crack morphology of tensile tested as shown in Figure.5a-c. SEM examination was recorded from the middle part of the fracture zone and characterizes the fracture morphology of the welded joints. Ultimately, FSW joint fracture occurred from the weaker region of the joints during the tensile test. Depending on the state of joints, the position of the soft region is found to differ from the TMAZ/SZ interface or HAZ. As welded joint separated from HAZ (115 HV) on the advancing side and the fracture position of the heat-treated joint was found in SZ/TMAZ interface. With deeper and larger dimples showing the as the weld joint fracture surface (Figure.5b). In the heat-treated joint (Figure.5c), which showed enlarged dimple surface, similar behaviour was also found; under tensile testing, these elongated dimples suggested fracture mode was ductile. The width of the dimples in the heat-treated joint was wide, it proposes that there is a large stretch area at the tip of the crack, leading to a massive plastic area ahead of the crack [8].

Table 5 Fracture region of FSW lap joints

Sl. No	Condition	Joints	Fracture Region
1.	As-weld joint	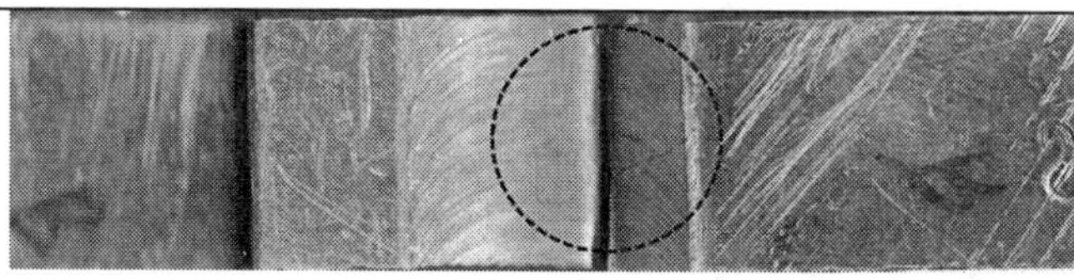	HAZ
2.	Solution treatment +artificial ageing treated joint	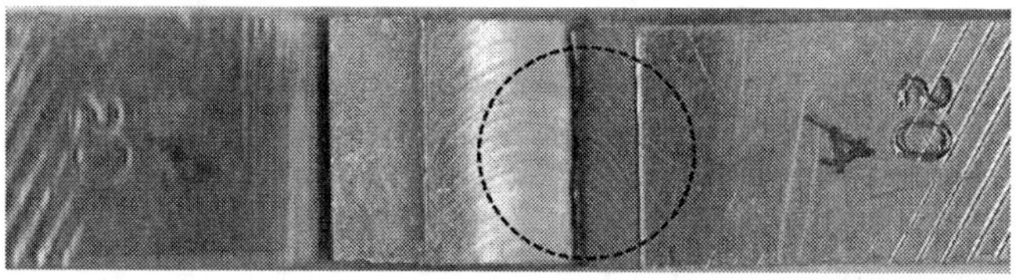	SZ/TMAZ

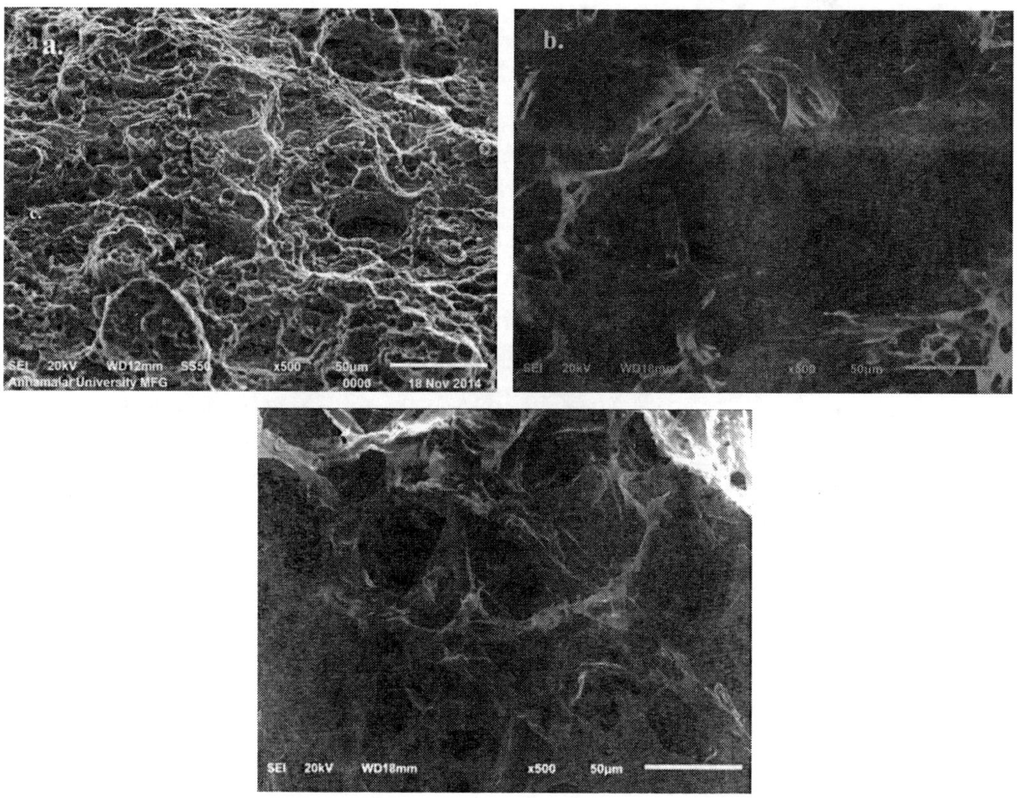

Figure.5 Fracture morphology of a) As received material b) As weld joint, c) Heat treated joint

IOP Conf. Series: Materials Science and Engineering 1059 (2021) 012037 doi:10.1088/1757-899X/1059/1/012037

3.5 TEM Micrograph

Two forms of precipitates (coarse and fine) were seen in the TEM micrograph of material in the obtained state (shown in Figure.6a). Rough precipitates differ in length from 60 to 110 nm, while fine precipitates differ in length from 20 to 60 nm with two distinct morphologies, including needle-like and spherical. Based on the EDS analysis, it is conformed that the fine precipitate was formed by Al, Cu and Mg elements. It suggests that the formation of Al_2Cu (Figure. 5d) and Al_2CuMg (Figure. 5e) precipitates. However, these precipitates were completely solutionized in the aluminium due to sliding heat in the weld cycle [12] except few coarse precipitates. The SZ of as welded joint showed complete fragmentation of coarse and fine precipitates (Figure. 6b) and SZ of heat treated joint revealed fine dissolution of strengthening precipitate except few coarse precipitates (Figure.6c).

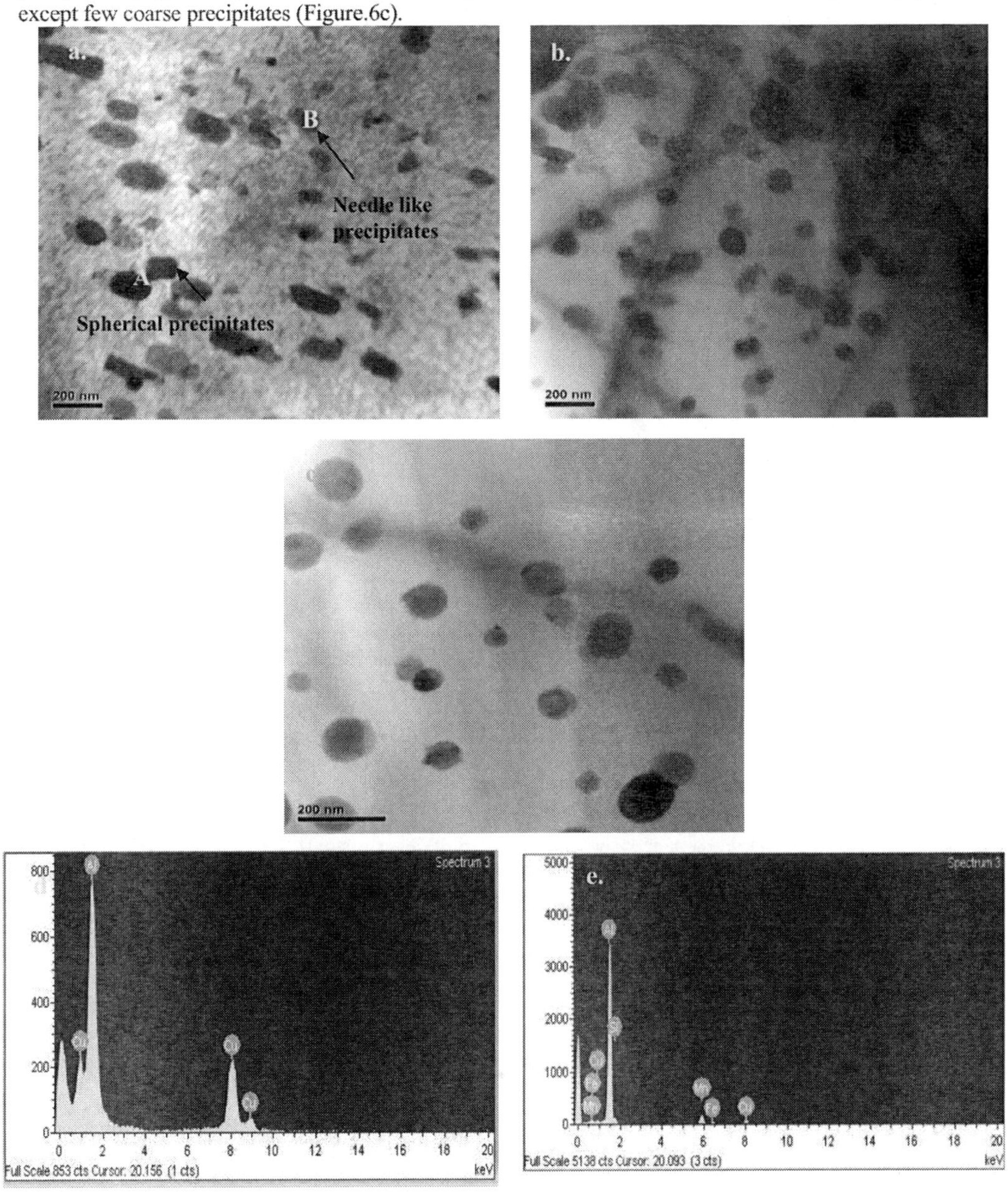

Figure.6 TEM micrograph of a) Parent material, b) As welded joint c) Heat treated joint, EDS analysis of d) Particle A and e) Particle B.

It can be concluded from the analytical outcomes, the morphology of strengthening particles were uniform compared to the as-weld joint. That is the explanation for heat-treated joints having a higher fracture load. The temperature of the solution is another significant factor in the regulation of grain stability in the FSW joints.

4. Discussions

The properties of FSW joints as a function of the process parameters, tool geometry, thermal cycle, and precipitate distribution and volume rather than grain size [13]. The strength of aluminum alloys has improved by many techniques after welding. SZ of FSW joints experienced by severe heat input in the range of 300 C to 480 C [14,15]. During this period, the precipitates were dissolved in the aluminum matrix completely. Whereas, in the HAZ on both retreating and advancing sides experienced by partial heat input. Moreover, this heat input makes the precipitates coarse in size, even though the PM grains were not altered during thermal cycle [14]. Hence, this coarsening action is the weakest region in as weld condition. Ageing treatment is the most effective method of re precipitation of precipitate, which increases strength compared to the as-weld joint. Moreover, in the FSW process, the precipitates become rough or dissolve and this result reduces the joint's strength. The heat treatment significantly improved the properties of the lower hardenss in the as welded joint. The heat-treated joints offered better strength (15.27kN) compared to the as welded joint. Solution treatment and artificial ageing process enhanced the size of grains in all regions, resulting in the size of grains increases in aluminium matrix. The formation of new grains in larger size is caused by the heat treatment process [16]. Dense and thick distribution of precipitates were observed in the weld region. Hence, the heat-treated joints yielded higher lap shear strength than as welded joints. Abnormal grain growth is the other phenomena in all FSW joints that have occurred in the weld due to inhomogeneous plastic deformation during the thermal cycle. Due to abnormal grain growth, it causes a) antistrophic in mobility and grain boundary. b) reduces the anchoring effect of precipitates in the grain by the solutioning process. The solutioning temperature is another factor, which will promote larger grains when the temperature is high. In this investigation, the solution process was performed at 496 C and 1 h holding time.

As per the Hall- Petch relationship, the weld strength depends on the grain size and precipitates distribution. Although, the hardness distribution in the weld is direct relationship with the welding strength. All the FSW joints are invariably formed by different microstructure in the weld region. This heterogeneous property was formed by thermal cycle during FSW. SZ region experienced by severe heat input and whereas HAZ experienced by lower heat input (250 C). This difference in heat input is the reason for as symmetric flow pattern in the FSW joint. The lowest hardness region of as welded joint was observed at the HAZ on both advancing and retreating sides. The soft region was formed by the dissolution or coarsening of precipitates during FSW cycle. Because the SZ of welded joints was homogenized by thermal cycle between 420C and 480C and causes the dissolution of strengthening precipitates (Al_2Cu). Precipitate dissolution is the cause for the formation of lower hardness at HAZ in as welded joint. Heat treated (Solution treatment followed by ageing process) joints possess peak hardness in HAZ, TMAZ and SZ. Heat treatment is associated with the increase size fraction of dissolved particle that enhances the presence of alloying element for re precipitation during ageing treatment, the lap shear strength of FSW joints in the as-welded condition offers inferior strength than heat treated joint. The reason for the existence of high lap shear strength in heat-treated joints is abnormal grain growth. The abnormal grain growth is a metallography transformation of which few grains progress abnormally at the expenditure of fine alpha matrix grains and it occurs when the regular grain progress stationary [17]. The formation of abnormal grain growth in the SZ has a substantial effect on lap shear strength of FSW joints further heat treatment. The formation of precipitate in as welded joint has been shown in coarse in size by agglomerating precipitate during thermal cycle. The reduction in dislocation density and coarsening of precipitates causes lower hardness and increasing lower hardness width as welded joint. The width of the lower hardness on HAZ in heat-treated joints was reduced and moved toward the weld center due to artificial ageing in the heat treatment process. The ageing process enhances the re precipitation of precipitates (θ') [18,19].

5. Conclusions

The friction stir lap-welded joints were successfully heat-treated using two techniques namely solution treatment followed and artificial aging process. The lap shear strength of the joints was correlated with heat-treated joints in systematically. The heat-treated joints conceived higher lap shear strength than the as welded joints, due to the recovery of lost strength in the lowest hardness region (soft region). The solution process was performed at 496 C for soaking period of 1 h make complete solubility of solute (Al2Cu) and retains the same by rapid cooling. The ageing process was performed at 160 C and soaking period of 12 h. The artificial ageing process promotes re-precipitation of precipitate in the weld region. Hence, the lowest hardness region in the heat-treated joints was minimized. The abnormal grain growth has substantial effect in lap shear strength of FSW joints after heat treatment.

References

1] ZhouC. YangX., LuanG.2006 Effect of root flaws on the fatigue property of friction stir welds in 2024-T3 aluminum alloys. *Mater.Sci. Engg. A*, **418,**155-60.

[2] SquillanceA. De RenzoA, Giorleo G, Bellucci F, 2004 A comparison between FSW and TIG welding techniques: modification of microstructure and pitting corrosion resistance in AA2024-T3 butt joints, J Mater. Process. Tech, **152,**97-105.

[3] Genevosis C, Deschamps A, Denquin A, Doisneau-coottignies B,2005, Quantities investigation of precipitation a mechanical behavior for AA2024 friction stir weld, Acta Mater, 53, 2447-58.

[4] Liu.H. J, Zhan.L. L, 2011, Effect of welding speed on microstructures and mechanical properties of underwater friction stir welded 2219 aluminum alloy, *Mater. Design.* **32**,1548-1553.

[5] Zang Z, Xiao,Z. Ma Y,2014, Influence of water cooling on microstructure and mechanical properties of friction stir welded 2014 aluminum alloy, *Mater.Eng.A,***614,** 6-15.

[6] Wang Ji, Ruidong Fu, Yiiun Li, Jianfeng Zang,2014, Effect of deep cryogenic treatment and low temperature aging on the mechanical properties of friction-stir-welded joints of 2024-T351, *Mater.Sci. Eng A,***609**, 147-153.

[7] RajendranC,Srinivasan K, Balasubramanian V, Balaji H, Selvaraj P,2019, Effect of Post Weld Heat Treatment on Strength and Microstructure of Friction Stir Welded Lap Joints of AA2014-T6 Aluminum Alloy. *Metal Science and Heat Treatment* **61**.5-6,305-310.

[8] Rajendran C, Srinivasan K, Balasubramanian V, H.Balaji, P.Selvaraj, 2016, Influence of post weld heat treatment on tensile strength and microstructural characteristics of friction stir welded butt joints of AA2014-T6 aluminium alloy, *J Mech Behav Mater,* **25**(3-4): 89–98.

[9] Chaitanya Sharma, dheerendra Kumar Dwivedi, Pradeep Kumar 2013 Effect of post weld heat treatment on microstructure and mechanical properties of friction stir welded joints or Al-Zn-Mg alloy by AA7039, *Mater. Design.***43**, 134-143.

[10] SrivatsanT S, Vasudevan S, Park L, 2007, Tensile deformation and fracture behavior of friction stir welded aluminum alloy 2024. *Mater.Sci.***456**, 235-45.

[11] Suvillian, A, Robson J D,2008, Microstructural properties of friction stir welded and post welded heat treatment 7449 aluminum alloy thick plate, *Mater. Sci. Eng A*, **478** (1-2), 351-60.

[12] Rajendran, C.Srinivasan K, Balasubramanian V, Balaji H, and Selvaraj P, 2019 Evaluation of load-carrying capabilities of friction stir welded, TIG welded and riveted joints of AA2014-T6 aluminium alloy. *Aircraft Engineering and Aerospace Technology*.

[13] Attallah, M Moataz., and G Hanadi. Salem. 2005 Friction stir welding parameters: a tool for controlling abnormal grain growth during subsequent heat treatment. Materials Science and Engineering: A **391**, 1-2, 51-59.

[14] Rhodes,C. G, MahoneyM. W., BingelW. H., SpurlingR. A., and BamptonC. C 1997 Effects of friction stir welding on microstructure of 7075 aluminum. *Scripta materialia* **36,** 1, 69-75.

[15] MahoneyM. W, RhodesC. G., FlintoffJ. G., BingelW. H., and SpurlingR. A.1998, Properties of friction-stir-welded 7075 T651 aluminum. *Metallurgical and materials transactions* A **29**, 7, 1955-1964.

[16] Rajendran, C., K. Srinivasan, V. Balasubramanian, H. Balaji, and P. Selvaraj. 2019 Effect of tool

tilt angle on strength and microstructural characteristics of friction stir welded lap joints of AA2014-T6 aluminum alloy. *Transactions of Nonferrous Metals Society of China***29,** 9, 1824-1835.

[17] Safarbali, B., M. Shamanian, and A. Eslami. 2018, Effect of post-weld heat treatment on joint properties of dissimilar friction stir welded 2024-T4 and 7075-T6 aluminum alloys." *Transactions of Nonferrous Metals Society of China* **28**, 7, 1287-1297.

[18] Rui-dong fu, Jian-Feng Zhang, Yi-jun Li, Ju Kang, Hui-Jie-Liu, Fu-chengZhang,2013, Effect of heat input and post welding natural aging on the hardness of stir zone for friction stir welded 2024-T3 aluminum alloy thin sheet, *Mater.SciEng A,***599,**319-324.

[19] Ji Wang, Ruidong Fu, Yijun Li, and Jianfeng Zhang. 2014, Effects of deep cryogenic treatment and low-temperature aging on the mechanical properties of friction-stir-welded joints of 2024-T351 aluminum alloy, *Materials Science and Engineering: A* **609,**147-153.

IOP Conf. Series: Materials Science and Engineering 1059 (2021) 012038 doi:10.1088/1757-899X/1059/1/012038

Fabrication of Pineapple Peeling Machine Using Pneumatic Solenoid Valve

Madhankumar S[1*], Suryakumar H[2], Sabarish R[2], Suresh M[2], Ummer Farook A[2]

[1]Assistant Professor, Department of Mechatronics Engineering, Sri Krishna College of Engineering and Technology, Coimbatore, Tamil Nadu, India - 641008.
[2]UG Scholar, Department of Mechatronics Engineering, Sri Krishna College of Engineering and Technology, Coimbatore, Tamil Nadu, India - 641008.

[*]Corresponding author, Email Id: madhankumars@skcet.ac.in

Abstract. Pineapple peeling device which used for slicing the pineapple to create a cylindrical pulp. The key goal of this research is to implement a pineapple peeler system to overcome the challenges encountered by small and medium-sized businesses, in which the system produced will minimize its time needed for the pineapple operation. From an engineering viewpoint, ergonomic concerns relevant to peeling have been discussed. For this project, the concept of the pineapple peeler has a cylindrical blade being used to strip the pineapple flesh. The benefits of this pineapple peeler established throughout this venture are that it can remove the leaves as well as the core of the pineapple, and also can peel the outer surface of that same pineapple. Before starting the device, the pineapple is installed in the holder of the machine. So, the first cutter is to remove the ends as well as the base of the pineapple. Following that, the pineapple surface can be extracted by using a cylindrical knife. The mechanism within that device has been used to monitor the entire function of the pineapple peeler system and therefore is interfaced with dual pneumatic tubes to operate the tool. The mechanical and electrical elements are the essential parts of this unit.

1 Introduction

The pineapple (Ananas comosus) is really the third biggest native fruit in the world behind banana & citrus. This is a tasty crop with a strong orientation as well as a strong nutritious value. That was one of India's largest commercial tropical fruits [1]. Pineapple has been a strong great source of vitamin A and B and quite high in folic acid and nutrients such as potassium, magnesium, calcium and titanium. It is indeed a number of quality of polysaccharides [2], a proteolytic juice [3]. Besides being consumed raw, the fruit might well be dried and cooked in various ways. Pineapples are eaten as a fruit or included in the processing of jams, jelly and lettuce [4]. The pineapple peeler has become a device that might eventually cut along with slice through pineapple it into tubular pineapple pulp [5]. Pineapple appears to have been the internal substance of the pineapple between both the external skin as well as the middle. The outer layer over its pineapple is harsh as well as thick to expel instantly [6]. The soul of a pineapple seems to be the strongest or perhaps the major ingredient of said pineapple. The core including its pineapple is sturdy and robust. The automatic pineapple filet knife and drilling device can remove the whole outer layer of its pineapple yet at the same time scrape the core of pineapple [7]. Pineapple pulp is being used for the production of syrup [8], flavoring, pineapple gin and tonic, pineapple mash as well as pineapple packaged mush [9]. The biggest benefit of the automated pineapple slices and colonizing system is that the unit could handle a pineapple through full mode. A fully automated slicer and the colonizing pineapple device would do all the necessary technology to convert the pineapple pulp [10], which implies that perhaps the external surface, as well as the pineapple central, could be erased by the device [11]. This same product performance but instead employment expenses can be minimized using the instantaneous pineapple peeler as well as drill piece of equipment. The autonomous pineapple filet knife as well as drill device is used in small and medium-sized enterprises.

The main objective of this project is to fabricate an automated pineapple peeler for overcoming abovementioned issues. For this project, the concept of the pineapple peeler has a cylindrical blade being used to strip the pineapple flesh. The benefits of this pineapple peeler established throughout this venture are that it can remove the leaves as well as the core of the pineapple, and also can peel the outer surface of that same pineapple.

2 Construction

There are two pneumatic cylinders mounted on a mild steel frame, where one cylinder is vertically mounted under the frame which is used to lift the pineapple. This cylinder is connected with the frame using U shaped clamp. Another pneumatic cylinder is placed in a round ram at the end of the piston ram. There is a blade mounted vertically. There is another conical blade which is used to peel the pineapple. Two 5/2 solenoid valves which are used to control the pneumatic cylinder. A tray is used to collect the pineapple.

The whole system is controlled using an electronic circuit consisting of a micro controller which is programmed using Arduino. It also consists of a relay circuit which runs the sequenced program. A step-down transformer is used to reduce the voltage to 12 volts. There is also another circuit used to reduce the voltage from 12 volt to 5 volts. The whole circuit is controlled by a push button switch. The whole system works on the principle of pneumatics which requires air compressor ranging from 8 bar to 12 bars. Fig. 1a shows the components of pineapple peeling device for construction and Fig. 1b shows the fabricated image of pineapple peeling machine.

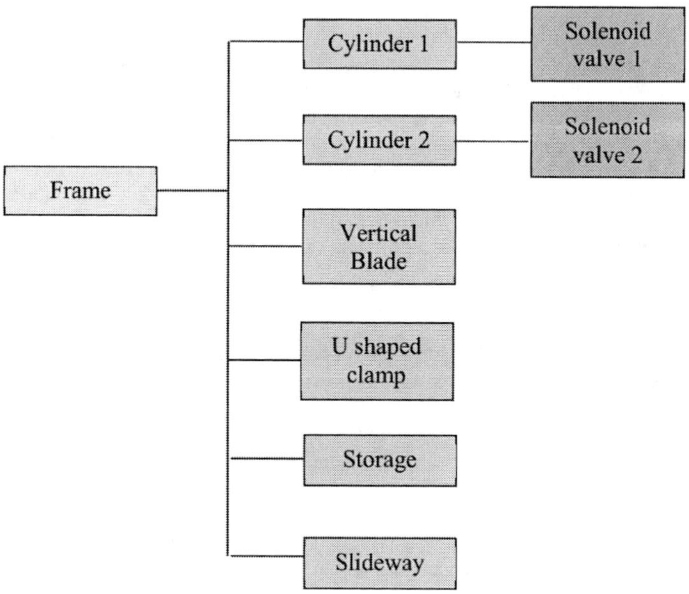

Figure 1. Peeling machine components

3 Working

First, there is an electronic push switch which is used to start the electronic sequence, the sequence is already programmed into the microcontroller. The step-down transformer is used to reduce the current from 230 volts to 12 volts which are further reduced to 5 volts when passed to the Arduino. The pineapple is placed on a U- shaped holder which is welded to a vertical pneumatic cylinder. There are two vertical blades that are used to slice the leaf and root of the pineapple. There is another horizontal pneumatic cylinder which is controlled by a 5/2solenoid valve which acts as a ram to slide the pineapple into the conical blade which will peel off the outer rough skin of the pineapple. The skin will be dropped off the edge of the table, the pulp of the fruit will be collected in the tray placed under the frame.

4 Design calculations

The design calculations were done for determining the dimensions, features, materials and so on of pineapple peeling machine [12], [13]. The beam and beam calculations of device as well as the design of cylinder and piston are presented below.
The reaction forces are at the supports of the peeling machine ends are 112.45 N. This will be calculated using shear force and bending moment diagrams [14]. The maximum bending moment is at center of the device, the value is 61.25 Nm. The dimensions the column is calculated using bending equation (1),

$$\sigma_b = \left(\frac{\sigma}{y} = \frac{E}{R} = \frac{Mb}{I} \right) \tag{1}$$

The diameter and breath values are 97.2 cm and 194.4 cm respectively. The factor of safety value for this machine is 3.4. Then, slenderness ratio relation is used to identify the column is short or long [15]. From calculations, identifying that the given column is long column. For long column crippling load formula is used,

$$P = \frac{n\pi^2 EA}{(\frac{L}{K})2} \tag{2}$$

From the above equation (2), the piston rod length has been calculated. The value got as 19 cm which is standardized using PSG design data book page number 7.20 [16]. Next the cylindrical blade dimensions are calculated. The radius of the blade is 4.5 cm. Finally, the dimensions of cylinder are calculated with the help of pressure terms. The standardized diameter of the cylinder and stoke length are 40 mm and 130 mm respectively. Based on these design calculations, the peeling device was fabricated which has been shown in fig. 1b.

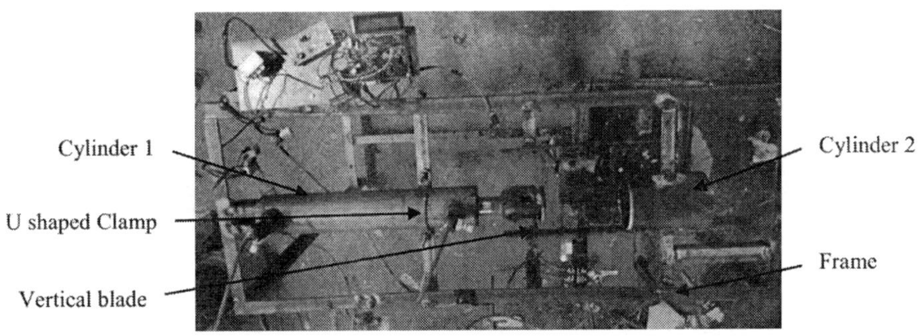

Figure 2. Fabricated image of Pineapple Peeling Machine (Video Link)

5 Conclusions

The pineapple peeling machine was fabricated successfully with the help of some components and the system can be improved with a help of a safety circuit and sensors which could be used to improve safety when operating. Furthermore, improved slideways can be developed to move the pulp into the collecting tray. Conveyors can be used to input the pineapple in the semi-circular clamp. It is noted how much time is needed to slice and core the pineapple by operating effectively. The full description is provided down. Installation i.e. putting the pineapple mostly on tray takes roughly 5sec. The time necessary for upstroke is approximately around 2 sec. as well as the time necessary for utilizing two additional cylinders is almost 5 sec. The overall time duration is therefore 12-15 seconds. Given the device and the ergonomics of the device, the device should be quite simple to run. The one pineapple requires fifteen seconds, such as although in 60 minutes, 300 pineapples could be worked and also in 7 hours, roughly 2100 pineapples could be used.

Acknowledgments

This work was supported by the Sri Krishna College of Engineering and Technology, India - 641008 as part of the motivation towards research and development of modern technologies.

IOP Conf. Series: Materials Science and Engineering 1059 (2021) 012038 doi:10.1088/1757-899X/1059/1/012038

References

[1] M. Asim *et al.*, "A review on pineapple leaves fibre and its composites," *Int. J. Polym. Sci.*, vol. 2015, pp. 1–16, 2015.

[2] T. E. Aruna, "Production of value-added product from pineapple peels using solid state fermentation," *Innov. Food Sci. Emerg. Technol.*, **57**, 102193, 2019.

[3] P. Rodsamran and R. Sothornvit, "Preparation and characterization of pectin fraction from pineapple peel as a natural plasticizer and material for biopolymer film," *Food Bioprod. Process.*, **118,**198–206, 2019.

[4] N. Delchier, A. L. Herbig, M. Rychlik, and C. M. G. C. Renard, 2016 "Folates in Fruits and Vegetables: Contents, Processing, and Stability," *Compr. Rev. Food Sci. Food Saf.*, **15**, 506–528.

[5] P. E. L. Jones and A. D. Jeffery, 1973" Pineapple Peeler,".

[6] J. G. de Oliveira and A. P. Vitória, 2011 "Papaya: Nutritional and pharmacological characterization, and quality loss due to physiological disorders. An overview," *Food Res. Int.*, **44,** 1306–1313,.

[7] P. E. A. Morse, "Method for Preparing Pineapple,".

[8] J. T. Casabar, R. Ramaraj, S. Tipnee, and Y. Unpaprom, 2020 "Enhancement of hydrolysis with Trichoderma harzianum for bioethanol production of sonicated pineapple fruit peel," *Fuel*, **279**, 118437.

[9] H. Dai *et al.*, 2020 "Direct fabrication of hierarchically processed pineapple peel hydrogels for efficient Congo red adsorption," *Carbohydr. Polym.*, **230**, 115599.

[10] R. C. Omar, R. Roslan, H. Taha, W. A. Wahab, and I. N. Z. Baharuddin, 2019 "Slope Improvement using Combination of Coconut Husks, Coconut Flakes, Luffa, Pineapple Leaves and Vege-Grout," *Test Eng. Manag.*, **81**, 3212–3218.

[11] K. C. Gross, C. Y. Wang, and M. E. Saltveit, 2016 *The commercial storage of fruits, vegetables, and florist and nursery stocks*, 66.

[12] and R. S. Madhankumar S, Vignesh T, Anandraj P, Anirudh Varadarajan, Arul Praveen T, 2020 "Design and Modelling of L-type Bi-directional Roller Conveyors for Glass Hauling," in *Lecture Notes in Electrical Engineering*, **626**, 269–279.

[13] S. Madhankumar, M. Jishnu, J. Karthick Prithiv, S. Gowrishankar, S. Rajesh, and R. Balamurugan, 2019 "Design and modelling of disaster relief vehicle using rocker bogie mechanism," *Int. J. Innov. Technol. Explor. Eng.*, **8**, 1274–1276.

[14] T. A. Selvan, A. Viswanathan, S. Madhankumar, and S. Sneha, 2019 "Design and fabrication of automated powder coating system," *Int. J. Eng. Adv. Technol.*, **8**, 899–902.

[15] T. A. Selvan, A. Viswanathan, S. Madhankumar, and S. Sneha, 2019 "Design and fabrication of automatic spring type chip conveyor," *Int. J. Recent Technol. Eng.*, **8**, 917–920.

[16] D. Padmanaban, *PSG Design Data Book*. 2020.

IOP Conf. Series: Materials Science and Engineering 1059 (2021) 012039 doi:10.1088/1757-899X/1059/1/012039

An algorithmic approach for minimizing test power in VLSI circuits

J. Poornimasre [1*], **HariKumarRajaguru** [2]
[1] Assistant Professor-II, Bannari Amman Institute of Technology, Sathyamangalam
[2] Professor, Bannari Amman Institute of Technology, Sathyamangalam

poornimasrej@bitsathy.ac.in,harikumarr@bitsathy.ac.in

Abstract: Testing is an approach to check the function of the circuit or device under the test after its fabrication. To test the device, test patterns are required. The test patterns can be generated with the help of EDA tool. Those patterns having huge number of unfilled bits. The unfilled bits need to be assigned with certain logical value. This paper proposed two new algorithms. First algorithm assigns the unfilled bits with effective logical value and second algorithm change the order of test set and both algorithms aims to reduce the test power. Investigation on benchmark circuits shows that, the results of proposed algorithms effectively reduces the test power.

1 Introduction

Very-large-scale integration (VLSI) is the process of creating an integrated circuit (IC) by combining millions of transistors in order to get a single chip. The complexity of the circuit has been increased over years with the advancement of technology and testing of those circuits become complex and more important in order to prove the quality of product. The testing is done to check the functionality of the circuits after the verification process.

An IC may contain 2^n test patterns which is harder to generate test patterns and to check its quality when the value of n increases. The time consumption for this testing will be very much higher. So when an IC chip is manufactured, its fault detection is a bit harder. This could make IC tests much easier and very much faster. Test pattern generator generates the test patterns. The generated test patterns are applied to devices or circuits that need to be tested. The output of the circuit for the given test set will be compared by the comparator. The comparator gets the stored response from memory and compares it with the output of CUT. If a match occurs for the entire test set then the respective CUT will be considered as PASS otherwise fail. The following figure -1 illustrates the basic testing process.

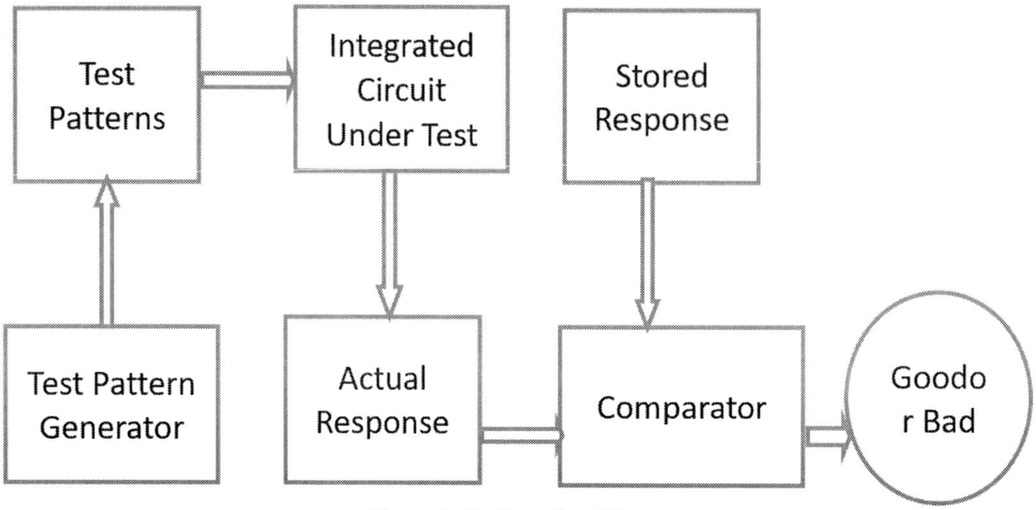

Figure 1. Testing of an IC

Due to the development of integrated circuits in the past few decade, the size of electronic component getting reduced and introduced the complexity in testing of those electronic devices. During Testing, the unique test patterns stored in the ATE are applied to the electronic device to evaluate its goodness in their functionality. Now a days the traditional techniques like ATE (Automatic Test Equipment) are becoming not suitable. Cost of ATE mainly depends on its memory size and its decoder complexity. The need of a minimum number of test set is required to reduce the ATE memory and its cost. Minimal test set also reduce the TAT.

The cost of an electronic device or an IC indirectly depends on test cost. The test cost again depends on test power. The reason is during testing, test vectors stored in ATE is applied to the Circuit. These test vectors contain logic 0, logic 1, and unfilled bits. The unfilled bits may be assigned with either logic 0 or 1. If any test vectors that contains more successive logic 0s and 1s may increase the power dissipation. It is similar to switch on and off the lightbulb continuously. The excessive heat that may damage the IC during its test. This affects the yield. So cost of the IC gets increased. The conclusion is test power is the major cause for test cost. So, the power dissipation during the test should be minimized.

2 Existing techniques

The test power is directly depending on logic value of unfilled bits because the unfilled bits occupies the major parts of entire test set. Thus, it is important to fill the logic value for unfiled bits. Assigning proper values to the unfilled bits may reduce the power dissipation. There are number of bit filling techniques exist. In Random bit filling, the unfilled bits are filled randomly with logical value of zeros and ones. But this technique increases the test power. Logic zero filling technique and Logic 1 filling technique fills the 'X' bit with logical value zero and one respectively. The 'X' bit is filled with adjacent bit in adjacent filling technique. Column Bit Filling (CBF) is based on hamming distance between successive test vectors [1][2]. Based on hamming distance test vectors are reordered and unfilled bits are filled column wise. Some code-based techniques also exist to reduce the test power. The proposed algorithm reducing the test power effectively compare to the above existing methods. The two categories of test power are maximum test power that exist during the test application and average of the all test pattern power. Both maximum (peak)power and average power are calculated based on weighted transition metric.

2.1 Equations

The Weighted transition metric can be represented as follows.

$$WTM = \sum_{i=1}^{n-1}(n-i)(t_{i,j} \oplus t_{i,j+1}) \quad [1]$$

Average power:

$$P_{avg} = \frac{\sum_{j=1}^{N} \cdot \sum_{i=1}^{n-1}(n-i)(t_{i,j} \oplus t_{i,j+1})}{N} \quad [2]$$

Peak power:

$$P_{peak} = \max_{j\in\{1\ldots N\}} \cdot \sum_{i=1}^{n-1}(n-i)(t_{i,j} \oplus t_{i,j+1}) \quad [3]$$

N= Number of test vectors chosen for a particular circuit.
$t_{i,j}$, $t_{i,j+1}$ are successive test bits of test vector.
n= total test bits in a single test vector.

3 Proposed algorithms

Algorithm-1

Test power is mainly depending on the position of unfilled bit also switching activity between successive test bits in the test vector. The proposed algorithm fills the unfilled test bit to a specific logic value based on the position. It combines the concept of 0 filling, 1filling and adjacent filling based on the position of 'X' bit. The algorithm steps are follows.

Algorithm Steps:

Step1: If the unfilled bit 'X' is LSB: then filled with adjacent bit. If its MSB then filled with previous bit.
Step2: If previous bit of 'X' bit is zero and adjacent bit is one then fill it with previous bit.
Step 3:If previous and adjacent bit are same then fill it with previous bit.
Step4: If 'X' bit occurs repeatedly then the length of 'X bit is calculated and filled based on step 1 to 3

Algorithm-2

The proposed arithmetic approach is very simple and does not require and complex hardware to generate the test. This approach utilizes the existing test set and modify it effectively. It's always wise to improve the test set instead of generating new test set. Mintest vectors are taken as reference and they are modified by using the arithmetic approaches. Hence by doing this, different test patterns will be obtained. The Mintest vectors of ISCAS benchmark has taken as reference vectors[6],[7],[8]. The don't care bits are filled with logic 0 initially. Logic 1 may increase the test power. Then test patterns are reordered in effective way to perform arithmetic approach. Then arithmetic operations are performed with successive test patterns

4 Experimental Results

The proposed algorithm is applied for reducing the test power with single scan chain architecture. This algorithm is implemented using Python program. The ISCAS 89Benchmark circuits are taken as reference circuits to carry out the research.

Table 1. ISCAS'89 Benchmark circuits and test patterns [9][10][11]

ISCAS'89 Circuits	No. of bits/Vector	Total bits	No of X bits
S-9324	247/159	39273	28672
S-13207	700/236	1,65,200	153887
S-15850	611/126	76986	64329
S-35932	1763/16	28208	9957
S-38417	1664/19	1,64,736	112154
S-38584	1464/136	1,99,104	163817

Table 2. Peak power Reduction- Algorithm 1

ISCAS'89 Circuit	ORIGINAL TEST SET	0 FILLING	1 FILLING	ADJACENT FILLING	CBF &DV	Proposed Algorithm1
S-5378	-	10127	10309	12516	12822	11430
S-9324	17494	12994	14471	20431	17169	14016
S-13207	135607	101127	113980	100896	125392	94619
S-15850	100228	81832	80453	99537	96452	70722
S-35932	683765	505295	446608	572169	660096	437619
S-38417	572618	531321	547921	583614	-	480557

IOP Conf. Series: Materials Science and Engineering 1059 (2021) 012039 doi:10.1088/1757-899X/1059/1/012039

Table 3. Average power Reduction -Algorithm 1

ISCAS'89 Circuit	ORIGINAL TEST SET	0 FILLING	1 FILLING	ADJACENT FILLING	CBF &DV	Proposed Algorithm1
S-5378	-	3336	3085	5851	10344	3469
S-9324	14630	5692	5870	14749	13492	4025
S-13207	122301	12416	16319	74854	103400	8028
S-15850	90899	20742	26503	102154	64275	13754
S-35932	601840	172665	174770	337658	443030	118935
S-38417	535875	136634	149640	610803	-	86707

Table 4. Peak power Reduction- Algorithm 2-Xor

ISCAS'89 Circuit	ORIGINAL TEST SET	0 FILLING	1 FILLING	ADJACENT FILLING	CBF &DV	Proposed Algorithm2
S-5378	-	10127	10309	12516	12822	10223
S-9324	17494	12994	14471	20431	17169	9772
S-13207	135607	101127	113980	100896	125392	121228
S-15850	100228	81832	80453	99537	96452	87787
S-35932	683765	505295	446608	572169	660096	537854
S-38417	572618	531321	547921	583614	-	190334

Table 5. Average power Reduction -Algorithm 2 -xor

ISCAS'89 Circuit	ORIGINAL TEST SET	0 FILLING	1 FILLING	ADJACENT FILLING	CBF &DV	Proposed Algorithm 2
S-5378	-	3336	3085	5851	10344	3806
S-9324	14630	5692	5870	14749	13492	3183
S-13207	122301	12416	16319	74854	103400	11374
S-15850	90899	20742	26503	102154	64275	23996
S-35932	601840	172665	174770	337658	443030	198004
S-38417	535875	136634	149640	610803	-	31025

Table 6. Peak power Reduction- Algorithm 2-or

ISCAS'89 Circuit	ORIGINAL TEST SET	0 FILLING	1 FILLING	ADJACENT FILLING	CBF &DV	Proposed Algorithm 2
S-5378	-	10127	10309	12516	12822	10505
S-9324	17494	12994	14471	20431	17169	10410
S-13207	135607	101127	113980	100896	125392	116711
S-15850	100228	81832	80453	99537	96452	90420
S-35932	683765	505295	446608	572169	660096	546382
S-38417	572618	531321	547921	583614	-	203141

Table 7. Average power Reduction -Algorithm 2 -or

ISCAS'89 Circuit	ORIGINAL TEST SET	0 FILLING	1 FILLING	ADJACENT FILLING	CBF &DV	Proposed Algorithm 2
S-5378	-	3336	3085	5851	10344	4419
S-9324	14630	5692	5870	14749	13492	4735

S-13207	122301	12416	16319	74854	103400	16036
S-15850	90899	20742	26503	102154	64275	29099
S-35932	601840	172665	174770	337658	443030	239145
S-38417	535875	136634	149640	610803	-	41646

Table 8. Peak power Reduction- Algorithm 2-and

ISCAS'89 Circuit	ORIGINAL TEST SET	0 FILLING	1 FILLING	ADJACENT FILLING	CBF &DV	Proposed Algorithm 2
S-5378	-	10127	10309	12516	12822	10322
S-9324	17494	12994	14471	20431	17169	5060
S-13207	135607	101127	113980	100896	125392	90342
S-15850	100228	81832	80453	99537	96452	49975
S-35932	683765	505295	446608	572169	660096	327437
S-38417	572618	531321	547921	583614	-	44054

Table 9. Average power Reduction -Algorithm 2 -and

ISCAS'89 Circuit	ORIGINAL TEST SET	0 FILLING	1 FILLING	ADJACENT FILLING	CBF &DV	Proposed Algorithm 2
S-5378	-	3336	3085	5851	10344	3003
S-9324	14630	5692	5870	14749	13492	1652
S-13207	122301	12416	16319	74854	103400	6193
S-15850	90899	20742	2650f3	102154	64275	8310
S-35932	601840	172665	174770	337658	443030	115689
S-38417	535875	136634	149640	610803	-	11009

The following figure 2 figure gives the comparative analysis on both algorithms. The algorithm 1 use the bit filling approach and second algorithm fills the all unfilled bit with logic 0 and ordering the test vectors based on its binary weights.The peak power and average power are cause the major impact in testing[12][13][14]. Results of the figure indicates that Peak power and average power effectively reduced with algorithm 2 with And arithmetic approach

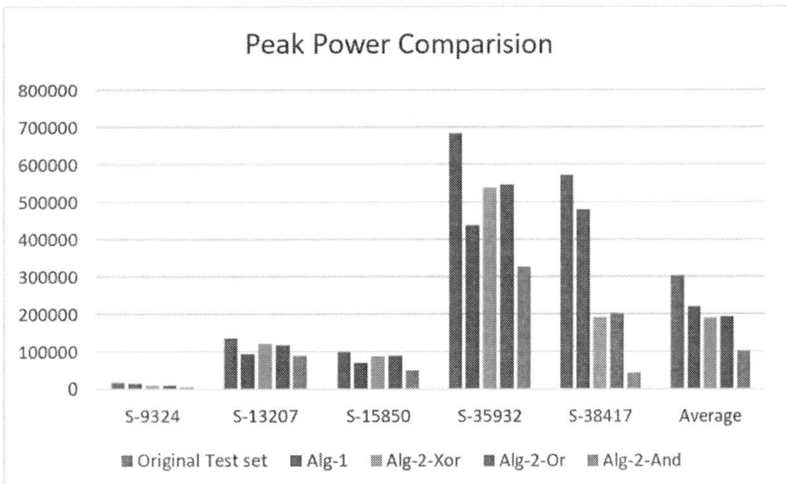

Figure 2. Peak power comparisons of Proposed Algorithms

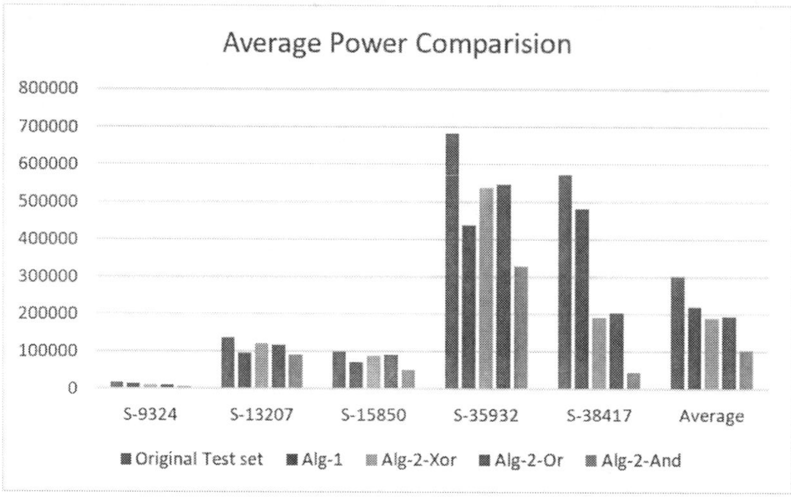

Figure 3. Average power comparisons of Proposed Algorithms

5 Conclusion:

The proposed X-filling algorithmic method reduces switching activity and thus reducing both peak test power and average test power. This algorithm improves the reduction of peak power 64%,10%,8%& 20% percentage over original test vectors ,0 filling, 1 filling and adjacent filling respectively. The reduction of average test power of the proposed algorithm is 91%,33%,37% and 79% over mintest vectors, zero filling, one filling and adjacent filling respectively. The second proposed algorithm indicates that the average of total peak power and mean power of benchmark circuits greatly reduced with binary And approach. Test data compression is used to reduce the size of test bits. Several compression techniques exist for test data compression [15]. The future work will be carried on data compression using arithmetic technique.

References:

[1] Mehla U S, Dasgupta K S and Devashrayee N M 2010 Hamming Distance Based Reordering and Columnwise Bit Stuffing with Difference Vector: A Better Scheme for Test Data Compression with Run Length Based Codes. In *VLSI Design International Conference on* (pp. 33-38). IEEE Computer Society.

[2] Mehta U S, Dasgupta K S, Devashrayee N M and Parmar H 2011Artificial intelligence based scan vector reordering for capture power minimization. In 2011 Nirma University *International Conference on Engineering* (pp. 1-6). IEEE.

[3] Paramasivam Kuppusamy, K Gunavathi Hamming distance-based test vector reordering techniques for minimizing switching activity during testing of VLSI circuits International AMSE Journal-Modelling A **80** (1) 1-13

[4] Paramasivam Kuppusamy, K Gunavathi A Arulmozhivarman 2017 Functional distance based test vector reordering for low power testing of VLSI circuits Journal of Advanced Research in Dynamical and Control Systems

[5] P Girard 2002 Survey of low-power testing of VLSI circuits *IEEE Design & test of computers*

[6] HariKumar R and Manjurathi B 2016 Test Data Compression and Power Reduction Using Similarity Based Reordering Technique for Wireless Systems. *Wireless Personal Communications* **90**(2) 713-728.

[7] Thilagavathi K and Sivanantham S 2018 Two-stage low power test data compression for digital VLSI circuits *Computers & Electrical Engineering* **71** 309-320.

[8] Butler K M, Saxena J, Jain A, Fryars T, Lewis J and Hetherington G 2004 Minimizing power consumption in scan testing: Pattern generation and DFT techniques. In *2004 International Conference on Test* pp. 355-364. IEEE.

[9] Mehta U S, Devashrayee N M and Dasgupta K S 2010 Hamming distance based 2-D reordering with

power efficient don't care bit filling: optimizing the test data compression method. In *2010 International Symposium on System on Chip* pp. 1-7. IEEE.

[10] Chandra A and Chakrabarty K 2003 Test data compression and test resource partitioning for system-on-a-chip using frequency-directed run-length (FDR) codes *IEEE transactions on computers* **52**(8) 1076-1088.

[11] Hamzaoglu I, Patel JH 1998 Test set compaction algorithms for combinational circuits *IEEE/ACM international conference on computer-aided design, digest of technical papers* (IEEE Cat. No. 98CB36287)

[12] Girard P, Nicolici N and Wen X 2010 *Power-aware testing and test strategies for low power devices*. Springer Science & Business Media.

[12] Chandra A and Chakrabarty K 2003 A unified approach to reduce SOC test data volume scan power and testing time *IEEE transactions on computer-aided design of integrated circuits and systems* **22**(3) 352-363.

[13] Nourani M and Tehranipour M H 2005 RL-Huffman encoding for test compression and power reduction in scan applications *ACM Transactions on Design Automation of Electronic Systems (TODAES)* **10**(1) 91-115.

[15] Yuan H, Guo K, Sun X and Ju Z 2016 A power efficient test data compression method for SoC using alternating statistical run-length coding. *Journal of Electronic Testing* **32**(1) 59-68.

Analyzing Properties of Air as a natural refrigerant using MCDM methodology

Mohammed Hameeduddin Haqqani, Mohammed Azizuddin, Syed Shuibul Qarnain, S. Bathrinath*

Department of Mechanical Engineering, Career Point University, Kota- 325003
Department of Mechanical Engineering, Deccan College of Engg &Tech, Hyderabad - 500001
Department of Mechanical Engineering, Kalasalingam Academy of Research and Education, Krishnankoil-626126

*Corresponding Author: bathri@gmail.com

Abstract. Refrigerant selection is a challenging task because it often requires compromising desirable properties. Since we have so many refrigerants in the market to choose from and all are either pure or mixtures. Achieving all desirable properties for a refrigerant for a specific application is very rare and a perplexing task, which takes months if not years to come to a conclusion to select the right refrigerant. Among these conflicting and challenging circumstances, the selection of a refrigerant would be easy if properties of a desirable refrigerant are listed and analyzed for their importance. There exist a literature gap providing information on prioritizing the properties required for an ideal refrigerant which is pure, natural and abundant. This proposed research provides a list of properties of air as a natural refrigerant and prioritizes the refrigerant properties based on weightage. Multiple Criteria Decision Making (MCDM) based DEMATEL methodology is used to provide weights for the refrigerant properties. The results show Ozone depleting potential among effect group and Cost of the refrigerant among because group are the top properties. The managerial application is also discussed.

Keywords: Air, Natural refrigerant, Refrigerant properties, MCDM, DEMATEL

1 Introduction

In the early18th century when vapor compression cycle was introduced the choice of refrigerants were solely classified into either flammable or toxic. The common refrigerants during early 1920's were Ammonia, Carbon Dioxide, Sulphur Dioxide, Hydrocarbons, Methyl Chloride and Water. At that time, (i.e. before ODP and GDP era) challenges to find refrigerants were based in Toxicity, Stability and Flammability [1]. So, the challenge was to identify a refrigerant which was more applicable for the use of commercial and industrial refrigeration machines.

Common refrigerants used earlier were [2]:

CFC: Available in the market for 50 Years and now it is stopped.

HFCF: Available since 25 Years – 10 more years to go.

HFC: Available since 50 Years – 35 years to go.

HFO: Available since begin life for the next 50 years

HC: Available 100 years the first time-now reborn

A lot of research and review is available in search of lower Global Warming potential refrigerants available in the market [3]. Most of the existing literature on refrigerants focus on HCF, CFC and synthetic mixtures, But the literature on natural refrigerants in literature is scare. This paper tries to fill the gap exist in the field of selection of refrigerants based on the physical and chemical properties of air. There are numerous papers on the performance of each of the refrigerant and its characteristics, but the simple selection criteria on the basis of cause and effect analysis is still not available. This paper addresses these simple criteria and its impact on the selection process of the right refrigerant for the right application.

2 Motivation for the research

This research is significant because in the present scenario we have so many refrigerants in the market and so many in the research laboratories under testing, it was deemed necessary to have an analogy to identify a refrigerant property which suites the application keeping in mind the trends of the industries and the International regulations. As all the industries are going towards banning of ODP refrigerants and minimizing GWP refrigerants year over year, the authors feel it was necessary to come up with a simple model to segregate the important elements in selecting a refrigerant for an Air conditioning system. This paper highlights basic key factors to select a refrigerant based on accepted models and MCDM analysis.

This research tries to answer the following research questions: -

a) What are the properties of Air as a natural refrigerant?

b) Which property of Air is most desirable and least desirable as a refrigerant?

The objectives of this research are: -

a) To list down the refrigerant properties of Air.

b) To prioritise the refrigerant properties of air and rank them according to the weights obtained.

This research answers the research hypothesis questions that; Are there properties of Air that can make Air to act as a as a natural refrigerant?

This research is novel because the properties of refrigerant as air is being evaluated based om MCDM methodology by taking into consideration the value of the experts from the mainstream professionals. The interactions of properties are studied through DEMATEL methodology with the help of cause and effect diagram provided in fig.2.

3 Materials and Methods

There are many MCDM methods used to study the interaction among characteristics of a system [4, 5]. DEMATEL is one of the versatile and widely accepted MCDM methodology to study the interaction among various variables in a system. DEMATEL methodology was proposed by Battle memorial institute in Geneva by Gabus and Fontella between 1972 and 1976. It transforms the experiences and inputs from the experts into a visual model with the help of a causal diagram [6]. It removes the vagueness and bias while decision making process [7, 8]. The detailed step by step process in DEMATEL methodology is as follows: -

Step 1: Using a Likert scale given in Table 1 a direct comparison matric is achieved by n number of experts. All the matrices obtained by n number of experts are averaged to one direct relation matrix. Equation (1) is used to compute the direct relation matrix R

$$D = \begin{bmatrix} 1 & f_{12} & f_{13} & \cdots & f_{1(n-1)} & f_{1n} \\ f_{21} & 1 & f_{23} & \cdots & f_{2(n-1)} & f_{2n} \\ \cdots & \cdots & \cdots & \cdots & \cdots & \cdots \\ \cdots & \cdots & \cdots & \cdots & \cdots & \cdots \\ f_{(n-1)1} & f_{(n-1)2} & f_{(n-2)3} & \cdots & 1 & f_{(n-1)n} \\ f_{n1} & f_{n2} & f_{n3} & \cdots & f_{n(n-1)} & 1 \end{bmatrix} \quad (1)$$

The Likert scale used in this research is with five scores as represented in Table 1

Table. 1 Likert Scale

Linguistic terms	Influence score

Extreme effectivity	4
High effectiveness	3
Medium effectiveness	2
Low effectiveness	1
No effectiveness	0

Step 2: Normalization, for the matrix D, obtain the sum of each row and each column respectively, Apply equation (2) to the highest value among the respective sum of rows and columns.

$$C = \frac{1}{\max_{1 \le i \le n}} \sum_{j=1}^{n} D_{ij} \qquad (2)$$

Step 3: Obtain Normalized Matrix 'N' by applying equation (3).

$$N = C \times D \qquad (3)$$

Step 4: Obtain the total relation matrix T by using equation (4),

$$T = F(I - F)^{-1} \qquad (4)$$

Step 5: Now add each row and column respectively.

To compute the summation of rows and columns by using equations (5) and (6) and are depicted as D and C correspondingly.

$$D = \left[\sum_{j=1}^{n} g_{ij}\right]_{n \times 1} \qquad (5)$$

$$C = \left[\sum_{i=1}^{n} g_{ij}\right]_{1 \times n} \qquad (6)$$

Step 6: Vector 'D' is obtained by adding sum rows and Vector 'R' is obtained by adding columns, construct a causal diagram by mapping the dataset D+C in Y axis and D-C in X axis.

4 Application of the methodology to the case study

The experts that contributed to Delphi technique and pairwise comparisons of Best worst method were 20 Decision makers. All 20 experts, working in South India with experience of more than a decade in refrigerant and Air conditioning sector and with a minimum qualification of a bachelor's degree in HVACR related field. Among the experts, three were university Professors with PhD, five experts were energy design consultants, five experts were utility maintenance managers for HVACR equipment's, four experts were energy design managers and remaining three were MEP project managers in Construction and contracting field.

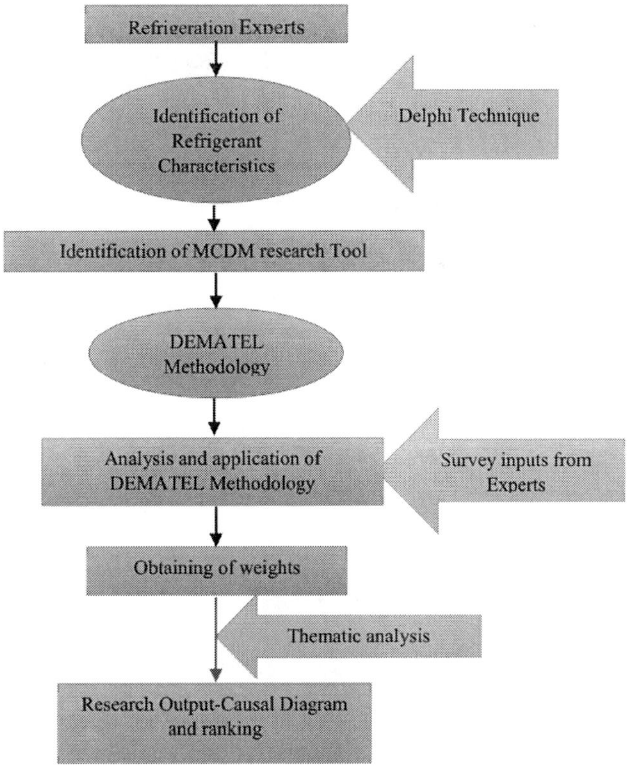

Figure 1 Research design

Figure 1 provides the research flow diagram and depicts the different stages of research steps followed in this paper. This survey resulted in 25 comprehensive refrigerant characteristics. From among the list of 25 driving factors many rounds of Delphi technique was applied with a panel of decision makers to arrive at the 15 most important refrigerant properties. To classify the factors on different themes, thematic analysis was done before applying the DEMATEL Methodology. The weights obtained for each refrigerant characteristic by the application of DEMATEL Methodology is provided in table 3 and 4.

5 Results and Discussion

From the analysis of this model the prominent vector (D+R) value results from highest to lowest showing C11 (ODP) as the number 1 property of concern and C6 (OS) being the least concern in selecting the refrigerant. The cause and effect diagram in Fig 2 shows 7 recommendations in effect group and 8 recommendations in cause group. Effect group recommendations are easily influenced by cause group elements and has the potential to change the entire system. The recommendation HEP (High Evaporator Pressure) which is the highest rank among the group has a major influence in selection the type of refrigerant. The highest recommendation among the list is C11 which is ODP (Ozone Depletion Potential) is 6.818 and C3 (HEP) which is 6.782 both are in effect group recommendations. This behaviour states that these are the two most important features of any refrigerant what needs to be considered while selection. Whereas the C6 (OS) takes the last rank, which mean it bears less impact on the selection criteria.

Critical Discussion: The essential factors in this model are cause group factors, which is the epicentre of all other factors around it as seen in Fig 2, Table 2 and 3. Which shows the highest potential by C11 (ODP) and C14 (Availability) to cause a change in the other group elements. From the Fig 2, among the Cause group the top two recommendations are C11(ODP) and C14 (A) Availability. Since the No1 items for everyone is sustainability and protecting the environment, ODP is the biggest, as we have seen earlier the R11 and R12 and now R22 will not be available.

The C12(GWP) which stands on third position is another key factor which influence in selecting the refrigerants, as we have seen in the recent times, R407C and now R410A which is used widely are to be replaced soon with new refrigerant due to its GWP potential. The C15(Life) of the refrigerant also plays an important factor and needs to be considered as it falls in the cause group. Over the years we have moved from CFC (Chloro Fluoro Carbons) refrigerants to HCFC (Hydro Chloro Fluoro Carbon) and then to HFC (Hydro Fluoro Carbon) and now we are in HFO (Hydro Fluoro Olefin) era, tomorrow we might go for synthetic refrigerants and so on. It is high time we start looking at the natural refrigerants as the most suitable alternative before we harm our environment.

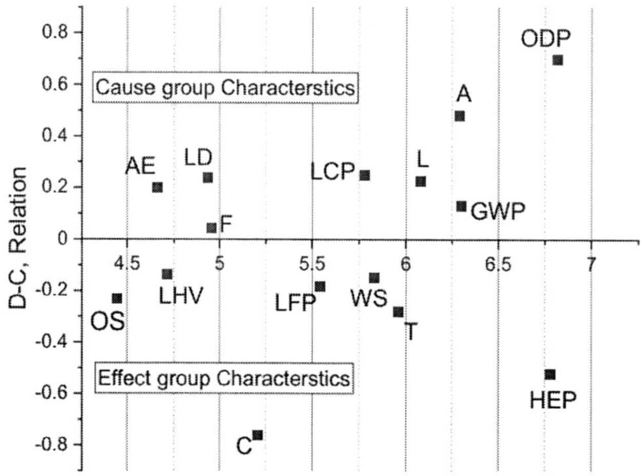

Figure. 2 Causal Diagram

6 Conclusions

From the results obtained from table-4, we can conclude that the most important recommendation that should be given priority for selecting refrigerant is C11(ODP and C13(C) which would play a key role going forward. The refrigerants where classified into Non-Flammable, Good Stability, Low Toxicity and moderate boiling point refrigerants. It was found from the above analysis that CFC or HCFC or HFC or HFO all of them have some degree of global warming effect and ozone depleting potential. From the above methodology it is very clear that the world needs to adopt refrigerants which are natural in availability and sustainable to the environment. Air being one of an abandon resource is the most viable option and more research needs to be taken place on it.

7 Limitation, Further Scope and Managerial Application

This research takes a holistic approach towards the critical segments in selecting a refrigerant which are the common practice followed by the industry today. One of the limitation of the research as we are progressing A forward there are lot of other factors being developed which could have impact on other sub factors and that could result in changes in ranking in the Cause and Effect part of this paper. The result of this paper can be used to look at the parameters for future refrigerants that are to be developed based on the guidelines of this paper. An automated APP can be developed backed by a logical program for the selection of future refrigerant. Using natural refrigerants is a challenging task, specially due to their thermodynamic properties, air as a refrigerates comes little closer than others in the line. More research is required in this section to come up with sustainable solution.

Acknowledgments

None.

References

[1] McLinden, M.O., Seeton, C.J. and Pearson, A., 2020. New refrigerants and system configurations for vapor-compression refrigeration. *Science*, **370** (6518), pp.791-796.

[2] McLinden, M. O., & Huber, M. L. (2020). (R) Evolution of refrigerants. Journal of Chemical & Engineering Data.

[3] Siddiqui, M.U., Owes, A., Al-Amri, F.G. and Saeed, F., 2020. Recent Developments in the Search for Alternative Low-Global-Warming-Potential Refrigerants: A Review. *International Journal of Air-Conditioning and Refrigeration*, **28**(03), 2030004.

[4] Qarnain, S.S., Muthuvel, S. and Bathrinath, S., 2020. Analyzing factors necessitating conservation of energy in residential buildings of Indian subcontinent: A DEMATEL approach. *Materials Today: Proceedings*.
[5] Qarnain, S.S., Muthuvel, S., Bathrinath, S. and Saravanasankar, S., 2020. Analyzing factors in emerging computer technologies favoring energy conservation of building sector. *Materials Today: Proceedings*.
[6] Qa Qarnain, S.S., Muthuvel, S. and Bathrinath, S., 2020. Modelling of driving factors for energy efficiency in buildings using Best Worst Method. *Materials Today: Proceedings*.
[7] Bhalaji, R.K.A., Bathrinath, S., Ponnambalam, S.G. and Saravanasankar, S., 2019. A Fuzzy Decision-Making Trial and Evaluation Laboratory approach to analyse risk factors related to environmental health and safety aspects in the healthcare industry. *Sādhanā*, **44**(3), 55.
Bathrinath, S., Charan, V.S., Ponnambalam, S.G. and Saravanasankar, S., 2019. Identification and evaluation of criteria of agile manufacturing using DEMATEL: A case from an indian metal fabrication industry. *Journal of Modern Manufacturing Systems and Technology*, *2*, 61-74.

Modelling the patient flow in an Out Patient Department (OPD) of a hospital using simulation techniques

S D Kulkarni[1]*, V Roshni[2], S Varshitha[2], M V Sandeep[2], T Monish[2], V Venkataraman[1]

[1] Assistant Professor, Department of Industrial Engineering and Management, Ramaiah Institute of Technology, Bengaluru – 560 054, Karnataka, India

[2] UG Student, Department of Industrial Engineering and Management, Ramaiah Institute of Technology, Bengaluru – 560 054, Karnataka, India

E-mail: sudhir@msrit.edu

Abstract. Hospitals are integral and vital entities of healthcare systems across the globe. Reduction of patient wait time is of utmost importance for the overall smooth functioning of any hospital. This paper discusses the analysis of the patient flow in the Out Patient Department (OPD) of a hospital. Data was collected with respect to patient arrival and service times in minutes. Input analyser was used to determine the distribution of the patient flow data collected at each section. A simulation model using Arena software was developed to study the patient flow in various sections of the OPD. The average waiting time and utilization of resources were determined for each sections. The results showed that there was a longer waiting time in billing section and Obstetrics and Gynaecology (OBG) section. Also there was high utilization of resources in these sections compared to the other sections of the OPD.

Keywords: Outpatient Department; Simulation model; Waiting time; Service time; Patient flow.

1. Introduction

Patient waiting time is a very serious concern for all the hospital administrations. Industrial engineering techniques can be effectively applied to reduce the waiting time and improve the system efficiency and enhance patient satisfaction. Several researchers have worked on studying the effect of patient wait time on the overall efficiency of the hospital patient flow. Simulation modelling has been effectively used by researchers to study the patient flow in various hospitals across the globe and has emerged as a powerful tool in identifying the bottle necks in a system.

A study on the use of mathematical-computer models in developing operating policies for a university-health-service outpatient clinic was done. A comparison on subsequent real-world results with those predicted by the models was made. An analysis of daily arrival patterns was used to schedule more number of appointment patients during periods of low walk-in demand in order to smoothen the overall daily arrivals. A Monte Carlo simulation model showed the effects of alternative decision rules for scheduling appointment periods during the day to increase patient throughput and physician utilization [1]. Total cash flow was proposed as a performance measure in addition to throughput, time in system, and queue times and lengths for analyzing a healthcare system. Scenario analysis and comparison of expenses were done to validate the results [2].

Discrete event simulation is one of the most widely used operations research tools. This encompasses selecting proper methods of collection and analysis of data, application of analytical techniques, verification of models, design and development of simulation experiments [3]. Arena is one of the most popular software tool used to build a simulation model. A systematic procedure to build a simulation model is explained and illustrated [4]. A discrete event simulation model was developed to examine doctor schedule mixes and various appointment schedules. An optimization program was developed to reduce the patient waiting time without adding extra resources [5]. Sampling techniques which are an integral part of statistical analysis is thoroughly explained. The detailed description of determining the sample size based on relevant parameters is illustrated [6].

Modelling and simulation of the capacity of utilities was done to show the importance of using advanced control techniques in enabling intelligent scheduling and streamlining the patient flow. Optimal pathway for patients can be designed to reduce queuing length in the hospitals [7]. The authors studied the effect of various arrival patterns on the ability of emergency department to treat patients. Arena 10.0 simulation model was used to estimate the requirement of additional resources [8]. A mismatch between demand on outpatient services and clinic capacity was observed. A simulation based operation management was proposed to provide future visibility in OPDs for all stakeholders [9].

The behaviour of the patients and their relatives arriving at hospital were studied. A simulation model was developed and the results highlighted the counters with maximum queue lengths and waiting times [10]. OPD facilities were classified into two types; primary and support facilities. Both were hosted in separate buildings which resulted in reduction in congestion in the floor area of OPD clinics and capacity optimization in both the buildings [11]. The impact of clinical disciplines on wait times was studied using the patient flow analysis. Results revealed that a significant amount of patient's total stay in the system was related to waiting time for physical examination [12]. Fundamental concepts of simulation modelling and its application in several cases is demonstrated [13]. Application of various probability and statistical methods to engineering scenarios is shown [14].

Literature review shows that simulation modelling is an important tool being used in solving the problems related to patient flow management. Resources in a typical hospital scenario mainly consists of doctors and supporting staff. The focus of the study was to understand the patient waiting time and utilization of the above resources associated with each section. High utilization of certain resources infers that they are occupied during the shift. Hence imbalance in resource utilization value indicates uneven distribution of work load among the resources, which leads to bottle necks and hence increased queue length and patient waiting time.

In this paper, sections of OPD, namely admission, billing, initial test, orthopaedics, Ear-Nose-Throat (ENT), OBG, ophthalmology and prescription were considered for study, as they had high patient flow. In each station the distributions of arrival time, waiting time and service time in minutes were determined using input analyzer. Important factors affecting queues in the system were recorded from the output of Arena simulation model. The waiting time was observed to be more in the billing section and OBG consultation section compared to the other departments or sections in the OPD of the hospital.

2. Data and Methods

The movement of patients as they entered the OPD was traced and recorded. Patient flow chart was prepared by observing the entire sequence of movement of patients in the OPD. Data for parameters required for building a simulation model were recorded during the entire working shift. Figure 1 illustrates the methodology adopted in approaching and solving the stated problem. The sections of OPD included in this study were admission, billing, initial test, orthopaedics, ENT, OBG, ophthalmology and prescription.

The sample size for data collection was determined using equation 1 which is based on sample size determination theory, where n is the number of samples required to achieve an accuracy (K) of 10%, $\bar{X}$ is the mean, σ is standard deviation and α=significance level (5%).

$$n = \left(\frac{Z_{\frac{\alpha}{2}}\,\sigma}{K\bar{X}}\right)^2 \tag{1}$$

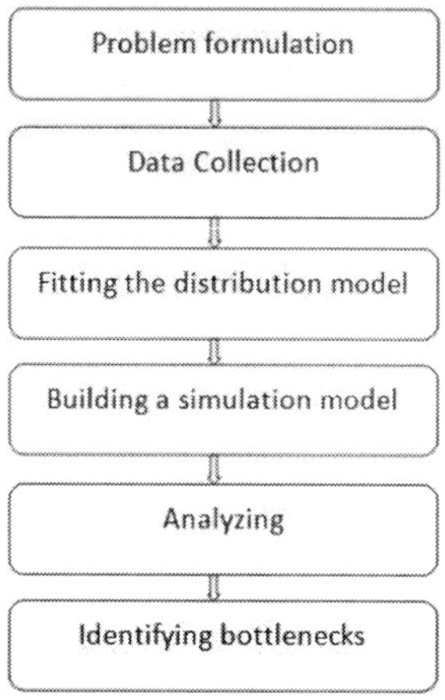

Figure 1. Methodology

Table 1. Number of data points required to achieve an accuracy of 10%

Section	Variable Considered	Mean (minutes)	Standard Deviation (minutes)	Required Number of Data
Admission	Inter-arrival time	4.87	4.22	289
Admission	Service time	2.7	1.39	102
Billing	Service time	2.37	1.22	102
Prescription	Service time	4.5	5.8	232
Initial Test	Service time	2.27	0.907	62
Orthopaedics	Service time	9.2	6.43	188
ENT	Service time	2.27	0.907	62
OBG	Service time	7.03	3.71	107
Ophthalmology	Service time	1	0.947	345

The entire process is simulated based on the inter-arrival time at the admission section, which is the input station and service time at successive stations. Table 1 shows the values of mean, standard deviation and number of samples required to achieve an accuracy level of 10%. Based on this, adequate data was collected and recorded for parameters such as arrival time, time between arrivals and service time in minutes. This data was fed in the input analyser to determine the distribution which fits the best for each section. Histograms plotted using MINITAB software shown in figures 2- 6 depict the frequency distribution of the inter-arrival/service times for all the sections considered for this study.

IOP Conf. Series: Materials Science and Engineering 1059 (2021) 012041 doi:10.1088/1757-899X/1059/1/012041

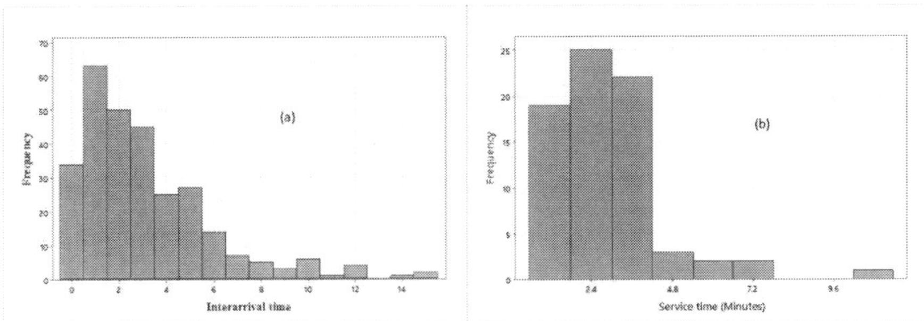

Figure 2. Histogram for admission section for (a) inter-arrival time and (b) service time in minutes

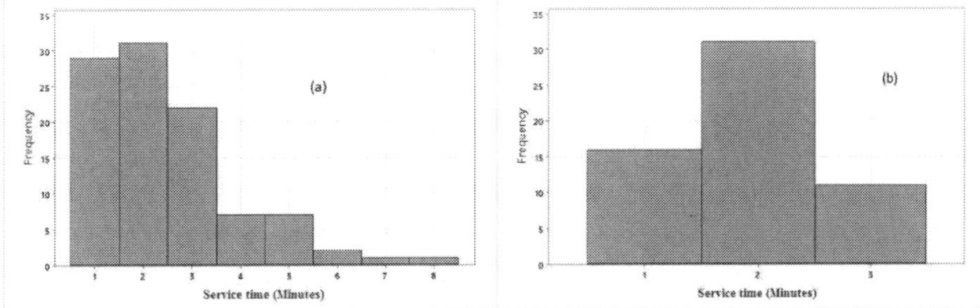

Figure 3. Histogram for service time in minutes (a) billing section (b) initial testing section

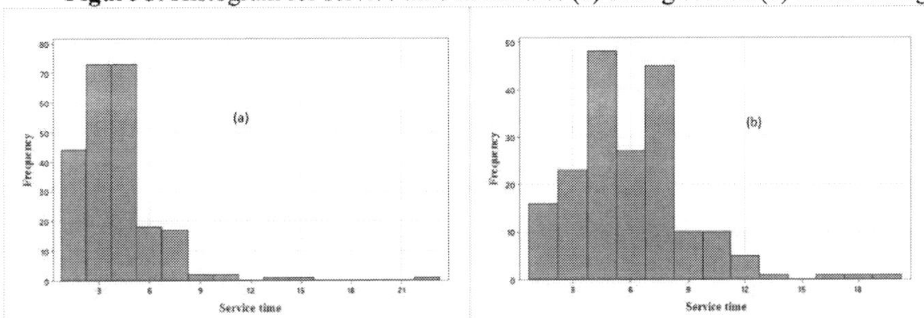

Figure 4. Histogram for service time in minutes (a) Orthopaedic section (b) ENT section

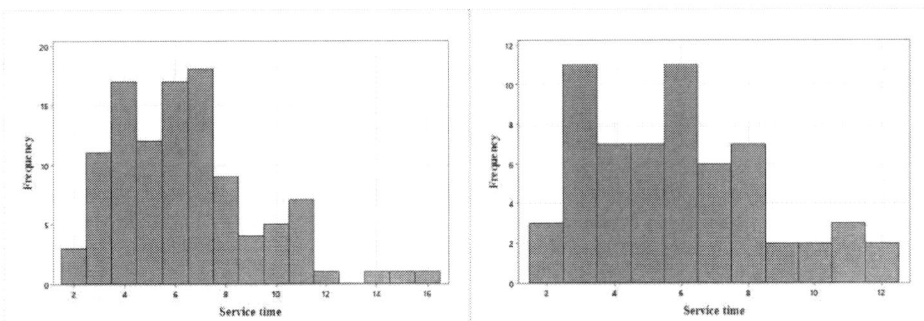

Figure 5. Histogram for service time in minutes (a) OBG section (b) Ophthalmology section

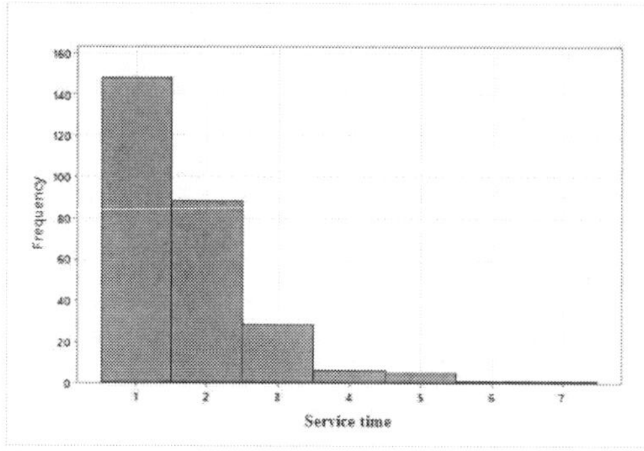

Figure 6. Histogram for service time in minutes for prescription section

The probability density functions of service time in minutes for the concerned departments are depicted in figures 7-11. These are obtained from the results of input analyser and give the type of probability distribution which best fits for the given set of data. The type of the best fitting distribution to be considered for further analysis along with the expression to be used are shown in table 2.

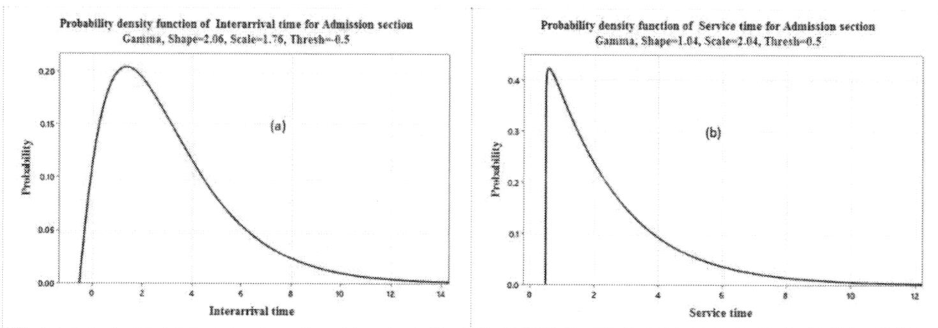

Figure 7. Probability density function for admission section considering (a) Inter-arrival time in minutes (b) Service time in minutes

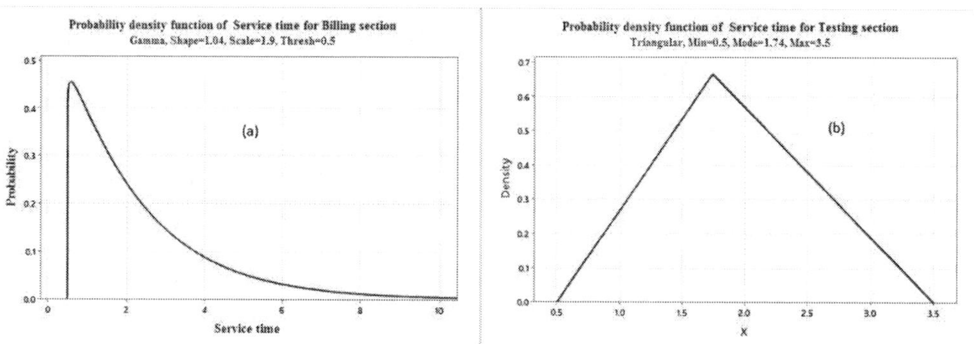

Figure 8. Probability density function of service time in minutes for (a) Billing section (b) Initial testing section

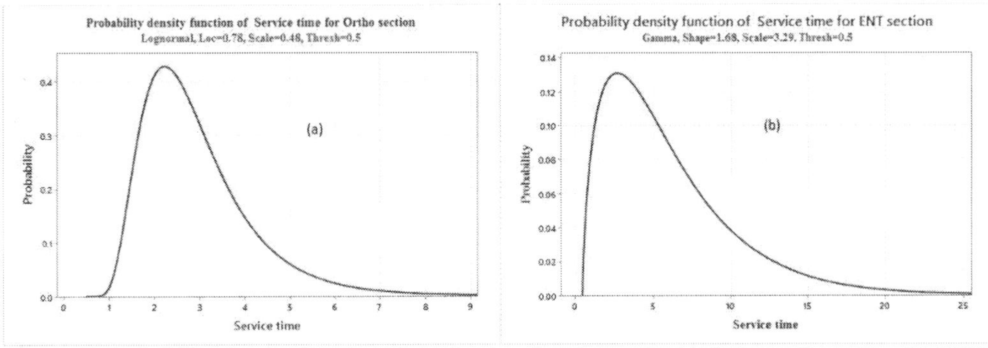

Figure 9. Probability density function of service time in minutes for (a) Orthopaedic section (b) ENT section

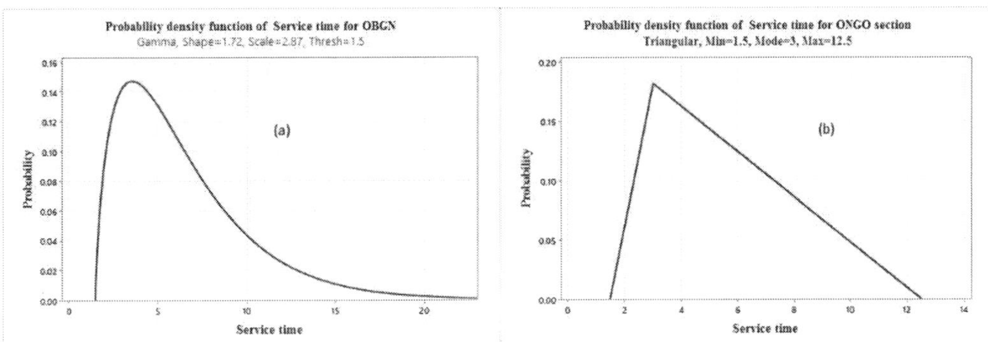

Figure 10. Probability density function of service time in minutes for (a) OBG section (b) Ophthalmology section

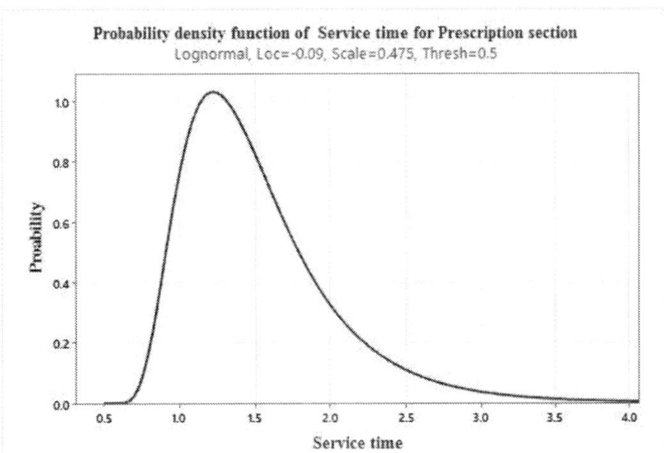

Figure 11. Probability density function of service time in minutes for Prescription section

Data for various sections was collected based on the sample size determination formula. The purpose was to ensure that a sound probability distribution is obtained for the various variables. In order to conduct the study, a histogram needs to be obtained and based on the histogram a chi-square test was conducted and accordingly the distribution was obtained. Input analyser software was used to assess the type of distribution. The choice of the appropriate distribution was based on the shape of distribution obtained using histogram, mean square error, chi-square test statistics and the p value.

Figures 2 to 6 show the histograms for the various variables considered. It can be seen that for the inter-arrival time at admission section there is variability as evidenced by a long tail, however service time has lesser variability. Likewise it can be seen that billing, testing, prescription section have moderate variability. This is partly because of the automation process involved, and if more automation can be considered then there is further possibility of reducing the variability at these sections. The area where more variability can occur would be at the doctors section. It can be seen from figure 4a and 9a that there is less variability for the orthopaedic section and large variability for the remaining three sections. The variability at these sections can create more waiting time for patients in the system.

In order to conduct the simulation study, it would be necessary to obtain an appropriate distribution for the data set. Various distributions were obtained using input analyser based on mean square error, shape of the histogram, chi-square value and p value. There were cases were the input analyzer would provide a low mean square error, however the chi-square value was large and the p value was small. In such cases the chi-square value with a smaller value coupled with a large p value was chosen. MINITAB software was used to generate these plots.

Table 2. Distribution, Chi Square, P value and Parameters for each section

Section	Variable Considered	Type of Distribution	Chi Square Statistics	P value	Parameters
Admission	Inter-arrival time	Gamma	6.33	0.402	$\alpha = 2.06$, $\beta = 1.76$, Threshold =-0.5
Admission	Service time	Gamma	2.43	0.49	$\alpha = 1.04$, $\beta = 2.04$, Threshold =0.5
Billing	Service time	Gamma	1.57	0.224	$\alpha = 1.04$, $\beta = 1.9$, Threshold =0.5
Prescription	Service time	Lognormal	0.756	0.413	$\mu = 1.17$ $\sigma = .928$ Threshold =0.5, Scale = -0.09, Shape =0.475
Initial Test	Service time	Triangular	0.0176	0.577	Minimum = 0.5, Mode = 1.74 Maximum = 3.5
Orthopaedics	Service time	Lognormal	6.67	1.69	$\mu = 3.58$, $\sigma = 2.25$, Threshold =0.5, Scale = 0.78, Shape =0.48
ENT	Service time	Gamma	6.59	0.477	$\alpha = 1.68$, $\beta = 3.29$ Threshold =0.5
OBG	Service time	Gamma	5.57	0.238	$\alpha = 1.72$, $\beta = 2.87$ Threshold =1.5
Ophthalmology	Service time	Triangular	2.93	0.575	Minimum = 1.5, Mode = 3 Maximum = 12.5

It is also to be noted that there were quite minimal differences between the mean square values of some of the closest distributions. Based on the above mentioned criteria the corresponding distributions were obtained. Table 2 provides the details of the various distributions fitted with the corresponding parameters. Figures 7-11 provide the corresponding probability density function for the various variables. It can be seen that density function plots are similar in shape to the histogram from figure 2-6, thereby justifying the choice.

Table 3. Expressions used in order to build the simulation model (time in minutes)

Section	Variable Considered	Expression
Admission	Inter-arrival time	-0.5 + GAMM(2.06, 1.76)
Admission	Service time	0.5 + GAMM(1.04, 2.04)
Billing	Service time	0.5 + GAMM(1.04, 1.9)
Prescription	Service time	0.5 + LOGN(1.17, 0.928)
Initial Test	Service time	TRIA(0.5, 1.74, 3.5)
Orthopaedics	Service time	0.5 + LOGN(3.58, 2.25)
ENT	Service time	0.5 + GAMM(1.68, 3.29)
OBG	Service time	1.5 + GAMM(1.72, 2.87)
Ophthalmology	Service time	TRIA(1.5, 3, 12.5)

Once the distribution is created the values are fed to the model using the expression given by the input analyzer. Table 3 provides details of the expression used in order to create the model. The simulation model using Arena software was built based on the distribution of data, constraints in the system such as departments, number of resources, resource availability and flow of patients. The output of the simulation model was analyzed to obtain the values pertaining to queue length, average waiting time, average service time and resource utilization. The results obtained for various sections of the OPD were compared and the sections with high queue length and those that were either over or underutilized were identified.

3. Results and Discussions

Simulation model was built as per the patient flow process and the expressions shown in table 3. Relevant modules were used for each service point encountered during the patient movement in OPD. Decision modules were created based on the data at sections such as at admission / registration, initial testing, department allocation and prescription. The simulation model developed in shown in figure 12.

The focus of this study was to understand the waiting time and utilization of resources such as doctors and supporting staff associated with each section. In order to build a simulation model for this process, the following time measurements are considered based on the literature. Inter-arrival time (i.e. time between arrival of successive patients) is measured at the beginning of the process, i.e. at admission section. Service time (time required to attend a patient by the resource stationed at that section) is measured for all other successive sections.

The model was run for 30 numbers of replicates for a single shift of OPD operation. The results obtained were recorded. Significant parameters such as patient waiting time and resource utilization were studied. Patient waiting time is an important aspect which has to be addressed by the hospital management as it impacts the satisfaction of the patient. Longer waiting times may have adverse effect on the patient satisfaction.

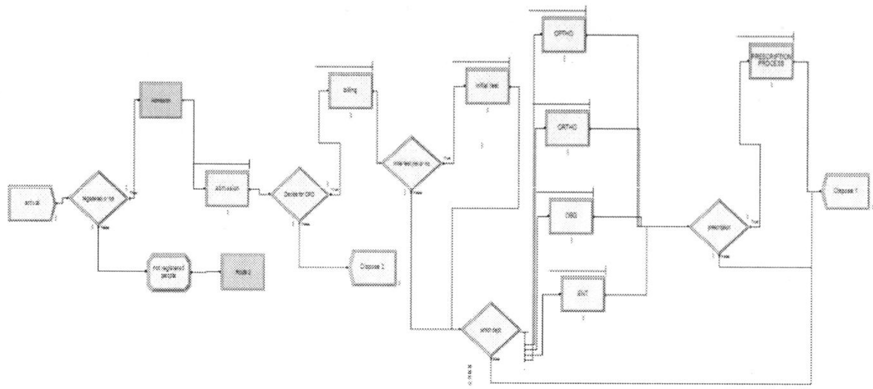

Figure 12. Arena simulation model of patient flow in OPD

Imbalance in resource utilization leads to uneven distributing of queue lengths and variation in patient waiting time across the sections. This leads to overall drop in system performance and hence induce stress in resources.

Also resource utilization needs to be addressed as over utilization may lead to fatigue in doctors and other staff which in turn may reduce their efficiency. Under utilization indicates wastage of resources and loss to the organization.

Table 4. Average waiting time, half width, minimum and maximum average times

Section	Average waiting time (minutes)	Half Width	Minimum Average Time (minutes)	Maximum Average Time (minutes)
Admission	0.3119	0.06	0.1081	0.6904
Billing	4.5067	0.63	1.9673	8.5607
Prescription	0.3661	0.06	0.1304	0.7913
Initial Test	0.2774	0.04	0.1667	0.5660
Orthopaedics	0.8760	0.19	0.2701	2.5690
ENT	1.6822	0.42	0.2105	4.9511
OBG	3.8111	1.09	0.2893	13.7036
Ophthalmology	0.8758	0.28	0	3.4124

Table 4 shows the average waiting time, half width, maximum and minimum average time for the various section of the OPD. In Arena, output variability is reported via the half width, which is half of the 95th percentile confidence interval for the metric. It can be seen that in case of billing and OBG sections the average waiting time of patients is significantly higher than other sections. Also the half width at these sections is higher which indicates that there exists a high variability in the data.

Table 5 shows the utilization of resources at these sections. It can be observed that billing and OBG sections have high resource utilization compared to others. Resources which are not optimally utilized are to be streamlined to avoid over stressing of the doctors and other staff members. Figure 13 illustrates the comparison of resource utilization in various sections of the OPD.

Table 5. Resource utilization at various sections of the OPD

Section	Average Utilization	Half Width	Minimum Average Time (minutes)	Maximum Average Time (minutes)
Admission	0.4109	0.01	0.3569	0.4551
Billing	0.7681	0.02	0.6385	0.8478
Prescription	0.3373	0.01	0.2696	0.3977
Initial Test	0.4088	0.01	0.3393	0.4831
Orthopaedics	0.2967	0.02	0.1672	0.3977
ENT	0.3595	0.02	0.1938	0.5088
OBG	0.5055	0.04	0.2279	0.7092
Ophthalmology	0.2562	0.02	0.1312	0.3816

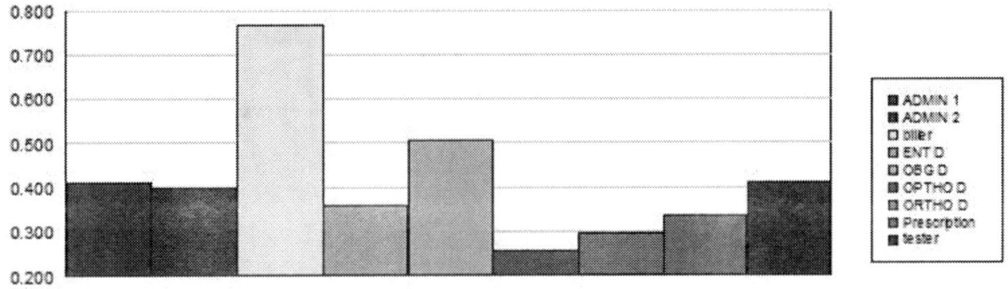

Figure 13. Comparison of Resource utilization in various sections of the OPD obtained from Arena results

4. Conclusions

This paper considered the Out-Patient Department of a hospital for studying the patient flow and issues related to waiting time and service time at various sections. Simulation model of the patient flow across the OPD was built considering the best fitting probability distribution. The results obtained the simulation model was analysed. It was observed that there was a higher waiting time at billing and OBG sections. Also the resource utilization at these sections was comparatively higher. This study was for the entire process covering all the sections and considering it as one system. The entire process was simulated based on the inter-arrival time at the admission section, which is the input station and service time at successive stations. To understand the variation within a section and from section to section, further investigation needs to be done by considering different scenarios and analyzing the obtained results. Reduction in patient waiting time will directly impact the satisfaction of the patients. Resource optimization to improve the efficiency of the doctors and staff can be further done after validating the model. The findings from further investigations can be used to improve the quality of service to the patients and improve the overall service of the hospital.

References

[1] Rising, E.J., Baron, R. and Averill, B., 1973. A systems analysis of a university-health-service outpatient clinic. *Operations Research*, ***21***(5), 1030-1047.

[2] Weng, M.L. and Houshmand, A.A., 1999, December. Healthcare simulation: a case study at a local clinic. In *Proceedings of the 31st conference on Winter simulation: Simulation---a bridge to the future-*, ***2***, 1577-1584).

IOP Conf. Series: Materials Science and Engineering 1059 (2021) 012041 doi:10.1088/1757-899X/1059/1/012041

[3] Banks, J., Carson, I.I. and JS, N., 2001. BL, and Nicol, DM. *Discrete Event System Simulation.*

[4] Kelton W D, Sadowski R P and Sturrock DT 2004 *Simulation with Arena* (McGraw-Hill: India)

[5] Wijewickrama, A.A. and Takakuwa, S., 2006, December. Simulation analysis of an outpatient department of internal medicine in a university hospital. In *Proceedings of the Winter Simulation Conference*, 425-432. IEEE.

[6] Cochran W G 2007 *Sampling Techniques* (Wiley: India)

[7] Zhao, L. and Lie, B., 2010. Modeling and Simulation of Patient Flow in Hospitals for Resource Utilization. *Simul. Notes Eur.*, ***20***(2), 41-50.

[8] Joshi A J and Rys M J 2011 Study on the effect of different arrival patterns on an emergency department's capacity using discrete event simulation *Int. J. Ind. Eng.,* **18**, 40-50

[9] Choi B K, Kang D, Kong J, Kim H, Jammoom A A, Mogbil A M and Alghandi T A 2013 Simulation based operation management of outpatient departments in university hospitals *Proc. Winter Simul. Conf.* 2287-98

[10] Ngowtanasuwan G and Ruengtam P 2013 Quality of life in the built and natural environment: applied simulation model for design of improving medical record area in outpatient department of a Governmental hospital *Proc. – Soc. Behav. Sci.* **101**, 147-58

[11] Kritchanchai D and Hoeur S 2018 Simulation modeling for facility allocation of outpatient department *Int. J. Healthcare Manage.* **11**,193–201

[12] Aeenparast A, Farzadi F, Maftoon F and Yahyazadeh H 2019 Patient flow analysis in general hospitals: how clinical disciplines affect outpatient wait times *Hosp. Pract. Res.* **4**, 128-33

[13] Law A M, Kelton W D and Kelton D M 2000 *Simulation modelling and analysis* (McGraw-Hill: USA)

[14] Montgomery D C and Runger G C 2011 *Applied statistics and probability for engineers* (Wiley: USA)

IOP Conf. Series: Materials Science and Engineering 1059 (2021) 012042 doi:10.1088/1757-899X/1059/1/012042

Study on plastic pet bottles characteristics to develop eco-friendly plastic paver blocks

Loganayagan [1], Keerthana [2], Abishek [2],Amudhan Vetrivel [2]
[1] Assistant professor, Department of Civil Engineering, Bannari Amman Institute of Technology, Sathyamangalam-638 401, India.
[2] UG student, Department of Civil Engineering, Bannari Amman Institute of Technology, Sathyamangalam-638 401, India.

Email ID: loganayagans@bitsathy.ac.in

Abstract: Production of plastic has been increased nowadays, nearly 16 tonnes are produced per day meanwhile the degradation of plastic is very difficult because of slow process degradation. This study focuses on an eco-friendly paving product. Here plastic waste has taken for the investigation to develop a product by replacing cement with plastic waste in paver blocks. The main objective of this investigation is to recycle plastic waste into pavers and to analyze its characteristics based on the recent developments in industrial needs. Nowadays plastic is more accustomed around the world. Hence this work paves the way to reduce plastic waste accumulation in the surrounding environment. The plastic wastes were collected from the nearest surrounding and it is mixed with M-sand of an optimum percentage to cast paver blocks. The plastic paver block reduces the plastic waste, controls the pollution, and reduces the overall cost of production. The application of pavement blocks is perfect on pathways and street ways as it is a simple way of finishing and laying, also cost-effective compared to conservative paving blocks. Here the strength properties of pavement blocks incorporating PET bottles are presented.

1.0. Introduction.

Pavement is used for surface coverings. Their construction materials include concrete and stones such as flagstone, Cobble stone etc., This pavement technology in India is being in practice for past decades for footpath and parking areas and now they have different uses. In this project properties such as compression test and oven test are carried out. The convention paver block consists of a raw material which includes cement, aggregates which are easily available but this paver block contains raw materials such as PET waste and quarry dust which are easily available and cost effective. From the observations of test results, PET can be reused with 50% of quarry dust and 25 % of fly-ash in Plastic Paver block [1].These paver blocks are used in places where there is no action of heavy loads since the compressive strength is low. Concrete paver block is a better option in road construction when compared to the conventional road which is made by bitumen and gravel [2]. One of the largest components of plastic waste is polyethylene terephthalate (PET). In order to reduce the plastic waste in environment paver blocks are made incorporating those PET waste and M sand and the strength properties are discussed in this project. Diversity of plastics applications is related with their specific properties, low density, easy processing, good mechanical properties, good chemical resistance, excellent thermal and electrical insulating properties and low cost (in comparison to other materials)[3].

A plastic as its behavior has impact on the environment at every extent of its stages from manufacture, usage and disposal. Recycling of plastics has lots of usage where it supplies energy economizing and declining greenhouse gas emissions. Plastic Waste management in respect to plastic can be done by recycling. If they are not recycled then they will become big pollutant to the environment [4]. Renewing it rescues the non-renewable sources such as gas and oil. In recent days many developing countries are focusing on recycling plastics because of the increase in consumption of plastics all over the world. Nowadays plastic wastes are utilized in construction fields. By the research of U.S. Department of energy, in each year 60million barrels of oil and another type of insulation would be safe by exploiting of plastic foam insulation in buildings [5]. NPL has successfully manufactured waste plastic into 3-60mm thick tiles. It is considered to be mechanically strong enough and tested for flammability and water absorption. The performance characteristics of concrete paving blocks make it suitable for the heaviest duty applications, able to support substantial loads and resist shearing and braking forces [6]. In Indian countries plastic has been used in highway construction, it is also economically beneficial as it lasts for 10 years. Main advantage of using plastics in road constructions is that the melting points of conventional roads were 55°c but melting points of plastic roads were around 66°c.

Nowadays, numerous waste materials and industrial byproducts such as glass, ceramic etc., Compared to glass and ceramic, plastic is low in cost, high strength to weight ratio and easy to shape it in any desired shape. Researchers planned to use plastic waste in form of concrete ingredient as the concrete is second most sought material by human beings after water [7]. The use of wastage plastic in concrete paver blocks is aimed at reducing aggregates content and leading to better economy and durability [8]. Concrete paving blocks (cement: quarry dust: sand = 1:1:2) by weight or volume were produced to serve as Control having tested the compressive strength and water absorption properties [9]. From the observations of test results, PET can be reused with 50% of quarry dust and 25 % of fly-ash in Plastic Paver block [10]. Wastes related to resin and PET wastes are used more in concrete constructions than other wastes.

1.1. PET:

Polyethylene terephthalate is considered to be a polyester material which is routinely used for making containers for food and beverages, pharmaceuticals and fibres. It is cited as PET or PETE and PETP or PET-P in earlier. It is one of linear thermoplastic resin where its formation is based on condensation of terephthalic acid and ethylene glycol. Its application is worldwide because it is having high tensile strength, great resistance towards chemicals, transparency, fabricability and stoutness against fire. One such factor should be considered it is completely non-biodegradable that's the reason its production rate is increased.

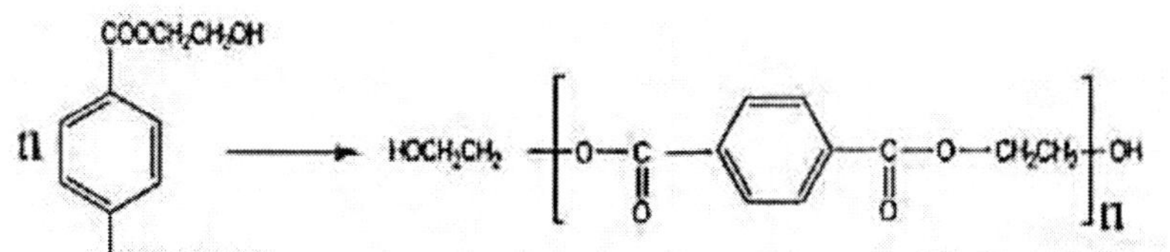

Figure 1: Polycondensation of BHET yields PET

Polyethylene terephthalate (PETE) is one of the polymerized forms of ethylene terephthalate (C10H8O4) monomer units. It possesses distinctive chemical and physical properties. It seems to be like amorphous glass like product. Additional heat treatment is given to the molten polymer or enhancing it by addictive to amend its crystallinity and the rigidity. When PET is heated above 72°c it leads to semi-crystalline state which tends to change into rubbery elastic form, easy to stretch or line up into desired shape. Between 255 and 265°c PET tends to melt. As it a monomer Bis (2-

IOP Conf. Series: Materials Science and Engineering 1059 (2021) 012042 doi:10.1088/1757-899X/1059/1/012042

Hydroxyethyl) terephthalate- BHET can be conglomerate by the esterification reaction between terephthalic acid and ethylene glycol with water as an end product. Another way of synthesis is transesterification reaction between ethylene glycol and dimethyl terephthalate with end product. Generally, Polymerization means large number of monomer molecules react together to form a polymer. Here it is through polycondensation reaction between the monomers (happen instantaneously after esterification or after transesterification) with end product of water.

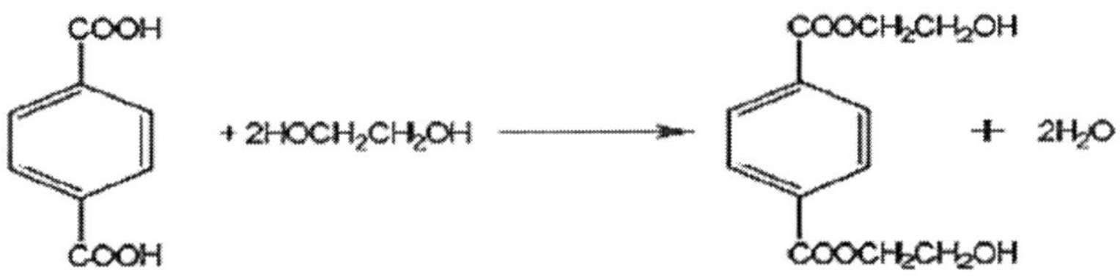

Figure 1.1: Chemical reaction between ethylene glycol and terephthalic acid yields BHET

2.0. Material properties.

2.1. Waste plastic bottle (PET):

Plastic waste used in making paver block was collected from nearby places. The type of plastic waste collected is polyethylene terephthalate (PET) and its properties are given in Table 1. It includes bottles used for soft drinks. The plastic used has thermoplastic properties, which means it can be molded and remolded recursively when heated.

Table1: Properties of collected PET bottles.

S.NO	PARTICULARS	VALUE
1	Melting point	>150 C
2	Density	1.38 g/cm^3
3	Tensile strength	55 Mpa
4	Boilingpoint	>230C

2.2. Quarry dust:

Crushed sand from quarry of size less than 4.75 mm were used. The advantages of quarry dust are cost effective, easily available and are used as a effective replacement of river sand. The properties for quarry dust are acquainted inTable 2.

Table 2: Properties of quarry dust.

S.NO	PARTICULARS	VALUE
1	Specific gravity	2.6
2	Water absorption	2.2
3	Density	15.1 KN/m^3

Figure 2. Melting of plastic waste.

2.3. Preparation of testspecimens:

The full plastic bottle is shredded into small plastic straps shown in (Img.3). The shredded plastic straps are heated in pan (img2) at a temperature >150 C, the plastic straps melts and become molten, once it becomes molten the quarry dust is added to it as per the design mix and are mixed together in the molten state. The cubes of size 70 X 70 X 70mm were cleaned and fixed and the mix were poured into the cubes in the molten state and let it to dry for 24 hours so that they become a hard mass. After drying they are demolded and tested for compression test.

Figure 3. Shredded plastic waste for processing.

2.4. MixRatio:

Block 1: plastic waste = 0.4, Quarry dust = 0.6;
Block 2: plastic waste = 0.5, Quarry dust = 0.5;
Block 3: plastic waste = 0.6, Quarry dust = 0.4.

IOP Conf. Series: Materials Science and Engineering 1059 (2021) 012042 doi:10.1088/1757-899X/1059/1/012042

2.5. Testing of specimen.

Compressivetest:

The compressive strength of the specimen is found using UTM machine. The cube size of 70 X 70 X 70mm were casted and the test was carried out to determine the compressive strength. The specimens were tested one by one and average value is taken as the compressive strength of specimen. UTM used in shown inimg 4.

Figure 4: Universal Testing Machine – compressive test.

Oven test:

As paver blocks were made from melting of plastic waste, it is necessary to know the melting point so oven test is performed.

3.0. Result and discussion.

3.1. Compressive test result:

Table 3shows the compressive strength of mix 2 with equal plastic waste and quarry dust is effective compared to other mixes.

Table 3: Compressive strength of individual paver test blocks.

Paver block	PET waste	Quarry dust	Compressive strength
Paver block 1	0.4	0.6	1.42 N/mm^2
Paver block 2	0.5	0.5	6.32 N/mm^2
Paver block 2	0.6	0.4	2.04 N/mm^2

Figure 5. Compression test cube.

3.2. Oven test result:

The heat resistance is an essential parameter to be known asthepaverblockshavebeencastedusingwasteplastics. Table4 states behavior of each specimen in different temperature maintained in a controlled system(oven).

Table 4: Oven test results.

Specimen	Temperature in C°	Remarks
Specimen 1	50	No change
	100	No change
	>140	Melts
Specimen 2	50	No change
	100	No change
	>140	Melts
Specimen 3	50	No change
	100	No change
	>140	Melts

3.3. Discussion:

Comparatively normal and plastic paver blocks have similar properties and durability characteristics that can give a good performance in paved areas, though not showing more improved parameters than normal paver blocks it appears to be a way to reuse plastic wastes (PET bottles).Mix 2 shows optimum strength and higher heat bearing ability among other two mixes making it perfect combination for making blocks.

4.0. Conclusion.

The results obtained from the tests clearly indicate thatthe block having equal ratio of plastic waste and quarry dust possess high compressive strength than that of other mixtures and all the test blocks melts again at a temperature greater than 140°C. Since these blocks have low compressive strength,they can be used in walking pavements where there is no action of heavy loads. The plastic paver blocks are economical than conventional paver blocks. This method of using recycled plastic waste in paver block serves the best way to reduce the impact of non-biodegradable plastic in the environment.

5.0. References:

[1] Nivetha, C., Rubiya, M., Shobana, S. and Vaijayanthi, R.G., 2006. "Production of plastic paver block from the solid waste (Quarry dust,fly ash &PET). *ARPN Journal of Engineering and Applied Sciences*, **11**,(2), 1819-6608.

[2] Dinesh, S., Dinesh, A. and Kirubhakaran, K., 2016. "Utilisation of waste plastic in manufacturing

of bricks and paver blocks". *International Journal of Applied Engineering Research*, **2(4)**, 364-368.

[3] Pandey, A.N., Yadav, A. and Chaudhary, D.,2019. "Waste Plastic Used In Paving Block: A Research". *International Journal of Scientific Research and Review,(7)* **4** ,279-543.

[4] Ghuge, J., Surale, S., Patil, B.M. and Bhutekar, S.B., 2019."Utilization of Waste Plastic in Manufacturing of Paver Blocks." *System*, *International Research Journal of Engineering and Technology*,(**6) 4**.

[5] Mittal, K., Agrawal, S., Bhadoria, O.P.S. and Pandey, M.,2020 "Sustainability and limitation of Plastic Modified Bituminous Interlocked Paving Block: A Supervision". *International Journal of Innovative Technology and Exploring Engineering*, ISSN: 2278-3075, **9(8),** July 2020

[6] Panimayam, S.A., Chinnadurai, P., Anuradha, R., Pradeesh, K. and Jaffer, A.U., 2017. "Utilisation of waste plastics as a replacement of coarse aggregate in paver blocks". *International Journal of ChemTech Research*.**10(8),** *ISSN*, 2455-9555

[7] Mohan, D.M.S., Vignesh, J., Iyyappan, P. and Suresh, C., 2018. "Utilization of plastic bags in pavement blocks". *Int. J. pure Appl. Math*, ***119*****(15)**,1407-1415.

[8] Pawar, S.S. and Bujone, S.A., 2017. "Use of Flyash and Plastic in Paver Block". *International Research Journal of Engineering and Technology (IRJET) e-ISSN*, 2395-0056.

[9] Agyeman, S., Obeng-Ahenkora, N.K., Assiamah, S. and Twumasi, G., 2019. "Exploiting recycled plastic waste as an alternative binder for paving blocks production". *Case Studies in Construction Materials*, ***11***, 00246.

[10] Shanmugavalli, B., Gowtham, K., Nalwin, P.J. and Moorthy, B.E., 2017. "Reuse of plastic waste in paver blocks". *Int. J. Eng. Technol*, ***6*****(02)**, 313-316.

Sanitization using Hexacopter Autonomous Drone

K Ramesh[1], **B Priya Dharshini**[1], **K Haridass**[1], **S Deepak Kumar**[1], **R Gokul Raj**[1], **V Hariprasad***[1]

[1] Department of Aeronautical Engineering, Bannari Amman Institute of Technology, Sathyamangalam, Tamilnadu, India

hari.aero007@gmail.com

Abstract. Multicopter are slowly finding its way into various industrial sector applications including security, oil & gas, public health, etc. Unmanned Aerial Vehicles provides faster and reliable solutions for many real time challenges. Spraying sanitizers using drone is one such solution which is effectively implemented during COVID-19 pandemic situation. Hexacopter is designed to carry up to 2 litres of sanitizer in its reservoir and the sanitizer spray system is designed to dispense the liquid as droplets at a constant flow rate. Hexacopter is made of carbon fibre and equipped with Pixhawk autonomous module as flight controller. The UAV can be operated in both manual and autonomous modes to increase the ease of operation for indoors as well as in outdoors sanitization process.

Keywords: Hexacopter; Sanitization; Spray; Autonomous

1 Introduction

Corona virus outbreak has a major impact on health, economy and daily life for the people around the world. Worldwide governments are taking various measures to overcome the effects caused by corona virus outbreak. Delay in the availability of vaccine and cyclic raise in the number of victims pose a challenge to the huge population to return to their routine. People who returned to their work are also being exposed to the infection. The current solution proposed by public health care officials is to frequently sanitize one's own self and their surroundings. In this aspect, disinfecting large hall, malls, shops, playground, parks, streets, etc is a challenging task. This requires number of people to be employed and lot many man hours to do it as a daily chorus. Drones can be effectively used to perform the sanitization process in a large area. Hexacopter drones (Figure 1) are capable for performing vertical take and landing which makes them most favourable choice of indoor as well as outdoor operation. In the outdoor environment, autonomous module installed in the drone can make the task easy for the pilot as it follows a predetermined path [1]. For the indoor operation, obstacle avoidance module can be most preferable for the safe flight. There are number of drones successful designed and tested for pesticide spraying in agriculture fields. The knowledge gathered from the same can be used to design a disinfection drone[2–4].

Pesticide spray drones are already employed in many agricultural field and successfully evaluated on its capability. The pesticide spray agri drones are used in open fields and provide greater flexibility in its operations [5–9]. It is easy to perform the pesticide spray process in an open field due to better GPS connectivity and less obstacles on its surrounding [10]. The drone available in the market for the agri-spray purpose can be redesigned to perform the sanitization process for a large area [11]. This will reduce the time required to sanitize a zone and requires less amount of sanitizer [12–14]. The commercially available drones are quite expensive, hence a cost effective autonomous drone is designed for the sanitization purpose.

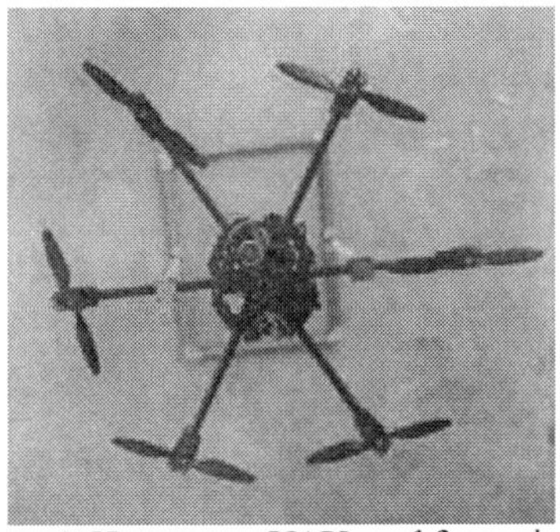

Figure 1. Hexacoptor UAV used for sanitation

2 Design and fabrication

The design and fabrication of sanitizer drone is divided into two sections. Section 2.1 explains the design and fabrication of multicoptor and section 2.2 explains the design and fabrication of the spay step-up. The drone was designed to carry 2 litre of sanitizer with endurance of 25 minutes and radio range up to 1.2 km. Each component was selected based on the mission profile and mission requirements. 3D printed components were also used in fabrication of the spray step-up.

2.1. Hexacopter Fabrication

Hexacoter which is capable of performing vertical take-off and landing is chosen for the sanitation purpose. The multicoptors is a classification of Unmanned Aerial Vehicle (UAV), which are capable of take-off and land vertically, unlike fixed wing UAV which requires a runway to take-off and land. The vertical take-off and landing characteristic of the multicoptor is crucial for the spray application. Multicopter UAV's have number of brushless DC (BLDC) motor which together produces the required take-off thrust. The designed hexacopter has 6 brushless DC motor coupled with 6 light weight 10* 4.5 plastic propellers and 30 amp electronic speed controller which can able to generate a total thrust up to 90 N. A light weight carbon fibre frame is used to house the various components including the BLDC motors and the spray set-up. Provision to mount the sanitizer reservoir and other allied components were provided at the bottom of the carbon-fibre airframe. The entire multicoptor is powered using single 11.1V 8000 mAh Lithium polymer battery. The Lithium polymer batteries has high power to weight ratio and high power density when compared with other batteries which make it ideal battery to power the entire drone. The power to operate the spay step-up is also drawn from the lithium polymer battery. Power from the battery is distributed to various components in the multicoptor using a dedicated power distributor and voltage regulator is used for components which operates in low voltage such as the flight controller.

Table 1. Hexacopter Specification

Component	Specification
BLDC Motor	920 kv
Propeller	10*4.5
Frame Material	Carbon Fibre
Battery	8000 mah
Flight controller	Pixhawk 2.8

The hexacopter is controlled manually using 6-channel transmitter and receiver which operates in 2.4 GHz ISM band. Real time flight data such as altitude, airspeed, Vertical speed, heading, etc. are transmitted to the ground control station (GCS) through on-board telemetry module connected to the flight controller. The flight controller is a bag of electronics which is used for automated flights. The flight controller has 3- axis MEMS based accelerometer and gyroscope for stabilization and dual redundant Global Positioning System (GPS) receiver to identify the current location. The flight controller uses ARM cortex processor, has high computing power with less energy consumption. Flight controller uses I2C, UART, CAN communication protocols to communicate with the other peripherals. The speed of the pump in the spray set-up is controlled using the flight controller.

2.2. Spray set-up

The spray set-up is used to deliver the sanitizer on a target area. The set-up consists of sanitizer reservoir, pump, flow tube(s) and nozzle(s). The custom designed 2 litre sanitizer reservoir is manufactured using High Density Polyethylene (HDPE) through injection moulding process. A high speed pump is designed to pump the sanitizer from the reservoir to the nozzle(s). The pump and its components were designed on solid works software (Figure 2 and Figure 3) and printed using 3D printer (Figure 4).

Figure 2. Impeller designed in solid works

Figure 3. Motor pump designed in solid works

The flow tubes are connected with pump in one end and to the nozzle on the other end. Mass flow rate in the nozzle and the sanitizer pump pressure is controlled in order avoid flow tube bulging and subsequent leakage due to pressure build in the flow tube. The spray set-up has two nozzles one mounted on the left arm and other on the right arm of the hexacopter and the nozzles are mounted at an equal distance from the sanitizer reservoir which is mounted at the centre of gravity of the hexacoptor.

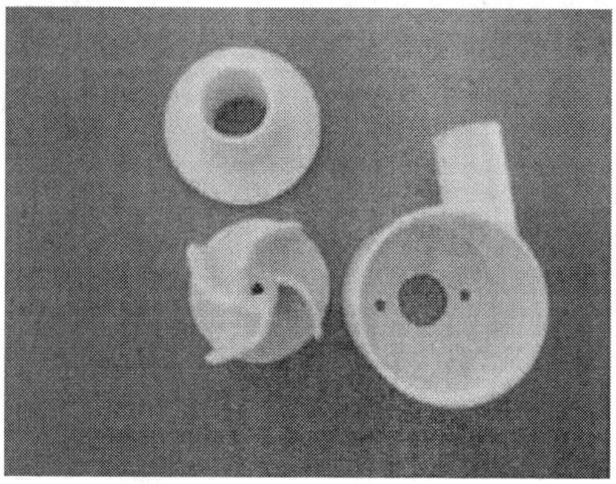

Figure 4. 3D printed components

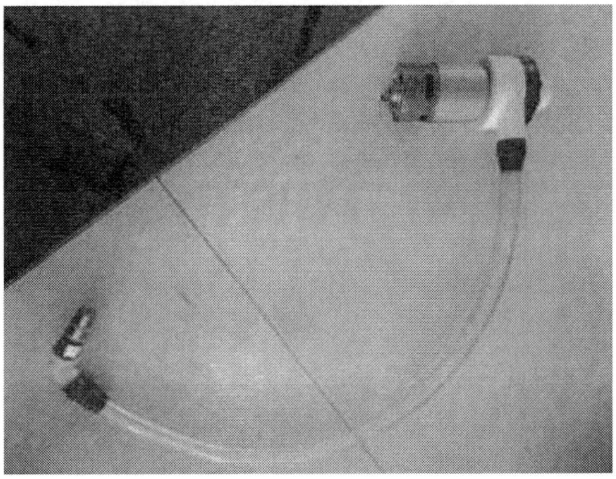

Figure 5. 8008 nozzle connected to pump

The 8008 flat fan nozzles are classified as very course nozzles with low drift potential factor and has a very large diameter droplet [15]. The 8008 nozzle has maximum spray angle of 80 degrees, can have a maximum mass flow rate 2.8 litre/minute and operating pressure of 2.5 bar (Figure 5). The entire dry set-up weight around 1.2 Kg and can be mounted on the bottom of the multicoptor.

3 Method of Sanitation Using Hexacopter

A pre-flight inspection on the hexacopter is to be performed on the model to ensure the working of all components and installed subsystem. Crucial components such as propellers, batteries, sanitizer reservoirs and flight controller are fastened tightly to the frame. Sanitizer is filled in the 2 litre container and prepared for spraying. The sanitizer need to be filled in the reservoir before uploading the flight envelope into the flight controller. The hexacopter drone using the Pixhawk flight controller is calibrated during the pre-flight inspection. The calibration process ensure the steady flight while change in weight experienced by the drone during the discharge of sanitizer.

The location need to be disinfected is fed into the drone as a flight envelop using the ground control unit. The flight envelope can be created using Real-Time Kinematic (RTK) device or by giving Global Positioning System (GPS) co-ordinates to the multicoptor using software like mission planner or Universal Ground Control (UGC), these software has a graphical user interface which allows the user to choose GPS co-ordinates manually. Since the mission planner software is more compatible with Pixhawk, it is chosen as the ground control unit for this application. The mission planner software uploads the flight envelope data using telemetry to the flight controller in the hexacopter where the information is processed and a virtual flight path is created. This virtual flight path can be viewed in mission planner software. The auto-pilot flight does not require any human intervention during the entire flight. All the flight telemetry data are transmitted to GCS, using on board telemetry module operating at 433 MHz which is received by the ground based receiver. The 8008 nozzle is coarse type nozzle which can able to spread the sanitizer on the target region precisely even with a crosswind (Figure 6). The dual nozzle configuration provides large coverage. The sanitizer pump speed which influence the pressure inside the nozzle is controlled by the flight controller, constant pressure inside the nozzle will ensure the even sanitizer deposition on the target area.

Figure 6. 8008 nozzle spray test

The reservoir pressure, sanitizer level, tube pressure and mass flow rate of the nozzle can be monitored using the on-board telemetry module which is used for flight telemetry data transmission. Once the marked areas are covered the multicoptor will automatically land where it took-off. If the sanitizer in the reservoir ran off before covering the given area, the multicoptor is programmed to land and sanitizer can be refilled in the reservoir and begins the spraying process for the remaining region from the point of return. The hexacopter drone can effectively sanitize an area of 200 m^2 for a single refill.

4 Conclusion

The following conclusion can be stated from the design and operating of sanitizer drone.

1. Hexacopter drone can effectively carry 2 liters of sanitizer and fly for 25 minutes with 8000 mAh battery
2. Large area can be disinfected with smaller work force and in a short duration
3. Improved safety for the personnel working in the disinfection process

References

[1] Radoglou-Grammatikis P, Sarigiannidis P, Lagkas T and Moscholios I 2020 A compilation of UAV applications for precision agriculture *Computer Networks* **172** 107148

[2] Faiçal B S, Freitas H, Gomes P H, Mano L Y, Pessin G, de Carvalho A C P L F, Krishnamachari B and Ueyama J 2017 An adaptive approach for UAV-based pesticide spraying in dynamic environments *Computers and Electronics in Agriculture* **138** 210–23

[3] Ivić S, Andrejčuk A and Družeta S 2019 Autonomous control for multi-agent non-uniform spraying *Applied Soft Computing* **80** 742–60

[4] Xue X, Lan Y, Sun Z, Chang C and Hoffmann W C 2016 Develop an unmanned aerial vehicle based automatic aerial spraying system *Computers and Electronics in Agriculture* **128** 58–66

[5] Guo Q, Zhu Y, Tang Y, Hou C, He Y, Zhuang J, Zheng Y and Luo S 2020 CFD simulation and experimental verification of the spatial and temporal distributions of the downwash airflow of a quad-rotor agricultural UAV in hover *Computers and Electronics in Agriculture* **172** 105343

[6] Abd. Kharim M N, Wayayok A, Mohamed Shariff A R, Abdullah A F and Husin E M 2019 Droplet deposition density of organic liquid fertilizer at low altitude UAV aerial spraying in rice cultivation *Computers and Electronics in Agriculture* **167** 105045

[7] Ling W, Du C, Mengchao Z, Yu W, Ze Y and Shumao W 2018 CFD Simulation of Low-attitude Droplets Deposition Characteristics for UAV based on Multi-feature Fusion *IFAC-PapersOnLine* **51** 648–53

[8] Guo S, Li J, Yao W, Zhan Y, Li Y and Shi Y 2019 Distribution characteristics on droplet deposition of wind field vortex formed by multi-rotor UAV *PLOS ONE* **14** e0220024

[9] Mukherjee A, Misra S, Sukrutha A and Raghuwanshi N S 2020 Distributed aerial processing for IoT-based edge UAV swarms in smart farming *Computer Networks* **167** 107038

[10] Wang G, Han Y, Li X, Andaloro J, Chen P, Hoffmann W C, Han X, Chen S and Lan Y 2020 Field evaluation of spray drift and environmental impact using an agricultural unmanned aerial vehicle (UAV) sprayer *Science of The Total Environment* **737** 139793

[11] Xu Y, Xue X, Sun Z, Chang C, Gu W, Chen C, Jin Y and Peng B 2019 Online spraying quality assessment system of plant protection unmanned aerial vehicle based on Android client *Computers and Electronics in Agriculture* **166** 104938
[12] Yang K, Yang G Y and Huang Fu S I 2020 Research of Control System for Plant Protection UAV Based on Pixhawk *Procedia Computer Science* **166** 371–5
[13] Ling W, Du C, Ze Y, xindong N and Shumao W 2018 Research on the prediction model and its influencing factors of droplet deposition area in the wind tunnel environment based on UAV spraying *IFAC-PapersOnLine* **51** 274–9
[14] Ahmad F, Qiu B, Dong X, Ma J, Huang X, Ahmed S and Ali Chandio F 2020 Effect of operational parameters of UAV sprayer on spray deposition pattern in target and off-target zones during outer field weed control application *Computers and Electronics in Agriculture* **172** 105350
[15] Van de Zande J C, Huijsmans J F M, Porskamp H A J, Michielsen J M G P, Stallinga H, Holterman H J and de Jong A 2008 Spray techniques: how to optimise spray deposition and minimise spray drift *Environmentalist* **28** 9–17

Design of experiment approach to identify optimal parameters for boring operation and tool life improvement for piston pins

V Venkataraman[1], P Raghavendra[2], M D'Souza[2], N Holla[2], N Jagadish[3]

[1]Assistant Professor, Industrial Engineering & Management, Ramaiah Institute of Technology, MSRIT Post, Bangalore 54
[2]U. G Student, Industrial Engineering & Management, Ramaiah Institute of Technology, MSRIT Post, Bangalore 54
[3]Proprietor, Durga Tech Industries, Peenya, Bangalore 58

Corresponding author email : vivekanand@msrit.edu

Abstract. Piston pins play a crucial part for providing support to the connecting rod which plays a pivotal role in functioning of the automobile. The current work focusses on identifying the factors which affect the inner bore diameter for piston pins during the boring process and also finding an optimal setup which can reduce tool life. In order to identify the parameters which, have an effect, factors such as speed, feed and depth of cut were considered and a full factorial design of experiments were conducted with 3 replicates. The response variable used while conducting include mean inner diameter and range. The experiments were conducted using randomization principle and experimental conditions were run according to random number generator using Minitab software. Analysis of variance (ANOVA) was conducted using Minitab and based on the ANOVA results, main effects and interaction plot, optimal parameters values were chosen which would provide optimal inner bore diameters. In addition, optimal parameters where accordingly chosen which would provide longer tool life.

Keywords: Boring operation, design of experiments, Analysis of variance, Main effects and Interaction Plots, Statistics, Piston Pins

1. Introduction

Drilling and Boring are two major operations and play a crucial role in ensuring quality of component or part satisfies the customer requirement. In order to conduct an effective boring process there need to be sound application of six sigma principles. Define, Measure Analyze, Improve and Control (DMAIC) approaches are effective approaches for achieving sound process [1]. The current work focusses on applying design of experiment approach to a boring process for providing an optimal set up parameters which would provide a suitable working range for achieving. Design of experiments (DOE) provide a sound approach to achieve an optimal setting for a particular process. The general guidelines for conducting design include defining a problem statement, selecting response variable, selecting factors and levels, type of design, conduction of experiment followed by statistical analysis and final concluding based on the results [2,12]. Several literatures on understanding the factors affecting drilling operation have been address and few literatures have been addressed on understanding the factors affecting boring operations. Some important literatures which discusses about factors affecting drilling and boring performance have been addressed in the current work. One of the earliest conduction of design of experiment methodology for boring process was conducted to understand the effects of using feed, speed on hole quality. Four different experiments with different feed vales and speed values were chosen with two levels each for conduction of DOE. The various responses which were measured include hole diameter, roundness error, hole location error. In their work the choice of the levels for first two experiment were within 25% range of the operating conditions based on the results obtained it was found the speed and feed effects were not significant on roundness error, however effects were seen for increase in hole diameter as feed increases, similar results were seen for second experiment. Due to these reason third experiment was conducted with large difference in levels and it was found speed and feed have significant effect. The above work laid a strong foundation for analyzing machining parameters using design of experimental approach [3]. A taguchi based DOE to understand the effects of several variables or factors (Speed, feed, Nose radius) on surface roughness, amplitude and volume of metal removed. It was found that factors considered played a crucial role in affecting the various response [4]. Later several authors have applied different types of design of experimental approach for evaluating boring performance based on speed, feed depth of cut [5-7]. The current work focusses on obtaining an optimal parameter for achieving a target value of bore diameter and also identifying the optimal parameter for improvement in tool life. The experiment was carried out using CNC machine of horizontal lathe type machine with spindle speed range of 1000 to 3000 rpm. The piston specification should be 10.02±.05 mm and length should be 47.5±.5 mm. The part diameter were measured using digimatic Vernier caliper which has an accuracy of .01 mm.

2. Methodology

Figure 1 provides the methodology. There are several phases which include data collection, data analysis, designing the 2^3 experiment, conduction of experiment based on the principle of randomization and replication, analysis of experiments using ANOVA and determining the optimal setting based on the main effects, interaction effects plot and statistical test.

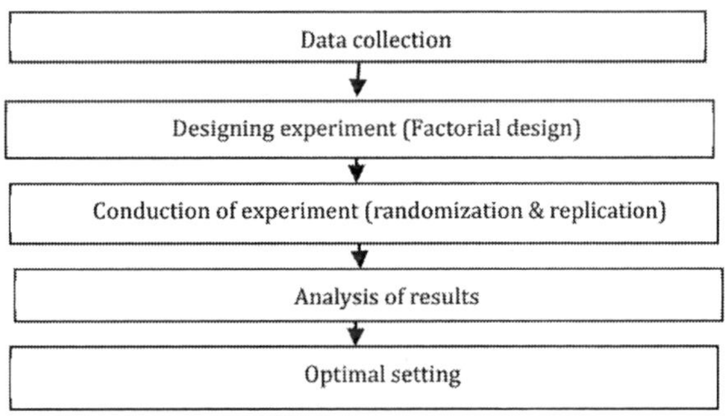

Figure 1 Methodology

2.1. Data collection

Data was initially collected for 30 samples of inner bore diameter and a line chart was plotted by considering specification limits, figure 2 provides the chart. It is seen from the above data that there are few parts which fall outside of the specification limits and the process is seen to be in the upper range of the specifications limits, since there was around 10% of parts outside of the limits. It would be worth understanding the process to understand which factors play a role and accordingly select a suitable set of parameters. Thus the objective of the work is to ensure that process was within the specification limits for which a design of experiment needs to be conducted

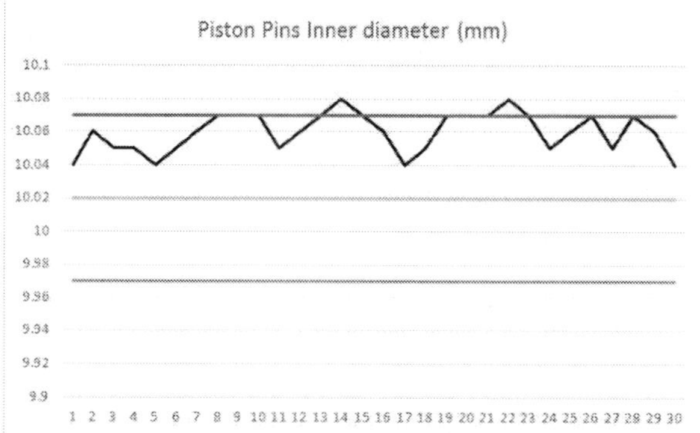

Figure 2 Bore Inner diameter for 30 samples

2.2. Design of experiment

Based on the various literature studied and with discussion with process experts, it was concluded that experiment would be conducted by varying three factors which include speed, feed and depth of cut. The levels of the factors were chosen as high and low based on the current operating condition (which is the center point). A 2^3 full factorial design with 3 replicates were chosen. Table 1 provides the details of the levels. Based on the above design with 3 replicates, would result in 24 runs. The response variable was inner bore diameter. Three reading of the inner bore diameter were recorded and the average value of the inner bore diameter was considered as the response variable. Range (maximum – minimum) of bore diameter for each treatment combination was considered.

Table 1 Factors and levels for Design of Experiments

PARAMETERS	Low (-)	High (+)
Speed (Rpm)	1800	2300
Feed (mm /rev)	0.10	0.15
Depth of cut (mm)	0.40	0.60

2.3. Conduction of experiments

Based on the design the randomization order was obtained using minitab software, Table 2 provides the list of randomized order (indicated by run order). The standard order indicates the standard order based on Yates design. CNC machine was used to conduct boring operation and the tool used was high speed steel. Based on the randomized order, the experiment was conducted and accordingly reading were recorded.

3. Results and Analysis

Two response variables are analysed which includes the Average Bore inner diameter and Range of bore diameter for each part. Box plot of data as shown in figure 3 indicates that there is minimal variability of the data set. For average inner bore diameter there is symmetricity of the data set

Table 2 Randomized order for conduction of experiment

STD. ORDER	RUN ORDER	SPEED (rpm)	FEED (mm/rev)	DEPTH OF CUT(mm)
4	1	2300	0.15	0.4
1	2	1800	0.1	0.4
23	3	1800	0.15	0.6
20	4	2300	0.15	0.4
24	5	2300	0.15	0.6
2	6	2300	0.1	0.4
19	7	1800	0.15	0.4
13	8	1800	0.1	0.6
12	9	2300	0.15	0.4
8	10	2300	0.15	0.6
10	11	2300	0.1	0.4
18	12	2300	0.1	0.4
5	13	1800	0.1	0.6
9	14	1800	0.1	0.4
11	15	1800	0.15	0.4
6	16	2300	0.1	0.6
16	17	2300	0.15	0.6
7	18	1800	0.15	0.6
15	19	1800	0.15	0.6
21	20	1800	0.1	0.6
17	21	1800	0.1	0.4
14	22	2300	0.1	0.6
22	23	2300	0.1	0.6
3	24	1800	0.15	0.4

indicating that data set might be normally distributed. This is also validated by considering the normal probability plot shown in figure 4 of average bore diameter. Based on the plot and p value for the test it is seen that data set follows normal distribution. Thus it can be seen there are no outliers in the data, hence it can be easy to analyse the data.

The next step involves conducting the ANOVA using Minitab software, The ANOVA results are shown in Table 3. It is seen that most of the factors are not playing a significant role in affecting the Average bore diameter, however it is seen that Speed and Feed interaction seems to play a role, due to a smaller p value. It is to be noted that the purpose of DOE also is to find an optimal set of values which can result in an optimal bore diameter and also to ensure that tool life is increased. For this purpose the main effects and interaction effects are analyzed. Figure 5 provides the main effects plot, it is seen that higher speed and lower feed results in lower value.

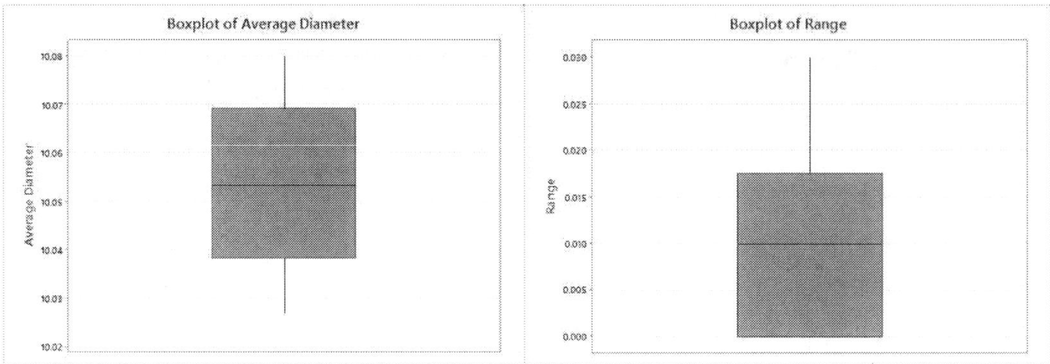

Figure 3 Box Plot for Average and Range for Bore Inner Diameter

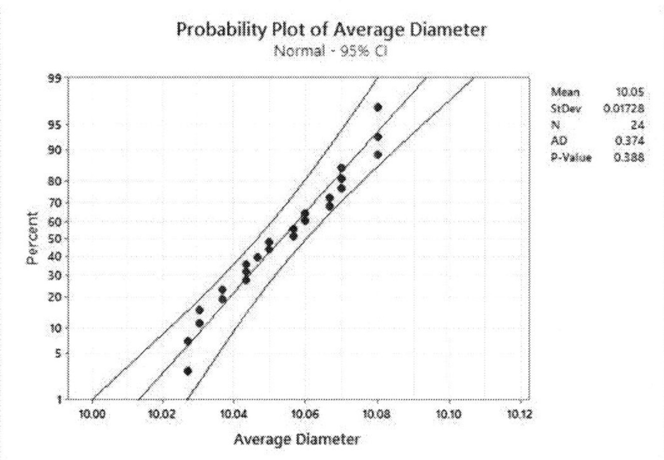

Figure 4 Normal Probability Plot of Average Bore Diameter

As seen from the graph it is seen that these changes are not large and the p values obtained from ANOVA table are not small enough hence it can be concluded that main effects do not play a role. It was already noted from ANOVA table that speed and feed interaction play a role and from figure 5 it is seen that there are two lines intersecting for speed and feed interaction, indicating there is an interaction effect. In addition, there is drastic change of bore diameter when feed is changed from a low value of .1 mm to .15 for a fixed lower speed of 1800 rpm. From the plots it is seen that a low speed with a high feed would be a suitable combination. In order to justify the choice of this combination individual plots (figure 6) and two sample t test were conducted for the various combination of speed feed interaction. From figure 5 it can be seen that the combination of 1800 rpm and a feed of .15 mm/rev has a low mean value. Table 4 provides the comparison for various combinations, it can be seen there is a significant difference between the combination of 1800 rpm with feed of .15 mm/rev and 1800 rpm with feed of .1 mm/rev, however the p value was large for other treatment combinations. Based on t test and individual plots it can be concluded that the optimal combination can be with a speed of 1800 rpm and a feed of .15 mm/rev.

The R^2 value was low for the model, since the objective was to find an optimal setting R^2 may not be useful in this context. The model adequacy is conducted by checking the residual plots. It can be seen from figure 7 that the probability values of the residues are in a straight line thereby indicating that it is normally distributed which is further validated by a large p value using Anderson darling test. It can also be seen that there is no pattern seen when residue and fitted values are plotted. Thus the model seems to be adequate.

Table 3 ANOVA for Factors affecting Bore Diameter

Term	Effect Coeff	Coef	SE Coef	T	P
Constant		10.053	0.003891	2583.5	0
Speed	-0.0022	-0.0011	0.003891	-0.29	0.78
Feed	-0.0056	-0.0028	0.003891	-0.71	0.49
Depth of cut	-0.0006	-0.0003	0.003891	-0.07	0.94
Speed*Feed	0.0111	0.0056	0.003891	1.43	0.17
Speed*Depth of cut	-0.0006	-0.0003	0.003891	-0.07	0.94
Feed*Depth of cut	0.0039	0.0019	0.003891	0.5	0.62
Speed*Feed*Depth of cut	0.0006	0.0003	0.003891	0.07	0.94

Table 4 Two sample t test of various treatment combination

Treatment combination comparison	T value	P value	Significant
1800 * .15 Vs 1800*.1	-1.83	.098	Yes for α =10%
1800 * .15 Vs 2300*.1	.30	.744	No
1800 * .15 Vs 2300*.15	.84	.419	No

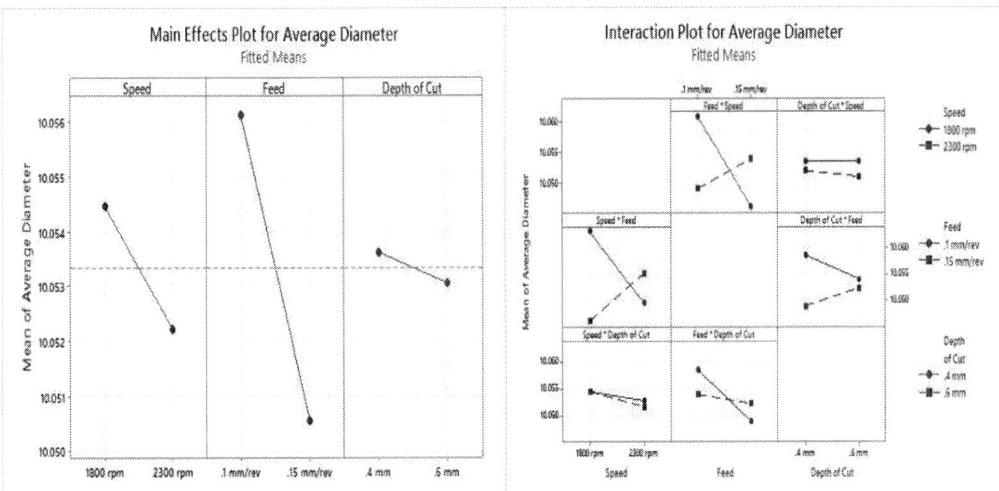

Figure 5 Main and Interaction Effects Plot for Bore Diameter

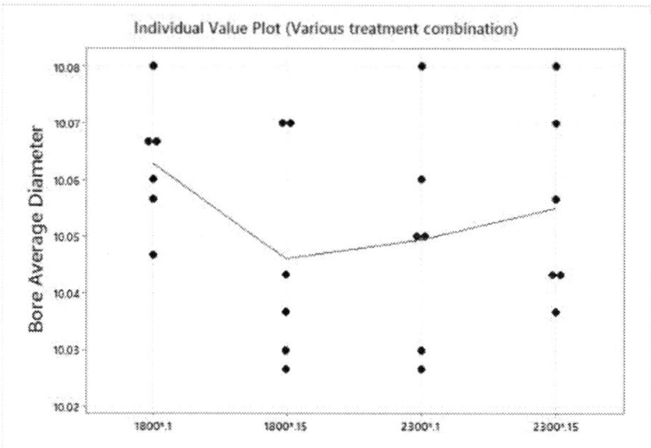

Figure 6 Individual Plots for Treatment Combination

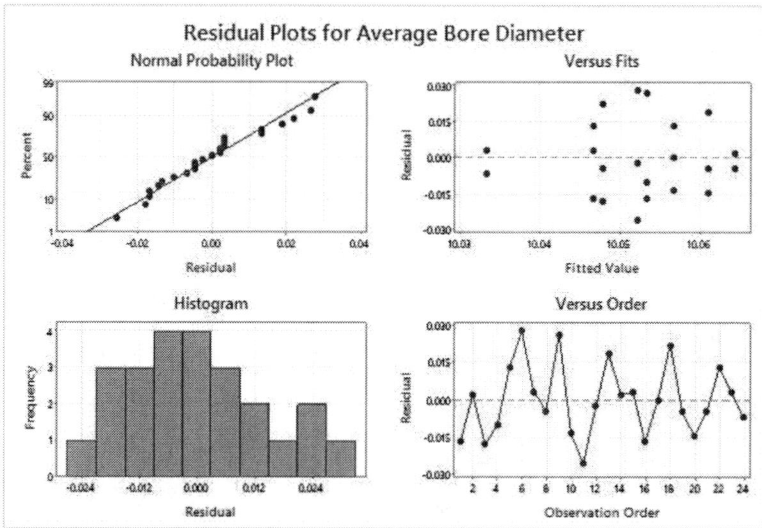

Figure 7 Residual Plots for Mean Diameter

Range was used a measure for determining the variation of bore diameter. ANOVA was conducted for the data set; Table 5 provides the details. It is seen that none of the factors are significant, the values of the range indicate that process is in control. In addition, the main effects and interaction plots were examined, it was seen from figure 8 main effects plot that when depth of cut or feed is changed from low to high value the range increases from lower to higher value. Based on the interaction plot the combination of Low speed with low feed produces low deviation. But it can also be seen from the individual value plot (figure 9) that the changes are not significant, to validate this a two sample t test was conducted. From the two sample t test (table 6) it can be seen that there isn't any significant difference between the combination which produces the lowest range value and the other treatment combination. It can therefore be concluded that any combination would be suitable for optimum range value. The residual plots shown in figure 10 were checked for model adequacy and it is seen from the plots that residues follow normal distribution and there is no pattern between the residue and the fitted value.

Table 5 ANOVA for Factors affecting Range

Term	Effect Coeff	Coef	SE Coef	T	P
Constant		0.009167	0.002041	4.49	0
Speed	0	0	0.002041	0	1
Feed	0.003333	0.001667	0.002041	0.82	0.426
Depth of cut	0.003333	0.001667	0.002041	0.82	0.426
Speed*Feed	-0.001667	-0.000833	0.002041	-0.41	0.689
Speed*Depth of cut	0.001667	0.000833	0.002041	0.41	0.689
Feed*Depth of cut	0.001667	0.000833	0.002041	0.41	0.689
Speed*Feed*Depth of cut	0.001667	0.000833	0.002041	0.41	0.689

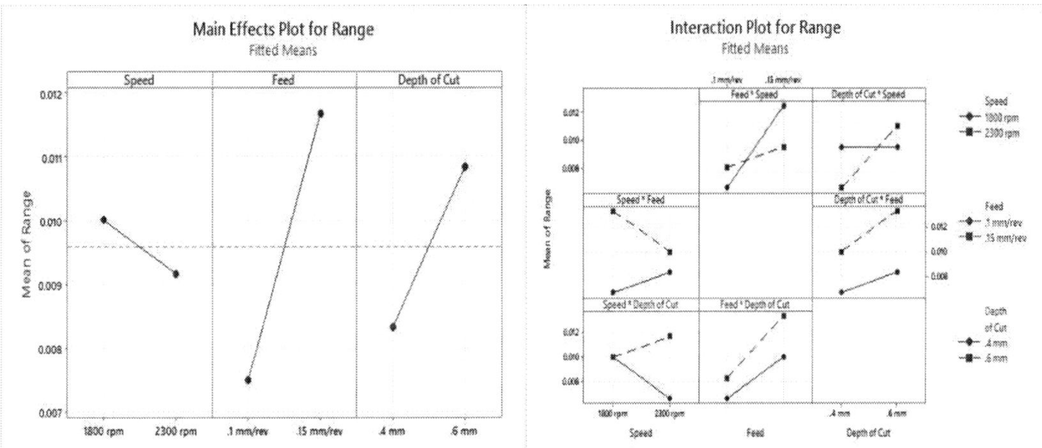

Figure 8 Main and Interaction Effects Plot for Range

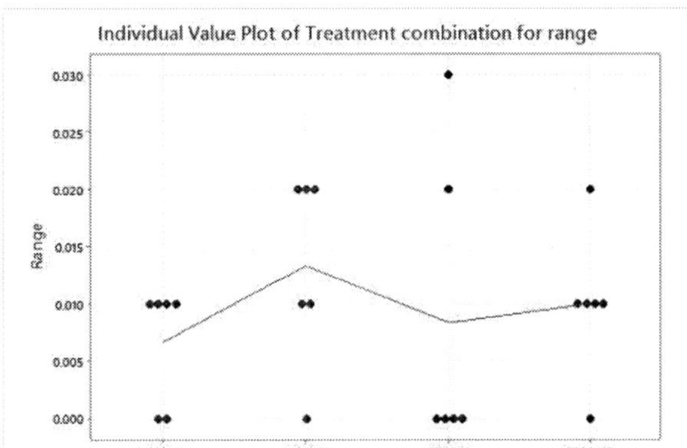

Figure 9 Individual Plots for Treatment Combination for range

Table 6 Two sample t test of various treatment combination for range

Treatment combination comparison	T value	P value	Significant
1800 * .1 Vs 1800*.15	-1.10	.296	No
1800 * .1 Vs 2300*.1	-1.00	.341	No
1800 * .5 Vs 2300*.15	-.29	.78	No

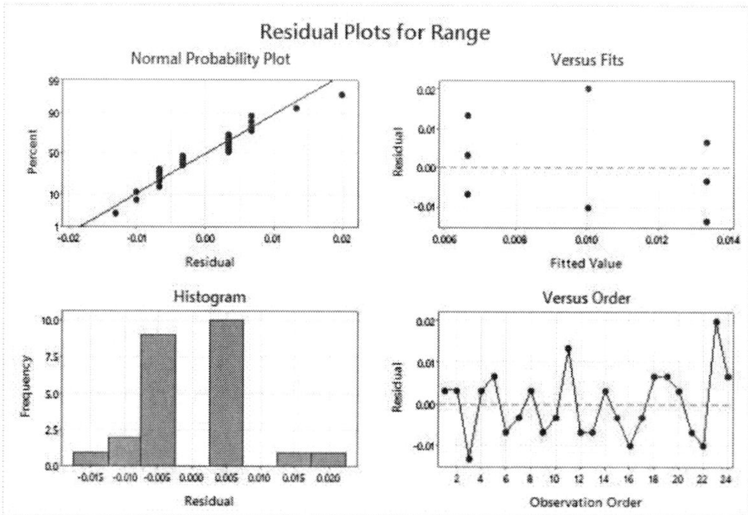

Figure 10 Residual Plots for Range

Since none of the treatment combination of speed and feed have a significant effect on the range it becomes easy to set an optimal parameter for the mean bore diameter. Therefore, a low speed combination of 1800 rpm with a feed of .15 mm/rev and a depth of cut of .4mm is chosen. This combination needs to be validated, a sample of 10 parts were bored using the above mentioned combination. It is seen from figure 11 that there is a decrease in bore diameter when compared to figure 1. Also seen is that process is well within the specification limits, to further validate it a two sample t test was conducted. The hypothesis framework in this case was $\mu_1 > \mu_2$, where μ_1 represents the mean based on speed of 2000 rpm, feed of .12 mm/rev, depth of cut of .5 mm and μ_2 represents the setting based on the design of experiments (speed 1800 rpm, feed .15 and depth of cut .4 mm). Based on the box plot in figure 12, it is seen that bore diameter is lower and the range is smaller for the parameters obtained from the designed experiment. From Table 7, it can be seen that there is a significant difference between current set up and designed experiment set up hence it can be concluded that $\mu_1 > \mu_2$. Thus it is seen that designed experiment set up produces a lower diameter and well within the specification limit.

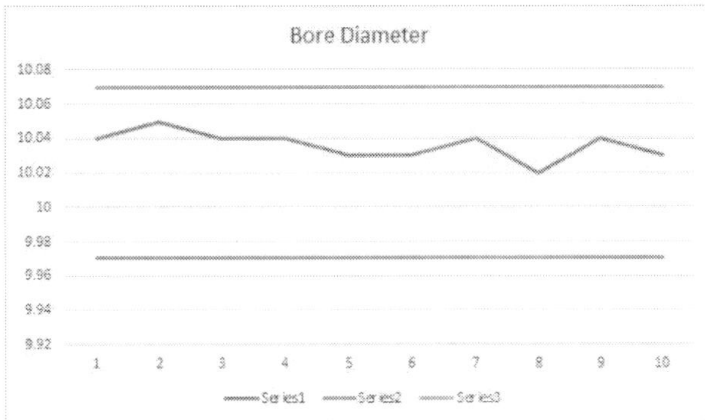

Figure 11 Bore Diameter for optimal set up

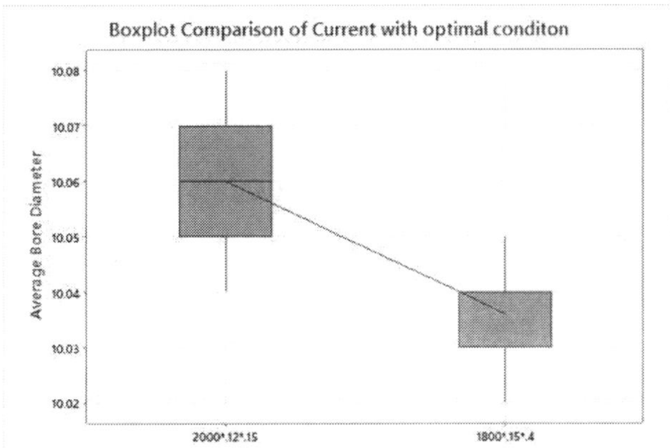

Figure 12 Box Plot for Bore diameter for current and optimal setup

Table 7 Two sample t test for comparison between optimal and current setup

Hypothesis	Treatment combination comparison (Speed, feed, Depth of cut)	T value	P value	Significant
$\mu_1 > \mu_2$	2000 rpm .12mm/rev .5 mm Vs 1800 rpm .15mm/rev .4 mm	6.95	.00	Yes

Tool life plays a vital role in affecting productivity therefore it would be worthwhile looking at the impact of the above mentioned parameters on tool life. Equation 1 provides the Extended Taylor modified tool life equation and various authors have dealt with tool life calculation [8-11]. In the equation V represents the speed, T represents tool life, d represents depth of cut and f the feed. Depending upon the tool types, n, x and y are accordingly chosen. Based on various authors recommendation for HSS n is chosen as .17, x as .77 and y as .37 [11]. Since tool life improvement is considered the ratio of tool life would be crucial, in such cases the units of the parameters will not matter. The ratio of tool life for current condition and the experimental set up is given in equation 2. Table 8 provides the details of the value for the various parameters for various experimental conditions. It is seen there is a slight decrease in tool life when the optimal condition is chosen. Thus there is a slight trade off which exist when choosing the optimal condition. It is also to be noted that depth of cut of .6mm for a combination of 1800 rpm and a feed of .15mm/rev was insignificant when compared to same combination of speed, feed with a depth of cut of .4. Based on this the tool life ratio was once again found and is shown in table 9. It is seen that there is a significant increase in tool life for this combination. Thus there is choice of either choosing a speed of 1800 rpm, feed of .15 mm/rev and a depth of cut of .5mm or a speed of 1800 rpm, feed of .15 mm/rev and a depth of cut of .6mm.

$$VT^n d^x f^y = C \quad (1)$$

$$\frac{T_2}{T_1} = \left(\left(\frac{V_1}{V_2}\right) * \left(\frac{d_1}{d_2}\right)^x * \left(\frac{f_1}{f_2}\right)^y\right)^{\frac{1}{n}} \quad (2)$$

Table 8 Tool life ratio of Current and optimal condition

Set up	Speed (rpm)	Feed (mm/rev)	Depth of cut (mm)	Ratio of $\frac{T_2}{T_1}$ (Ratio of tool life)
Current Condition	2000	.12	.5	$\left(\left(\frac{1800}{2000}\right) * \left(\frac{.15}{.12}\right)^{.77} * \left(\frac{.4}{.5}\right)^{.37}\right)^{\frac{1}{.17}}$=.909
Optimal Condition	1800	.15	.4	

Table 9 Tool life ratio of Current and Experimental condition

Set up	Speed (rpm)	Feed (mm/rev)	Depth of cut (mm)	Ratio of $\frac{T_2}{T_1}$ (Ratio of tool life)
Current Condition	2000	.12	.5	$\left(\left(\frac{1800}{2000}\right) * \left(\frac{.15}{.12}\right)^{.77} *\right.$
Experimental Condition	1800	.15	.6	$\left.\left(\frac{.6}{.5}\right)^{.37}\right)^{\frac{1}{.17}}$=2.19

4.0 Conclusion

The current work brings in the importance of design of experiment approach in boring process diameter improvement. It was seen that based on the experimental set up an optimal speed of 1800 rpm with a feed of .15mm/rev ensures that process can be brought well within the range of upper and lower specification limit. It was also seen that additional combination of varying depth of cut such as .4 mm or .6 mm with a speed of 1800 rpm with a feed of .15mm/rev also provides reduced bore diameter. It can also be noted that there only a slight decrease in tool life for the combination of 1800 rpm with a feed of .15mm/rev with a depth of cut of .4 mm. However, a depth of cut of .6 mm provided a larger tool life for a combination of speed of 1800 rpm and feed of .15 mm/rev. Thus depending upon the situation and condition either of these combination can be chosen to ensure the process is well within control.

5.0 Author's contributions and Acknowledgements

The authors would like to provide their contribution for the paper as described below
V. Venkataraman: Conceptualization, Methodology, Formal Analysis, Investigation (Conducting a research and investigation process), Validation, Writing – Original Draft, Supervision, Project administration, Visualization, Resources (Study materials)

P. Raghavendra: Formal Analysis, Investigation (Conducting a research and investigation process, performing the experiments, data collection), Validation, Visualization

M. D'Souza: Formal Analysis, Investigation (Conducting a research and investigation process, performing the experiments, data collection), Validation, Visualization

N. Holla: Formal Analysis, Investigation (Conducting a research and investigation process, performing the experiments, data collection), Validation, Visualization

N. Jagadish: Resources, Supervision, Project administration

The authors would like to thank Mr. Sudheer Kulkarni and Dr.R. Shobha of Ramaiah Institute of technology for their suggestions.

References

[1] Juran J M and Godfrey A.B 1998 *Juran's Quality Handbook* (Mcgraw Hill, New York)

[2] Montgomery D C 2013 *Design and analysis of experiments*, eight ed., (John Wiley & Sons, New York)

[3] Furness R J Wu CL Ulsoy A G 1996 Statisitcal Analysis of the Effects of Feed, Speed, and Wear on Hole Quality in Drilling, *J. MANUF. SCI.E-T. ASME.* **118** (3), 367-75

[4] Venkata Rao K Murthy B S N, Mohan Rao N 2013 Cutting tool condition monitoring by analysing surface roughness, work piece vibration and volume of metal removed for AISI 1040 steel in boring *Measurement* **46** (10), 4075-84

[5] Vaishnav M P Sonawane S A 2013 Optimization of Boring Process Parameters by Using Taguchi Method *Int. J. Engg. Res. Tech* **3** (8), 717-721

[6] Bansal G Sodhi H S Singh J 2014 Optimization of Machining Parameters for MRR in Boring Operation Using RSM *IJRMET* **4** (2), 127-31

[7] Kumar P et al 2014 Analysis and Optimization of Parameters Affecting Bore Deviation in Boring Process *IJITR* **2**(3), 967-72

[8] Taylor F W 1907 On the Art of Cutting Metals, *ASME Trans*. **28**, 31-350

[9] KRONENBERG, M., 1940. *Machining with single point tools*. Cincinnati Milling Machines, **8.**

[10] Woldman N E Gibbons, R C 1951 *Machinability and Machining of Metals*, (McGraw-Hill, New York), **2** 29-54

[11] Stephenson D A Agapiou J A 2016 *Metal Cutting Theory and Practice*, 3rd Edition, CRC Press, Boca Raton, **9**, 549-73

[12] Montgomery D C Runger G C 2003 *Applied Statistics and Probability for Engineers* third ed (John Wiley & Sons, New York)

Design Enhancement and Thermal Analysis of Disc Brake

B Roshan Rino, M Pranesh Kumar and S Velmurugan

Mechanical Engineering, Bannari Amman Institute of Technology, Erode,India

roshanrino.me19@bitsathy.ac.in, praneshkumar.me19@bitsathy.ac.in, vsvmurugan@gmail.com.

Abstract. In automobiles, the function of disc brakes is controlled by jamming the brake pads opposed to a rotary disc that is usually attached to a wheel. A composite material set is preferred to manufacture the brake pads. This braking process produces frictional forces which cause deceleration and eventually stops the rotation of the disc. The indulgence of the generated heat due to friction is important for successful braking. Changes in temperature of the brake cause radial and axial bend; this variation in shape, in turn, distresses the proper alignment of the pads and the disc. The aim of this paper is to design a brake disc and perform the thermal stress analysis by applying five different materials namely Gray cast iron (GCI), reinforced Ti-composite (TMC),Ti-alloy (Ti-6Al-4V), reinforced Al-Cu alloy(AMC 2), and reinforced Al-composite(AMC 1).The modelling and analysis of the disc brake are achieved with the aid of SOLIDWORKS and ANSYS software.

1 Introduction

A disc brake is used as calipers to jam the pads set contrary to a disc to create friction, this action reduces the speed of a wheel, either to decrease its rotational speed or to grasp it as motionless. [1,2 &3] The energy of motion is transformed into excess heat which is dispersed into the air. [4,5&6] The disc is designed to be ventilated which allows heat to dissipate more efficiently. When compared to drum brakes disc brakes operates much more efficiently and with better capabilities if dissipating heat. These brakes have applications in the automotive industry for almost all types of vehicles. The increase in production of these types of brakes for commercial purpose was in the year 1955 the same model sold over 1.5 million pieces over the course of 20 years without a single modification being made. The same type of brakes is used in motorcycles bi-cycles trucks as well as other heavy automotive vehicles such as train and serves the purpose quite efficiently and no other better options have been yet found. This paper concentrates on the disc brakes which are used in a commercial four-wheeler. A ventilated disc brake model was designed using SOLIDWORKS software and the effects on the model under influence of high temperature was studied for the five materials and the results were compared for all the materials.

2 Properties of materials

Grey cast iron has high thermal conductivity and specific heat capacity as well as good frictional properties this material generally used in disc brakes [7]. Ti-alloy, Ti-6Al-4V, otherwise named as TC4, Ti64, or ASTM Grade 5, is an alpha- beta titanium alloy with a good strength-to-weight ratio and exceptional resistance to corrosion, [8] When compared with steel and nickel-base materials Titanium matrix composites (TMCs) offer stiffness and high specific strength. High temperature TMCs can propose up to 50% weight reduction compared to monolithic superalloys which compensate the strength and stiffness [9&10]. Hardening alluminium metals by adding other elements is common method, one of these methods is to add copper to the metal to greatly increase the hardness of the material currently the most widely used composition is addition of 6.7% of copper. Aluminium metal matrix composites which are reinforced with silicon carbide have much significance as an engineering material (AMC 1), having high specific strength and stiffness play a major role in situations where the material is expected to last for long durations without any replacement in which saving weight is an important feature [11].

Table 1 shows the physical properties of the selected materials. The conventional material GCI is chosen because of its high frictional coefficient and reasonable specific heat capacity. However, the other materials have similar properties.

Table 1. Physical properties of the selected materials [5].

Properties / Materials	1 Compressive Strength (Mpa)	2 Frictional Coefficient (µ)	3 Wear Rate ($x10^{-6}mm^3/N/m$)	4 Specific heat Capacity (kJ/kg.K)	5 Specific gravity (kg/m^3)
GCI	1293	0.41	2.36	0.46	7.2
Ti-6Al-4V	1070	0.34	246.6	0.58	4.42
TMC	1300	0.31	8.19	0.51	4.68
AMC 1	406	0.35	3.25	0.98	2.7
AMC 2	761	0.44	2.91	0.92	2.8

3 Modelling and Analysis

3.1 Modelling

The disk break of the car is designed in solidworks 2020. The disk break is shown in the figures 1&2.

Figure 1. Modelling of Disk Break

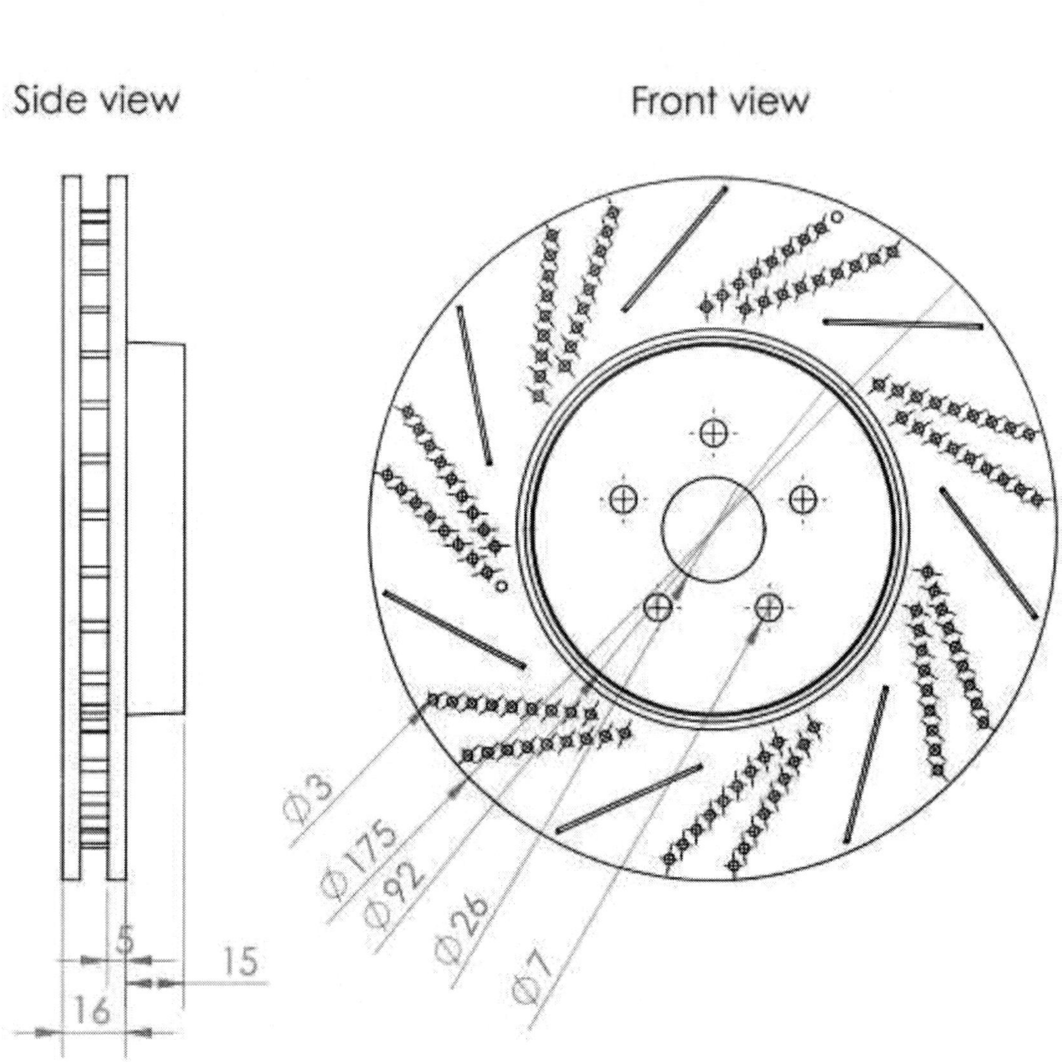

Figure 2. Drawing of disc

3. 2 Analysis

For the purpose of analyzing the models Ansys R2 is used, the solidworks file is imported into IGS format. Equivalent stress and the deformation is calculated for the disk break using different materials Gray cast iron (GCI), reinforced Ti-composite (TMC), Ti-alloy (Ti-6Al-4V), reinforced Al-Cu alloy (AMC 2), and reinforced Al-composite (AMC 1) to find the which material is best for disk break. Generally, the maximum temperature a disc brake reaches is from 300°C – 800°C depending upon the maximum speed of the vehicle since the analysis is not for supercars an average temperature of 500°C was used the convection rate was given as "Stagnant air simplified case" after applying these condition in the next step for static structural analysis a pressure of 2MPa was added on either sides (as shown in figure 3) to simulate the pressure applied by the calipers and the results were obtained as follows [12].

Figure 3. Application of pressure and fixtures

IOP Conf. Series: Materials Science and Engineering 1059 (2021) 012045 doi:10.1088/1757-899X/1059/1/012045

4 Result and Discussion

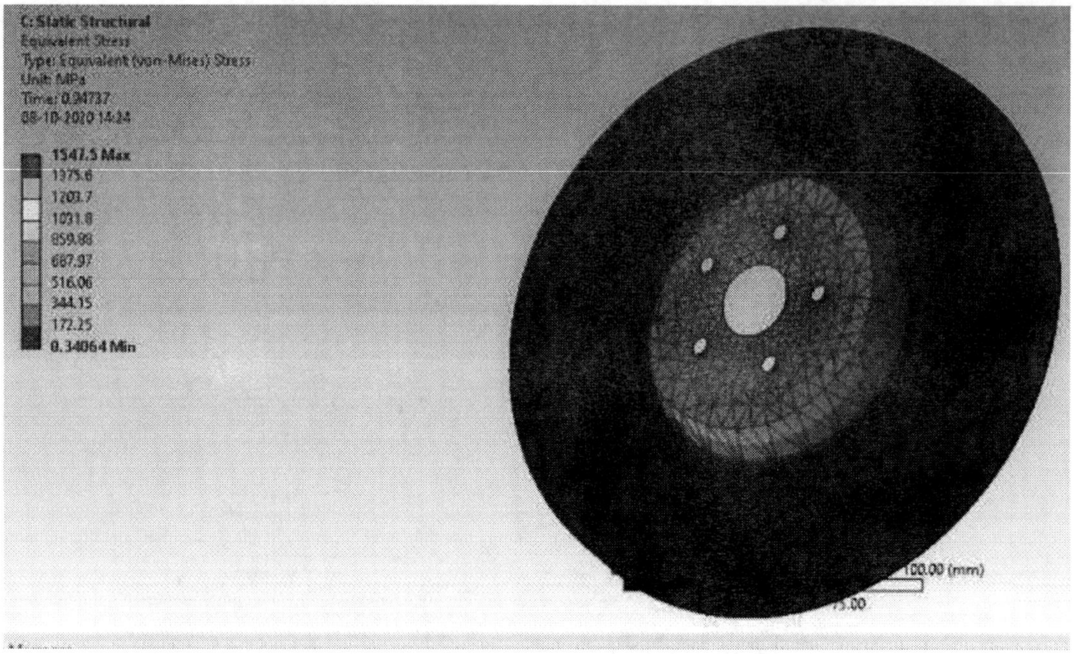

Figure 4. Equivalent stress (Grey Cast Iron)

Figure 4 shows the equivalent stress that acts upon the disc (Grey Cast Iron) after dissipation of heat and action of pressure from the calipers is found to be a minimum of 0.34 and a maximum of 1547 MPa. We can also observe that the stress levels are highest in the bolt holes due to the fixed geometry. Since Grey Cast Iron is the conventionally used material for these disc brakes these results will be taken as reference for all the other 4 materials.

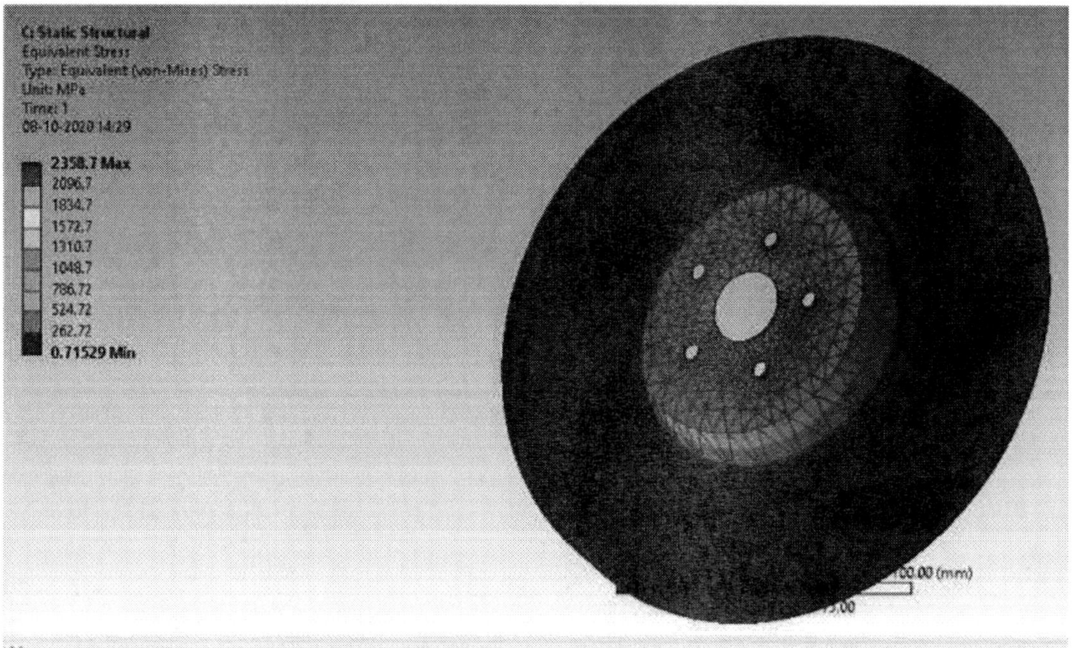

Figure 5. Equivalent stress (Al-Cu alloy)

Figure 5 shows the equivalent stress that acts upon the disc (Al-Cu alloy) after dissipation of heat and action of pressure from the calipers is found to be a minimum of 0.715 and a maximum of 2358 MPa. We can also observe that the stress levels are highest in the bolt holes due to the fixed geometry. The stress level of Al-cu alloy is higher when compared to Grey Cast Iron as well as all other materials taken for this study therefore this material is least good at handling stress.

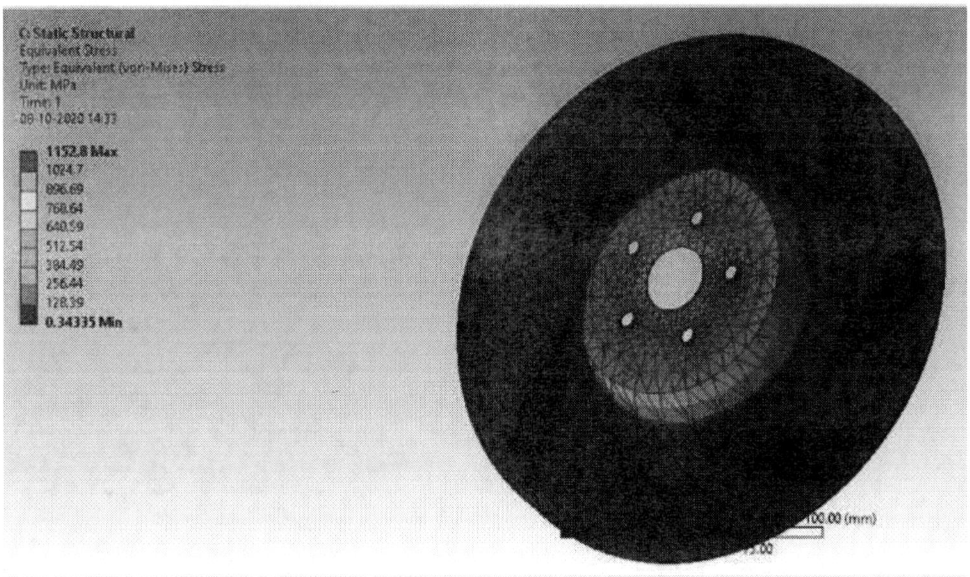

Figure 6. Equivalent stress (Ti alloy)

Figure 6 shows the equivalent stress that acts upon the disc (Ti alloy) after dissipation of heat and action of pressure from the calipers is found to be a minimum of 0.343 and a maximum of 1152 MPa. We can also observe that the stress levels are highest in the bolt holes due to the fixed geometry. Here the stress level is reasonable when compared to all other materials it has lower stress than Grey Cast Iron only the metal matrix materials have better results.

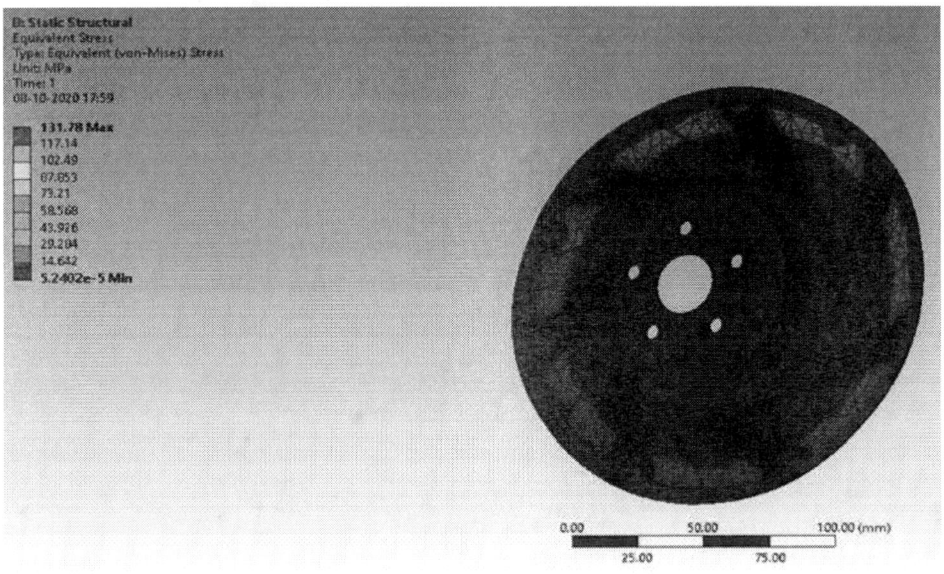

Figure 7. Equivalent stress (Ti-composite)

IOP Conf. Series: Materials Science and Engineering 1059 (2021) 012045 doi:10.1088/1757-899X/1059/1/012045

Figure 7 shows the equivalent stress that acts upon the disc (Ti-composite) after dissipation of heat and action of pressure from the calipers is found to be a minimum of $4.2e^{-5}$ and a maximum of 131 MPa. We can also observe that the stress levels are highest in the bolt holes due to the fixed geometry. The Titanium is the second best at handling stress. It can handle stress much better than the conventional material Grey Cast Iron.

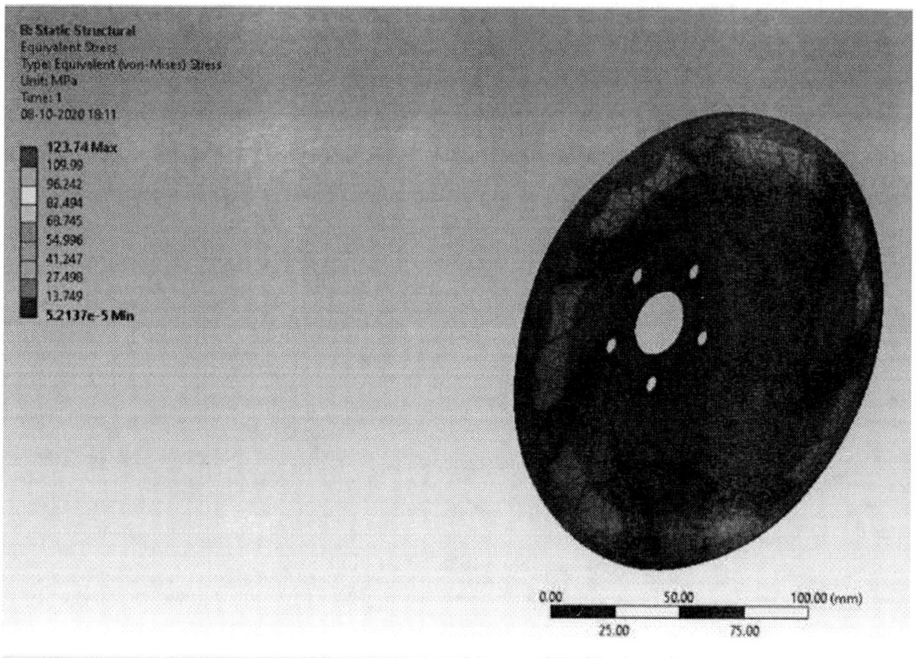

Figure 8. Equivalent stress (Al composite)

Figure 8 shows the equivalent stress that acts upon the disc (Al composite) after dissipation of heat and action of pressure from the calipers is found to be a minimum of $5.2e^{-5}$ and a maximum of 123 MPa. We can also observe that the stress levels are highest in the bolt holes due to the fixed geometry. The aluminium metal matrix material has the lowest stress level among the five chosen materials it significantly lower than the conventional material that is GCI. Therefore as far as thermal stress is concerned Al composite is the best material.

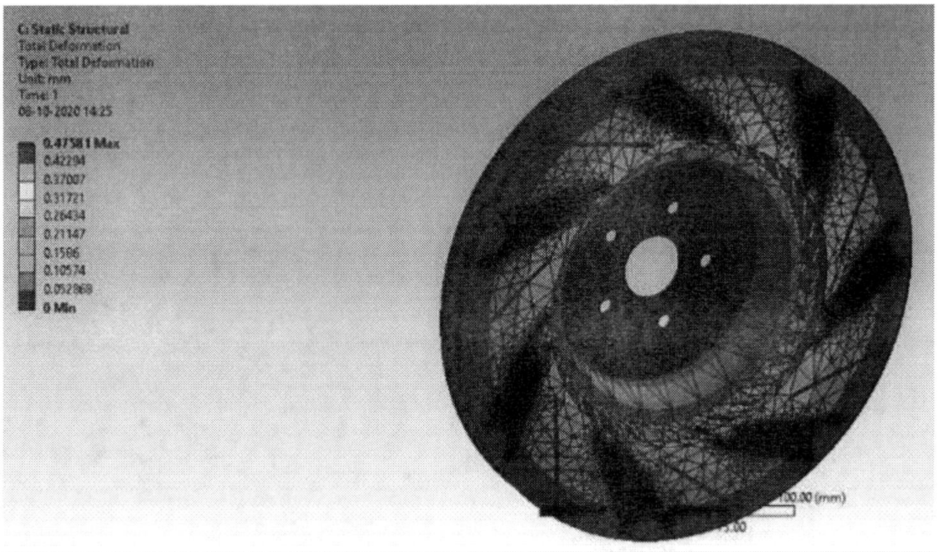

Figure 9. Deformation (Grey Cast Iron)

Figure 9 shows the deformation suffered by the disc when material Grey Cast Iron was added and subjected to condition mentioned above the values given are the total deformation in mm. the minimum deformations value is 0 and the maximum deformation is 0.47. The deformation occurs more at the outer edges of the disc due to the pressure from the calipers. The deformation of Grey Cast Iron for reference to compare with other chosen materials

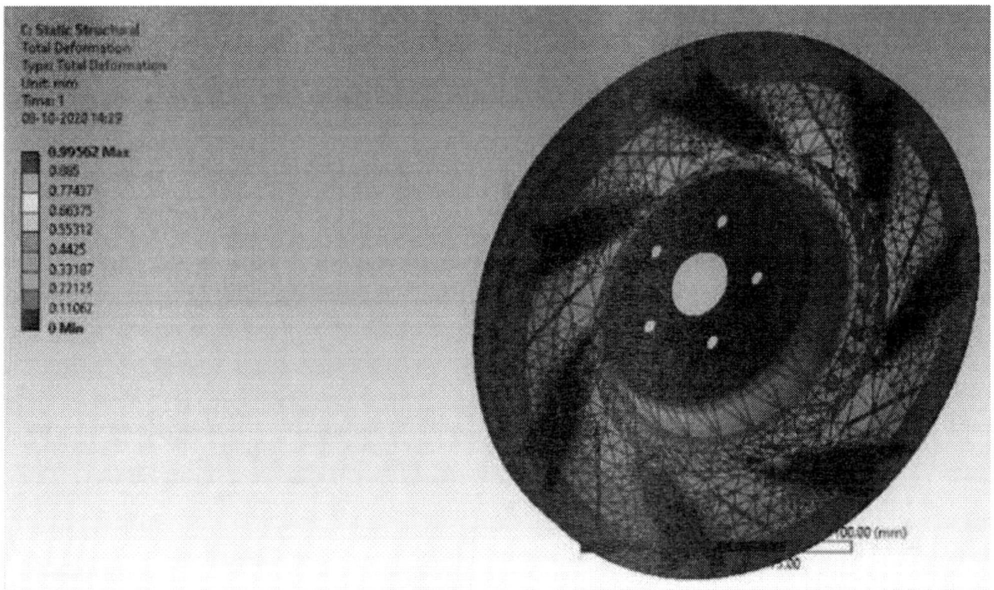

Figure 10. Deformation (Al-Cu alloy)

IOP Conf. Series: Materials Science and Engineering 1059 (2021) 012045 doi:10.1088/1757-899X/1059/1/012045

Figure 10 shows the deformation suffered by the disc when material Al-Cu alloy was added and subjected to condition mentioned above the values given are the total deformation in mm. the minimum deformations value is 0 and the maximum deformation is 0.995 The deformation occurs more at the outer edges of the disc due to the pressure from the calipers. The deformation of this material is much better than the compared with the matrix materials however when compared to the conventional GCI its deformation is greater.

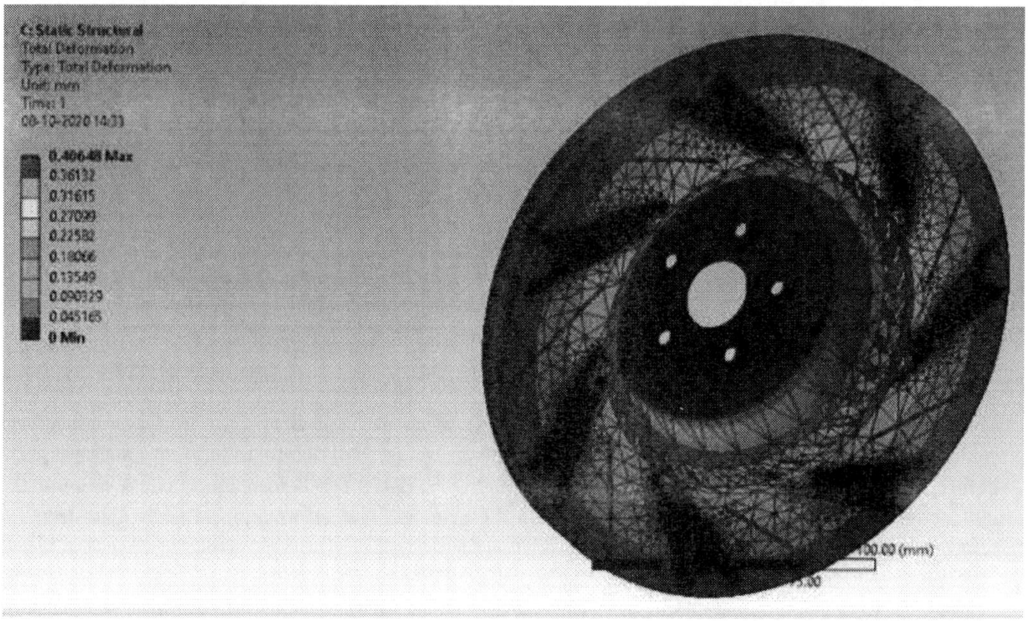

Figure 11. Deformation (Ti alloy)

Figure 11 shows the deformation suffered by the disc when material Ti alloy was added and subjected to condition mentioned above the values given are the total deformation in mm. the minimum deformations value is 0 and the maximum deformation is 0.406. The deformation occurs more at the outer edges of the disc due to the pressure from the calipers. The deformations occurring when this material is the least among the 5 chosen materials therefore this material could be a good alternative for GCI.

IOP Conf. Series: Materials Science and Engineering 1059 (2021) 012045 doi:10.1088/1757-899X/1059/1/012045

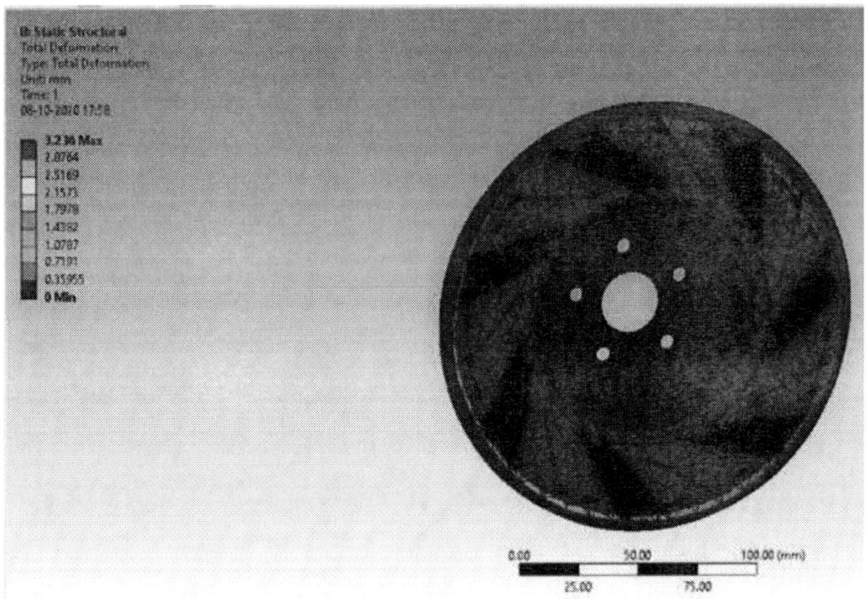

Figure 12. Deformation (Ti-Composite)

Figure 12 shows the deformation suffered by the disc when material Ti-composite was added and subjected to condition mentioned above the values given are the total deformation in mm. the minimum deformations value is 0 and the maximum deformation is 3.2. The deformation occurs more at the outer edges of the disc due to the pressure from the calipers. This material deforms much greater than GCI which would suggest that this material is not a proper replacement for Grey Cast Iron.

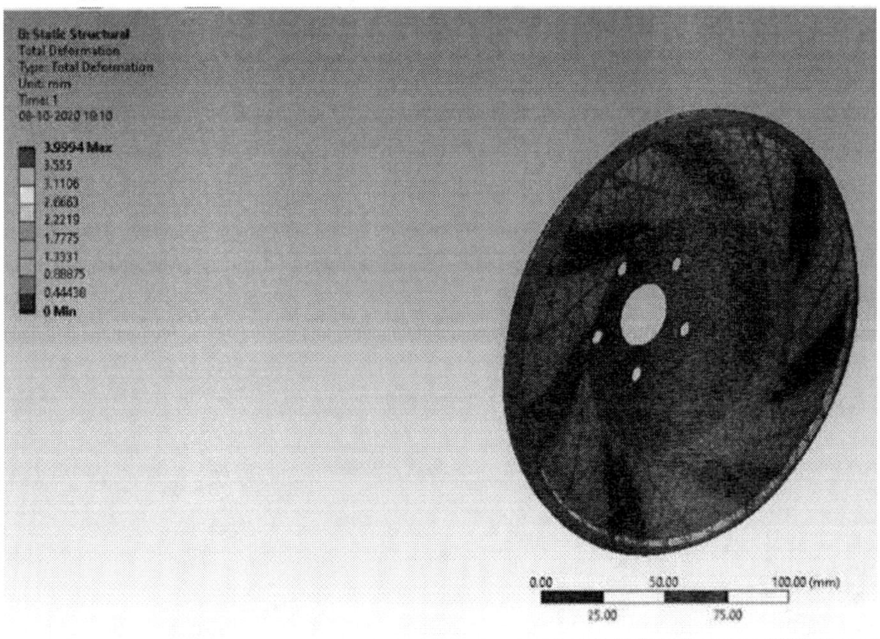

Figure 13. Deformation (Al composite)

Figure 13 shows the deformation suffered by the disc when material Al composite was added and subjected to condition mentioned above the values given are the total deformation in mm. the minimum deformations value is 0 and the maximum deformation is 3.9. The deformation occurs more at the outer edges of the disc due to the pressure from the calipers. Both the metal matrix composites have significantly higher deformation when compared to all the other materials.

After the analysis process was complete the results were put into 2 graphs for the purpose of easier comparison one of the graphs compares the maximum stress suffered by each material and the other compares the maximum deformation suffered by each of the chosen material

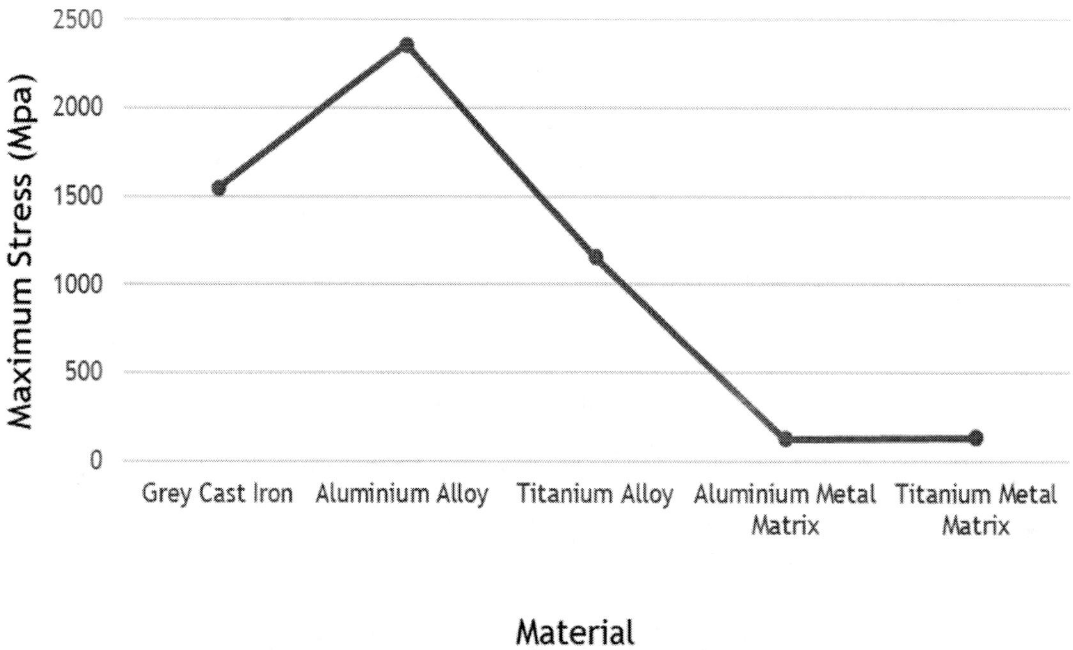

Figure 14. Comparison of Maximum stress

The Figure 14 shows the maximum stress suffered by each material when subject to functioning conditions. Here it is observed that aluminium composite and titanium composite and aluminium has the lowest stress titanium alloy is the next best thing and Grey Cast Iron also has some reasonable levels of stress.

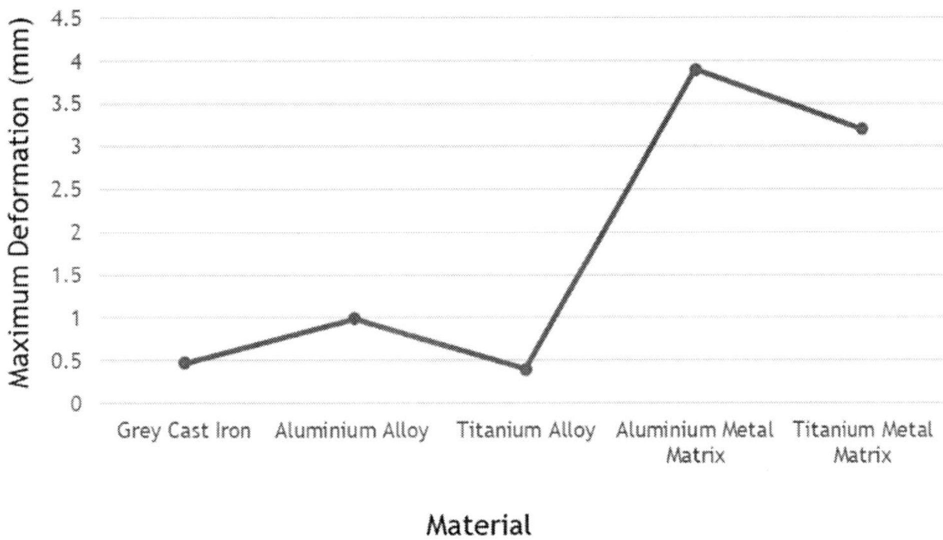

Figure 15. Comparison of Maximum Deformation

The Figure 15 shows the maximum deformation of the disc when subjected to functioning conditions it is observed that the materials which suffers the least amount of deformation are both Grey Cast Iron and Titanium alloy.

5 Conclusion

From the above graphs and figures we can observe that both Titanium alloy and GCI (Grey Cast Iron) show significantly better results than the other materials. Grey Cast Iron deforms only 0.5 mm at the given conditions and titanium alloy deforms nearly 0.4 mm which is less when compared to the other selected materials which have over 3mm of deformation. When it comes to stress levels it is observed that both the metal matrix materials have significantly lower stress level below 500 MPa whereas both Grey Cast Iron and Titanium alloy have stress levels ranging from 1000 to 1500 MPa, however when both stress and deformation is taken into consideration only GCI and Titanium alloy seem to be a viable option from this, we can conclude that both titanium alloy and Grey Cast Iron can be used effectively for disc brakes.

References

[1] Grieve D G Barton D C Crolla D A and Buckingham J T 1998 Design of a lightweight automotive brake disc using finite element and Taguchi techniques *Proceedings of the Institution of Mechanical Engineers, Part D: Journal of Automobile Engineering* **212**(4) 245-254.

[2] Ghadimi B Sajedi R and Kowsary F 2013 3D investigation of thermal stresses in a locomotive ventilated brake disc based on a conjugate thermo-fluid coupling boundary conditions *International communications in heat and mass transfer* **49** 104-109.

[3] Hwang P and Wu X 2010 Investigation of temperature and thermal stress in ventilated disc brake based on 3D thermo-mechanical coupling model *Journal of mechanical science and technology* **24**(1) 81-84.

[4] Jiguang C and Fei G 2015 Temperature field and thermal stress analyses of high-speed train brake disc under pad variations *The Open Mechanical Engineering Journal* **9**(1).

[5] Saiz C B Ingrassia T Nigrelli V and Ricotta V 2015 Thermal stress analysis of different full and ventilated disc brakes *Frattura Ed Integrita Strutturale* **9**(34).

[6] Akop M Z Kien R Mansor M R and Rosli M M 2009 Thermal stress analysis of heavy truck *Journal of Mechanical Engineering and Technology (JMET)* **1**(1) 43-51.

[7] Cueva G Sinatora A Guesser W L and Tschiptschin A P 2003 Wear resistance of cast irons used in brake disc rotors *Wear* **255**(7-12) 1256-1260.
[8] Blau P J Jolly B C Qu J Peter W H and Blue C A 2007 Tribological investigation of titanium-based materials for brakes, *Wear* **263** 7-12 1202-1211.
[9] Muhammad D Hayat Harshpreet Singh Zhen He and Peng Caob 2019 Titanium metal matrix composites: An overview.
[10]Ahmad F Lo S J Aslam M and Haziq A 2013 Tribology behaviour of alumina particles reinforced aluminium matrix composites and brake disc materials *Procedia engineering* **68** 674-680.
[11]Maleque M A Dyuti S and Rahman M M 2010 Material selection method in design of automotive brake disc *In Proceedings of the world congress on engineering* **3**.
[12]Hwang P and Wu X 2010 Investigation of temperature and thermal stress in ventilated disc brake based on 3D thermo-mechanical coupling model *Journal of mechanical science and technology* **24**(1) 81-84.

Effect of Titanium Di-Boride/Copper (TiB_2-Cu) Electrode on the Machinability of OHNS Steel by EDM Process

Jayahariharapranav P [1], Dinesh D[2*], MathanKumar P*[3], Rajamurugan.G[4], Jayakumar N[5]

[1] UG Student, Department of Mechanical Engineering, Bannari Amman Institute of technology, Sathyamangalam, Erode-638401, TamilNadu, India

2,3,5 – Assistant Professor ,Department of Mechanical Engineering, Bannari Amman Institute of technology, Sathyamangalam,Erode-638401,TamilNadu,India

4- Associate Professor, Vellore Institute of Technology, Vellore -632014, Tamilnadu, India.

Corresponding author: dhinesh1583@gmail.com.

Abstract: The technological challenges faced by any industry in machining OHNS steel in the use of tools and die by conventional processing techniques point to tool wear rate and poor surface finish. To overcome these, Electrical Discharge Machine (EDM) process, an inevitable Unconventional Machining process is used in the precision and micro component manufacturing industries. Here an attempt has been made to machine OHNS material with a lately developed copper titanium diboride (Cu-TiB_2) of 16% wt of TiB2 is prepared through powder metallurgy route which enhances the shortcomings obtained in conventional techniques. The experiment models are made to optimize through statistical Response Surface Methodology (RSM). The experiments were conducted by considering the effect of pulsating direct current, pulse time ON, and pressure flushing. Then, ANOVA is executed to predict the impact of each parameter and its significance concerning the output responses of roughness (SR) and Volume of Material removed (MRR). The equation developed for each output response and then the optimum process parameters obtained in the desirability based multi-objective optimization using RSM. The minimum surface roughness and maximum material removal rates are attained at 14 A, 16(μs) and 0.6 MPa. Then the confirmation experiments were carried to validate the model and it is witnessed that the error percentage attained was within the satisfactory range.

Keywords: Electrical Discharge Machine (EDM), Roughness (SR), Volume of material removed (MRR), Response Surface Methods (RSM).

IOP Conf. Series: Materials Science and Engineering 1059 (2021) 012046 doi:10.1088/1757-899X/1059/1/012046

1. Introduction:

The Spark Machining was a thermo-electric material removal process thoroughly used applied unconventional machining processes to machine hard materials with complex geometry. This method involves a controlled erosion of materials that are electrically conductive exercising initial repeated electric sparks produced between the conductive work and tool electrode that are separated by a dielectric medium.

The Powdered Metallurgy (PM) process facilitated the development of the electrode tool which was acquired by the combination of cobalt (Co) and Chromium (Cr) powders which were ultimately used in the Electric Discharge Machining process (EDM).The image generated by Energy Dispersion (EDaX) had analyzed spectra which had a surface containing elements of O, Al, Cr, Fe, Co and C at its zenith. The alloying of mild steel with the electrode was established through this data[1]. Modification of the high carbon steel surface was progressed using an electrode consisting of a combination of Copper-Chromium-Nickel by the method of powdered metallurgy through EDM. This particular electrode tool produced, that was utilized in the EDM process had produced better micro hardness for the machined surfaces of the workpiece[2] .The deposition of the hard layers throughout the Electric Discharge Machining process begins accompanied by the erosion of the compact electrode tool which was produced by the process of powdered metallurgy resulting in the generation of borides and hard carbides. The zenith of Titanium, carbon, and boron make the presence of tool material affirmative[7].

Copper is one of the most widely and efficiently used earth elements attributing to its various material properties such as electrical conductivity, ductility, malleability, elasticity, and strength. Copper is also said to have very good potential for oxidation. This being included in the production of the Cu-SiC composite electrode tool by powdered metallurgy which in turn is used on a workpiece through EDM to obtain a copper matrix [8]. Selective Laser Melting (SLM), a conventional tool electrode manufacturing process was incorporated to deliver a tungsten carbide tool electrode which was depended on the volume energy density to account for the multiple varieties of microstructures. Three phases of factors namely build direction processing stage and parameter set were driven into consideration for the tool composition. As the presence of the cobalt element is minimized in the electrode, the further procedure to be undergone by the electrode for cobalt decrease is also neglected[3,6]. Electric discharge machining of oil hardening on shrinking tool steel where the fabrication of metal matrix composite electrode by combining Al_2O_3 with copper powder was operated where the Voltage and current are the established parameters in which the wear rate of the tool, removal rate of material, percentile wear rate, the roughness of surface were established as output parameters. Various examinations that were accounted for Cu-TaC(Copper tantalum) electrodes generated by the process of powdered metallurgy were conducted with two different compositions (Cu with 30% and 50% wt Tac). Using the Taguchi method to correlate the usefulness of the electrodes that were produced through the process of powdered metallurgy that is being compared to the conventional copper electrode[4,5].

Another challenging parameter of the spark machining is the Volume of Material Removed (MRR). The electrode formed by the powdered metallurgy will have a different composition of the copper with chromium and molybdenum to replicate a fair volume of material removed. Both the electrode tool upon being fused onto the workpiece has ascertained material deposits onto the workpiece[9].

To further improvise the surface hardness and the abrasive wear resistance of the workpiece, an electrode tool using the combination of copper powder and boron carbide powder is manufactured by the process of powder metallurgy. Only through this process, the weight percentage of the boron carbide powder (B_4C) is about 20% in composition with the copper powder electrode. The workpiece being EDMed using this particular electrode and analyzed produced hardness 4 times better than the other workpiece, and colossal reduction up to 65% - 78% in the abrasion wear [10]. In this paper the focus on the Copper- Titanium diboride of 16% wt electrode–prepared through Powder Metallurgy route effects on the OHNS in EDM machine was studied.

2. Experimental procedure:

The material used for this investigation is an OHNS material of 25×25×5mm with a chemical composition listed in Table1.

Table 1. Chemical composition of High OHNS Steel (wt %)

Element	C	Mn	Cr	V	Ni	Mo	Iron
Composition	0.95	1.15	0.49	0.19	0.05	0.13	Balance

It is widely used in tool and dies material, milling cutters, and gauging tools. The Copper-Titanium diboride electrode is prepared in the different composition of 4, 8, 12, 16, 18 % wt is prepared using powder metallurgy route by mixing copper powder (44 μm) and titanium diboride (44 μm) in a V-Blender for a time period of 10 hours then the blended powder of 2.5g is transferred to the die contains 12mm in diameter and pressed using a hydraulic pressing machine with a force of about 14000kg. The prepared powder metallurgy tool is then sintered using a tubular furnace at a temperature of 1100°C in a closed environment, in continuous of that brazing is performed to mount to the Sparking Machine. It is found that the 16%wt produces an optimum wear rate of tool which is significant in roughness [11]. Present investigation, focus on tool electrode of TiB_2 –Cu is used as shown in Figure.1

Figure 1. Copper –Titanium diboride electrode

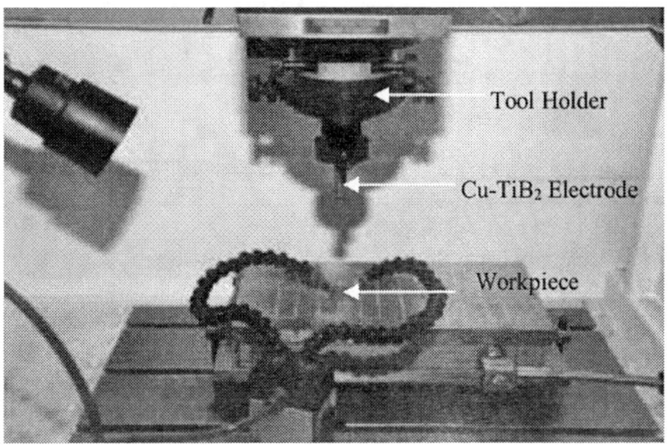

Figure 2. Set up for EDM Experiments.

Table 2. EDM operating condition

Working condition	**Description.**
Electrode material	Copper –titanium diboride
Spark Gap	0.1mm
Dielectric fluid	EDM oil
Specimen material	OHNS Steel Material
Voltage	50 Volts
Pulse Off time	20 μs

Twenty experiments are carried out using an Electra 5535 EDM machine set up as shown in Figure. 2 and its operating condition are shown in Table 2. The input parameters like Pulse Direct current (Ip) Pulse time ON (T_n), and Pressure flushing (Fp) at three levels are considered as shown in Table 3. The specimens are machined using EDM in which EDM oil is used as a dielectric medium. The roughness (SR) is measured using Roughness tester SJ-210 Mitutuyo as shown in Figure.3 three readings of each specimen were noted on various points on the surface undergone spark machining. Also, Volume of Material removed (MRR) is intended by the mass loss of the workpiece by a digital weighing scale with 0.001g accuracy and the time is taken by each trial to a depth of 1mm.

$$\text{MRR} = \frac{(Wi - Wf)}{t} \text{ g/min} \quad (1)$$

Where Wi –weight of the Work in g before Machining ,Wf- weight of the work in g after machining , t- time taken to complete machining in minutes.

Table 3. Level of EDM Machining Parameters

Factors	**Variables**	**Units**	**Levels**				
			-1.68	**-1**	**0**	**1**	**1.68**
A	Pulse Direct Current (I)	Amperes	2	5	10	15	18
B	Pulse Time ON- T_n	Microsec onds	10	14	20	25	28
C	Pressure Flushing (F)	Mpa	0.3	0.6	1	1.4	1.7

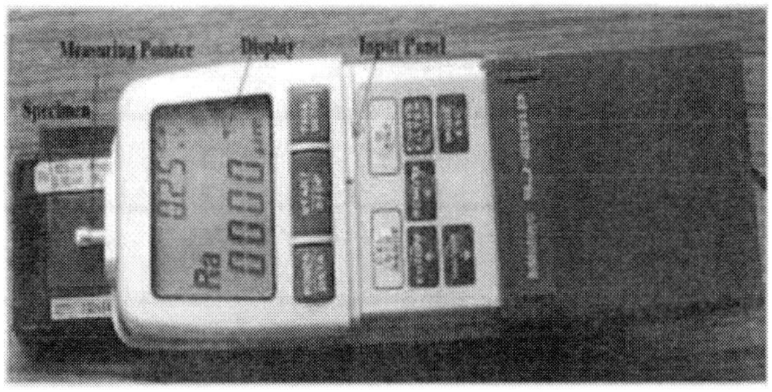

Figure 3. Surface roughness tester

3. Experimental design:

Statistical response surface methods using rotatable centred central composite approach is utilized in design of experiments. The location of the optimum is unknown pair to running the experiment, it make sense to use a design that provides equal precision of estimation in all directions. A central composite is made rotatable by the choice of α.

3.1 Developing the design Matrix:

Pulse Direct current (I) Pulse time ON (T_n), and flushing Pressure (F) are considered as the influencing inputs in the machining of OHNS Steel. The Design of experiments is a noteworthy statistical tool that describes the series of steps to perform experiments Sequentially. It is designed based on three input parameters at three levels which contains ten star points (+1,-1), Axial points and centre Points of six each with 20 trials with a value of α =1.68 are carried out based on the run order and output responses obtained is shown in Table 4.

3.2. Development of mathematical models:

Using the experimental result which is obtained from the central composite rotatable DoE and applying regression analysis, the modelling of the selected response with a small quantity of independent input process parameter can be gained. The response surface exists expressed as follows.

$$y = a_0 + \sum_{i=1}^{n} a_i x_i + \sum_{i=1}^{n} a_{ii} x_i^2 + \sum_{i<j}^{n} a_{ij} x_i x_j + \varepsilon \quad (2)$$

In the above equation,

y – Surface Roughness (response parameter)

x_i and x_j - input variables

a_i, a_{ii}, and a_{ij} – Coefficient of the linear terms, quadratic terms and interaction effects respectively.

Table 4. Design of experiments and its results

Run	Pulse Direct Current (I)	Pulse time ON (T_n)	Pressure Flushing (F_p)	SR (μm)	MRR (g/Min)
1	10	28	1	8.08	0.108
2	2	20	1	4.758	0.021
3	10	10	1	3.78	0.035
4	10	20	1	7.192	0.134
5	10	20	1	8.177	0.162
6	15	25	0.6	8.237	0.216
7	15	25	1.4	8.57	0.22
8	10	20	1	7.5	0.127
9	5	14	0.6	5.251	0.072
10	5	25	1.4	6.596	0.036
11	15	14	0.6	6.026	0.134
12	10	20	1	7.166	0.126
13	15	14	1.4	10.019	0.225
14	18	20	1	10.21	0.225
15	10	20	1	6.457	0.124
16	10	20	1	7.239	0.106
17	10	20	1.6	6.208	0.149
18	5	14	1.4	5.129	0.031
19	5	25	0.6	7.269	0.039
20	10	2	0.3	6.208	0.149

4. Statistical Analysis:

Statistical analysis is performed using ANOVA to know about the influencing parameters of selected process and their output responses. Table 5 and 6 shows the ANOVA result of roughness parameter and volume of material removed with 95% confidence level.

Table 5. Analysis of Variance for Surface Roughness.

Source	Sum of Squares	df	Mean Square	F-value	p-value	Status
Model	43.47	9	4.83	6.39	0.0039	significant
A-Pulse Current (I_p)	23.14	1	23.14	30.62	0.0002	
B-Pulse On(T_{on})	9.65	1	9.65	12.77	0.0051	
C-Flushing pressure(F_p)	0.9129	1	0.9129	1.21	0.2975	
AB	0.9268	1	0.9268	1.23	0.294	
AC	3.28	1	3.28	4.34	0.0639	
BC	2.22	1	2.22	2.93	0.1176	
A^2	0.603	1	0.603	0.7979	0.3927	
B^2	1.71	1	1.71	2.27	0.163	
C^2	0.8762	1	0.8762	1.16	0.3069	
Residual	7.56	10	0.7558			
Lack of Fit	6.01	5	1.2	3.87	0.082	not significant
Pure Error	1.55	5	0.3105			
Cor Total	51.02	19				

4.1. Developing the final model:

The Mathematical regression model was developed for the output responses of the material removal rate and was evaluated by F-test . It was found that the model F-value for the surface roughness is 6.39 and the lack of fit is 3.87 .Also ,the model f-value for Material removal rate is 15.79 and the lack of fit is 2.38 . This result implies that the mathematical model was statistically evaluated and it's proven the significance of the developed model. ANOVA was done to check the adequacy of the model. Insignificant factors are eliminated by a backward elimination method to improve the significance of the model. The developed model for SR and MRR is shown in Equation (3) and (4)

$$SR = -7.89824 + 0.01972 *A + 0.960470 *B + 5.19353 *C - 0.012377(A*B) + 0.320062(A*C) - 0.239261 (B*C) + 0.008182 A^2 - 0.011400B^2 - 1.54109 *C^2 \quad (3)$$

$$MRR = -0.127073 - 0.003361 *A + 0.024696 *B - 0.129203 C + 0.000477 A*B + 0.008687A*C - 0.002784 A*C - 0.000029 A^2 - 0.000626 B^2 + 0.052977 C^2 \quad (4)$$

Where A –Pulse Direct Current B-Pulse time ON, C- Pressure Flushing.

Table 6. Analysis of Variance for Volume of Material Removed

Source	Sum of Squares	df	Mean Square	F-value	p-value	Status
Model	0.0807	9	0.009	15.79	< 0.0001	significant
A-Pulse Current (I_p)	0.0675	1	0.0675	118.92	< 0.0001	
B-Pulse On(T_{on})	0.0022	1	0.0022	3.81	0.0796	
C-Flushing pressure(F_p)	0.0002	1	0.0002	0.3356	0.5752	
AB	0.0014	1	0.0014	2.43	0.1502	
AC	0.0024	1	0.0024	4.26	0.0661	
BC	0.0003	1	0.0003	0.5288	0.4838	
A^2	7.39E-06	1	7.39E-06	0.013	0.9114	
B^2	0.0052	1	0.0052	9.09	0.013	
C^2	0.001	1	0.001	1.82	0.2066	
Residual	0.0057	10	0.0006			
Lack of Fit	0.004	5	0.0008	2.38	0.1811	not significant
Pure Error	0.0017	5	0.0003			
Cor Total	0.0863	19				

4.2. Consequence of input Variables on the surface roughness:

Roughness is an important influencing parameter on machining of OHNS Steel. Figure 4, 5 and 6 depicts the relationship between the roughness with three input parameters and its interactions represented in a 3D plot surface. It is observed from the Figure 4. Increase in Pulse direct current 5 A to 15 A increases in roughness occurs due to generation of large discharge energy between the tool and workpiece forms a crater wear surface on the workpiece [17] and Pulse time ON increases tends to increases the roughness. At a minimum pulse direct current (5A) and pulse time ON (14μs) produces a minimum roughness. Figure 5 represents that the increases in roughness causes the pulse direct current increasing but it has attained maximum surface roughness value at minimum Pulse direct current of 5A and 0.6 Mpa pressure flushing when compared to Figure 4.

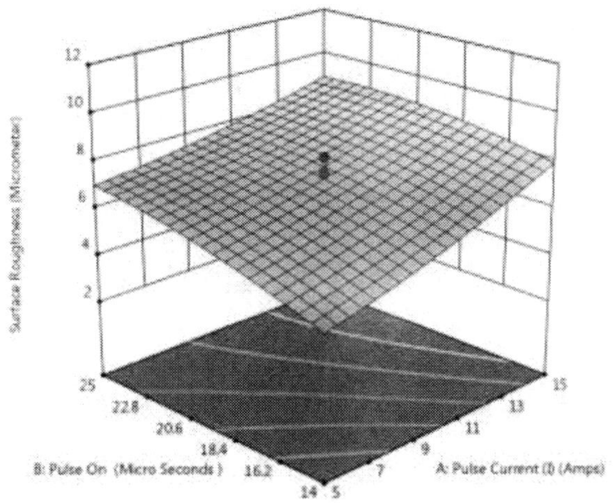

Figure 4. Effect on Roughness With Pulse Direct current and Pulse Time ON

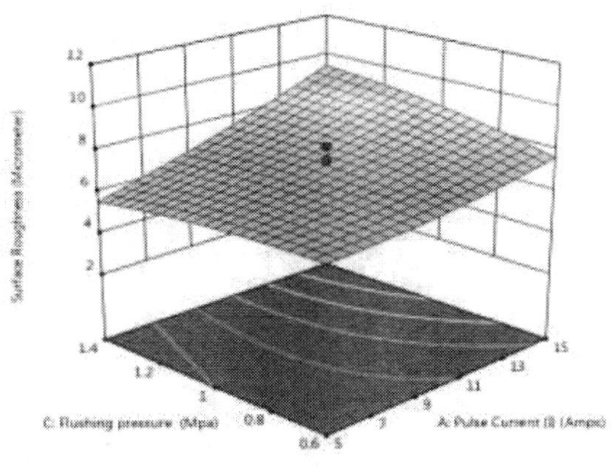

Figure 5. Effect of Roughness with Pulse Current and Pressure Flushing

Due to the increase the number of sparks with respect to time [12-13]. The pressure flushing has negligibly lesser effect on roughness. Figure.6 demonstrates the increase of pulse current increase the surface roughness that forms a stable arc in the spark gap [11,14] and increase of pressure flushing slightly increases the roughness.

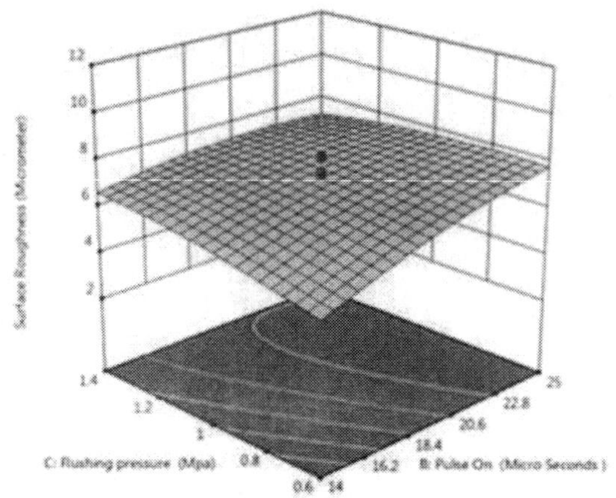

Figure 6. Effect of Roughness with Pulse time ON and Pressure Flushing

4.3 Consequence of input variables on the Volume of Material removed:

The Volume of material removed is a parameters that is to be considerably important in EDM to the three input parameters and their interaction is illustrated in 3D Plot surface as shown in Figure 7,8 and 9. Figure.7 indicates upon further increase in pulse direct current and pulse time ON the Volume of Material removed is increases it implies that higher discharge energy .When the pulse current (5A) and pulse on time (14μs) produce a maximum Volume of material removed. Figure 8. Implies that increase of pulse direct current increases the volume of material removed is slightly higher than that as discussed in Figure.7.

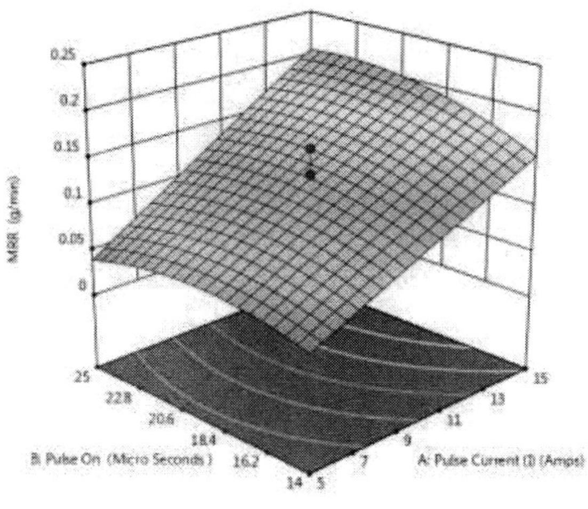

Figure .7 Effect of MRR with Pulse Direct current and Pulse time ON

The pressure flushing increases Volume of material removed increases slightly Figure.9 represents that the volume of Material removed increases on increasing the pulse time ON. From both the cases it is observed that pressure flushing have a less significant factor either it increases on MRR or decreases on SR. It has been confirmed in the ANOVA result of F-ratio of 3.81 has a lesser value compared to other two input parameters.

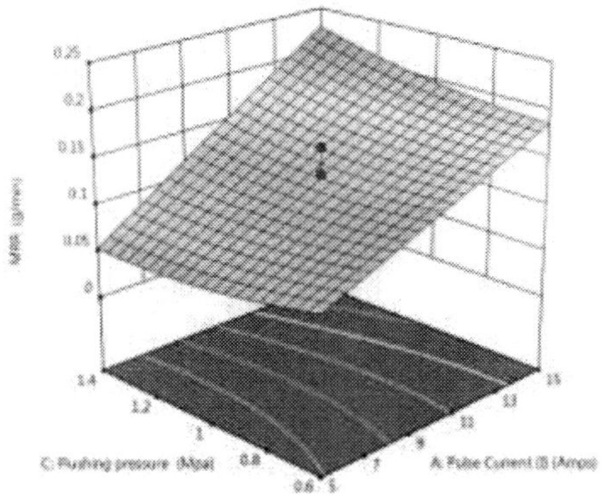

Figure. 8. MRR Vs Pulse current and pressure flushing.

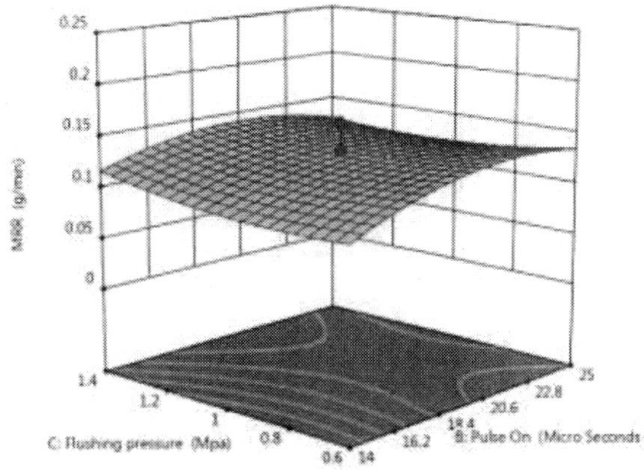

Figure. 9. MRR with Pulse time ON and pressure flushing

5. Confirmation experiments:

In order to verify the developed surface roughness and MRR regression model has been valued with the aid of Design experts 12.0. The desirability value falls between zero and one .The solution with higher desirability is considered as the optimum parameter is considered as shown in table 7. The experiments were conducted again with three trials for the conformity of experiments the same is feed into the developed equation to find the predicted value .Then experimental values are compared with the predicted value as shown in Table 8.The errors values calculated are in satisfactory level of less than 10% .

Table .7 Optimum process parameters

Parameters	**Optimum Value**
Pulse Current	14 A
pulse on time	15 **μs**
Flushing Pressure	0.6 Mpa

Table 8. Predicted and observed value of OHNS Steel

Output Response	**Objective**	**Calculated Values**	**Obtained values**	**Error %**
SR (μm)	Minimum	6.52968	6.828	-4.3690
MRR (g/min)	Maximum	0.147959	0.139	6.4453

6. Conclusion:

An experimental study is performed in EDM and studied the influencing input parameters in relation to $Cu-TiB_2$ electrode on OHNS Steel workpiece the following conclusion is observed:

i. The Pulse direct current is influenced more on roughness (SR) and Volume of material removed (MRR).
ii. Pulse time ON increases has major influence on the other two input parameters.
iii. Pressure flushing increases or decreases with pulse Direct current and pulse time ON interaction with output responses of MRR and SR has considerably less effect on both.
iv. Pulse Direct Current is the major influencing factor that is predicted using ANOVA with a F-value of 30.62 and 118.92 in Surface roughness and MRR respectively.
v. The optimum parameters obtained using face centred central composite design using RSM with high desirability value is Pulse direct current of 14 A, Pulse time ON of 15μs and Flushing pressure of 0.6 Mpa .

References

[1] Ahsan Ali Khan, Mohammed Baba Ndaliman and Mohammad Yeakub ali et.al. 2012 Surface Modification by EDM using Co-Cr Sintered Powder Metallurgy Electrode, *Advance Material Research,* **576** 56-59.

[2] Amoljit singh Gill, Sanjeev Kumar 2015 Surface Alloying by Powder Metallurgy Tool Electrode Using EDM Process *Materials today proceedings* **2** 1723-1730.

[3] Uhlmann, Eckart, Bergmann, André, Bolz, Robert and Gridin Witalij 2018 Application of additive manufactured tungsten carbide tool electrodes in EDM *Procedia CIRP* **68** 86-90.

[4] Naveen beri, Sachin Maheswari, Chitra Sharma and Anil kumar 2010 Technological advancements in electrical discharge Machining with Powder metallurgy processed electrodes A review *materials and manufacturing process* **25** 1186-1197.

[5] Mathalai Sundaram C, SivasubramanianR and Chandrasekaran K 2016 Experimental investigations of new powder metallurgy electrode in electrical discharge machining of OHNS using taguchi method *International journal of advanced engineering technology* **7** 559-564

[6] Naveen beri. Maheswari S, Sharma C, and Anil Kumar 2008 Performance Evalation of powder metallurgy electrode in electrical discharge machining of AISI D2 steel using taguchi method *International journal of mechanical,aerospace ,industrial ,Mechatronics and manufacturing engineering* **2** 225-229.
[7] Afzaal ahmed 2015 Deposition and analysis of composite coating on aluminium using Ti-B_4C powder metallurgy tools in EDM *Materials and manufacturing processes* **31** 467-474
[8] Nalin somani,Naman Sharma,Yash Kumar Gautam, Paras Khatri ,et.al 2018 Fabrication of Cu-sic composites using powder metallurgytechnique *Materials today proceeding* **5** 28136-28141.
[9] Azar Equabal, Md.Israr Equabal and Anoop Kumar Sood 2019 An investigation on the feasibility of fused deposition modelling process in EDM electrode manufacturing *CIRP Journal of Manufacturing science and technology.* **26** 10-25.
[10] Can cogun, ziya esen, asim genc, Ferah cogun et.al 2015 Effect of powder metallurgy Cu-B_4C electrodes on workpiece surface characteristics and machining performance of electrical discharge machining" *Journal of engineering Manufacture* **230** 2190-2203.
[11] Kumar Mathan P Sivakumar K and Jayakumar N 2018 Multiobjective optimization and analysis of copper–titanium diboride electrode in EDM of monel 400TM alloy Materials Manufacturing Process **33** 1429-1437.
[12] El-Taweel T A 2009 Multi-Response Optimization of EDM with Al–Cu–Si–TiC P/M Composite Electrode *Journal of Advance Manufacturing Technology* **44** 100-113.
[13] Singh G A and Kumar S 2016 Surface Roughness and Microhardness Evaluation for EDM with Cu-Mn Powder Metallurgy Tool *Materials Manufacturing Processes* **31** 514-521.
[14] Patowari P K Mishra U K Saha P and Mishra P K 2011 Surface Integrity of C-40 Steel Processed with WC-Cu Powder Metallurgy Green Compact Tools in EDM Materials and Manufacturing Processes 26 668-676.
[15] Patel K M, Pandey P M and Rao P V 2009 Determination of an Optimum Parametric Combination Using a Surface Roughness Prediction Model for EDM of Al2O3/SiCw/TiC Ceramic Composite *Materials Manufacturing Processes* **24** 675-682.
[16] Das A Prakash J M 2012 Experimental Investigation on Surface Modification of Aluminum by Electric Discharge Coating Process Using Tic/Cu Green Compact Tool-Electrode *Machining Science and Technology* **16** 601-623.
[17] RahangM and Patowari P K 2016 Parametric Optimization for Selective Surface Modification in EDM Using Taguchi Analysis*Materials and Manufacturing Processes* **31** 422-431.

Analysis of Shrinkage Defect in Sand Casting by Using Six Sigma Method with Taguchi Technique

M A Omprakas[1]*, M Muthukumar[1], S P Saran[1], D Ranjithkumar[1], C M Shantha kumar[1], S Thiruppathi Venkatesh[1] and M Sengottuvelan[2]

[1] Department of Mechanical Engineering, Nandha Engineering College, Erode – 638052, Tamilnadu, India.
[2] Department of Mechatronics Engineering, Bannari Amman Institute of Technology, Sathyamangalam - 638401, Tamilnadu, India

omprakasangappan@gmail.com

Abstract. Casting industries play a major role in the field of manufacturing. The production of complex shape and size products is manufactured in a single process, which cannot be produced in other manufacturing processes. Because the other process needs more than one step to convert a raw material into a product. When producing the casting, the quality of the casting should be maintained without defects. This is not possible as we can't produce a cent percentage of accuracy. But the percentage of defects can be reduced with the help of certain quality control tools and techniques. In this paper, the main focus is to reduce the shrinkage defect which occurs in the External Bearing Ring of ductile cast iron which is produced in the leading casting industry in Coimbatore. The data have been collected from the industry for the six months and the defects have been identified with the help of the Six-Sigma DMAIC (Define, Measure, Analyze, Improve, Control) technique. The quality control tools are applied in different stages of the DMAIC technique for identifying and controlling the defects. Also, the Taguchi technique is applied for creating the L9 orthogonal array from the Minitab software. Finally, the best possible solution is obtained and it is suggested to the industry for defects reduction.

Keywords: Casting defects; Six-sigma; DMAIC process; Taguchi; ductile iron.

1 Introduction

Casting is one of the oldest techniques in the days around 4000 B.C. The casting process was used for manufacturing of gold ornaments. Few years later, it was used for the production of weapons and tools with metals like copper as materials. Then after, the casting has been used for the production of products which were having different shapes from small to complex shape and size of products and different materials like cast iron and ductile iron, etc. Due to its major benefits and needs, the casting production plays a major role in the field of manufacturing. The occurrence of defects in the casting affected the casting industry economically. So, the defect occurrences should be reduced and the quality of the casting should be improved for which the application of some techniques like six-sigma and quality tools can be used. The defect in the particular area is identified and some changes are made to reduce the defects. In this paper, the External Bearing Ring made of ductile cast iron which is the one of the main components in the windmill is considered for analysis. The leading casting industry in Coimbatore are facing this shrinkage defect on this particular product as high in numbers. So, this shrinkage defects are reduced by applying the six-sigma technique and quality control tools, with the help of Minitab software.

The Six-Sigma DMAIC technique is used, for the reduction of defect. In a metal casting industry, the casting production is made by using green sand-casting process. Here the misrun defect is reduced from 31% to 4% [1]. The Six-Sigma DMAIC technique is used in the small-scale industry as a case study. To reduce the rejection and rework of leaf spring manufacturing process. As a result, the overall rejection rate is reduced from 48.33% to 0.79% [2]. The Six-Sigma DMAIC technique is used to improve the efficiency and profitability of all the operations as a business strategy. The Blow holes, Misrun, Slag inclusion, Rough surface defects are reduced in the sand casting process. The percentage of rejection is reduced from 6.98 % to 3.10 % and made a profit of Rs.2.35 lakhs appx [3]. The performance level and efficiency of the casting process was improved by adopting the Six-Sigma technique. DMAIC approach is applied here. The process parameters of the green sand casting are optimized here, which contributes to diminish the casting defect, this is the conclusion of this case study. The optimized parameter levels for melt shop process are ferromanganese (3.5 kg), ferrosilicon (5 kg), ferrochrome (1.2 kg), pig iron (195 kg), steel (48 kg) and cast iron (350 kg) [4]. It is aimed to review the research work made by several researchers and an effort is made to give the technical solution for minimizing various casting defects and to improve the entire process of casting manufacturing [5].

The steel casting manufacturing process is affected by shrinkage defect and gas defect, it is reduced by using the Taguchi's DoE (Design of Experiments). Totally nine experiments are made and four parameters are chosen. The four parameters are sand-binder ratio, mould moisture, pouring temperature, and amount of de-oxidant material added to the melt. In this the experiment six is selected as a best parameter [6]. The Six-Sigma technique is applied to the flywheel manufacturing casting industry, to reduce the defect percentage. The process map, cause-and-effect matrix and the failure mode effective analysis are used as a primary tool. The working parameters, namely moisture content, green strength, permeability, and loss on, ignition on sand preparation are measured. The experimental results were analyzed and modelled through response surface methodology (RSM). The optimized process parameters are taken from the experiment [7]. The Design of Experiments (DoE) and Computer Simulation Techniques are combined together and used for analyze the moulding related defect and sand related defect of green sand-casting process [8]. The foundry and casting manufacturing is mainly focused. The causes and remedial measures are given to each casting defect by using the cause and defect analysis, concept [9]. The minimization of cold shunt defect in an automobile cylinder block of gray cast iron grade FG150. The seven quality control tools, check sheet, pareto analysis, cause effect diagram, flow chart, scatter diagram, histogram and control chart are used here. The defects are reduced by controlling the pouring temperature and alloy composition [10].

2 Materials and Methods

Six-sigma is one of the best quality control techniques, for improving the quality of the casting by reducing the defects. In six-sigma, we can produce 3.4 defects per million opportunities. The Define, Measure, Analysis, Improve, and Control are the different stages of the DMAIC technique, which is used for the identification of the defect and the root cause of the defects. This is to reduce the defect with the help of Taguchi technique by forming the orthogonal array.

The first phase of Six-Sigma (DMAIC) is the Define phase. In this phase, a thorough study is carried out in the leading foundry industry located in Coimbatore, India where the ductile cast iron products are produced. The data was collected from the industry for the last six months and are tabulated in Table 1. From the number of products and quantity of rejection in the data, the percentage of good casting and the percentage of rejection are calculated.

IOP Conf. Series: Materials Science and Engineering 1059 (2021) 012047 doi:10.1088/1757-899X/1059/1/012047

Table 1. Data collected from the Casting Industry

S. No.	Product Name	Total Quantity of Product	Quantity of Rejection	Good Casting Produced	% of Good Casting	% of Rejection
1	HUB	1577	23	1554	98.54%	1.45%
2	FAN HUB	2341	112	2229	95.21%	4.78%
3	BASE TRUNNION	564	8	556	98.58%	1.41%
4	CLAW BEAM ASSEMBLY	1945	45	1900	97.68%	2.31%
5	STEERING BRACKET	2014	32	1982	98.41%	1.58%
6	MOTOR COVER	854	20	834	97.65%	2.34%
7	COLLAR	1245	26	1219	97.91%	2.08%
8	SUSPENSION LEG	1120	61	1059	94.55%	5.44%
9	EXTERNAL BEARING RING	3099	400	2699	87.09%	12.90%
10	PISTON	1743	23	1720	98.68%	1.31%
11	SUSPENSION	865	26	839	96.99%	3.00%
12	BEARING COVER	991	40	951	95.96%	4.03%
13	RETAINING RING	1121	45	1076	95.98%	4.01%
14	PLANETARY FIN	1467	71	1396	95.16%	4.83%
15	AXLE CASING	743	36	707	95.15%	4.84%

The second phase in Six-Sigma (DMAIC) is the Measure phase. In this phase, the maximum number of defects occurred on the particular product is identified with the help of a histogram as shown in Figure 1 using Table 1.

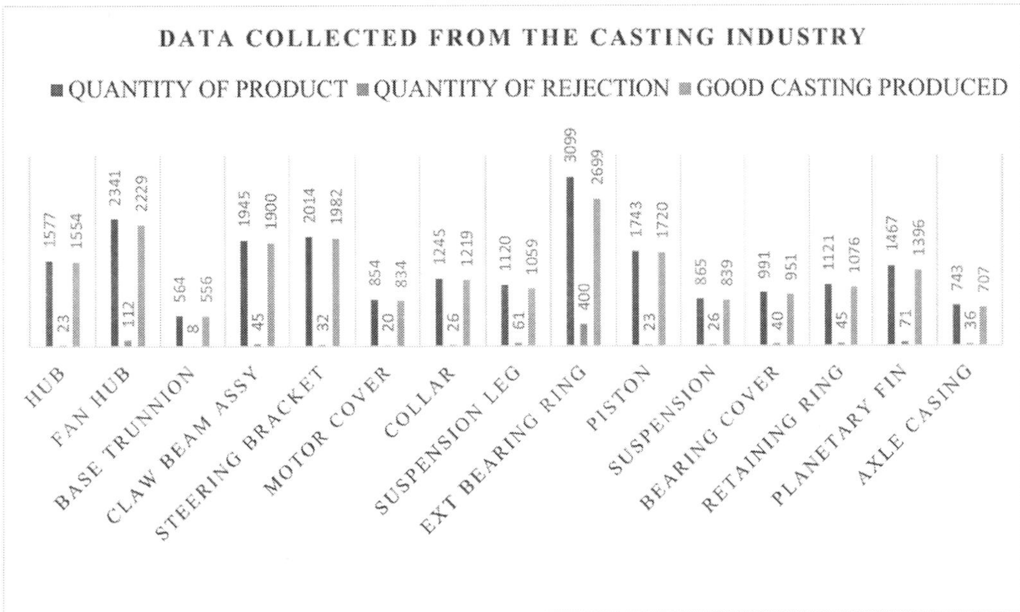

Figure 1. Data collected from the casting industry

In Figure 1, it is clearly shown that the External Bearing Ring product is having a higher rejection rate of 400 out of 3099 products. So, the External Bearing Ring product is taken and the different defects in it are tabulated in Table 2.

Table 2. Products with different kind of defects

S. No.	Product Name	Defects				
		Shrinkage	Sand Drop	Pin Hole	Sand Inclusion	Scabbing
1	HUB	7	4	0	3	1
2	FAN HUB	18	25	4	5	2
3	BASE TRUNNION	1	1	0	0	3
4	CLAW BEAM ASSEMBLY	0	14	9	0	4
5	STEERING BRACKET	24	0	0	8	5
6	MOTOR COVER	2	7	3	0	6
7	COLLAR	16	4	2	2	7
8	SUSPENSION LEG	5	7	18	2	8
9	EXTERNAL BEARING RING	187	27	18	23	9
10	PISTON	0	8	9	0	10
11	SUSPENSION	4	9	6	2	11
12	BEARING COVER	7	8	12	0	12
13	RETAINING RING	9	12	3	5	13
14	PLANETARY FIN	0	21	18	0	14
15	AXLE CASING	7	6	3	5	15

When compared to all other products, the External Bearing Ring is only having 400 rejections out of 3099 products as already discussed in the define phase. In Figure 2, the External Bearing Ring product is having 187 shrinkage defect, 27 sand drop defect, 18 pinhole defect, 23 sand inclusion defect and 9 scabbing defect. The major defect in this product is shrinkage, as it is having 187 out of 400 defects. When compared to all other defects, its rate is high. So, the shrinkage defect is alone taken and an attempt is made to reduce the percentage of shrinkage defect.

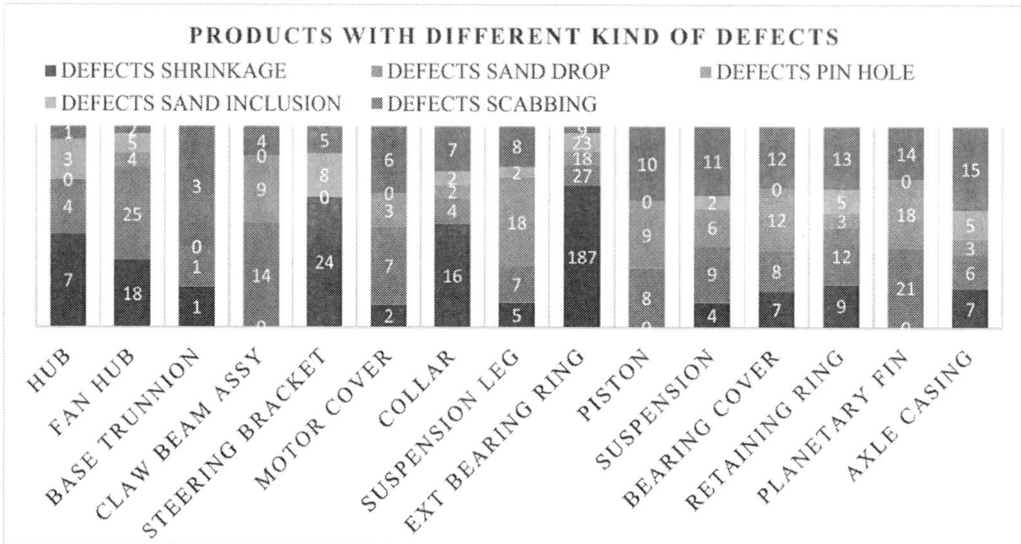

Figure 2. Products with different kind of defects

The third phase in Six-sigma (DMAIC) is the Analyze phase. In this phase, the possible causes for the occurrence of shrinkage defect are identified using the cause and effect diagram as shown in Figure 3. The six different areas in the cause and effect diagram are measurement, materials, machine, environment, manpower, and method. Using this, the major parameter for the occurrence of defect can be identified in that particular area. This is the brainstorming process which is carried out with the help of company employees and standard values.

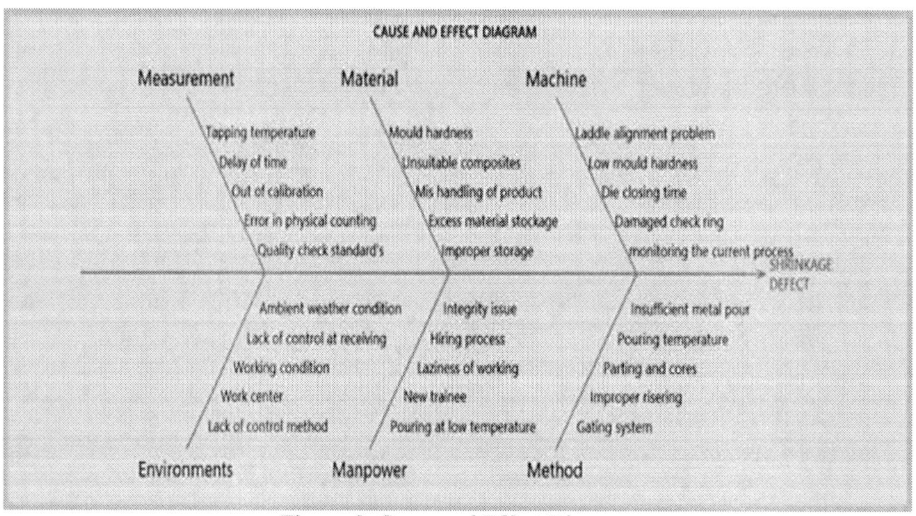

Figure 3. Cause and Effect Diagram

In Figure 3, the different parameters for the occurrence of the shrinkage defect are identified. The majorly affecting parameter is taken with the help of experienced persons from the casting industry and also referred the standard conditions for the shrinkage defects. Three parameters are taken from the Figure 3 namely, mould hardness, tapping temperature and pouring temperature.

The fourth phase in Six-sigma (DMAIC) is the Improve phase. In this phase, the parameters are taken from Figure 3, along with their range of values from high to low, are tabulated in Table 3. Three level of values for each parameter are considered. Taguchi technique is applied for the formation of L9 orthogonal array as shown in Table 4 using the Minitab software. In Table 5, the three-level range of values are applied in the L9 orthogonal array.

Table 3. Parameters with three levels

Factors Designation	Control Factors	Range	Level 1	Level 2	Level 3
A	Mould Hardness (nu)	80 - 50	75	65	55
B	Tapping Temperature (℃)	1570 – 1560	1570	1565	1560
C	Pouring Temperature (℃)	440 - 1380	1430	1400	1370

Table 4. L9 Orthogonal Array

Trial No.	A	B	C
1	1	1	1
2	1	2	2
3	1	3	3
4	2	1	2
5	2	2	3
6	2	3	1
7	3	1	3
8	3	2	1
9	3	3	2

Table 5. Substitution of Three-Level values in Orthogonal Array

Trial No.	A	B	C
1	75	1570	1430
2	75	1565	1400
3	75	1560	1370
4	65	1570	1400
5	65	1565	1370
6	65	1560	1430
7	55	1570	1370
8	55	1565	1430
9	55	1560	1400

From Table 5, for each combination 30 castings produced, the output (or) result is based on the number of defects occurred during the casting process. Shrinkage defect is majorly focused and other defects are neglected. Also, there are no other defects are occurred majorly during this identification. In Table 6, the result of the produced casting is tabulated in percentage, for each combination of values the percentage of defect is shown in result column. The Taguchi analysis is made for the output with the parameters A, B, C shown in Table 7, taking 'smaller is better'. Figure 4, is the signal to noise ratio graph where 'smaller is better' is taken.

Table 6. Result for the combination of Orthogonal Array

Trial No.	A	B	C	Result (O/P)
1	75	1570	1430	13.33
2	75	1565	1400	20.00
3	75	1560	1370	33.33
4	65	1570	1400	26.67
5	65	1565	1370	40.00
6	65	1560	1430	20.00
7	55	1570	1370	46.67
8	55	1565	1430	20.00
9	55	1560	1400	26.67

Taguchi Analysis: O/P versus A, B, C:

Smaller is better

Table 7. Response Table for Signal to Noise Ratios

Level	A	B	C
1	-29.31	-28.33	-31.96
2	-28.86	-28.03	-27.69
3	-26.33	-28.13	-24.85
Delta	2.98	0.31	7.11
Rank	2	3	1

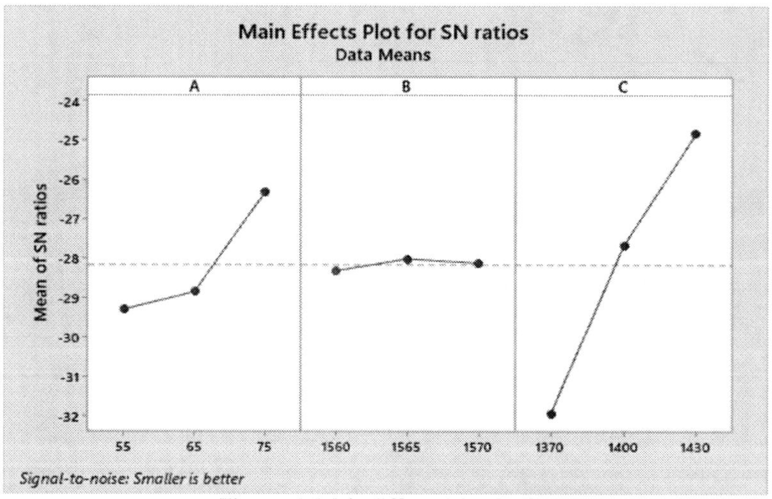

Figure 4. Main Effects Plot for SN Ratios

Table 8 shows the Response table for Means. In Figure 5, the mean effect plot for mean graph is made and 'smaller is better' is chosen. Because the defect percentage only needs to be reduced, from the Figure 4, Figure 5, Table 7 and Table 8 based on the rank the combination of parameters chosen which have the best possible solution for shrinkage defect reduction. Using regression test, it is checked whether the obtained combination is correct with the help of regression analysis from the output value verses the parameters A, B, C. Table 9 shows the coefficients. The R-sq (adj) obtained in Table 10, is 89.36% which is more than the satisfactory condition of 85%. So, the obtained value combination is

IOP Conf. Series: Materials Science and Engineering 1059 (2021) 012047 doi:10.1088/1757-899X/1059/1/012047

the best possible solution. The obtained combination is Mould Hardness of 75 nu, Tapping Temperature of 1570℃ and Pouring Temperature of 1430℃.

Table 8. Response Table for Means

Level	A	B	C
1	31.11	26.67	40.00
2	28.89	26.67	24.44
3	22.22	28.89	17.78
Delta	8.89	2.22	22.22
Rank	2	3	1

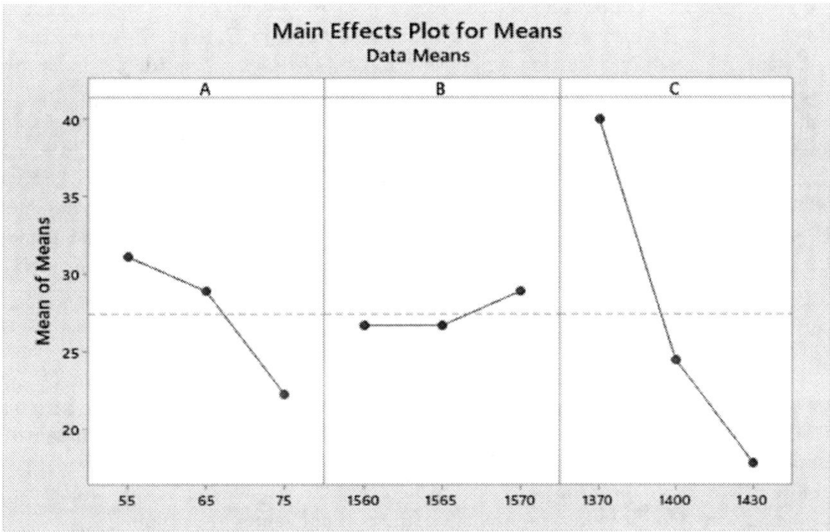

Figure 5. Main Effects Plot for Means

Regression Analysis: O/P versus A, B, C

Regression Equation

O / P = 227 – 0.444 A + 0.222 B – 0.3704 C

Table 9. Coefficients

Term	Coef	SE Coef	T-Value	P-Value	VIF
Constant	227	454	0.50	0.638	
A	-0.444	0.143	-3.10	0.027	1.00
B	0.222	0.287	0.77	0.474	1.00
C	-0.3704	0.0478	-7.75	0.001	1.00

Table 10. Model Summary

S	R-sq	R-sq(adj)	R-sq(pred)
3.51364	93.35%	**89.36%**	77.97%

The fifth and final phase of the Six-sigma (DMAIC) is Control phase. In this phase, the obtained combination of Mould Hardness of 75 nu, Tapping Temperature of 1570℃ and Pouring Temperature of 1430℃ from the improve phase for the External Bearing Ring product, are once again checked, by making 50 castings. From the pouring of 50 castings, it is found that the shrinkage defect count is only 3 and remaining defects are also minimal when compared to the previous defect count. Hence, in this control phase, the defects are controlled and maintained with the obtained combination values. The obtained percentage of rejection is 6%, which is less when compared to initial percentage of 12.90%. So, in the upcoming days the defect percentage will be in control.

3 Conclusion

The optimum combination of value obtained with the help of Six-Sigma (DMAIC) technique. In define phase, data are collected for six months and found that the good casting is 95.54% and the amount of rejection rate is high as 4.46% out of the total production count of 21689. Data are tabulated and histogram is plotted from that values, the most affected product is identified which is the External Bearing Ring product. While producing the External Bearing Ring product in casting, the amount of rejection is 12.90%. In measure phase, the different defects in External Bearing Ring product are identified and the majorly occurring defect is taken. Majorly occurring defect is shrinkage defect. Because the defect count is 187 out of 400 rejections out of 3099 total products produced. In analyze phase, the cause and effect diagram is used to identify the different parameters which is the possible one for the occurrence of the shrinkage defect. The chosen parameters are Mould Hardness in nu, Tapping Temperature in℃, Pouring Temperature in℃. These parameters are identified with the help of journals and experienced persons from the industry. In the improve phase, the parameter ranges are identified and three level of values are chosen with the help of Taguchi technique and Minitab software. The L9 (3^3) orthogonal array is formed and the chosen range values are tabulated in it. For this combination the output is taken like 30 casting in each combination. With the help of Signal to Noise ratio, the smaller value is chosen and it is checked with the Regression equation. The obtained optimum solution is Mould Hardness of 75 nu, Tapping Temperature of 1570℃ and Pouring Temperature of 1430℃. Also, the Regression Value R-sq (adj) is 89.36%. It satisfied the condition because the R-sq (adj) value must be above 85%. In the final phase, the combination of values obtained from the improve stage is tested again for 50 products and the got the number of rejections as 3. The obtained percentage of rejection is 6%, which is less when compared to initial percentage of 12.90%. With the help of Six-sigma, it is possible to reduce the defect percentage and the Taguchi technique which is used to form the orthogonal array and from this combination, the optimal values are obtained.

References

[1] Chen, J.C. and Buddaram Brahma, A.R., 2014. Taguchi-based Six Sigma defect reduction of green sand casting process: an industrial case study. *Journal of enterprise transformation*, **4**(2), 172-188.

[2] Gijo Shreeranga Bhat E V, Jnanesh N A 2014 Application of Six Sigma methodology in a small-scale foundry industry *International Journal of Lean Six Sigma,* **5**,193-211

[3] Amit sharma, Virender verma and Deepak juneja 2014 Utilization of Six-Sigma (DMAIC) approach for reducing casting defects *International Journal of Engineering Research and General Science,* **2**,1065–1075

[4] Kumar S, Satsangi, P S and Prajapati D R 2013 Improvement of Sigma level of a foundry: a case study *The TQM Journal*, **25**, 29-43.

[5] Hemant Thakkar and Sunil Chaudhari 2014 Review on analysis of foundry defects for quality improvement of sand casting *International journal of engineering research and application,* **4**, 615-618

[6] Achamyeleh A Kassie and Samuel B Assfaw 2013 Minimization of Casting Defects *International organization of Scientific Research Journal of Engineering,* **3**,31-38

[7] Kumaravadivel A and Natarajan U 2013 Application of Six-Sigma DMAIC methodology to sand casting process with response surface methodology *International Journal of Advanced Manufacturing Technology*, **69**,1403-1420

[8] Uday A Dabade, and Rahul C Bhedas gaonkar 2013 Casting defect analysis using desing of experiments (DoE) and computer aided casting simulation technique *Forty sixth CIRP conference on manufacturing system,* **7**, 616-621
[9] Avinash Juriani 2015 Casting defects analysis in foundry and their remedial measures with industrial case studies *International Organization of Scientific Research Journal of Mechanical and Civil Engineering,* **12**,43-54
[10] Jadhav B R and Santosh J Jadhav 2013 Investigation and analysis of cold shut casting defect and defect reduction by using 7 quality control tools *International Journal of Advanced Engineering Research and Studies,* **2**, 28-30

Mechanical Concept on Design and Development of Irrigation System to Help Rural Farmers for Their Agriculture Purpose During Unavailability of External Power

Ramesh Chandra Nayak[1], Manmatha K. Roul[2], Abhisek Sarangi[3], Asish Sarangi[4], Abinash Sahoo[5]
Mechanical Engineering Department, Synergy Institute of Engineering & Technology, Dhenkanal, Odisha, India
Mechanical Engineering Department , Gandhi Institute for Technological Advancement (GITA), Bhubaneswar, Odisha, India
Mechanical Engineering Department, IGIT, Sarang, Dhenkanal, Odisha, India
Mechanical Engineering Department, CET, Bhubaneswar, Odisha, India
Mechanical Engineering Department, Bhubaneswar college of Engineering, Bhubaneswar

Email : rameshnayak23@gmail.com

Abstract: Odisha is one of the states in India, where almost 70% families depend on agriculture. Odisha is a common destination for tropical cyclones that generally initiate in the Bay of Bengal. These cyclones destroy the total agriculture system and electricity infrastructure of Odisha every year. Agriculture is the backbone of the economy of any state. Economy can be boosted by the adoption of modern technology and advanced irrigation methods. Continuous irrigation is only possible by using pumps which is operated by diesel, petrol, kerosene or electric power connected with tube well. For such type of arrangement farmers generally depend on electricity, petrol or diesel. But due to continuous electric failure for number of days as a result of frequent cyclones, farmers suffer a lot. With continuous increase in the prices of petrol, diesel, kerosene or electricity, poor farmers find it difficult to bear the cost. The aim and purpose of this work is to develop an irrigation system to help farmers for irrigation purpose to obtain better yields in agriculture with minimum expenses. In the present work mechanical concepts such as Mechanical Advantage (MA), Velocity Ratio (VR), Efficiency, energy storage concepts are used. The system consists of gears, flywheel, handle, bearings and other assembly unit. This arrangement is placed on a frame which is made up of mild steel angles having dimension 40 x 40 x 5 mm. A shaft having 25 mm diameter is used with a flywheel and four numbers of bearings. Driver and driven gears having 200 and 40 number of teethes in simple gear train arrangement are taken in this work. The concept is very simple and its manufacturing process is also very easy. There will be no issue for such type of product as it is environmental friendly and the main purpose of this system is to save the environment and reduce pollution. The idea can be easily implemented because here the input materials are easily available. So by using this system there is no need of any electric supply for irrigation purpose. The work presented here is innovative in nature and has not been developed by any other researchers in the past. By using this system even weak and physically disabled persons can lift water for the purpose of agriculture without any difficulty.
Keywords: Continuous irrigation; farmers; environmentally friendly; design and development

1. Introduction

Agricultural yields depend on availability of continuous irrigation method. Now a day the governments of several countries are providing various facilities to their farmers for enhancing agricultural yields and thus boosting the economy of the country. In some of countries, the government is also working for utilizing water from big rivers for irrigation purpose by constructing canals. The government is also spending lot of money for irrigation purpose like drilling of bore wells and repairing the old ones. The revenue generation of a country depends largely on the agricultural yields. In India almost 70% families depend on small and large scale agriculture. Due to job crisis the educated youths are also choosing agriculture as a prime profession. Even though the government has implemented various provisions for irrigation, the farmers still face problems every year and lose nearly 30% to 40% of their net profit due to unavailability of suitable methods of irrigation for the purpose of agriculture. Odisha is one such state in India where 70% families depend on agriculture. Almost all farmers use bore well at their farms where lifting of water from their bore well is done by using pumps operated by electricity, petrol or diesel. Odisha, being a state prone to natural calamities in the forms of cyclone and high flood, suffers from prolonged electricity failures which linger for 15 to 20 days after each disaster. During these periods the electricity supply is disrupted due to damage of transmission lines. So, during these critical periods, the farmers suffer the most. During COVID 19 pandemic period from March to July -2020, the farmers faced huge problems due to restriction in mobility imposed by the government. This system has been developed to solve the problem of irrigation faced by the farmers due to non-availability of electricity or petroleum products to run their pumps at the time when they need them the most.

The designed product has been used in one of the farm with an area of five acres where the farmers were producing vegetables. Earlier, the farmers were spending nearly ten thousand rupees every month for providing irrigation to the land. Moreover, they also faced problems in operating the pumps due to failure in electricity and poor means of transporting the petroleum products. The application of this system with their bore well has almost sorted out their problem. The farmers are now using this system without spending a single penny. Even more than five physically disabled farmers are using this system successfully for irrigating their farms. The arrangement of the system that consists of number of gears, bearings, shafts, fly wheels is shown in Figure 1. All the arrangements are placed on a frame and the dimension provided in the design may be changed for future works.

A numbers of works have been done by researchers on irrigation methods for various environmental conditions. Deepa *et al.* [1] studied on intelligent agriculture by provision of IOT breakthrough and found that the automated agriculture system is anticipated to benefit the farmers in managing an irrigation system. Sirimorok *et al.* [2] studied on knowledgeable watering systems established on framework of low bandwidth divided applications (LBDA) in cloud computing breakthrough in knowledgeable systems and computing and found that the irrigation scheduling model can effectively enhance the prognostic accuracy for the watering system. Gorguner *et al.* [3] studied on modelling effects of future climate alteration on reservoir repository and irrigation water requirement in a Mediterranean basin and found under all eight etched climate alteration projections growing trends for the annual water requirements are anticipated throughout 21st Century. Sandhu and Irmak [4] studied on performance evaluation of Hybrid-Maize stereotype for rain fed, finite and all-encompassing irrigation conditions and found that the highest dissimilitude in the middle of simulated and calculated data were detected for rain fed ministration, indicating incongruity of the archetype in dry/rainfed suburb and the necessity for archetype enhancements in water stress conditions. Mao *et al.* [5] studied on modelling the salt aggregation and leaching procedures in arid agrarian areas with a latest mass balance model and found that the developed archetype can decipher homogeneous and heterogeneous soil water fluctuation and solute transmission effectively.

Scheiber *et al.* [6] studied on hydrochemical adjustment of irrigation groundwater origin in an alluvial artesian basin and found that this adjustment of groundwater origin will allow managers, irrigators and modellers to evaluate the long term feasibility of groundwater utilisation in alluvial watersheds. Shtull Trauring *et al.* [7] studied lessening salinity of cured waste water with broad scale desalination and found that superior TWW with a consequential portion of desalted base water can allow groundwater salinity remediation. Filgueiras *et al.* [8] studied L.P. Soil water fulfilment and real evapotranspiration prognostication using remote sensing data along with regression algorithms and found that between the regression archetypes tested cubist regression, gradient boosting machine and random forest fit the SWC and ET data. Zhou [9] studied an amalgamated irrigation strategy for water preservation and quality enhancement of cash crops and found that such strategy would inspire the metamorphosis of irrigation strategies from the traditional field irrigation postulation to a new chapter of irrigation superintendence. Huang [10] studied on agricultural water augmentation with a divided eco hydrological archetype in a mountain plain concavity and found that such augmentation in a exemplary year would result in 0.4% enhancement of crop creation for the whole concavity.

Liu *et al.* [11] studied on endowment of clay aquitard to sinkhole iron agglomeration and water quality and found that intrusion, sediment chemistry and hydraulic gradients affect the iron distribution in aquitard pore water. Massa *et al.* [12] studied on lessening nutrient and water losses from soilless harvesting in southern Europe and found that employing advanced recirculation blueprints have made it possible for repurposing the solution under saline conditions with optimal utilisation and minimal expulsion of nutrients. Dillis *et al.* [13] studied cannabis water-use practices and its impact on water resources and the environment and found that extraction of water from groundwater wells mostly occur during the summer dry season and showed their impact on connected surface water. Surya *et al.* [14] studied in-situ soil moisture sensors (SMS) for quantification and detection of soil moisture using advanced materials and found that Molybdenum disulfide (MoS2) offers the highest sensitivity than rest materials. Computing water use and groundwater rejuvenate under flood irrigation in the Heihe River Basin of north western China was done by Li. [15] He found that 50%irrigation quota improved water productivity for spring wheat and spring maize, improved utilization of crop water and protected ground water resources. Limon [16] studied index-based hydrological drought coverage for irrigated agriculture in southern Spain and found that by applying high ordinary deductibles the cost of marketable premiums could be reasonably priced for irrigators. Jimenez *et al.* [17] studied on intelligent agents and multi-agents for irrigation scheduling and found that by including artificial intelligence growth in crop yield and water savings could be achieved. Challenges regarding execution of transition from conventional irrigation to fertigation in countries having moderate climate using regained waste water was discussed by Chojnacka *et al.* [18].

The incorporation of soil moisture sensor measurements, root zone simulation forecasts and remotely sensed NDVI time-series was studied by Fontanet *et al.* [19] to improve crop performance and irrigation scheduling. Liu et al. [20] studied the accumulation of toxic metal lead (Pb) in irrigated water, wheat plant and soil samples and found that high levels of Pb could affect local animal and human population along with soil and wheat plant contamination. Pressure drop caused by two-phase flow of oil/water emulsions through sudden expansions and contractions was studied by Roul and Dash [21]. Edwards *et al.* [22] studied American Indian reservation farmer's problems for investing in irrigation and found that trust ownership, a lack of access to capital using land as a collateral for on-farm investment in eastern Utah created the income gap between adjacent communities and Indian farmers. Zhao *et al.* [23] discovered a novel double-layer ditch system (DDS) with a fibrous stuffing to increase the removal of pollutant in agricultural drainage ditches and its effective in both in terms of performance and annual average removal efficiencies than single-layer ditch system. Irrigation performance evaluation was done by Benavides *et al.* [24] by using relative irrigation supply (RIS) as indicator. By using ANCOVA model they found that on-demand distribution systems and current forced on-farm irrigation systems improved RIS significantly. The main objective of this work is to help farmers for continuous irrigation purpose. It can be easily handled and operated. A physically disabled person can also use it. Multiple activities are possible by using this system. A woman farmer can easily access it. Dimensions provided in this work can also be varied as per requirement. It can enhance agricultural yielding and thus boost economy.

2. Experimental Set Up

The fig.1 shows the experimental setup of the irrigation system which consists of spur gears, bearings, shaft, handle and fly wheel. The system is connected with the plunger rod of tube well in eccentric way. Gears are provided in simple gear train arrangements.

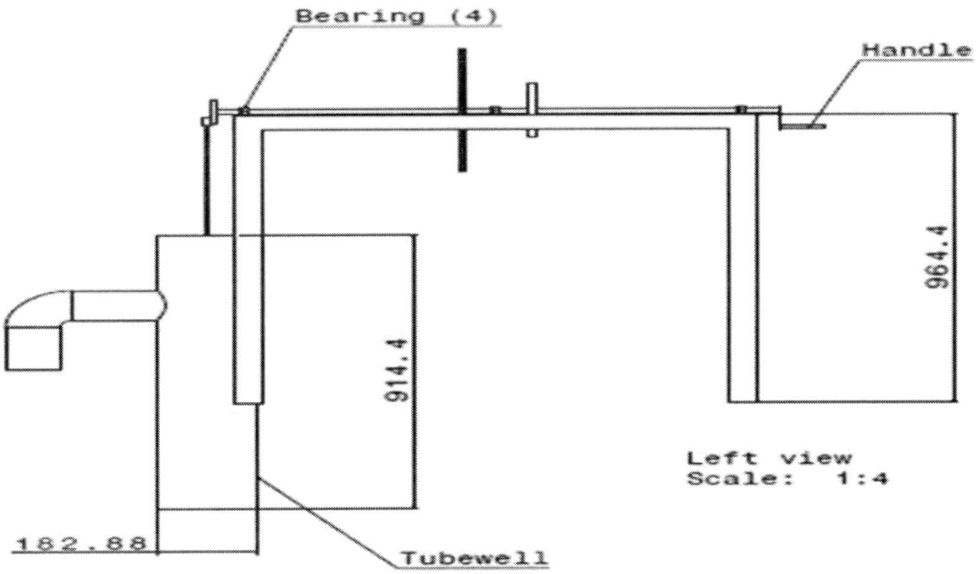

Figure 1. Design layout of the continuous irrigation system

The number of teeth on driver and driven gears are 200 and 40 respectively. Two shafts having length 640 mm and 600 mm are considered in this system. Driver and driven gears are mounted on these two shafts. Handle is connected at the end of one shaft and flywheel is connected at the end of other shaft. The shaft that carries flywheel at one end contains a pulley on its other end. That pulley is connected with plunger rod of the tube well. The shaft to which handle is connected carries driver gear having 200 numbers of teeth and other shaft to which flywheel and pulley are connected, carries driven gear having 40 numbers of teeth. Both gears are meshed with each other. Bearings are provided for smooth rotation of shafts. The experimental setup is constructed in the workshop with the help of welding machine, cutting machine, and gear hobbing machine. Spur gears are manufactured by using gear hobbing machine. After the manufacture of gears, black oxide finish method was used for preventing rust. The dimensions taken for this purpose may be varied as per requirement.

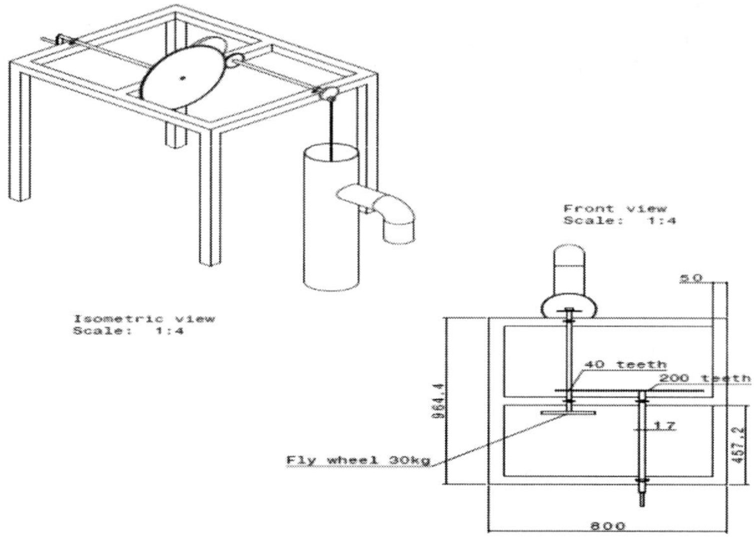

Figure 2. Isometric view and Front view of the continuous irrigation system

The figure 2 shows the isometric view and front view of the set up used for continuous irrigation system. Flywheel of mass 30 kg is provided for energy storage and smooth out put power deliberation purpose. When handle is rotated, the driver shaft and driver gear also rotate. As the driven gear is meshed with the driver gear, so the driven gear also rotates which causes driven shaft to rotate. One end of driven shaft is attached with flywheel and the other end is attached with the pulley. Due to rotation of pulley, the plunger rod, which is connected with the pulley, moves up and down. As a result, water is lifted in the upward direction with the motion of the plunger rod. All such arrangements are provided on a frame made up of mild steel angle of cross section 40 x 40 x 5 mm. The figure 3 and figure 4 show the pump body with motion of plunger rod.

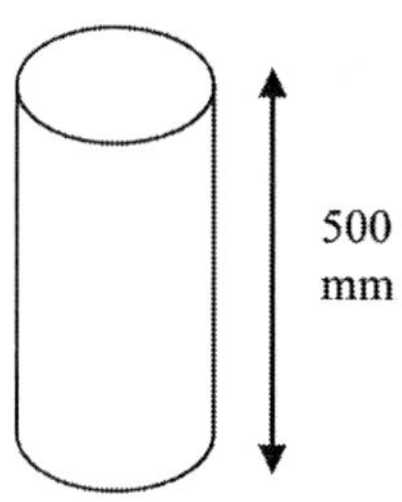

Figure 3. Up and down motion of plunger rod

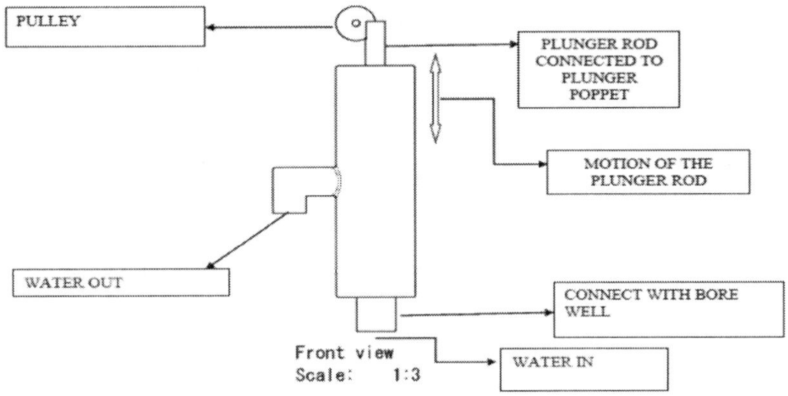

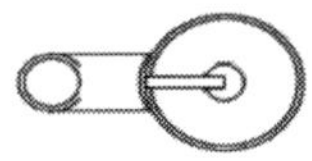

Figure 4. Motion of plunger rod in pump body

After water is discharged from the tube well exit, discharge of water is measured by provision of V-notch. Discharge through V-notch is given by the formula [25].

$$Q = \frac{H^{2.48}}{21900} \quad (1)$$

Q = Discharge

H = Head

Discharge from the pipe (Bore well) is calculated by using the relation.

$$Q = \frac{0.0174\, D^2 x}{\sqrt{y}} \quad (2)$$

Where
x = x- coordinate in cm
y = y-coordinate in cm
D = Pipe diameter in cm
Q = Discharge in litre per second.

3. Results and Discussion:

A survey was conducted in one of the district named cuttack in the state of odisha, where almost 80% families depend on agriculture. The data provided here corresponds to an agricultural farm having 3 acre area. Earlier that farmer was using the diesel pump, petrol pump and sometimes electric operated pump for irrigation purpose. The total expenditure on such cases is given in table1. The initial cost of the pump, running and maintenance cost are also shown in this table.

Table 1. Irrigation Cost per acre

Sl. No	Variables	Diesel Pump (2 HP)	Petrol Pump (2 HP)	Electric Pump (2 HP)	Present arrangement
1	Purchase Cost (Rs)	17000/-	12000/-	10,000/-	6,000/-
2	Expenditure per day	450/-	500/-	200/-	0
3	Maintenance per week	300/-	200/-	100/-	0

From table 1, it is observed that farmers are spending more for irrigation purpose and also there is the chance of unavailability of petrol, diesel and electric supply. To solve this, issue our system was used on the same farm and it was observed that without incurring any running and maintenance cost, same output was obtained. Its initial cost is also very less as compared to the cost of pumps. This is a new concept and innovative in nature. This arrangement can be used for solving irrigation problem and it can supply continuous flow of water with less effort and with reduced cost. This system is very easy to operate and can be easily installed with the existing tube well. This system is suitable for agriculture as well as for household purposes. It can also be used for supplying water for constructional work.

Time line to develop the product is given as follows. For product development it takes about 1 month. For scaling up to commercialization of the setup it takes about 2 months. Break - even point is about 2 months. There is a large market for this system for continuous supply of water from the bore well for the purpose of agriculture. There is no issue for the development of such type of product as it can save the environment and reduce air pollution.

The increase in Mechanical Advantage results in increase in efficiency. This concept is applied here in order to improve the efficiency of the system. When efficiency increases the effort applied to lift the given amount of water from the well for the purpose of irrigation decreases. The concept of energy storage by flywheel has been concerned here, due to which by less effort the lever of tube well operates smoothly.

For the operation of this system there is no need of electric supply and it does not require any petrol or diesel for its operation. The cost of this product is also very less and hence it can be easily installed by farmers. The weight of this product is also very less as a result it will be handy for the farmers to install it in their field. Soil property for agriculture purpose will not be affected by using this product. Continuous irrigation can be provided with the help of this system. Installation and maintenance are not major factors for such system. Zero emission is the main advantage of such system. Any types of farmers can use this method comfortably with less effort. This system can be used to solve the problem of irrigation faced by the farmers due to non-availability of electricity or petroleum products to run their pumps at the time they need these the most.

4.Conclusions:

The following conclusions can be drawn from this study.

i) The increase in Mechanical Advantage results in increase in efficiency as a result of which the effort applied to lift a specified amount of water from the well for irrigation purpose reduces.
ii) The concept of energy storage by flywheel has been applied here for smooth operation of the system and continuous supply of water for irrigation purpose with minimum effort.
iii) The cost of this product is very less as compared to other alternative available methods.
iv) The weight of this product is also less and therefore it can be easily taken by farmers to their field.
v) Soil property will in no way be affected by using this method as irrigation on the farm at any time is possible by using this method.
vi) Installation and maintenance are easy for such system.

vii) Zero emission is a major advantage of such system as this system works without any external energy sources.
viii) Any types of farmers like women and disabled persons can use this system.

References

[1] Deepa B, Anusha C and Chaya Devi, P 2020 Smart agriculture using iot Advances *in Intelligent Systems and Computing*, **1171**, 11-19.
[2] Sirimorok N, As M, Yoshida K and Köppen M 2021 Smart watering system based on framework of low-bandwidth distributed applications (LBDA) in cloud computing Advances *in Intelligent Systems and Computing*, 1263 AISC 447-459.
[3] Gorguner M, Kavvas M.L 2020 Modeling impacts of future climate change on reservoir storages and irrigation water demands in a Mediterranean Basin *Science of the Total Environment*, **748**, 141246.
[4] Sandhu R, Irmak S 2020 Performance assessment of Hybrid-Maize model for rainfed limited and full irrigation conditions Agricultural *Water Management*, **242**, 106402.
[5] Mao W, Zhu Y, Wu J, Ye M, and Yang J 2020 Modelling the salt accumulation and leaching processes in arid agricultural areas with a new mass balance model *Journal of Hydrology*, **591**, 125329.
[6] Scheiber L, Cendón D.I, Iverach C.P, Hankin S.I, Vázquez-Suñé E and Kelly B.F.J 2020 Hydrochemical apportioning of irrigation groundwater sources in an alluvial aquifer *Science of the Total Environment*, **744**, 140506.
[7] Shtull-Trauring E, Cohen A, Ben-Hur M, Tanny J and Bernstein N 2020 Reducing salinity of treated waste water with large scale desalination Water *Research*, **186**, 116322.
[8] Filgueiras R, Almeida T.S, Mantovani E.C, Dias S.H.B, Fernandes-Filho E.I, da Cunha F.F and Venancio L.P 2020 Soil water content and actual evapotranspiration predictions using regression algorithms and remote sensing data *Agricultural Water Management*, **241**, 106346
[9] Zhou H, Chen J, Wang F, Li X, Génard M, and Kang S 2020 An integrated irrigation strategy for water-saving and quality-improving of cash crops: Theory and practice in China *Agricultural Water Management*, **241**, 106331.
[10] Huang F, Mo X, Hu S, and Li L 2020 Agricultural water optimization coupling with a distributed ecohydrological model in a mountain-plain basin Journal *of Hydrology*, **590**, 125336.
[11] Liu Y, Ma T, Chen J, Xiao C, Liu R, Du Y, and Fendorf S 2020 Contribution of clay-aquitard to aquifer iron concentrations and water quality *Science of the Total Environment*, **741**, 140061.
[12] Massa D, Magán J.J, Montesano F.F and Tzortzakis N 2020 Minimizing water and nutrient losses from soilless cropping in southern Europe *Agricultural Water Management*, **241**, 106395.
[13] Dillis C, McIntee C, Butsic V, Le L, Grady K, and Grantham T 2020 Water storage and irrigation practices for cannabis drive seasonal patterns of water extraction and use in Northern California *Journal of Environmental Management*, **272**, 110955.
[14] Surya S.G, Yuvaraja S, Varrla E, Baghini M.S, Palaparthy V.S, and Salama K.N 2020 An in-field integrated capacitive sensor for rapid detection and quantification of soil moisture *Sensors and Actuators, B: Chemical*, **321**, 128542.
[15] Li D 2020 Quantifying water use and groundwater recharge under flood irrigation in an arid oasis of northwestern China *Agricultural Water Management*, **240**, 106326.
[16] Gómez-Limón J.A. 2020 Hydrological drought insurance for irrigated agriculture in southern Spain *Agricultural Water Management*, **240**, 106271.
[17] Jimenez A.F, Cardenas P.F, Canales A., Jimenez F, and Portacio A 2020 A survey on intelligent agents and multi-agents for irrigation scheduling *Computers and Electronics in Agriculture*, **176**, 105474.
[18] Chojnacka K, Witek-Krowiak A, Moustakas K, Skrzypczak D, Mikula K, and Loizidou M 2020 A transition from conventional irrigation to fertigation with reclaimed wastewater: Prospects and

challenges, *Renewable and Sustainable Energy Reviews*, **130**, 109959.

[19] Fontanet M, Scudiero E, Skaggs T.H, Fernàndez-Garcia D, Ferrer F, Rodrigo G and Bellvert J 2020 Dynamic Management Zones for Irrigation Scheduling *Agricultural Water Management*, **238**, 106207.

[20] Liu W, Zafar A, Khan Z.I, Nadeem M, Ahmad K, Wajid K, Bashir H, Munir M, Malik I.S, and Ashfaq A. 2020 Bioaccumulation of lead in different varieties of wheat plant irrigated with wastewater in remote agricultural regions, *Environmental Science and Pollution Research*, **27** (22), 27937-27951.

[21] Roul, M.K. and Dash, S.K. 2009 Pressure Drop Caused by Two-phase Flow of Oil/Water Emulsions Through Sudden Expansions and Contractions: A Computational Approach, *International Journal of Numerical Methods for Heat & Fluid Flow*, **19**, 665-688.

[22] Ge M, Edwards E.C and Akhundjanov S.B 2020 Irrigation Investment on an American Indian Reservation American *Journal of Agricultural Economics*, **102** (4), 1083-1104.

[23] Zhao Y, Shu X, Tu Q, Yang,Y, Liu C, Fu D, Li W, and Duan C 2020 Pollutant removal from agricultural drainage water using a novel double-layer ditch with biofilm carriers *Bioresource Technology*, **310**, 123344.

[24] Benavides J, Hernández-Plaza E, Mateos L and Fereres E 2020 A global analysis of irrigation scheme water supplies in relation to requirements Agricultural *Water Management*, **243**, 106457.

[25] Henderson F.M 1966 *Open Channel Flow*. MacMillan Company, New York, USA

Effects of Longitudinal and Transverse Direction Opening in Reinforced Concrete Beam: The State of Review

Dinesh KannaM[1], Arun M[2]

[1]UG Student and [2]Associate Professor,

Department of Civil Engineering, PSG Institute of Technology and Applied Research,Coimbatore - 641 062, Tamil Nadu, India.

arun@psgitech.ac.in

Abstract The implementation of hollow cores in beams is getting popular nowadays because of its practical advantages. The reduction of disadvantages in the beams with an opening is based on the engineers' knowledge and practical experiences too. Most of the researchers have started to learn about the hollow beams and also, the strengthening of these beams. Based on the reviews on various journals in the field of hollow beams, this article investigates the effect of shape, size, and position of opening and reviews the strengthening of hollow beams. Finally, it brings a judgment with the suggestions to provide hollow beams and the future scope for the research in the area of hollow beams.

Keywords size of opening; shape of opening; longitudinal and web opening; reduced stiffness; strengthening of beams.

1.INTRODUCTION

Millions of Reinforced Concrete (RC) components are being built in the construction industry all over the world every year. The basic amenities and additional facilities of water supply and sewage services, electricity, telephone and air conditioning vents are to be provided in the structures. In the modern construction era, the provision of opening is unavoidable to accommodate these essential utilities in RC beams at longitudinal and/or transverse directions. The provision of opening in beams will result in the reduction of self-weight of beams. Consequently, the prerequisites size of other structural members that support these beams will be reduced, leading to the saving in the construction materials and cost (Arun Murugesan and ArunachalamNarayanan [37,38]. Many researchers have reported on the behavior of steel beams, composite beams, and pre-stressed beams with openings in the transverse direction. A lot of research works have been taken in the area of RC beams with openings in the longitudinal and transverse direction. The majority of the work focuses on the strength characteristics and behavior of beams with openings in terms of Flexure, Shear, Flexure-Shear, Pre-cracking, and post-cracking deflection, crack width, external and internal strengthening separately with various types of loading condition. This paper explores the effect of opening, optimum shape, size, and position of the opening in reinforced concrete shallow beams without strengthening, various strengthening techniques preferred by different authors.

IOP Conf. Series: Materials Science and Engineering 1059 (2021) 012049 doi:10.1088/1757-899X/1059/1/012049

2.EFFECTS ON SHAPE OF THE OPENING

M.A. Mansur and Kiang-Hwee Tan [35] considered the most common simple shapes as openings from the study of Prentas. H. Madkour [31] has found that the reduction in ultimate loads and increased deflection, when the openings of large rectangular shapes located in shear zones. BengiAykac et al., [10] have compared their experimental results with another research consisting of triangular openings. The results indicated that the beams with circular openings have more load carrying capacity than square and triangular openings under all types of tensile reinforcement groups (lightly, moderately, and heavily reinforced). J.H. Ling et al., [30]said that the beams with circular openings offer 8% higher strength than the square openings. SoroushAmiri and Reza Masoudnia [9] analyzed the effects of opening on the behavior of concrete beams and their results showed that the circular openings had achieved 9% more ultimate load carrying capacity while compared with equivalent square opening. They reported that because of the existing corners in the square openings, the reduction of ultimate load carrying capacity is increased more than the circular openings. Saeed Ahamed Al-Sheikh [7] compared the reduction in ultimate loads between the various shapes of openings in beams and said that the reduction of the circular, square and rectangular openings are about 1%, 19%, and 23% respectively, and the reduction in load capacity between circular and equivalent square opening is about 8%. When the rectangular and square opening of the same height is compared, the results show that the Rectangular opening had a reduction in ultimate load of 3 - 4% more when compared to square opening.

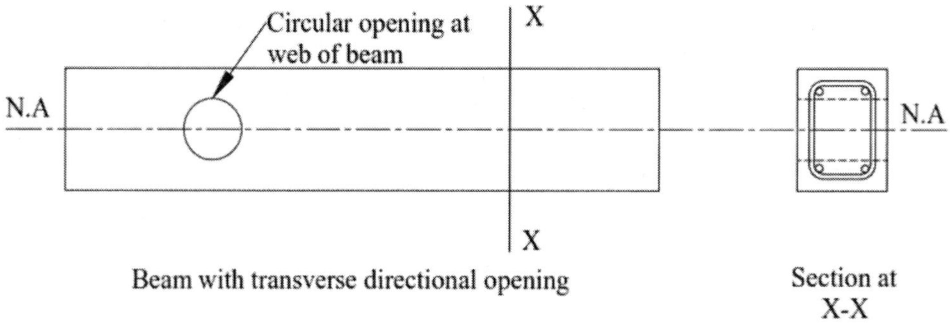

Figure 1. Beam with transverse opening

G. Balaji and R. Vetturayasudharsanan [11] investigated RC beams with circular openings and concluded that the Hollow beams had better ductility and less deflection at yield stage than control beams. Jain Joy and Rajesh Rajeev [28] recommended that the beams having longitudinal circular hollow core sections carried more loads because it avoids sharp edges. Jeyakumar et al., [27] said that the beams with circular openings had 9.58% more strength than beams with square openings. S. Manikandan et al., [33] experimentally investigated and observed that the hollow beams with circular openings had the same similarities of control beams in the ductility behavior, deflection at yield and flexural strength characteristics. Saif A Hassan et al., [24] concluded that the failure mode changes with respect to the position and number of openings because of the provision of circular opening.

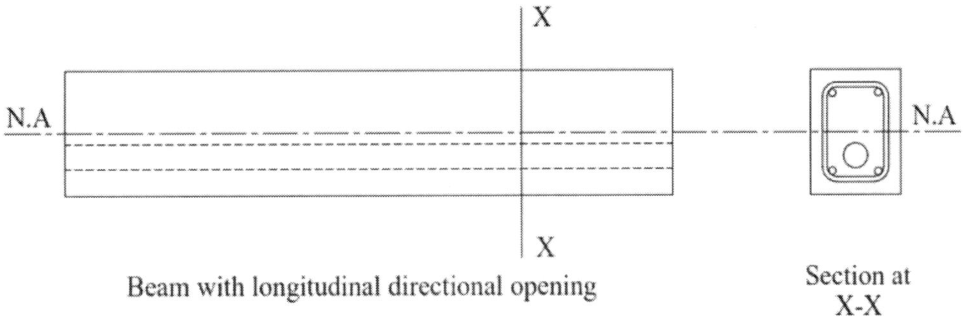

Figure 2. Beam with longitudinal opening

3.EFFECTS ON SIZE OF THE OPENING

M. A. Mansur and Kiang-Hwee Tan[35] and Saeed Ahamed Al-Sheikh[7] haveclassified the beams which are having the opening depth/diameter less than or equal to 0.4D (D – overall depth of beam) as termed as small openings. JavadVaseghiAmiri and MortezaHosseinAlbygie[8] said that when the opening diameter exceeds 1/3rd the depth of the member, it causes greater reduction in the serviceability of the beam. Eventually J.H. Ling et al., [30] reported that the reduction of beam strength was about proportion to the opening size. The beam lost approximately one-third of its strength when the opening size reached one-third of beam height. The beam lost one-sixth of the strength as the opening size was one-sixth.

M. A. Mansur [36] categorized with the type of failure that the size of the openings which allows the beams to maintain the beam type behavior are termed as small openings and others are large openings. When the size of opening lies between 30% and 40% of the overall depth of the beam, there will be a shift from Beam type behavior to frame type behavior. Therefore, it is recommended to determine the size of the opening between the above-mentioned ranges to conveniently apply the simple beam theory.

J.H. Ling et al., [30] recommended that when the opening height should not exceed 0.2 times the beam depth and the opening length should be limited to 0.05 times the beam length. While exceeding this limit, the beam strength was more severely affected. They also concluded that the opening size should not exceed 0.25D maintain at-least 80% beam performance. Ihsan A.S. Al-shaarbaf et al., [6] had concluded that the increase in the opening length and depth of opening results in a substantial decrease in the ultimate load. As well as H. Madkour [31] said that, under non-linear characteristics of the hollow beam, increasing the opening height causes the increased deflection and decreased ultimate load. SoroushAmiri and Reza masoudnia[9] concluded the results more distinctly that there is no effect when the beam had the opening with a diameter of less than 0.48D. Subsequently, and S. Naganathan et al., [3] have reviewed and suggested that the depth of openings can be taken up to 0.5D and further recommended that when the openings become multiple openings can be used with a minimum space of 0.5D between two adjacent openings.

For hollow beams with longitudinal openings, Jeyakumar et al.,[27] have reviewed and reported that there was no effect on the ultimate load capacity when the circular opening size is less than 0.44D. The ultimate load capacity is reduced to 34.29% when the circular opening diameter is more than 0.44D. AhamadJabbarHussainAlshimmeri and Hadi Nasir Ghadhban Al-Maliki[1] have made a statement that the presence of hollow recess. In the RC beams had reduced the load carrying capacity and increased the deflection with respect to the increased the size of the square and rectangular openings up to 58% and 75% respectively when compared to solid beam for same applied load. Nasr. Z. Hassan et al., [24]has reported that the use of PVC pipe in the tension zone had no significant influence if the diameter of the pipe is less than 0.25b at two studied locations (160 mm and 180 mm from the top of the beam) where (b) is the breadth of the beam.

Arun Murugesan and Arunachalam Narayanan [38] and H. Madkour (2009)concludedthe provision and increment of opening will reduce the moment of inertia of the beam which leads to the reduction of lever arm distance between T_u and C_u. This reduction ultimately causes the decrement in first crack load and ultimate load.SherinKurikesu and Abhirami S [29]had analysed the geo-polymer concrete beams with longitudinal openings and concluded that the geopolymer beams with a hollow-core of 32 mm ϕ have more load carrying capacity, bending stress, shear force, less deflection than Geopolymer concrete solid beams. G. Balaji and R. Vetturayasutharsanan [11] has investigated the flexural hollow beams with 63.5mm ϕ (for single opening), 32.5mm ϕ (for double openings) circular PVC pipes and G.I. pipes, and concluded that the flexural behavior of hollow beams with double openings had considerable reductions in the behavior when compared to the beams with single opening and the concrete in beams can be removed upto 17.5% in the tension zone. Sallal R. Abid et al., [2]had dealt with the flexural behavior of HSC beams with different size reductions and said that the size reduction of 16% in the beams recorded the strength more than that of solid beam.

4.EFFECTS ON POSITION OF THE OPENING

JavadVaseghiAmiri and MortezaHosseinalibygie [8]observed that the position of opening has an effect which cannot be considered for the serviceability of the beam. But the openings can be placed in the mid of shear span, not close to the supports. Saeed Ahamed AL-Sheikh's [7] report opposes the results of JavadVaseghiAmiri and MortezaHosseinalibygie [8]by reporting that the beams with openings at shear zone had more reductions than the openings located above the supports and at the flexure zone. M. A. Mansur [35]recommends that the provision of openings entirely within the tension zone causes no change in the load carrying mechanism under pure bending. Fouad el Ame et al.,[16] reported that the provision of openings in the bottom chords with a diameter of circular opening less than 40% of the beam's effective depth brings no significant changes in terms of ultimate load capacity or first crack instant. When the openings are located in compression fibers, the behavior of the beam becomes more brittle with less plastic deformation. Besides the concrete in the upper fibers undergo extensive damages and the beam cracks more than usual. When the depth of the openings is equal to or greater than 0.4d, the type of failure changes from a beam type failure to frame type failure that occurs suddenly without significant previous signs.

J.H. Ling et al., [30]recommended for the shallow beams, openings should avoid the regions with maximum displacement and shear. Also, the opening affects the beam stiffness more significantly when it is placed at the mid-span, where the moment is predominant. For a higher beam capacity, the opening should avoid the beam mid-span and the support, where the maximum moment and shear occurs. It should avoid the potential pathway of the first crack and the critical shear crack. Thus the opening could be placed somewhere between the mid-span and shear regions.When increasing the opening size at support affected the beam ductility more significantly than the mid-span. S.C. Chin et al., [13,14]has investigated for a large square opening and placed the opening in the shear region (0.5d from support) and observed a brittle shear failure. Ihsan A.S. Al-shaarbaf et al., [6] concluded that the load capacity is decreased when the applied load to the nearest end of the opening decreases. SoroushAmiri and Reza masoudnia [9]have recommended that the adequate concrete in compression chord with proper shear reinforcement is required to obtain an ultimate flexural strength.

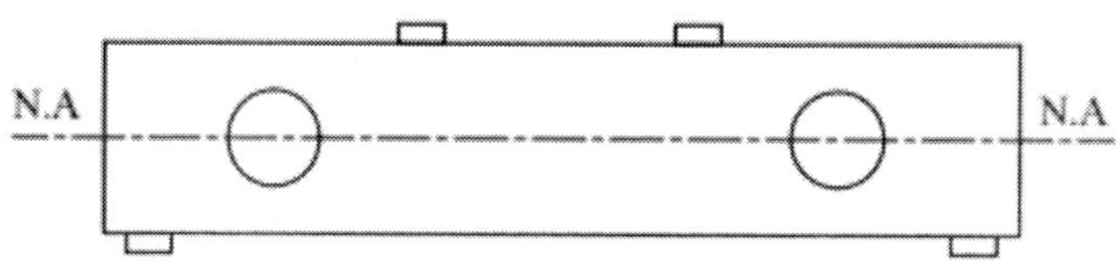

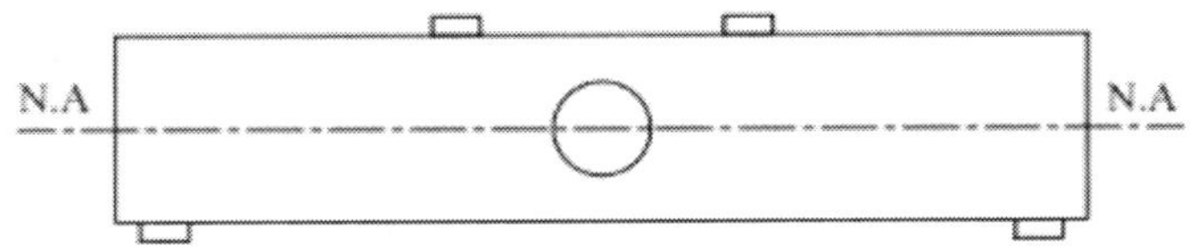

Figure3. Optimum position of opening in web of beams

Nilesh H. Saksena and Patel P G [43]concluded that the provision of opening at L/4 distance had more reductions than the openings provided at L/2 distance, which has no effect. Jain Joy and Rajesh Rajeev [28]said that the presence of a hollow circular PVC pipe instead of concrete in the low stressed zone has not caused a significant reduction in strength of reinforced concrete beams SoroushAmiri and Reza Masoudnia[9]have reviewed researches from various authors and said that the opening position should not be positioned not more than half of the depth of beam from the supports or from the load point.

Kamal GhamryMetwally et al., [42]discussing the position of rectangular web opening, as the relative influence of bending moment compared to that of shearing force increases as the opening is shifted away of the nearest support, the first crack appeared in beams with the opening located near the support faster than that in beams with the opening located away from the support. End openings resulted in an average reduction of 80% and 82% in the beam's cracking failure loads respectively relative to the corresponding beams without opening. This reduction slightly decreased when opening shifted away from the support and became about 70% to 65% respectively, and they also said that for the beams with a central opening, the crack pattern was similar to that of beams without opening. YaarubGatiaAbtan and HussainDhafer Abdul Jabbar [22] havestudied BLTO models (Beams with longitudinal and transverse opening) and concluded that the hollow core at 0.5D had no effect and it is considered as the optimum hollow core section. Stefee S.R and Ravi .C [46]has concluded with the optimum depth of web opening at the below neutral axis is 100 mm (from top) for the beam depth of 150 mm.

For longitudinal openings, Manikandan S et al., [33]have observed that introducing opening at a tension zone of size up to 25% had non-considerable reductions. Arun Murugesan and Arunachalam Narayanan [37,38]have concluded thatit is better to provide the openings below the stress block than other positions and when the distance between the center of opening and centroid of beam decreases the moment of inertia and leads to the reduction of cracking moment of resistance. Dhinesh N. P and Satheesh V. S [15]had reported that when the depth of opening is increased, it affects the ultimate load carrying capacity and deflection. Parthiban N and Neelamegam M[39]have observed that the presence of PVC pipes instead of concrete in the low stressed zone, increases the strength of RC beams up to 21%. Nasr. Z. Hassan et al., [24]has concluded that the best location to install PVC pipes in a reinforced concrete beam is 0.53d from top of sectionaway from the tension reinforcement to avoid brittle shear failure.

SherinKurikesu and Abhirami S[29] concluded that increasing the size of opening with pipe at neutral axis had not much difference in the flexural capabilities. YaarubGatiaAbtan and HussainDhaferAbdulJabbar [22] have studied with BLTO models and concluded that the presence of hollow-core led to decreased load resistance which also had lower value when hollow-core approaches tension or compression zone. They observed that more depreciation in the ultimate load and first crack load if the hollow core was found in the tension zone or compression zone. Saif A Hassan et al., [25]had concluded that the openings should not be placed near the supports and the failure mode changes with respect to the position and number of openings.

5.EFFECT OF OPENING IN RC BEAMS

M. A. Mansur [35] reported that under service loads, provision of opening causes reduction in stiffness which led to the excessive deflection, redistribution of internal forces and moment in continuous beams. Ihsan A.S. Al-Shaarbaf et al., [6] concluded that openings in a RC beam alters the beam behavior from simple to complicated. H. Madkour[31], Ibrahim

H. Elkersh et al., [19] and S. Naganathan et al., [3]observed that the diagonal cracks will be formed at earlier stage when the openings provided in the web of the beams and it reduces the shear capacity of the beam, due to the discontinuity created in the flow of stresses because of the openings. BengiAykac et al., [10] predicted that if the diagonal reinforcement is not provided then the top and bottom chord will act separately due to the opening. Saif A Hassan et al., [25] have observed that provision of opening in the existing beams is not recommended because it leads to the reduction in beam stiffness and the section will fail without indication.

Luis Herrera et al., [26]concluded that the specimen with web opening had comparable ultimate capacities to that of the specimen without opening with only a 15% decrease in compression and a 21% decrease in tension capacity. Fouad el ame et al., [16]reported that beams with opening size of less than 0.4d (d-effective depth of beam) exhibit a ductile behavior and beams with 0.4d or greater openings exhibit a brittle behavior. J.H. Ling et al.,[30] concluded that the stiffness was significantly affected by the size and the position of opening. If the size of the opening is about d/3 will affect the 36 % of stiffness of the beam. Stefee S.R et al., [46]have observed that the crack width is larger in beams with openings than solid beams and exceeds limitations of crack width irrespective of the size of openings and location. The opening reduces the cost of construction, providing opening does not require any skilled laborers, it also reduces the height of the ceiling, which leads to saving the overall height of the building. J.H. Ling et al., [30]reported that the shear crack occurred at around the same time regardless of the size, location of the opening. From Gajipara J.M et al., [20]under pure cyclic torsion, the torsional flexibility of the beam is governed by the longitudinal reinforcement and the post cracking deflection is governed by both the longitudinal and shear reinforcements. When most of the authors recommend hollow RC beams for the beams under torsion, this conclusion from Gajipara J.M et al., [20]improves the usage and improvement of RC hollow beams under torsion.

For longitudinal openings in the RC beam, A.S. Alnauimi et al., [4]have experienced that the failure load is closer to design loads with ductile nature for the tested beams. Ali Said Alnuaimi et al., [5]observed that the hollow beams had more displacement measured because the solid beams are likely to carry more loads. They recommend hollow beams when the torsion is predominant for the same reinforcement. Nibin Varghese and Anup Joy [47]have observed from the results of their research that all partially replaced beams (concrete in various zones are partially replaced by PVC pipes) had less deflection behavior than control beams at the safe load. Manish Kumar et al., [34]has observed that when increasing the difference in grade of concrete in hollow heterogeneous beams, the deflection is reduced. (Beams partially replaced by a different grade of concrete, below neutral axis).

6.SUGGESTIONS PROVIDED TO AVOID OR REDUCE THE EFFECT OF OPENING

- A separate simplified, empirical design procedure can be adopted for the design of beams with openings. If this fails, either the FEM or Strut &Tie procedure can be adoptedas an alternative practical solution. (Kamal GhamryMetwally et al., [42])
- In the case of openings in the existing beams, the strengthening can be done by,
  - Steel plates
  - Fiber reinforced polymer materials
  - CFRP sheets
  - Additional reinforcement (Nazar K. Oukali [41])

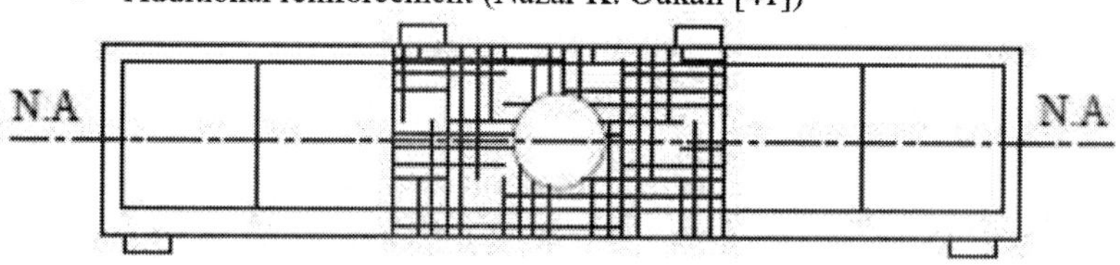

Figure 4. Provision of FRP sheets around the opening

- In order to reduce the effect of opening under strengthening category, it is recommended to use CFRP or FRP laminations to regain its half of design structural capacity (S.C Chin et al., [13,14]).
- FRP provides resistance against failure and increases the load carrying capacity of the beams(K.M. El-Sayed et al., [18]). CFRP sheets provide durability against corrosion to the concrete (Nazar K. Oukali [41]).
- The orientation of AFRP (Aramid fiber reinforced polymers) enhances the shear strength of the RC beams and the efficient angle of installation is 60° to the axis of the beam. (Bashir H. Osman et al., [40])

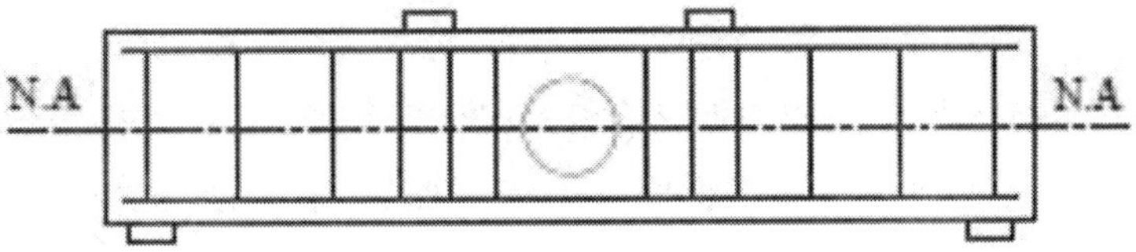

Figure 5. Provision of additional shear reinforcement

- Based on the usage of materials,
  1. Adopt an increased amount of shear reinforcement (Hadi Nasir Ghadhban Mohammed Al-Miliki [23])
  2. Special type of reinforcements can be provided around the opening in the form of
     - additional stirrups above and below the opening.
     - extra stirrups at both sides of the opening
     - longitudinal reinforcement in the top and bottom chords. (Kamal GhamryMetwally et al., [42])
  3. To reduce the area of concrete in the tension zone upto 25%, it is recommended to use G.I. pipes instead of PVC pipes (G. Balaji and R. Vetturayasudharsanan. [11])
  4. The flexural strength of RC beams having opening depends upon either the yielding of steel in tension or crushing of concrete in compression zone (Marta Slowik [45])
  5. When the opening is away from the load point, the additionally provided reinforcements will take the stress concentration around the opening. (Siddhalingesh B Kudatani et al., [44])
  6. Usage of diagonal reinforcement provides not only the shear resistance but also the flexural strength, stiffness and ductility to the RC beams with openings. (BengiAykac et al., [10], JavadVaseghiAmiri and MortezaHosseinalibygie [8]and J.H. Ling et al., [30])
  7. The increase in strength of concrete is more effective to improve the serviceability of the beams with larger openings. (JavadVaseghiAmiri and MortezaHosseinalibygie [8])
  8. Providing the reinforcement at the opening led to an increase of the load to initiate the shear crack by 20% and a substantial increase in the load for the crack to reach the opening.
  9. Strengthening using External Prestressing Technique (EPT) enhances the torsional capacityand also, improves the ductility behavior. (Abd El-HakimKhalil et al., [17])

7.CONCLUSION

The effects of opening either transverse direction or longitudinal direction in reinforced concrete beams by various authors have been studied. Based on the vigorous effort made in this area the following conclusionsare drawn:

1. It could be concluded that the usage of circular-shaped openings is more effective than other shapes of openings since it avoids sharp edges which are subjected to high-stress concentration while loading.
2. It is recommended to have a circular opening less than 0.55D of the beam for web opening. For longitudinal opening; the size of the opening to be restricted within 0.5D of the beam. The other shapes of web opening; Increase in size of opening decreases the mechanical properties of the beam, and it leads to the change in the mode of failure.
3. It can be concluded that, for web opening in RC beams, the optimum position to have a opening at mid-depth and the center between the support and load point to reduce the changes made in the beam due to opening.
4. When the opening location moves towards the point of application of load or towards the support, there will be an occurrence of reduction in its load-carrying capacity.
5. The optimum position to provide a longitudinal opening in the RC beam is just below the neutral axis. If the opening moves into the compression zone or tension zone, then the reduction in the structural capacity of RC beams will occur.
6. It is proposed to select the strengthening materials based on the purpose, availability, cost, availability of skilled labor and time.
7. It is desirable to work on the following field to elevate this work to the next level.
    - Performance of different types of concrete in the field of hollow beams
    - Investigation of different strengthening materials and techniques to reduce the deflection
    - Behavior of hollow RC elements under earthquake forces.

REFERENCES

[1] A.J. Alshimmeri, H. N. Al-Maliki. 2014*Structural Behavior of Reinforced Concrete Hollow Beams under Partial Uniformly Distributed Load. J Eng.* **20(7)**:130-145.
[2] Abbass A A, Abid S R, Arna'otd F H, Al-Ameri R A, Özakça M. 2020*Flexural response of hollow high strength concrete beams considering different size reductions. Structures.* **23**, 69-86. DOI:10.1016/j.istruc.2019.10.001
[3] Ahmed A, Fayyadh M M, Naganathan S, Nasharuddin K. 2012 *Reinforced concrete beams with web openings: A state of the art review. Mater Des.*40:90-102. DOI:10.1016/j.matdes.2012.03.001
[4] Alnauimi AS, Bhatt P. 2004 *Direct design of hollow reinforced concrete beams. Part II: an experimental investigation. StructConcr.* 5**(4)**:147-160. DOI:10.1680/stco.5.4.147.56911
[5] Alnuaimi AS, Al-Jabri KS, Hago A. 2008*Comparison between solid and hollow reinforced concrete beams. Mater Struct Constr.*41(2):269-286. DOI:10.1617/s11527-007-9237-x
[6] Al-Shaarbaf IAS, Al-Bayati NAMJ, Al-Kaisy DIA. 2007*Nonlinear finite element analysis of reinforced concrete beams with large opening under flexure. EngTechnol J.* **25(2):**210-228.
[7] Al-sheikh SA. 2014 *Flexural behaviour of RC beams with opening. Concr Res Lett.* **5(2)**:812-824.
[8] Amiri JVA, Hosseinalibygie M. 2004 *Effect of Small Circular Opening on the Shear and Flexural Behavior and Ultimate Strength of Reinforced Concrete Beams Using Normal and High. J Sch Eng.* **16**,37 - 52.
[9] Amiri S, Masoudnia R. 2011*Investigation of the opening effects on the behavior of concrete beams without additional reinforcement in the opening region using the FEM method.Aust J Basic Appl Sci.* **5(5):**617-627. DOI:10.13140/RG.2.1.2108.1041
[10] Aykac B, Kalkan I, Aykac S, Egriboz YE. 2013 *Flexural behavior of RC beams with regular square or circular web openings.EngStruct.* **56**:2165-2174. DOI:10.1016/j.engstruct.2013.08.043
[11] Balaji G, Vetturayasudharsanan R. 2020 *Experimental investigation on flexural behavior*

of RC hollow beams.Mater Today Proc. **21**(XXXX):351-356. DOI:10.1016/j.matpr.2019.05.461
[12] Bernardo LFA, Lopes SMR. 2011 *High-strength concrete hollow beams strengthened with external transverse steel reinforcement under torsion. J CivEngManag.* **17(3):**330-339. DOI:10.3846/13923730.2011.589204.
[13] Chin S C, Shafiq N, Nuruddin M F. 2012 *Strengthening of RC beams with large openings in shear by CFRP laminates: Experiment and 2D nonlinear finite element analysis. Res J ApplSciEng Technol.***4(9):**1172-1180.
[14] Chin SC, Shafiq N, Nuruddin MF. 2016 *Behaviour of RC beams with CFRP-strengthened openings. StructConcr.* **17(1):**32-43. DOI:10.1002/suco.201400111
[15] Dhinesh N P, Satheesh V S. 2017 *Flexural Behaviour of Hollow Square Beam.*(3):1-7. http://ijseas.com/volume3/v3i3/ijseas20170321.pdf
[16] El Ame, Fouad&Mwero, John &Kabubo, Charles. 2020 *Openings Effect on the Performance of Reinforced Concrete Beams Loaded in Bending and Shear.Engineering, Technology and Applied Science Research.* **10**. 5352-5360.
[17] El-Hakimkhalil A, Etman E, Atta A, Fayed S. 2015 *Torsional Strengthening of RC Box Beams Using External Prestressing Technique.IOSR J Mech CivEngVer VII.* 12(2):2320-2334. DOI:10.9790/1684-12273041
[18] El-kareimShoeib A, El-sayedSedawy A. 2017 *Shear strength reduction due to introduced opening in loaded RC beams. J Build Eng.* 13:28-40. DOI:10.1016/j.jobe.2017.04.004
[19] Elkersh IH, Lotfy E, Ellatif EA, Aboshosha MR. 2017*Flexural Behavior of Reinforced Concrete Beams With Nano-Metakaolin. Int J Latest Trends Eng Technol.* 8(2):25-36. DOI:10.21172/1.82.031
[20] Gajipara J M, Patel P.V, Raiyani S.D. 2015 *Behavior of Rc Hollow Beam Under Pure Cyclic Torsion.Int J Res Eng Technol.* 04(25):407-412. DOI:10.15623/ijret.2015.0425060
[21] Gamal AA. 2015*Strengthening Techniques in Reinforced Concrete Hollow Core. International Journal of Engineering and Innovative Technology (IJEIT)* Vol. **4**, Issue 8, **4(8):**159-173.
[22] GatiaAbtan Y, Dhafer Abdul-Jabbar H. 2019*Experimental Study to Investigate the effect of Longitudinal and Transverse Openings on the Structural Behavior of High Strength Self Compacting Reinforced Concrete Beams. J Eng Sustain Dev.* 66-79. DOI:10.31272/jeasd.23.1.5
[23] Hadi L, Ghadhban N. 2013*Experimental Behavior of Hollow Non-Prismatic Reinforced Concrete Beams Retrofit With CFRP Sheets.J Eng Dev.* **17(5)**:1813-7822.
[24] Hassan NZ, Ismael HM, Salman AM. 2018 *Study Behavior of Hollow Reinforced Concrete Beams. Int J CurrEng Technol.***8(6)**:1640-1651.
[25] Hassan SA, Hemzah SA, Alyhya WS. 2020 *Experimental investigation of the structural behaviors of self-compacted reinforced concrete hollow beams with in-place circular openings. IOPConfSer Mater Sci Eng.* 671(1). DOI:10.1088/1757-899X/671/1/012117
[26] Herrera L, Lemnitzer A. 2014 *Cyclic Performance of RC Beams With Web Openings. Second European Conference on Earthquake Engineering and Seismology,Istanbul* ***August 25-29***, *pg*1-11.
[27] Jeyakumar A, Ravikumar MS, Nalanth N. 2019 *Hollow beams strengthened with steel fibers and FRP wrapping. Int J RecentTechnol Eng.*8(2):1129-1133. DOI:10.35940/ijrte.B1661.078219
[28] Joy J, Rajeev R, Student PG. 2014*Effect of Reinforced Concrete Beam with Hollow Neutral Axis. IJSRD -International J Sci Res Dev.* 2(10online):2321-2613. www.ijsrd.com
[29] Kurikesu S. 2018 *Analytical Study on Geopolymer Concrete Beam With Hollow Space Below Neutral Axis.Int J AdvEng Res Dev.* 5:332-342.
[30] Ling JH, Tang HS, Leong WK, Sia HT.2020 *Behaviour of Reinforced Concrete Beams with Circular Transverse Openings under Static Loads. Journal of Science and Applied Engineering (JSAE)*, **3**(May):1-16.

[31] Madkour H.2009 *Non-linear analysis of strengthened RC beams with web openings. ProcInstCivEngStruct Build.*162(2):115-128. DOI:10.1680/stbu.2009.162.2.115
[32] Mahmoud AM. 2012*Strengthening of concrete beams having shear zone openings using orthotropic CFRP modeling. Ain Shams Eng J.*3(3):177-190. DOI:10.1016/j.asej.2012.02.005
[33] Manikandan.S, Dharmar.S, Robertravi.S. 2015*Experimental Study on Flexural Behaviour of Reinforced Concrete Hollow Core Sandwich Beams.International Journal of Advance Research In Science And Engineering IJARSE*, Vol. No.**4**, Special Issue (01),8354(4):937-946.
[34] Manish Kumar, G.P.Khare, PukhrajSahu. 2019*Flexural Behaviour of RCC Beam with Partially Replaced Concrete below Neutral Axis.International Research Journal of Engineering and Technology.* 619-624.
[35] Mansur M A, 2006 *Design of Reinforced Concrete Beams with Web Openings. Proceedings of the 6th Asia-Pacific Structural Engineering and Construction Conference (APSEC 2006),*104-120.
[36] Mansur M.A. and Kiang-Hwee Tan.1999 *Concrete beams with openings: analysis and design. Choice Rev*36(11):36-6307-36-6307. DOI:10.5860/choice.36-6307
[37] Murugesan A, Narayanan A. 2018 *Deflection of Reinforced Concrete Beams with Longitudinal Circular Hole.Pract Period Struct Des Constr.* 23(1):1-15. DOI:10.1061/(ASCE)SC.1943-5576.0000356
[38] Murugesan A, Narayanan A. 2017*Influence of a longitudinal circular hole on flexural strength of reinforced concrete beams. Pract Period Struct Des Constr.* 22(2):1-10. DOI:10.1061/(ASCE)SC.1943-5576.0000307
[39] N ParthibanMN. 2017*Flexural Behavior of Reinforced Concrete Beam With. Int Res J Eng Technol.* 4(4):2263-2274. https://www.irjet.net/archives/V4/i4/IRJET-V4I4573.pdf
[40] Osman BH, Wu E, Ji B, Abdulhameed SS. 2017 *Repair of Pre-cracked Reinforced Concrete (RC) Beams with Openings Strengthened Using FRP Sheets Under Sustained Load. Int J ConcrStruct Mater.* **11(1)**:171-183. DOI:10.1007/s40069-016-0182-3
[41] Oukaili N, Al-Shammari A. 2014*CFRP strengthening of RC beams with multiple openings subjected to static and impact loads. AdvStruct Eng.*17(12):1747-1760. DOI:10.1260/1369-4332.17.12.1747
[42] Ramadan O M, Metwally K G, Shaban W M. 2015*Proposed recommendations for the design of reinforced concrete beams with openings.*
[43] Saksena N H, Patel P G. 2013 *Effects of the circular openings on the behaviour of concrete beams without additional reinforcement in the opening region using FEM method.Int J AdvEng Technol.* **4(2)**:40-42.
[44] Siddhalingesh B Kudatini, H Eramma. 2016 *Experimental Investigation on Internally Strengthened RC Beam with Rectangular.*15410-15417. DOI:10.15680/IJIRSET.2016.0508175
[45] Słowik M. 2019 *The analysis of failure in concrete and reinforced concrete beams with different reinforcement ratio. Arch Appl Mech.*89(5):885-895. DOI:10.1007/s00419-018-1476-5
[46] Stefee SR and Ravi C. 2020*Experimental Investigation of Flexural Member of Beam Opening in Web by using PVC Pipe. International Research Journal of Engineering and Technology (IRJET).*3737-3741.
[47] Varghese N and Joy A. 2015*Flexural Behaviour of Reinforced Concrete Beam with Hollow Core at Various Depth. Int J Sci Res.***5(5)**:741-746. DOI:10.21275/v5i5.nov163467

IOP Conf. Series: Materials Science and Engineering 1059 (2021) 012050 doi:10.1088/1757-899X/1059/1/012050

Study of Solar Panel Efficiency and Estimation at an Institution Edifice

Mohammed Baba Sharfuddin Ali Khan [1] **, Dr.Syed Mujahed Hussaini** [2]**, Abrar Hussain** [3]
[1] M.tech (Scholar), JNTU, Hyderabad
[2] BITS Pilani, Hyderabad
[3] Assoc Prof , Hyderabad
E-mail: babasharfuddinkhan@gmail.com

Abstract. A photovoltaic cell is an electrical device tool that transforms the vitality of light to other form with an aid of pv effect, an actual and chemical occurrence. An arrangement of pv cell is elucidated as a device where voltaic attributes, as current, voltage, resistance, exposure to light. Singly solar cells are often electrical structure blocks of photovoltaic modules, known casually as solar panels. A single junction silicon based solar type can make a limited open-circuit potential of nearly 0.5- 0.6 volts. Solar cells are generally set out as photovoltaic, ignoring of whether a source is from sun or an unnatural light. In computing to making energy, [1]it can be pre-owned as a photon-detector (ex. infrared detectors), perceive light, electromagnetic radiations near to visible span, or quantifying light intensity. [2]The operation of a photovoltaic (PV) stall needs basic attribute: Engagement of light, causing either electron-hole pair or excitants. The disconnection on charge conveyor for adverse types. The disconnection on uprooting of those bearers to an outer circuit. In variance to a solar thermal collector supplies heated by absorbed sun lighting, for a purpose of either straight heat or deviated electrical power generated from a heat source. A "photo electrolytic cell" on the other hand, refers straight to a kind of photovoltaic cell (as of developed since Edmond Becquerel, new dye-sensitized solar blocks), to a device that splits water directly into hydrogen and oxygen using only solar radiance.

1 Introduction

The site we have selected is an institution building; the institution is established on a large piece of land measuring 5 acre's whereby different blocks vary in sizes between 2000-4000sq.yrds totaling to required space of open land thereby leaving the open space vacant for a garden and an open plot space for parking. [3]Here parking space can also be used for installation of solar panels using the total requirement for installation was 5000sq.yrds, solar panel varying between 100-200v can be used in order to make a setup of required amount of power source the site selected has a hefty range of electric bill varying between 60,000-70,000pm and the units consumed by the structure is around 6000 units and the subsidy given by the MNRE is around 25% and the rest 75% is to be taken care by the institution's responsibility which can easily be reimbursed in near future as per the requirements. The general set up is shown in Figure 1.

Figure 1. General set up

2 Materials and Methods

The pv cells selected were of silicon based also anti-solar plate [4]conductors which also generated energy during off times of rays, total subsidy given by MNRE is 20%, dual circuit flow selected, inverter, hassle-free battery system and general tools of material used as advised also the readings of irradiance has been noted in Fig 2.

1. Hassle free battery system- ranging a life span of almost 15 years.

2. Silicon based pv cells with a life span of 25 years.

3. Inverter with batteries attached aiding in reaching maximum life of desired form.

4. Aid from the government in bagging a subsidy of around 20% of the total cost incurred on the whole project completion.

5. Clear grounded system for good variance in order to avoid any type of back ends or losses.

6. Temporary stand's were made of tensile iron to hold the weight of solar panels and be able to withhold it for long duration of time till it gets calculated.

7. On the other side of the stand we took dependency from the side wall's of roof top in order to get good angle of inclination and also maintain a good posture for a long time without any disturbance's and get good behavior in case of earthling.

8. The area selected was away from unwanted shades from any other surfaces.

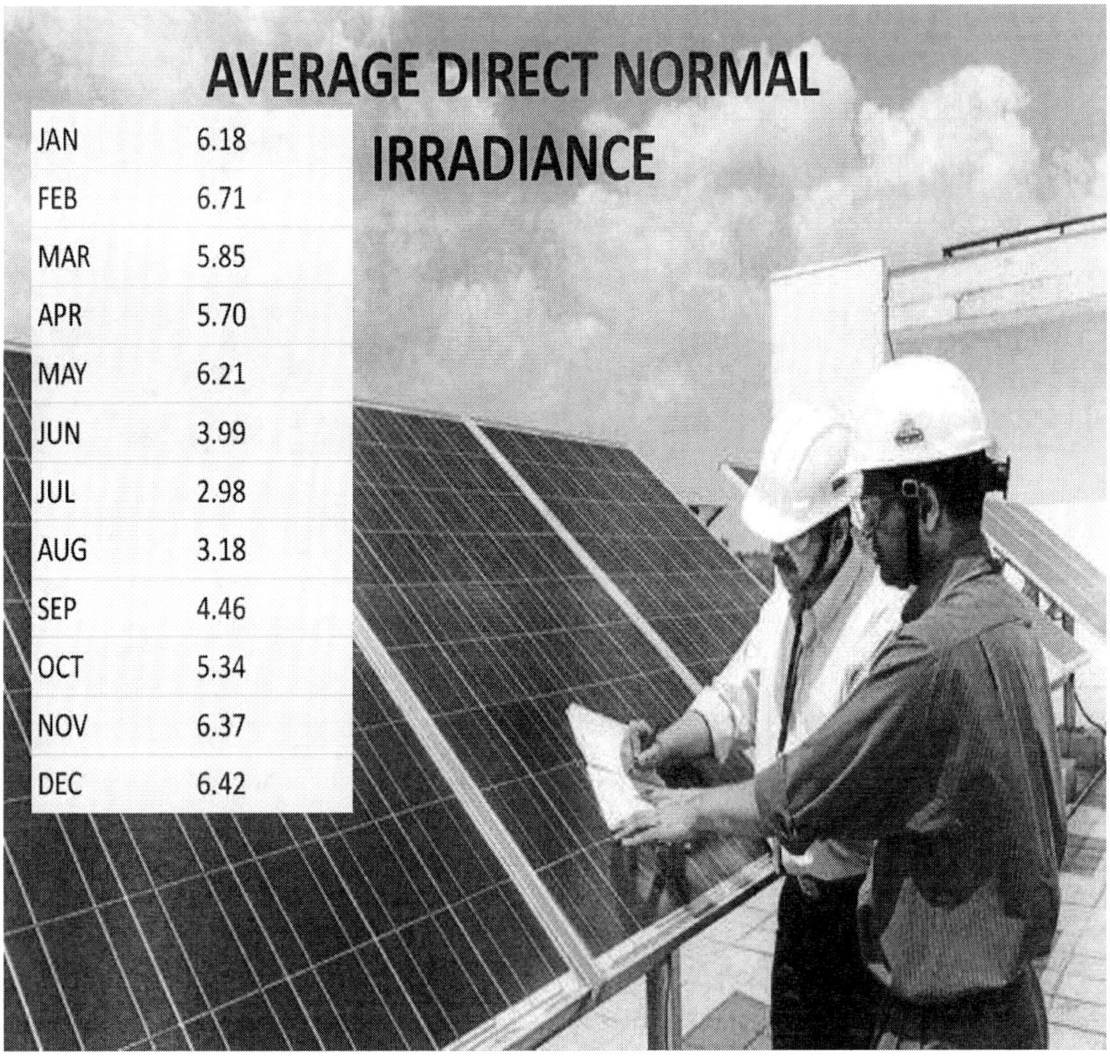

Figure 2. Average Direct Normal Irradiance

The irradiance experienced from the city of Hyderabad were noted down as per the calendar year with different months whereby the highest noted month was between Feb-may and the lowest one was maintained between July-august (Figure 2).

The whole data was [5]maintained simultaneously as per the requirements of the site and given in accordance to the data noted and average 100v panel was under constant observation from timely based manner.

3. Results and Discussion

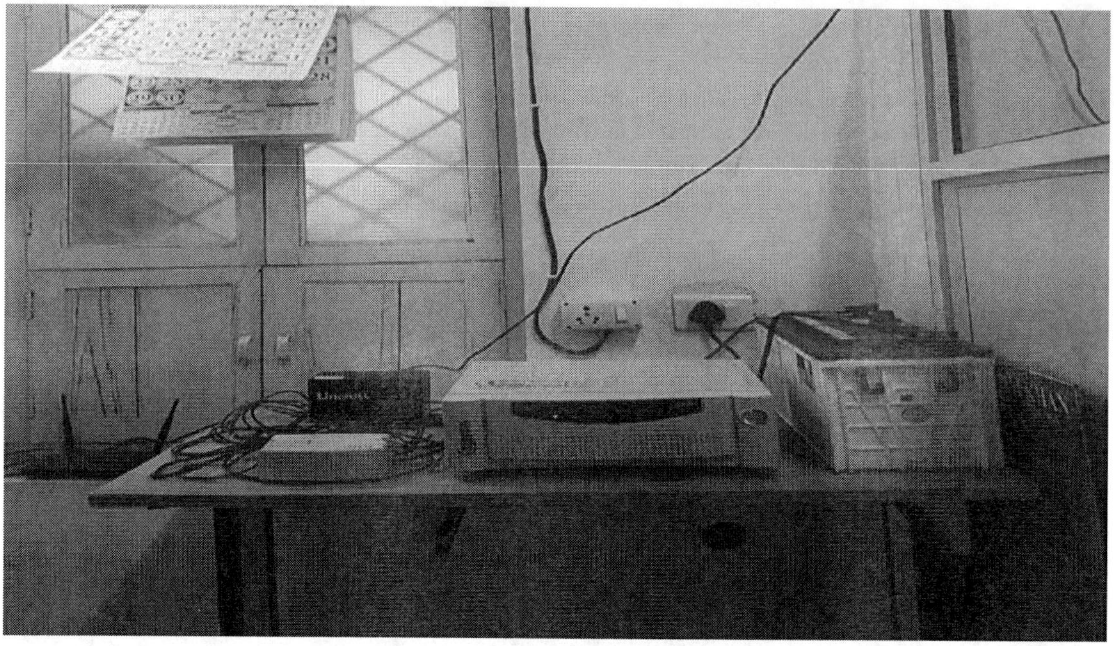

Figure 3. Inverter with Battery setup.

The complete experimental setup included an inverter with life support from a hassle free battery system along with single jointed wire in order to avoid power losses with an aid from dual circuit flow to avoid fluctuations from the system working on it from timely manner based voltage generation (Figure 3). As the internal [6]system is kept under shade in order to keep constant temperature of the internal system and avoid any type of losses due to increment in temperature which increases the capacity of losses during power generation thereby decreasing its life and efficiency also the maintenance cost increases inexorable losses from the whole system of setup generation.

Thus, the readings were calculated and the system power generation was ended from consistent Endeavour from various sites of the whole system either from the output or the input against the system in a moderate and even way out.

4 Conclusions

Therefore, the whole readings were noted using the data generated and it has been concluded that the environment of the institution selected needs a large area for the generation of power so as to meet the requirements and hereby conclude that owing to the shades from nearby places like buildings, plants, residential places increases the [7]maintenance costs, reducing the power generation, increase in the amount of solar collectors in order to reach the consumption limit which indirectly increases the overall cost of the plant thus making it overrated,. Inefficient, time-consuming, lower life time for the overall plant size.

IOP Conf. Series: Materials Science and Engineering 1059 (2021) 012050 doi:10.1088/1757-899X/1059/1/012050

References

[1] https://www.aeee.in/wp-content/uploads/2017/10/AEEE-ECBC-Report-Final-for-NITI-Aayog-BEE-UNDP-GEF.pdf

[2] Alliance for Energy Efficient Economy. Available at: http://www.aeee.in/wp-content/uploads/2017/05/ECBC-Overview.pdf

[3] Alliance for Energy Efficient Economy. Available at: http://www.aeee.in/wp-content/uploads/2017/05/ECBC-Implementation-Approaches.pdf

[4] Scenario of Energy Conservation Building Code (ECBC) in India

[5] GD rai book on Non-conventional Energy Sources

[6] Scenario Human Resources and Management by GD Birla

[7] Energy technology by Khanna publisher's

IOP Conf. Series: Materials Science and Engineering 1059 (2021) 012051 doi:10.1088/1757-899X/1059/1/012051

Multi-response optimization of the Thermal properties of Bio fluids using Taguchi-Grey Analysis

Selvalakshmi [1], Karthikeyan I [2], Immanual R [3], Paul Vinofer A[4], Varadha E[5] Chokkalingam B[6]

[1]Department of Electrical and Electronics Engineering, RMK Engineering college, Thiruvallur, India

[2,3,6]Department of Mechanical Engineering, Sri Ramakrishna Institute of Technology, Coimbatore, India

[4]Department of Mechanical Engineering, Anna University Regional campus Tirunelveli, India,

[5]Department of Mechanical Engineering, Holycross Engineering college, Tuticorin, India,

E-mail: selvalakshmibe@gmail.com

Abstract. In order to study and investigate the overall performance of bio fluids for solar thermal applications, this work consists of integration of Taguchi method with Taguchi Grey Relational Analysis. The preparation of bio fluids has three levels of input control parameters such as Type of oil, Percentage of concentration and Stirrer speed. Output optimization indicators for the experiment were chosen as Thermal conductivity, Density and Viscosity. The comparative study made consists of coconut oil, sunflower oil and castor oil in the bio fluids preparation. These oils are available readily in the market and are quite inexpensive. The bio fluids are prepared by mixing the above oil with water in appropriate concentration. The experiment is run for the optimal solution using Taguchi analysis using L9 orthogonal array. The influential parameters which are significant in contributing the performance of bio fluids were studied using ANOVA table. After multi objective parametric optimization, the control parameters optimization level shows the use of Sunflower oil with 15% concentration at 15000 RPM stirrer speed produces the optimal performance. Based on the Ranking and Sum of squares from ANOVA table, it is evident that the most influential parameter is the type of oil followed by Percentage of concentration and Stirrer speed respectively.
Keywords: Taguchi Grey Relational Analysis, Type of oil, Percentage of concentration, Stirrer speed, Thermal conductivity, Density, Viscosity

1. Introduction

In solar thermal applications various modification are made to improve the heat transfer and overall performance of the collector. Among those modifications the change of heat transfer fluid plays a vital role. Uses of Nano fluids are increasing due to their ability to improve their respective system performance. [1] Use of Aluminum oxide (Al_2O_3) with water at various proportions like 0.1%, 0.4%, and 0.8% in the heat exchanger results in enhanced performance. Also there is an improvement of 18%, 25%, 30% of heat transfer coefficient with the respective proportion of nano fluids and water. Properties like thermal conductivity and pressure drop also shows higher percentage of improvement. Thermos fluids are also used to improve the cavitation flow [2]. Liquid nitrogen is used as a thermal fluid and it causes a temperature drop, which outcomes in a smaller cavity. Nano fluid properties are measured and analyzed by Letícia Raquel Oliveira et.al in which Titanium Iso propoxide was used as a nano fluid and led to Increase in thermal conductivity and viscosity of about 6% and 4% respectively. C. Selvam et.al considered the thermo physical properties of silver nano fluids used with ethylene glycol and water. This led to increase in the thermal performance and also had significant effect on various other properties such as thermal conductivity, viscosity, density and specific heat and showed an increase of about 12%, 35%, 2.16% and 7% for the respective properties when measured by different devices and processes. The use of water with copper nano fluid in a separate layer of a heat exchanger was investigated by Joanna Wilk et.al also the evaluation of the thermo physical properties obtained was carried out. Nano fluids are used as a heat transfer fluid in solar collector [6]. A stable ionic nano fluid improves the optical properties where it acts as a superior absorption material. The stability of this nano fluids found to be good since this fluid can be used for 20 days without any settlement. [7] The thermo physical properties of Nano fluids and various contributing parameters in thermal applications were reviewed. Metals with good thermal conductivity are used as a nanofluids and this goes to a next level where Bio fluids were developed. Bio fluids are derived from some biological plant parts and it is used in the nano form. The properties of brominated vegetable oil which is a bio derived additive is measured and compared with natural, unmodified oil [8]. The results provide increase in density and viscosity and also show a good solubility. Also study on Bio fluids derived from weed and the thermal conductivity compared with various Nano fluids were investigated [09]. Two such different weeds are taken and the enhancement of the thermal conductivity was found to be around 8%.to 103% at different temperature levels. The performance of biofluids also shows some good improvement in the thermal properties, Hence the various bio fluid properties were measured and estimated in this research. Bio fluids can be prepared at various level of concentration and stirrer speed but the best combination is needed to be established so that there is a maximum gain in the performance level. A best optimization technique is required to get better results with various combinations in the bio fluids. Many have employed optimization techniques in them researches to generate better results. In the case of multi-response optimization, Input Parameters of electric discharge machining were optimized for the optimum condition by grey relational analysis [10]. Conditions such as input voltage, capacitance, resistance, feed rate and spindle speed were optimized and this condition provided a micro hole of about 40μm average diameter with 10 aspect ratio. From Taguchi Method analyzation it was Found out that Squeeze pressure is an important parameter which improvises the mechanical property in squeeze casting process [11]. The optimized parameters completely eliminate the defects that were produced before using optimal condition.

Grey Taguchi method is also used to optimize the emulsion process parameters of water in diesel emulsion. Properties like surfactant concentration, stirrer speed and water concentration are optimized for a good output response [12]. Deformation temperature and friction coefficient affect the forging process of helical gear and an optimal value for the above parameter is estimated using Taguchi Method [13].

Optimal parameters are used by [14] in diesel engine and the performance of brake specific fuel consumption and brake thermal efficiency were improvised to 13.3% and 15.1% respectively. [15] 10% of water concentration in diesel was optimized by Taguchi method and the optimized concentration showed reduced NOx emission by 32.5%. The present study deals with the preparation of bio fluids using water and selective oils by stirring action and optimizing its vital input and output parameters. The parametric optimization of input parameters which influence the overall performance of prepared bio fluids such as type of oil, Percentage of concentration, Stirrer Speed are individually optimized for high thermal conductivity, High density and minimum absolute viscosity using normal Taguchi method. The collective overall results were optimized by integrating Normal Taguchi and Grey Taguchi analysis. The main novelty of this work lies in the use of normal Taguchi and grey analysis with a combination of Anova technique to investigate and determine the overall performance of bio fluids and its effective application in solar applications.

2. Materials and Properties

2.1 Materials

Table. 1 Properties of Oils

Properties	**Coconut oil**	**Sunflower oil**	**Castor oil**
Density	924 Kg/m^3	918.8 Kg/m^3	961 Kg/m^3
Boiling point	177°C	232°C	313 °C
Viscosity	80 Centipoise	49.14 Centipoise	650 Centipoise
Thermal conductivity at 25°c	0.1665 W/mK	0.168 W/mK	0.180 W/mK

The purpose of preparing the bio fluids is to find the feasibility as a Heat transfer fluid in solar thermal applications. The comparative study made consists of coconut oil, sunflower oil and castor oil in the bio fluids preparation. These oils are available readily in the market and are quite inexpensive. The bio fluids are prepared by mixing the above oil with water in appropriate concentration. The properties of the different oil used for experiments is listed in the Table 1. The research was carried out using the Coconut oil (Parachute brand), Sunflower oil (Saffola brand) and Castor oil (Generic brand) respectively.

2.2 Bio fluids preparation

The preparation of bio fluids is carried out in room temperature, using an electric variable speed Stirrer (0-15000 RPM), The experimental setup is shown in figure 1. Water is employed as the base fluid for the preparation of bio fluid. Both oil and water were mixed in desired proportions in a transparent glass container and agitated for nearly 30 minutes in order to obtain blending. Photographs were taken at every stage of the sample preparation for documentation purpose and is shown in figure 2. Similar process is repeated for preparation of all the biofluids sample. The prepared bio fluid blends have enhanced properties compared to its base fluids, water and oil.

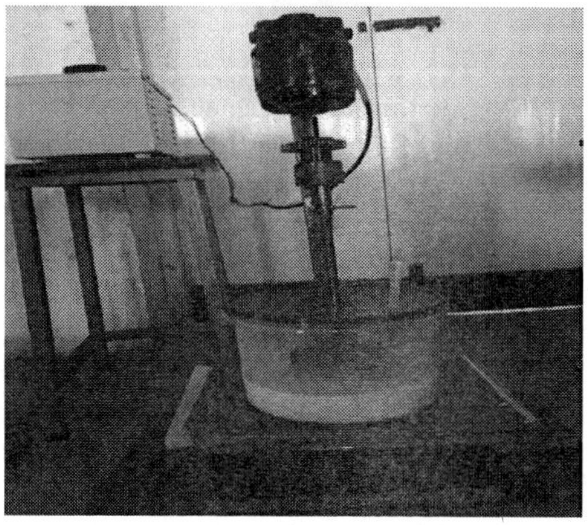

Figure.1 Experimental setup of the BioFluid preparation process

Figure. 2 Bio fluids preparation stages

2.3 Measurement of Bio Fluid Thermal properties

2.3.1 Thermal Conductivity

The thermal conductivity of the prepared fluids was measured by a KD2 Pro Thermal Properties Analyzer device shown in figure 3. For measurement of thermal conductivity, 50 ml of prepared sample were utilized. For accurate results, the experiment is conducted after 20 minutes to allow the system to attain thermal equilibrium. The liquid sample was poured into the measuring cylinder kept inside the chamber to avoid heat convection losses. The needle and the probe of the analyzer were dipped into the test sample and attached to the monitor respectively. Thus the thermal conductivity of the test samples displayed in the monitor is recorded.

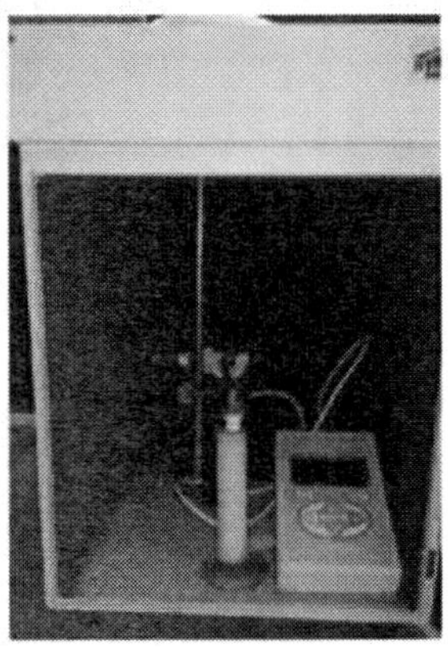

Figure. 3 Thermal Conductivity Measuring Instrument

2.3.2 Density

For measuring density, the hydrometer, an instrument used to measure relative density is utilized. The measuring cylinder is filled with 500 ml of the prepared sample and the measurement is done as shown in figure 4. The hydrometer is carefully inserted into the cylinder. When the hydrometer takes its position, the density of the sample is recorded. The experiment was repeated for accurate results as the sample has to stay still without vibrations.

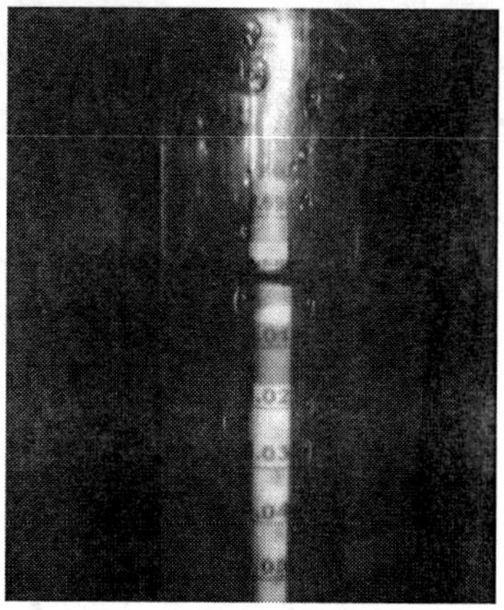

Figure. 4 Density Measurement

2.3.3 Viscosity

The viscosity measurement of the prepared sample is carried out by using Redwood Viscometer as shown in figure 5. For measurement of viscosity, 50 ml of prepared sample is used. Initially, the viscometer cup was cleaned using ether and dried. The sample was poured into viscometer cup and the outer bath was filled with water. A clean thermometer is inserted into the

Figure. 5. Redwood Viscometer

cup along with the stirrer. Heater is switched on until the temperature of the sample reaches 30 °C. Now the sample is made to flow through the orifice to measure Redwood seconds. Time taken for collection of 50 ml samples in the volumetric beaker was measured using a stopwatch. Using the above recordings, absolute viscosity of the samples is calculated.

2.4 Taguchi-Grey analysis

The Design of experiment using the Taguchi orthogonal array approach is the mostly sought and very economical one. It is a type of general fractional factorial design that is based on a design matrix proposed by Dr. Genichi Taguchi and considers a selected subset of combinations of multiple factors at multiple levels. and satisfies the needs of solving problem and products design optimization projects. By applying the Taguchi's method one can reduce the time, resources and money required for little experimental investigation. The DOE provides the relation existing between the input and output variables. In order to determine the factors that influence the performance criteria, Taguchi's orthogonal array L9 (3^4) is used. Also it is possible to determine the factors that are more important than others using this array. The Taguchi technique uses an orthogonal array to deliver the core of input factors with a least number of experiments [16]. In the outdated Taguchi technique, individual response only is optimized in the experimental domain [17-19]. To solve multi- objective optimization problems, the Taguchi method is combined with grey relational analysis [20-22]. The current research involves a Taguchi L9 orthogonal array for evaluation of performance. In the preparation of bio fluid process, the recognize is made for three control variables that act as input parameters: (i) Type of oil (ii) Percentage of Oil Concentration and (iii) Stirrer speed. Table 2 shows the Preparation of bio fluid process parameters and their levels.

Table .2. Bio Fluid Preparation Variables and Levels

VARIABLES	LEVELS		
	1	2	3
A Type of oil	Coconut oil	Sunflower oil	Castor oil
B Percentage of Concentration	5%	10%	15%
C Stirrer speed	5000	10000	15000

The type of oil for different levels was chosen based on the oil's heat transfer efficiency [23-24]. Percentage of oil concentration is selected based on the economic point of view therefore it is arbitrarily chosen as 5 %, 10% and 15% and similarly the fixed Speed levels are 5000, 10000 and 15000 RPM. For solar thermal applications the heat transfer fluids having higher thermal conductivity, High density and low viscosity were recommended [25, 26].The thermal conductivity and density of the bio fluids has been used with "the larger-the-better," characteristic which is calculated as follows:

$$x_{(i)}p = \frac{|y_i(p) - \min y_i(p)|}{\max y_i(p) - \min y_i(p)} \quad (1)$$

The Viscosity of the bio fluids has been considered as "the smaller-the-better," characteristic which is calculated as follows:

$$x_{(i)}p = \frac{\max y_i(p) - y_i p}{\max y_i(p) - \min y_i(p)} \quad (2)$$

Here y_i (p) is the reference order (original), x_i (p) = the comparison order, i = 1, 2,……, m, p = 1, 2,…., n, where n= number of experiments and retorts along with k.

where max $y_i(p)$ are the maximum computed of $y_i(p)$ and min $y_i(p)$ are the minimum computed values of $y_i(p)$, respectively.

The grey relational coefficient $\varphi_i(p)$ can be calculated as follows:

$$\varphi_{(i)}p = \frac{\Delta_{min} + \epsilon\Delta_{max}}{\Delta_{oi}(kp) + \epsilon\Delta_{max}} \quad (3)$$

$$\Delta_{0i}(p) = \|x_0(p) - x_i(p)\| \quad (4)$$

Where Δ_{0i} = difference value which is absolute.

Δ_{max} = maximum value of differences of all compared series (absolute)

Δ_{min} = minimum value of differences of all compared series (absolute)

The main aim of distinguishing coefficient € (0 ≤ € ≤ 1) is to slightly modify the dominant effect of Δ_{max} whenever it is higher. Hence in our present work a value of 0.5 is set for €. The value of y_0 (grey relational grade) can be determined as follows:

$$y_0 = \sum_{i=1}^{n} \Phi_i(p)\delta y_i \ldots\ldots \sum \delta = 1 \qquad \text{Eq (5)}$$

where δ is the weighting factor.

y_0 is the nominal value of the equivalent response.

While converting the multiple grey relation grades, a high weightage is assigned for thermal conductivity rather than for density and viscosity because thermal conductivity directly influences the efficiency of solar thermal applications. The set of following weightage values for different responses for made: Thermal Conductivity = 0.5, Density = 0.2, and Viscosity = 0.3. In orthogonal array, more than one parameter can be varied at a time hence it makes more suitable for the present study. Based on the design variables and levels chosen, L9 orthogonal array carries out the data which is shown in Table 3.

Table 3. Array (L9) of the experimental results.

Trails	Input Process Paramenters			Output Responses			S/N Ratio		
	Type of oil	Percentage %	Speed (rpm)	Thermal Conductivity W/mK	Density Kg/ m³	Absolute Viscosity (Centipoise)	Thermal Conductivity	Density	Absolute Viscosity
1	Coconut Oil	5%	5000	0.525	998	1.0382	-5.59681	59.98	-0.326
2		10%	10000	0.506	1015	1.0032	-5.91699	60.12	-0.028
3		15%	15000	0.326	1024	0.9883	-9.73565	60.206	0.102
4	Sunflower	5%	10000	0.687	990	0.9963	-3.26087	59.	0.0

	oil							91	32
5		10%	15000	0.606	1005	1.0132	-4.35055	60.04	-0.114
6		15%	5000	0.82	1020	1.0482	-1.72372	60.17	-0.409
7	Castor oil	5%	15000	0.459	1005	1.0482	-6.76375	60.04	-0.409
8		10%	5000	0.533	1018	1.0082	-5.46546	60.15	-0.071
9		15%	10000	0.404	1050	1.0032	-7.87237	60.42	-0.028

2.5 Results and Discussions

2.5.1 Determination of optimal Preparation parameters to provide maximum Thermal Conductivity by Taguchi method

To find the Optimum thermal conductivity, the loss function value is modified into a signal to noise ratio (S/N Ratio). S/N ratios are the log functions of the predicted results which would be the as main aim of the optimization problem. Then Larger-the-Better (LTB) characteristic for S/N Ratio is used to calculate deviation of quality function from the expected value. The S/N ratio for Thermal conductivity is calculated using the equation 6,

$$\frac{S}{N} = -10\log\frac{1}{N}\sum_{u=1}^{N_i}\frac{1}{y_u^2} \qquad \text{Eq (6)}$$

Where I denotes the number of measurements indicates the trial number and N_i reflects the number of trials for the experiment

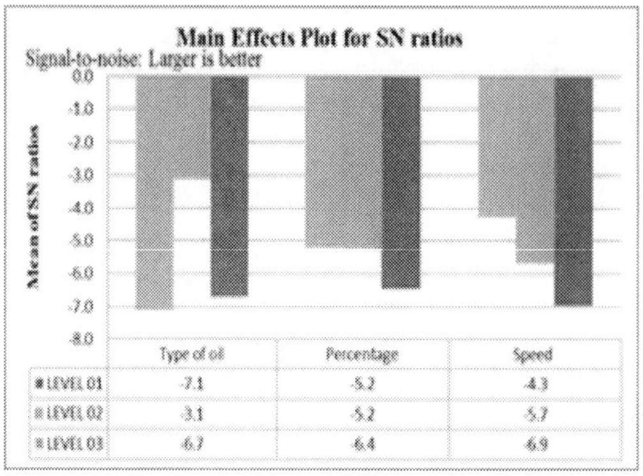

Figure.7 S/N Ratio Plot for individual Parameter at various levels for Optimum Thermal Conductivity

It is necessary to produce the biofluids with maximum thermal conductivity for best performance. From table 3, It is observed that the maximum thermal conductivity (0.82 W/mK) is obtained for the parameter of Type of oil = sunflower oil, Percentage of Concentration= 15% and Stirrer speed= 5000 in the 6th trial. Second and third highest thermal conductivity is also obtained from the same Sunflower oil. Hence the selection of oil plays a vital role in influencing thermal conductivity. On arranging the oil in the order of thermal conductivity, it is seen that Sunflower oil > Castor oil > Coconut oil. Hence it is evident that, other parameters such as % of concentration and Stirrer speed comes into play only after the type of oil. Minimum thermal conductivity (0.326 W/mK) is obtained for the parameter of Type oil=Coconut oil, Percentage of Concentration = 15% and Stirrer speed =15000 in the 3rd trial. Now by applying the Taguchi method to evaluate the best possible solution based on the thermal conductivity values. The plot for the SN ratio obtained (figure 6), larger the S/N ratio, larger is the influence of the parameter in the performance. From the graph, it is observed that the Sunflower oil, with 5% of concentration in 5000 RPM gives the optimal solution for best performance.

2.5.2 Determination of optimal Preparation parameters to provide maximum Density by Taguchi method

To find the Optimum Bio fluid preparation parameters for maximum density, S/N ratio is normalized according to Larger the Better. S/N Ratio is calculated and it is shown in figure 7. Similar to thermal conductivity, the biofluids with high density is analyzed in the trials from table3

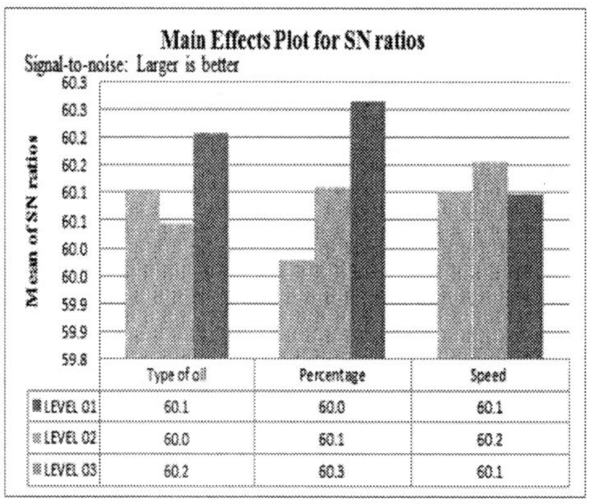

Figure.8 S/N Ratio Plot for Each Parameter at Different levels for Optimum Density

From table 3, It is observed that the maximum Density (1050 Kg/m^3) is obtained for the parameter of Type of oil=Castor oil, Percentage of Concentration=15% and Stirrer speed= 10000rpm in the 9th trial. Now Taguchi method is applied to evaluate the best possible solution based on the density values. Second and third highest density is also obtained from the same 15% of the concentration. On arranging the percentage of concentration in terms of density, it is seen that 15% > 10 % > 5%. Hence the percentage of concentration plays a vital role in influencing density. Hence it is evident that, other parameters such as type of oil and Stirrer speed comes into play only after the percentage of concentration. Minimum density (990 kg/m^3) is obtained for the parameter of Type oil=Sunflower oil, Percentage of Concentration = 5% and Stirrer speed =10000 in the 4rd trial. Now by applying the Taguchi method to evaluate the best possible solution based on the density values. The plot for the S/N ratio obtained (figure 7), is larger the S/N ratio, larger is the influence of the parameter in the performance. From the graph, it is observed that the Castor oil, with 15% of concentration in 10000 RPM gives the optimal solution for best performance.

2.5.3 Determination of optimal Preparation parameters to provide Minimum Viscosity by Taguchi method

To find the Optimum Bio fluid preparation parameters for Minimum Viscosity, smaller, the Better S/N Ratio is calculated and it is shown in figure 9.

$$\frac{S}{N} = log_{10} \sum_{u}^{N_t} y_u^2/n$$

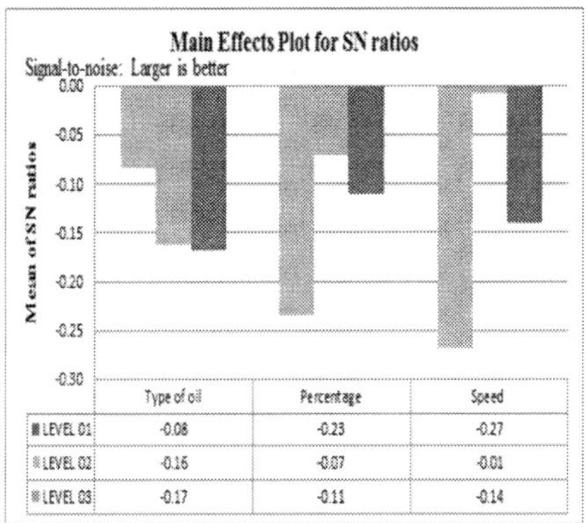

Figure.9 S/N Ratio Plot for Each Parameter at Different levels for Optimum Viscosity

For best performance, the biofluids with low viscosity are analyzed. It is observed that the Minimum Viscosity (0.9883 Centepoise) is obtained in the 3rd run for the parameter; Type of oil =Coconut oil, Percentage of Concentration =15% and Stirrer speed = 15000 rpm. Unlike thermal conductivity and density, no parameter directly dominates in viscosity evaluation. Now by applying the Taguchi method to evaluate the best possible solution based on the viscosity values. The plot for the S/N ratio obtained (figure 8), larger the S/N ratio, larger is the influence of the parameter in the performance. From the graph, it is observed that the Coconut oil, with 10% of concentration in 10000 RPM gives the optimal solution for best performance.

2.5.4 Taguchi Grey analysis

In order to optimize the process parameters for combined optimized outputs, then regulation of the experimental data for determining grey relational grade. Different optimal combinations for various optimal cases are carried out. The S/N ratio for the overall grey relational grade is calculated. Table 4 shows the grey relational coefficients, grey relational grades, and S/N ratios of each run and ranks the runs based on grey relational grade. R.No.6, has the greatest response among all cases showing the Type of oil for level 2, Percentage of contribution for level 3, Stirrer speed for level 1 (Sunflower oil ,15% concentration at a stirrer speed of 5000 rpm).This is found to have a good output response during the L9 orthogonal array experiments and is believed to be almost precise to the optimal value.

Table.4 Grey relational grade, S/N Ratio and Rank table.

Trail	Output Responses			Grey relational coefficient			Grey relational grade	S/N ratio	Rank
	Thermal Conductivity	Density	Absolute Viscosity	Thermal Conductivity	Density	Absolute Viscosity			
	W/mK	Kg/ m3	(Centipoises)	0.5	0.2	0.3			
1	0.525	998	1.0382	0.456	0.366	0.375	0.41	-7.66	8
2	0.506	1015	1.0032	0.440	0.462	0.668	0.51	-5.80	5
3	0.326	1024	0.9883	0.333	0.536	1.000	0.57	-4.82	4
4	0.687	990	0.9963	0.650	0.333	0.789	0.63	-4.03	2
5	0.606	1005	1.0132	0.536	0.400	0.546	0.51	-5.81	6
6	0.82	1020	1.0482	1.000	0.500	0.333	0.70	-3.09	1
7	0.459	1005	1.0482	0.406	0.400	0.333	0.38	-8.33	9
8	0.533	1018	1.0082	0.463	0.484	0.601	0.51	-5.87	7
9	0.404	1050	1.0032	0.373	1.000	0.668	0.59	-4.63	3

To predict the significance of process parameters, the total irregularity of grey relational grades must be distinguished by calculating the mean sum of squares (SS). SS can be found by using:

$$SS = \sum_{i=1}^{K}(y_i - y_m)^2 \qquad \text{Eq (7)}$$

Where y_i is the mean response for the i_{th} experiment and y_m is the grand mean of the response. Depending on the DOF and SS, value of mean sum of squares is computed. Table 5 lists the S/N ratio responses and the significance of the process parameters on overall performance. All three input parameters factors ,Type of oil: 67.80%, stirrer speed: 13.84% and Percentage of concentration: 18.36% significantly contribute to increased performance as shown in Table 5. Moreover, the optimum process parametric setting is A2B3C1, in which Type of Oil = sunflower oil, Percentage of concentration = 15% and stirrer speed = 5000 rpm. Grey Analysis has been made for the above three tests. It has been noticed that the proportions Sunflower oil+ 15%+ 15000 RPM gives the Optimum output.

Table 5. Analysis of Variance (ANOVA) Table

Level	Type of Oil	% of Concentration	Stirrer Speed
1	0.5081	0.5428	0.5959
2	0.653	0.5072	0.534
3	0.489	0.6	0.5202
Delta	0.164	0.0927	0.0757
Rank	1	2	3
Optimum level	A2	B3	C1
SS	0.048	0.013	0.0098
DOF	2	2	2

The optimal sample has been prepared for the proportions of Sunflower oil+ 15%+ 5000 rpm

3. CONCLUSION

A statistical approach of design of experiments by Taguchi techniques was used to minimize the number of experiments and for getting optimal results. The influence of processing parameters for the thermal conductivity properties location was analyzed. The optimum parameters were obtained from Design of Experiments using L9 array. Thermal Conductivity, Density, Viscosity of various proportions of liquids were observed. From the properties of the bio fluids the Taguchi Analysis has been made.

Key results are summarized as follows;
The results are identified for maximum thermal conductivity, maximum density and minimum viscosity.

1. Maximum thermal conductivity of 0.82 mK is obtained for Type of oil=Sunflower oil, Percentage of concentration= 15% and Stirrer speed =5000 RPM. The optimal thermal conductivity is identified in Type of oil=Sunflower oil, Percentage of concentration= 15% and Stirrer speed =5000 RPM.
2. Maximum density of 1050 kg/m^3 is obtained for Type of oil=Castor oil, Percentage of concentration = 15% and Stirrer speed =10000 RPM. The optimal density is identified in Type of oil=Castor oil, Percentage of concentration=15% and Stirrer speed =10000 RPM
3. Minimum viscosity of 0.9883 centipoise is obtained for Type of oil=Coconut oil, Percentage of concentration = 15% and Stirrer speed =15000 RPM. The optimal viscosity is identified in Type of oil=Castor oil, Percentage of concentration= 10% and Stirrer speed =10000 RPM
4. The best possible solutions are identified by the Grey Analysis and optimal combination was determined as Sunflower oil with 15% concentration at 5000 rpm.

Nomenclature
Rpm- Revolution per minute

S/NRatio – Signal to Noise ratio
SS – Sum of Squares
DOF – Degree of Freedom

References

[1] P. C. MukeshKumar J, Kumar, R. Tamilarasan, S. Sendhil Nathan and S. Suresh, 2014 Heat transfer enhancement and pressure drop analysis in a helically coiled tube using Al2O3 / water nanofluid, *Journal of Mechanical Science and Technology* 28 (5) 1841-1847.
[2] Tairan Chen, Biao Huang, Guoyu Wang and Kun Wang, 2015 Effects of fluid thermophysical properties on cavitating flows, *Journal of Mechanical Science and Technology* **29** (10) 4239~4246.
[3] Letícia Raquel Oliveira, Anielle Christine Almeida Silva, Noelio Oliveira Dantas, Enio P and BandarraFilho, 2017 Thermophysical properties of TiO2-PVA/water nanofluids, *International Journal of Heat and Mass Transfer* **115** 795–808.
[4] C. Selvam, D. Mohan Lal, and Sivasankaran Harish,2016 Thermophysical properties of ethylene glycol-water mixture containing silver nanoparticles, *Journal of Mechanical Science and Technology* **30** (3) 1271-1279.
[5] Joanna Wilk, Robert Smusz and Sebastian Grosicki, 2017 Thermophysical properties of water based Cu nanofluid used in special type of coil heat exchanger, *Applied Thermal Engineering* S1359-4311(17)30088-1.
[6] Wenjing Chen, ChangjunZou and Xiaoke Li, 2017 An investigation into the thermophysical and optical properties of SiC/ionic liquid nanofluid for direct absorption solar collector, *Solar Energy Materials & Solar Cells* **163** 157–163.
[7] MunishGuptaa, VinaySingha, Rajesh Kumara and Z. Saidb, 2017 A review on thermophysical properties of nanofluids and heat transfer applications, *Renewable and Sustainable Energy Reviews* **74** 638–670.
[8] Matthew J. Thomas, Katlyn A. Bramblett, Blane D. Green and Kevin N.West, 2015 Thermophysical and absorption properties of brominated vegetable oil, *Journal of Molecular Liquids* **211** 647–655.
[9] Meher Wan, RashmiParashar, Neeraj Kumar, R.R. Yadav, Rajiv Prakash, Jane Catherine Ngila and VyomParashar,2015 Heat transfer biofluids: A novel approach towards weed management, *Ecological Engineering* **84** 492–495.
[10]Jong Hyuk Jung and Won Tae Kwon, 2010 Optimization of EDM process for multiple performance characteristics using Taguchi method and Grey relational analysis, *Journal of Mechanical Science and Technology* **24** (5) 1083-1090.
[11] P. Senthil,and K. S. Amirthagadeswaran, 2012 Optimization of squeeze casting parameters for non symmetricalAC2Aaluminium alloy castings through Taguchi method, *Journal of Mechanical Science and Technology* **26** (4) 1141~1147.
[12] Suresh Vellaiyan,and K. S. Amirthagadeswaran, 2016 Taguchi-Grey relational-based multi-response optimization of the water-in-diesel emulsification process, *Journal of Mechanical Science and Technology* **30** (3) 1399~1404.
[13] Wei Feng and Lin Hua, 2011 Multi-objective optimization of process parameters for the helical gear precision forging by using Taguchi method, *Journal of Mechanical Science and Technology* **25** (6) 1519~1527.
[14] Suresh Vellaiyan and Koduvayur S. N. Amirthagadeswaran, 2017 Multi-Response Optimization of Diesel Engine Operating Parameters Running with Water-In-Diesel Emulsion Fuel, *Thermal Science*, **21**, 427-439.
[15] Suresh Vellaiyan and K S. Amirthagadeswaran Emission characteristics of water-emulsified diesel fuel at optimized engine operation condition, *Petroleum Science and Technology*, **35** 13, 1355-1363
[16] P. Senthil and K. S. Amirthagadeswaran,2012 Optimization of squeeze casting parameters for non symmetricalAC2Aaluminium alloy castings through Taguchi method, *Journal of Mechanical Science and Technology*, **26** (4) 1141-1147.
[17] W. H. Yang and Y. S. Tarng, 1998 Design optimization of cutting parameters for turning operations based on the Taguchi method, *Journal of Materials processing Technology*, **84** (1- 3) 122-129.
[18] D. C. Chen and C. F. Chen,2007 Use of Taguchi method to study a robust design for the sectioned beams curvature during rolling, *Journal of Materials processing Technology*, **190** (1) 130-137.

[19] C. P. Fung and P. C. Kang,2005 Multi-response optimization in friction properties of PBT composites using Taguchi method and principal component analysis, *Journal of Materials Processing Technology*, **170** (3) 602-610.

[20] W. Feng and L. Hua, 2011 Multi-objective optimization of process parameters for the helical gear precision forging by using Taguchi method, *Journal of Mechanical Science and Technology*,**25** (6) 1519-1527.

[21] J. H. Jung and W. J. Kwon,2010 Optimization of EDM process for multiple performance characteristics using Taguchi method and Grey relational analysis, *Journal of Mechanical Science and Technology*, **24** (5) 1083-1090.

[22] H. Aydin, A. Bayram, U. Esme, Y. Kazancoglu and O.Guven, 2010 Application of grey relational analysis(GRA) and Taguchi method for the parametric optimization of friction stir welding (FSW) process, *Materials and Technology*, **44** (4) 205-211.

[23] Putri WA, Fahmi Z, Sutjahja IM, Kurnia D, Wonorahardjo S. Thermo physical parameters of coconut oil and its potential application as the thermal energy storage system in Indonesia. *In Journal of Physics: Conference Series, 739*, 012065. IOP Publishing.

[24] Castro MP, Andrade AA, Franco RW, Miranda PC, Sthel M, Vargas H, Constantino R, Baesso ML. 2005 Thermal properties measurements in biodiesel oils using photo thermal techniques. *Chemical Physics Letters* **5**, 18-22.

[25] Vignarooban K, Xu X, Arvay A, Hsu K, Kannan AM.2015 Heat transfer fluids for concentrating solar power systems–a review. *Applied Energy*, **146,** 383-96.

[26] Hoffmann, J.F., Vaitilingom, G., Henry, J.F., Chirtoc, M., Olives, R., Goetz, V. and Py, X., 2018. Temperature dependence of thermo physical and rheological properties of seven vegetable oils in view of their use as heat transfer fluids in concentrated solar plants. *Solar Energy Materials and Solar Cells*, **178**, 129-138.

IOP Conf. Series: Materials Science and Engineering 1059 (2021) 012052 doi:10.1088/1757-899X/1059/1/012052

Comparative study of Al6061 and AlSi10Mg produced by Selective Laser Melting Process

A Sabarinivas[1] and Dr R Subramanian[2]
[1] PG Scholar, PSG College of Technology, Coimbatore, India
[2] Professor, PSG College of Technology, Coimbatore, India

E-mail: sabari.creater@gmail.com

Abstract. Alloy 6061 has an excellent corrosion resistance property. The main advantage of the Aluminium alloy is lightweight. The alloy may be high in strength and suitable for heat treatment. This article discusses the comparison between two different materials namely As-received Al6061 and AlSi10Mg of the component produced by Selective Laser Melting process (SLM). Selective Laser Melting is one of the techniques of Additive Manufacturing or 3D printing. In recent years, Additive manufacturing is increasingly used in several industries including healthcare, automotive, aerospace and defence.

1 Introduction

As-received component and SLM 3D oriented materials are subjected to the heat-treatment process. The cast components in an aluminium alloy, when processed conventionally, often require heat treatment to improve mechanical properties. Heat treatment includes T6 cycle of solution annealing, quenching and age hardening. For applications which require a combination of good thermal properties and low weight, parts in EOS AlSi10Mg serve the purpose [3]. Solution heat treatment applies to heat-treatable alloys. It involves a heat treatment process in which alloying constituents are taken into solution, followed by rapid quenching [4]. Additive manufacturing is a process which can produce complex shapes which are quite difficult through traditional casting and machining methods. The process has the flexibility in material selection and can give high accuracy of final parts after fabrication of the component [6].

Some of the most common processes in additive manufacturing are Selective Laser Melting (SLM), Ballistic Particle Manufacturing (BM), Liquid Thermal Polymerization (LTP), Fused Deposition Modelling (FDM) etc. Here, the Additive Manufacturing technique was used. Selective laser melting, an additive manufacturing technique can melt the metal fully and convert into a solid three-dimensional part [4,6]. Some of the alloys used in the process are 15-5 stainless steel, cobalt-chromium, titanium Ti6Al4V, aluminium AlSi10Mg [1]. The material used here is aluminium alloy AlSi10Mg (Silicon - 10%, Magnesium - 0.20-0.45%, Aluminium – balance).

2 Experimental Work

Square shape rod of 10mm length and 5.8mm height of samples were produced using an SLM 280HL machine. In the build operation, 3 samples were produced. Then comparison was done with As-received Al6061 rod with a 25mm length and height of 30mm diameter. For the study of mechanical properties (strength, hardness), a tensile test was performed. Then microstructure was analysed for both materials of Al6061 and AlSi10Mg of heat-treated using Vertical muffle furnace. The furnace had its maximum temperature up to 950°C with a chamber size of 180mm×200mm under non-heat-treated conditions. The following parameters (scanning speed, laser power, hatch angle) were used in the processing of SLM AlSi10Mg alloy.

- Layer thickness-30μm
- Laser power
  - Standard contour-320W,560mm/s(laser scan speed)
  - Down contour-80W
  - On part contour-320W,50mm/s(laser scan speed)
- Infill stripe type exposure with a stripe of 7mm and an overlap of 0.2mm with a laser power of 370W and a scan speed of 1300mm/s.Hatch spacing of 0.13mm
- Up-skin -no exposure pattern,with laser power 360W and scan speed 1000mm/s
- Hatch spacing 0.21mm
- Down skin- no exposure pattern, with laser power 360W and scan speed 1500mm/s. Hatch spacing 0.1mm.
- Hatch angle rotation of 67° every layer
- The laser power of 370W and scan speed of 3100mm/s, unidirectional scanning. Layer thickness is 60μm

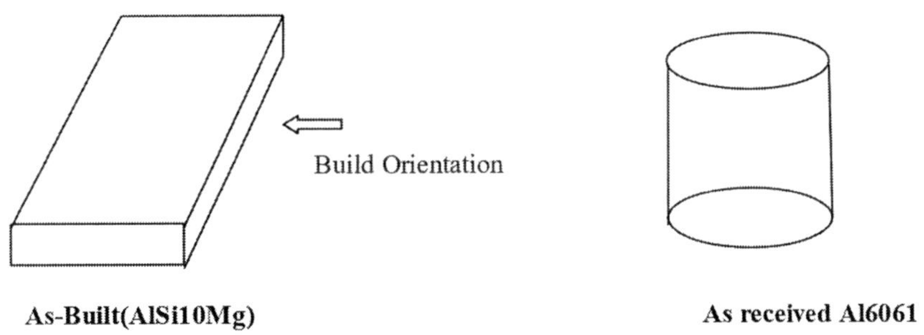

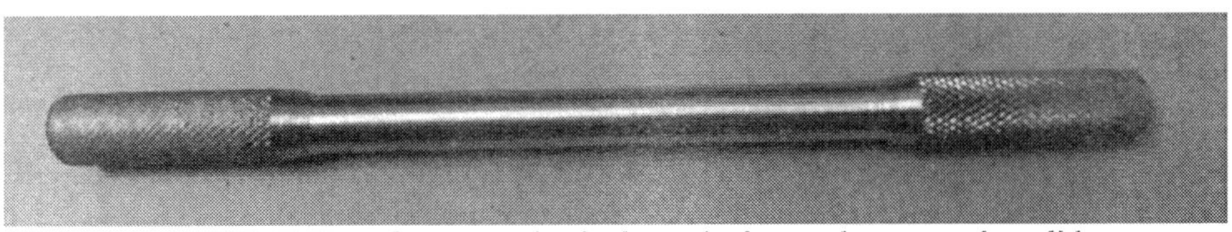

Tensile test specimen for As-received Al6061 in the non-heat treated condition

2.1 As received Al6061 and 3D printed AlSi10Mg

The material chosen for the research is Al6061 and AlSi10Mg.The chemical composition for Al6061 and AlSi10Mg was analysed by using Optical Emission Spectroscopy [1,2].

Table 1. Elemental composition of As-received Al6061

Si %	Fe%	Cu%	Mn%	Mg%	Cr%
0.698	0.595	0.726	0.326	1.657	0.145
0.740	0.561	0.751	0.325	1.779	0.138
0.719	0.578	0.738	0.326	1.718	0.142

Ni%	Zn%	Ti%	Ag%	Al%
0.035	2.237	0.025	<0.00050	93.38
0.030	2.446	0.021	<0.00050	93.13
0.033	2.386	0.023	<0.00050	93.25

Table 2. The elemental composition of 3D printed AlSi10Mg

Element	Composition%
Al	Balance
Si	9-11
Fe	0-0.055
Cu	0-0.1
Mn	0-0.45
Mg	0.20-0.45
Ni	0-0.05
Zn	0-0.10
Pb	0-0.05
Sn	0-0.05
Ti	0-0.15

3 Results and Discussions

3.1 Microstructure (Comparison between Al6061 and AlSi10Mg)

The microstructure of the samples was observed through optical microscopy [1,2,4,7].

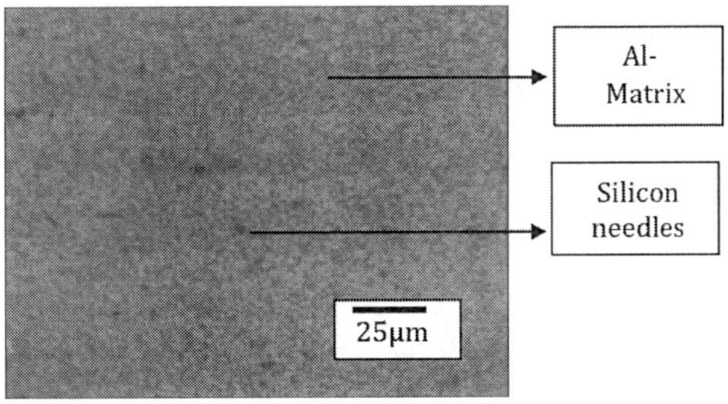

Figure 1a. The microstructure of As-received (Al6061)

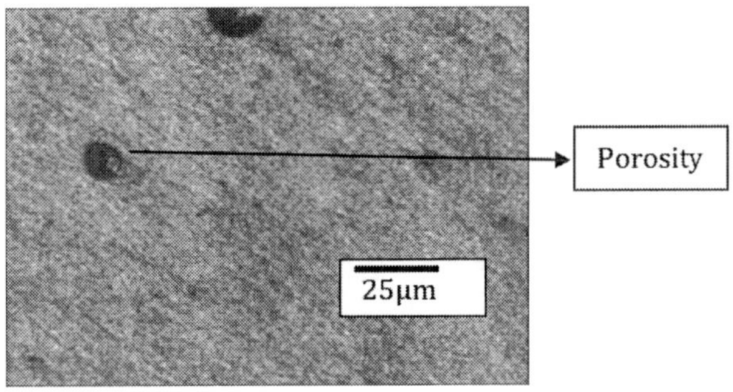

Figure 1b. The microstructure of SLM of AlSi10Mg

Optical microscope Image of Fig1a As-received Al6061 showed the presence of Si needles in an Al-matrix at Non-Heat treated condition. Fig1b AlSi10Mg shows the presence of porosity in an Al-matrix at Non-Heat Treated Condition.

3.2 Tensile Test

The tensile test was performed For As received Al6061[Fig2] and the result was as illustrated

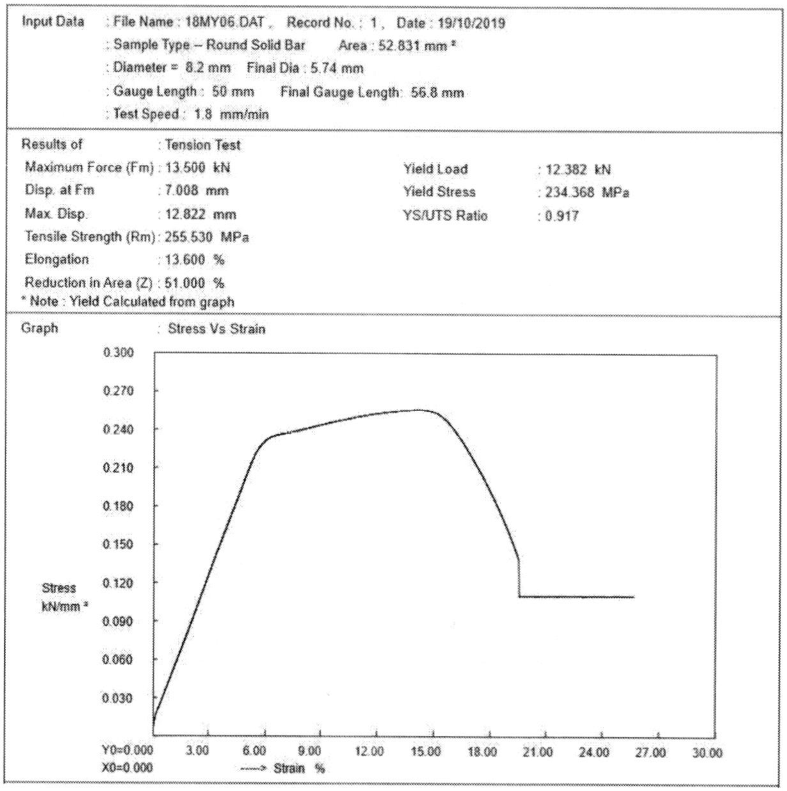
Input Data : File Name : 18MY06.DAT , Record No. : 1 , Date : 19/10/2019
: Sample Type -- Round Solid Bar Area : 52.831 mm ²
: Diameter = 8.2 mm Final Dia : 5.74 mm
: Gauge Length : 50 mm Final Gauge Length: 56.8 mm
: Test Speed : 1.8 mm/min

Results of : Tension Test
Maximum Force (Fm) : 13.500 kN
Disp. at Fm : 7.008 mm
Max. Disp. : 12.822 mm
Tensile Strength (Rm) : 255.530 MPa
Elongation : 13.600 %
Reduction in Area (Z) : 51.000 %
* Note : Yield Calculated from graph
Yield Load : 12.382 kN
Yield Stress : 234.368 MPa
YS/UTS Ratio : 0.917

Graph : Stress Vs Strain

Figure 2. Tension test result graph

3.3. Fracture Surface

Figure 3a. SEM Image of a tensile test sample

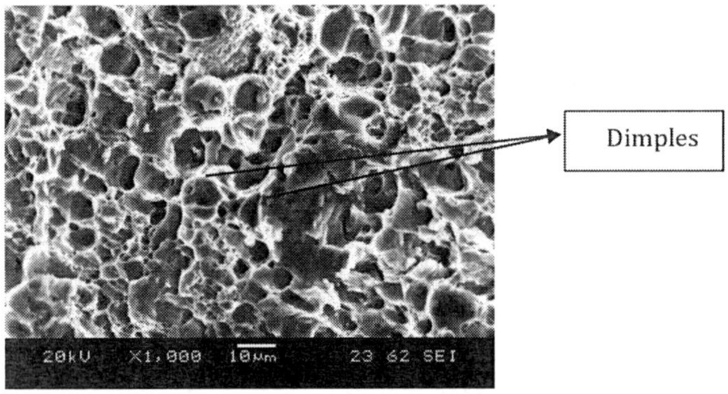

Figure 3b. SEM image of a fractured surface of As-received Al6061

SEM images (Fig 3a, 3b)of the fracture surface of As-received Al6061 and Engineering stress-strain curve (Fig 2) showed that the As-received Al6061 material is in a ductile manner in non-heat treated condition. The SEM image (Fig 3b) represented the elongated dimples which were seen in non-heat treated Al6061 and also enhanced ductility. It is hard to analyse the material as ductile or brittle as it is based on elongation present only [1,6,7]. The fractography (Fig 3a) indicated that the material breaking point occurred on high load=12.382 KN.

3.4. Heat treatment process

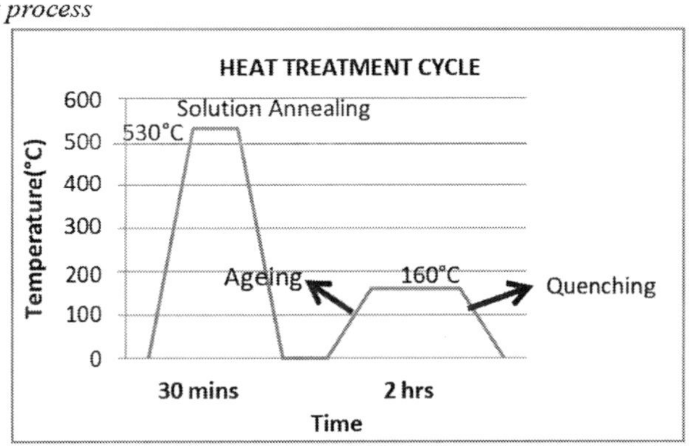

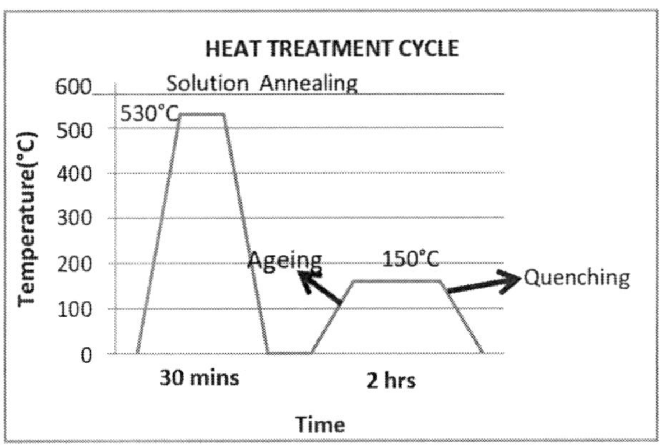

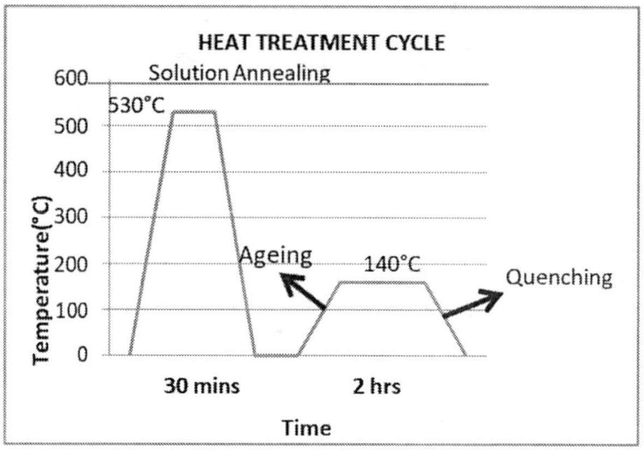

Figure 4. Heat treatment cycles

The samples As-received Al6061 and AlSi10Mg were heat-treated following a T6 condition by solution heat treatment (SHT) for 30minutes at 530°C [8,9] (figure 4). After this quenching was done, followed by artificial ageing at three different temperatures 140°C,150°C,160°C for 2 hours. Here the temperatures did not change the morphology of the Al6061 and AlSi10Mg samples. But the temperatures showed an effect in the hardness of the materials. After ageing, the hardness levels were 71.66Hv at 160°C,67.8Hv at 150°C and 66.2Hv at 140°C respectively for As-received heat-treated Al6061 sample. In AlSi10Mg SLM sample, the hardness level was 72.4Hv at 160°C. Better hardness was observed when compared to that of Al6061. The storage time between quenching the samples and start of the ageing process was kept to a minimum of 1 hour[9].

3.5. Microstructure (Al6061 vs AlSi10Mg with heat treatment)

The microstructure of the samples subjected to the heat-treatment process, followed by T6 condition was observed.

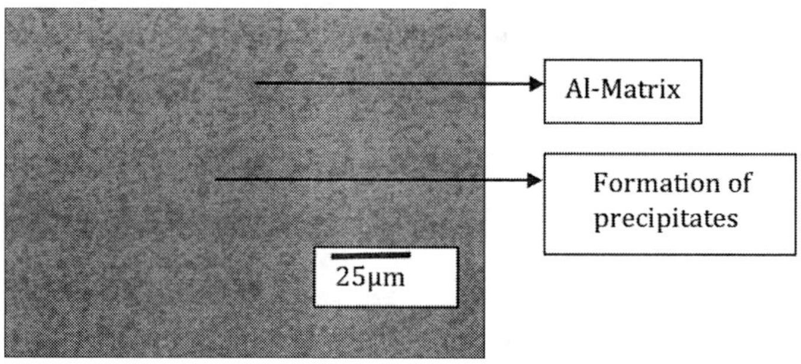

Figure 5a. Solution heat treatment of As-received Al6061 at 530°C

The Image(Fig5a), shows, Precipitates Formed and uniformly distributed in Al phases.

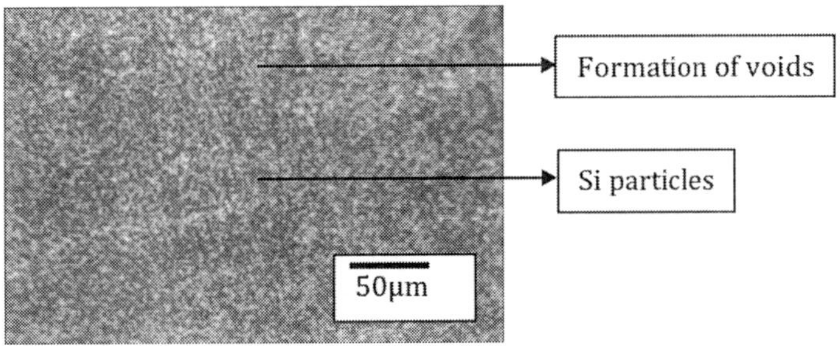

Figure 5b. Solution heat treatment of As-built AlSi10Mg at 530°C

The Image(Fig 5b), it was observed that the voids nucleated at the Si particles. This void affect the hardness, of the material

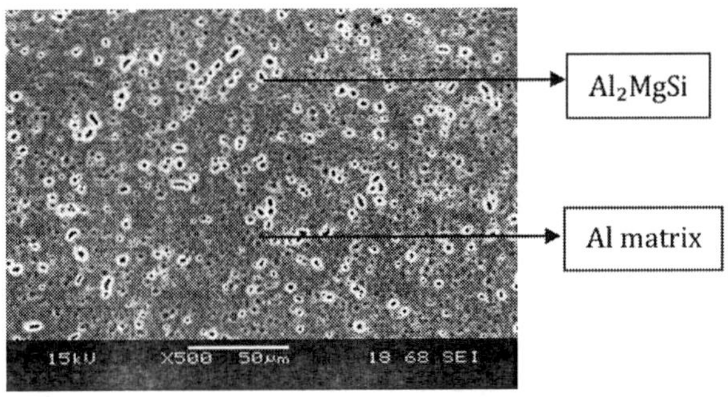

Figure 6. SEM image of T6 condition at 160°C (As received Al6061)

The SEM image(Fig6) shows the silicon and magnesium were dispersed in an Aluminium matrix, coarser Silicon particles were uniformly distributed.

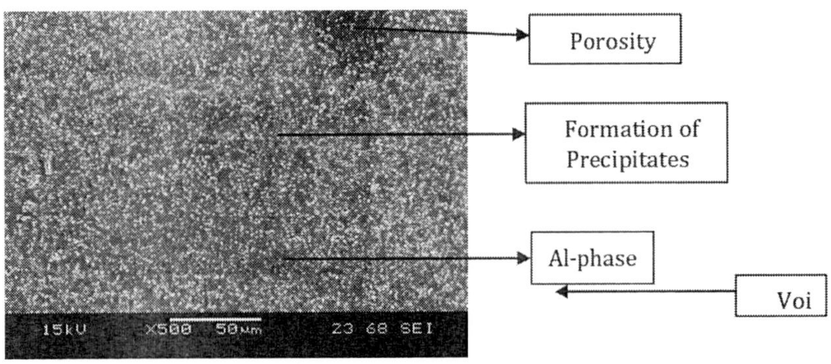

Figure 7. SEM image of T6 condition at 160°C (As-built of SLM of AlSi10Mg)

The SEM image(Fig7) showed fine microstructure, which was achieved after heat treatment of solution Annealing and Ageing.

3.6. Hardness

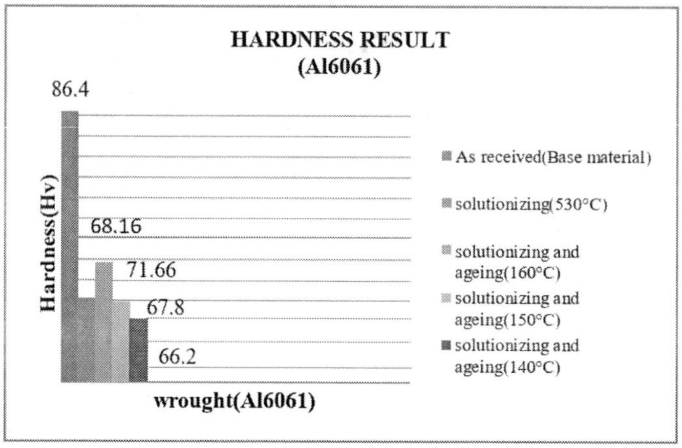

Figure 8. Hardness result for As-received Al6061 (Non-Heat treated vs Heat-treated)

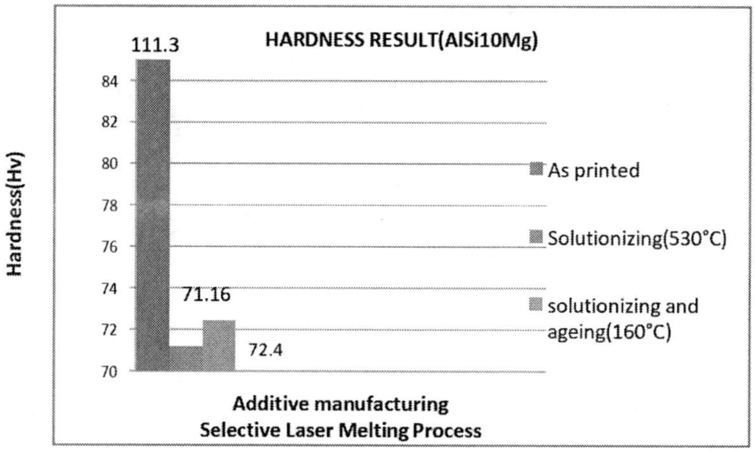

Figure 9. Hardness result for As-built AlSi10Mg (Non-Heat treated vs Heat treated)

The hardness test was carried out to evaluate the influence of heat treatment on the hardness of the specimen[7]. From the test conducted(Vickers hardness), it was observed that hardness of heat-treated specimen was lower than that of As-built specimen. The higher hardness value indicates a fine dispersion of Si particles in the Al matrix. Hence, the As- built specimen held higher hardness value. After solution treatment, a decrease in the hardness value was observed in As-built Al6061. The hardness value decreases on increasing solution treatment temperature. In this research, the microstructure of As-received Al6061 and AlSi10Mg in heat-treated and Non- heat-treated were studied. As received microstructure showed that precipitates were formed in the aluminium matrix at Non-heat treated condition. The SLM of AlSi10Mg contained defects like porosity in Non-heat Treated condition.

The pores in the SLM parts are classified into three types namely fusion errors, gas pores and shrinkage pores. The formation of pores would result in poor metallurgical bond [4]. Various mechanical properties of aluminium alloys have been reported in the literature, such as tensile behaviour, hardness. Aluminium 6061 is tested, and the stress-strain curve shows that material is in ductile condition and better ductility based on elongation [1].

Proceeding heat treatment T6 condition, Si particles are distributed uniformly along with the alpha aluminium matrix and Si particles up to 0.719%. For the heat-treated material, voids nucleate at Si particles by de-cohesion. After treatment, the microstructure has no fine grains. Also, there is a reduction of contributing grain boundaries to strengthening [1,2,6,7].

4 Conclusion

In the study of research, Aluminium alloys have better strength when compared to other materials. In research work, it is clearly explained that As-received Al6061 is ductile and its good strength is shown through stress versus strain curve.

On comparing As-received Al6061 and SLM of AlSi10Mg, AlSi10Mg is better Hardneabilty than that of As-received, and Fine microstructure was achieved at the Heat Treatment condition(Ageing)

This research finds out problem identification in these components through the works conducted. After the heat treatment process, the hardness is decreased within an ageing time of 2 hours. After ageing, the hardness of the material is slightly reduced because of Time, Temperatures, In Al6061 Defects are lesser than that of Asbuilt AlSi10Mg, But Better Hardenability is Achieved in AlSi10Mg. Selective laser Melting process of AlSi10Mg Process have less time to produce the material compared to conventional Process and Low in cost

In future ultrafine grain refinement can be achieved for increased better strength, hardness and reducing porosity by using severe plastic deformation technique (SPD) of Equal Channel Angular Pressing (ECAP) method.

References

[1]Zrinik J, Dobatkin S V, Mamuzic I 2008 Processing of metals by severe plastic deformation (SPD)-the structure and mechanical properties responds *Metalurgija* **47(3)** 213-215

*[2]*Benjamin Fulcher A, David Leighk, Trevor Watt J 2012 Comparison of AlSi10Mg and Al6061 processed through DMLS *Proceedings of the Solid Freeform Fabrication (SFF) Symposium, Austin* **46** 405-406

[3]Surya Prakash Allampalli 2013 Influence of swaging on subsequent annealing behaviour of Aluminium 6061

[4]Kunze Olakanmi EO 2013 Selective laser sintering/melting (SLS/SLM) of pure Al, Al–Mg, and Al–Si powders: Effect of processing conditions and powder properties *Journal of Materials Processing Technology* **213(8)** 1387-1405

[5]Theresa Hanemann, Luke Carter N, Moritz Habschied, Nicholas Adkins J E, MoatazAttallah M, Martin Heilmaier 2019 *Journal of Alloys and Compounds* **795** 13-15

*[6]*Caplar John Martin H, Brennan Yahata D, Jacob Hundley M, Justin Mayer A, Tresa Pollock M 2017 *3D printing of high-strength aluminium alloys*

[7]Wei Li, Shuai Li, Ang Zhang, Yan Zhou 2016 Effect of heat treatment on AlSi10Mg alloy fabricated by selective laser melting: Microstructure evolution, mechanical properties and fracture mechanism *Material Science & Engineering A* **663** 116-125

[8]Heat Treating of Aluminium Alloys *ASM Handbook* **4** Heat Treating 841-879

*[9]*Jacobs M H 1999 *Precipitation Hardening (Training in Aluminium Application Technologies, Interdisciplinary Research Centre in Materials)* 26-32

[10] Nesma Aboulkhair T, Chris Tuck, Ian Ashcroft, Ian Maskery 2016 *Materials Science and Engineering A* **84** 15-18

[11] Zhang J, Wei Q, Bourell D, Shi Y 2019 *Journal of Materials Science and Technology* **35** 270-84

IOP Conf. Series: Materials Science and Engineering 1059 (2021) 012053 doi:10.1088/1757-899X/1059/1/012053

Performance Optimization of Thermoelectric Generators using Taguchi Method

P. Ragupathi[1], Debabrata Barik[2], S. Aravind[2], G. Vignesh[2]

[1]Research Scholar, Department of Mechanical Engineering, Karpagam Academy of Higher Education, Coimbatore, India, 641 021.
[2]Department of Mechanical Engineering, Karpagam Academy of Higher Education, Coimbatore, India, 641 021.

ponragupathi@gmail.com

Abstract. The thermoelectric generators (TEGs) are devices that are utilized to convert the heat energy into electrical energy directly and the working principle of this device is based on See beck effect. Thermoelectric power production is a smart method for the direct translation of heat energy into an electrical one. This work explores a method to get the optimum process parameters on the performance of various TEGs by finding the conversion efficiency to recover the waste heat and converts it into electricity. For this purpose, an experimental setup was designed and fabricated to determine the performance of TEGs. The TEGs made by Bismuth Telluride (Bi_2Te_3), Lead Telluride (PbTe), and Aluminium Oxide (Al_2O_3) were taken for the performance analysis. The process variables are heat input, TEG material and temperature difference. The experiments were conducted by using Taguchi's L9 orthogonal array to reduce the number of experiments. The results found that the heat input of 90W, TEG material of Bi_2Te_3 and the temperature difference of 75°C gives the maximum conversion efficiency of 2.45% from thermal to electrical energy. The statistical analysis of variance (ANOVA) showed that the most influential parameter on the performance of TEGs was heat input. The R^2 and R^2 (adj) values were found to be 93.25% and 86.50%, this shows that the developed model is significant and can predict the optimal solution.

1 Introduction

All the equipments and machines which are used to do work and all other energy transformations produce heat along with the desired output power. This heat is directly allowed to pass in the environment and wasted, so it is called as waste heat. This wasted heat after being passed in the atmosphere causes the rise in the average temperature of the surroundings. This has unpleasant result on the environment [1-3]. In general, the heat energy produced is not fully converted into useful work and some amount of that energy is wasted to the surroundings. Practically there is no potential to transfer all the input energy into work [4, 5]. In this condition, increasing the efficiency of that equipment and machines directly is somewhat difficult. By utilizing the waste heat for other purposes will indirectly shows the increasing of efficiency of that machine. Electricity generation is an alternative and effective method to utilize this waste heat. The direct conversion of heat energy into electricity is done by TEGs [6-8]. The TEGs are working based on the Seebeck effect. This effect is a phenomenon in which the voltage will be generated if there is a temperature gradient between two dissimilar electrical conductors or semiconductors [9, 10]. TEGs comprise of three main elements: Thermoelectric (TE) systems, TE materials and TE modules. Temperature gradient from heat is converting into electric voltage and generates electricity by TE materials. These TE materials should have the properties of higher electrical conductivity (σ_e) and lower thermal conductivity values (κ_t)

[11]. The lower thermal conductivity value makes sure that if one side of TEG is made as hot then another side remains as cold, which tends to cause a great temperature gradient. The Seebeck coefficient (S) is the measure of the degree of electrons flow with respect to the temperature gradient. The figure of merit zT ($S^2\sigma T/k$) decides the performance of a given material to generate the TE power [12, 13]. Bi_2Te_3, PbTe, and Al_2O_3 are largely used as TE materials as they have the properties of low thermal conductivity and high electrical conductivity [14]. These TE materials contain unusual elements which make them very costly. The TE module is made up of units of TE materials and is joined electrically in series and thermally in parallel [15]. The negatively and positively charged (n-type and p-type) semiconductor are joined to form a couple. There are several couples in the module. If there is a temperature gradient exists between the two materials then the electricity will flow in the circuit. The TE system produces electric power by absorbing heat from a source like a hot exhaust waste heat using TE modules [16]. Maintaining of high temperature difference in the actual applications is very difficult. Air or water can be utilized to cool the cold side of the TEGs. Heat exchangers are also can be used to supply the heating effect and cooling effect on both sides of the TE modules [17]. Designing of a consistent TEG system which will operate at high temperature is a very difficult task. Balancing of heat transfer through the TE modules and increasing the temperature difference across those modules requires a wide engineering design to attain the maximum efficiency of the system. For attaining this condition, the heat transfer methodologies should be properly designed in TEG field. The minimization of thermal losses at many places because of interfaces between TE materials is also done. Elimination of large drop in pressure between the hot and cold sources is another tough work [18].

Lot of advantages makes the TEGs as a good power producing source. TEGs are devices which have no moving components so that they are free of maintenance. TEGs are pollution free devices because there is no chemical reaction takes place inside them [19]. They are effectively used in space investigation probes due to their low weight and noiseless operations. There is no need of any external power supply for the functioning of TEGs. Because of that they are called as passive devices. Various areas like space investigation, automobiles, remote unmanned locations and cathodic shield etc. are using the TEGs as a power supply [20]. Lower in efficiency and higher surface temperatures are some of the limitations of TEGs. While using TE devices the heating will be increased by high electrical output resistance and it is not suitable for some applications due to the low thermal conductivity. The material which one has less thermal conductivity and more electrical conductivity is best suited for TEGs [21].

Taguchi method is one of the optimization methods to analyze the function of significant process variables to attain a best possible result for an objective function [22, 23]. This technique can be successfully utilized to help in designing trials and get the best possible conditions by using orthogonal array, so that the time and attempt can be minimized [24, 25]. The Taguchi method can give an organized approach for easing the experimental plan. Meanwhile, this method is widely used to find the best possible conditions of a system [26, 27].

In this paper the power at the output and the efficiency of conversion of various TEGs comprised of different materials such as Bismuth Telluride (Bi_2Te_3), Lead Telluride (PbTe) and Aluminium Oxide (Al_2O_3) is experimentally investigated and compared. The details on energy transfer, potential difference, electricity flow and performance are found to give a comparative analysis for various TE materials. The optimum performance of TEGs were found using Taguchi method.

2 Methodology

2.1. Experimental Setup

TEGs are semi-conductor devices that are used to convert the temperature differences and heat energy into a DC power supply. Seeback effect is utilized in TEG semiconductor devices to produce voltage. This produced voltage forces electrical current and generates useful power. TEGs comprised of Bi_2Te_3, PbTe and Al_2O_3 materials were used in this experiment. A voltmeter and an ammeter are used for measuring electric potential difference and the current flow between two points in the circuit. Dimmer stat is utilized to adjust the output potential/voltage to electrical circuits. Heater is used to provide a heat input to TEG and is controlled by the dimmer stat. Water flow is allowed to TEG to maintain the temperature in the cold side. The Figure 1 and 2 shows the experimental methodology and experimental setup.

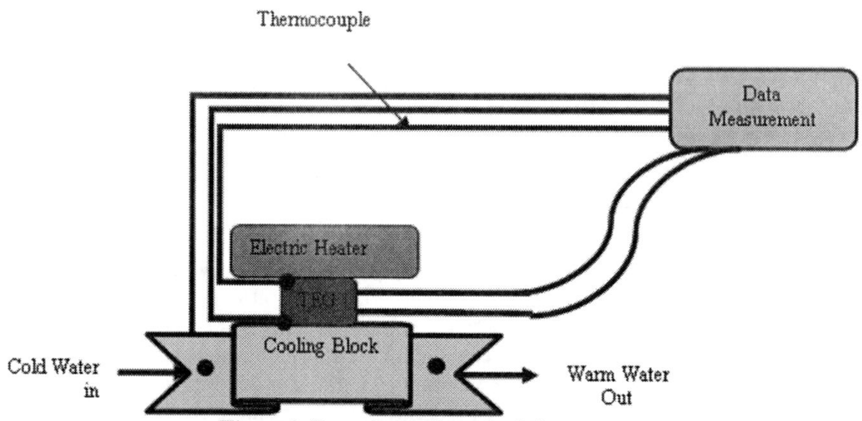

Figure1. Experimental methodology

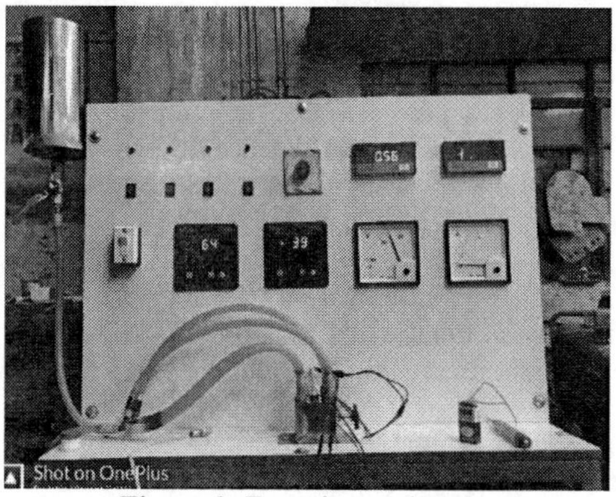

Figure 2. Experimental setup

2.2 Experimental Procedure

The performance optimization trial was carried out to find the efficiency of conversion of the TEG from thermal energy to electrical energy with process variables. This experiment was done by inserting the TEG between an electrical heating source and water cooling source. Heaters served as heating devices and TEGs with a dimension of 40 mm x 40 mm x 4 mm were used to produce electric power due to the temperature gradient across its junctions. The variation in the heat input value to the electric heater was made and the temperature measurements were noted using thermocouples that were attached to a data measurement system. At this condition, the TEG surface temperature and cooling water temperature were noted down. The output terminal of the TEG is attached to the voltmeter and the ammeter in order to monitor the power produced. If the heater is switched on then simultaneously the cold water is fed to the duct. Then the TEG became hot in one side and cold in another side. This temperature difference produced electricity. The efficiency of energy conversion of the TEG was found by using the following relation.

$$\eta_{TEG} = P_o / Q_i \quad (1)$$

Where, P_o is the output electrical power produced by the TEG and Q_i is the rate of heat input to the hot side of the TEG.

The performance optimization experiments were conducted by using the following process variables with 3 levels which are mentioned in Table 1 Taguchi L9 orthogonal array was utilized to find the optimum process variable.

Table 1. Process variables and their levels

Factors	Process variables	Symbol	Unit	Level 1	Level 2	Level 3
I	Heat Input	Q	W	30	60	90
II	TEG Material	M	-	Bi_2Te_3	PbTe	Al_2O_3
III	Temperature Difference	ΔT	^oC	25	50	75

3 Results and Discussion

Results were obtained and described for the above mentioned trial conditions with the help of the following figures and tables.

3.1 η_{TEG}

Figure 3 indicated the relationship between mean value of SNR to the process variables heat input, TEG material and temperature difference with the method of larger is better because high η_{TEG} essential for energy conversion from thermal to electrical energy. The plot for main effect of means was showed in figure 4. The maximum η_{TEG} would gain at 90W heat input with Bi_2Te_3 material and 75^oC of temperature difference. The Table 2 shows the corresponding design in 7 th experiment process combinations.

Table 2. L9 Array for experiment

Experiment No.	Q	M	ΔT	η_{TEG}	SNR
1	30	Bi_2Te_3	25	1.98	5.93330
2	30	PbTe	50	1.95	5.80069
3	30	Al_2O_3	75	1.90	5.57507
4	60	Bi_2Te_3	50	2.10	6.44439
5	60	PbTe	75	2.05	6.23508
6	60	Al_2O_3	25	2.02	6.10703
7	90	Bi_2Te_3	75	2.45	7.78332
8	90	PbTe	25	2.31	7.27224
9	90	Al_2O_3	50	2.24	7.00496

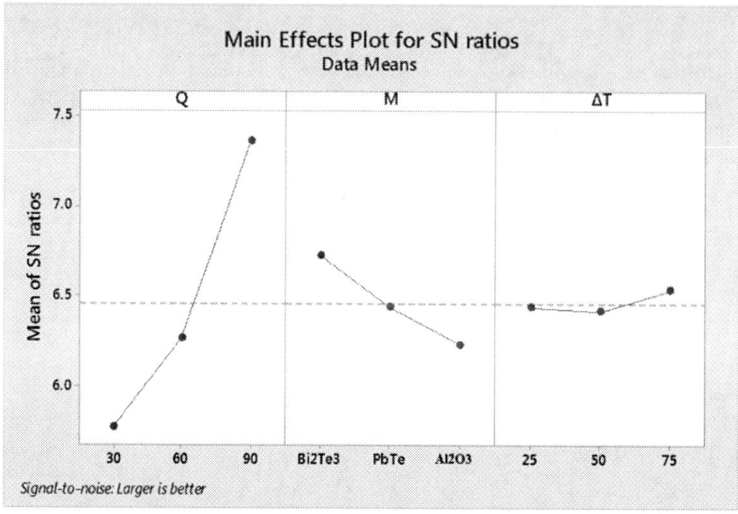

Figure 3. Main effects plot for SNR

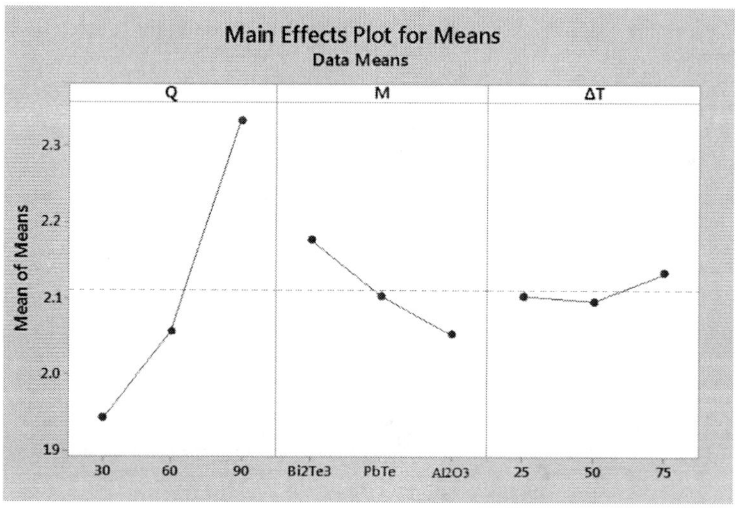

Figure 4. Main effects plot for means

The interaction plot for η_{TEG} with regard to heat input, TEG material and temperature difference was shown in figure 5. All the relation indicated that the η_{TEG} were directly relative to variables taken. The highest η_{TEG} was produced at higher value of heat input and higher value of temperature difference. Figure 6 indicated the relation of residual vs percentage in the normal probability plot, fitted value vs residuals in the versus fits, residual vs frequency in the histogram and observation order vs residual in the versus order plot for η_{TEG}. It seems that the values in the normal probability plot were almost close to the mean line. In the meanwhile, versus plot also showed that the some points are close to the mean line. In the versus order plot, it showed that the acceptable outcome of 3 values were crossed over the line of zero.

In Table 3 and 4 ranks for the process variables were indicated. The main significant variable was primarily heat input, second material and finally temperature difference for SNR response table and table of response for means. Hence the heat input and the temperature difference provided additional contribution in η_{TEG} which was indicated in Figure 7. Table 5 shows the response values and the equation of regression and Table 6 shows the ANOVA results.

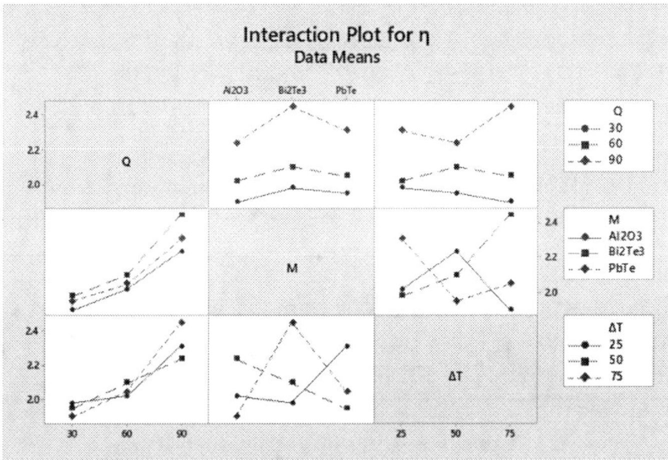

Figure 5. Interaction plot for η_{TEG}

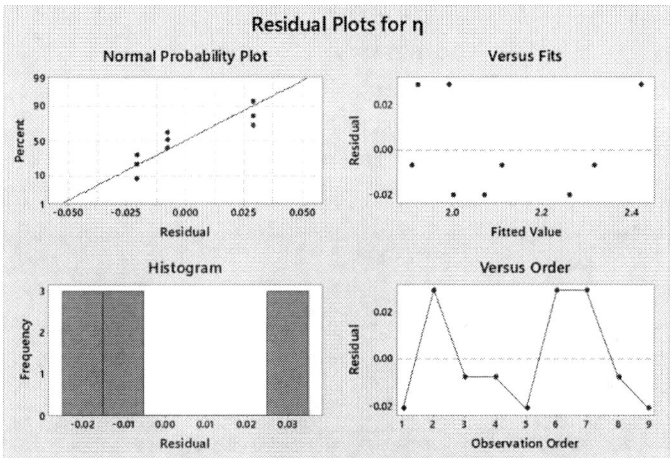

Figure 6. Residual plots for η_{TEG}

Table 3. η_{TEG} – Table of Response for SNR

Level	Q	M	ΔT
1	5.694	6.693	6.324
2	6.192	6.130	6.149
3	7.107	6.170	6.519
Delta	1.413	0.563	0.371
Rank	1	2	3

Table 4. η_{TEG} - Table of Response for Means

Level	Q	M	ΔT
1	1.927	2.170	2.073
2	2.040	2.030	2.033
3	2.270	2.037	2.130
Delta	0.343	0.140	0.097
Rank	1	2	3

Table 5. Response Values and Equation of Regression

Response	Regression Equation	S	R^2	R^2 (adj)
η_{TEG} (%)	2.0789- 0.1522 Q_30-0.0389 Q_60+ 0.1911 Q_90 + 0.0911 M_Bi2Te3-0.0489 M_PbTe-0.0422 M_Al_2O_3- 0.0056 ΔT_25-0.0456 ΔT_50+ 0.0511ΔT _75	0.0218131	93.25%	86.50%

Table 6. Analysis of Variance (ANOVA)

Source	DF	Adj SS	Adj MS	F-Value	P-Value
Heat Input	2	0.183622	0.091811	24.52	0.039
Material	2	0.037422	0.018711	5.00	0.167
Temp Difference	2	0.014156	0.007078	1.89	0.346
Error	2	0.007489	0.003744		
Total	8	0.242689			

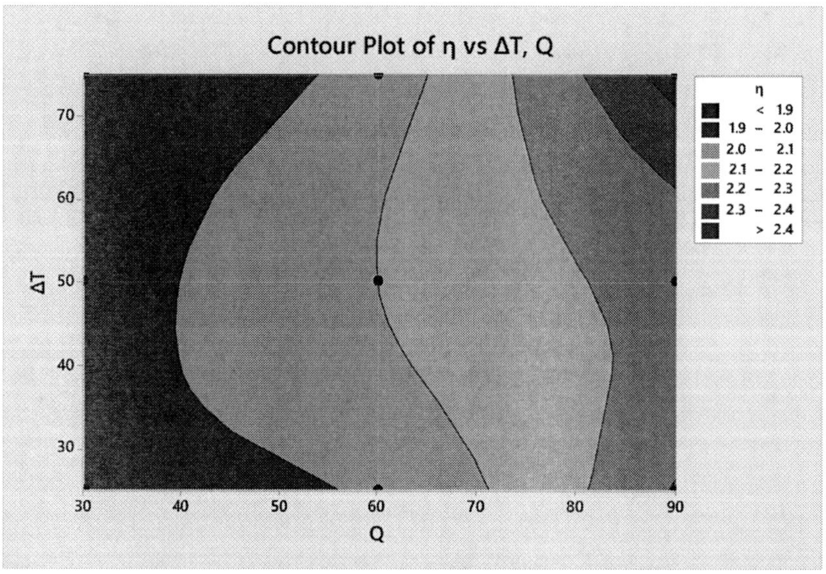

Figure 7. Contour plot for η_{TEG}

4 Conclusions

The experimental investigation on the performance optimization of TEGs using process variables such as heat input, TEG material and temperature difference was done and the conclusions of this work were stated below

1. The main significant parameter for η_{TEG} was heat input and the temperature difference.
2. The maximum value of η_{TEG} was 2.45% for the optimum process variables such as 90W heat input, TEG material of Bi_2Te_3 and the temperature difference of 75°C.
3. The lowest value of η_{TEG} was 1.90% for the optimum process variables such as 30W heat input, TEG material of $Al_2O_3$and the temperature difference of 75°C.
4. The ANOVA showed that the most influential parameter on the performance of TEGs was heat input. The R^2 and R^2 (adj) values were found to be 93.25% and 86.50% respectively.

Overall the TEG material of Bi_2Te_3 indicated the best heat to electric energy conversion efficiency compared to other TEG materials operated under same ranges of temperature. This TEG can be used for efficient electricity generation from waste heat of internal combustion engines, industrial hot chimneys, silencer of on road vehicles and cooking stoves.

Abbreviations

Al_2O_3	Silicon Germanium
Bi_2Te_3	Bismuth telluride
k_t	Thermal Conductivity
η_{TEG}	TEG-Efficiency
σ_e	Electrical Conductivity
ΔT	Temperature Difference
ANOVA	Analysis of Variance
DC	Direct current
DF	Degree of freedom
M	TEG Material
^oC	Degree Celsius
PbTe	Lead Telluride
P_o	Output electrical power
Q	Heat Input
Q_i	Rate of heat input
S	Seeback Coefficient
SNR	Signal to Noise Ratio
TE	Thermoelectric
TEG	Thermoelectric Generator
W	Watt

References

[1] Hewawasam LS, Jayasena AS, AfnanMMM, Ranasinghe RACP and WijewardaneMA 2020 Waste heat recovery from thermo-electric generators (TEGs) Energy Reports **6** 474-479.

[2] Li W, PaulMC, Siviter J, Montecucco A, Knox AR, Sweet T, Min G, Baig H, Mallick TK, Han G and Gregory DH 2016 Thermal performance of two heat exchangers for thermoelectric generators Case studies in thermal engineering **8** 64-175.

[3] Elghool A, Basrawi F, Ibrahim H, Ibrahim TK, Ishak M, Yusof TM and Bagaber SA 2020 Enhancing the performance of a thermo-electric generator through multi-objective optimisation of heat pipes-heat sink under natural convection Energy Conversion and Management **209** 112626.

[4] Barma MC, Riaz M, Saidur R and LongBD 2015 Estimation of thermoelectric power generation by recovering waste heat from Biomass fired thermal oil heater Energy Conversion and Management **98** 303-313.

[5] Dhass AD, Krishna R and Sreenivasan M 2020 Numerical analysis of a variety of thermoelectric generator materials Materials Today: Proceedings.

[6] Li G, Zhu D, Zheng Y and Guo W 2020 Mesoscale combustor-powered thermoelectric generator with enhanced heat collection Energy Conversion and Management **205** 112403.

[7] Addanki S and Nedumaran D 2019 Simulation and fabrication of thermoelectric generators for hand held electronic gadgets Materials Science and Engineering: B **251** 114453.

[8] Arsie I, Cricchio A, Marano V, Pianese C, De Cesare M and Nesci W 2014 Modeling analysis of waste heat recovery via thermo electric generators for fuel economy improvement and CO_2 reduction in small diesel engines SAE International Journal of Passenger Cars-Electronic and Electrical Systems **7** 246-255.

[9] Bensaid S, Brignone M, Ziggiotti A and Specchia S 2012 High efficiency Thermo-Electric power generator International journal of hydrogen energy **37** 21385-1398.

[10] Chen WH, Huang SR and Lin YL 2015 Performance analysis and optimum operation of a thermoelectric generator by Taguchi method Applied Energy **158** 44-54.

[11] DateA, Date A, Dixon C, Singh R and Akbarzadeh A 2015 Theoretical and experimental estimation of limiting input heat flux for thermoelectric power generators with passive cooling Solar Energy **111** 201-217.

[12] Ahmed S, Mousa MG and Hegazi AA 2018 Performance analysis of a passively cooled thermoelectric generator Energy Conversion and Management **173** 399-411.

[13] Elghool A, Basrawi F, Ibrahim TK, Habib K, Ibrahim H and Idris DMND 2017 A review on heat sink for thermo-electric power generation: Classifications and parameters affecting performance Energy conversion and management **134** 260-277.

[14] Jia X and Guo Q 2020 Design study of Bismuth-Telluride-based thermoelectric generators based on thermoelectric and mechanical performance Energy **190** 116226.

[15] Lundgaard C and Sigmund O 2019 Design of segmented off-diagonal thermoelectric generators using topology optimization Applied Energy **236** 950-960.

[16] Lv S, He W, Jiang Q, Hu Z, Liu X, Chen H and Liu M 2018 Study of different heat exchange technologies influence on the performance of thermoelectric generators Energy Conversion and Management **156** 167-177.

[17] Kim CN **2018** Development of a numerical method for the performance analysis of thermoelectric generators with thermal and electric contact resistance Applied thermal engineering **130** 408-417.

[18] Shareef MH, Sajid A, Majeed AAand Adnan MAB Efficiency Calculation of a Thermoelectric Generator International Journal of Science and Research **5** 1520-1522.

[19] Wu Y, Zuo L, Chen J and Klein JA2016 A model to analyze the device level performance of thermoelectric generator Energy **115** 591-603.

[20] Ragupathi P, Barik D, Vignesh G and Aravind S 2020 Electricity Generation from Exhaust Waste Heat of Internal Combustion Engine Using Al_2O_3 Thermoelectric Generators Journal of Applied Science and Engineering **23** 55-60.

[21] Ragupathi P, Barik D, Pradeep S and Lakshan L 2018 A Review on Waste Heat Recovery Technologies in Internal Combustion Engines International Journal of Research in Advent Technology 270-276.

IOP Conf. Series: Materials Science and Engineering 1059 (2021) 012053 doi:10.1088/1757-899X/1059/1/012053

[22] Chen WH, Huang SR and Lin YL 2015 Performance analysis and optimum operation of a thermoelectric generator by Taguchi method Applied Energy **158** 44-54.

[23] Kishore RA,SanghadasaM and Priya S2017Optimization of segmented thermoelectric generator using Taguchi and ANOVA techniques Scientific reports **7** 11-15.

[24] Rezania A, Atouei SA and Rosendahl L 2020 Critical parameters in integration of thermoelectric generators and phase change materials by numerical and Taguchi methods Materials Today Energy **16** 100376.

[25] Ji D, Wei Z, Mazzoni S, Mengarelli M, Rajoo S, Zhao J, Pou J and Romagnoli A 2018 Thermoelectric generation for waste heat recovery: Application of a system level design optimization approach via Taguchi method Energy Conversion and Management **172** 507-516.

[26] Kishore RA, Kumar P and Priya S 2018 A comprehensive optimization study on Bi 2 Te 3-based thermoelectric generators using the Taguchi method Sustainable Energy & Fuels **2**1 175-190.

[27] Anant Kishore R, Kumar P, Sanghadasa M and Priya S 2017 Taguchi optimization of bismuth-telluride based thermoelectric cooler Journal of Applied Physics **122** 2025109.

Mathematical Modeling and Characteristic Analysis on the Dynamics of Slider Flapping Wing Mechanism

Sudeesh Arputham D[1], Yuvan Nethaji T[2], Sujith Kumar V[3], Mohaideen S[4], Dr.G.Suganya Priyadharshini[5]

[1,2,3,4] Student, B.E. Mechanical Engineering, Coimbatore Institute of Technology, Coimbatore, Tamil Nadu, India

[5] Assistant Professor, Department of Mechanical Engineering, Coimbatore Institute of Technology, Coimbatore, Tamil Nadu, India

E-mail: sudeesharputham@gmail.com

Abstract: This paper addresses the modelling of a distinct flapping wing mechanism sans any gears deploying only the primitive slider and hinge type joints. The proposed contraption is meticulously designed so as to hamper vibrational flutter and enhance durability. Initially, equations of forces enacting at significant motion transformation junctions are articulated in terms of Differential Algebraic Equations (DAE) and then these DAEs culminate as the characteristic mathematical model for the mechanism. The derived mathematical model is then harnessed to optimize the performance characteristics of the system. Further, results attained from the mathematical formulation are assessed and validated with graphical analysis.

Keywords: Flapping Wing Mechanism, Differential Algebraic Equations, Mathematical Model

1. Introduction

A seamless flapping wing mechanism for Micro Wing Aerial Vehicle (MAV) is designed to accomplish an unhindered flapping action with enhanced durability. The mechanism is designed with the considerations of the implicational and geometrical conditions regards to the application of Micro-Aerial Vehicle (MAV) Flight. Neglecting the use of gears, the links and hinge joints are equipped. The mechanism transforms the rotational motion from a motor into a periodic symmetric flapping action. The Links implied as wing of the MAV mechanism are enforced with a slider motion besides the hinge joints. This is done in order to bring out the proper inline flapping motion and eliminating the phase lag between both the links. In order to optimize the performance of the mechanism, optimum flapping angle and suitable dimensions of varied linkages have to be chosen.

A mathematical model in the form a Differential Algebraic Equation is derived for arriving at a controlled stability analysis for the proposed mechanical system. Conventionally, the Euler Lagrange method has been followed by many authors to arrive at their mathematical formulation. But this method yields a very complex solution which is very disadvantageous in scenarios where we have to compute a unique solution for the derived model. In this paper, we adopt a finite element approach where we break our mechanism into sub mechanisms at the nodes of motion transformations and analyze them individually and then combine them as complete formulation. The relation between the different linkages are assessed and ultimately the coherence between optimum flapping angle, minimum auxiliary displacement and maximum torque is established in a graphical representation. A unique solution for the optimum flapping angle is finally attained which aids in extrapolating the other characteristic performances like flapping frequency.

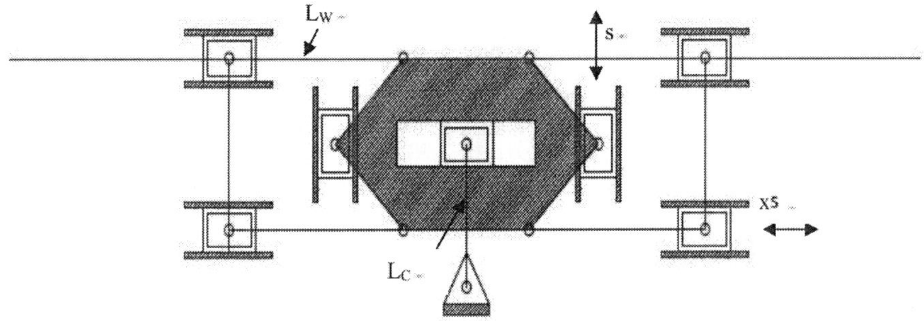

Figure 1. Free-Body Diagram of Slider Flapping Wing Mechanism

In the aforementioned proposed mechanism two primary motion transformations takes place –

a. Rotational Motion of the Crank Link (denoted by L_C in figure) is transformed into Vertical Oscillatory Linear Motion of the Body Plate (the hexagonal shaped linkage)
b. Vertical Oscillatory Motion of the Body Plate is transformed into a combination of Sliding and Rotational Motion of the Wing Link (denoted by L_W)

Adopting a Finite elemental approach for the purpose of reducing complexity, the three Rigid bodies (Crank Link, Body Plate and Wing Link) are separately assessed.

1.1 Geometrical Transformation (2-Translation Variables_2-Rotational Angles)

Let s = Vertical Displacement of the Body Plate from its initial position at time t
x = Horizontal Displacement of the Hinge Point of the Wing Link at the slider from its initial position at time t

θ = Angle made by the Crank Link from the horizontal at time t

φ = Angle made by the Wing Link from the horizontal at time t

The crank link has mass of m_C and length L_C. The mass of the body link is denoted by m_B. The mass and length of the wing link is represented by m_W and L_W respectively.

From the mechanism assembly, the relationship between the crank rotation angle(θ) and vertical displacement of the body plate(s) is given by

$$s = L_C \sin\theta \qquad \text{------------ (1)}$$

Therefore, $$\dot{s} = L_C \cos\theta(\dot{\theta}) \qquad \text{-------------(2)}$$

$$\ddot{s} = -L_C \sin\theta(\ddot{\theta}) \qquad \text{-------------(3)}$$

From the mechanism assembly, the relationship between the wing rotation angle(φ) and vertical displacement of the body plate(s) is given by

$$s = L_W \sin\phi \qquad \text{------------ (4)}$$

Therefore, $$\dot{s} = L_W \cos\phi(\dot{\phi}) \qquad \text{-------------(5)}$$

$$\ddot{s} = -L_W \sin\phi(\ddot{\phi}) \qquad \text{-------------(6)}$$

From the mechanism assembly, the relationship between the horizontal displacement of the hinge joint along the slider(x) and vertical displacement of the body plate(s) is given by

$$x = L_W - \sqrt{L_W^{\,2} - s^2} \qquad \text{--------------(7)}$$

$$\dot{x} = \left(\frac{s}{\left(L_W{}^2 - s^2\right)^{\frac{1}{2}}} \right) \dot{s} \quad \text{--------------(8)}$$

$$\ddot{x} = \left(\frac{s}{\left(L_W{}^2 - s^2\right)^{\frac{1}{2}}} \right) \ddot{s} + \left(\frac{L_W{}^2}{\left(L_W{}^2 - s^2\right)^{\frac{3}{2}}} \right) \dot{s}^2 \quad \text{--------------(9)}$$

1.2 Crank Link
The Crank link exhibits a Pure Rotational Motion rotating about the z-axis as

Force Equation
Initially a Torque is applied to the crank link of length L_c which initiates the rotary motion of the crank with a particular angular acceleration

Therefore Input Force is given by , $F_{In} = \dfrac{\tau}{L_C}$ --------------- (10)

In order to sustain the rotational motion, a centripetal force is also generated at the crank link. This force complements the vertical linear movement of the Body Plate.

Hence the Force generated at the Crank Link, $F_C = m_C L_C \dot{\theta}^2$ --------------- (11)

Body Plate:

The Body Plate exhibits a Pure Translational Motion moving linearly along the y-axis

Force Equation
The Body Plate exhibits a vertical oscillatory motion due to the influence of the crank and geometry of the set up.
Therefore Force experienced by the Body Plate, $F_B = m_B \ddot{s}$ --------------- (12)

1.3 Wing Link
The Wing link is imparted a Combined Translational and Rotational Motion along x-axis and about z-axis respectively to acquire the flapping action.
Force Equation
The wing link undergoes combined translational and rotational motion. Hence two types of forces act on the wing link. The first kind is due to Linear acceleration of the joint in the slider. The second kind is because of the angular acceleration of the wing link.
For the Force due to Rotational Motion Inertia term is to be included

Since the wing link is of rectangular cross section, $I = \dfrac{m_W L_W{}^2}{12}$

Therefore, Force generated by the Wing Link, $F_W = m_W \ddot{x} + \dfrac{1}{L_W} I \ddot{\phi}$

$$F_W = m_W \ddot{x} + \frac{m_W L_W \ddot{\phi}}{12} \quad \text{-------------- (13)}$$

2. Mathematical Formulation:
Transformation 1: Rotary Motion of Crank to Linear Vertical Motion of Body Plate
Equations (8), (9) and (10) constitute to this Transformation.
The equilibrium force equation for Transformation 1 is given by,

$$F_{In} = F_C + F_B$$

$$F_{In} = \frac{\tau}{L_C} = m_C L_C \dot{\theta}^2 + m_B \ddot{s} \quad \text{-------------------- (14)}$$

IOP Conf. Series: Materials Science and Engineering 1059 (2021) 012054 doi:10.1088/1757-899X/1059/1/012054

Transformation 2: Linear Vertical Motion of Body Plate to Combined Sliding and Rotational Oscillation of Wing Link

Equations (3) and (4) constitute to this Transformation.

The equilibrium force equation for Transformation 2 is given by,

$$F_B = F_W$$

$$m_B \ddot{s} = m_W \ddot{x} + \frac{m_W L_W \ddot{\phi}}{12} \quad \text{-----------------------} (15)$$

Substituting Equation (6) in Equation (5), we get

$$F_{In} = m_C L_C \dot{\theta}^2 + m_W \ddot{x} + \frac{m_W L_W \ddot{\phi}}{12} \quad \text{-----------------------} (16)$$

Representing Equation (16) as a function of s,

$$F_{In} = m_C L_C \frac{\dot{s}^2}{L_C^{\ 2}\cos^2\theta} + m_W\left(\left(\frac{s}{\left(L_W^{\ 2}-s^2\right)^{\frac{1}{2}}}\right)\ddot{s} + \left(\frac{L_W^{\ 2}}{\left(L_W^{\ 2}-s^2\right)^{\frac{3}{2}}}\right)\dot{s}^2\right) - \frac{m_W L_W}{12}\frac{\ddot{s}}{L_W \sin\phi}$$

On Further Simplification,

$$F_{In} = m_C \frac{\dot{s}^2}{L_C\cos^2\left(\sin^{-1}\left(\frac{s}{L_C}\right)\right)} + m_W\left(\left(\frac{s}{\left(L_W^{\ 2}-s^2\right)^{\frac{1}{2}}}\right)\ddot{s} + \left(\frac{L_W^{\ 2}}{\left(L_W^{\ 2}-s^2\right)^{\frac{3}{2}}}\right)\dot{s}^2\right) - \frac{m_W}{12}\frac{\ddot{s}}{\sin\left(\sin^{-1}\left(\frac{s}{L_W}\right)\right)}$$

----------------------------------- (17)

The above equation depicts the relation between Input Force and vertical displacement s. Since we need a relation involving the flapping angle φ and Torque τ to optimize parameters, the above equation is further modified as

$$F_{In} = m_C \frac{\dot{s}^2}{L_C\cos^2\theta} + m_W\left(\left(\frac{s}{L_W\cos\phi}\right)\ddot{s} + \left(\frac{L_W^{\ 2}}{L_W^{\ 3}\cos^3\phi}\right)\dot{s}^2\right) - \frac{m_W}{12}\frac{\ddot{s}}{\sin\phi}$$

$$F_{In} = m_C \frac{L_C}{L_C^{\ 2} - L_W^{\ 2}\sin^2\phi}\dot{s}^2 + m_W\left(\left(\frac{s}{L_W\cos\phi}\right)\ddot{s} + \left(\frac{1}{L_W\cos^3\phi}\right)\dot{s}^2\right) - \frac{m_W}{12}\frac{1}{\sin\phi}\ddot{s}$$

$$F_{In} = m_C \frac{L_C L_W^{\ 2}\cos^2\phi}{L_C^{\ 2} - L_W^{\ 2}\sin^2\phi}\dot{\phi}^2 + m_W L_W\left((\sec\phi)\dot{\phi}^2 - (\sin\phi\tan\phi)\ddot{\phi}\right) + \frac{m_W}{12}L_W\ddot{\phi}$$

$$\tau_{In} = \frac{m_C L_C^{\ 2} L_W^{\ 2}\cos^2\phi}{L_C^{\ 2} - L_W^{\ 2}\sin^2\phi}\dot{\phi}^2 + m_W L_C L_W(\sec\phi)\dot{\phi}^2 - m_W L_C L_W(\sin\phi\tan\phi)\ddot{\phi} + \frac{m_W}{12}L_W L_C\ddot{\phi}$$

$$\boxed{\tau_{In} = \frac{m_C L_C^{\ 2} L_W^{\ 2}}{L_C^{\ 2} - L_W^{\ 2}}(\cot^2\phi)\dot{\phi}^2 + m_W L_C L_W(\sec\phi)\dot{\phi}^2 - m_W L_C L_W(\sin\phi\tan\phi)\ddot{\phi} + \frac{m_W}{12}L_W L_C\ddot{\phi}}$$

--------------------------------(18)

Equation (18) is the required mathematical model that establishes a relation between input torque and output flapping angle.

Since the same kind of material is employed for the whole mechanism and the density of the links are very low when compared to the length dimensions ($m_C \ll L_C$ and $m_W \ll L_C$), we can neglect m_C and m_W.

Let $A = \dfrac{L_C^2 L_W^2}{L_C^2 - L_W^2}$, $B = L_C L_W$, $C = \dfrac{L_W L_C}{12}$

Therefore, $$\boxed{\tau_{In} = A(\cot^2\phi)\dot{\phi}^2 + B[(\sec\phi)\dot{\phi}^2 - (\sin\phi\tan\phi)\ddot{\phi}] + C\ddot{\phi}} \qquad \text{------------------(19)}$$

Equation (19) is the mathematical model of the proposed Flapping Wing Mechanism with an input of F_{In}

3. Stability Analysis

For the proposed mechanism, three combinations for the Crank Link (L_C) and the Wing Link (L_W) are possible. In order to deduce the optimum flapping angle, the following cases are considered.

(i) $L_C=L_W$ (ii) $L_C> L_W$ (iii) $L_C < L_W$

Case (i) $L_C=L_W$

From Equation (18), $A = \dfrac{L_C^2 L_W^2}{L_C^2 - L_W^2}$

At $L_C= L_W$ the value of A becomes infinity and hence the system will be unstable.

If the Length of the Crank Link is equal to the Length of the Wing Link, then during Transformation 1 (Rotary Motion of Crank to Linear Vertical Motion of Body Plate), the angular acceleration of the Crank Link will be discontinuous and the Crank Link will cease to rotate. Thus, the Flapping action will be restricted and unstable.

Case (ii) $L_C > L_W$

From Equation (7), $x = L_W - \sqrt{L_W^2 - s^2}$

At $\theta = 90°$, $s = L_C$ $\quad x = L_W - \sqrt{L_W^2 - L_C^2}$

Since $L_C > L_W$ $\quad x = L_W - i\sqrt{L_C^2 - L_W^2}$

Thus, the system has a discontinuity at $\theta = n\pi/2$ and hence the system will be unstable.

If the Length of the Crank Link is greater than the Length of the Wing Link, then during Transformation 2 (Linear Vertical Motion of Body Plate to Combined Sliding and Rotational Oscillation of Wing Link), the sliding motion of the Wing Link will be hindered due to geometry when the Crank Link is rotated to an angle of $n\pi/2$. Hence a Singularity occurs at that point and the mechanism ceases to exhibit the desired Flapping action.

Case (iii) $L_C < L_W$

If the Length of the Crank Link is lesser than the Length of the Wing Link, all the derived equations satisfy. Thus, the drawbacks of the above mentioned geometries can be overhauled and both the Transformations 1 & 2 will be executed seamlessly resulting in a stable mechanism with uniform Flapping action. Hence, we can infer that the length of the wing link should always be greater than that of the crank link in order for the mechanism to produce the desired flapping action.Further analysis is made taking into consideration the stability criterion and the characteristic parameters are assessed and deduced by means of graphical analysis utilizing the mathematical formulation. The iterative results are plotted to attain optimum parameters.

4. Graphical Analysis

Considering the stable case of $L_W > L_C$, a graph is plotted for Flapping Angle versus Vertical Body Plate with the derived mathematical model for different combinations of L_C and L_W and a similar periodic pattern is obtained.

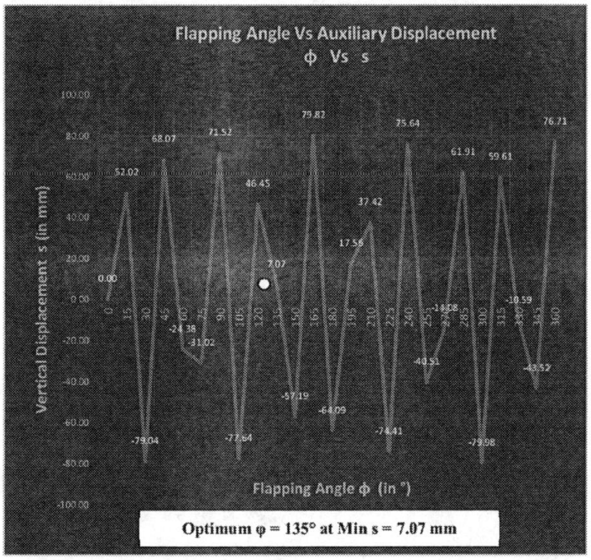

Figure 2. Flapping Angle Vs Auxiliary Displacement of Body Plate when $L_W = 2L_C$

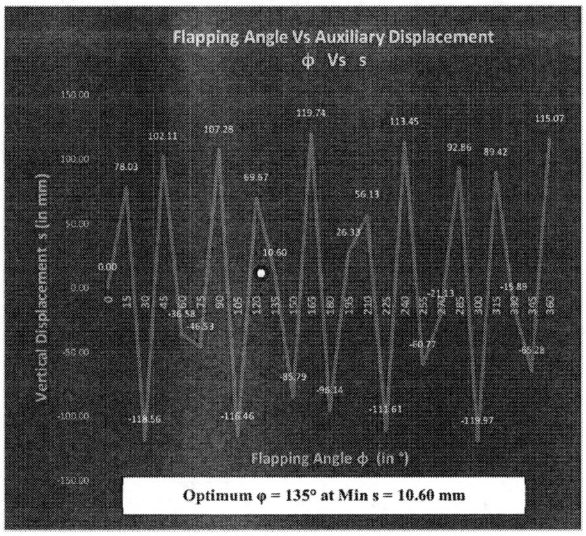

Figure 3. Flapping Angle Vs Auxiliary Displacement when $L_W = 3L_C$

From the results, it is found that the minimum body displacement (s_{min}) occurs at a flapping angle of 135°.

Hence 135° is the optimum flapping angle which can provide the required flapping frequency with reduced body displacement which in turn minimizes the overall vibrations that incur during the dynamic motion.

From the derived mathematical model, a graph is plotted between Flapping Angle and Torque and the observed results are as below.

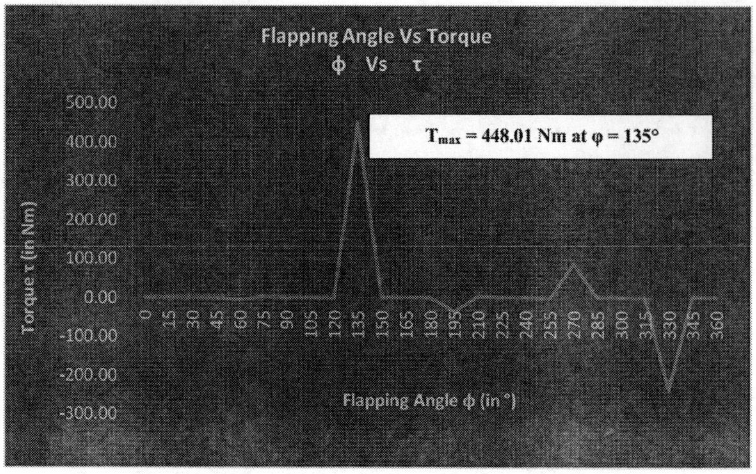

Figure 4. Flapping Angle Vs Torque when $L_W = 2L_C$

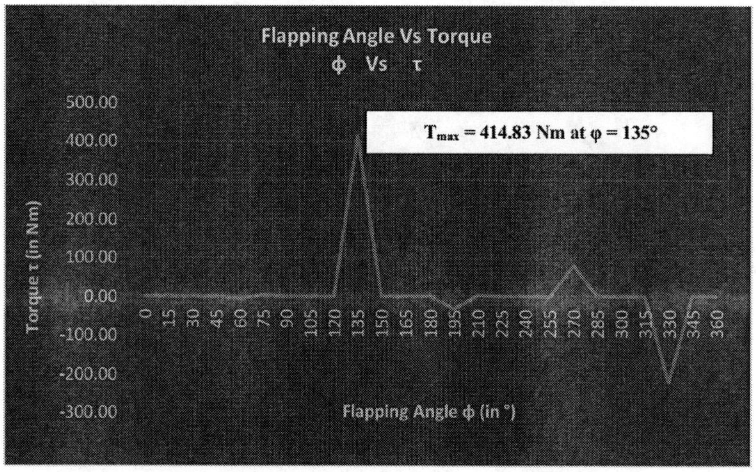

Figure 5. Flapping Angle Vs Torque when $L_W = 3L_C$

The maximum torque can be generated at the optimum flapping angle. In our case the optimum flapping angle is at 135° which yields a maximum torque of 448.01 Nm and 414.83 Nm respectively in both the observed cases.

It is pertinent for the Flapping Wing Mechanism to generate maximum torque at a minimum auxiliary displacement in order to minimize the vibrational flutter, reduce the wear rate and enhance the flapping frequency. Thus, from the graphical analysis, we deduce that a flapping angle of 135° serves the purpose of optimization.

5. Results and Discussions

The relations between the lengths of the crank link L_C and wing link L_W are attained. It is found out that the stability of the mechanism completely depends upon the relation between the two links. The mechanical system exhibits a stable periodic and symmetric flapping action if the length of the Wing Link is designed in such a manner that its dimensions are greater than the length of the Crank Link (i.e., $L_W > L_C$). In order to obtain optimum performance characteristics like high flapping frequency and less vibrations with reduced wear, it is necessary that the mechanism generates high torque with least possible link displacement. By employing the formulated mathematical model, the different torque values for various flapping angles are computed for different combinations of Crank Link L_C and Wing Link L_W dimensions. Since the flapping angle has an explicit relation to the auxiliary displacement of the body plate, the different range of values for auxiliary displacement is also computed. When the results are plotted in a graph the optimum Flapping angle for the proposed contraption is found out to be 135° for which auxiliary displacement is minimum and torque generated is maximum.

6. Conclusion

Mathematical Modelling Approaches like Denavit-Hartenberg matrix transformations, Euler-Lagrange and Hamiltonian Approaches can only yield effective results in case of simple mechanisms. But for a more complex mechanism such as the proposed slider flapping wing mechanism, the conventional approaches cannot be applied to conclusively predict the performance parameters as all of these approaches yield only general equations without any unique solutions. Adopting the proposed methodology yields an effective mathematical formulation which aids in optimizing the mechanism and enhancing its efficiency by obtaining a unique solution unlike the other methodologies. Since all the linkages are explicitly connected to the flapping angle, the design dimensions of interconnected links L_C and L_W can be determined. The flapping angle deduced as the optimum one can be utilized to acquire the flapping frequency of the mechanism which is root parameter around which a Micro Aerial Vehicle design revolves upon.

References

[1] Ali Selk Ghafari, Abbas Ebrahimi, Ali Meghdari, Zahra Jahanbin, 2015 Multi-Body Simulation of a Flapping-Wing Robot Using an Efficient Dynamical Model, Journal of the Brazilian Society of Mechanical Sciences and Engineering, **38** 133-149.

[2] Ansari S A, Zbikowski R, Knowles K, 2006 Aerodynamics Modelling of Insect-Like Flapping Flight for Micro Air Vehicles, *Progress in Aerospace Sciences*, **42** 129-172.

[3] Cem Civelek, 2006 Mathematical Modelling of Rotational Mechanical Elements of Higher Order and Their Characteristics, *Mathematical and Computer Modelling,* **43** 957–964.

[4] Cesar Francisco Bogado Martinez, Max Suell Dutra, Luciano Santos Constantin Raptopoulos, 2018 Proposed Control for Wing Movement, Type Flat Plate, For Ornithopter Autonomous Robot, Journal of the Brazilian Society of Mechanical Sciences and Engineering, **40** 456 1-10.

[5] Chao Wing, Chaoying Zhou, Cong Liu, Xingwei Zhang, 2010 An Optimization on Single-Crank-Double-Rocker Flapping Mechanism, *Fourth International Conference on Genetic and Evolutionary Computing,*978-1-4244-8891-9.

[6] Elena S Gebel, 2015 Mathematical Modeling of Dynamics of Multi-Lever Linkages, *25th DAAAM International Symposium on Intelligent Manufacturing and Automation, DAAAM, Procedia Engineering,* **100** 1562 – 1571.

[7] Halit Kaplan, 2014 Mathematical Modeling and Simulation of High-Speed Cam Mechanisms to Minimize Residual Vibrations, *Proceedings of the Institution of Mechanical Engineers, Part C: Journal of Mechanical Engineering Science,* **228** 2402-2415.

[8] Harijono Djojodihardjoa , Alif Syamim Syazwan Ramlib, Surjatin Wiriadidjajac, 2012 Kinematic and Aerodynamic Modelling of Flapping Wing Ornithopter, *International Conference on Advances Science and Contemporary Engineering,* **50** 848-863.

[9] Jithin M P, Jude Olivero, Naveen M Nair, Sarath R, Vinod Yeldho Baby 2018 Design of Ornithopter Flapping Wing Mechanism, *International Research Journal of Engineering and Technology,* **5** 4 2395-0056 4297-4300.

[10] Krzysztof Augustynek, Andrzej Urbaś, 2018 Mathematical Modelling of Spatial Linkages with Clearance, Friction and Links' Flexibility Effects, *The 5th Joint International Conference on Multibody System Dynamics,* **1**-14.

[11] Phan H V, Kang T, and Park H C, 2017 Design and Stable Flight of A 21 G Insect-Like Tailless Flapping Wing Micro Air Vehicle with Angular Rates Feedback Control, *Bioinspiration & Biomimetics,* **12** 3 036006.

[12] Rajkiran Madangopal, Zaeem A Khan, Sunil K Agrawal, 2005 Biologically Inspired Design of Small Flapping Wing Air Vehicles using Four-bar Mechanisms and Quasi-steady Aerodynamics,*Journal of Mechanical Design,* **127** 4 809 -816.

[13] Ramesh Sattenapalli, Ravikanth Raji P, 2015 Kinematic Analysis of Micro Air Vehicle Flapping Wing Mechanism, *International Journal of Scientific Engineering and Technology Research,* 33 2319-8885.

[14] Silbilski K, 2004 Dynamics of Micro-Air-Vehicle with Flapping Wings, *Acta Polytechnica,* **44** 2 15-21.

[15] Sriyulianti Widhiarini, Ji Hwan Park, Bum Soo Yoon, Kwang Joon Yoon Il-Hyun Paik, Jong Heon Kim, Chan Yik Park, Seung Moon Jun Changho Nam, 2016 Bird-mimetic Wing System of Flapping-wing Micro Air Vehicle with Autonomous Flight Control Capability, *Journal of Bionic Engineering,* **13** 458–467.

[16] Vukobratović M K, Filaretov V F and Korzun A I, 1994 A Unified Approach To Mathematical Modelling Of Robotic Manipulator Dynamics, *Cambridge University Press,* **12** 5 411 – 420.

[17] Xinyan Deng, Schenato L, Wei Chung Wu, Sastry S, 2006 Flapping flight for biomimetic robotic insects: part I-system modeling, IEEE Transactions on Robotics, **22** 4 776 -788.

Computational study of heat-transfer in extended surfaces with various geometries

Mohankumar D, Shashank, Pazhaniappan Y, Nithesh Kumar RA, Ragul R, Manoj Kumar P, Nikhil Babu P

Department of Mechanical Engineering, KPR
Institute of Engineering and Technology,
Coimbatore.
dmohankumar94@gmail.com

Abstract

In this work, an analysis has been done on extended surfaces with various geometries to identify the better heat transfer rate. The geometries of extended surfaces namely – bar fins, stepped fins, stepped fins with indentations have been modelled, investigated and studied. Based on the design identification and numerical analysis the extended surfaces with increases surface area has increased heat transfer. From numerical data the fins with indentations has increased surface area and heat transfer. The indentations have been designed with varying diameter used in the fins are 4mm, 3mm, 2mm respectively and analysed using a fixed computational method. It is found that the spherical notches with varying diameter were more effective than the other fins and making the notches increases the heat transfer was much effective than increasing the diameter of the indentation in the stepped fin. This work shows the design, analysis, and result of some extended surface geometries with fixed input boundary condition.

Keywords: Fins, Notches, Heat Transfer, Varying Shape, CFD, Spherical Notches, Indentation.

1. Introduction

An extended surface called as fin is attached with a heated body to increase the convective heat transfer. Fins find applications in electro and mechanical components like internal combustion engines, compressors, heat exchangers, motors, etc. In electronics industry fins are used to avoid overheating or burning like in computers and laptops where hardware's are placed in a compact and small space as microstructures. The main objective of using the fins is, to increase convective heat transfer between the hot surface where fin is attached and its environment. The optimization of heat transfer rate is based on the following two approaches: i) to minimize the volume or mass for a given body and ii) to increase the surface area of the body. The rectangular geometry is mostly used because it is simple and inexpensive to manufacture.

From the analysis carried out on rectangular fins [1], it was observed that high fins have optimum heat transfer that occurs within a narrow range of spacing between the fins, whereas in shallow fins, optimum heat transfer occurs over a wider range of fin spacing. In an experiment conducted [2] on perforated fins, it was observed that along the perforated fin length the temperature drop is higher than that for non-perforated fin. According to [3] observation on heat transfer by natural convection with perforated pin, fin array implementation of notch and holes in extended surfaces dramatically increase the surface area and heat dissipation rate. Review of [4] heat transfer enhancement methods to maximize TEG (Thermo Electric Generator) performance in automotive applications showed that the diamond shaped pin fins provided the best heat transfer characteristics compared to other types [5]. The analysis also resulted in heat transfer enhancement features with different fin height to the channel height ratio while keeping a steady hydraulic diameter [6]. Fin geometry investigation [7] on heat exchanger efficiency by simulation and optimization methods for the application of diesel exhaust shows that fins arranged longitudinally are more efficient than circular or annular ones due to lower back pressure and improved heat transfer [8]. Experimentation on natural convection and surface heat transfer [9] using a cavity with vertically arranged fins shows that by increasing the number of vertical fins results in increased fluid temperature, increased convective and radiative heat transfers for $N < 7$ and stabilize beyond $N \geq 9$ fins. Heat transfer optimization and analysis of the engine fins with varying surface roughness [10] shows that, when the roughness of the body increases, the heat flux and the rate of heat dissipation increases. The scope of the work is to analyse all the three types fin geometries for maximum heat transfer rate.

2. Geometric modelling and material selection

2.1 Geometric modelling

Three types of fins i.e. normal rectangular fins, rectangular stepped fins, stepped fins with circular indentation are to be investigated here [11],[12], the basic dimensions of different types of fins are shown in Table 1.

Table 1: Dimensional parameters of various fin geometries

Fin Type	Normal Fin	Stepped Fin			Stepped Fin with indentations		
		Step from bottom			Step from bottom		
		1	2	3	1	2	3
Length (mm)	70	23	23	24	23	23	24
Breadth (mm)	50	50			50		
Width (mm)	5	5	4	3	5	4	3
Total Area (mm^2)	7250	7500			8799.4		
Material Used	Aluminium 6061						

2.2 Material selection

According to the analysis of materials [13] Aluminum 6061 is an Al-alloy hardened through precipitation, Aluminum 6061 consist of 97.9% Al, 0.6% Si, 1.0%Mg, 0.2%Cr and 0.28%Cu. It is best suitable material for stepped fin with circular indentations. The Alloy plate is one of the most versatile alloys which is used in welding assemblies, marine frames, chemical equipment's, electronics parts, fasteners, heat exchangers [14] and heat sinks. It is extremely popular for its medium to high strength requirements, good toughness, corrosion resistance and excellent thermal properties [15]. The physical properties of Al 6061 are mentioned in Table 2 below.

Table 2: Physical Properties of Aluminum 6061

PROPERTY	MAGNITUDE
Density	2710 kg/m3
Thermal Conductivity	152 – 202 W/m K
Specific Heat Capacity	897 J/kg K
Specific Gravity	2.7
Tensile yield strength	276 MPa
Shear strength	207 MPa
Fatigue strength	96.5 MPa

2.3. CFD modelling - 3D Physical model

The 3D model of the fins, created using SolidWorks 2016 modelling software and their isometric view are shown in Figure 1. The size of indentations of the fins with indentations are fixed in way that the surface won't become porous and accordingly the stepped structures will have decreasing size of indentations when going from from the base of the fins.

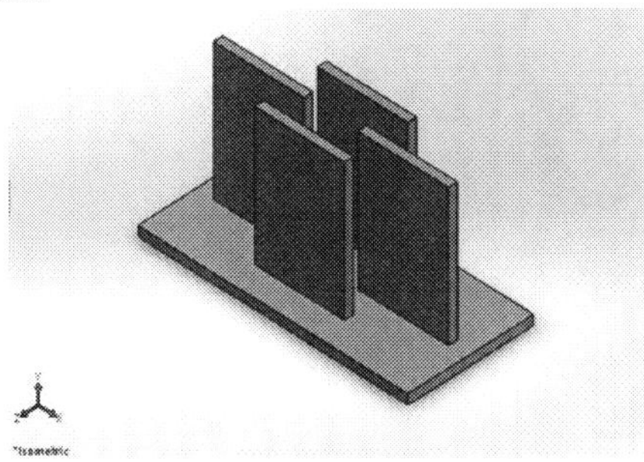

(a)

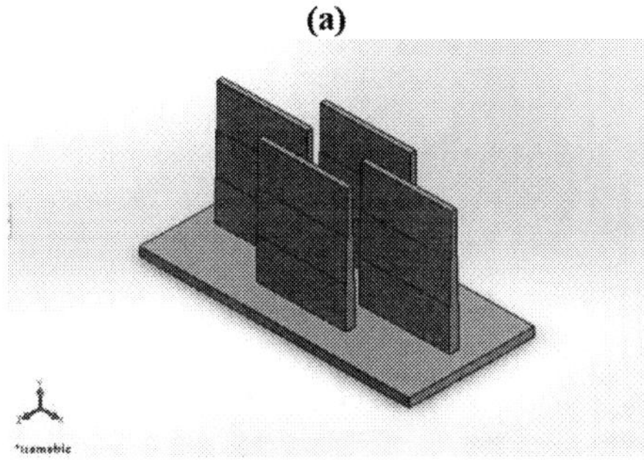

(b)

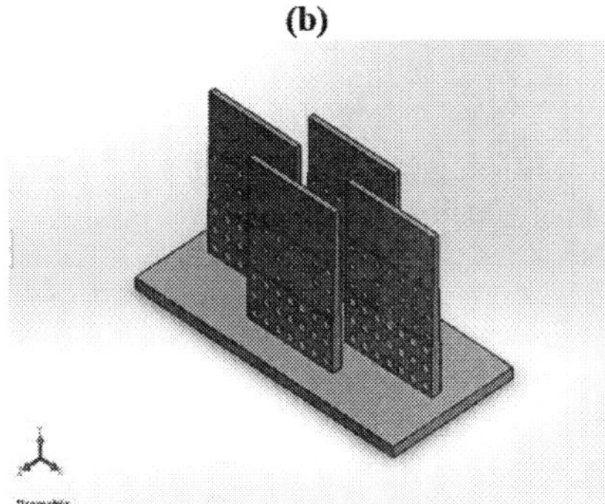

(c)

Figure 1: 3D models of (a) Rectangular normal fin

(b) Rectangular stepped fin (c) Rectangular stepped fin with indentations.

2.4. Meshing

The fine meshed images of all the three models taken into study are shown in the below figure 2.

Table 3: Mesh Independency test results

S.No	Element Size	No. of Elements	T_{tip} (°C)	ΔT (°C)
1	Very Coarse	142084	456	194
2	Less Coarse	250863	424	226
3	Coarse	436076	412	238
4	Fine	1341528	370	280
5	Very Fine	1892045	366	284

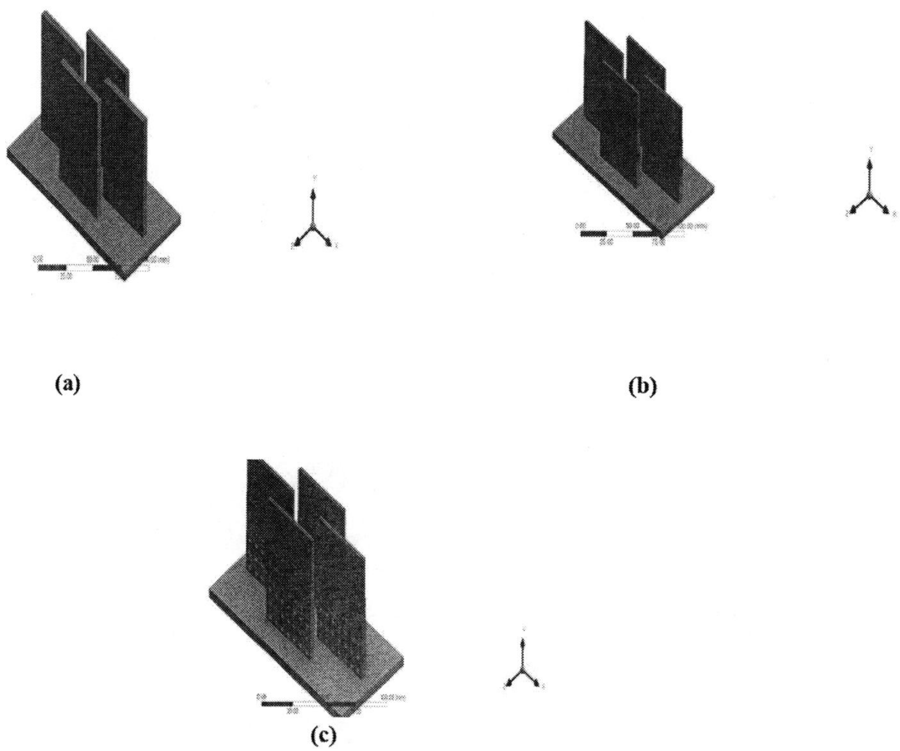

(a) (b) (c)

Figure 2: Mesh models of (a) Rectangular normal fin (b) Rectangular stepped fin (c) Rectangular stepped fin with indentations.

2.4.1 Mesh Independency Test. The Mesh Independency test for Stepped Fin with Indentation has been carried out at very coarse, less coarse, coarse, Fine and very fine element size. As seen in result Table 2 and Figure 3 the difference between ΔT for fine and very fine mesh is only 1.40%. So, fine mesh is selected.

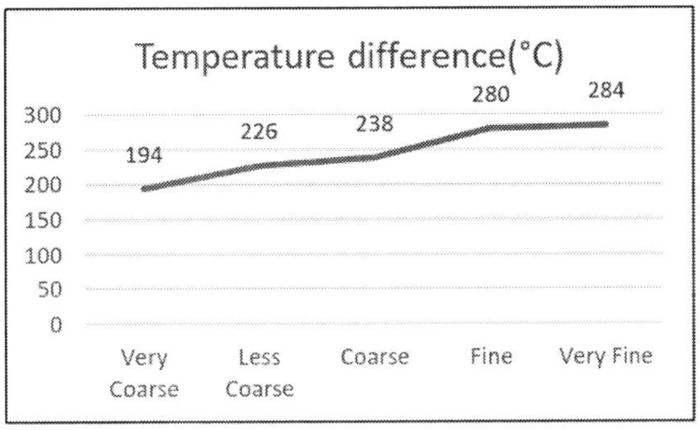

Figure 3: Graph Representing the mesh independency test for stepped fin with indentation

3. Mathematical methodology

The heat transfer rate [16] from the application at the base temperature (Tb) to the surrounding atmospheric temperature (T_∞) is given by Newton`s law of cooling [17], [18].

$$Q = \eta_f h A_f (T_b - T_\infty)$$

Where,

η_f – total fin efficiency, $\eta_f = (\tanh(mL)) / mL$
A_f – Surface area of the fin in m^2, $A_f = 2wL$
h – Convective heat transfer coefficient in W/m^2K
m – dimensional function
L – Length of the extended surface
w – Width of the extended surface

4. Boundary conditions

The boundary conditions applied to the solid and fluid domain during the analysis in ANSYS FLUENT are as follows,

Material	–	Air
Material	–	Solid
Density	–	1.1768 Kg/m^3
Inlet velocity of air	–	10 m/s
Specific heat	–	C_p=1005 J/kg.K
Inlet air temperature	–	300 K
Thermal conductivity	–	K= 0.0262 W/m^2
Operating condition	–	1 bar

5. Results and Discussions

5.1 Temperature Contour

After the analysis, the result from CFD are shown by means of vectors and contour plots, the following images in Figure 4 shows the temperature contour of normal fins, stepped fins and fins with indentations. From these results it is clear that the temperatures at the tip of the stepped fins with indentations have less temperature than the normal and stepped fins and highest temperature difference.

T_{tip} (With indentations) < T_{tip} (Stepped) < T_{tip} (Normal)

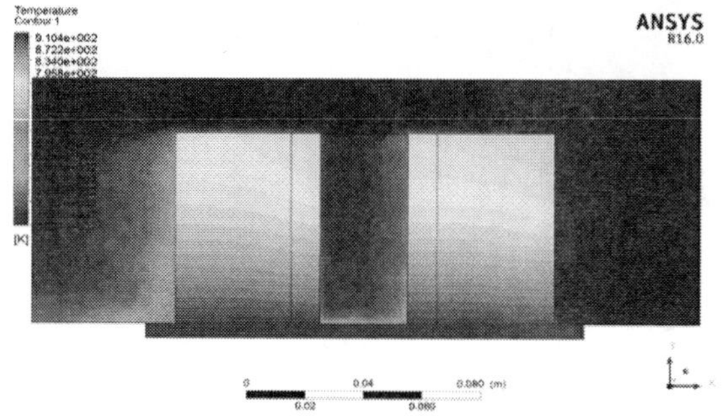

(a)

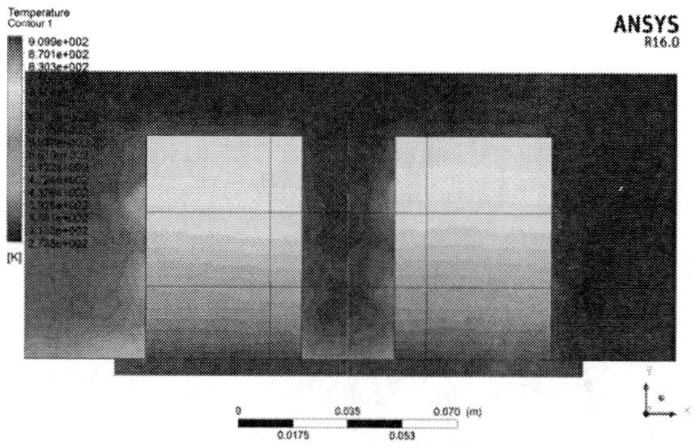

(b)

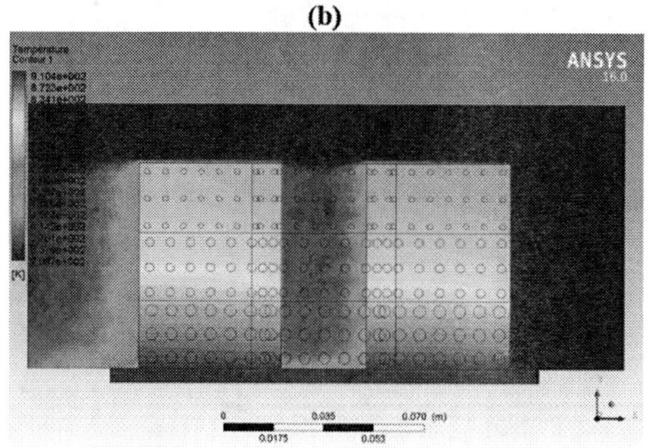

(c)

Figure 4: Temperature Contour of (a) Normal Fin (b) Stepped Fins (c) Stepped Fin with Indentations

5.2 Velocity contour

The below Figure 5 shows the velocity streamline of different types of fins, from this it is clear that the average velocity of air around the rectangular stepped fin is highest among all three fins and rectangular fins with indentation least. But the velocity of air near the surface of the fin is highest in rectangular fins with circular indentation due to increased drag. Since drag is more in rectangular fin with notch, the time of contact between air and fin surface is more which improves heat transfer rate.

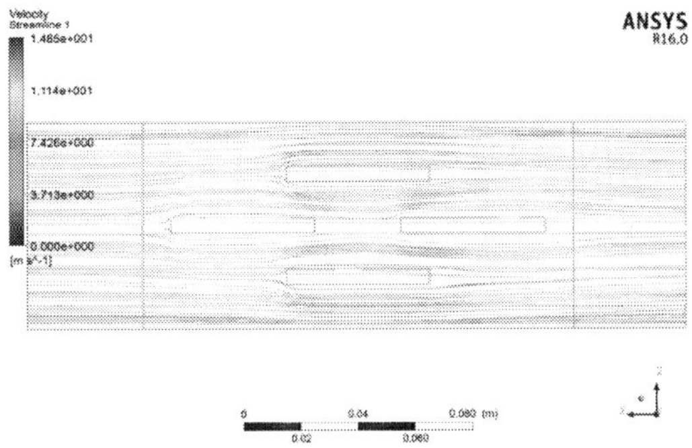

(a)

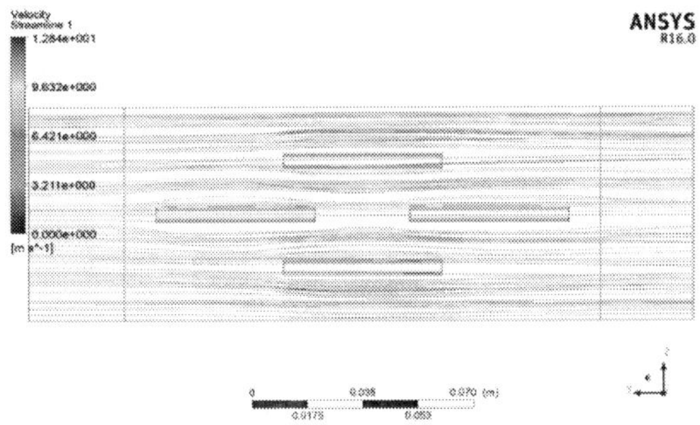

(b)

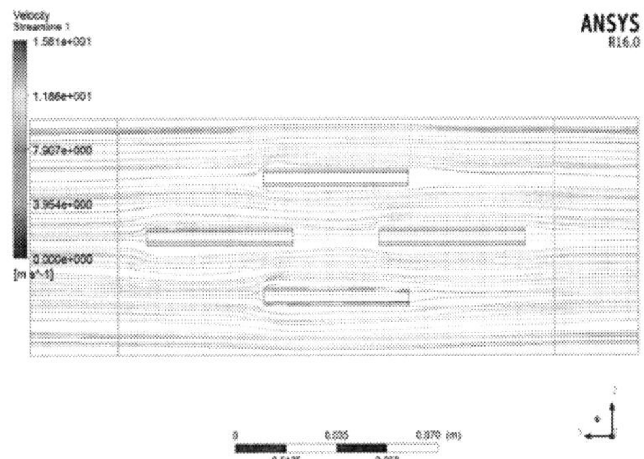

(c)

Figure 5: Streamline Velocity (a) Normal Fin (b) Stepped Fins (c) Stepped Fin with Indentations

5.3 Pressure contour

Figure 6 shows the pressure contour of different types of fin developed due to impact of flowing fluid medium particles. From the observation, the pressure generated on rectangular normal fin is the highest due to the higher impact surface area and the pressure generated on rectangular stepped fin with indentation has the lowest due to small impact surface.

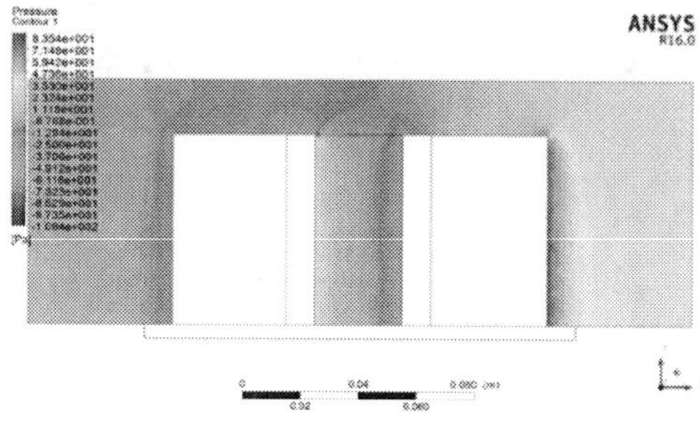

(a)

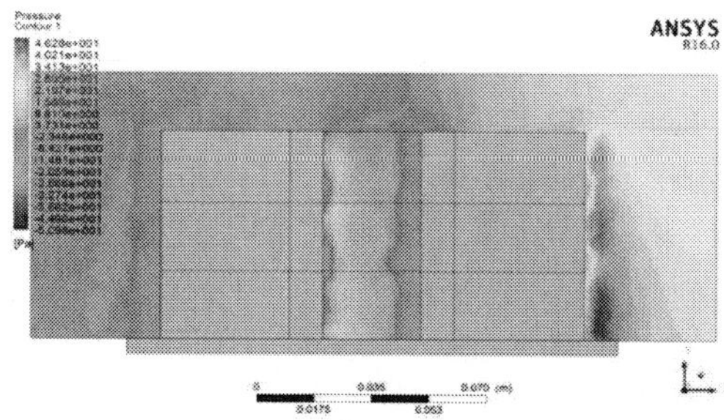

(b)

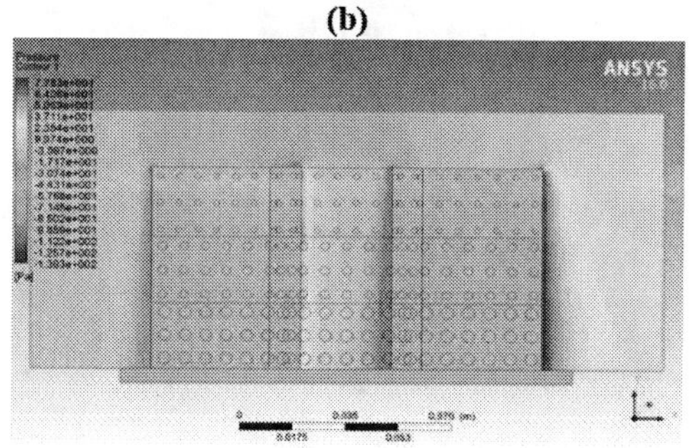

(c)

Figure 6: Pressure contour for (a) Normal Fin
(b) Stepped Fin (c) Stepped Fin with Indentations

5.4 Comparison of parameters of different geometry of fins

Table 4: Comparison of surface area, volume, heat transfer and heat transfer per unit volume when fin base temperature and surrounding temperature is constant

S. No	**Type of Fin**	**A (cm^2)**	**η_f %**	**V (cm^3)**	**Q (W)**	**Q/V (KW/m^3)**
1	Normal Fin	0.725	84	0.183	189.7	10466
2	Stepped Fin	0.75	81	0.151	186.9	12418
3	Fin with indentation	0.88	80	0.142	219.3	14598

Where,

Surface area	–	A
Volume	–	V
Heat transfer	–	Q
Heat transfer per unit volume	–	Q/V
Efficiency	–	η_f

According to the Table 3 data the efficiency of the rectangular fin is 3.51% better than the stepped fin and 5% better than the stepped fin with indentation but the volume of the normal fin is 16.96% more than stepped fin and 99.84 % more than stepped fin with indentation. The ratio of heat transfers and volume shows that stepped fin with indentation have better geometry for improved heat transfer capability.

Table 5: Comparison of fin base temperature, fin tip temperature, actual heat dissipation rate and efficiency according to actual data

S. No	Variable	Normal Fin	Stepped Fin	Fins with Indentations
1	**T_b (°C)**	650	650	650
2	**T_{tip} (°C)**	447	428	370
3	**ΔT (°C)**	203	222	280
4	**Q_{fin} (W)**	172.55	185.63	243.31
5	**η**	51%	53%	60%

Where

Temperature at base	–	T_b
Temperature at tip	–	T_t
Temperature Difference	–	ΔT
Heat Transfer	–	Q_{fin}
Efficiency	–	η

All the fins are subjected to similar base temperature of 650°C but the highest temperature difference is seen in rectangular stepped fin with indentation of 280°C and lowest in rectangular normal fins of 203°C. Due to high temperature difference rectangular stepped fin with indentation has highest heat transfer rate of 243.303W.

6. Conclusion

The thermal analysis of fins by modifying its geometry has been completed. By observing the analysis, the results of aluminum alloy 6061 fins with circular notches is more efficient, as the temperature difference and rate of heat transfer are more than the normal bar fins and the stepped fins of same material.

- Weight and volume of the material used can be reduced with the increase in surface area of the fins, so that number of fins for fixed heat transfer rate can also be reduced wherever needed.
- The usage of less material and resources results in the improved efficiency. Also, in the fins with indentations wake zone is reduced and drag is increased, which improves the time of contact between the heated surface of the fin and the surrounding atmospheric air, which increases the overall efficiency of the extended surface.

Acknowledgements

The authors would like to thank the Simulation and Analysis section, Department of Mechanical Engineering, KPR Institute of Engineering and Technology, Coimbatore – India, for providing the system facilities and their continuous support.

References

[1] Adhikari C, Wood D H and Pahlevani M 2020 Optimizing Rectangular fins for natural Convection cooling using CFD *Therm. Sci. Engg. Prog.* **17** 100484.

[2] Ibrahim Tamir K., Mohammed Marwah N, Mohammed Kamil Mohammed and Najafi G 2018 Experimental Study on the effects of perforations shapes on vertical (heated fins performance under forced convection heat transfer *IJHMT* **118** 832-846.

[3] Shitole Pankaj, Bhosle Santosh, Kulkarni Kishor, Joshi Sarang 2018 Experimental Investigation of Heat Transfer by Natural Convection with Perforated Pin Fin array *Procedia Manufacturing* **20** 311-317.

[4] Jaideep Pandit, Megan Thompson, Srinath Ekkad V and Huxtable Scott T 2014 Effect of pin fin to channel height ratio and pin fin geometry on heat transfer performance for flow in rectangular channels *Int. J. Heat Mass Tran.* **77** 359-368.

[5] Kumar P M, Saravanakumar P T, Mylsamy K, Kishore P and Prakash K B 2019 Study on thermal conductivity of the candle making wax (CMW) using nano-TiO_2 particles for thermal energy storage applications *AIP Conf. Proc.* **2128** 1 020027.

[6] Manoj Kumar P, Anandkumar R, Sudarvizhi D, Prakash K B and Mylsamy K 2019 Experimental investigations on thermal management and performance improvement of solar PV panel using a phase change material *AIP Conf. Proc.* **2128** 1 020023.

[7] Hatami M, Ganji D D and Gorji Bandpy M 2015 Investigation of fin geometry on heat exchanger performance by simulation and optimization methods for diesel exhaust application *Neural Comput & App.* https://doi.org/10.1007/s00521-015-1973-1

[8] Manoj Kumar P, K Mylsamy, Karthick Alagar and K Sudhakar 2020 Investigations on an evacuated tube solar water heater using hybrid-nano based organic phase change material *Int. J. Green Energy* **17** 13 872-883. doi.org/10.1080/15435075.2020.1809426

[9] El Moutaouakil, Bounkendil M, Zrikem Z and Abdelbaki A 2020 Natural Convection and surface radiation heat transfer in a cavity with vertically oriented fins *Int. J. Thermophys.* **41**.

[10] Pulkit Sagar, Puneet Teotia, Akash Deep Sahlot and Thakur H C 2017 Heat Transfer analysis and optimization of engine fins of varying surface roughness Mat. Today-Proc. 4 8565-8570.

[11] Saurabh Pathak, Om Parkash, Ravikant, "Thermal analysis of fins with Varying Geometry of different materials, *IJSRD*, 2321-0613.

[12] Kumar P M, Anandkumar R, Mylsamy K and Prakash K B 2020 Experimental investigation on thermal conductivity of nanoparticle dispersed paraffin (NDP) *Mater. Today-Proc.* doi.org/10.1016/j.matpr.2020.02.798

[13] Obula Reddy Kummitha and Reddy B V R 2017 Thermal Analysis of cylinder block with fins for different materials using ANSYS *Mat. Today-Proc.* **4** 8142-8148.

[14] Karthick A, Manokar Athikesavan M, Pasupathi M K, Manoj Kumar N, Chopra S S and Ghosh A 2020 Investigation of inorganic phase change material for a semi-transparent photovoltaic (STPV) module *Energies* **13** 14 3582. doi.org/10.3390/en13143582

[15] Kumar P M, Saravanakumar P T, Mylsamy K, Kishore P and Prakash K B 2019 Study on thermal conductivity of the candle making wax (CMW) using nano-TiO_2 particles for thermal energy storage applications *AIP Conf. Proc.* **2128** 1 020027.

[16] Yunus A. Cengel, "Heat Transfer A Practical Approach", Tata McGraw Hill, 2010

[17] Frank P. Incropera and David P. Dewitt, "Fundamentals of Heat and Mass Transfer", Wiley & Sons, 1998.
[18] Kothandaraman, C.P., "Fundamentals of Heat and Mass Transfer", New Age International, New Delhi, 1998.

Exergy Analysis on Performance of Groundnut Solar Dryer with Forced Convection

Subramaniyan C[1*], Prakash KB[2], Kalidasan B[3], Bhuvanesh N[4], Amarkarthik A[5]

[1,2,3,4] *Assistant Professor, Department of Mechanical Engineering, Bannari Amman Institute of Technology, Sathyamangalam-638401, India*

[5] *Professor, Department of Mechanical Engineering, Bannari Amman Institute of Technology, Sathyamangalam-638401, India*

Email ID: *subra.csm@gmail.com*

Abstract. The moisture present in the farming yields and in seeds is one of the greatest affecting factors which are playing a major role in storing it for longer days. The quality of products can be increased by preventing the formation of bacteria, yeast, etc. The moisture content present in the groundnut must be between 10 to 15% (weight basis). In this study a numerical analysis was investigated on a solar dryer, which is used to diminish the moisture content present in the groundnut to preserve it for longer days. The performance analyses of forced convection solar dryer with dryer compartment were analysed based on local ambient conditions at sathyamangalam for various required design and operating parameters such as moisture removal rate, air mass flow rate, air inlet, and outlet temperature. The proposed basic solar collector aperture area is about 2 m^2 which is made of the copper plate as an absorber plate coated with black powder coating for better absorption. The moisture content present in the freshly harvested groundnut varies from 25-30 % (weight basis). The analysis was done for drying groundnut to remove the moisture content to the required moisture level of 10% by utilizing solar energy. The results from the study illustrate that the Efficiency of dryer is a maximum of 39.2% for the optimized mass flow rate of 0.022 kg/s and also it was found that the dryer overall efficiency and average collector temperature is gradually decreasing for increasing the mass flow rate.

Keywords: Energy; uncertainty; Efficiency; Moisture; Heat gain

Nomenclature

t	Time, hour
I	Radiation, W/m^2
T_a	Ambient temperature, K
T_{a1}	Inlet air temperature , K
T_{ab}	Average absorber temperature, K
Q_l	Heat loss, W
Q_{in}	Heat absorbed, W
T_{a2}	Outlet air temperature, K
Q_u	Heat gain, W
E_s	Total input energy, W
ΔE_{loss}	Energy loss from the dryer to the ambient air, W
w_b	Work done for blower, W
ρ_a	Density, kg/m^3
η	Efficiency, %

$m_a^{\cdot}$	Mass flow rate of air, kg
$C_{pa}^{\cdot}$	Specific heat of air, kJ/kg K
$m_c^{\cdot}$	Mass of copper plate, kg
$C_{pc}^{\cdot}$	Specific heat of copper plate, kJ/kg K
M_m	Moisture, kg
M_i	Initial Moisture, kg
M_f	Final Moisture, kg
ρ	Reflectivity
α	Absorptivity
τ	Transmissivity
A	Collector area, m^2
U_T	Uncertainty in Thermocouple, K
U_{RH}	Uncertainty in Relative humidity thermo hydrometer, kg
U_V	Uncertainty in Anemometer, m/s
U_I	Uncertainty in Solar power meter, W/m^2
U_W	Uncertainty in Weighing machine, kg
U_η	Uncertainty in dryer Efficiency, %

1. Introduction

Drying was the first method for preserving food items used by humans. It is the process in which farm products are preserved by removing the moisture content with the help of solar power. This manner of preserving helps us to store the items for a longer period of time. In earlier days food items are dried in open sun light by placing them over mats, roofs or ground. This open drying has many disadvantages like rain, heavy wind, sand dust, etc. Also wastage of food due to insects, birds, etc. These climatic and infectious risk causes deterioration and decomposition of food particles. Also it needs a vast area of land and more time duration. Tray dryers are equipment that is helpful in drying grains in spite of rain, wind, insects, birds and dust. Only minimum space is required for the installation of the dryer and also relatively less expensive when compared to traditional solar drying. The drying rate depends up on numerous constraints such as radiation from sun, surrounding temperature, speed of wind, humidity ratio and initial moisture content, crop varieties and product that is exposed to the radiation. Analyzing the performance of solar dryer can be done for the different operating and design parameters such as collector area, mass flow rate of air and tray design. The air temperature at entrance of drying chamber is increased by increasing the length of the drying chamber [1]. A newly designed solar dryer for drying the maize was utilized for analysing the performance of solar dryer at different angle of flat plate solar collector. The result shows that the more radiation input can be obtained at an angle of 45^0 in the solar dryer. Peak value of solar radiation can be occurred at 11am and then afterwards it gradually decreases [2].

Estimating the drying time of the green pea at different temperatures was analysed to obtain the required final moisture content. Maximum of up to 25% of moisture content has been removed from the green pea after 1 hour of drying at the temperature of 348K. From the result it was found that the drying time increases when the inlet temperature of drying chamber decreases [3]. Numerical analysis of the solar dryer was done for the different psychometric properties of air for the required output parameter dry food crops. The results shows that the dryer gives the better performance at optimized operating parameter of input temperature of about 32^0C and relative humidity of 80%, the output temperature of about 45^0C and relative humidity of 40% [4]. The solar dryer with mixed-mode was fabricated and an experimental study also been done on the dryer in which the results shows that the efficiency and product drying rate was 57.5% and 0.62 kg/h respectively [5]. Design and development of the setup of solar dryer to analyse the performance of dryer at different temperatures by drying potato slice was done. The performance analyse of the dryer was done for the moisture content in the potato slice at different temperatures such as 40^0C, 50^0C, 60^0C and 70 ^{0}C at constant mass flow rate of air and at the efficient performance of the dryer the moisture content can be reduced from 85% to 21% [6]. Various design and fabrication models were discussed for effective performance of the solar dryer for drying applications. The collector area of the dryer for the mass of the drying products were also been compared for different required and ambient conditions [7].

A brief Discussion were also been made about different types of solar dryers in which the difference between the natural and forced convection of the solar dryer were elaborately discussed with the cost efficient types of solar dryers for farmers in rural areas [8]. The novel solar drying system is designed and the performance of the system was evaluated for drying seaweed for the moisture removal of 90% (wet basis). The results show for the optimized drying time, air flow rate, the efficiency of the collector was 35 % with the best payback period of 2.33 year [9]. An elaborated study also been done on various solar dryers with respect to the drying product, performance and economical availability of the dryer [10]. A new unsteady state analytical model was derived for an inclined multi-pass dryer in which a deep-bed dryer attached with in-built thermal storage. The performance of a solar air heater was evaluated for the rate of moisture rate and air humidity [11]. An experimental study was conducted on the performance of solar dryer with solar tracking system to examine the enhancement on the dryer performance with respect to the fixed and tracking position. The result shows that the dryer performance is enhanced to 75% from 52 % by tracking the sun instead of fixed in a position [12]. Performance analysis was done on the greenhouse solar dryer for ginger drying with parabolic roof structure enclosed by polycarbonate plates with the base area of 912.4 m2 [13]. A novel active indirect solar dryer was developed and performance study also been done for banana drying application. The dryer efficiency was found to be in the range of 13.85 to 31.84% [14]. A brief review on economic analysis and significance of structural factors on the design of solar dryers was discussed based on the various drying applications and technologies in the selected Asian countries [15].

The PV/T drying system was designed and fabricated for finding the performance of the closed-loop control unit assisted solar dryer for enhancing the drying rate [16]. The mathematical modelling also been done for finding the thermal buoyancy in the convection on solar dryer for various inlet opening to vent outlet opening ratios [17]. A novel hybrid solar-biomass dryer is designed and fabricated for finding the energy and exergy efficiency of the dryer. The results show that the collector efficiency is in range of 21% to 22 % for the effective design and operating parameters [18]. A novel solar indirect dryer is designed and developed which chimney and powered fan. The result shows the significance of the design and operating parameters on the maximum average temperature for the optimized value of 4.3^{0}C [19]. An experimental study on solar dryer was done to investigate the moisture deduction rate in three unlike chambers (chimney type, attic space type, thin tube chimney type). The results shows that the moisture removal rate in three different chambers were 44.5%, 33.3% and 58.9% for chimney type, attic space type and thin tube chimney type respectively [20]. Adaption of packed bed design of absorber plate were also been done on the solar dryer for enhancing the exergy. A novel thermal energy storage system supported solar dryer with was designed and developed for bitter gourd drying application. An exergy and environmental analysis was done for evaluating the thermal performance of solar dryer different design and operating parameters. The mass flow rate were also been optimised for the better moisture removal rate and efficiency [21-22].

In this present study, the design and development of solar flat collector dryer was done to find the enhancement of the thermal performance of the dryer system by adapting the black powder coating absorber copper plate for better absorption. The significance of the moisture content present in the fresh harvested groundnut on the thermal performance of the dryer system also been investigated. The experimental analysis were also been done for drying the groundnut to remove the moisture content to the required moisture level by utilizing the solar energy. The effect of mass flow rate on performance of dryer is also to be found for optimizing the dryer overall efficiency and average collector temperature at maximum.

2. Experimentation:

The design of solar dryer was based on the maximum requirement and utilization of the solar input and heating output energy. The dryer consists of a flat plate collector of aperture area of 2 m^2 (2 m x1 m) and forced convection system which is assisted with an air blower. The dryer set up consists of 3 major parts namely flat plate collector, tray drying unit and natural draft system. Figure.1 shows the drying compartment and tray setup (0.6 x 0.5 m, 3 No's) design which is entirely insulated with glass wool and the Figure .2 represents the design of solar collector area and blower setup with natural chimney draft. The collector setup is inclined to the horizontal surface at an angle of 25° which is attached in front of the drying unit for the maximum collection of solar radiation. The air blower is attached for the suction of air and to blow the air at the required velocity to the collector. A regulator is also fixed with the blower for regulating the flow rate of the air. The additional secondary reflector mirror setup (0.7 x 0.6 m) is provided to increase the received solar flux in the copper plate in the solar collector. The natural draft chimney is designed in a way to provide more pressure difference on the compartment tray for the flow of air which is shown in Figure.3.

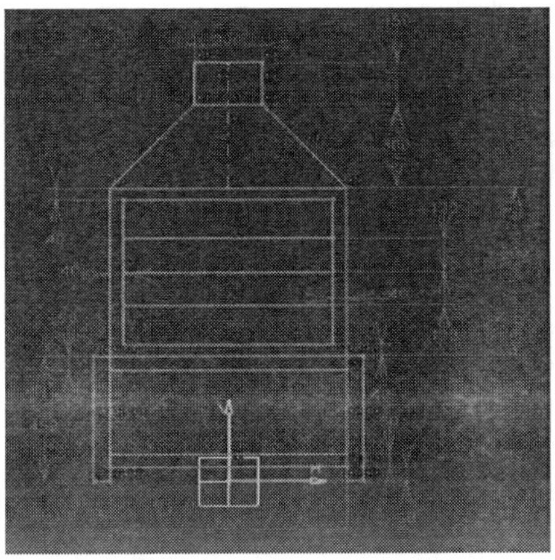

Figure.1 Design of tray compartment

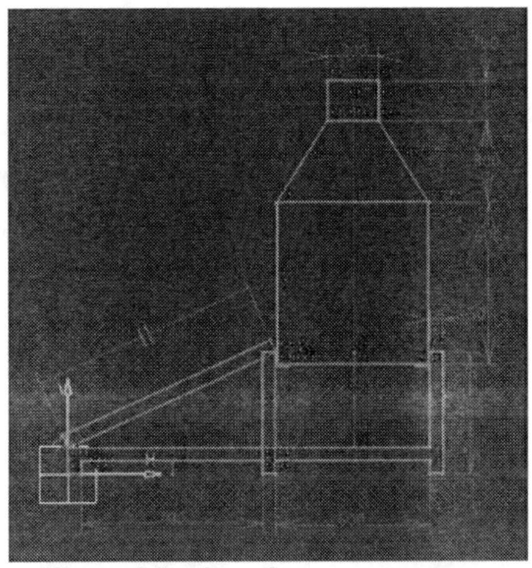

Figure.2 Design of tray compartment

To increase the absorptivity of the copper plate in the solar collector section, black powder coating (absorptivity of 0.95) was given to the plate. The aluminium foil is used as a secondary reflector (reflectivity of 0.9) at the vertical surface of the drying compartment tray for enhancing the optical efficiency of the solar dryer. The acrylic cover (transmissivity of 0.9) is provided at the solar collector as a transparent medium which replaces the glass materials for better economic of the solar dryer.

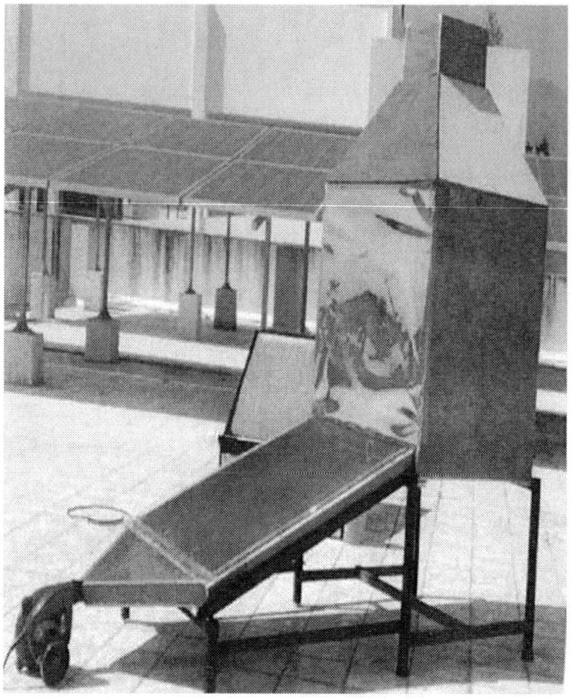

Figure.3 Experimental set up of solar dryer

The dryer operating parameters are confirmed for its steady state conditions while the experimentation study is started for finding the output parameters. The operating parameters were taken for a day from 10.00 AM to 03.00 PM within the interval of 10 minutes. The corresponding values of the direct radiation and temperatures at various nodes (atmospheric, copper plate, transparent cover, air inlet and outlet) were noted and it is shown in Table.1. All the readings were taken at the steady state conditions for various flow rate of air which is blown into the solar collector area.

Table. 1. Experimental parameters.

T	I	T_{a1}	T_{ab}	T_{a2}
(hr)	(W/m^2)	(K)	(K)	(K)
10	790	304	322	306.1
11	816	305	324	308.2
12	873	307	328	310.3
1	852	306	325	309.1
2	825	305	322	307.9
3	798	305	320	307.2

3. Analyzing methods:

3.1. Dryer efficiency:

Moisture content in 1 kg of ground nut is denoted by M_m and Moisture ratio (M_r) is given by following equation (Eq. (1)).

$$M_r = \frac{m_m}{m_i} \qquad (1)$$

Dryer instantaneous efficiency and moisture removed by the dryer are given by following equation (Eq. (2) & (3)) respectively.

$$\eta = \frac{Q_u}{E_s} \tag{2}$$

$$m_e = m_f - m_i \tag{3}$$

Dryer instantaneous total energy supply, solar flux received and optical efficiency of the collector are given by following equation (Eq. (4) & (5) & (6)) respectively.

$$E_s = (I \times A) + w_b \tag{4}$$

$$E_4 = \tau \times \alpha \times \varepsilon (I \times A) \tag{5}$$

$$\eta_{op} = \frac{E_4}{\tau \times \alpha \times \varepsilon (I \times A)} \tag{6}$$

Energy balance: Energy in to the collector = Energy out from the collector

$$E_4 = E_{in} = E_{out} = (mc_p\Delta T)_{plate} + (mc_p\Delta T)_{air} + \Delta E_{loss}$$

$$E_4 = [m_c c_{pc}(T_{ab2} - T_{ab1})] + [m_a c_{pa}(T_{a2} - T_{a1})] + \Delta E_{loss}$$

Dryer instantaneous energy given to the drying product for moisture removal and Dryer instantaneous efficiency are given by (Eq. (7) & (8)) respectively.

$$E_5 = E_4 - [(m_a c_{pa}(T_{a3} - T_{a2}))] \tag{7}$$

$$\eta_d = \frac{E_5}{E_1 + P_6} \tag{8}$$

Dryer instantaneous energy given to the drying product in the mode of convective heat transfer for moisture removal rate which is equal to the sensible heat loss by the air is given by (Eq. (9)).

$$Q = hA(T_a - T_p) \tag{9}$$

$$m_a c_{pa}(T_{a1} - T_{a2}) = hA(T_{ab} - T_a)$$

The average temperature and heat gain of the dryer are given by (Eq. (10) & (11)) respectively.

$$T_{avg} = \frac{T_{ab} + T_a}{2} \tag{10}$$

From Equation (9) & (10),

$$\Delta T_{ab} = \frac{0.729(I \times A) - m_a c_{pa}(\Delta T_a)_c}{m_c c_{pc}}$$

$$Q_u = \frac{\dot{m}_m \times h_{fg}}{t} \tag{11}$$

3.2. Uncertainty analyses in solar dryer:

The uncertainties on the measuring parameters are tabulated in table. 2 and the same values are used to find the uncertainty on the efficiency of the solar dryer which is given by (Eq. (12))

Table. 2. Uncertainty parameters.

Notation	Description	Specification
U_T	Thermocouple	±0.5 °C
U_{RH}	Relative humidity thermo hydrometer	±1%
U_V	Anemometer	±0.1 m/s
U_I	Solar power meter	±5 W/m²
U_W	Weighing machine	±0.001 kg

$$U_{\eta d} = \sqrt{\left(\frac{\partial \eta d}{\partial \dot{m}}\right)^2 U_{\dot{T}}^{\,2} + \left(\frac{\partial \eta d}{\partial T_4}\right)^2 U_{RH}^{\,2} + \left(\frac{\partial \eta d}{\partial T_1}\right)^2 U_{\dot{v}}^{\,2} + \left(\frac{\partial \eta d}{\partial T_1}\right)^2 U_{\dot{I}}^{\,2} + \left(\frac{\partial \eta d}{\partial T_1}\right)^2 U_{\dot{W}}^{\,2}} \quad (12)$$

4. Result and Discussion

The average temperature of the solar dryer was calculated for different flow rate of air from the blower. Figure.4 shows that variation of average temperature of the collector with respect to different mass flow rate of air. Initially for the fully opened blowing condition of 0.044 kg/s, the collector average temperature is 312.4 K. whereas diminishing the mass flow rate of air leads to increasing the average temperature of the collector. The maximum average temperature is around 322.3 K at the optimized mass flow rate of 0.022kg/s.

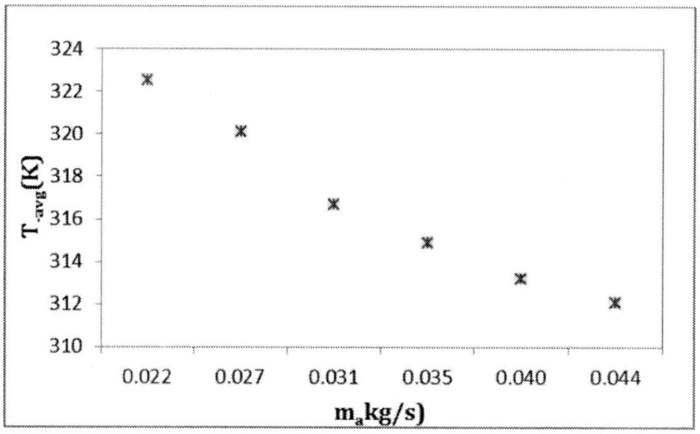

Figure.4 Variation on dryer average temperature with different mass flow rate

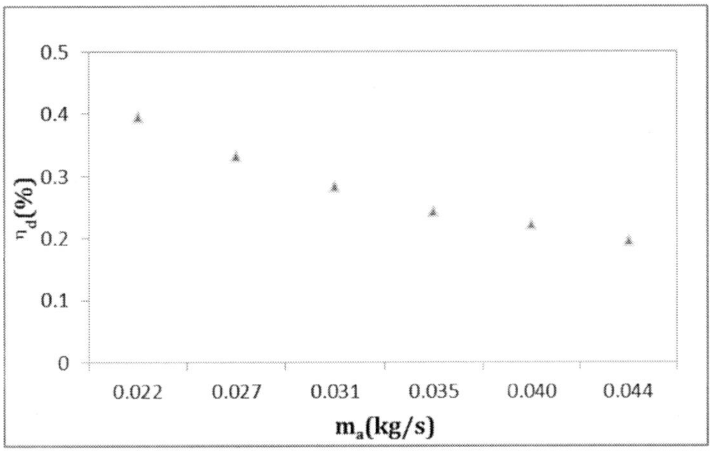

Figure.5 Variation on dryer efficiency with different mass flow rate

Figure.5 shows that variation on efficiency for different mass flow rate of air in the dryer. The maximum efficiency of about 40% is obtained at the mass flow rate of 0.022kg/s. The efficiency is further drops when mass flow rate increases and it reaches to 20% for the mass flow rate of 0.044kg/s and the efficiency is also decreases when the mass flow rate less than 0.022kg/s.

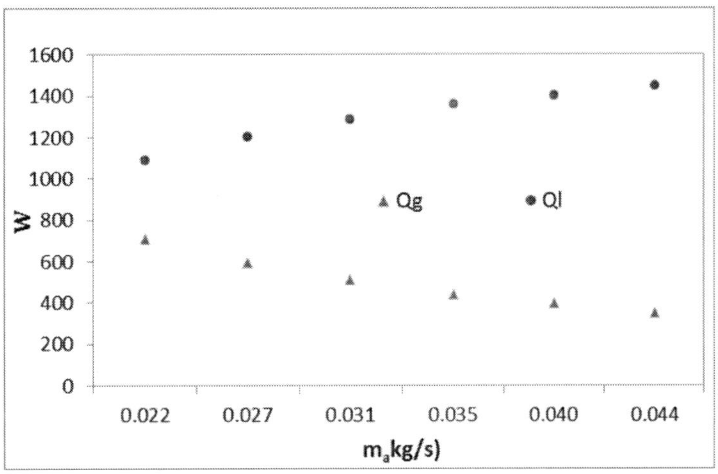

Figure.6 Variation on heat loss and heat gain for different mass flow rate

Figure.6 shows that heat loss and heat gain for various mass flow rate of air in the dryer. The dryer heat gain is about maximum, for the mass flow rate of air at 0.022kg/s and the value of the maximum heat gain of the dryer is about 725W. The heat gain is further decreases to 425W for the mass flow rate of air at 0.044kg/s. The heat loss is about maximum for mass flow rate of 0.044kg/s due to availability of time is very less which is not enough to conduct the heat from air to grains. Figure.7 shows the amount of moisture content removed for various mass flow rate of air at dryer. At the mass flow rate of 0.022kg/s, the reduction on moisture content is more when compare to the mass flow rate of 0.044kg/s for the same time interval. The amount of reduced moisture content is decreasing at increasing time interval from initial to 7 hours.

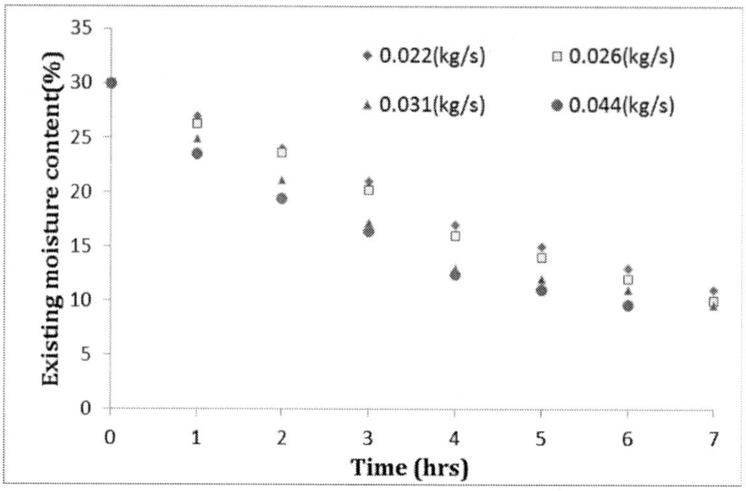

Figure.7 Reduction on moisture content with respect to drying time

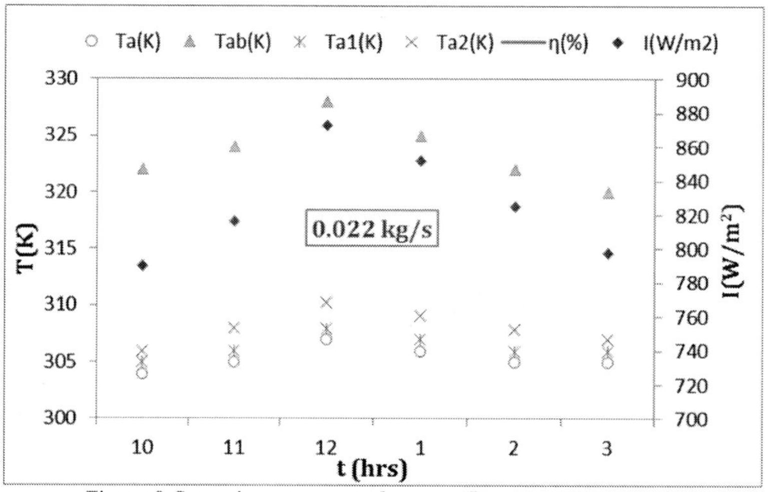

Figure.8 Operating parameter for mass flow rate of 0.022kg/s

Figure.8 shows the temperatures at various nodes with respect to time. The Ambient Temperature at 10AM is 304K and it is gradually increases to the maximum temperature of 307K at 12 noon and then decreases to 305K at 3PM. Absorber plate Temperature at 10AM is about 322K and it is gradually increases to the maximum temperature of 328K at 12 noon and then it is gradually decreases to 319K at 3PM. Collector air inlet Temperature at 10AM is about 304K and it is gradually increases to the maximum temperature of 307K at 12 noon and then it is gradually decreases to 305K at 3 PM. Collector air outlet Temperature at 10AM is about 306K it is gradually increases to the maximum temperature of 310K at 12 noon and then it is gradually decreases to 307K at 3 PM. Radiation at 10AM is 782W/m^2 and it is gradually it increases to the maximum radiation of 870 W/m^2 at 12 noon and then gradually it is decreases to 798 W/m^2. Figure.9 shows that instantaneous efficiency of the dryer at different time for the fixed mass flow rate. The efficiency is about maximum of 44% at the noon due to the availability of maximum radiation and then efficiency is gradually decreases to 25% at 3PM.

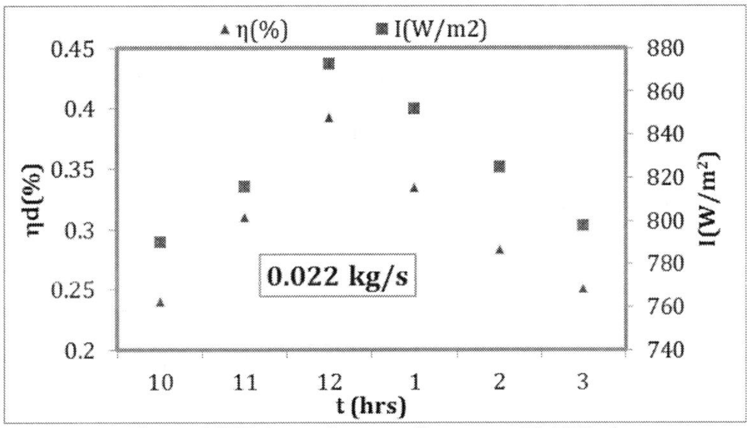

Figure.9 Efficiency for mass flow rate of 0.022kg/s

5. Conclusion

In this present study, the design and fabrication of secondary reflector assisted solar dryer was done with forced convection system and the performance of the solar dryer is analyzed for various mass flow rates of air, radiation and time interval. The mass flow rate of air across the dryer is varied from 0.022kg/s to 0.044kg/s and the maximum efficiency of dryer is obtained at the mass flow rate of 0.022kg/s at 12 noon. It was also found that the efficiency of the dryer is further decreases when the mass flow rate is increases from 0.022kg/s and the efficiency of the dryer is further decreases when the mass flow rate of air is decreases lower than 0.022 kg/s.

References

[1] Aboul-Enein, S., El-Sebaii, A. A., Ramadan, M. R. I., & El-Gohary, H. G. (2000). Parametric study of a solar air heater with and without thermal storage for solar drying applications. Renewable energy, 21(3-4), 505-522.

[2] Gatea, A. A. (2010). Design, construction and performance evaluation of solar maize dryer. Journal of Agricultural Biotechnology and sustainable development, 2(3), 39-46.

[3] Arunsandeep, G., Lingayat, A., Chandramohan, V. P., Raju, V. R. K., & Reddy, K. S. (2018). A numerical model for drying of spherical object in an indirect type solar dryer and estimating the drying time at different moisture level and air temperature. International Journal of Green Energy, 15(3), 189-200.

[4] Ayensu, A. (1997). Dehydration of food crops using a solar dryer with convective heat flow. Solar energy, 59(4-6), 121-126.

[5] Bolaji, B. O., & Olalusi, A. P. (2008). Performance evaluation of a mixed-mode solar dryer. AU Journal of Technology 11(4), 225-231.

[6] Mohan, V. C., & Talukdar, P. (2013). Design of an experimental set up for convective drying: experimental studies at different drying temperature. Heat and Mass Transfer, 49(1), 31-40.

[7] Ekechukwu, O. V., & Norton, B. (1999). Review of solar-energy drying systems II: an overview of solar drying technology. Energy conversion and management, 40(6), 615-655.

[8] El-Sebaii, A. A., & Shalaby, S. M. (2012). Solar drying of agricultural products: A review. Renewable and Sustainable Energy Reviews, 16(1), 37-43.

[9] Fudholi, A. H. M. A. D., Othman, M. Y., Ruslan, M. H., Yahya, M., Zaharim, A., & Sopian, K. (2011). Technoeconomic analysis of solar drying system for seaweed in Malaysia. In Proc. of the 7th IASME/WSEAS Int. Conf. on Energy, Environment, Ecosystems and Sustainable Development (EEESD, 11) (pp. 89-95).

[10] Fudholi, A., Sopian, K., Ruslan, M. H., Alghoul, M. A., & Sulaiman, M. Y. (2010). Review of solar dryers for agricultural and marine products. Renewable and sustainable energy reviews, 14(1), 1-30.

[11] Jain, Dilip, and Rajeev Kumar Jain. (2004)"Performance evaluation of an inclined multi-pass solar air heater with in-built thermal storage on deep-bed drying application." Journal of Food Engineering 65.4: 497-509.

[12] Daş, M., & Akpinar, E. K. (2020). Determination of thermal and drying performances of the solar air dryer with solar tracking system: Apple drying test. Case Studies in Thermal Engineering, 100731.

[13] Nimnuan, P., & Nabnean, S. (2020). Experimental and simulated investigations of the performance of the solar greenhouse dryer for drying cassumunar ginger (Zingiber cassumunar Roxb.). Case Studies in Thermal Engineering, 100745.

[14] Etim, P. J., Eke, A. B., & Simonyan, K. J. (2020). Design and development of an active indirect solar dryer for cooking banana. Scientific African, 8, e00463.

[15] Udomkun, P., Romuli, S., Schock, S., Mahayothee, B., Sartas, M., Wossen, T., & Müller, J. (2020). Review of solar dryers for agricultural products in Asia and Africa: An innovation landscape approach. Journal of Environmental Management, 268, 110730.

[16] Vengsungnle, P., Jongpluempiti, J., Srichat, A., Wiriyasart, S., & Naphon, P. (2020). Thermal performance of the photovoltaic–ventilated mixed mode greenhouse solar dryer with automatic closed loop control for Ganoderma drying. Case Studies in Thermal Engineering, 100659.

[17] Sekyere, C. K. K., Adams, F. W., Davis, F., & Forson, F. K. (2020). Mathematical modelling and validation of the thermal buoyancy characteristics of a mixed mode natural convection solar crop dryer with back up heater. Scientific African, e00441.

[18] Ndukwu, M. C., Simo-Tagne, M., Abam, F. I., Onwuka, O. S., Prince, S., & Bennamoun, L. (2020). Exergetic sustainability and economic analysis of hybrid solar-biomass dryer integrated with copper tubing as heat exchanger. Heliyon, 6(2), e03401.

[19] Demissie, P., Hayelom, M., Kassaye, A., Hailesilassie, A., Gebrehiwot, M., & Vanierschot, M. (2019). Design, development and CFD modeling of indirect solar food dryer. Energy Procedia, 158, 1128-1134.

[20] Islam, M., Islam, M. I., Tusar, M., & Limon, A. H. (2019). Effect of cover design on moisture removal rate of a cabinet type solar dryer for food drying application. Energy

Procedia, 160, 769-776.

[21] Vijayan, S., Arjunan, T. V., & Kumar, A. (2020). Exergo-environmental analysis of an indirect forced convection solar dryer for drying bitter gourd slices. Renewable Energy, 146, 2210-2223.

[22] Vijayan, S., Arjunan, T. V., Kumar, A., & Matheswaran, M. M. (2020). Experimental and thermal performance investigations on sensible storage based solar air heater. Journal of Energy Storage, 31, 101620.

DFMA analysis of front axle assembly of an excavator

Govindarajalu Venkatean[1], Sadagopan Palaniswamy[2]
[1]Associate Professor, Suguna College of Engineering, Coimbatore, India
[2]Professor, PSG College of technology, Coimbatore, India

Email : [1]govindarajalu@yahoo.com, [2]sadagopan.p@gmail.com

Abstract. The front axle assembly of a hydraulic excavator consists of front axle, stub axle and thrust bearing. The front axle carries the weight of the front part of the vehicle, as well as to facilitate steering and to absorb shocks due to road surface conditions. In addition, it absorbs torque applied on it due to the braking of vehicle. It is constructed with I-section in center and the ends are made either circular or elliptical. Stub axles are connected to the front axle by king pins. Front wheels are mounted on stub axle's arrangement for steering. A nominal assembly gap in the front axle assembly has to be maintained for better performance and for easy maneuvering of the vehicle during operation. The gap depends on the model and the assembly dimensions of the front axle, stub axle and thrust bearing. A typical value ranging from 0.1 and 0.5mm is required for one particular model being manufactured by the local industry. It is reported that there are more number of assembly rejections in maintaining the said gap which lead to higher cost and downtime due to rework or interchange of parts. To reduce this, DFMA analysis was carried out using the concept of stack up tolerance to reduce the rejections and improve the performance of front axle assembly, without affecting the manufacturing.

1 Introduction

Excavators are earth movers commonly used for digging, pulling and lifting in construction and agricultural fields. The front axle assembly of an excavator is the most important part in carrying the load. It carries the vertical load of the vehicle including the load it handles. The front axle assembly consists of front axle, stub axle and thrust bearing, which are connected with a king pin and is used for directional control of the vehicle. Similar to any assembly of parts, the front axle assembly components are designed with specific tolerances for manufacturing and clearances are provided based on its functional requirement viz. smooth and ease of steering, up and down movement of the wheels based on earth surface. Due to the cumulative tolerance effect, the gap in the assembly varies widely from a very large clearance to nil clearance or interference. These large bandwidths of the clearance due to the cumulative effect of the tolerance of components vary the performance and life of the assembly. Hence, based on either field study or from the historical data of similar assembly, the tolerance gap width is restricted to certain range. The tolerance gap due to the assembly part plays an important role in avoiding rejections at the assembly stage. The stub axle is assembled with a front axle and a thrust bearing as shown in figure 1. The opening in the stub axle is 177.5 mm (shown as Dim A in fig. 2) with an existing tolerance of +0.5/+0.2mm. It is also further provided with a constraint tolerance from the right and left end of the part as 88.0 ±0.2 and 89.5 ± 0.2 mm respectively.

The part that fits inside the stub axle is the thrust bearing with a width of 23.416 mm (shown as Dim B in fig. 3) with an existing tolerance of ±0.254 mm and front axle with a width of 154 mm. (shown as Dim C in fig. 4) with an existing tolerance of -0.1 / -0.3 mm.

The existing tolerance produced with high percentage of assembled parts with less than 0.1 mm gap and greater than 0.5 mm were high resulting in difficulty in assembly and usage of shim. This research work does a DFMA analysis on the existing tolerances using stack up method to target the assembly tolerance gap between 0.1 mm and 0.5mm to address the above problems identified.

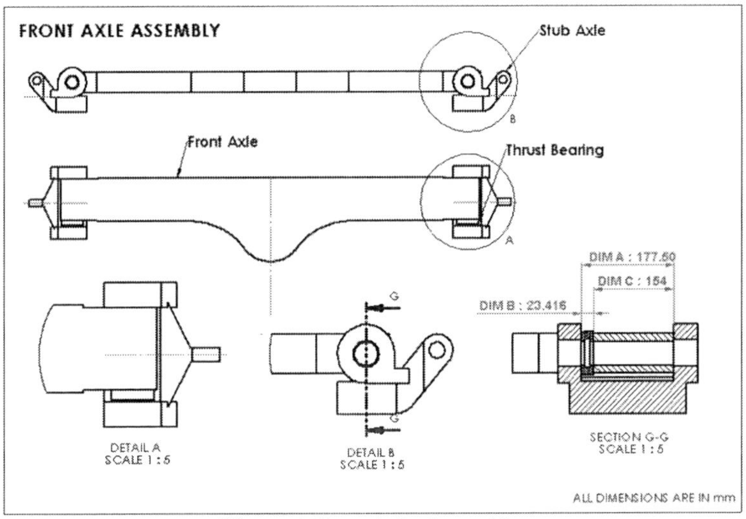

Figure 1 Front Axle Assembly

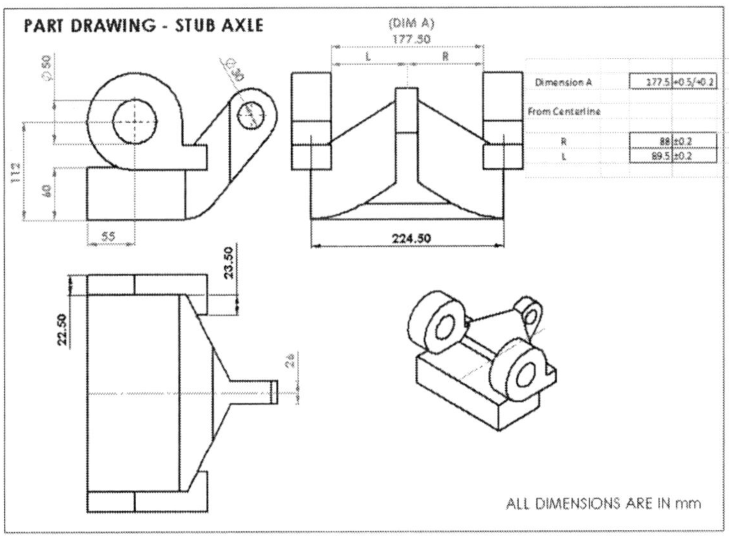

Figure 2 Stub Axle

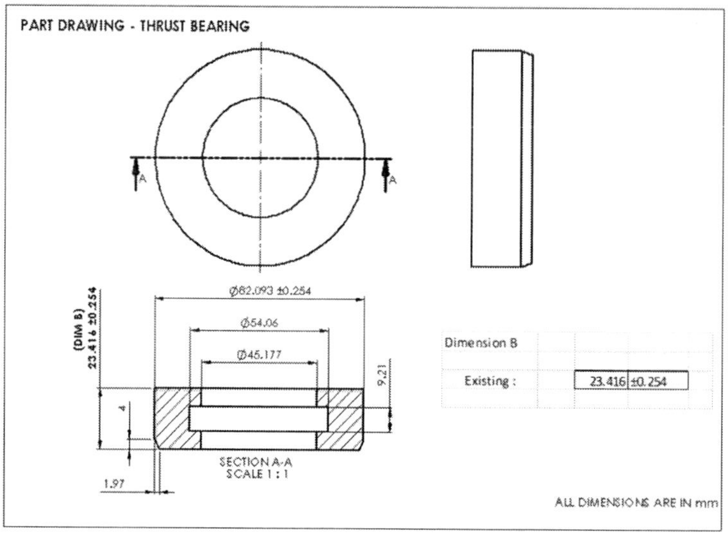

Figure 3 Thrust Bearing

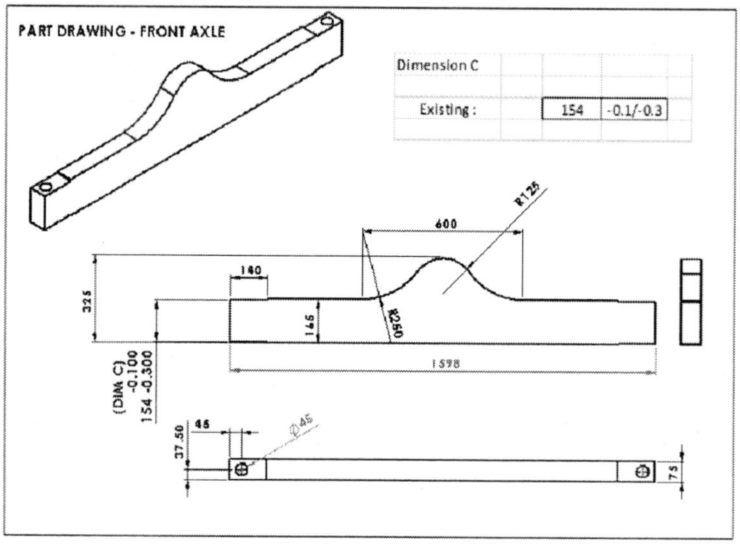

Figure 4 Front Axle

2 Review of Literature

K.W. Chase et al. (1988) discussed about applying worst case and statistical method tolerance analysis and allocation by proportional scaling factor, constant scaling factor and constant precision factor depending on size and using optimization techniques (cost vs tolerance function and lagrange multipliers). They also developed estimated mean shift model and concluded that for advanced analysis and optimization of tolerance allocation methods proper quality control techniques are needed to determine the capability of the process, cost analysis and to model and perform DOE. Fritz Scholz et al. (1995) discussed various tolerance stack up methods with mathematical analysis. He used inflation factor for various distributions and concluded with risk analysis stating to maintain a risk of 0.27% out of compliance assemblies.

Kenneth W chase (1999) discuss about the tolerance allocation by center of mean, limits for spread and reduction of spread. He also discussed about allocation by scaling, weight factor and least cost optimization. He summarized that the optimization reduced the cost and the modified tolerance after adjusting to conform to process limits, resulted in decrease of cost. Zhinua Zou et al. (2004) researched the gap based approach for mechanical assembly to find out the condition of fits. They developed an algorithm to find out the condition of fit inside the assembly. Each of the fitting condition was provided by a weight factor for linear sum of gap size and geometry parameters. Ragu.K. et al.(2007), analyzed a five stage pump mathematically with monte carlo simulation and probability. They also compared similar results for different number of stages and concluded stating the need to standardize the coupling. Selective assembly was recommended for non standard coupling length.

Mathieu Mansuy et al. (2011) in their research work proposed a new algorithm to express the relationship in linear form between the connecting tolerance and functional conditions with modification coefficient. He concluded that the relationship between a functional condition and the geometric tolerances on different surfaces was linear in worst case tolerance allocation. Jhy-Cherng Tsai et al. (2013) researched by redistribution of tolerance by negative tolerancing. They concluded that by providing negative tolerancing the accumulated tolerance can be constraint for assemblies with more number of parts instead of providing precision tolerance which result in high cost. Mathieu Mansuy et al. (2013) in their paper defined a generic algorithm for analyzing the tolerance of tri-dimensional mechanism. They developed an algorithm to satisfy the analysis of all kinds of tri-dimensional mechanism. Dinesh Shringi et al. (2013) discussed non traditional tolerance stack up conditions for worst case, RSS, spots model and mean shift model for a bearing and shaft assembly. They concluded that the cost for RSS model is less as compared to other models. Yang et al. (2013) analyzed straight built mechanical assemblies using probabilistic approach to analyse the variation on eccentricity of the assembly. They developed linear expressions and analyzed with direct, best and worst built assemblies. These expressions shall be used to rotate and select the orientation that minimize and maximize the eccentricity of the final assembly. Jhy-Cherng Tsai et al. (2015) investigated the effect of grouped random assembly. They concluded that the assembly tolerance are reduced with suitable grouping when the tolerance of set of components are close. They claimed 96% reduction is achieved when the tolerance of two component sets are equal.

Benjamin Schlesih et al. (2016) compared tolerance analysis approach for rigid mechanical models. They concluded that the skin model shapes predicted real assembly characteristics. This model is based on generation and sealing of the work parts which have been deviated from the tolerances and processed by computational geometry algorithm. Gaurav Ameta et al. (2018) analysed by tolerance map (T-Map) which uses the number of points from the size, position, orientation and form of the tolerance. They analyzed tolerance with size, form variation, parallelism and with manufacturing bias to produce probability density function (PDF) on T-maps to classify the possibilities of a feature in its tolerance zone. They concluded that the PDF can be used for tolerance allocation to reduce costs and loosen tolerance.

Satchit Ramnath et al. (2018) in their research compared various tolerance analysis methods applied to complex assembly. They concluded that the T-map model was precisely modeled as compared to Monte-Carlo and VisVSA Model for 3D variations and their interactions.

3 Objective

The existing tolerance resulted in high percentage of parts not falling within the required tolerance gap of 0.1 mm to 0.5 mm. Therefore the objective of this research is aimed to allocate tolerance by tolerance stack up analysis in DFMA to establish the tolerance gap and to increase the probability to fall between 0.1 mm to 0.5 mm such that 1) interference or difficulty in assembly is avoided due to less gap (less than 0.1mm) and 2) also to avoid use of shim due to too much gap (greater than 0.5 mm).

4 Methodology

Based on the literature survey and research objectives, the methodology to be adopted for the present research work is shown in Figure 5

1) The existing design and tolerances provided are detailed along with the need/objective of the research. The need raised as the tolerance gap was too low or too high in most of the parts.
2) DFMA analysis carried out on the existing parts to find out the distribution of tolerance with mean and standard deviation
3) With the results of DFMA, Z score and subsequently probability analysis carried out to find the percentage of parts at various values of gaps.
4) The target mean and range of tolerance is determined to fit within a probability of 5 % variation
5) From the target mean and range and with probability requirements, reverse calculation was done to find the modified tolerance
6) DFMA analysis carried out on the modified parts to find out the distribution of tolerance with mean and standard deviation
7) With the results of DFMA, Z score and subsequently probability analysis carried out to find the percentage of parts at various values of gaps
8) Above results were compared with the results obtained from existing tolerance analysis (Pt. 3) to confirm the target tolerance gap is met

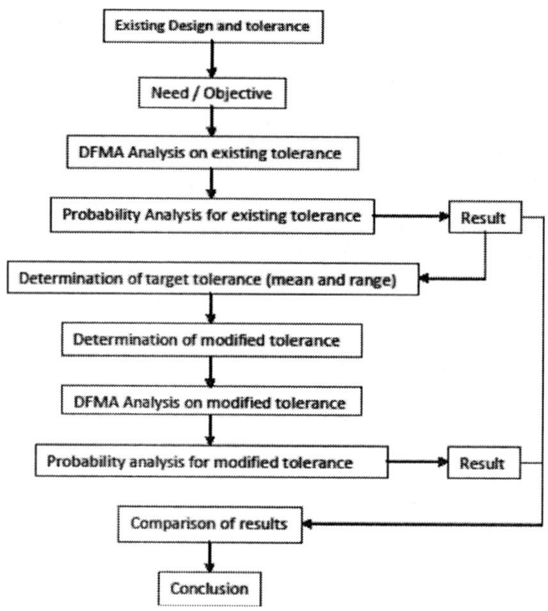

Figure 5 DFMA analysis on tolerance stack up of front axle assembly

5 DFMA analysis of the existing tolerance:

The original dimensions are :

Stub axle opening

Nominal	:	$177.5^{+0.5}/_{+0.2}$		
From Centerline	:	Right side	- 88±0.2	
		Left Side	- 89.5±0.2	
		Total	:	177.5±0.4
Range of Tolerance				
Nominal	:	177.7 to 178.0		
From Centerline	:	177.1 to 177.9		
Overall	:	177.1 to 178.0 (min. to max.)		

Mean (Existing Tolerance) (M_{ET}) : (177.1+178.0)/2 = 177.55
Standard deviation (Existing Tolerance) (SD_{ET}) : 0.150

Assembly

Front Axle (Nominal) : 153.8 ± 0.1 ($154^{-0.1}/_{-0.3}$)
Thrust Bearing : 23.416 ± 0.254
Nominal assembly dimension (M_A) : 153.8+23.416 = 177.216

Calculation of Assembly tolerance:

Arithmetic Tolerance:

Tolerance, u = $\sum u_i$ = 0.1+0.254 = 0.354

Normal distribution

Tolerance, u = $\sqrt{\sum u_i^2}$ = $\sqrt{(0.1^2+0.254^2)}$ = 0.273

Stack up tolerance was evaluated using arithmetic tolerance and normal distribution methods for comparison. The tolerance stack up analysis resulted in with a tolerance of 0.354 (sum of tolerances) using arithmetic tolerance method (worst case) and 0.273 using normal distribution method. The overall range of tolerance was from 171.1 to 178.0 mm with a mean of 177.55 mm. The standard deviation of the assembly (SD_A) is 0.091, calculated from the stack up tolerance of 0.273 mm from normal distribution.

The normal distribution on the existing stub axle and the assembly of front axle and thrust bearing is shown in figure 6. It is observed from the graph that the mean of assembly (front axle and thrust bearing) is offset to the left of the stub axle by 0.334 mm but the distribution is widely spread gap (standard deviation is high) and therefore it is obvious that the assembly rejections would be more as reported by the industry

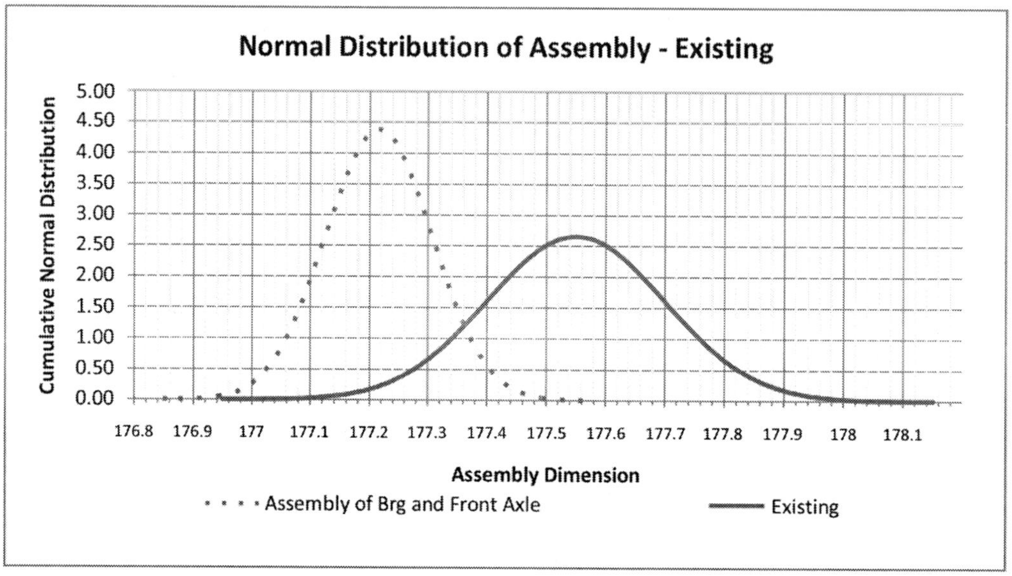

Figure 6 Normal distribution of Assembly – Existing

5.1 Analysis of the Existing Tolerance

Analysis was done to find out the probability at (1) No gap, (2) with 0.1 mm gap and (3) with 0.5 mm gap.

Mean of Existing tolerance (M_{ET}) : 177.550
Standard deviation (SD_{ET}) : 0.150

Difference in mean ($M_{D\text{-}ET}$) :

1) For No gap : $M_{D\text{-}ET\,(gap=0)} = M_{ET}\text{-}M_{A(gap=0)}$
$= 177.550\text{-}177.216$
$= 0.334$

2) For Gap at 0.1 mm : $M_{D\text{-}ET\,(gap=0.1)} = M_{ET}\text{-}M_{A(gap=0.1)}$
$= 177.550\text{-}177.316$
$= 0.234$

3) For Gap at 0.5 mm : $M_{D\text{-}ET\,(gap=0.5)} = M_{ET}\text{-}M_{A(gap=0.5)}$
$= 177.550\text{-}177.716$
$= -0.166$

Difference in standard deviation ($S_{D\text{-}ET}$):

For no gap, gap at 0.1 mm and gap at 0.5mm (Same for all 3 values)

$S_{D\text{-}ET\,(gap=x)} = \sqrt{((SD_{ET})^2\text{-}(SD_{A(gap=x)})^2)}$; (x=0,0.1,0.5)
$= \sqrt{(0.091^2+0.150^2)}$
$= 0.175$

IOP Conf. Series: Materials Science and Engineering 1059 (2021) 012057 doi:10.1088/1757-899X/1059/1/012057

The difference in mean was found to be 0.334, 0.234, -0.166 respectively for above gap levels. The standard deviation is same for all three gap levels which is 0.175.

5.2 Calculation of Probability for existing tolerance

The probabilities that the parts not meeting the above three levels of gap (g=0 mm, g=0.1mm and g=0.5mm) were calculated. The Z score ($Z_{ET(g=x)} = (0- M_{D-ET(gap=x)})/S_{D-ET(gap=x)}$) calculated resulted in values of -1.904, -1.334 and -0.946 for the above gaps respectively. The probability for the corresponding Z score from normal table was 0.028, 0.091 and 0.172 respectively.

The probability of tolerance at Z score from the standard normal distribution table for not maintaining a minimum 0.1 mm gap is 9.1% and for the tolerance exceeding 0.5 mm the probability is 17.2 % which requires usage of shim in the assembly. Also, it was observed that the probability of parts with no gap is 2.8 % (Interference) and therefore those parts are difficult to assemble.

The figure 7 shows the probability in y axis for the specified level of gap in mm in the x axis. We target to bring the probability below 5% for both at less than 0.1 mm and greater than 0.5 mm ie. 95% of the assembly parts shall maintain a minimum gap of 0.1 mm and also with a maximum gap of 0.5 mm.

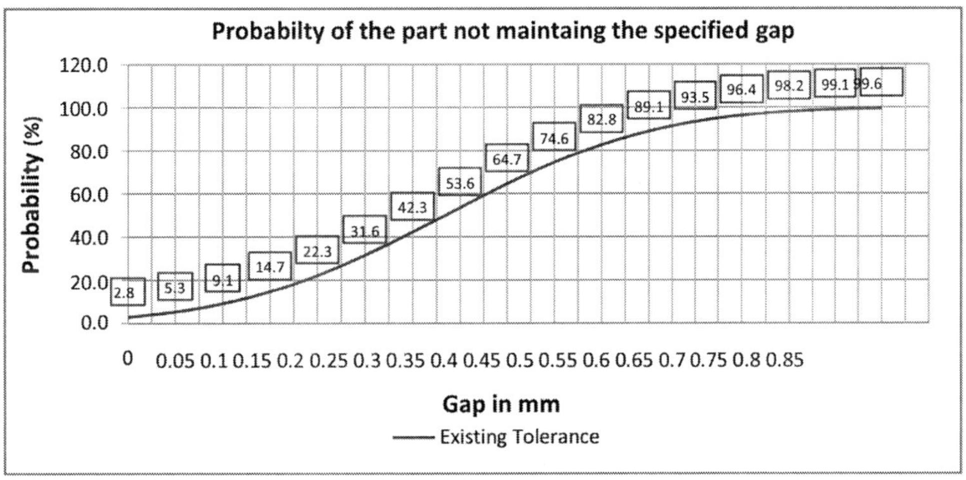

Figure 7 Probability of the part not maintaining the specified gap

6 Determination of Modified Tolerance

Mean Gap Suggested (mean of the target gap): (0.1+0.5)/2 = 0.3 mm nominal (to maintain a gap of 0.1 mm to 0.5mm)

Nominal Gap to be maintained shall be

=177.216+0.3 (Mean gap)

=177.516

New Nominal Dimension (rounded of for manufacturing) : 177.50

IOP Conf. Series: Materials Science and Engineering 1059 (2021) 012057 doi:10.1088/1757-899X/1059/1/012057

Tolerance to be worked out such that the probability should be <= 0.5 or 5% (ie. for gap < 0.1 mm and should be >=0.95 or 95% for gap > 0.5 mm)

Therefore Z score = -1.645 and 1.645 (from standard normal table) for above conditions

For less than 0.1 mm gap, the difference in mean is 0.184 (177.5-177.316) and the difference in standard deviation = (0-diff. in mean)/ Z score = 0.112. Similarly for greater than 0.5 mm gap, the difference in mean is -0.216 (177.5-177.716) and the difference in standard deviation is 0.131. Therefore limiting to the minimum spread, the lower value of 0.112 mm is taken for further calculation, which shall provide even better results for gap greater than 5 mm.

Standard deviation = $\sqrt{((0.112)^2-(0.091)^2)}$

= 0.065

Modified tolerance taken as 3*SD = 3*0.065 = 0.195 mm (rounded off to 0.2 mm)

Modified tolerance : ± 0.2 mm

Range : 177.300 to 177.700

Reworking new tolerance for Stub axle:

Existing : $177.4^{+0.5}/_{+0.2}$ ie Nominal at 177.85

Modified : $177.3^{+0.3}/_{+0.1}$ ie Nominal at 177.50

Range : 177.4 to 177.6

Both Nominal and Range falling within required calculation limits

From centerline – Nominal : to be maintained the same nominal as 177.5

Therefore centerline to nominal to be reworked for right and left side with a difference of 1.5 mm approximately

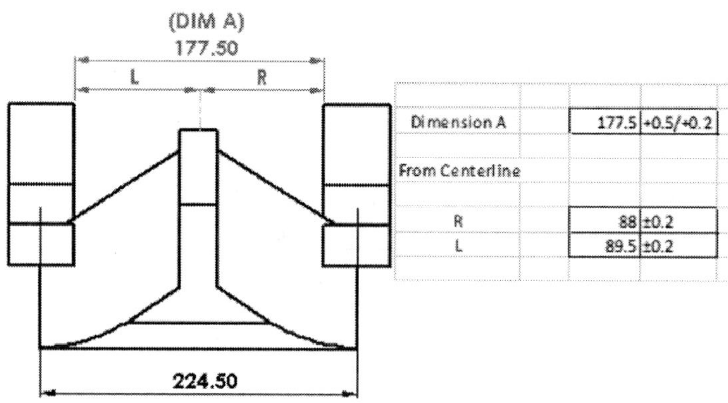

Figure 8 Stub axle with nominal and centerline dimensions

Figure 8 shows the existing nominal and right and left side dimensions from centerline. These centerline dimensions were reworked for right and left side with a difference of 1.5 mm between the two.

Right side = (177.5-1.5)/2 = 88.0

Left side = (177.5+1.5)/2 = 89.5

Total = = 177.5

Tolerance for Centerline deviations to be fixed such that the deviation should be between 177.3 to 177.7. Since the tolerance range shall be given to individual side, a minimum of 0.1 mm tolerance shall be provided for each half. ie ± 0.2 mm in total

Tolerance range for dimension from centerline is 177.3 to 177.7

7 The DFMA analysis with the modified dimensions are shown below

The modified dimensions are :

Stub axle opening

Nominal : $177.3^{+0.3}/_{+0.1}$

From Centerline : Left side - 88±0.1

Right Side - 89.5±0.1

Total : 177.5±0.2

Range of Tolerance

Nominal : 177.4 to 177.6

From Centerline : 177.3 to 177.7

Overall : 177.3 to 177.7 (min. to max.)

Mean (Modified Tolerance) (M_{MT}) : (177.3+177.7)/2 = 177.50

Standard deviation (Modified Tolerance) (SD_{MT}) : 0.067

Assembly

Front Axle (Nominal) : 153.8±0.1 ($154^{-0.1}/_{-0.3}$)

Thrust Bearing : 23.416 ± 0.254

Nominal assembly dimension. (M_A) : 153.8+23.416 = 177.216

Calculation of Assembly tolerance:

Arithmetic Tolerance:

Tolerance, u = $\sum u_i$ = 0.1+0.254 = 0.354

Normal distribution

Tolerance, u = $\sqrt{\sum u_i^2}$ = $\sqrt{(0.1^2+0.254^2)}$ = 0.273

Stack up tolerance was evaluated using arithmetic tolerance and normal distribution methods for comparison. The tolerance stack up analysis resulted in with a tolerance of 0.354 (sum of tolerances) using arithmetic tolerance method (worst case) and 0.273 using normal distribution method. The overall range of tolerance was from 177.3 to 177.7 mm with a mean of 177.50 mm. The standard deviation of the assembly (SD_A) is 0.091, calculated from the stack up tolerance of 0.273 mm from normal distribution.

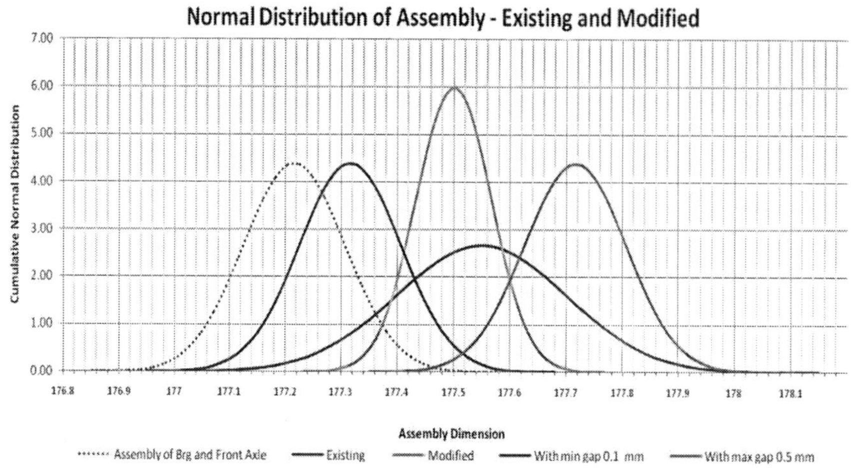

Figure 9 Normal distribution of Assembly – Existing and Modified

Figure 9 shows the normal distribution of assembly parts at 0.1 and 0.5 mm gap. Also, it shows the normal distribution of stub axle parts with existing tolerance and modified tolerance. It can be seen that the mean is close to the midway of the 0.1mm and 0.5mm curve as compared to the mean of the existing tolerance and also the standard deviation is much less than the existing tolerance which may lead to better performance of the assembly.

The above graph shows the normal distribution of assembly parts at 0.1 and 0.5 mm gap. Also it shows the normal distribution of stub axle parts with existing tolerance and modified tolerance. It can be seen that the mean is close to the midway of the 0.1mm and 0.5mm curve as compared to the mean of the existing tolerance and also the standard deviation is much less than the modified tolerance.

7.1 Analysis of the modified tolerance

Analysis was done to find out the probability at (1) No gap, (2) with 0.1 mm gap and (3) with 0.5 mm gap.

Mean of Modified tolerance (M_{MT}) : 177.500

Standard deviation (SD_{MT}) : 0.067

Difference in mean ($M_{D\text{-}MT}$) :

1) For No gap : $M_{D\text{-}MT\,(gap=0)} = M_{MT} - M_{A(gap=0)}$
$= 177.500 - 177.216$
$= 0.284$

2) For Gap at 0.1 mm : $M_{D\text{-}MT\,(gap=0.1)} = M_{MT} - M_{A(gap=0.1)}$
$= 177.500 - 177.316$
$= 0.184$

3) For Gap at 0.5 mm : $M_{D\text{-}MT\,(gap=0.5)} = M_{MT} - M_{A(gap=0.5)}$
$= 177.500 - 177.716$
$= -0.216$

Difference in standard deviation ($S_{D\text{-}ET}$):

For no gap, gap at 0.1 mm and gap at 0.5mm (Same for all 3 values)

$$S_{D\text{-}MT\,(gap=x)} = \sqrt{((SD_{MT})^2 - (SD_{A(gap=x)})^2)}\ ;\ (x=0,0.1,0.5)$$

$$= \sqrt{(0.091^2+0.067^2)} \quad = 0.113$$

The difference in mean is 0.284, 0.184, -0.216 respectively for above gap levels. The standard deviation is same for all three gap levels which is 0.113.

7.2 Calculation of Probability for existing tolerance

The probabilities that the parts not meeting the above three levels of gap (g=0 mm, g=0.1mm and g=0.5mm) were calculated. The Z score ($Z_{MT(g=x)} = (0 - M_{D\text{-}MT(gap=x)})/S_{D\text{-}MT(gap=x)}$) calculated resulted in the values of -2.518, -1.631 and -1.915 for the above gaps respectively. The probability for the corresponding Z score from normal table were 0.006, 0.051 and 0.028 respectively

8 Results and Discussion

The probability for the modified tolerance at Z, from the standard normal distribution table, for not maintaining a minimum of 0.1 mm gap is 5.1% and for the tolerance exceeding 0.5 mm the probability is 2.8 % which requires less usage of shim at present, also the probability of parts with less than 0 gap is almost zero which means none of the assembly parts will result in interference.
The probability for the existing tolerance for not maintaining a minimum 0.1 mm gap is 9.1% and for the tolerance exceeding 0.5 mm gap, the probability is 17.2 % which requires usage of shim in the assembly. Also, it was observed that the probability of parts with no gap is 2.8 % (Interference) and therefore those parts are difficult to assemble.
Comparing the results of the existing and modified tolerance it can be clearly noted that due to the modified tolerance the parts not falling within the required tolerance gap of 0.1mm to 0.5mm has reduced by 18.4%, which is a substantial reduction in rejection of parts

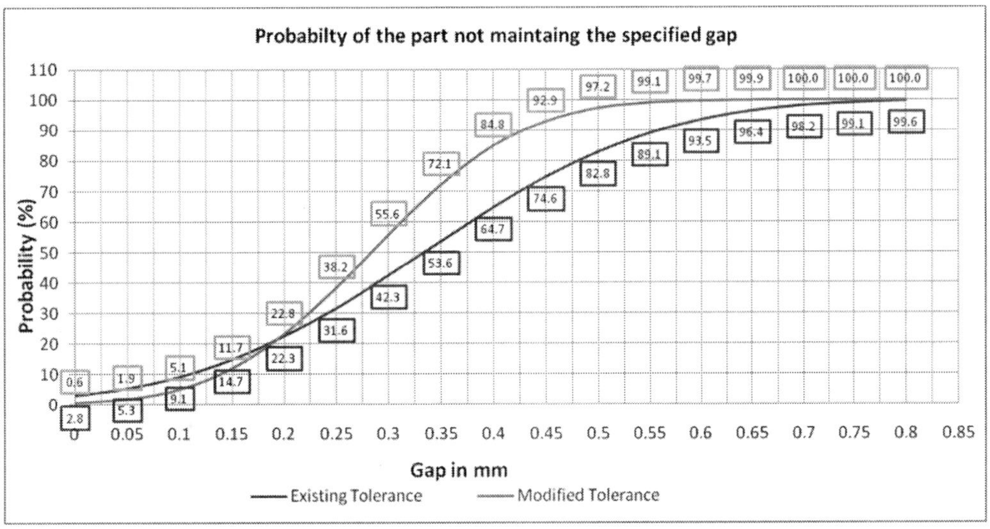

Figure 10 Probability of the part not maintaining the specified gap

Figure 10 shows the probability along the y-axis for the parts not maintaining the minimum gap specified along the x-axis. It clearly shows that the probability for the modified tolerance is much less than the existing tolerance.

9 Conclusion

By stack up tolerance on existing tolerance and by probability analysis if was deduced that the assembly rejection percentage was higher as reported by the industry. By applying the DFMA analysis to control within 5 percent probability and with reverse calculation the modified tolerance is calculated and with further analysis the following improvements are achieved.

1. The gap variation (maximum to minimum) in the assembly is reduced by 0.100 mm for the nominal dimension and by 0.400 mm for the dimension measured from the center line.
2. The probability for the existing parts that could not be assembled in the first attempt is 2.8%, whereas for the modified tolerance obtained in this research study it is almost reduced to zero percent which reduces improper assembly.
3. The probability of maintaining within the gap range of 0.1 mm to 0.5 mm is reduced by 18.4% which contributed to increased assembly efficiency.
4. The modified design has better interchangeability for parts, easy for assembly and does less usage of shim and in addition none of the randomly selected parts resulted in interference.
5. Based on the analyses, new modified tolerances were derived and it is found that the assembly of the parts is having 5.1 % probability to maintain a minimum gap of 0.1 mm and 2.8% probability exceeding the 0.5mm tolerance. The total probability of not maintaining within the range is 7.9% which is very less compared to the exiting tolerance specifications wherein the total probability is 26.3%. Also, it has a zero percent probability of parts with interference.

References

[1] Benjamin Schleich and Sandro Wartzack 2016 A Quantitative Comparison of Tolerance Analysis Approaches for Rigid Mechanical Assemblies *Procedia CIRP.* **43** 172-7,
[2] Dinesh Shringi and Kamlesh Purokit 2013 Analysis of new non traditional tolerance stack up conditions, *International Journal of Engineering Research and Application.* **13(2)** 1319-24
[3] Fritz Scholz 1995 Tolerance stock analysis methods *Boeing information and support services*
[4] Gaurav Ameta, Joseph K Davidson and Jami J Shah 2018 Statistical tolerance analysis with T-Maps for assemblies, *Procedia CIRP.* **75** 220-25
[5] Jhy-Cherng Tsai and Shang-Rung Wu 2103 Tolerance analysis and re-distribution of a machine tool spindle by negative tolerancing, *Procedia CIRP.* **10** 267-70
[6] Jhy-Cherng Tsai, Fa-Chih Chen and Jyun-Hao Dai 2015 Reduction of tolerance stack-up by grouped random assembly for components with uniform distributions *Procedia CIRP.* **27** 260-63
[7] Kenneth W Chase 1999 Tolerance allocation methods for designers, *ADCATS.* 99-6
[8] Kenneth W.Chase and W.H.Greenwood 1988 Design issues in Mechanical Tolerance Analysis, *ADCATS* **87.5** 50-59
[9] Mansuy, Mathieu & Giordano, Max and Hernandez, Pascal 2011 A new calculation method for the worst case tolerance analysis and synthesis in stack-type assemblies. *Computer-Aided Design.* **43** 1118-25
[10] Mansuy, Mathieu & Giordano, Max and Hernandez, Pascal 2103 A Generic Method for the Worst Case and Statistical Tri-dimensional Tolerancing Analysis, *Procedia CIRP.* **10** 276-82
[11] Ragu.K and Mohanram. P. V 2007 Tolerance design of multistage radial flow submersible Pumps *MECHANIKA.* Nr1.63
[12] Satchit Ramnath, Payam Haghighi, Aniket Chitale, Joseph K Davidson and Jami J. Shah 2018 Comparative Study of Tolerance Analysis Methods Applied to a Complex Assembly *Procedia CIRP.* **75** 208-13
[13] Yang Z, McWilliam S, Popov A, Hussain.T and Yang.H 2013 Dimensional variation propagation analysis in straight-build mechanical assemblies using a probabilistic approach *Journal of Manufacturing Systems*, **32** 348-56

[14] Zhihua Zou and Edward P.Morse 2004 A gap-based approach to capture fitting conditions for mechanical assembly *Computer Aided Design.* **36** 691-700
[15] Standard Normal Distribution Table, https://www.ztable.net/, retrieved on 8, September 2020

Progression in Fire Retardant Properties of Polymer Composites – A Review

Nithiyapathi.C[1*], Thirunavukkarasu.K[2], Daniel Das.A[3], Tamilvendan.D[4]
[1] Assistant Professor, Department of Mechanical Engineering, Karpagam Academy of Higher Education, Coimbatore-641021, Tamilnadu, India
[2] Professor, [3] Assistant Professor, Department of Mechanical Engineering, Karpagam Academy of Higher Education, Coimbatore-641021, Tamilnadu, India.
[4] Assistant Professor, Department of Aeronautical Engineering, Hindusthan College of Engineering and Technology, Coimbatore -641032, Tamilnadu, India.

nithiyapathi.c@kahedu.edu.in

Abstract: Emerging new composite materials from existing materials is the actual challenge for the greater part of the material engineers. In this way, there are enormous inspections, attempts rising in the field of composites to compose new materials with the updated mechanical, electrical and thermal properties. The available articles dependent on the thermo-mechanical investigation of epoxy polymers and other polymer matrix composites with various fibre reinforcement methods are deficient and are less useful. Therefore, this survey focuses on the use of thermal investigation tests and parameters to contemplate thermal properties of (Glass Fibre Reinforced Epoxy Polymer Composites) GREP composites. This survey additionally refreshes the exploration that has been acted in the ongoing years on the frontage of thermal behaviour of advanced materials for various applications.

1 Introduction

In recent days, the application of glass fibre based reinforcement in composite structures are seems to be wide and large due to its improved mechanical behaviour in polymer composites. Due to high mechanical properties, it is used in several developing industries such as: Automotive, Aircrafts, Ships, Electrical & Electronics appliances and Sporting products. The reason for GREP usage is, the most fascinating because of its sufficient consumption in various applications which fit in various structural components [1-3].

Epoxy Polymer (EP) resin has become the most used polymer matrix with glass fibre. EPs are generally utilized as matrices in advanced composites because of their great impregnation and adhesion to fiber strengthening, bringing about incredible mechanical execution; chemical and electrical resistance and low shrinkage on cure. Their utilization is specifically advocated in application zones where their specialized points of interest balance their greater expenses contrasted with other thermosetting polymers, for example in aviation industry. However, due to its restricted advantage of inherent lower thermal conductivity and stability, their wide range of application is restricted, especially where excellent heat dissipation and low thermal expansion are necessarily required [4-5].

In the recent years, scientists made intentional endeavours so as to change the properties of epoxy with the consideration of added substances or fillers to expand the framework commanded composite

properties. Anyway the flammability/combustibility of EPs despite everything speaks to a confinement in basic applications as an incidental fire occasion includes wellbeing dangers, yet in addition loss of mechanical properties.
So, the natural flammability of epoxy polymer matrices prompts fire risk for GREP composite products. It brings large concern of need for improving the fire retardant property of GREP.

2 Combustion of Polymers and Various Fire retardants

Combustion of polymers is a multi-staged process and it is due to its physical and chemical reactions. As shown in Figure-1, the three factors: Heat, Fuel and Oxygen are coexisting to initiate and sustain the combustion of polymers. So the interruption or retard of combustion process is also realized by its physical actions and chemical actions as in the same way.

The physical action of retardant can be applied by sustaining the combustion through (i) an endothermic process; (ii) formation of protective char layer; and (iii) addition of fillers/additives [6-7].

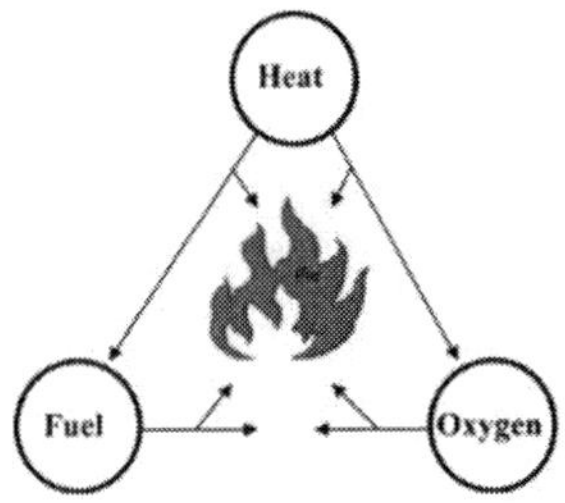

Figure 1 Fire Triangle

These methods are work as follows:

- Through an endothermic process, cooling is allowed to the underlying polymer to a certain amount of temperature below the critical temperature to sustain the combustion level.
- The protective char layer serves as heat shield to prevent the transfer of both combustible fuels and oxygen between the flame and pyrolosis zone.
- The addition of fillers and additives can change inert non-combustible gases to dilute the fuel.

Other than, addition of flame retardants, can hunt and terminate the active H• and HO• radicals in the gas phase [7-9]. There are six more fire retardant chemistries for selection which is having more advantage than each other individually, as listed in table-1.

Table 1. Various fire retardant categories and its characteristics

Flame Retardants (FR)	Additives	Advantage	Disadvantage	PMC used	Ref
1. Halogenated (Organobromine, organochlorine)	- Brominated FR	- soluble in polymer - tendency to lower T_g - ignition resistance and self-extinguishing behavior	- unacceptable losses in mechanical properties - Release large smokes and CO - Release corrosive gases	- Epoxy - Vinyl esters	[10]
2. Phosphorus	- intumescent additives	- based on polymer, chemical structure it will be act as vapor-phase and condensed-phase active flame retardants	- some phosphorus-based are plasticize the matrix	-	[8]

		single molecules	- DOPO (9,10-dihydro-9-oxa-10-phosphaphenanthrene-10-oxide)	- having a mix of vapor-phase and condensed-phase activity - very good hydrolytic stability	- difficult to process - higher heat leads to loss in mechanical properties - more vapor-phase active produces more smoke and corrosive gases - difficult to synthesize - more expensive	-	[7-8]
		oligomeric species	- Fyrol ™ PMP (Poly methylphenyl phosphinate),	- oligomeric phosphorus act as condensed-phase active flame retardants			
			- red phosphorus	- good hydrolytic stability - Proper encapsulation leads good flame retardancy at low loadings	- If not encapsulated, it leads to health and corrosion issues	- Epoxy	[11]
			- ammonium polyphosphate	- works through a char formation mechanism during burning	- provide protection in time-to-temperature - protect from structural failure types of fires	- Epoxy Polyurethanes	[11]
3.	Mineral fillers		- Metal hydroxides, carbonates	- good to face the most fire safety criteria - by use of non-flammable gases it will dilute total fuel at combustion - low smoke release rates - Effective fire resistant	- smoke release and fire performance shortcomings exist - limited use during fire - effectively resist fire at early stage but not after - cannot use alone, as fire retardant	- Epoxy	[7,12]
4.	Intumescent flame retardants		- Carbon based char - pentaerythritol	- protect the flame affected area - good ignition resistance, self-extinguishment - lowered heat release - yield good time-to-temperature	- Cannot blend directly, can apply at the surface only - Absorb water and durability of the coating is questionable. - corrosion and weight issues due to water absorption	-	[7,9]
5.	Inorganic/ceramic forming		- zinc stannate, zinc hydroxyl stannate, ammonium molybnenate,	- can form ceramic reinforced thermally stable chars on the top surface - produce ignition	- Since it is inorganic, are not phase-miscible with many polymers	-	[13,14]

	Zinc borates	resistance - reduced flame spread - improved heat release performance	- Problem in processings - Loss of mechanical properties - Sometimes act as co-cured fire protection barrier		
	- geopolymers	- superior fire performance - high level of fire safety performance	-	- FRP, aerospace applications	
6. Polymer nanocomposites	-	- Lower heat release rate - Provide anti-drip behavior to thermoplastics	- Some are have worse ignition resistance - Polymer viscosity cause challenge in fabrication - Chance for voids and dry-spots is usual	-	[15,25]

Although the fire property of composite is not only depending on the added filler, also by its own constituent materials that is fibre as well as resin, which is supposed to have a negative response to the heat.

2.1 Effect of Fibres in Fire Retardants

In general, even though the flammability in polymer matrix composites is more over controlled by the selection of polymer, the choice of fibre and the quantity can have the effects on the fire safety performance and flammability during various firing circumstances. Most reinforcing fibres are carbon, glass and aramids in PMCs for high performance industrial applications usually in aerospace, automotive applications. Exotic fibers such as metals, ceramic, basalt materials are also used but not frequently as compared to the usual one because of inadequate reach of end demands and level of performance. Eventually the fibre is used to attain the mechanical performance nevertheless flammability is not a design requirement for the choice of fibre in the development of PMC. The effect of kind of fibre on flammability in PMC may quite simple or difficult based on the type of test the flammability [7, 10].

Table 2. Flammability Characteristics of various fiber used with PMC

Type of Fiber	Flammability Characteristics	Ref
Glass	• Generally non-combustible • Poor thermal conductivity • At very high heat flux it can melt-decompose-degrade	[13]
Carbon	• Not fully non-combustible, but decreased heat release rate • Thermally conductive along its axis • At very high heat flux it will burn • Difficult to extinguish the fire flow • Cause inhalation and exposure hazards	[13]
Aramid	• fibers tend to char or ash when exposed to fire • Burn away faster than carbon fiber • Low peak Heat release rate • Burning time is large compared to carbon and glass fiber	[13]
Thermoplastic	• Rapid deformation due to melt and deform process when exposed to heat and flame • Worse flammability • Increased flame spread / fire flow	[14]

	• Will destroy the mechanical properties easily	
Natural (Cellulosic)	• fibers will burn when exposed to flame • negative contribution to flammability performance • addition of fire retardants is must required for better flammability performance	[17] [19]

2.2 Effect of Polymers in Fire Retardants

The chemical structure of the polymer is the basic key to resolute the combustion and heat release rate of polymer with its atoms whether it drips or chars during burning. For the most PMCs thermoset polymers are the dominated material over the thermoplastic polymers. The reason is thermoset polymers tend to hold in place when exposed to fire where as thermoplastic tend to deform and flow and in some cases drip, while exposed to fire. Meantime there is also some exceptional cases on both thermosets and thermoplastic polymers, like PEEK (Poly ether ether ketone), PVC (Poly vinyl chloride) etc, [10].

2.3 Effect of Fillers in Fire Retardants

Flame retardants (FR) are the chemicals that are added to composite materials focused to prevent or/and slower the further development of ignition by various physical and chemical procedures. There are three wide categories of flame retardant additives as listed in table-3.

Table 3. working mechanisms of various FR

Category	FR type		Mechanisms
I	vapor phase reactants	—	Includes FR chemistries that inhibit combustion through chemical reaction. eg. Halogenated additives, Phosphorus
II	endothermic decomposing diluents	—	Cools the decomposing polymer by means of releasing non-flammable gas that dilutes the fuel. eg. Mineral fillers that releases water and CO_2
III	condensed phase char formers	—	Comprise FR chemistries for cross linking the decomposing polymer into a more thermally stable char

Based on the working mechanism each category has its own advantage and limitation, however it is the need of time to propose and suggest eco-friendly form of fire retardants which may release no or less hazardous, corrosive gases and as well as decreased effect on mechanical property degradation.

3 Thermo mechanical Analysis of Recent Epoxy Polymer Blends

Thermo mechanical analysis is a technique to study the thermal behaviour of designed material using several tests with the application of heat/temperature. It is to be necessary to evaluate the fire behaviours of designed composites to improve its thermal behaviour. To evaluate the fire behaviors of composites, cone calorimetry test is performed. Several important detailed data such as time to ignition (TTI), peak heat release rate (pHRR), total heat release (THR), total smoke production (TSP), peak smoke production rate (pSPR), peak CO production (COP) and peak CO2 production (CO2P) were derived from the test. [16-18]

Some researchers used more than one filler to improve fire retardant properties and that is too blended with epoxy to result better properties during combustion. Tie Zhang and et Al., proved that the results obtained using synergism of more than one retardant fillers produces better LOI values compared with single filler added. Also it reveals organic compounds and metal oxides synergism for fire retardant effectively reduces the heat release rate and promotes char formation [12]. Another researcher recently experienced addition of grapheme oxide and titanium-di-oxide synergism result the improvement on mechanical and thermal behavior against fire which exposed to prove significantly. Focusing on thermo mechanical behavior is the compromising method adopted now days to improve the existing composite structure synthesized with polymers. By the effect of adding more than one additive/filler it disclose that existing polymers such as epoxy are increased twice in its mechanical behavior as well as peak load carrying ability at elevated temperatures [4, 11].

The following table intimates that the results of synergistic effects for epoxy polymers in recent years. All the results are showing as in table-4 is that additive materials with FR properties improved the fire retardancy in positive high direction but the limitation is to be considered under the releasing agents which are harmful to the environment while burning or firing with respect to table-1.

Table 4. Fire retardant behaviors of various combinations and their recent highlights

Epoxy composite blends	FR additives	Highlights in FR properties			Refs
		THR	PHR	LOI	
Epoxy with nano materials (octaMethyl-polyhedral oligomeric silsesquioxane, ammonium polyphosphate)	1 wt% OM-POSS +4 wt% APP	30% ↓	50% ↓	-	[8]
Epoxy with carbon nanotubes poly (phenylphosphonic-4,4'-diaminodiphenyl-methane	5 wt% CNT-PD	29% ↓	47% ↓	-	[9]
Epoxy with organically modified MMT; OM-POSS	2.5 wt% OMMT intercalated withDNA	31% ↓	21% ↓	32%	[10]
Epoxy with layered double hydroxides + melamine polyphosphate	5 wt% LDH + 15 wt% MPP	-	35%↓	24.4%	[16]
Epoxy with layered double hydroxides + butylene succinate	6 wt% LDH intercalated with cardanol-BS	19% ↓	62% ↓	29.2%	[20]
Epoxy with layered double hydroxides hydroxypropyl -sulfobutyl-beta-cyclodextrin sodium+ dodecylbenzenesulfonate	6 wt% LDH intercalated with sCD/DBS	35% ↓	66% ↓	26.8%	[21]
Epoxy with Graphene nanofillers	5 wt% FRs-rGO	-	35% ↓	19.5%	[22]
	4 wt% DPP-GO	35.6%	30.8%	25.2%	[23]
Epoxy with aluminum trihydroxide	2wt% ATH/EP		21%↓	-	[24]
	10wt% ATH/EP	6% ↓	50.7% ↓	-	
Epoxy with polydopamine	2wt% PDA/EP	16.6% ↓	53.5% ↓	-	
	10wt% PDA/EP	15% ↓	61.3% ↓	-	

Study of flammability properties through thermal gravimetric analysis (TGA) and Differential thermal analysis (DTA) curves along with cone colorimeter provide detailed behavior of designed composite under firing. TGA and DTA curves [26-27] will show the moss loss and residual weights during combustion which expose the characteristic of thermal decomposition.

4 Conclusion

The flame retardant property is additionally considered with the mechanical behaviour of new composites now a day. All the research findings are explicitly showing that the addition of fire retardant additives and the modified version of the same matrices revealed a positive increment in the thermal behaviour of existing constituents of the composite structure. There is a need to develop various synergistic approaches to design composites based on echo friendly and null effect on the environment. Therefore, it is concluded from this survey that FRs based on natural sources either via bio-based / mineral-based products such as metal hydroxides have a good impact on both thermal and mechanical behavioural conditions and less impact on the environment during design material fire. Still, there is no compromise with the degradation of the matrix and fibre property during combustion.

References

[1] Rajaei M, Kim N K, Bhattacharyya D, 2018 Effects of heat-induced damage on impact performance of epoxy laminates with glass and flax fibres, *Composite Structures.* **185** 515-23
[2] Wang X, Song L, Pornwannchai W, Hu Y, Kandola, B 2013 The effect of graphene presence in flame retarded epoxy resin matrix on the mechanical and flammability properties of glass fiber-reinforced composites. Composites *Part A: Applied Science and Manufacturing*, **53** 88–96.

[3] Kandare E, Chukwunonso AK, Kandola BK., 2011 The effect of fire-retardant additives and a surface insulative fabric on fire performance and mechanical property retention of polyester composites. *Fire Mater*; **35(3)** 143–55.
[4] Kavimani, V, Soorya Prakash K, Titus Thankachan, and Udayakumar R 2020 Synergistic improvement of epoxy derived polymer composites reinforced with Graphene Oxide (GO) plus Titanium dioxide (TiO2). *Composites Part B: Engineering.*
[5] Yang X, Liang C, Ma T, Guo Y, Kong J, Gu J, Zhu J, 2018 A review on thermally conductive polymeric composites: classification, measurement, model and equations, Mechanism and Fabrication Methods, *Advanced Composites and Hybrid Materials* **1(2)** 207–30
[6] Laoutid F, Bonnaud L, Alexandre M, Lopez-Cuesta J M, Dubois , 2009 New prospects in flame retardant polymer materials: from fundamentals to nanocomposites. *Mat Sci Eng R.* **63** 100-25.
[7] Morgan A B 2013 *Flame retardant fiber-reinforced composites* University of Dayton Research Institute, USA Woodhead Publishing Limited
[8] Fang Fang, Siqi Huo, Haifeng Shen, Shiya Ran, Hao Wang, Pingan Song, Zhengping, Fang, 2019 A Bio-based Ionic Complex with Different Oxidation States of Phosphorus for Reducing Flammability and Smoke Release of Epoxy Resins *Composites Communications.* Published by Elsevier Ltd.
[9] Gérard C, Fontaine G, Bellayer S, Bourbigot S 2012 Reaction to fire of an intumescent epoxy resin: Protection mechanisms and synergy. *Polym Degrad Stabil.***97** 1366-86
[10] Wang S, Xin F, Chen Y, Qian L, Chen Y 2016 Phosphorus-nitrogen containing polymer wrapped carbon nanotubes and their flame-retardant effect on epoxy resin *Polym Degrad Stabil.* **129** 133-41.
[11] Horold S, Clariant Produkte, Gersthofen 2014 Phosphorus-based and Intumescent Flame Retardants *Polymer Green Flame Retardants* Elsevier
[12] Tie Zhang, Weishi Liu, Meixiao Wang, Ping Liu, Yonghong Pan, Dongfa Liu 2016 Synergistic effect of an aromatic boronic acid derivative and magnesium hydroxide on the flame retardancy of epoxy resin *Polymer Degradation and Stability.*
[13] Mouritz A P, Gibson AG 2006 Fire properties of polymer composite materials *Springer.*
[14] Horrocks A R 2001 *Fire Retardant Materials.* Cambridage England Woodhead Publishing Limited
[15] Wentao He, Pingan Song, Bin Yu, Zhengping Fang, Hao Wang 2020 Flame retardant polymeric nanocomposites through the combination of nanomaterials and conventional flame retardants *Progress in Material Science*
[16] Zabihi O, Ahmadi M, Khayyam H, Naebe M.: Fish DNA-modified clays 2016 Towards highly flame retardant polymer nanocomposite with improved interfacial and mechanical performance. *Sci Rep.* **6** 38194
[17] Sumesh, K R, Kanthavel K, and Kavimani V 2020, Peanut oil cake-derived cellulose fiber: Extraction, application of mechanical and thermal properties in pineapple/flax natural fiber composites. *International Journal of Biological Macromolecules.* **150** 775-85
[18] Sumesh, K R, Kanthave Kl, and Kavimani V 2020 Machinability of hybrid natural fiber reinforced composites with cellulose micro filler incorporation *Journal of Composite Materials.* 0021998320918020.
[19] Senthilkumar Krishnasamy, Senthil Muthu Kumar Thiagamani, Chandrasekar Muthu Kumar, Rajini Nagarajan, Shahroze R M, Suchart Siengchin, Sikiru O. Ismail D, Indira Devi M P 2019 Recent advances in thermal properties of hybrid cellulosic fiber reinforced polymer composites *International Journal of Biological Macromolecules* **141** 1–13
[20] Kaul P K, Samson A J, Selvan G T, Enoch I, Selvakumar P M 2017 Synergistic effect of LDH in the presence of organophosphate on thermal and flammable properties of an epoxy nanocomposite *Appl Clay Sci.* **135** 234-43
[21] Wang X, Kalali E N, Wang D Y 2015, Renewable cardanol-based surfactant modified layered double hydroxide as a flame retardant for epoxy resin *ACS Sustain Chem Eng.***3** 3281-90.
[22] Kalali E N, Wang X, Wang D Y, 2015, Functionalized layered double hydroxide-based epoxy nanocomposites with improved flame retardancy and mechanical properties *J Mater Chem A.***3** 6819-26

[23] Qian X, Song L, Yu B, Wang B, Yuan B, Shi Y 2013 Novel organic-inorganic flame retardants containing exfoliated graphene: preparation and their performance on the flame retardancy of epoxy resins *J Mater Chem A*.**1** 6822-30
[24] Chen W, Liu Y, Liu P, Xu C, Liu Y, Wang Q 2017 The preparation and application of a graphene-based hybrid flame retardant contain a long-chain phosphaphenanthrene *Sci Rep*. **7** 8759-70
[25] Kiliaris P, Papaspyrides C D 2010 Polymer/layered silicate (clay) nanocomposites: An overview of flame retardancy *Prog Polym Sci*. **35** 902-58
[26] Wenmu Yang, Shuying Wu, Wei Yang, Anthony Chun-Yin Yuen, Yang Zhou, Guan Yeoh, Cyrille Boyer, Chun H. Wang 2020 Nanoparticles of polydopamine for improving mechanical and flame-retardant properties of an epoxy resin, *Composites Part B*. **186** 107828
[27] Jumahat A, N. Zamani R 2016, Thermogravimetry analysis of nanosilica-filled epoxy polymer, *Materials Research Innovations*.

Investigations on the Mechanical and Tribological Performance of Nickel Aluminum Bronze- $CaCO_3$ Composite

Vijay Bhardwaj, R Saravanan, M Govindaraju, R Vaira Vignesh

Department of Mechanical Engineering, Amrita School of Engineering, Coimbatore, Amrita Vishwa Vidyapeetham, India

r_saravanan@cb.amrita.edu

Abstract. In the present work, commercially available Nickel Aluminum Bronze (NAB) alloy was composited with $CaCO_3$ through casting route. The developed composite was characterized for microstructure, micro hardness, tensile strength (room temperature and high temperature), and wear rate. The results indicate an improvement in tensile strength by 10%, micro hardness by 2.5 times, and wear resistance by 29% than that of the as-received specimen. This is attributed to the reinforcement of $CaCO_3$ in NAB, which in turn resulted in refinement of the grains, reduction of the β'-phase and preferential formation of intermetallic precipitates of κ-phases (κ_{II}, κ_{III}, and κ_{IV}).

1. Introduction

Nickel Aluminum Bronze (NAB) is a copper-based alloy with the composition containing 4% to 5% of nickel, 8% to 10% of aluminum, with like iron, silicon, or manganese as a minor element. Conventionally, NAB alloys offer good mechanical strength, wear resistance, and corrosion resistance [1-4]. Hence, NAB alloys are cast to fabricate engineering components such as valves, parts of pumps, ship propellers, bearings, and gears [5-8]. However, bulk or surface modification of NAB alloys would augment their properties and hence improve the effective service life of engineering components.

The as-cast NAB alloy consists of the coarse copper-rich α phase, the β' phase (martensitic β), and various intermetallic precipitates (κ_{II}, κ_{III} and κ_{IV}), based on Fe_3Al or NiAl [9-14]. The coarse α phase weakens the tensile strength whereas the β' phase accelerates the fatigue crack growth rate. Besides, the β' phase and α phase undergo preferential corrosion attacks because of galvanic couple effects in the marine environment [10-12, 14]. Furthermore, the lamellar κ_{III} phase could make NAB alloy susceptible to crevice corrosion. Recently, introducing the uniform and refined microstructures into as-cast NAB alloy via different techniques has attracted considerable attention [6, 14-21]. In this study, the influence of CaO on the microstructural evolution, micro hardness, and Tribological performance of the NAB alloy was investigated. The major application of the composite is in the area where base metal wear out faster such as bearings, valve and also in the area where NAB used in high temperature application.

2. Materials and Methodology

In this study, commercially available NAB was used as the base material. The composition of the NAB alloy, which was determined using arc spectroscopy, is given in table 1.

Table 1. Composition of NAB alloy.

Element	Cu	Pb	Ni	Mn	Al	Fe
Composition	80.790	0.101	5.050	1.022	9.099	3.749

NAB alloy was melted in an electric furnace, as described in our earlier research work. The as-received NAB alloy was cut into small pieces and was placed in a graphite crucible. The furnace was gradually heated and was held at 1250°C for 30 minutes. The slag formed in the course of melting was removed. The reinforcement $CaCO_3$ (0.20 weight %) was added to melted NAB and stirred mechanically. Sequentially, the melt was poured into a mold (Φ 18mm × 150mm) and was air-cooled.

The microstructure of the developed composite was observed along the cross-section. The specimen was polished as per standard metallographic techniques and was etched using an etchant (solution of 100 ml H_2O, 5 ml HCl, and 5 grams $FeCl_3$) for 3 seconds. The micrographs were obtained using a Carl Zeiss microscope. The micro hardness of the developed composite was determined using Mitutoyo Vicker's micro hardness tester. The specimens were indented at a normal load of 100 g for 15 seconds at ten different regions, and the average micro hardness was reported. Specimens for the tensile test (room temperature and high-temperature) were prepared as outlined by the standard ASTM E8.

The hot-tensile test was performed at 600°C. The Tribological characteristics (dry-sliding condition) of the developed composite were assessed using a pin-on-disk tribometer. The tribology-test specimen preparation and tribology test conduction were in line with the standard ASTM G99. The tribology test parameters are given in table 2.

Table 2. Tribological Test Parameters.

Sl. No.	Parameters	Units	Value
1	Sliding distance	m	1500
2	Load	N	20
3	Velocity	m/sec	2.5

3. Results and Discussion

3.1 Microstructure

The base material had a coarse grain structure with Cu- rich α phase, β'-phase, and intermetallic precipitates of κ-phases. The microstructure of the fabricated NAB-0.2$CaCO_3$ composite is shown in Figure 1, which depicts the refinement of the grains, reduction of the β'-phase, and preferential formation of intermetallic precipitates of κ-phases (κ_{II}, κ_{III}, and κ_{IV}).

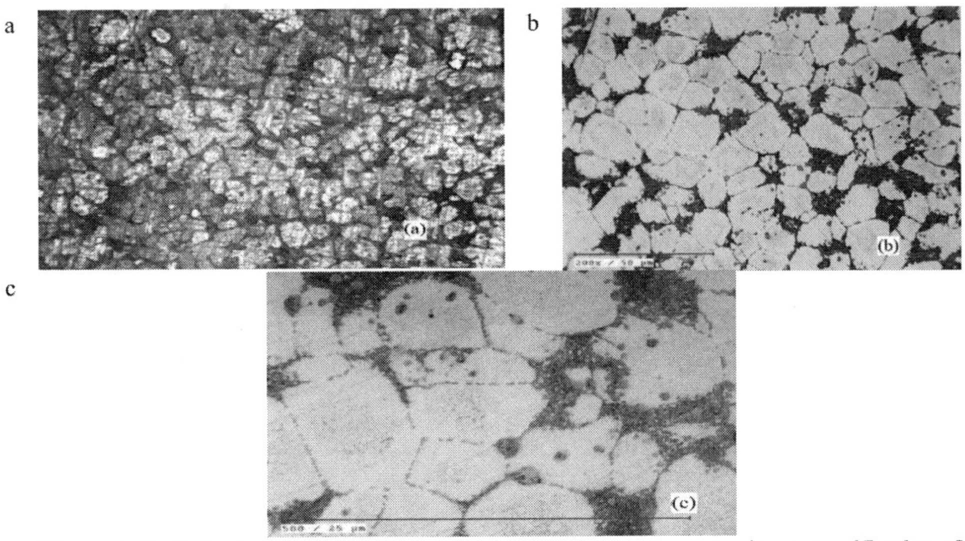

Figure 1. Optical micrograph of (a)Base Metal, NAB-0.2$CaCO_3$ composite at magnification of (b) 200x; (c) 500x

The microstructural evolution is attributed to the presence of $CaCO_3$ that dissociates as CaO and CO_2 at around 825°C. CaO is a ceramic phase and acted as potential sites for heterogeneous nucleation, and in turn, acted as a grain-pinning agent. This enabled refinement of microstructure in the NAB-$CaCO_3$

composite. The fine dispersion of CaO is evident from the microstructural analysis. The high-resolution scanning electron micrograph of NAB-0.2$CaCO_3$ composite depicting the phases is shown in figure 2.

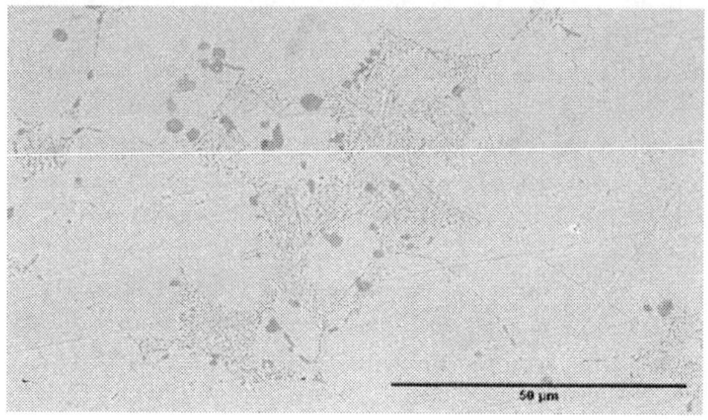

Figure 2. High-resolution scanning electron micrograph of NAB-0.2$CaCO_3$ composite depicting the phases.

The elemental composition of the developed composite was determined using energy-dispersive X-ray spectroscopy (EDS). The chosen area for the EDS analysis and the EDS graph is shown in Figure 3 (a) and Figure 3 (b) respectively. The elemental composition analysis confirmed the presence of the following elements in the developed composite: Cu, Ni, Al, Fe, Ca, and O. The XRD graph is shown in Figure 3 (c). The XRD analysis confirmed the normal phases of NAB alloy and CaO in the developed composite. No new intermetallic phases were observed.

a

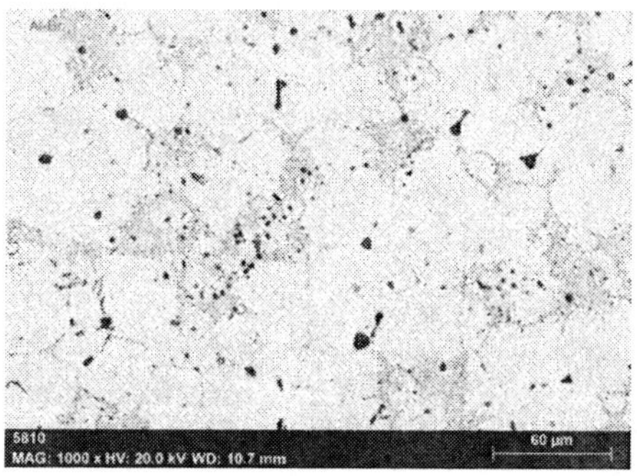

b

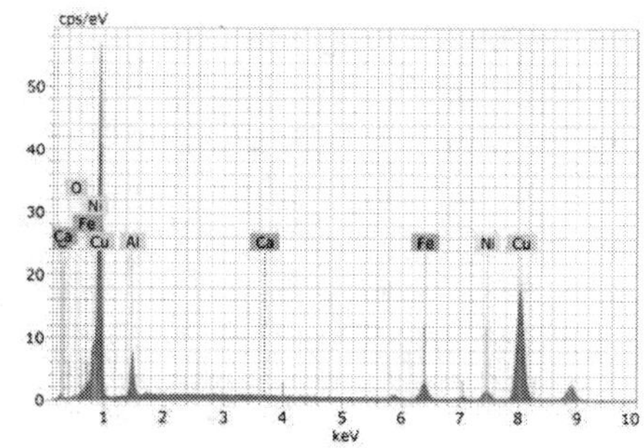

c

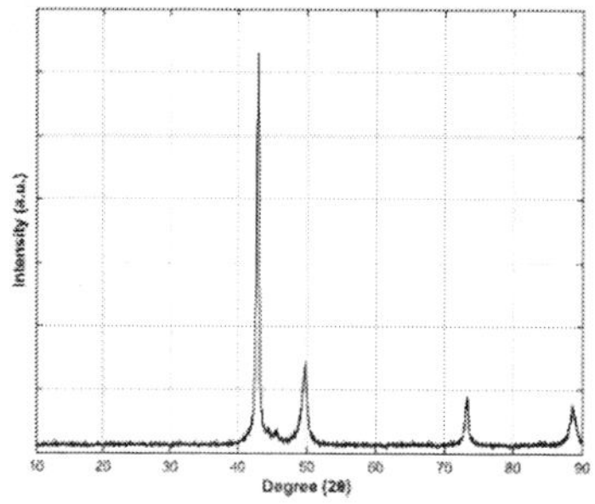

Figure 3. a) Chosen are for EDS analysis; b) EDS graph; c) XRD graph

3.2 Micro hardness

The average micro hardness of the as-cast base material was 210 HV. The microstructural refinement and presence of hard ceramic phase CaO in the developed composite resulted in an average micro hardness of 350 HV. The micro hardness was two and a half times higher than that of the base material. The results are shown in Table 3.

Table 3. Hardness of the developed composite

Sample	1	2	3	4	5
Hardness(HV)	330	362	345	338	375

3.3 Tensile Strength

The average tensile strength of the base material was 338 MPa. The developed composite exhibited an average tensile strength of 370 MPa, which was ~10% higher than that of the base material. The grain refinement augmented the tensile strength of the developed composite. The results are shown in Table 4.

Table 4. Tensile strength of the developed composite

Sample	1	2	3	4	5
Tensile Strength (MPa)	352	370	393	367	368

3.4 Hot Tensile Strength

The conventional NAB alloy had an average hot-tensile strength of 94MPa at 538°C [24]. The developed composite was subjected to hot-tensile test at 600°C. The average hot-tensile strength was 127 MPa. The results indicate that the hot-tensile strength of the developed composite is higher than that of the conventional NAB alloy. The results are shown in Table 5.

Table 5. Hot Tensile Strength of the developed composite

Sample	1	2	3	4	5
Hot-Tensile Strength (MPa) at 600°C	127	120	124	127	137

3.5 Wear Rate

The wear rate of the base material and the developed composite was determined using a pin-on-disc tribometer. The wear rate of the base material was 13.03×10^{-4} mm^3/m and that of the developed composite was 10.08×10^{-4} mm^3/m. The results explicit the improvement in wear resistance by 29%. The results are shown in Table 6.

Table 6. Wear rate of the developed composite

Sample	1	2	3	4	5
Wear Rate ($\times10^{-4}$ mm^3/m.)	10.1	10.3	9.8	10.1	10.1

4. Conclusion

NAB alloy was composited with 0.2%$CaCO_3$ successfully. The results demonstrated the following:

- Microstructural analysis confirms fine dispersion of CaO, refinement of the grains, reduction of the β'-phase, and preferential formation of intermetallic precipitates of κ-phases (κ_{II}, κ_{III}, and κ_{IV}) in NAB-0.2%$CaCO_3$
- The microstructural refinement enabled an increase in microhardness (2.5 times), tensile strength (~10%), hot-tensile strength, and wear resistance (~29%) of the NAB-0.2%$CaCO_3$ than that of the base material.

Reference

[1] Czyryca E 2000 *Fatigue Crack Growth Thresholds, Endurance Limits, and Design*: ASTM International)

[2] Luo Q, Qin Z, Wu Z, Shen B, Liu L and Hu W 2018 The corrosion behavior of Ni-Cu gradient layer on the nickel aluminum-bronze (NAB) alloy *Corrosion Science* **138** 8-19

[3] Shankar K V and Sellamuthu R 2016 An investigation on the effect of nickel content on the wear behaviour and mechanical properties of spinodal bronze alloy cast in metal mould *International Journal of Materials Engineering Innovation* **7** 89-103
[4] Thossatheppitak B, Suranuntchai S, Uthaisangsuk V, Manonukul A and Mungsuntisuk P 2013 Mechanical properties at high temperatures and microstructures of a nickel aluminum bronze alloy. In: *Advanced Materials Research*: Trans Tech Publ) 82-9
[5] Yamatogi T, Murayama H, Uzawa K, Mishima T and Ishihara Y 2011 Study on Composite Material Marine Propellers マリンエンジニアリング **46** 330-40
[6] Ilangovan S, Vignesh R V, Padmanaban R and Gokulachandran J 2019 Effect of composition and aging time on hardness and wear behavior of Cu-Ni-Sn spinodal alloy *Journal of Central South University* **26** 2634-42
[7] Carlton J 2018 *Marine propellers and propulsion*: Butterworth-Heinemann)
[8] Yamatogi T, Murayama H, Uzawa K, Kageyama K and Watanabe N 2009 Study on cavitation erosion of composite materials for marine propeller. In: *Seventeenth International Conference on Composite Materials (ICCM17), Edinburgh, UK, July,* 27-31
[9] Lv Y, Wang L, Xu X and Lu W 2015 Effect of post heat treatment on the microstructure and microhardness of friction stir processed NiAl Bronze (NAB) alloy *Metals* **5** 1695-703
[10] Michels H T and Kain R M 2003 Effect of Composition and Microstructure on the Seawater Corrosion Resistance of Nickel Aluminum Bronze. In: *CORROSION 2003*: NACE International)
[11] Qin Z, Zhang Q, Luo Q, Wu Z, Shen B, Liu L and Hu W 2018 Microstructure design to improve the corrosion and cavitation corrosion resistance of a nickel-aluminum bronze *Corrosion Science* **139** 255-66
[12] Wang C, Jiang C, Chai Z, Chen M, Wang L and Ji V 2017 Estimation of microstructure and corrosion properties of peened nickel aluminum bronze *Surface and Coatings Technology* **313** 136-42
[13] Wu Z, Cheng Y F, Liu L, Lv W and Hu W 2015 Effect of heat treatment on microstructure evolution and erosion–corrosion behavior of a nickel–aluminum bronze alloy in chloride solution *Corrosion Science* **98** 260-70
[14] Qin Z, Wu Z, Zen X, Luo Q, Liu L, Lu W and Hu W 2016 Improving corrosion resistance of a nickel-aluminum bronze alloy via nickel ion implantation *Corrosion* **72** 1269-80
[15] Abbasi-Khazaei B and Keshavarz S 2017 Nickel-aluminum-bronze/Al2O3 surface nanocomposite produced by friction-stir processing: Corrosion properties and microstructure *Materials and Corrosion* **68** 883-91
[16] Ilangovan S and Sellamuthu R 2012 An investigation of the effect of Ni content and hardness on the wear behaviour of sand cast Cu–Ni–Sn alloys *International Journal of Microstructure and Materials Properties* **7** 316-28
[17] Ilangovan S, Vignesh R V, Padmanaban R and Gokulachandran J 2018 *Progress in Computing, Analytics and Networking*: Springer) 559-71
[18] Nair S, Sellamuthu R and Saravanan R 2018 Effect of Nickel content on hardness and wear rate of surface modified cast aluminum bronze alloy *Materials Today: Proceedings* **5** 6617-25
[19] Thapliyal S and Dwivedi D K 2016 Microstructure evolution and tribological behavior of the solid lubricant based surface composite of cast nickel aluminum bronze developed by friction stir processing *Journal of Materials Processing Technology* **238** 30-8
[20] Venkatesan S, Jerald J, Asokan P and Harichandran R 2020 A study on the mechanical and corrosion behaviour of bronze–TiB2 metal matrix composites *Materials Today: Proceedings*
[21] Zhai W, Lu W, Zhang P, Wang J, Liu X and Zhou L 2018 Wear-triggered self-healing behavior on the surface of nanocrystalline nickel aluminum bronze/Ti3SiC2 composites *Applied Surface Science* **436** 1038-49

IOP Conf. Series: Materials Science and Engineering 1059 (2021) 012060 doi:10.1088/1757-899X/1059/1/012060

Single-step electrochemical synthesis of zinc oxide nanoflowers and multi-walled carbon nanotubes nanoconjugates

Ishu Singhal and B. S. Balaji

School of Biotechnology, Jawaharlal Nehru University, Delhi 110067, India

ishu_singhal@yahoo.com

Abstract. Zinc Oxide (ZnO) nanoparticles are investigated for use in various biomedical applications such as in the treatment of cancer because of their high targeting potential. Zinc oxide nanomaterials when conjugated with multi-walled carbon nanotubes (MWNT) show enhanced properties. In our present work, nano-conjugation between zinc oxide nanoflowers (ZnO NFs) and multi-walled carbon nanotubes has been successfully achieved. Here, we have used a two-electrode electrochemical setup for preparation as well as conjugation between ZnO nanoflowers and MWNT. The x-ray diffraction, Raman spectra, and transmission electron microscopy imaging for the ZnO NFs/MWNT nano-conjugate shows the successful synthesis of the nano-conjugate. The electrochemical method used here is a single step, economical, and without the use of any additional surfactant or capping agents. The only use of pure zinc electrodes as a precursor material favors high purity, easy and a less time-consuming purification process.

1. Introduction

Zinc Oxide (ZnO) nanomaterials are known to possess unique physical, chemical, optical and antimicrobial properties due to their high surface area to volume ratio. Owing to their remarkable properties, zinc oxide nanomaterials are used in many types of applications, like UV-absorbers in the textile industry due to their optical properties [1], in agriculture to improve the yield and quality of crops [2], as food additives and in the food packaging industry due to antimicrobial property [3], in sunscreens in the cosmetics industry as they not only scatter but also reflect ultraviolet A and B radiations [4], ZnO nanoparticles are investigated for use in various biomedical applications such as in the treatment of cancer because of their high targeting potential [5, 6] and so on. Zinc oxide nanomaterials when conjugated with multi-walled carbon nanotubes show enhanced properties. Decoration of ZnO quantum dots on multi walled carbon nanotubes were used as a good photosensitive UV photodetector due to their photo-elastic, high performance and photo-detection properties [7]. The ZnO nanomaterials when conjugated on multi walled carbon nanotubes (MWNT) protect the electronic properties of MWNT and can be used as a highly sensitive biosensor for the detection of carcinoma [8], malaria [9] etc.

Different methods have been used for the synthesis of ZnO/MWNT nano – conjugate in the past. Kaur et. al synthesized ZnO/MWNT conjugates by electrostatic interaction between ZnO functionalized with an amine group and MWNT functionalized with chlorine group [10]. Jiang et al. modified MWNT with sodium dodecyl sulphate (SDS) resulting in a negative charge on the MWNT surface, thereby enabling ZnO to be deposited via non – covalent interactions [11]. The sol-gel method has also been used for the synthesis of ZnO nanoparticles conjugated with MWNT by using amine-functionalized multi-walled carbon nanotubes [12]. Microwave-assisted synthesis has also been achieved by irradiating the mixture of zinc acetate dehydrate and MWNT [13]. Pulsed laser deposition was used by Gupta et al. for the preparation of ZnO nanocrystals which were then deposited on the MWNT surface using the hydrothermal method [14]. ZnO/MWNT conjugate was also prepared using atomic layer deposition, where ZnO was synthesized using zinc dust followed by coating on the outer and inner surface of MWNT [15].

All of these methods have resulted in the successful synthesis of zinc oxide nanomaterials conjugated with multi-walled carbon nanotubes. However, all of these synthesis methods are multistep and require costly materials such as precursor reagents, capping agents, surfactants, etc. and instruments. Here we are reporting for the first time, one-step electrochemical synthesis of zinc oxide nanoflowers (ZnO NFs) with multi-walled carbon nanotubes. In our developed electrochemical method, a pure zinc electrode is used as a precursor, therefore, no need for any additional metal salt and stabilizers. Additionally, the process can be optimized at a high current density to achieve high output yield. The conjugate prepared was characterized using transmission electron microscopy, X-ray diffraction, and Raman spectroscopy. All these characterization techniques showed and confirm the successful synthesis of ZnO NFs/MWNT nano-conjugate.

2. Material and methods

For the synthesis of the conjugated nanocomposite, a two-electrode electrochemical setup was used. The zinc electrodes, with a dimension of 70 mm length and 6 mm diameter, were spaced 40 mm apart and fixed. The electrolyte solution was prepared by adding 2 M sodium chloride (Merck) in 200 mL double distilled de-ionized water. 50 mg of MWNT is added in the prepared electrolyte solution and then further sonicated for 3 hours at room temperature. Here, no capping agent or additional metal salt was used. 200 ml of double-distilled de-ionized water was taken as the medium for the electrolyte cell to operate. The electrochemical cell was operated at a voltage of 3 V DC under a controlled and careful stirring rate of 400 RPM. The formation of gas bubbles is observed immediately. After a few minutes of operation, the electrolytic solution changed its color to an opaque grey precipitate. The experimental run was carried out for 5 minutes at a temperature of 60 °C. After the experimental run, the precipitate obtained was ultrasonicated for fifteen minutes. Further, the solution was kept separately for overnight stirring. The next day, the sample was washed and purified several times using ultracentrifugation and decantation method to achieve the pure sample in powder form for further morphological, structural and spectroscopic characterizations.

X-ray diffraction (XRD) was carried out using PANalytical X'Pert Pro X-ray diffractometer. Raman spectroscopy was carried using the combined confocal Raman AFM (atomic force microscope) system by WITec Denmark. Transmission electron microscopy (TEM) for all the samples was carried out using a Jeol 2100F microscope.

3. Results and discussion

The structural characterization of the as prepared ZnO NFs/MWNT nano-conjugate was carried out using X-ray diffraction. The graph is shown in Figure 1, and it can be observed that the zinc oxide nanoflowers are of wurtzite hexagonal in structure. The main peaks corresponding to the hkl values are (100), (002), (101), (102), (110), (103), (200), (112), (201), (004), and (202). All of these peaks correspond to the hexagonal structure of zinc oxide [16]. The peak for multi-walled carbon nanotubes (MWNT) is also present which corresponds to its (002) plane [17].

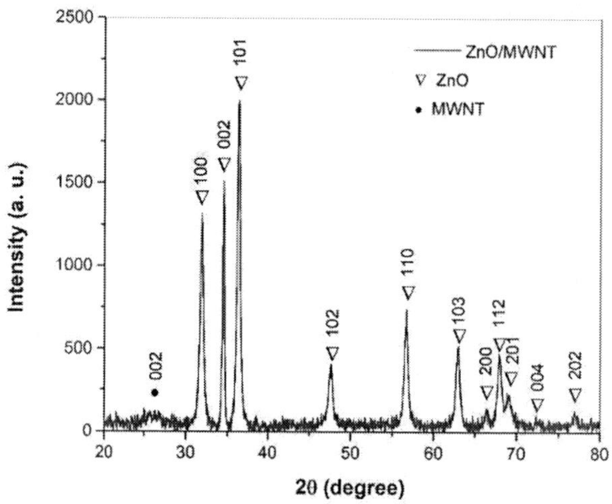

Figure 1. X-ray diffraction graph for the ZnO NFs/MWNT nano-conjugate.

Raman spectroscopy for the as synthesized ZnO NFs/MWNT was also carried out for studying the vibrational properties of the nano-conjugate. The Raman spectra obtained is shown in Figure 2. It can be observed from the spectra that the disorder-induced D band and G band are present at 1343 cm^{-1} and 1576 cm^{-1} respectively. The bands at 2690 cm^{-1} correspond to the 2D band found in multi-walled carbon nanotubes. A weak G+D peak is also present in the spectra [18, 19]. The peaks present at 434 cm^{-1} corresponds to the higher frequency mode (E_{2H}) of ZnO. This peak is the characteristic peak for the wurtzite structure of ZnO nanoparticles. The peak present 329 cm^{-1} is for the multi-photon mode E_{2H}-E_{2L}. The peak appearing at 569 cm^{-1} is assigned to the A_{1L}/E_{1L} mode for the hexagonal structure of ZnO. The acoustic combination of A_1 and E_2 modes can also be observed at 1101 cm^{-1} [20]. The presence of all the vibrational modes of zinc oxide and multi-walled carbon nanotubes confirms the synthesis of ZnO NFs/MWNT nano – conjugate.

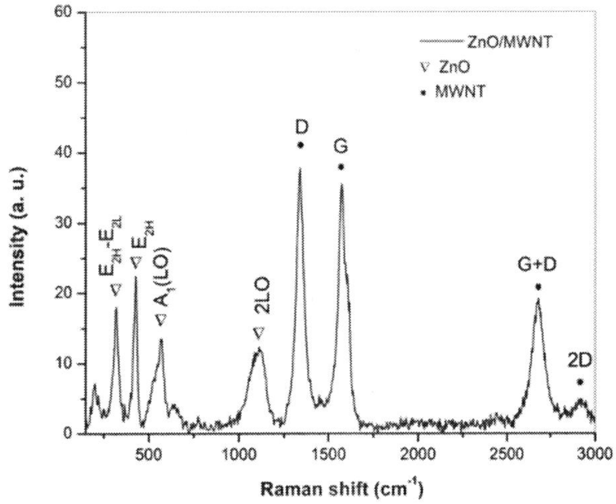

Figure 2. Raman spectra for the ZnO NFs/MWNT nano-conjugate.

The morphological characterization was carried out using transmission electron microscopy. It can be observed from Figure 3(a) that the flower-like nanostructures with an average size range of 500 – 700 nm have been obtained. The image shows that these structures have formed with nanocrumb growing away from the centre [19]. It can also be observed from the image that these fusiform like nanoflowers are conjugated with multi-walled carbon nanotubes. The HRTEM (High-Resolution-TEM) image in Figure 3(b) shows that the lattice fringes are 0.27 nm for the ZnO nanoflowers and 0.37 nm for multi-walled carbon nanotubes [21]. The selected area electron diffraction (SAED) pattern is shown in Figure 3(c). The planes corresponding to (100), (101), (102), (110), and (103) for zinc oxide nanoflowers can be seen in the image [22]. Also, the planes (100), and (002) for multi-walled carbon nanotubes were also obtained [23].

IOP Conf. Series: Materials Science and Engineering 1059 (2021) 012060 doi:10.1088/1757-899X/1059/1/012060

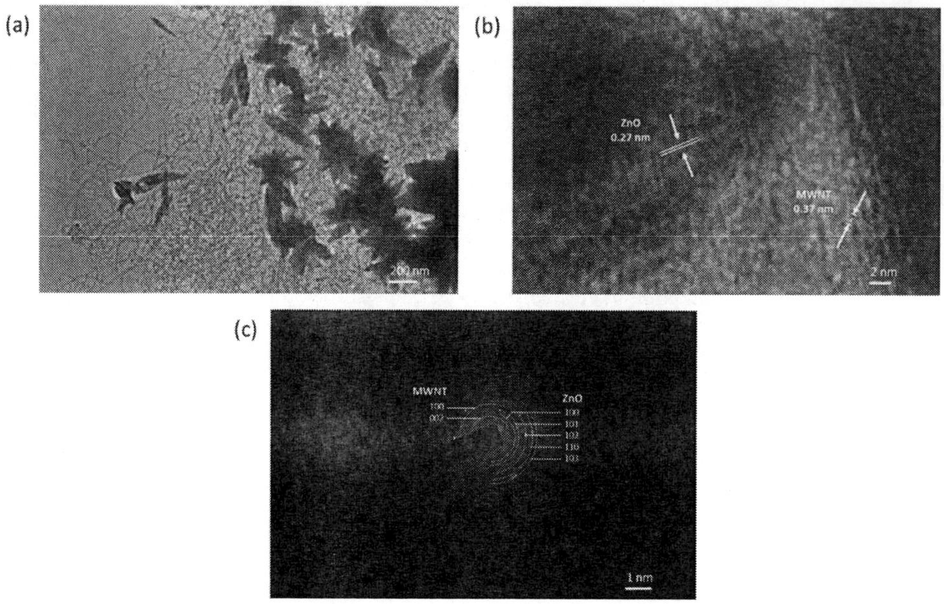

Figure 3. Transmission electron microscopy of ZnO NFs/MWNT nano-conjugate, (a) TEM image, (b) HRTEM, and (c) SAED pattern.

The different synthesis routes for the preparation of zinc oxide nanostructures and multi-walled carbon nanotubes nano-conjugate are listed in Table 1. All of these methods have some advantages and some disadvantages. However, the electrochemical synthesis method used in this study provides different types of zinc oxide nanostructures depending on the parameters. This method is a single step, time-saving, and economical with high yield. Therefore, this method can easily be scaled up for industrial production of nanomaterials and their conjugates.

Table 1. Different synthesis methods for ZnO nanostructures conjugated with MWNT, their advantages, and disadvantages.

Synthesis Methods	Nanomaterial	Advantages	Disadvantages	Ref.
Wet chemical method	Quantum dots, nanoflowers, nanorod, nanoparticles	A simple and inexpensive method, can be grown on the substrate easily	The density of the ZnO coating is not well controlled, difficult to maintain homogeneous ting films (due to cracks)	[24]
Electrostatic interaction	ZnO Quantum dots interaction with SDS treated MWNT	Densely self-assembled conjugates are synthesized, augmenting photocatalytic properties	Difficult to optimize the size and shape of the ZnO on the surface. Purification (The surfactant is very hard to remove)	[11]
Sol-gel method	Nanoparticles	Not much expensive and very simple method, uses only low temperatures	There is a shrinking of large volume and possibility to crack due to drying	[25]
Microwave irradiation	Nanosphere	Quicken the rate of the reaction, higher	Scalability is a problem. Need for expensive equipment	[26]

		yield, and shorter reaction time		
Hydrothermal method	Nanowire, and nanocrystals	Less temperature is enough for the synthesis. The composition of nanomaterials is under good control	Calcination at high temperature is required during the synthesis process	[27]
Ball milling	Nanoparticles	Scalable method of synthesis, no need of solvents	Chances for contamination due to the mechanochemical method of synthesis, Crystal defects occur Lounder noise due to the steel balls milling.	[24]
Atomic Layer Deposition	Nanoparticles	High control of the thickness of the deposition	The presence of functionalized groups modify the shape	[28]
Ultrasonic Spray Pyrolysis	Nanodots	Low cost, a high degree of control of size and shape can be accomplished.	Cannot be easily scaled up Hard to determine growth temperatures	[29]
Electrochemical	Nanoflowers	Low cost, high yield, economical, single-step, lesser experiment time, low temperature	Difficult to maintain the stoichiometric ratio.	Current work

4. Conclusion

In the present work, nano-conjugation between zinc oxide nanoflowers and multi-walled carbon nanotubes has been successfully achieved. Here, we have used a two-electrode electrochemical setup for preparation as well as conjugation between ZnO nanoflowers and MWNT. The XRD and Raman spectra, TEM imaging for the ZnO NFs/MWNT nano-conjugate shows the successful synthesis of the nano-conjugate. XRD reveals that the zinc oxide nanoflowers achieved are of wurtzite hexagonal structure type. The flower-like nanostructures has an average particle size ranging between 500 – 700 nm. The electrochemical method used here is a single step, economical, without the use of any additional surfactant or capping agents. The only use of pure zinc electrodes as a precursor material favors high purity, easy and a less time-consuming purification process. Further, the current demonstrated method can also be used for the synthesis of other wide range of metal oxide nanomaterials and their conjugates. Moreover, due to high surface area to volume ratio of zinc oxide NFs and its conjugate with MWNTs, they can be used as an electrode material for development of sensors, for sensing of analytes such as gas molecules, heavy metal ions and biomolecules. Their application in remediation of water resources can also be explored where the properties of both the counterparts of nano-conjugates i.e. ZnO NFs and MWNT can be exploited.

References

[1] Wang R, Xin J H, Tao X M and Daoud W A 2004 ZnO Nanorods grown on cotton fabrics at low temperature *Chemical Physics Letters* **398** 250-5

[2] Prasad T N V K V, Sudhakar P, Sreenivasulu Y, Latha P, Munaswamy V, Reddy K R, Sreeprasad T S, Sajanlal P R and Pradeep T 2012 Effect of Nanoscale Zinc Oxide Particles on the Germination, Growth and Yield of Peanut *Journal of Plant Nutrition* **35** 905-27

[3] Chaudhry Q, Scotter M, Blackburn J, Ross B, Boxall A, Castle L, Aitken R and Watkins R 2008 Applications and implications of nanotechnologies for the food sector *Food additives & contaminants. Part A, Chemistry, analysis, control, exposure & risk assessment* **25** 241-58

[4] Lu P J, Huang S C, Chen Y P, Chiueh L C and Shih D Y 2015 Analysis of titanium dioxide and zinc oxide nanoparticles in cosmetics *Journal of food and drug analysis* **23** 587-94
[5] Hanley C, Layne J, Punnoose A, Reddy K M, Coombs I, Coombs A, Feris K and Wingett D 2008 Preferential killing of cancer cells and activated human T cells using ZnO nanoparticles *Nanotechnology* **19** 295103
[6] Rasmussen J W, Martinez E, Louka P and Wingett D G 2010 Zinc oxide nanoparticles for selective destruction of tumor cells and potential for drug delivery applications *Expert opinion on drug delivery* **7** 1063-77
[7] Boruah B D and Misra A 2016 Conjugated assembly of colloidal zinc oxide quantum dots and multiwalled carbon nanotubes for an excellent photosensitive ultraviolet photodetector *Nanotechnology* **27** 355204
[8] Paul K B, Singh V, Vanjari S R K and Singh S G 2017 One step biofunctionalized electrospun multiwalled carbon nanotubes embedded zinc oxide nanowire interface for highly sensitive detection of carcinoma antigen-125 *Biosensors & bioelectronics* **88** 144-52
[9] Paul K B, Panigrahi A K, Singh V and Singh S G 2017 A multi-walled carbon nanotube-zinc oxide nanofiber based flexible chemiresistive biosensor for malaria biomarker detection *The Analyst* **142** 2128-35
[10] Kaur R, Paul A K and Deep A 2014 Conjugation of chlorinated carbon nanotubes with quantum dots for electronic applications *Materials Letters* **117** 165-7
[11] Jiang L and Gao L 2005 Fabrication and characterization of ZnO-coated multi-walled carbon nanotubes with enhanced photocatalytic activity *Materials Chemistry and Physics* **91** 313-6
[12] Baby T T and Ramaprabhu S 2011 Non-enzymatic amperometric glucose biosensor from zinc oxide nanoparticles decorated multi-walled carbon nanotubes *Journal of nanoscience and nanotechnology* **11** 4684-91
[13] Potirak P, Pecharapa W and Techitdheera W 2013 Microwave-assisted synthesis of ZnO/MWCNT hybrid nanocomposites and their alcohol-sensing properties *Journal of Experimental Nanoscience* **9** 96-105
[14] Gupta B K, Grover V, Gupta G and Shanker V 2010 Highly efficient luminescence from hybrid structures of ZnO/multi-walled carbon nanotubes for high performance display applications *Nanotechnology* **21** 475701
[15] Li X L, Li C, Zhang Y, Chu D P, Milne W I and Fan H J 2010 Atomic Layer Deposition of ZnO on Multi-walled Carbon Nanotubes and Its Use for Synthesis of CNT-ZnO Heterostructures *Nanoscale research letters* **5** 1836-40
[16] Jay Chithra M, Sathya M and Pushpanathan K 2015 Effect of pH on Crystal Size and Photoluminescence Property of ZnO Nanoparticles Prepared by Chemical Precipitation Method *Acta Metallurgica Sinica (English Letters)* **28** 394-404
[17] Huq M M, Hsieh C-T and Ho C-Y 2016 Preparation of carbon nanotube-activated carbon hybrid electrodes by electrophoretic deposition for supercapacitor applications *Diamond and Related Materials* **62** 58-64
[18] Feng J M and Dai Y J 2013 Water-assisted growth of graphene on carbon nanotubes by the chemical vapor deposition method *Nanoscale* **5** 4422-6
[19] Cai Q, Gao Y, Gao T, Lan S, Simalou O, Zhou X, Zhang Y, Harnoode C, Gao G and Dong A 2016 Insight into Biological Effects of Zinc Oxide Nanoflowers on Bacteria: Why Morphology Matters *ACS applied materials & interfaces* **8** 10109-20
[20] Sankara Reddy B, Venkatramana Reddy S and Koteeswara Reddy N 2013 Physical and magnetic properties of (Co, Ag) doped ZnO nanoparticles *Journal of Materials Science: Materials in Electronics* **24** 5204-10
[21] Zhang Z, Zhao Y, Sun L, Liu D, Shen J, Zhou W, Luo Q, Jin A, Yang H, Gu C and Xie S 2009 Growth and electrical properties of zinc oxide nanowires *Journal of nanoscience and nanotechnology* **9** 1119-22
[22] Chen D, Ai S, Liang Z and Wei F 2016 Preparation and photocatalytic properties of zinc oxide nanoparticles by microwave-assisted ball milling *Ceramics International* **42** 3692-6
[23] Kumar A, Singh K and Pandey O P 2014 One Step Synthesis and Growth Mechanism of Carbon Nanotubes *Journal of Materials Science & Technology* **30** 112-6
[24] Li X, Qin Y, Picraux S T and Guo Z-X 2011 Noncovalent assembly of carbon nanotube-inorganic hybrids *Journal of Materials Chemistry* **21** 7527
[25] Carter C B and Norton M G 2007 *Ceramic Materials*: Springer)

[26] Shi S and Hwang J-Y 2003 Microwave-assisted wet chemical synthesis: advantages, significance, and steps to industrialization *Journal of Minerals and Materials Characterization and Engineering* **02** 101-10
[27] Gan Y X, Jayatissa A H, Yu Z, Chen X and Li M 2020 Hydrothermal Synthesis of Nanomaterials *Journal of Nanomaterials* **2020** 1-3
[28] Johnson R W, Hultqvist A and Bent S F 2014 A brief review of atomic layer deposition: from fundamentals to applications *Materials Today* **17** 236-46
[29] Rane A V, Kanny K, Abitha V K and Thomas S 2018 Methods for Synthesis of Nanoparticles and Fabrication of Nanocomposites 121-39

IOP Conf. Series: Materials Science and Engineering 1059 (2021) 012061 doi:10.1088/1757-899X/1059/1/012061

Technological Advancement of Solar Thermal System Desalination Process – A Review

B Kalidasan[1], Divyabharathi R[2], AK Pandey[3], Subramaniyan C[1], and Mohankumar S[4]

[1] Assistant Professor, Department of Mechanical Engineering, Bannari Amman Institute of Technology, Sathyamangalam, Tamilnadu-638401

[2] Assistant Professor, Department of Agriculture Engineering, Bannari Amman Institute of Technology, Sathyamangalam, Tamilnadu-638401

[3] Associate Professor, Research Centre for Nano-Materials and Energy Technology (RCNMET), School of Science and Technology, Sunway University, No.5, Jalan Universiti, Bandar Sunway, Petaling Jaya, 47500 Selangor Darul Ehsan, Malaysia

[4] Assistant Professor, Department of Automobile Engineering, Kumaraguru College of Technology, Saravanampatti, Tamilnadu-641049

E-mail: kalidasancinna@gmail.com

Abstract. Increase in demand for drinking water due to growth in world population and industrialization has lead route for innovating new technologies for purifying sea water. Desalination by conventional energy source increases the emission of greenhouse gases and leads to global warming. As a solution for pollution and water scarcity, solar desalination techniques are incorporated worldwide for converting brackish water into portable water using solar power. Solar power helps water purification via solar thermal systems and also using solar photovoltaic panels which provide electric supply. This review article consolidate the solar thermal energy based small, medium and large scale solar desalination plants for extracting pure water. The environmental benefits and the futuristic challenges of utilizing solar thermal system for purification of water is also highlighted.

1. Introduction

Water being an essential source for human livelihood, it becomes a must needed factor for human survival. Earth is known as blue planet as it constitutes about 71% of water. But, the percentage of fresh water available on earth surface is only 1%. Advisable salt content for human drinking water is 1000 ppm whereas sea water has salt content of about 35000 ppm [1]. Distributions of saline and fresh water in earth surface are represented in Figure 1. It is also been observed that on an average about 40000 lakhs of human beings from India and China are suffer from lack of water every month. In order to meet the overall water supply worldwide, need for fresh water is at utmost alarm.

To provide a solution for the threating problem, desalination is the most recommended solution. Desalination is a process in which saline water is converted into fresh water. Desalination technique requires energy for its operation. During desalination saline or brackish water is heated till it reaches the evaporation point, at which the water gets evaporated to steam, on condensing back the steam we obtain water free from dissolved salts. Conventional fuel like petrol, kerosene and coal used for heating of water causes GHG emission and causes air pollution, energy in the form of electricity generated using thermal power plants also causes air pollution due to combustion of coal. Thus researchers and scientist experimentally work on techniques to convert saline water into drinking water using solar power. Solar power in the form of light is used in PV panels for electricity generation, which can be used in reverse osmosis and other purification techniques. As well solar power in the form of heat energy is used in solar thermal systems like solar stills and solar concentrators for directly evaporating the brackish water.

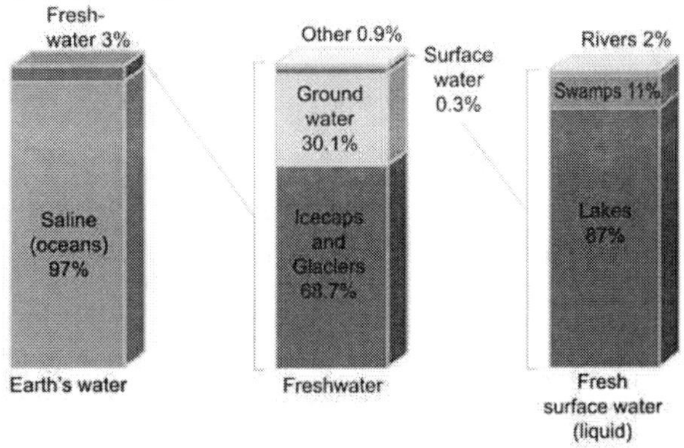

Figure 1. Saline and Fresh water Distribution

Solar energy is an abundant non-polluting energy source available free of cost. Solar thermal system like solar stills, solar chimney and solar concentrator techniques are widely used for purification of saline water [2]. Table 1 consolidated the recent review articles published on solar thermal system based desalination. From literature [3-7] it's quite evident that no review article has consolidated the advanced technologies opted with small scale, medium scale and large scale solar thermal system based desalination technique for purification of brackish water in order to make it fit for drinking. This review article summarizes advanced solar thermal system based desalination process along with the advanced technologies, environmental benefits and the futuristic challenges.

Table 1. Recent Review articles

Author & Year	Title	Remark
Aboelmaaref et al. 2020 [3]	Hybrid solar desalination systems driven by parabolic trough and parabolic dish CSP technologies: Technology categorization, thermodynamic performance and economical assessment.	The performance and economic benefits of desalination units operating using parabolic trough reflector and parabolic concentrators are reviewed in detail along with the challenges in implementation.
Omar et al. 2020 [4]	Pathway for integrated concentrated solar power – Desalination: A Critical Review	Critical review on performance and payback period for few selected concentrated solar desalination units analyzed and the benefits were summarized
Kalidasan et al. 2020 [5]	Phase change materials integrated solar thermal energy systems: Global trends and current practices in experimental approaches.	A detailed review on application of Phase change materials into solar thermal system for water and air heating with enhanced efficiency were consolidated with respect to low, medium and high temperature system.
Maia et al. 2019 [6]	An overview of the use of solar chimneys for desalination.	Focus was made on summarizing the numerical simulation of solar desalination plant operating using solar

		chimneys.
Jani & Modi et al. 2018 [7]	A review on numerous means of enhancing heat transfer rate in solar-thermal based desalination devices.	Heat transfer enhancement methods for improving the efficiency of solar stills for fresh water generation were reviewed in detail.

The review article has been divided into six section. A brief introduction indicating the need for purification of brackish water, and the novelty of the review article is consolidated in first section. Second section highlight the basic operation of desalination and the drawback of desalination process carried out using conventional methods. Section three elaborated the theme of the review article, all the solar thermal system based on low, medium and high temperature consolidated the recent advancements and performance of solar thermal systems for purification of water. In the fourth section the environmental benefits of using solar power for fresh water extraction are explained. Section five highlights the key challenges faced while using solar thermal system for desalination process. And the last section concludes the review article.

2. Desalination

Solar desalination is a process of extracting salt and dissolved solids from saline or sea water with the support of solar radiation to make the saline water fit for drinking. Evaporation and condensation are the basic principle of solar desalination. During this process solar radiation strike the top surface of the solar still, through which major amount of flux gets transmitted to the inner basin where water is stored. The basin area is made of black sheet in order to help higher absorptivity and reflected only a minor amount of radiation to the environment. Figure 02 represents the working of solar still. The solar radiation rises the temperature of the water thereby evaporation process takes place. The evaporated water precipitates gest condensed in the glass top cover, condensed water droplets are collected which produces fresh water. Based on the quantity of fresh water production, solar desalination system are subdivided into small scale solar thermal desalination system, medium scale solar thermal desalination system and large scale solar desalination system.

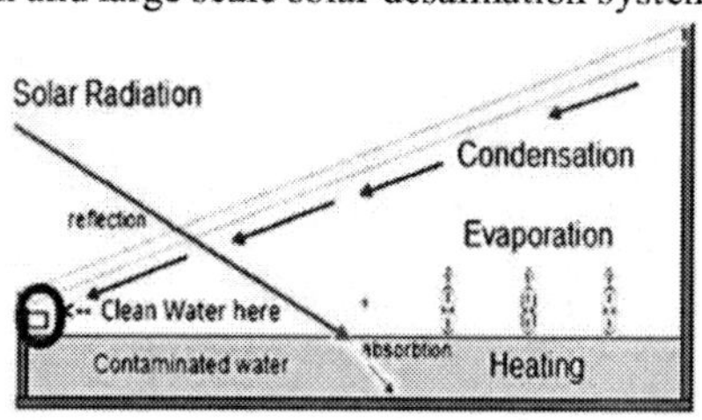

Figure 2. Working of Solar Desalination System

3. Advancement in Solar thermal system based Desalination

Solar energy is utilized as direct electric power using photovoltaic and in cooling and heating system as solar thermal system. Solar thermal system like solar water heater, solar parabolic trough and solar Fresnel collectors are used for heating water, similarly solar air heater and solar chimney are opted for air heating. In this section the advancement and technological developments carried out in solar desalination set up based on water production rate are classified and reviewed.

3.1 Small Scale Solar Thermal Desalination Technique

In this review section the advanced techniques tailed with small scale solar desalination systems are consolidated. Size of a solar desalination plant depends on the rate of fresh water produced in a day. Small scale solar still includes single slope solar still, double slope solar still, pyramid type solar still, spherical solar still, wick type solar etc. which tends to produce fresh water in 01-20 L/day. Table 2 shows the recent techniques made with small scale solar desalination plant.

Table 2. Advancement of Small Scale Solar Desalination Systems

Authors	Layout	Modification	Productivity
Pal et al. [8]		Double slope solar still with multi-wick	9.02 L/day
Hansen et al. [9]	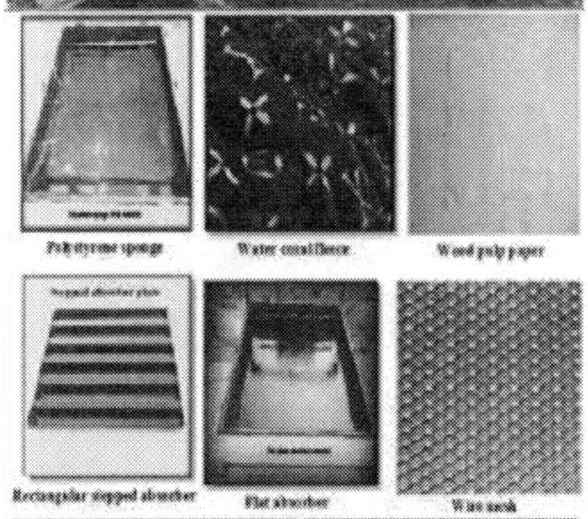	Inclined Solar Still with different wick	4.28 L/day
Modi et al. [10]	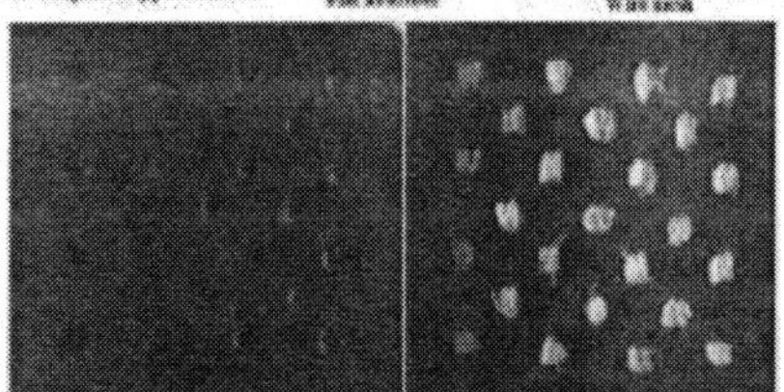	Single slop double basin area with wick	2.7 L/day
Muraleedharan et al. [11]		Solar still with frensel lens concentrator and Al_2O_3 thermic fluid in heat exchanger	12.190 L/day with Rs. 1.54
Terasaki et al. [12]	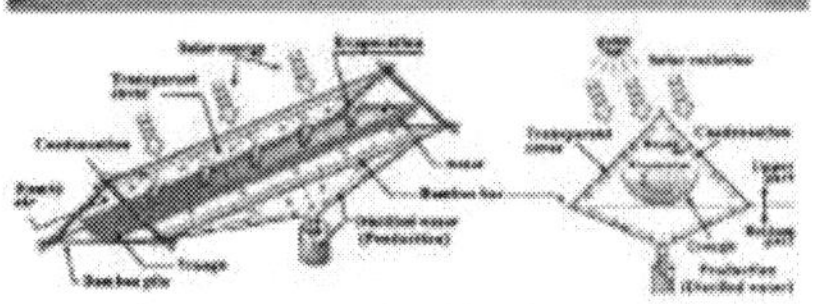	Triangular shaped Solar still with performance and economic analysis	3 L/day with Rs. 2.72/L
Kabeel [13]	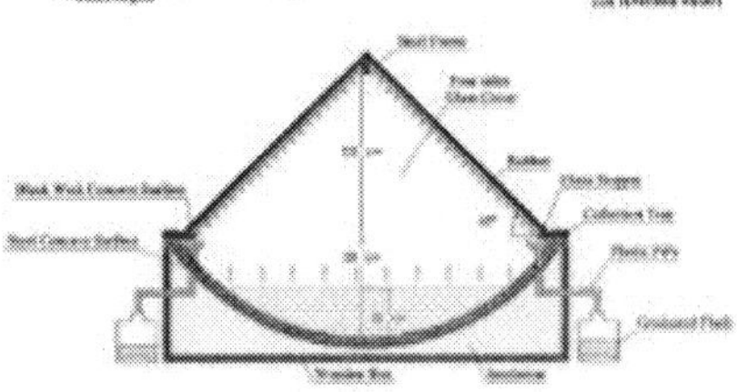	Concave solar still with efficiency 45 %	4.0 L/day with Rs. 4.85/L

Kabeel et al. [14]	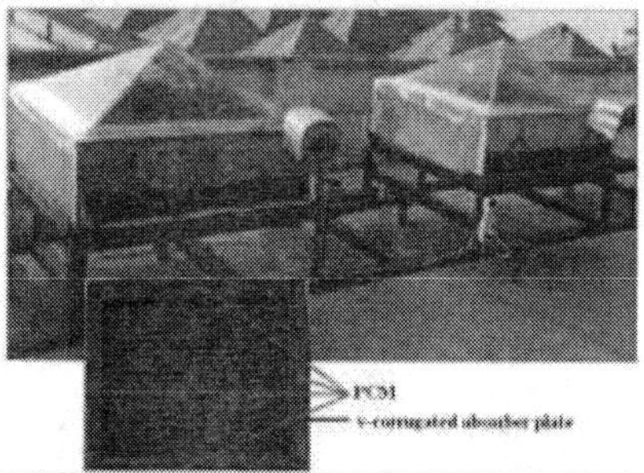	V-corrugated Absorber Plate with PCM	3.5 L/day with Rs. 17.59/L
Arunkumar & Kabeel et al. [15]		Compound parabolic concentrator concentric circular tubular solar still with PCM	5.78 L/day
Dashtban & Tabrizi et al. [16]	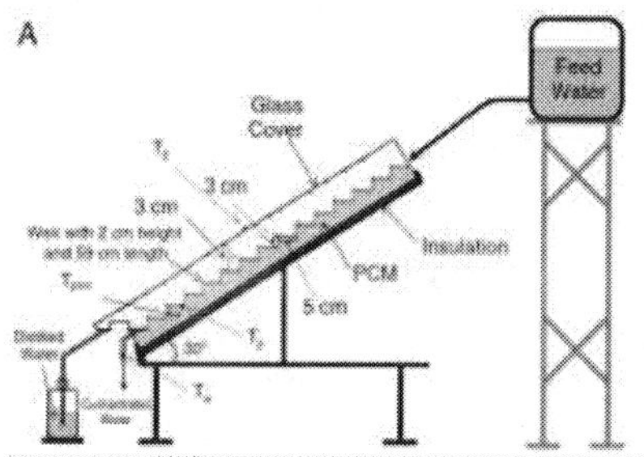	Cascade Solar Still with PCM	6.7 L/day
Rabhi et al. [17]	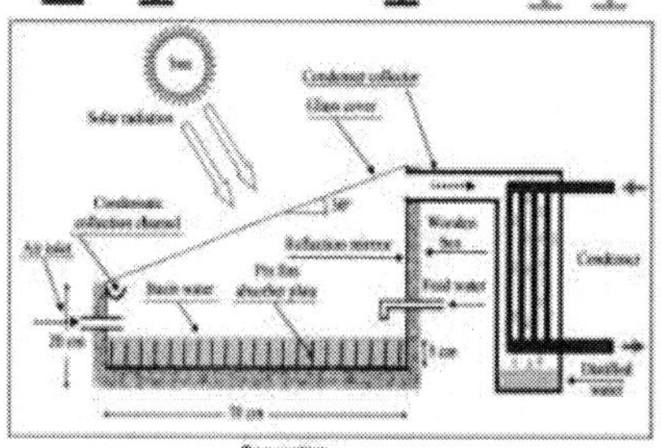	Single slope single basin solar still with pin fins	Water production improved by 41.2%
Shanazari & Kalbasi et al. [18]	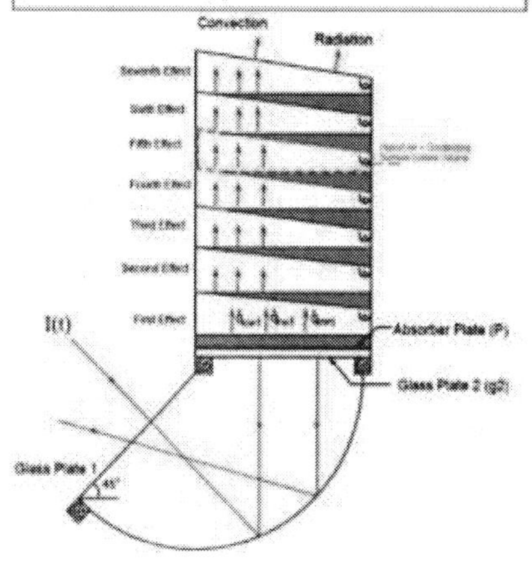	Inverted absorber multi effect solar still	Efficiency improvement by 407.3 %

IOP Conf. Series: Materials Science and Engineering 1059 (2021) 012061 doi:10.1088/1757-899X/1059/1/012061

3.2 Medium Scale Solar Thermal Desalination Technique

In this review section the advanced techniques tailed with medium scale solar desalination systems are consolidated. Medium scale solar with humidification and dehumidification type of solar desalination system are elaborated. In medium scale solar system the fresh water production rate is about 100-250 L/day. Table 3 shows the recent techniques made with medium scale solar desalination plant.

Table 3. Advancement of Medium Scale Solar Desalination Systems

Authors	Layout	Modification	Productivity
Sharshir et al. [19]	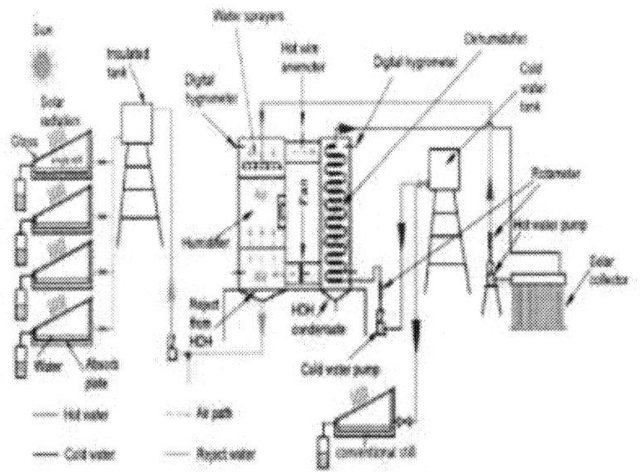	Solar hybrid desalination system with Humidification and Dehumidification with evacuated solar water heater.	66.3 L/Day With Rs. 36.53 to 2.53 per L
Tabrizi et al. [20]	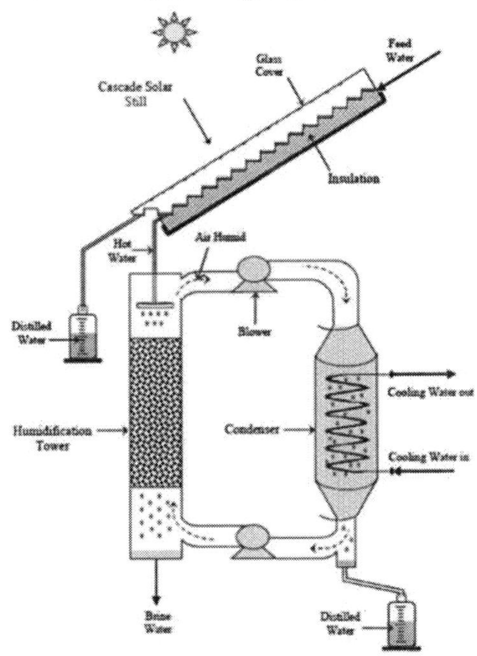	Cascade Solar Still Desalination with humidification and dehumidification	Efficiency Improved by 39%.

Elminshawy et al. [21]		Active solar desalination system with humidification and dehumidification process integrated with geothermal energy.	192 L/day with Rs. 0.22 per L
Al-Otoom et al. [22]	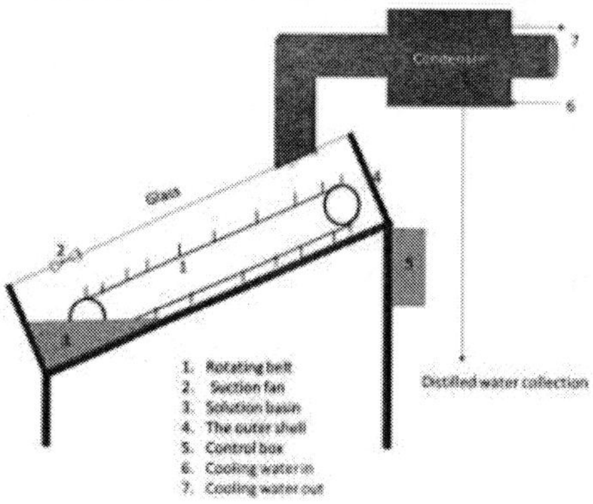 	Water desalination with continuous solar humidification and dehumidification	140 L/day

3.3 Large Scale Solar Thermal Desalination Technique

In this review section the advanced techniques tailed with large scale solar desalination systems are consolidated. Large scale solar with solar chimney type of solar desalination system are elaborated. In large scale solar system the fresh water production rate is about 1000 L/day. Table 4 shows the recent techniques made with large scale solar desalination plant.

Table 4. Advancement of Large Scale Solar Desalination Systems

Authors	Layout	Modification	Productivity
Zuo et al. [23]	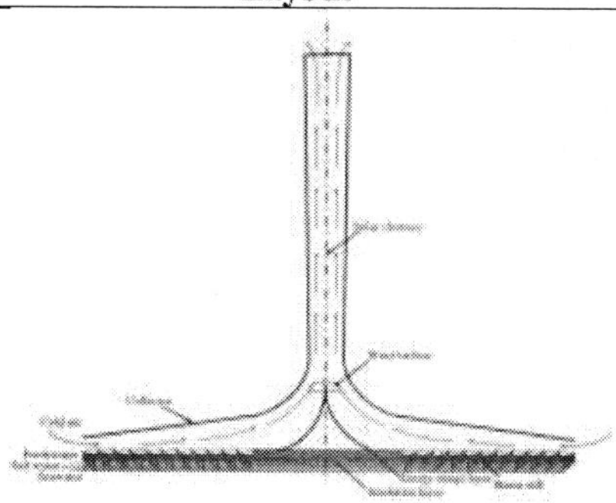	Solar Chimney with desalination of sea water.	190410 L/day

IOP Conf. Series: Materials Science and Engineering 1059 (2021) 012061 doi:10.1088/1757-899X/1059/1/012061

Asayesh et al. [24]	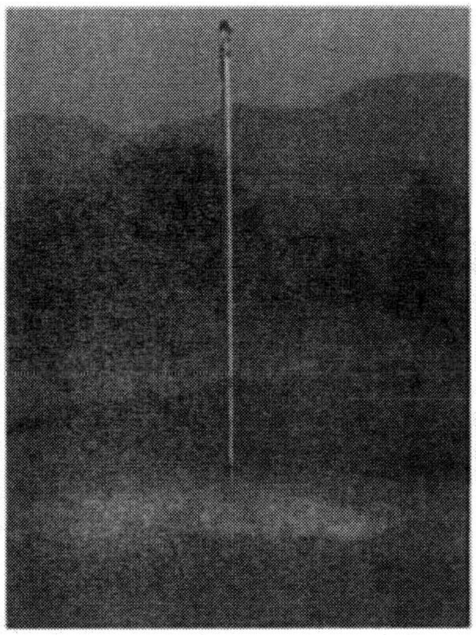	Combined solar chimney based solar desalination plant.	1000 L/day
Rahbar & Riasi [25]	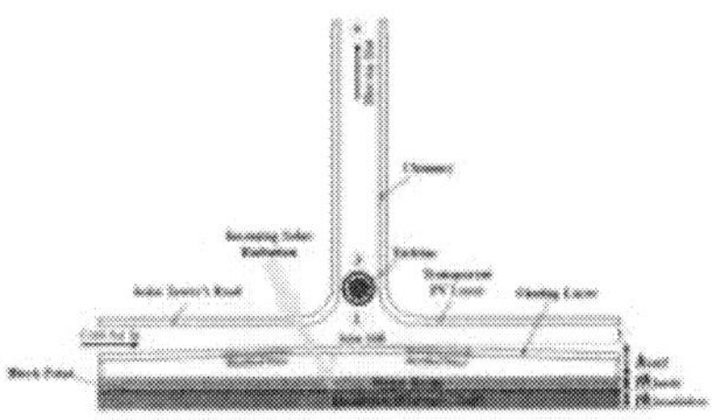	Solar chimney with solar Photovolatic	94848 L/day

4. Environmental Pros & Cons

Purification of brackish water using solar still is an efficient way to reduce emission of GHG involved in water purification techniques. Numerous research are carried out with idea to resolve the intermittent nature of solar power by using various thermal energy storage system. Integrating phase change material for thermal energy storage are a booming technology with nano additives. Special care has to be carried out to ensure that the PCM does not make any changes to the water, as well the disposal of PCM after its thermal stability is eco-friendly. Solar still with nanoparticles are also opted in large scale, nanoparticle beings a concern for environment, as the passes through membrane, special care has taken for its disposal and integrating with solar thermal desalination techniques.

5. Challenges faced by STS based Desalination

Solar stills, solar concentrators and solar chimney are few solar thermal systems classified based on different temperature level, operating to produce fresh water. Though the process is environmental friendly, certain consequence and challenges faced in real-time applications are mentioned below.

a) Thermal Storage - In order to overcome the intermittent nature of solar power, phase change materials are used as thermal batteries. Phase change material undergo the problem of low thermal conductivity and lack of thermal reliability with number of thermal cycles.

b) Economic – STS such as single slope and dual slope solar still are economically beneficial, with quick payback period. Whereas integrating fins, nano fluids and phase change materials tends to increase the payback period and overall cost. Economic evaluation for high temperature solar thermal system like solar chimney are beneficial only while operating as a dual or tri-generation cycle.

c) Man Power – Need for skilled labor is always required for solar power based desalination unit. Perfect operating conditions such as, glass temperature, solar tracking, free from dust, away from shadow etc. has to be always taken care of for better fresh water production rate.

d) Quality of Water – Technique to ensure the quality of fresh water produced using solar

IOP Conf. Series: Materials Science and Engineering 1059 (2021) 012061 doi:10.1088/1757-899X/1059/1/012061

thermal system has to be monitored regularly has the system involves metals, nano fluids and phase change materials.

6. Conclusion and Future Scope

Growth in population has led to scarcity of drinking water worldwide. Researchers and scientist are working on desalination process to convert sea water into fresh drinking water. Use of solar radiation for evaporating and condensing of sea water is environmental friendly, and does not require electric power generated using conventional fuel. Solar still are low temperature solar desalination system operating with improvement in performance using nano fluids, phase change materials, extruded fins, tracking system and concentrators. Integrating fins for better improvements of PCM fins based solar still increases the charging rate as well also increases the discharging rate which has to be reduced for better performance. Solar concentrators and solar chimney requires huge investment during initial set up and is economically feasible only with a dual or tri-generation system. Quality of fresh water produced using solar thermal system are of higher concern. Researches are focusing to increase the fresh water production rate in the existing techniques. Solar thermal based desalination process is an environment-friendly method, which reduces the emission of greenhouse gas. This research articles also highlights the research gap and drawbacks of STS based desalination process.

References

[1] Ahmed F E, Hashaikeh R, and Hilal N 2019 Solar powered desalination–Technology, energy and future outlook. *Desalination* (**453**) 54-76.

[2] Reif J H, and Alhalabi, W 2015 Solar-thermal powered desalination: Its significant challenges and potential. *Renewable and Sustainable Energy Reviews* (**48**) 152-165.

[3] Aboelmaaref M M, Zayed M E, Zhao J, Li W, Askalany A A, Ahmed M S, and Ali E S 2020 Hybrid solar desalination systems driven by parabolic trough and parabolic dish CSP technologies: Technology categorization, thermodynamic performance and economical assessment. *Energy Conversion and Management* (**220**) 113-103.

[4] Omar A, Nashed A, Li Q, Leslie G, and Taylor R A 2020 Pathways for integrated concentrated solar power-Desalination: A critical review. *Renewable and Sustainable Energy Reviews* (**119**) 109-609.

[5] Kalidasan B, Pandey A K, Shahabuddin S, Samykano M, Thirugnanasambandam M, and Saidur R 2020 Phase change materials integrated solar thermal energy systems: Global trends and current practices in experimental approaches. *Journal of Energy Storage* (**27**) 101-118.

[6] Maia C B, Silva F V, Oliveira V L, and Kazmerski L L 2019 An overview of the use of solar chimneys for desalination. *Solar Energy* (**183**) 83-95.

[7] Jani H K., and Modi K V 2018 A review on numerous means of enhancing heat transfer rate in solar-thermal based desalination devices. *Renewable and Sustainable Energy Reviews* (**93**) 302-317.

[8] Pal P, Yadav P, Dev R, and Singh D 2017 Performance analysis of modified basin type double slope multi–wick solar still. *Desalination* (**422**) 68-82.

[9] Hansen R S, Narayanan C S, and Murugavel K K 2015 Performance analysis on inclined solar still with different new wick materials and wire mesh. *Desalination* (**358**) 1-8.

[10] Modi K V, and Modi J G 2019. Performance of single-slope double-basin solar stills with small pile of wick materials. *Applied Thermal Engineering* (**149**) 723-730.

[11] Muraleedharan M, Singh H, Udayakumar M, and Suresh S 2019 Modified active solar distillation system employing directly absorbing Therminol 55–Al_2O^3 nano heat transfer fluid and Fresnel lens concentrator. *Desalination* (**457**) 32-38.

[12] Terasaki H, Umemura T, Akao T, Ito M, Fukuhara T, and Islam K M S 2016 Estimation of Production Cost of a Triangular Solar Still for Remote Area of Bangladesh.

[13] Kabeel A E 2009 Performance of solar still with a concave wick evaporation surface. *Energy*

34(10) 1504- 1509.

[14] Kabeel A E, Teamah M A, Abdelgaied M, and Aziz G B A 2017 Modified pyramid solar still with v-corrugated absorber plate and PCM as a thermal storage medium. *Journal of Cleaner Production* (**161**) 881-887.

[15] Arunkumar T, and Kabeel A E 2017 Effect of phase change material on concentric circular tubular solar still-Integration meets enhancement. *Desalination* (**414**) 46-50.

[16] Dashtban M, and Tabrizi F F 2011 Thermal analysis of a weir-type cascade solar still integrated with PCM storage. *Desalination* **279** (1-3) 415-422.

[17] Rabhi K., Nciri R, Nasri F, Ali C, and Bacha H B 2017 Experimental performance analysis of a modified single- basin single-slope solar still with pin fins absorber and condenser. *Desalination* (**416**) 86-93.

[18] Shanazari E, and Kalbasi R 2018 Improving performance of an inverted absorber multi-effect solar still by applying exergy analysis. *Applied Thermal Engineering* (**143**) 1-10.

[19] Sharshir S W, Peng G, Yang N, Eltawil M A, Ali M K A, and Kabeel A E 2016 A hybrid desalination system using humidification-dehumidification and solar stills integrated with evacuated solar water heater. *Energy conversion and management* (**124**) 287-296.

[20] Tabrizi F F, Khosravi M, and Sani I S 2016 Experimental study of a cascade solar still coupled with a humidification–dehumidification system. *Energy Conversion and management* (**115**) 80-88.

[21] Elminshawy N A, Siddiqui F R, and Addas M F 2016 Development of an active solar humidification-dehumidification (HDH) desalination system integrated with geothermal energy. *Energy conversion and management* (**126**) 608-621.

[22] Al-Otoom A, and Al-Khalaileh A T 2020 Water desalination using solar continuous humidification– dehumidification process using hygroscopic solutions and rotating belt. *Solar Energy* (**197**) 38-49.

[23] Zuo L, Zheng Y, Li Z, and Sha Y 2011 Solar chimneys integrated with sea water desalination. *Desalination* **276** (1-3), 207-213.

[24] Asayesh M, Kasaeian A, and Ataei A 2017 Optimization of a combined solar chimney for desalination and power generation. *Energy Conversion and Management* (**150**) 72-80.

[25] Rahbar K, and Riasi A 2019 Performance enhancement and optimization of solar chimney power plant integrated with transparent photovoltaic cells and desalination method. *Sustainable Cities and Society* (**46**) 101-441.

IOP Conf. Series: Materials Science and Engineering 1059 (2021) 012062 doi:10.1088/1757-899X/1059/1/012062

Automotive Brake Part Inspection and Fault Localization using Deep Learning

Santhiya Rajan[1*], R. Rameswari[2], Suresh Gunasekaran[3]

[1] PG Scholar, Embedded Systems, Dept of EIE, Bannari Amman Institute of Technology
[2] Assistant Professor, Dept. of EIE, Bannari Amman Institute of Technology
[3] Assistant Professor, Dept. of ECE, Bannari Amman Institute of Technology

immanuelihs@gmail.com

Abstract. Inspection of brake components is very essential to detect the damaged manufactured parts before it is assembled in any vehicle. Manual inspection of brakes is extremely difficult since most of defects are very minute and cannot be identified by human eyes. Therefore, automatic inspection of manufactured brakes is indispensible to prevent failure of brakes and accidents. Previously, various research articles perform inspection of brake through conventional image processing and traditional image processing algorithms. However, these techniques are capable of identifying a single fault only and are less robust to detecting numerous faults. Further, the existing techniques hardly localize the exact location of faults in the surface of brake. In order to over these drawbacks, in this research we utilize deep learning object detection algorithms namely Single Shot Detector and Faster RCNN to identify and localize the exact location of fault on the brake surface. Furthermore, the proposed system is capable to detect different types of faults in a single algorithm and is robust to brake's material surface, environmental and lightening factors. The deep learning algorithms are trained using transfer learning on custom collected dataset. The proposed algorithms deliver an accuracy of 95.64% and mAP of 73.2% on cylindrical grey shade brakes.

1 Introduction

Brakes are the very important part in any transportation vehicle and it is highly important that the developed brake parts are of good standard to stop any accidents. According to the recent survey, it is saddening to know that largest number of accidents in the world occurs in India. In 2017 alone, there were 147913 accidents on road, in which 48764 were on 2 wheelers, 26869 happened on cars, 20457 happened to people walking on roads and 3559 were on cycles. Approximately, 400 people die every day on our Indian roads due to different types of accidents. Based on the analysis done by the National Highway Traffic Safety Administration (NHTSA), it is observed that 22% of fatal accidents happen due to brake failure. This shows that one of the primary reasons for road accidents is failure of Brakes. Therefore, it is very crucial that all the developed brakes are of good quality without any defects and damages.

Failure of Brake is primarily due to damages found on the brake surface. The most commonly occuring damages on the brake surface are dents, rust, cracks, etc. In many cases, these defects are very small and it is extremely difficult to spot those with human eyes without any gadgets, so it is important to detect the faulty pieces by some other means. In this regard, one of the largely used techniques is computer vision. The surface image of the manufactured brake is captured using a camera and computer vision algorithms are used to examine the surface of the brake to detect any possible defects present in them.

Most of the existing computer vision methods use conventional image processing along with machine learning algorithms to detect dents and scratches present on the surface of the brake. But, these methods are subject to low performance accuracy and demand extensive manual labor work. Further, each algorithm is designed to capture a specific type of fault and also the developed with numerous constraints like type of base metal surface, lightening condition, etc. For example a separate model is designed to detect dent, another model to detect scratches and so on. When these different models are combined together, their performance decreases drastically and it is very difficult to develop a single robust model that is versatile enough to detect different types of faults in a single shot. Therefore, there is a necessity to develop a single versatile model that is capable to detect multiple faults. Henceforth, in order to build a reliable automotive inspection system, in this article, we have proposed a deep learning based computer vision system to detect the dents and scratches present on the brake. Further, the proposed system is capable to detect, localize and classify different types of faults on the surface of the brakes.

Our proposed system utilizes object detection algorithms that are prevalent in deep learning applications to detect different types of faults on the surface of the brake. Since fault detection in brakes is a very specific problem. There is very limited dataset present for this application. Therefore, the object detection models used for this task are trained used transfer learning technique with pretrained weights from COCO 2017 dataset. Further, Single Shot Detector and faster RCNN object detection algorithms are experimented in this work. The pretrained weights are retrained with annotated brake dataset.

2. Literature Survey

There are different research works that were previously implemented for defect detection in metal and fabric surfaces. Table 1 shows the gist of previously completed different works.

Table 1. Survey of Existing Research Works

Author/Title of the paper	Method Description	Percentage of Detection(%)	Name of the Defects
Jagdish Lal Raheja, Sunil Kumar, Ankit Chaudhary / "Fabric defect detection based on GLCM and Gabor filter: A comparison" [1]	Gabor and GLCM based approach	98.33 % for gabor based approach. 95% for GLCM based approach	Fabric defect detection: including missing yards, broken yards, and changes in tension during production in a textile LOOM
T. Aarthi, M. Karthi, M. Abinesh / "Detection and analysis of surface defects in metals using wavelet transform" [2]	Wavelet Transform - HAAR, Daubechies		Surface defects in metals includes corrosion, mechanical wear
Guan ming lei and hao lul a steel surface defect recognition algorithm based on improved deep learning network model using feature visualization and quality evaluation [3]	VGG19, Resnet, DeVGG19, Decision tree and VSD	87.84 % from resnet 61.842% from VGG19	Surface defects of the hot-rolled steel are collected, such as rolled-in scale (Rs), patches (Pa), crazing (Cr), pitted surface (PS), inclusion (In), and scratches (Sc)
Jagdish Lal Raheja [4]	GLCM	98.33%	Metal Surface Inspection including crack, pinhole, Spot, corner, edge and scratches

Mostafa Sadeghi [5]	Gabor wavelet	90-98%	
K.N.Sivabalan [6]	Gabor and Gaussian Filter	85-93%	
Rashmi S Deshmukh [7]	Feature extraction and Segmentation	90-95%	

From the table 1 it is observed that GLCM and gabor based techniques are the widely used ones. Gabor filter based defect detection has offered good performance in spatial and frequency domain. Usually, an even symmetric gabor filter is quite good enough at detecting blob- structured fabric defects, whereas, an odd symmetric one is works well at detecting edge shaped fabric defects. Gabor filter approach is computationally expensive than GLCM. GLCM operate only in the invariant environmental condition. Apart from this, Harr wavelet is discrete in nature so, it's not differentiable and it is suitable only discrete data. Additionally, harr wavelet is used to detect edges in an image and analysis of signals with impulsive transitions. In article [3], it is found that VGG converges faster than the other net like VSD. VSD has offered high accuracy than VGG with higher number of iteration. Articles from [8] – [15] are other few significant articles that have utilized deep learning based approaches for performing similar tasks. In all the above mentioned previous works it is observed that either traditional image processing based techniques or CNN architectures were only used for classifying the type of fault. However, these techniques had rarely focused on localization of these faults. But localization of these faults is highly important during manufacturing process in order to determine the exact position of the defects. Henceforth, in this research, we solve this problem by utilizing deep learning based object detection algorithms for detection and classification of defects on the surface of brakes.

3. Materials and Methods

3.1 Dataset:

The parts of the brake that were inspected were plunger, master cylinder and piston. The surface of these three components was examined to ensure quality of the brakes and prevent any possible brake failure. The most widely occurring defects in these parts are cracks, dents, scratches, rust and smears. The brake components required for analysis are sponsored by Brakes India Private Limited as a part of Brake India Hackathon, 2020. With insights from industry experts and engineers from Brakes India Private Limited, the component surfaces were exposed cracks, dents, scratches, rust and smears possible damages that could occur during preprocessing, manufacturing and post processing. Subsequently, defects were captured using Canon EOS 800d DSLR camera having 65 mm extender. (aperture = 8.0, iso = 200). 1000 images in each class were collected to form the dataset. The defect inspection proposed in this paper is implemented using object detection techniques in deep learning. Figure 1 (a) shows the overall steps involved in preparing dataset for object detection. Figure 1(b) and (c) shows few sample images collected in surface of plunger for scratch and dent defects.

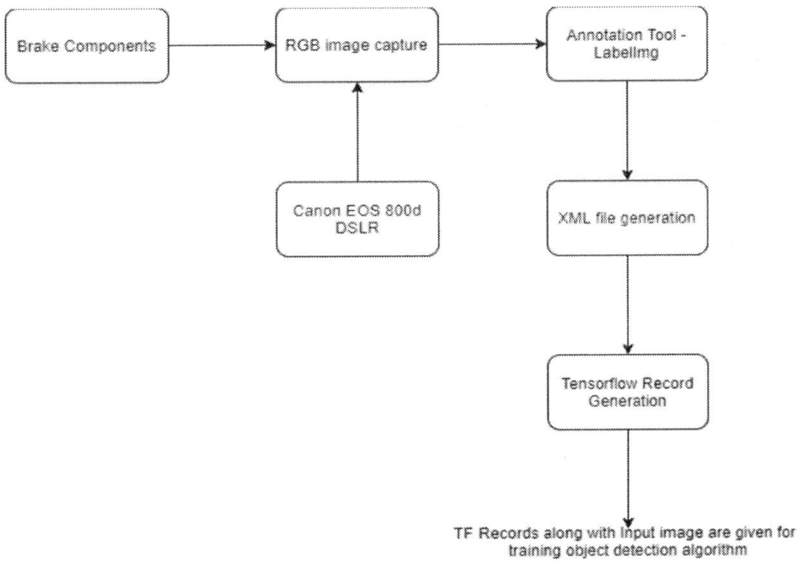

Figure 1(a). Steps involved in dataset preparation

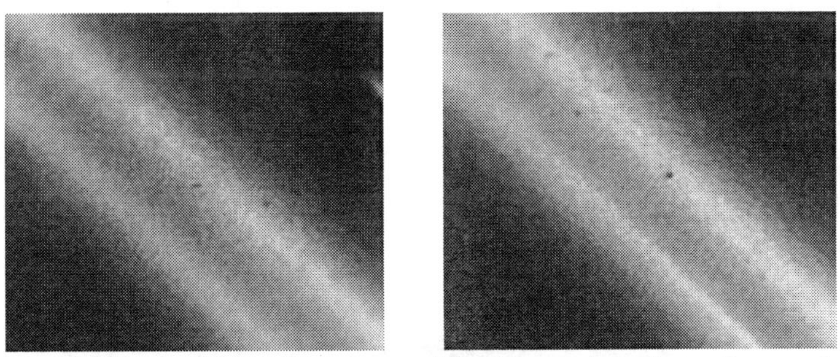

Figure 1 (b), (c). Sample images from dataset. The figures given shows the sample captured images on plunger surface

3.2 Object Classification vs. Object Detection dataset preparation:

Once the dataset is collected the next step is annotation of dataset. Figure 2 shows the comparison between object classification and object detection in deep learning. In classification, an entire image is mapped to a single class, whereas in object detection multiple classes present within a single image are located using bonding box and classified according to their classes. Since, in our proposed work, we identify faults using object detection techniques, it is important that the training set also contains these bounding boxes along with class labels for each class.

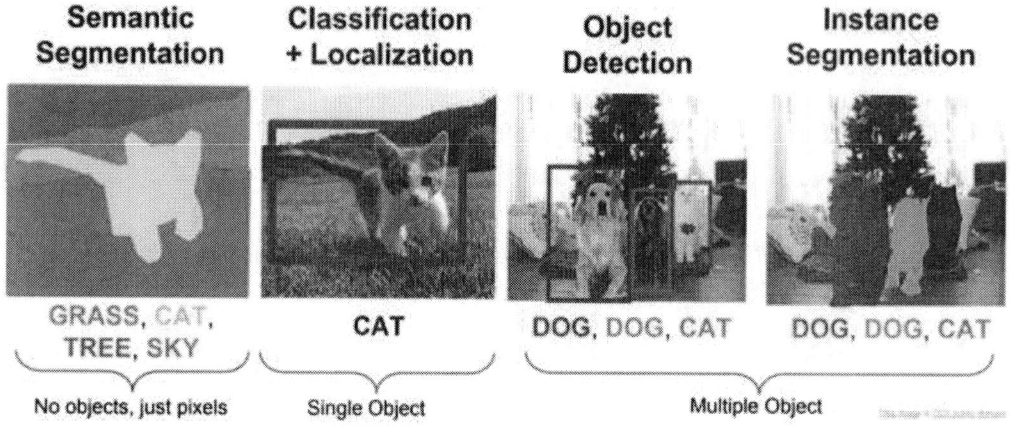

Figure 2. Comparison between classification and detection in deep learning

3.3 Label Img - Annotation Tool:

This process of obtaining a dataset with bounding box attributes along with class label is called annotating the dataset and the tool used for this purpose is called annotation tool. Annotation tool is a software program that allows the user to manually draw rectangular bounding box around the object to be detected, here in this case around the defect region on the surface of the brake. Following this, the annotation tool automatically determines the height, width, diagonal length of each bounding box and the annotation tool generates a comma separated value (csv) or eXtensible Markup Language (XML) file with geometric details of each bounding box along with class name. The entire image dataset directory can be completely loaded onto the annotation tool and the csv file for the entire dataset is generated. The generated csv file is then used to train the object detection algorithms using transfer learning technique. There are numerous types of annotation tools available for annotating the dataset. In this work, we have utilized LabelImg annotation tool. Figure 3 shows a screenshot of labelimg annotation tool. In the figure 2, the green bounding boxes are the manually drawn on the surface of the plunger for dent defect on brakes.

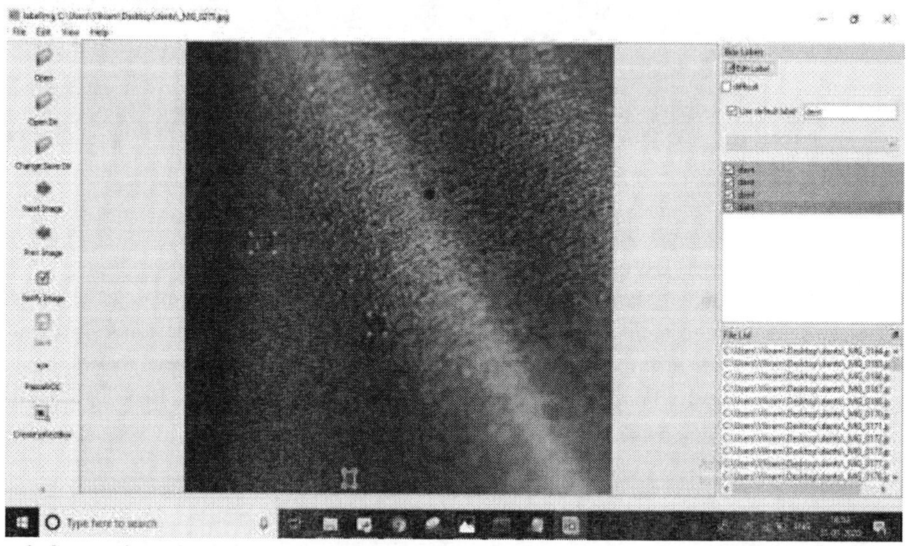

Figure 3. Screenshot of Label Img Annotation tool with bounding box drawn for dent defect on surface of plunger.

4. Methodology:

The defects on the surface of the brake identified using deep learning based object detection algorithms. These object detection algorithms overcome numerous disadvantages of traditional object detection based algorithms.

4.1 Traditional Object Detection Algorithms Vs Deep learning based Object Detection Algorithms:

Before the advent of deep learning, conventional image processing and machine learning based techniques were utilized for object detection. In conventional image processing, feature detectors were used to extract specific image features according to the object to be detected and the extracted features are processed by any one of the machine learning algorithms. In this type of approach, for each type of object to be detected, the feature to be extracted varies hugely. For a texture based objects, wavelet / GLCM features are extracted, similarly for shape based objects histogram of oriented gradients features used. In this regard, as different types of objects are included for detection, the type of features keeps on adding linearly. Therefore, for detecting 10 different objects, 10 different types of feature extractors have to be used in order to ensure high performance accuracy and in many cases for each type of feature detector, a separate machine learning algorithm is required. Therefore, developing a generic machine learning algorithmic for different objects with different feature detectors is highly hectic and even if such a versatile model is developed the accuracy of the developed model will again be unreliable and less efficient.

In such case, CNN and CNN based object detector comes to rescue. With the evolution of convolutional neural network, multiple object classification with more than 1000+ classes became practically possible with an accuracy equivalent to that of human level recognition. CNN are type of neural networks that mimic the performance of human brain and are capable of extracting edge to abstract in different types of objects. Such high level performance of CNN is possible due to the availability of huge dataset and high computational resources.

Any CNN architecture has two stages namely: feature extraction and classification. From edge to complex features are extracted in a single framework in feature extraction stage and the extracted features are same for all types of objects. These CNN based features are the base for deep learning object detection. Figure 4 shows the comparison between traditional and deep learning based object flow.

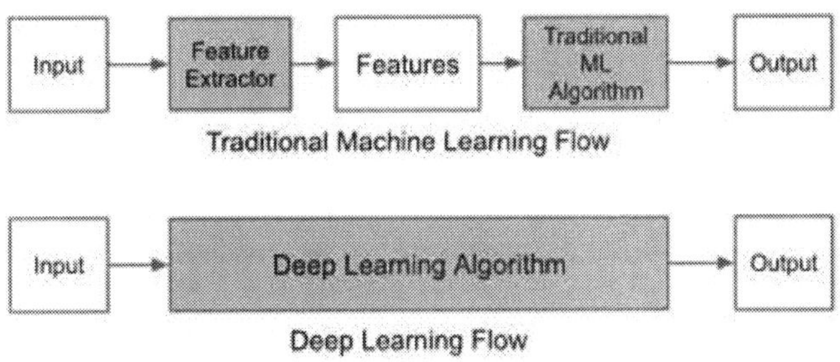

Figure 4. Comparison between traditional and deep learning based methods

4.2 Deep Learning Based Object Detection:

Deep learning based object detection is predominantly performed with CNN as feature extractor followed by either region proposal network or single shot detector based network. Regional proposal and single shot detector are the two major types detection algorithms that are widely prevalent in deep learning localization tasks. Therefore, we have selected one type of single shot detector and regional proposal network for analysis. Under single shot method we examine single shot detector (SSD) [16] network and under regional proposal technique we examine Faster Region – Convolutional Neural Network (FRCNN) [17].

IOP Conf. Series: Materials Science and Engineering 1059 (2021) 012062 doi:10.1088/1757-899X/1059/1/012062

4.2.1 Single Shot Detector:

Single shot is a single network object detection algorithm that is used object detection. Unlike other region proposal based frameworks, SSD has a single architecture for both detection and classification. In SSD, object localization and classification are performed during the forward pass itself in a single shot, therefore the architecture gets its single shot. SSD architecture is built of three major parts namely – initial CNN architecture for feature extraction, multibox detector network for detection and finally non maximum suppression.

VGG 16 architecture is used for extracting features from input image since it has a rich feature extraction framework than the other existing CNN models. VGG architecture was modified by deducting the final fully connected layers and including layers from conv6 layers. The extracted feature maps from VGG 16 are sent to multibox framework. Multi box is a regressor architecture that is utilized for obtaining the bounding box coordinates. This multi box technique is inception style architecture with 1x1 convolutions. Figure 5(a) shows the multi box regressor used in single shot detector. These multi boxes return location and confidences of predicted fault present on the surface of the brake. The output predictions are then processed by non maximum suppression block to determine the bounding box with highest confidence value. Following this output with highest confidence score is given as output. Figure 5(b) shows the overall SSD architecture.

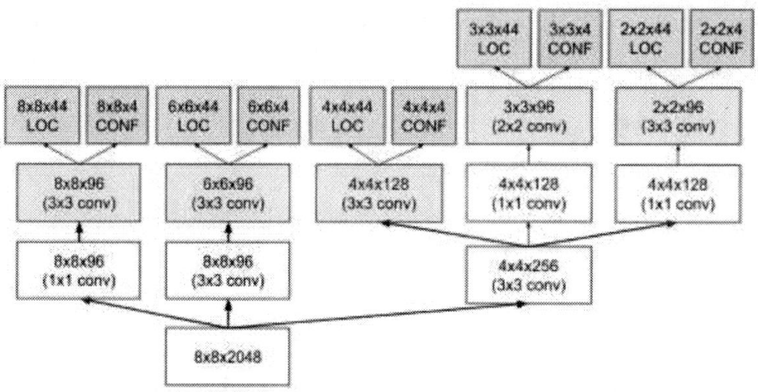

Figure 5(a). Multi-scale convolution prediction architecture for obtaining location and confidence of mutlibox

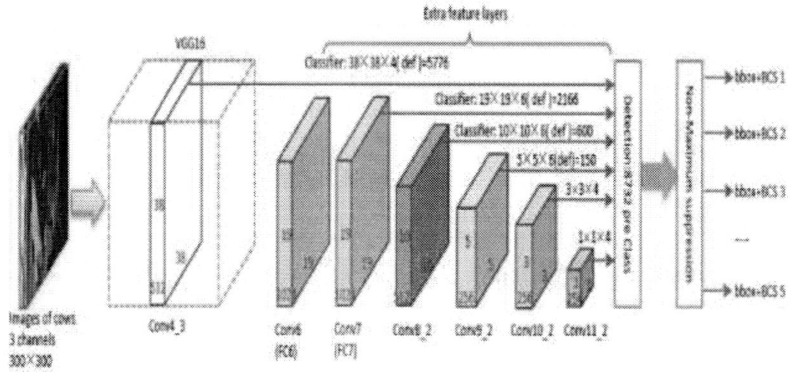

Figure 5(b). Single Shot Detector Algorithm

4.2.2 Faster RCNN:

Fastern RCNN is an architecture that evolved from Region-Convolutional Neural Networks (R-CNN) and fast RCNN networks. Unlike SSD, that has a single one shot architecture for detection, localization and classification, faster RCNN has separate frameworks to perform classification, localization and detection. Faster RCNN has four major steps namely – feature extraction using CNN architecture, obtaining a list of bounding boxes, assigning label to each bounding box and finally determining the probability of each label and bounding box. The initial feature extraction is performed using any state of the art CNN models like inception, VGG, etc. The original paper had pre-trained ZF net and VGG models. The feature maps obtained from the CNN feature extractors are sent to the region proposal network. These region proposal networks generate region proposals of approximately locations of where objects would be. These generated region proposals only perform localization but do not classify the object present within the bounding box. The region proposals obtained from this step are sent to a classifier framework that classifies the object present within the bounding box. Subsequently, a regressor checks the possibility of object occurrence within the bounding box coordinates. Figure 6 shows the overall faster – RCNN architecture.

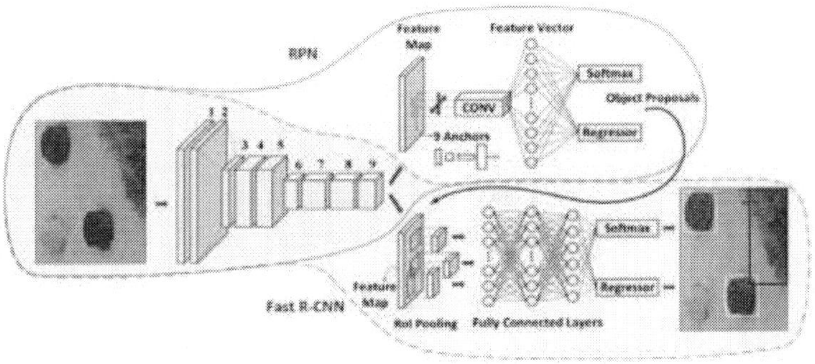

Figure 6. Faster RCNN Algorithm

In single shot detector algorithm, one single network is used for localization and classification tasks together, whereas in FRCNN region proposal networks are used for localization along with extracted features from CNN based architecture. SSD utilizes anchor boxes at different aspect ratio and unlike other algorithms SSD learns offset rather learning the box itself. Faster RCNN is an evolution from fast RCNN networks wherein selective search algorithms were replaced with region proposal networks to obtain the region of interests. In order to tackle the differences in aspect ratio and scale of objects, faster RCNN introduced the concept of anchor boxes for three different scales.

Further, these algorithms are trained using transfer learning techniques with annotated dataset collected previously. For transfer learning, trained weights from COCO dataset are used for initialization of the algorithm. Generally, object detection algorithms are evaluated using accuracy, error and mean Average precision (mAP). Apart from this, another important parameter for consideration while selecting object detection algorithm is speed vs accuracy tradeoff while deploying in real time application environment. Usually, faster RCNN models have little higher accuracy than SSD networks on the other hand SSDs are faster than FRCNN networks.

5. Results and Discussion:

5.1 Training Experimental Setup

The selected algorithms were trained for 500 epochs with 1000 images in each class. Cracks, dents, scratches, rust and smear the classes taken for consideration. The training and test dataset are split in the ratio 7:3 for testing purposes. Further, after training, the models were evaluated were accuracy, mAP and error. Very small amount of data augmentation was done to improve the performance of the model. All annotated images were converted to TF record format to be trained in tensorflow library. Tensorflow and python based programming platform were used to perform this analysis.

5.2 Factors to Analyzed in Evaluating the Performance of Object Detection Algorithms:

Generally, there are numerous factors to be considered to ensure that deep learning based object detection models delivers the best performance. Some of them are selecting the best feature extractors, correctly determining the output stride of feature extractor, selecting the most appropriate input image resolution, fixing the threshold values in non-max suppression (non max suppression is used to select the best bounding box from all suggestions offered by the model), usage of data augmentation and selection of best hyperparameters for the training scenario. Among these, in this research work, we examine the most important factors namely – feature extractors and input image resolution.

5.3 Experimental Results:

Table 2 shows the training and testing performance of the model on the dataset. The training accuracy for SSD is 95.4% and faster RCNN accuracy approximates to 97.10%. Similarly, training error for SSD and faster RCNN is 6.17% and 4.23% respectively. Our evaluation on test dataset gave an accuracy of 95.64% with an error rate of 5.6 for faster RCNN model and an accuracy of 94.16% with an error of 7.1 for SSD model. Table 2 shows the comparison between faster RCNN and SSD. Figure 7 shows the output image obtained with bounding box correctly detected around the faulty regions. Further, it can be observed that faster RCNN is slightly slower than SSD model for processing a single test image. This shows that there is a tradeoff between faster RCNN and SSD regarding accuracy and test time for processing a single test image. Faster RCNN is found to deliver a higher accuracy performance at a slightly slower speed than SSD. Though SSD is faster than faster RCNN, still its performance accuracy is slightly lesser than faster RCNN. Also, Fastern RCNN is the best choice when the faults are difficult to be identified by naked human eyes. Similarly, SSD is more preferred when real time processing speed is the most important factor for consideration.

5.4 Analysis of Different CNN feature Extractors

The performance of object detectors in deep learning is largely influenced by the type of convolutional neural network used for extracting features from the raw input image. Therefore, the selected models were also investigated with different CNN models as feature extractors along frame per second performance of those. Both faster RCNN and SSD models were experimented with VGG, Inception, Resnet, inception resnet and mobile net CNN models for feature extraction.

From the tests, it was found that faster RCNN having inception resnet as feature extractor with gave the highest accuracy at 1 frame per second. When the number of proposals was decreased by half, the accuracy of the model increased further. On the other hand, SSD using mobilenet as feature extractor gave the highest mAP for real time processing based applications.

In terms of memory consumption, SSD occupies the least memory footprint among all types of object detection models. It requires lesser than 1 Gb of total memory for storage and operation purposes.

5.5 Analysis of Image Resolution:

Image resolution plays a vital role in determining the performance of object detection. When the resolution of input image increases, the accuracy of detection also increases at the cost of higher inference time. When the resolution of input image increases, the accuracy of detection for smaller objects increases. This increase in resolution as benefits higher objects present in input image. Therefore, in case of fault detection on metal surface, it is very important that the input image have a considerably high resolution to ensure high performance accuracy. In all our analysis, the input image resolution was maintained at 512x512.

Further, the resolution of input image was decreased by half along the width and height and our analysis was repeated again. It was found that accuracy decreased by 15.88% on test dataset on average when analyzed with both the models. The decreased resolution had also impacted the inference of models and it was found that the inference time got reduced by 27.4% on an average on both the models.

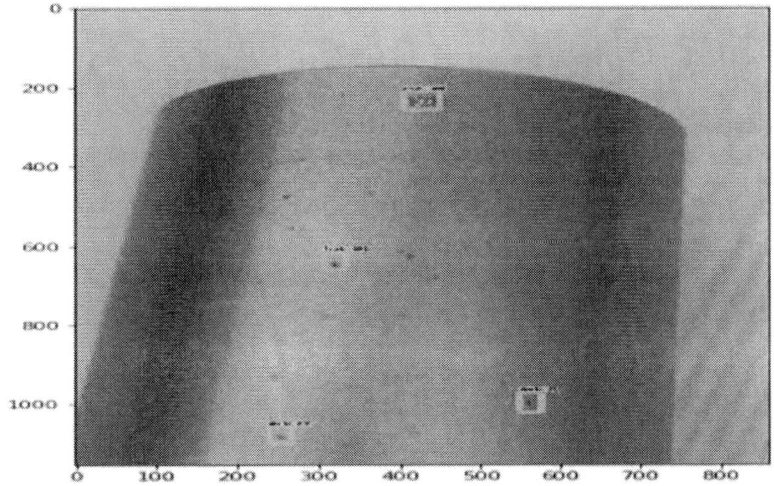

Figure 7. Detected defect portion with bounding box

Table 2. Performance Analysis of Faster RCNN and SSD networks

	Training Accuracy	Training Error	Test Accuracy	Test Error	Time taken to obtain output (sec/image)
SSD	95.4%	6.17%	94.16%	7.1%	0.36
Faster RCNN	97.10%	4.23%	95.64%	5.6%	0.42

6. Conclusion and Future Work:

Therefore, in this work we have presented deep learning object detection based automated brake inspection system for brake parts. Faster RCNN and SSD deep learning algorithms were utilized to detect the faulty regions on the surface of components. The developed model was trained on our custom collected dataset and is found to overcome the drawbacks of conventional image processing based solutions offering high reliability and also exactly localizes the presence of defects on the surface of the brake. The developed system is found to deliver an accuracy of 95.64% with an error rate of 5.6 and the model takes approximately 0.42 seconds to process a single test image. Therefore, through this research work, inspection of brake components was automated using deep learning based algorithms.

In future, we plan to execute the analysis of defect surface of brakes using semantic and instance segmentation algorithms. Therefore, through segmentation based approach, we aim to determine the exact area measurement of defect region and thereby aid brake designers to rectify those faults and defects.

Reference

[1] Raheja J L, Kumar S and Chaudhary A 2013 Fabric defect detection based on GLCM and Gabor filter: A comparison. *Optik, 124*(23) 6469-6474.
[2] Aarthi T, Karthi M and Abinesh M 2013 Detection and analysis of surface defects in metals using wavelet transform *International Journal of Scientific and Research Publications 3*(6) 147-158.
[3] Sadeghi M and Memarzadehzavareh F 2013 Flaws detection in steel plates Using Gabor Wavelet. *Life Science Journal 10*(2s).
[4] Sivabalan K N and Gnanadurai D 2011 Efficient defect detection algorithm for gray level digital images using Gabor wavelet filter and Gaussian filter *International Journal of Engineering Science and Technology 3*(4) 3195-3202.
[5] Deshmukh R S, Deshmukh P R and Taley A M 2012 Comparison analysis for efficient defect

detection algorithm for gray level digital images using median filters Gabor filter and ICA *International Journal of Advanced Research in Computer Science and Software Engineering* **2**(1).

[6] Guan S, Lei M and Lu H 2020 A Steel Surface Defect Recognition Algorithm Based on Improved Deep Learning Network Model Using Feature Visualization and Quality Evaluation *IEEE Access* **8** 49885-49895.

[7] Kim S, Kim W, Noh, Y K and Park F C 2017 May Transfer learning for automated optical inspection. In *2017 International Joint Conference on Neural Networks (IJCNN)* 2517-2524. IEEE.

[8] Song K and Yan Y 2013 A noise robust method based on completed local binary patterns for hot-rolled steel strip surface defects *Applied Surface Science* **285** 858-864.

[9] Wu Y, Qin Y, Wang Z and Jia L 2018 A UAV-based visual inspection method for rail surface defects. *Applied sciences*, **8**(7) 1028.

[10] Cen Y G, Zhao R Z, Cen L H, Cui L H Miao Z J and Wei Z 2015 Defect inspection for TFT-LCD images based on the low-rank matrix reconstruction. *Neurocomputing* **149** 1206-1215.

[11] Lei J, Gao X, Feng Z, Qiu, H and Song M 2018 Scale insensitive and focus driven mobile screen defect detection in industry *Neurocomputing* **294** 72-81.

[12] Li Y, Zhao W and Pan J 2016 Deformable patterned fabric defect detection with fisher criterion-based deep learning *IEEE Transactions on Automation Science and Engineering* **14**(2) 1256-1264.

[13] Chondronasios A, Popov I and Jordanov I 2016 Feature selection for surface defect classification of extruded aluminum profiles *The International Journal of Advanced Manufacturing Technology* **83**(1-4) 33-41.

[14] Gibert X, Patel V M and Chellappa R 2016 Deep multitask learning for railway track inspection *IEEE transactions on intelligent transportation systems* **18**(1) 153-164.

[15] Liu W, Anguelov D, Erhan D, Szegedy C, Reed S, Fu C Y and Berg A C 201 Single shot multibox detector In *European conference on computer vision* 21-37.

[16] Ren S, He K, Girshick R and Sun J 2016 Faster r-cnn: Towards real-time object detection with region proposal networks *IEEE transactions on pattern analysis and machine intelligence* **39**(6)1137-1149.

IOP Conf. Series: Materials Science and Engineering 1059 (2021) 012063 doi:10.1088/1757-899X/1059/1/012063

DESIGN OF HVAC SYSTEM FOR BEVERAGE INDUSTRY

Pradeep A D[1]*, Vivekanandhan R[2], Rameshkumar T[3], Sangaravadivel P[4]

Department of Mechanical Engineering, Bannariamman Institute of Technology, Sathyamangalam, Tamilnadu - 638401, India

adpradeep@live.com

Abstract.The objective of this project is to design HVAC system for beverage industry. Firstly, the requirements and Design Basic Report (DBR) of the HVAC system are defined. DBR gives a clear picture of filtration level, required conditions, Cost-effectiveness, Energy conservation, Fire safety, Indoor air quality (IAQ) and Ease of maintenance. Air Flow diagram makes you understand the Air Handling Unit (AHU) zoning for the HVAC system. Beverage industry zoning areas include pulp cutting, blending and homogenizer, filler, anteroom, CIP room and pasteurization. Design calculation provides the data for recirculation and 100% fresh air supply and return system. Processing line is having sequence process rooms and each room have its own individual cooling load calculations. Design calculation gives the value of Tonnage of refrigerant, CFM and Dehumidification capacity of the system. Ducting layout issue the best routing for the site installation to distribute the air from the supply and return system. It includes the details of Damper, Ducting, Duct Insulation and Instrumentation to use in the facility. Supply and return system placement are also covered in this layout drawing and the Bill of Quantity is prepared.

1 Introduction

Heating, ventilation, and air conditioning (HVAC) is the technology developed for environmental comfort both indoor and vehicular spaces. Environmental comfort in spaces aimed at providing thermal comfort and acceptable air quality. Based on the principles of thermodynamics, fluid mechanics and heat transfer, HVAC system design is a sub discipline of mechanical engineering.

HVAC is a part of residential structures such as hotels, single family homes, senior living facilities, apartment buildings and, medium to large industries, hospitals and office buildings. Transportation facilities such as cars, trains, airplanes, ships and submarines, and in marine environments, where safe and healthy building conditions are regulated with respect to humidity and temperature, using fresh air from outdoors.

Ventilating or ventilation (the V in HVAC) is the natural or mechanical/forced process of exchanging or replenishing air in any confined space. Ventilation provides high indoor air quality which involves temperature control, humidity control, oxygen replenishment, and removal of moisture, odors, smoke, heat, dust and airborne bacteria. Ventilation in addition to introducing outside air, maintains circulation in the interiors and prevents stagnation of interior air. Ventilation includes both the exchange

IOP Conf. Series: Materials Science and Engineering 1059 (2021) 012063 doi:10.1088/1757-899X/1059/1/012063

of air to the outside as well as circulation of air within the building. It is one of the most important factors for maintaining acceptable indoor air quality in buildings.

2 Scheme Design and DBR

The purpose of Design Basis Report (DBR is to specify the requirements and general scheme of HVAC system for proposed Tetra line project. The HVAC system design scheme is based on a) Standards and Guidelines, b) Basic Concept Design and c) System Description. The HVAC system design shall conform / comply with the requirements of SMACNA, ASHRAE, ISHRAE and NBC.

Basic concept design is done based on Metrological data of location, occupancy level, internal heat generation due to lighting load, equipment load, etc.. Input data for air conditioning system and ventilating system were provided based designated temperature to be maintained, occupancy, volume and thermal load data. Lighting and equipment load, ACPH, volume and area data, temperature are used as input parameters for AHU, Air-conditioning and ventilating system [1].

System description provides the AHU configuration needed based on input design parameters. Filtration level, type of blower, chilled water cooling coil requirements, operating pressure, electrical requirement and electrical motor details define the selection of AHU for HVAC system.

2.1 Standards and Guidelines

The HVAC system design shall conform / comply with the requirements of the following Standards and Guidelines as mentioned below:

a. Sheet Metal & Air Conditioning contractors' National Association (SMACNA)
b. American Society of Heating, Refrigerating and Air Conditioning Engineers (ASHRAE)
c. Indian Society of Heating, Refrigerating and Air Conditioning Engineers (ISHRAE)
d. National Building code of India (NBC)

2.2 Basic Concept Design

Concept design of HVAC system is developed using the above mentioned guidelines stated for the purpose. Concept design part need data pertaining to location, climatic conditions, conditions to be maintained in particular area and number of people working in that area [2].

2.2.1 Basic Service conditions

The HVAC system designing will be done considering the environmental conditions such as temperature, humidity etc, as applicable around the facility, during all seasons. The design condition for Bidadi (Place near Bangalore) is as mentioned below (Reference Considered nearest City Bangalore from ISHRAE HVAC Handbook / Metrological Data).

Table 1. Metrological Data at 12° 79'' 89' (N) 77° 38'' 64' (E) and752.7 m. above Mean Sea Level

Season	Dry bulb temperature [°F]	Wet bulb temperature [°F]	Relative Humidity [%]
Summer	96.0 (35.6°C)	78.0 (25.6°C)	45
Monsoon	82.0 (27.8°C)	78.0 (25.6°C)	82
Winter	58.0 (14.4°C)	54.0 (12.2°C)	78

The selection of HVAC equipment shall be governed by safety, reliability and compatibility with respect to specified future expansion, design margins, suitability for environment, economic considerations and past service history.The SI & FPS system of units shall be used.English language shall be used for all drawings, texts and communications.For the purpose of electrical grounding calculations (soil electrical resistivity) and cable rating calculations (soil thermal resistivity); the soil resistivity data indicated in the Geotechnical investigation report for provided by Client shall be used.

2.2.2 Occupancy Level

IOP Conf. Series: Materials Science and Engineering 1059 (2021) 012063 doi:10.1088/1757-899X/1059/1/012063

Maximum no. of people working in a given area will be considered for HVAC designing. The amount of physical human effort needed to do the work is proportional to heat generated while doing the work. Moderate physical working shall be considered for calculating latent and sensible load from the personnel.

2.2.3 Fresh Air Selection

The amount of fresh air taken into the system is decided on process happening in the given area. More fresh air intake will reduce the efficiency of HVAC system. Hence, fresh air will be selected on the following basis

a. Minimum fresh air 10% of CFM (As per industrial standard practice.)

b. Ex-filtration through doors

2.2.4 Internal Load

In addition to above discussed parameters, process happening in the area will have a major effect in designing HVAC system. Usage of electrical equipment and process equipment will generate heat due to energy conversion and this is to be considered while deciding on conditions to be maintained in processing area.

Table 2. Internal Load

Type of Load	Values
Lighting Load	2 Watts/sq.ft.
Heat Dissipated from theoccupancy Sensible Latent	265 / 365 Btu/hr
Heat Dissipated fromelectrical motor	Inefficiency of the motor. For calculations 20% considered for all motors.
Heat Dissipated From AHU\motor	Motor HP calculated based on the dehumidifier. CFM and same is considered for heat load calculations.
Heat gain return air duct	20 CFM leakage per 100 sq.ft. of duct.
Convection + radiation heat	At actual
Heat Dissipated from UPS System	On Full load of UPS system, for calculations 8% considered
Heat Dissipated from Battery System	On Full load of Battery system, for calculations 2% considered

3 Process Description

Beverage industry in which HVAC system to be installed consists of a tetra-pack line processing. The line moves through the following rooms. Room description and processes involved in those rooms are to be provided as input for detailed design of HVAC system.

Tetrapack line processes start at pulp cutting room and moves to blending and homogenizer room. Filling room, PM store and compactor are the areas down the process line that are to be considered for HVAC system design. Partial installation of HVAC system in a building without isolation of rooms will make the system inefficient. Office rooms, both existing and new office are considered in HVAC system design and necessary conditions are maintained in office area accordingly. Battery room and UPS room which normally gets heated are included, while designing the system.

Table 3. Input Data for Air conditioning System

S.No.	Description	Area (sq. ft.)	Lighting Load (Watts)	Equip Load (kW)	Occupancy	Design Temperature (°C)	ACPH
1	Pulp Cutting	627	1254	2	2	$24^0 \pm 2^0$	4 to 6
2	Blending & Homogenizer Room	1899	3798	28	2	$24^0 \pm 2^0$	4 to 6
3	Filling	1305	2610	12	2	$22^0 \pm 2^0$	4 to 6
4	PM Store	870	1740	0	2	$28^0 \pm 2^0$	4 to 6
5	Compactor	1423	2134	0	2	$28^0 \pm 2^0$	4 to 6
6	Existing Office	236	472	1	4	$24^0 \pm 2^0$	4 to 6
7	Office	86	172	0.5	2	$24^0 \pm 2^0$	4 to 6
8	UPS Room	241	482	32	---	$22^0 \pm 2^0$	4 to 6
9	Battery Room	146	392	8	---	$22^0 \pm 2^0$	4 to 6

Table 4. Input Data for Air Washer, Exhaust and Ventilation System

S.No.	Description	Length (m)	Width (m)	Area (m^2)	Height (m)	Volume (m^3)
1	Air Washer System					
1.1	CIP Room	11.25	9.7	109	4.8	524
1.2	Pasteurizer Room	6.02	9.7	58	4.8	280
1.3	Considering 6 nos. pot cooling points in at1000 CFM per point					
2	Exhaust Air System					
2.1	CIP Room	11.25	9.7	109	4.8	524
2.2	Pasteurizer Room	6.02	9.7	58	4.8	280
3	Ventilation System					
3.1	CIP Tank Area Exhaust	5.4	2.5	13.5	4.8	64.8
3.2	CIP Tank Area Fresh Air	5.4	2.5	13.5	4.8	64.8

The various data inputs required for HVAC system, Ventilation air washer system and exhaust air system from various rooms and areas are given in the tables 3 and 4. Based on the inputs, Air handling system configuration suitable for beverage industry is given in table 5.

IOP Conf. Series: Materials Science and Engineering 1059 (2021) 012063 doi:10.1088/1757-899X/1059/1/012063

Table 5. AHU System configuration

AHU Configuration	Recommendations
Unit Construction	Double Skin, Single Blower.
Filtration level	Supply through Pre filter (G-4), Fine Filter (F6 & F9) in Plenum. Fresh air filter G-4
Type of Blower	Plug type fan with flow measuring device comprising of Calibrated nozzle and differential pressure switch. Supply air fan shall be provided with variable speed Drive and shall be compatible with VFD.
Chilled Water cooling coil	Minimum 4-6 row deep
AHU Bleed	Bleed air exhaust to atmosphere through G-4 Filter
Dampers	At AHU return and supply air cut-outs, fresh air and bleed air cut-out. Motorized Fire Damper inSupply and Return, mounted on AHU.
Pressure	Controlled air flow to protect the product throughPressure differential.
Magnallic Gauge	To be installed across the Filter
Construction arrangement	All door shall be hinged type and Limit Switch
Electrical Requirement	AHU shall have remote start stop arrangement andStart-stop Pushbutton with ON-OFF-Trip indication Control station to be considered in Room
Electrical Motor	415 V 3 Phase , IE 2 Type Motor suitable to operate through VFD

4 System Description

HVAC system design is concerned with heat removal and maintaining comfort conditions, allied systems like ventilation air washer system and exhaust air systems are also to be designed properly for better and durable functioning of system.

4.1 HVAC System

The HVAC systems shall be designed to produce an environment, 24 hours a day, entire year that will have the appropriate air change rates, temperatures, and achievable humidity levels as providing comfort conditions in the following areas

4.1.1 Pulp cutting Room, Blending and Homogenizer Room

The pulp cutting Room, blending and homogenizerRoom shall be fed with one recirculation AHU which is located in the loft area inside the pasteurizer room. The supply air from the AHU is passed through rectangular duct above the false ceiling area. It is suggested to have the jet nozzle diffuser for supply air in the pulp cutting and blending and homogenizer room.

The return air shall be collected through the return air grills which are fixed at the false ceiling and connected to the Return air duct. The return air duct running above the false ceiling to reach mixing box of AHU.

The supply air duct shall be provided with thermal insulation with 19mm thick nitrile rubber insulating material and return air ducts shall be provided with thermal insulation with 13mm thick nitrile rubber insulation. The fresh air is drawn from the outside to the mixing box through the fresh air duct at ceiling level. The fresh air duct shall be with louver fixed at the external wall, damper and filter. The cooling coil shall be connected with chilled water supply and return pipes with necessary valves and fittings. The chilled water temperature shall be around 7 °C and 12 °C.

The suggested air treatment process is as follows. Fresh air will be sent through Pre-filter (G4) and then supplied to Mixing Box. Return air is mixed with 10% fresh air in mixing chamber and passed through Pre-filter (G4), Intermediate filter (F6), Cooling Coil, Fan and after passing through fine filter (F9) in Plenum, it is supplied to rooms through Jet Nozzles mounted at false ceiling.

4.1.2 Filling area

The Filling area shall also be fed with samerecirculation AHU which is located in the loft area inside the pasteurizer room. The supply air from the AHU is passed through rectangular duct above the false ceiling area. It is suggested to have the jet nozzle diffuser for supply air in the filling area from the ceiling and part of supply air is supplied through the return air grille fixed in the panel and level of the return air grille from the floor level maintained is around 500mm.

The return air shall be collected through the return air grills which are fixed at the false ceiling and connected to the Return air duct. The return air duct running above the false ceiling to reach mixing box of AHU.The fresh air is drawn from the outside to the mixing box through the fresh air duct at ceiling level. The fresh air duct shall be with louver fixed at the external wall, damper and filter. The supply air duct shall be provided with same amount of thermal insulation as specified above.The ducts running above the false ceiling area shall be rectangular in shape and ducts running in non-false ceiling areas shall be with circular in shape. The cooling coil is connected with chilled water supply and return pipes.

The suggested air treatment process is as follows.Fresh air will be sent through Pre-filter (G4) and then supplied to Mixing Box. Return air is mixed with 10% fresh air in mixing chamber and passed through Pre-filter (G4), Intermediate filter (F6), Cooling Coil, Fan and after passing through fine filter (F9) in Plenum, it is supplied to rooms through Jet Nozzles mounted at false ceilingand few supply grills mounted at wall panels at floor level and is around 500mm.

4.2Ventilation Air Washer System

Two stages Air washer system is adopted to feed the CIP, Pasterization and spot cooling to Tetra Line Areas. The inside condition will be few degrees less than the ambient.

Air washer unit shall be with double skin construction and comprising of Blower section, sensible heat exchanger, adiabatic Heat exchanger, filters and recirculation pumps. Filters shall be G4, F6 and F9 and pumps shall be submersible type.Common unit is considered to feed the CIP, Pasteurization and spot cooling requirement for Tetra line area.

The Air washer unit is placed outside near to the feeding areas. The air washer system unit shall be provided water connection and arrangement shall be made for adequate water quantity and required water quality.The duct from the air washer unit shall be running above the false ceiling and air supplied through the grills. The supply air duct shall be thermally insulated with 19mm thick nitrile rubber insulating material.

Fresh air is passed through pre-filter (G4) and Intermediate filter (F6). Using spray arrangement and fan it is again passed through fine filter (F9) in Plenum before supplying to rooms through grills mounted at false ceiling.

4.3Exhaust Air System

Exhaust air system shall be provided for CIP and Pasteurization areas. The Common exhaust fan is placed outside and near to the feeding areas.The duct from the Exhaust air unit shall be running above the false ceiling and air drawn from exhaust grill which are fixed in the false ceiling. Used air is filtered using pre-filter (G4) before its exhaust to the atmosphere.Spot cooling supplied to the Tetra line Area shall escape through the turbo ventilator provided at the roof level.

5 Detail Engineering

Detail engineering or detailed design engineering is a development of all required construction documents and drawings up to AFC (Approved for Construction) Stage for the construction and detailed bill of materials (BOM) for the bulk material procurement based on the basic or front end engineering design (FEED) package. The detailed Design and engineering is limited to verifying design basis but producing all construction drawings after incorporating vendor information.

5.1 Air Flow Diagram

Air Flow Diagram consisting of details air handling unit parts, dimensions and zoning. Areas that covered with one air handling unit called AHU zoning. This will give you the parts information AHU size, blower, filter, heating coil, cooling coil, louvers, dampers, and the number of rooms covered.

Air Flow diagram makes you understand the AHU zoning for the HVAC system. Tetra line process areas pulp cutting, blending and homogenizer, filler, anteroom, CIP room, Pasteurization and tetra line area zonings are included. Design calculation provides you a data for recirculation and 100% fresh air supply and return system. Tetra packs line having sequence process rooms, each room having individual cooling load calculation. Calculation gives the value of Tonnage of refrigerant, CFM, Dehumidification and capacity of the systems.

5.1.1 Recirculation System

Recirculation system is providing the fresh air considering space and occupancy level according ISHRAE standard. People to breathe and space ventilation. Thermal comfort zones are recommended to recirculation system for better efficiency of operation.AFD for Recirculation system schematic drawing is prepared for understanding the flow.

5.1.2 Air Washer Unit

Air washer is a piece of equipment that is designed to improve air quality by scrubbing the air that moves through it and adjusting humidity levels keep the environment consistent. As per the process requirement of the tetra pack line we go with the air washer unit with 100% exhaust system.It is the healthier system since it works on 100% fresh air. Low installation cost and low running cost. As per the process requirement of the tetra pack line we go with the air washer unit with 100% exhaust system.AFD for Air Washer Unit schematic drawing is also prepared for understanding the flow.

5.2Design Calculation for heat generation/gain

Design calculation gives you a full study of heat gain from all part of the room also from the equipment. Heat gain from the wall, roof, and processing equipment areused to calculatethe overall heat load, based on which, the ton of refrigerant (TR) and dehumidified CFM are decided [3,4].In beverage industries, pulp cutting room, filler room and ante room are recommended to recirculation system as per ASRHAE standard and sample calculation of one room is given below.

5.2.1 Pulp Cutting Room

The pulp contains the juice of the fruit. The pulp usually removed from fruit juice by filtering it out. The colour of the pulp can change depending on the species and the ripening stage. Usually it has the colour of the fruit's peel. This is process room for juice flavour where all the fruit getting processed and sized for soft drink preparation. The next process is to blend and homogenise the drink. Design conditions for pulp cutting room is given in table 6.

5.2.2 Filler Room

Filling describe the process whereby a receptacle in tetra pack is filled with a viscous and pasteurized product.Filler machine need thermal comfort for better performance so we are recommending the recirculation system.

5.2.3 Ante Room

Ante room acts as a store and they do not get wet due to condensation while unloading for dispatch.

IOP Conf. Series: Materials Science and Engineering 1059 (2021) 012063 doi:10.1088/1757-899X/1059/1/012063

5.2.4 Design calculation for Pulp cutting room

Temperature, Humidity, Occupancy of people, etc.. are given as input to calculate total air change required per hour. Table 7 provides the heat load calculation for pulp cutting room. Total heat generated from the room is calculated which includes heat emitted from walls, roofs, occupants, equipment, latent heat, sensible heat and heat from outside air [3,5].

Table 6. Design conditions for pulp cutting room

Design Conditions (°F)					
	DB	WB		RH%	GR/LB
Outside	96	78		45	134
Room	72			55	72
Diff.	24	Daily range - 22			62
Ventilation:		CFM/per			
People	2	X	15		
1 air change/Hr					235
Outside air		40		CFM	
ADP				57	
Dehumidified rise				13.5	
Dehumidified CFM				**3962**	
Check figures					
CFM/TR				544	
Sq.ft/person				352	
Watts/sq.ft				2.0	
Sq.ft/TR				97	
Dehum. CFM/sq.ft				5.62	
Total air change/Hr				9.9	

Table 7. Heat load calculation for Pulp cutting room

JOB NAME :	TETRA PACK LINE		**AREA**		704.534	Sq.ft
SPACE USED FOR :	Pulp cutting		**VOLUME**		14096.3	Cu.ft
Length 5.745	Breadth	11.393			Height	20.008
Item	Area/ Qty		Sungain		Temp. diff factor	BTU/HR
SOLAR GAIN - GLASS						
Glass	0	Sq.ft		X	0.56	0
Glass	0	Sq.ft		X	0.56	0
SOLAR & TRANS. GAIN - WALLS & ROOF						
Wall	0	Sq.ft		X	0.366	0
Wall		Sq.ft		X	0.366	0
Wall		Sq.ft		X	0.366	0
Wall		Sq.ft		X	0.366	0
Roof		Sq.ft		X	0.12	0
TRANS.GAIN - EXCEPT WALLS & ROOF						
All glass	0	Sq.ft	24	X	1.13	0
Partition (wall)	674.31	Sq.ft	19	X	0.375	4804
Partition (panel)	1376.66	Sq.ft	19	X	0.1	2616
Partition (glass)	135.63	Sq.ft	19	X	0.446	1149
Ceiling	704.53	Sq.ft	19	X	0.1	1339
Outside air	400	CFM	24	X	1.08	1037
INTERNAL HEAT						
People	2	People		X	265	530
Appliances		BTU/Hr		X		0
Lights	1409.07	Watts	1.08	X	3.413	5194
Equipment	12	kW	1	X	3413	40956
System gain & safety factor -5%						2750
Total room sensible heat						57760
Room latent heat						
Outside air	400	CFM	62	X	0.68	1686
People	2	People		X	365	730
Steam		lb/Hr		X	1076	0
System gain & safety factor -5 %						121
Total room latent heat						2537
Room total heat						60297
Outside air heat						
Sensible	400	CFM	24	X	1.08	9331
Latent	400	CFM	62	X	0.68	15178
System gain & safety factor - 3 %|						2544
Return air heat gain						0
GRAND TOTAL HEAT			**TR =**		**7.28**	87350

5.3 Design Calculation for Ventilation System

Two stages Air washer system is adopted to feed the CIP, pasteurization and spot cooling to Tetra Line Areas. The inside condition will be few degrees less than the ambient.Air washer unit shall be with double skin construction and comprising of Blower section, sensible heat exchanger, adiabatic Heat exchanger, filters and recirculation pumps. Filters shall be G4, F6 and F9 and pumps shall be submersible type.

Common unit is considered to feed the CIP, Pasteurization and spot cooling requirement for Tetra line area. The Air washer unit is placed outside near to the feeding areas. The air washer system unit shall be provided water connection and arrangement shall be made for adequate water quantity and required water quality.

Table 8. Ventilation area for CIP, Pasteurizer and tetra line

Sl.No.	Description	Length (m)	Width (m)	Area (m^2)	Height (m)	Volume (m^3)	Volume (cu.ft)
I	**2 Stage Air Washer System**						
1a	CIP ROOM	6.3	11.5	72	5.5	398	14060.98
1b	CIP ROOM	3.42	11.5	39	4	157	5551.35
2	Pasteurizers Room	5.72	11.04	63	5.5	347	12255.82
3	Considering 6 Nos spot cooling points in Tetra Line Area at 1000 cfm per point						

Table 8 provides volumetric area for ventilation in CIP, Pasteurization and Tetra line. Table 9 provides ventilation calculations to find total air supply and exhaust from the system [4,6,7].

Table 9. Ventilation calculation for CIP, Pasteurizer and tetra line

Sl. No.	Description	Volume (cu.ft)	ACPH	Supply Air CFM	Rounded Supply Air CFM	Exhaust Air CFM (90% of Supply Air)	Rounded Exhaust Air CFM
I	2 Stage Air Washer System						
a	CIP ROOM	14060.98	40	9373.99	9400	8460	8500
b	CIP ROOM	5551.35	40	3700.90	3700 13100	3330	3500 12000
2	Pasteurizer Room	12255.82	40	8170.54	8400	7560	7500
3	Considering 6 Nos spot cooling points in Tetra Line Area at 1000cfm per point				6000		0
				Total	**27500**		**19500**

5.4 Ducting layout

Ducts are used in heating, ventilation, and air conditioning (HVAC) to deliver and remove air. The needed airflows include, for example, supply air, return air, and exhaust air. Ducts commonly also deliver ventilation air as part of the supply air. As such, air ducts are one method of ensuring acceptable indoor air quality as well as thermal comfort.

Ducting layout issue the best routing for the site installation to distribute the air from the supply and return system. It consisting of the details of Damper, Ducting, Duct Insulation and Instrumentation to use.A duct system is also called ductwork. Planning (laying out), sizing, optimizing, detailing, and finding the pressure losses through a duct system is called duct design.

6 Conclusion

Design calculation provides the data for recirculation and 100% exhaust System. Tetra packs line having sequence process rooms and for each room, individual cooling load calculation. Calculation gives the value of Tonnage of refrigerant, CFM, Dehumidification and capacity of the systems. Ducting layout issue the best routing for the site installation to distribute the air from the supply and return system. Details like damper requirements, ducting, duct insulation and instrumentation to use in the facility will be available. Supply and return system placement are also covered in the layout drawing. Bill of quantity provides the detailed equipment requirements for HVAC system installation.

References

[1] Mohd Zaheen Khan 2018 Influence of Thermal Parameters on the Heat Load Calculation of the Building Using HAP Software *Journal of energy technologies and policies* **8** (4) 17-26.

[2] Chen Huang, JingYue, Tianyu Bai, Liugen Lv, Qianru Zhang and Jianchang Chen 2017 A Preliminary Research on Load Calculation Method of Stratified Air Conditioning System with Low-sidewall Air Inlets and Middle-height Air Outlets in Large Space Building *Procedia Engineering* **205** 2561-2568.

[3] Down P G, Billington N S and Over E 2014 Heating and Cooling Load Calculations – International Series of Monographs *In: Heating, Ventilation and Refrigeration* 5th edition Pergamon Turkey.

[4] Eng. Essa S and Ben-Essa, 2018 Cooling and heating loads in residential buildings in Kuwait *Journal of Engineering Research and Application* **8**(5) 01-14.

[5] Imal M 2016 A New System Design for Energy Management in HVAC Control Systems for Textile Plants *Acta Physica Polonica Series A* **130**(1) 245-248.

[6] Joppolo C M and Romano F 2017 HVAC System Design in Healthcare Facilities and Control of Aerosol Contaminants: Issues, Tools, and Experiments *Indoor Air Quality in Healthcare Facilities Springer Briefs in Public Health Springer, Cham* https://doi.org/10.1007/978-3-319-49160-8_8.

[7] Elnour M and Meskin N 2017 Multi-zone HVAC control system design using feedback linearization *5th International Conference on Control, Instrumentation, and Automation* Shiraz 249-254, doi: 10.1109/ICCIAutom.8258687.

IOP Conf. Series: Materials Science and Engineering 1059 (2021) 012064 doi:10.1088/1757-899X/1059/1/012064

Fabrication of Paddy Transplanter Machine

Arun Kumar Rajamanickam[1*], Uvaraja V C[2], Selvamuthukumaran D[3], Surya K[4]

[1,4] Department of Mechanical Engineering, Sri Krishna College of Engineering and Technology, Coimbatore, Tamilnadu, India
[2,3] Department Mechanical Engineering, Bannari Amman Institute of Technology, Sathyamangalam Tamilnadu, India

E-mail: arunkumarr@skcet.ac.in

Abstract. Wheat and Paddy are the new focuses in farming where still, relatively few scientists and makers take an interest. This field faces a few issues, for example, how to boost the benefit, how to build efficiency and how to diminish the expense. In India, two sorts of horticultural gear are utilized, manual strategy (ordinary technique) and motorized sort. Automation includes the utilization of a gadget between the work and power source. Transplantation is the on the extensive labor processes in agriculture, where in cultivation field the workers need to work in stooping position which creates the physical ailments to labors. In this work, to eliminate the difficulties in rice cultivation design of semi-automatic rice transplanting machine by cost effective method was discussed. This machine helps the farmer to improve the productivity and efficiency. Machine was designed and fabricated by using the available materials. Compare to the commercial machines the project has more economic. Investigation of the machine performance indicate that it significantly reduces the effort of labors and translation time.

Keywords: transplanting machine, cultivation, agricultural, farmers, plantation

1. Introduction

Rice cultivation is one of the commercial food crops and a staple diet for around half the world's population. India is among the top producers of rice and contributes about 20% of world production. Agriculture is the backbone of India's economy and over 58% of the rural sector still depends on agriculture as their major source of livelihood [1]. Annual yield worldwide is approximately 535 million tons. However, a recent survey has shown a significant decline in the production of paddy in rural areas because of the cost of cultivation and labour problems.

To compound the problem, most of the farmers in rural areas still follow traditional methods of planting and harvesting which leads to several postural issues for the farm-hands. It takes 2 weeks for 40 people to plant 35 acres of land with a work-hour of 7.5 hr /day. To compound to the long hours, employees work in a slouching position in the fields [2].

Figure 1. Manual Transplantation

The repeated strain from this process leads to physical ailments like back pain, neck pain, musculo-skeletal disorder (MSD) etc. Motivated to address this durdgery, we have designed a low cost machine with a planting mechanism to make an affordable prototype for common farmers that helps reduce the labour effort, and increases productivity by adhering to the Legowo row rice planting system.

Figure 2. Mechanized Transplantation

2. Literature Survey

Pradhanand S etal, investigate the conventional transplantation by involving 16 labours in varies age group. By considering varies parameter they have analysed two row, three row and four row transplanter methods. The investigation results hows that two row rice plantation is the excellent method for rice cultivation.

G.Singh et al. conducted experimentation on manual rice transplantation at the International Rice Research Institute farm. The results states that at this field the transplanter complete the process with capacity of 3.4 ha/h in different water depth. Rajvir Yadav et al investigate the performance of manual rice transplanter and found the pulling force required for the transplanter. Heighnt at which the force applied is thr important parameter which affect the output of the process.

Sharma and Singh developed a mat type nursery-raising device for rice planter. They reported that 72.02 and 33.33% saves the time and labour respectively, over the manual method of raising type nursery. The cost of raising mat type nursery for 1 ha was Rs. 299.50 with the developed device where as it was Rs. 1608.75 with manual method. Sivaswami and Anie John found that nearly 85% of labour cost was reduced by the introduction of paddy transplanters compared to manual transplanting.

3. Preparation Process

3.1 Land preparation process.

The process of rice cultivation includes several steps beginning with land preparation in which the land is prepared by tilling with a tractor in order to make it soft. Then the field is filled with water and again using a tractor with a cage well, the land is ploughed. A leveller is used to level the ground.

IOP Conf. Series: Materials Science and Engineering 1059 (2021) 012064 doi:10.1088/1757-899X/1059/1/012064

3.2 Seedling preparation process:

In seed preparation the seeds are scattered in the nursery. After the seedlings mature they are transplanted. During transplantation, seedlings from the nursery are taken after 24 days and planted again with proper spacing in between them for better yield. Organic insecticides are also sprayed to prevent insects. After 140 days of growth, the crops are ready to get harvested. The harvesting process could be manual. Sometimes the farmers also use machines to reduce the effort and time involved in the process.

3.3 Traditional / hand transplanting:

Hand transplanting is mostly practiced in India. For manual transplanting a person has to stand in puddled field and bend for long time for putting rice seedlings into soil. This process is highly labor intensive, tedious and tiresome. For manual transplanting (nursery raising, uprooting of the seedling and transplanting) nearly 250- 300 man-hours are required for 1 hectare. The transplanting operation involves the uprooting of 20 to 25 days old seedlings and transplanting them (2-3 seedlings per hill) in well puddled and leveled field. For manual transplanting the workers change their posture very frequently. This may cause suffering from musculoskeletal disorder.

3.4 Mechanized method:

Looking towards the labor shortage in the farm operations, government promotes mechanization in all the possible way to make the farming profitable. The farmers have come forward to adopt transplanting with the help of self-propelled rice transplanter. Due to small land holding and weak economic position, farmers are not in a position to purchase the machine individually, but on hiring basis the technology should be adopted.

4. Transplantation Methods and Types

4.1 Legowo method:

Legowo system is a way of planting rice, if there are several lines and then interspersed with 1 blank line. But at the very edge of the row, 2x seeds are planted more than the middle row. For this way of planting, usually in the agricultural world it is called legowo (2: 1), (3: 1), (4: 1), (5: 1), (6: 1) and so on.

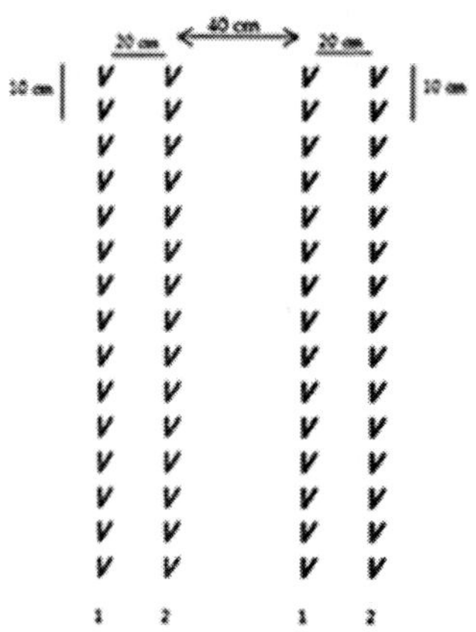

Figure 3. Legowo method

Figure 4. System of Rice Intensification

4.2 System of Rice Intensification (SRI):

System of Rice Intensification is not a secure technological method. Different set of methodologies are used to increase productivity by changing the seeds, nutrients, water, land, and human labour.

5. Design Process

Paddy seedling, planting unit, power transmission and attachments are the vital mechanisms of Rice Transplanting machines.

Planting Unit:

In mechanism design the following parameters are plays the vital role.

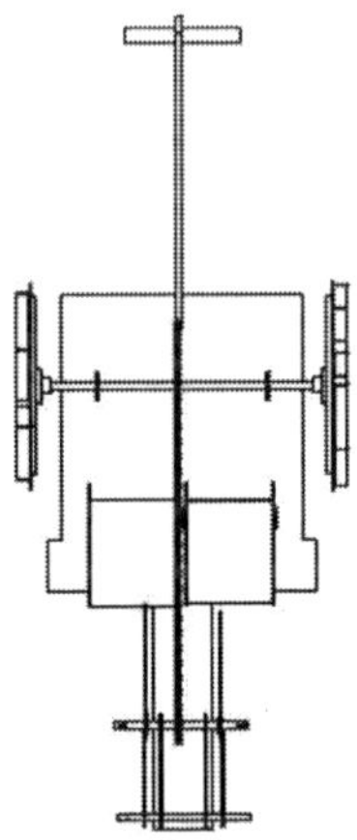

Figure 5. Top View

Depth of planting -. Planting depth is the important parameter in root growth. In this machine the depth was set 50 millimeter below the ground level.

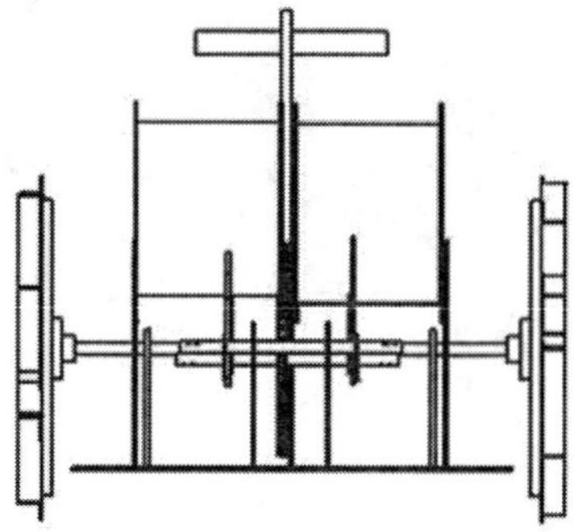

Figure 6. Front View

Plant catching mechanism - Four-bar linkage mechanism adapted to obtain the required motion. While designing the mechanism there are several parameters were considered (location of catching, distance, No. of plants, angle of the tray).
Moving pathway- The required pathway was obtained by four bar mechanism
Designing of Tray: Speed of tray movement, width and length are the varies factors considered in designing of tray mechanism. To make constant feeding of the seed mat to the planting arm it should come down to the end of the try by gravity.

6. Fabrication Work

There are different types of rice transplanting machine that are fabricated based on the sand. Here we have a rice transplanting machine that can be affordable by all farmers. We have designed and fabricated a special type of wheel that could be able to move in wet land. The wheels are attached to the shaft which makes the wheel to rotate in same direction. The shaft is attached to the base plate through the shaft support plate.

The driver sprocket is fixed at the center of the shaft. The tray is fabricated in such a way that, in which the four bar mechanism pick and place the rice in the wet mud. The tray is attached to plate through the tray support. The tray support is welded to the base plate. The tray is kept in 60-degree angle.

Figure 7. Transplanting machine

We have designed a special planting arm and the planting arm are connected by the shaft. The driven sprocket is fixed at the center of the sprocket. The chain is used to connect both the sprocket. When the wheel rotates the power transferred from driver sprocket to driven sprocket. The driven sprocket makes the planting to pick the rice and place it in wet mud.

7. Conclusion

In paddy field to improve the effectiveness and efficiency of rice cultivation there is huge demand for automation. Cost and complex design of the mechanism are the major drawback of existing transplanter. It was difficult to do the service and other maintenance in the conventional shop. It leads to develop the rice transplanter for the farmers who are affected with health related issues. Mechanized rice transplanter helpful to the farmer possess land of small size which reduce the involvement of labour in transplantation.

References

[1] M Darmawan,, "Analysis Of Legowo Row Planting System and system Of Rice Intensification (Sri) Of Paddy Field (Oryza Sativa L.) Toward Growth And Production,2016, *Agrotech Journal*, **1**, 14-18.

[2] Vineet Sharma Shashank Singh Rakesh Kumar, "Comparative study on mechanical transplanting and direct seeding of rice," 2014, Institute of Agricultural Engineering, *Scientific Journal of Agricultural Engineering*, **4,** no. 2.

[3] S. Pradhan and S.K. Mohanty (2014), ErgoEconomical Analysis of Different Paddy Transplanting Operations in Eastern India, *IOSR Journal of Agriculture and Veterinary Science* ,**6**, 23-2

[4] G. Singh, T. R. Sharma, C.W. Bockhop (1985), Field Performance Evaluation of a Manual Rice Transplanter, *Journal of agricultural Engineering Research*,**32**, 259268

[5] RajvirYadav, Mital Patel, S.P. Shukla and S. Pund (2007), Ergonomic evaluation of manually operated six-row paddy transplanter, *International Agricultural Engineering Journal* , **16** (3-4),147-157

[6] Sharma, S. C. and T. P. Singh (2004). Pantnagar developed a mat type nursery raising device for rice transplanter. *Agril. Engng. Today*, **28**(1-2): 70-72.

[7] M. Kamaraj, Akshay Kumar Chhabria, Kartick Kumar, Nishant Kumar,2017 "Design and Fabrication of MultiPurpose Farming ToolsEquipped", *International Journal of Innovative Research in Advanced Engineering (IJIRAE)*,05, **4**, ISSN: 2349-2163.

[8] M.V.Achutha, Sharath Chandra, Nataraj.G.K.,2016 "Concept Design and Analysis of Multipurpose Farm Equipment", International Journal of Innovative Research in Advanced Engineering (IJIRAE) 02, **3,** ISSN: 2349- 2763.

[9] Girish and Srihari,2015 "Design and fabrication of multipurpose farm equipment", *International Journal for Scientific Research & Development*, **4**, 06.

[10] Dr. C.N.Sakhale, Prof. S.N.Waghmare, "Multipurpose Farm Machine", *International Research Journal of Engineering and Technology (IRJET)* **03**, 09, e-ISSN: 2395 -0056.

IOP Conf. Series: Materials Science and Engineering 1059 (2021) 012065 doi:10.1088/1757-899X/1059/1/012065

A comprehensive review on solar cooker tracking techniques for performance enhancement

V Gokul[1], S Dhinesh[1], K Kalaiyarasu[1], B Kalidasan[2] and M Ravikumar[3]

[1] Department of Mechanical Engineering, Bannari Amman Institute of Technology, Sathyamangalam, Tamilnadu-638401

[2] Assistant Professor, Department of Mechanical Engineering, Bannari Amman Institute of Technology, Sathyamangalam, Tamilnadu-638401

[3] Professor, Department of Mechanical Engineering, Bannari Amman Institute of Technology, Sathyamangalam, Tamilnadu-638401

E-mail: kalidasancinna@gmail.com

Abstract. The paper provides a review on solar tracking system for solar cooker and its potential in increasing its efficiency. This paper mainly discusses the solar trackers for increasing the work output. This technique assists in increasing the amount of radiation falling on solar cookers. The types of solar tracking based on their degree of freedom are single axis solar tracking system and dual axis solar tracking system and its classification are discussed in this paper. The solar tracker types based on the actuating systems are also discussed. The tracking system suitable for the solar cooking applications is mentioned in this paper.

1. Introduction

In the modern world, the energy requirement for cooking is increasing dramatically. Majority of the staple foods need cooking before consumption. According to the Food and Agriculture Organization (FAO) of the United Nations, of all the world's total energy consumption, the Food sector contributes around 30 %. Cooking using biomass accounts for indoor air pollution resulting due to inefficient cooking practices. According to the World Health Organization (WHO), the impact of human health due to indoor air pollution includes Pneumonia, Stroke, Heart diseases and lung cancer.

Increasing global warming and change in climate has encouraged the use of renewable energy sources. It provides clean energy without harming the environment. Renewable energy has impacted on various fields like transportation, energy generation etc. Of all the renewable energy sources available, solar energy is considered to be the predominant source of all energy resources. According to the Centre of Climate and Energy Solutions, of total renewable energy generation in the United States, Solar energy generation is projected to increase from 11 % in 2017 to 48 % by 2050.

Both light and heat energy of the sun can be extracted to do useful work. For heat energy extraction, numerous devices are available [1] and the efficiency depends on the type of heat collector. Generally, concentrated heat collectors have higher efficiency than non- concentrated heat collectors

[2]. Due to the intermittent nature of the sun, the energy utilization is possible only during the solar hours. To rectify this disadvantage, thermal energy storage systems have been introduced. Different types of thermal energy storage have been proposed [3].

Concentrated heat collectors should be tracked according to the position of the sun to extract maximum solar radiation. Daily solar hours vary according to the latitude and longitude of the Earth. Types of tracking based on degree of freedom and actuating mechanism is discussed in this paper. Modern research works have been carried out in integrating solar cells on heat collectors [4]. Previous review articles on solar cooker are consolidated in Table 1.

Table 1. Recent Review articles

Author & Year	Journal Title	Remarks
Tian and Zhao 2013 [1]	A review of solar collectors and thermal energy storage in solar thermal applications	This review discusses solar collectors, thermal storage materials and heat transfer enhancement technologies.
Omara et al. 2020 [3]	Improving solar cooker performance using phase change materials: A comprehensive review	This review focuses on the different phase change material for application of solar cookers.
Cuce et al. 2013 [6]	A comprehensive review on solar cookers	This review discusses the design, thermodynamic analysis and performance enhancement of solar cookers.
Muthusivagami et al. 2010 [7]	Solar cookers with and without thermal storage—a review	This review paper discusses types of solar cooker.
Nsengiyumva et al. 2018 [11]	Recent advancements and challenges in Solar Tracking Systems (STS): A review	This paper briefly discusses the advancements in tracking systems.
Hafez et al. 2018 [18]	Solar tracking systems: Technologies and trackers drive types–A review	This review focuses on the tracking system for all PV and thermal storage systems.

The intention of this study is to review the solar tracking methods and to decide the optimum type for the specific application of solar cooking since all types of tracking mechanism cannot be used in solar cooking due to design constraints. Overview of solar cookers has been discussed in section 2. Different types of Solar tracking has been discussed in section 3 and suitable tracking for solar cookers are discussed in section 4. Finally, the conclusion is summarized in section 5.

2. Overview of Solar Cooker

A solar cooker uses the energy of direct sunlight to heat or cook the food materials. The principles of solar cooker were proposed by Horace Benedicts de Saussure in 1767. The working of solar cooker and solar cooker characteristics are discussed by Löf, G. O. [5]. There are different types of solar cookers available. There are numerous configurations of solar cookers and is constantly enhanced by researchers and scientists [6]. Therefore the solar cookers were broadly classified into solar cookers with storage and storage cookers without storage [7] as mentioned in Figure 1.

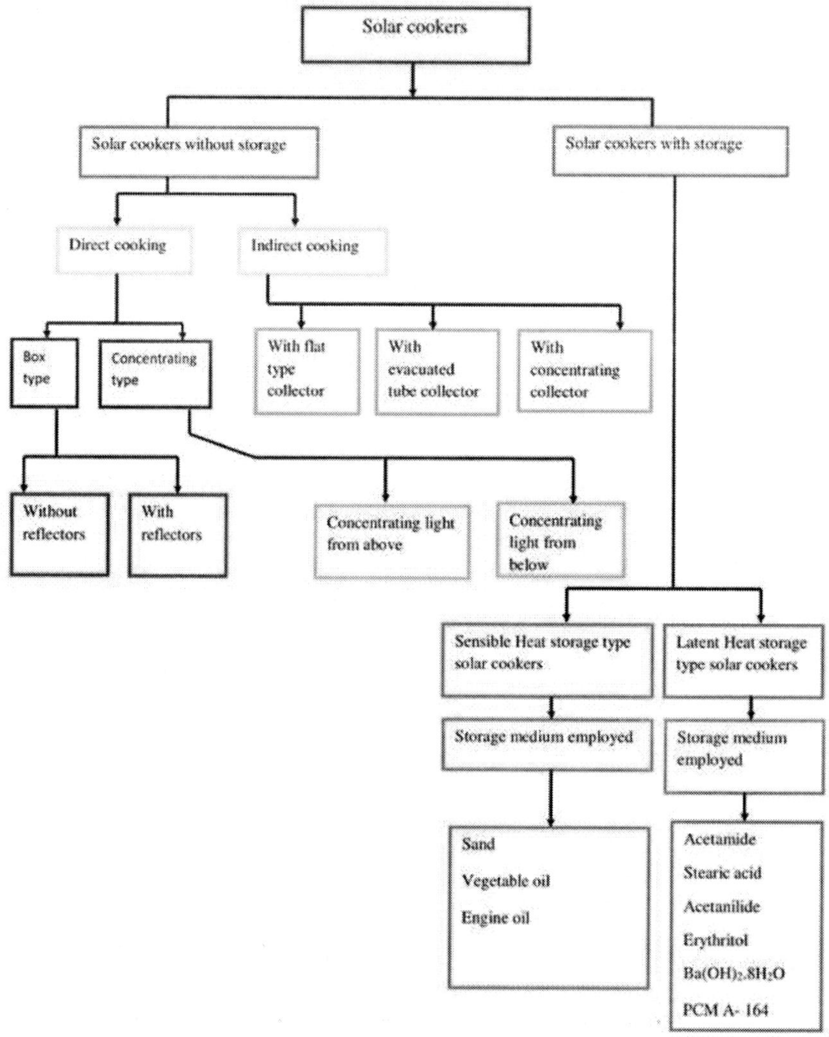

Figure 1. Classification of Solar Cookers [7]

2.1 Solar cookers without Storage

This type of cookers does not contain materials to store thermal energy. The solar radiation incident on these cookers got converted into heat energy for use at the instance. These cookers were divided further into direct and indirect solar cookers depending upon the heat transfer mechanism [8].

2.1.1 Direct solar cookers

Direct solar cookers use the solar radiation without the aid of fluids for heat transfer. Solar radiation is directly used to cook food. The commonly used solar cookers of this type are Box type collectors and concentrating collectors. Box type solar cookers are simple in construction and suitable for limited cooking, since they have relatively low heat capacity, while concentrating solar cookers are highly efficient and reach higher temperatures than box type solar cookers.

Meanwhile concentrating solar collectors face a disadvantage of orienting the apparatus to adjust the solar cooker in position of the focus of the concentrators. Therefore tracking mechanism is needed to continuously track the position of the sun and has been discussed in the section 3.

2.1.2 Indirect Solar cooker

Indirect solar cookers use thermal fluid which is heated by solar radiation and the heat is transported to the place of cooking [9]. Indirect cooking is explained in Figure 2.

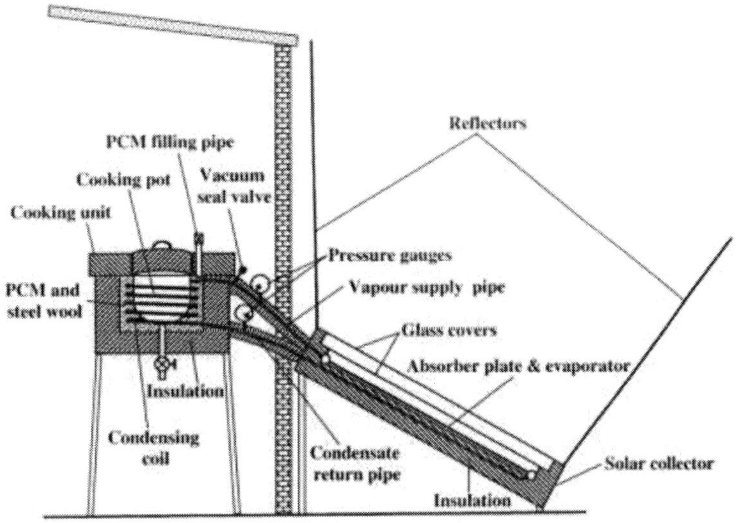

Figure 2. Indirect Solar cooking [9]

Indirect cooking can be accomplished using flat plate collectors, evacuated tube collectors or concentrated collectors.

2.2 solar cooker with thermal storage

The main disadvantage of a solar cooker is that it produces effective work output only during the solar hours. The output is minimized after the solar hours or during the cloudy days. Therefore a solar cooker with thermal energy storage was introduced. These materials store the thermal energy during the solar hours and can be utilized after solar hours. The thermal energy storage system for solar cookers can be further classified into sensible heat storage type solar cookers and latent heat storage type solar cookers. Latent heat energy storage system stores more heat compared to the sensible heat energy storage system [10]. Therefore latent heat energy storage systems have a significant effect on heat storage.

3. Tracking Methods of Solar Cooker

Generally, a solar tracking system adjusts the face of the radiation falling surfaces to follow the movement of the Sun. This increases the efficiency of solar cookers. Classification of solar tracking systems are shown in Figure 3.

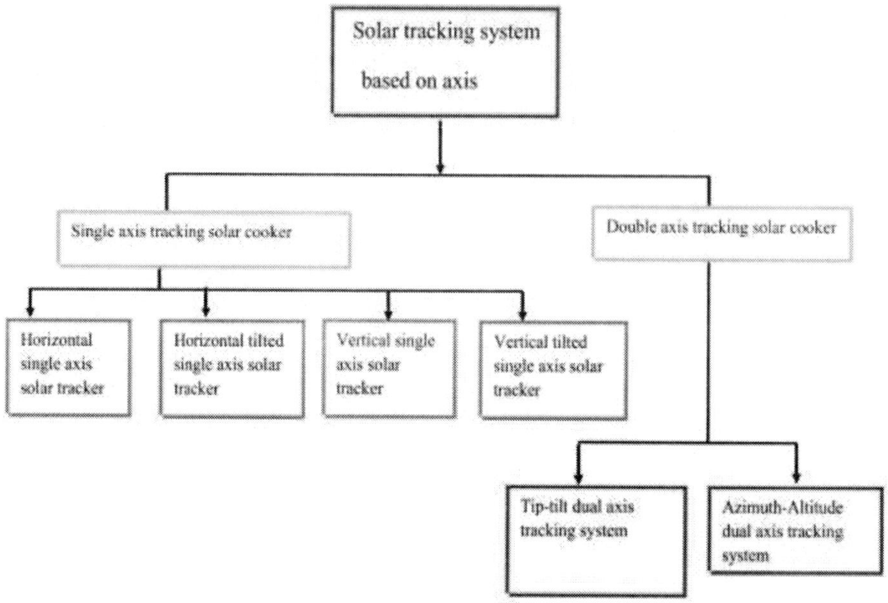

Figure 3. Types of solar trackers based on degree of freedom

3.1 Solar trackers based on number of axis

Based on the number of axes, solar trackers are classified into Single axis tracking and Dual axis tracking [11]. Single axis trackers have one degree of freedom whereas dual axis trackers have two degrees of freedom.

3.1.1 Single axis tracking solar cooker

When movement of the surface happens by rotation about one axis, it is called single-axis tracking. Further, the single axis tracker is classified into Horizontal Single-Axis Solar Tracker (HSAT), Horizontal Tilted Single-Axis Solar Tracker (HTSAT), Vertical Single-Axis Solar Tracker (VSAT), and Vertical-Tilted Single-Axis Solar Tracker (VTSAT) [11]. Zhao et al. [12] simulated the effects of absorber tube installation errors and reflector tracking errors on a single axis parabolic trough collector using a combined method of coordinate transformation and MCRT method. Results showed that with increase in the incident angle, tracking error has slightly more influence on absorber tube flux distribution. As the Concentration ratio increases, the errors become smaller. Sallaberry et al. [13] analysed the parabolic trough collector and discussed a procedure for the estimation of the optical losses due to the positioning angle error in a single-axis solar tracker.

3.1.1.1 Horizontal single axis Solar Tracker

Horizontal single-axis solar tracker rotates on a fixed axis which is parallel to the ground. It tracks the sun throughout the day. This tracker is considered to be the most cost effective method as it requires less material for construction. Wang et al. [14] developed a PLC based automatic Sun tracker for parabolic trough concentrator. The experimental results show that the tracking system has high accuracy and tracking error of the system is less than 0.6°

3.1.1.2Horizontal Tilted Single-Axis Solar Tracker

A similar tracker to HSAT but installed at certain tilt. Tilted-axis tracking systems are complex than HSAT, concrete foundation may require. Since mechanical components are not shared between the trackers, cost per panel cannot be reduced in larger arrays [15].

3.1.1.3 Vertical Single-Axis Solar Tracker

VSAT rotates on a fixed axis which is perpendicular to the ground. These trackers maintain a consistent angle of solar incidence when the Sun is lower in the sky because the profile of VSAT is not parallel to the ground. Farooqui [16] developed power free single axis tracking for solar cookers. The setup consists of stretched spring, water container, bearing and intravenous drip set for draining water. Due to pressure difference, the stretching of spring occurs, therefore tracking occurs. Figure 4 shows the mechanism for vertical single axis tracking system

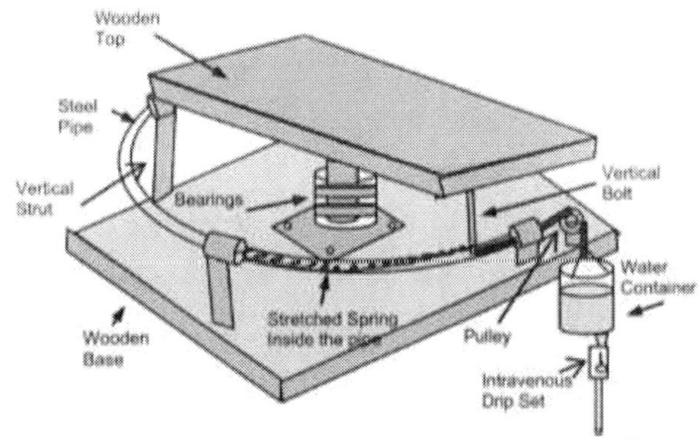

Figure 4. Vertical Single axis tracking system [16]

3.1.1.4 Vertical-Tilted Single-Axis Solar Tracker

These types of trackers are similar to horizontal tilted single-axis trackers. The only difference is that the tilt is parallel to the horizontal position and rotates about the vertical axis [17].

3.1.2 Dual axis tracking solar cooker

When movement of the surface happens by rotation about two axis, it is called dual-axis tracking. Dual axis tracker system has primary axis and secondary axis. Dual-axis tracking can accurately orient the solar device towards the sun. They are more expensive than a single Axis tracking system. Generally there are two types of dual Axis tracking system [18]. They are tip-tilt (TTDAT) and azimuth-altitude (AADAT). Abouziyan [19] investigated the paraboloid and box solar cookers with dual axis trackers on both cookers and showed that paraboloid cookers perform better than box type cookers. Attalage and Reddy et al. [20] proposed a correlation of annual collectible energy of two axis tracking flat plate solar collectors for 26 US locations using simulated results based on typical meteorological year data.

3.1.2.1 Tip-tilt dual axis tracking system

TTDAT has horizontal primary axis and the secondary axis normal to the primary axis and parallel to the solar surface. The primary axis can be shared with other arrays hence reducing installation cost.

3.1.2.2 Azimuth- Altitude dual axis tracking system

AADAT has its primary Axis vertical to the ground. Secondary axis is called as elevation axis. It is normal to the primary axis. The advantage of these type of tracking is that the weight is equally distributed around the pole. Gaafar et al. [21] designed and developed dual axis tracking system for parabolic trough collector and tested under two conditions: with and without solar tracker. The average percentage of the thermal gain increase of the Receiver with Two-Axis Tracking is 50.59% during the Flow Test and 54.25% during the No-Flow test. Chong and Wong [22] derived a General formula for on-axis sun-tracking system using coordinate transformation method to improve the tracking accuracy.

3.2 Types of drive in Solar Tracking

There are five classifications based on their tracking technology [18]. They are active tracking, passive tracking, semi-passive tracking, manual tracking, and chronological tracking.

3.2.1 Active Tracking

Active trackers use motors and gear to direct the tracker towards the sun by a controller responding to the solar direction. Parthipan et al. [23] designed one axis three position solar tracking system for paraboloidal dish collector. Tracking system with single-axis freedom can increase output by approximately 20%, whereas the tracking system with double axis freedom can increase the output by 40%. The system was designed to lower the cost of tracking. Grass et al. [24] compared the optics of non-tracking and novel types of tracking solar thermal collectors for process heat applications. Salawu et al. [25] proposed the method of microprocessor controlled solar tracking system and eliminated hunting, and usual problem with appropriate motor circuit design as shown in Figure 5.

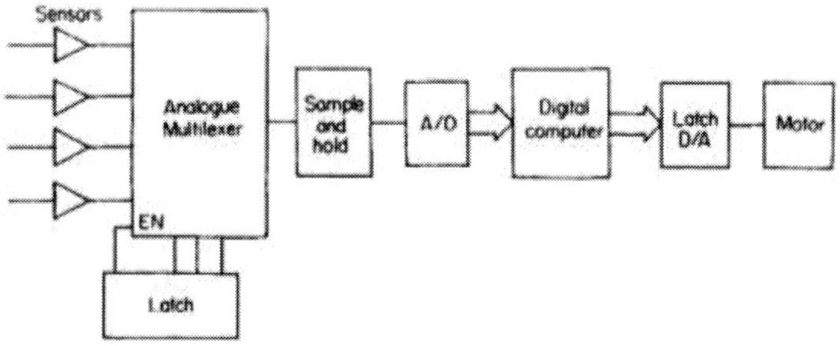

Figure 5. Block diagram of solar tracker [25]

3.2.2 Passive Tracking

A passive solar tracker follows the principles of thermal expansion and uses a compressed gas fluid with low boiling point that rotates to the other side to make the tracker move in response to an imbalance in pressure. Farooqui [26] designed a mechanism for one-dimensional tracking along the azimuth of box type solar cookers to eliminate the need of usual manual solar tracking using gravitational potential energy stored in springs.

3.3.3 Semi Passive Tracking

Semi passive tracking system uses a micro heliostat array, fresnel lens and receiver to track the position of the sun with minimal movement and mechanical efforts. Noel et al. [27] discussed the design process of a semi- passive tracking system. Tracking is performed by the rotating prisms arrays, which are lightweight individual prisms instead of big heliostat.

3.3.4 Manual Tracking

Manual solar tracker is a method where the tilt angle is adjusted manually. This method is a cost effective method and simple in construction. One of the significant applications is that the manual tracking is applied to the secondary axis of a tracker which requires seasonal tilting compared to the primary axis which requires daily tilting. Mwithiga et al. [28] designed a solar dryer and manually adjusted the angle according to the position of the sun. Results showed that Drying of parchment coffee could be reduced to 2–3 days as compared to sun drying which takes 5–7 days.

4. Tracking Techniques of Solar Cooker

4.1 Tracking method used for solar cooker based on axis

Since various tracking methods are available, horizontal single Axis tracking is widely used in single axis solar cookers, azimuth altitude trackers are used in dual axis solar cookers. Akayleh et al. [17] fabricated a parabolic trough collector working on a clockwork mechanism consisting of a gearbox and control unit. Results showed that the ability of the tracking system to produce bread is 3-5 times higher than the fixed system. Further, the proposed system costs 970 Euros while without tracking costs 380 Euros. Theban et al. [15] compared non-tracking and HSAT parabolic solar cooker and achieved operational efficiency of 31% with tracking mechanism involved. Figure 6 shows the solar cooking dish with and without tracking system

Figure 6. Non-tracking (left) and HSAT (right) [15]

Wang et al. [14] used PLC based Automatic Sun tracker for parabolic trough collector and achieved tracking error less than 0.6° This system is reliable and does not depend on weather during operation. The overview structure of PLC is shown in Figure 7.

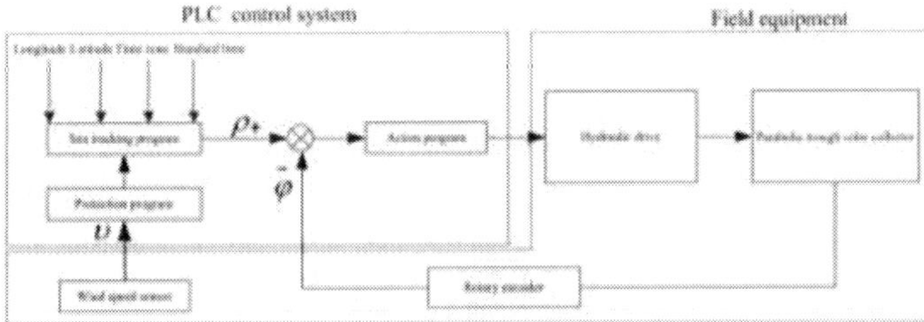

Figure 7. Overall Structure of PLC [14]

Bhave et al. [29] developed a solar cooker with thermal energy storage system using salt hydrate. The system consists of dual axis azimuth altitude trackers with manual control. Nuwayhid et al. [38] fabricated a simple dual axis azimuth altitude tracking system parabolic solar cooker for educational purpose.

Agarwal [30] designed and fabricated azimuth altitude dual axis solar tracker using four bar mechanism. The tracker consists of motor and worm gear drives for actuation. Figure 8 shows the tracking mechanism with four bar linkage.

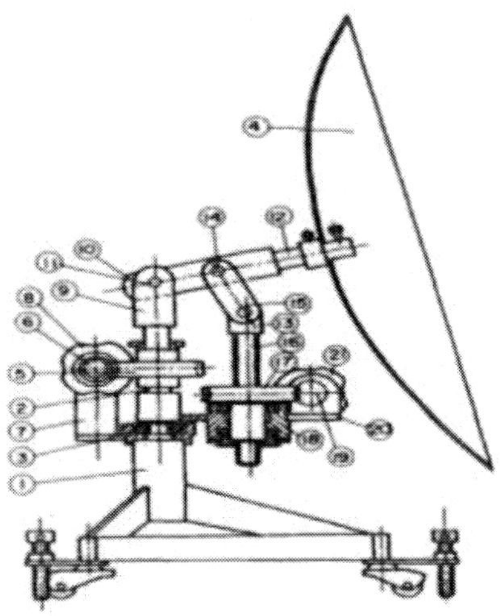

Figure 8. Tracking mechanism consisting of four bar linkage [30]

4.2 Tracking mechanism based on actuating mechanism for solar cookers.

Since all the discussed mechanism can be applicable for solar cookers, most experiments were carried out using active trackers. Skouri et al. [31] used LDR sensors, PIC microcontroller and motors to orient the parabolic solar cooker towards the sun. Figure 9 shows the parabolic solar dish with tracking system.

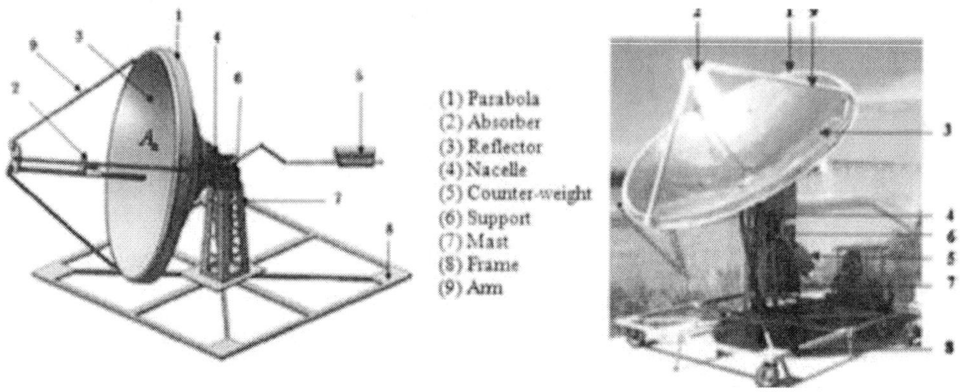

Figure 9. Parabolic tracking [31]

Li et al. [34] developed a semi passive type solar cooker, which uses a lens to concentrate the sunlight to the cooking surface. The system possesses two motors one for controlling the lens position and other for controlling the azimuth. The experimental set up of solar cooker is shown in Figure 10.

Figure 10. Tracker using concentrating lens [34]

Misra and Aseri [35] new approach for thermal performance improvement of solar cookers. He compared the natural and forced heat transfer to the utensil. In his experiment a box type cooker with manual tracking is used which is depicted in Figure 11.

IOP Conf. Series: Materials Science and Engineering 1059 (2021) 012065 doi:10.1088/1757-899X/1059/1/012065

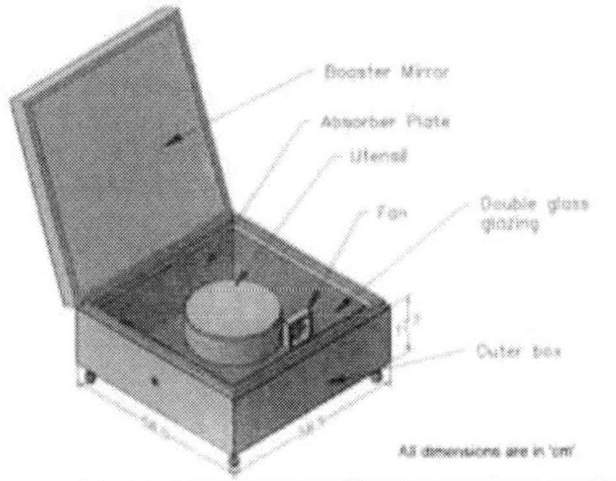

Figure 11. Box type solar cooker with manual tracking [35]

Different tracking system and the type of solar cooker used by researchers are consolidated in Table 2.

Table 2. Types of tracking system Opted for Solar Cookers

Authors	Year	Country & Location	Solar tracking method	Solar tracking mode	Solar tracking type	Solar Cooker Type
Bhave et al. [29]	2018	India	Dual	Manual	Azimuth altitude	Parabolic solar cooker
Ali S. Akayleh [17]	2018	Tafila, Jordan	Single	Active	Clockwork mechanism	Parabolic trough
Mbodji and Hajji [36]	2016	Morocco	Dual	Active	Azimuth altitude	Parabolic solar cooker
Parthipan et al. [23]	2016	India	Single, Dual	Active		Paraboloidal dish
Theebhan et al. [15]	2016	Malaysia	Single	Active	Horizontal	Parabolic dish
Wang et al. [14]	2016	China	Single	Active	Horizontal	Parabolic trough collector
Skouri et al. [31]	2016	Tunisia	Dual	Active	Azimuth altitude	Parabolic solar concentrator
Gaafar and Zobaa [21]	2016	UK	Dual	Active	Azimuth and altitude	Parabolic trough collector
Farooqui [16]	2015	Pakistan	Single	Passive	Vertical	Box type solar cooker

Sallaberry et al. [13]	2015	Spain	Single	Active	Horizontal	Parabolic trough collector
Lwin and win [32]	2014	Myanmar	Dual	Active	Azimuth-tilt	Parabolic dish
Okafor [38]	2013	Nigeria	Dual	Manual	Azimuth altitude	Parabolic solar cooker
Li et al. [34]	2013	USA	Dual	Semi passive	Azimuth altitude	Box type
Farooqui [26]	2013	Pakistan	Single	Passive	Novel one dimensional	Box type solar cooker
Misra and Aseri [35]	2012	India	Single	Manual	Vertical	Box type solar cooker
Abu-Malouh et al. [33]	2011	Jordan	Dual	Active	Two axis tracking	A parabolic solar cooker
Nuwayhid et al. [37]	2001	Lebanon	Dual	Active	Azimuth altitude tracking	Parabolic solar cooker
Abou-Ziyan [19]	1998	Egypt	Single, Dual	Active	Azimuth elevation tracking	Parabolic dish solar cooker
Agarwal [30]	1992	India	Dual	Active Manual	Altitude Azimuth	Parabolic solar cooker

5. Conclusion

The current review article consolidates the research work carried out to improve the performance of solar cookers by tracking technology. Among the tracking technologies, based on axis, dual axis tracking provides better efficiency. For cost is concerned, single axis Tracking proves to be effective. Based on actuating mechanism, active trackers for primary axis and manual trackers for secondary axis is the efficient way. Though numerous works have been carried out, the effect of wind in performance of tracking based solar cookers are yet to be explored. Performance improvement of solar cooker would tend to replace the existing conventional method of cooking, thereby would reduce the emission of polluting gas.

References

[1] Tian Y and Zhao C Y 2013 A review of solar collectors and thermal energy storage in solar thermal applications *Applied energy* **104** 538-53.

[2] Zambolin E and Del Col D 2010 *Experimental analysis of thermal performance of flat plate and evacuated tube solar collectors in stationary standard and daily conditions*. Solar Energy, **84** (8) 1382-96.

[3] Omara A A, Abuelnuor A A, Mohammed H A, Habibi D and Younis O 2020 Improving solar cooker performance using phase change materials: A comprehensive review *Solar Energy* **207** 539-563.

[4] Bergene T and Løvvik O M 1995 Model calculations on a flat-plate solar heat collector with integrated solar cells *Solar energy* **55 (6)** 453-62.

[5] Löf, G O 1963 Recent investigations in the use of solar energy for cooking *Guide to the Records of the Solar Energy Applications Laboratory and Solar Village*.

[6] Cuce E, and Cuce P M 2013 A comprehensive review on solar cookers *Applied Energy* **102** 1399-421.

[7] Muthusivagami R M, Velraj R and Sethumadhavan R 2010 Solar cookers with and without thermal storage-a review, *Renewable and Sustainable Energy Reviews* **14(2)** 691-701.

[8] Negi B S and Purohit, I 2005 Experimental investigation of a box type solar cooker employing a non-tracking concentrator, *Energy Conversion and Management* **46(4)** 577-604.

[9] Hussein H M S, El-Ghetany H H and Nada S A 2008 Experimental investigation of novel indirect solar cooker with indoor PCM thermal storage and cooking unit, *Energy conversion and management* **49(8)** 2237-46.

[10] Padmaraju S V, Viginesh M and Nallusamy N 2008 Comparative study of sensible and latent heat storage systems integrated with solar water heating units. *Renewable Energy & Power Quality Journal*, **1(6)** 55-60.

[11] Nsengiyumva W, Chen S G, Hu L and Chen X 2018 Recent advancements and challenges in Solar Tracking Systems (STS): A review. *Renewable and Sustainable Energy Reviews*, **81** 250-79.

[12] Zhao D, Xu E, Wang Z, Yu Q, Xu L and Zhu L 2016 Influences of installation and tracking errors on the optical performance of a solar parabolic trough collector *Renew Energy* **94** 197–212.

[13] Sallaberry F, Pujol-Nadal R, Larcher M, Rittmann-Frank M H 2015 Direct tracking error characterization on a single-axis solar tracker *Energy Convers Manag*. **105** 1281–90.

[14] Wang J, Zhang J, Cui Y and Bi X 2016 Design and implementation of PLC-based automatic sun tracking system for parabolic trough solar concentrator *MATEC Web Conf.* **77** 1–4.

[15] Theebhan M, Long C Y and Tiong T T 2016 Design and development of a solar thermal collector with single axis solar tracking mechanism *MATEC Web of Conferences* **78** 01002.

[16] Farooqui, S Z 2015 An improved power free tracking system for box type solar cookers *Solar Energy* **120** 100-103.

[17] Akayleh A S, Al-Soud M S and Al-Jufout S A 2020 Design and development of a solar-based cooker with a mechanical sun tracking system. *International Journal of Ambient Energy*, **41(7)** 808-12.

[18] Hafez A Z, Yousef A M and Harag N M 2018 Solar tracking systems: Technologies and trackers drive types–A review *Renewable and Sustainable Energy Reviews* **91**, 754-782.

[19] Abou-Ziyan HZ 1998 Experimental investigation of tracking paraboloid and box solar cookers under Egyptian environment *Appl Therm Eng*, **18** 1375–94.

[20] Attalage R A, Reddy T A1992 Annual collectible energy of a two-axis tracking flat-plate solar

collector *Solar Energy* **48** 151–5.

[21] Gaafar, Amr E and Ahmed F Zobaa 2016 Economical design of a two-axis tracking system for solar collectors 39-6.

[22] Chong K K, and Wong CW 2009 General formula for on-axis sun-tracking system and its application in improving tracking accuracy of solar collector *Sol Energy* **83** 298–305.

[23] Parthipan J, Nagalingeshwara Raju B, Senthilkumar S 2016 Design of one axis three position solar tracking system for paraboloidal dish solar collectors *Mater Today Proc* **3** 2493–500.

[24] Grass C, Schoelkopf W, Staudacher L, Hacker Z 2004 Comparison of the optics of non tracking and novel types of tracking solar thermal collectors for process heat applications up to 300 °C. *Sol Energy* **76** 207–15.

[25] Salawu R I and Oduyemi TA 1987 A microprocessor controlled solar tracking system *J Microcomput Appl* **10** 55–62

[26] Farooqui S Z 2013 A gravity based tracking system for box type solar cookers *Sol Energy* **2** 62–8.

[27] León N, García H and Ramírez C 2014 Semi-passive solar tracking concentrator *Energy Procedia* **57** 275-84.

[28] Mwithiga, Gikuru and Stephen Njoroge Kigo 2006 Performance of a solar dryer with limited sun tracking capability *Journal of food engineering* **74** 247-52.

[29] Bhave A G, and Thakare K A 2018 Development of a solar thermal storage cum cooking device using salt hydrate. *Solar Energy* **171** 784-9.

[30] Agarwal A K 1992 Two axis tracking system for solar concentrators *Renewable energy* **2(2)** 181-2.

[31] Skouri S, Ali A B H, Bouadila S, Salah M B and Nasrallah S B 2016 Design and construction of sun tracking systems for solar parabolic concentrator displacement. *Renewable and sustainable Energy Reviews*, **60** 1419-29.

[32] Lwin M M and Win N N 2014 Microcontroller based smart solar tracking system *IntJSci Eng Techno Res*. **3** 1845–50.

[33] Abu-Malouh R, Abdallah S and Muslih IM 2011 Design, construction and operation of spherical solar cooker with automatic sun tracking system *Energy Convers Manag.* **52** 615–20

[34] Li P W, Kane P and Mokler M 2013 Modeling of solar tracking for giant Fresnel lens solar stoves. *Solar energy.* **96** 263-73.

[35] Misra R, and Aseri T K 2012 Thermal performance enhancement of box-type solar cooker: a new approach. *International Journal of Sustainable Energy* **31(2)** 107-18.

[36] Mbodji N and Hajji A 2016 Performance testing of a parabolic solar concentrator for solar cooking. *Journal of Solar Energy Engineering* **138 (4)** 1-10.

[37] Nuwayhid R Y, Mrad F and Abu-Said R 2001 The realization of a simple solar tracking concentrator for university research applications *Renewable Energy* **24 (2)** 207-22.

Investigation on Boiler Grade Tubes and Tube Plate Using FWTPET Process assisted with PCA

A Daniel Das[1]*,K Thirunavukkarasu[2]

[1]Assistant Professor, Karpagam Academy of Higher Education, Coimbatore – 641 021, Tamilnadu, India
[2]Senior Professor, Karpagam Academy of Higher Education, Coimbatore – 641 021, Tamilnadu, India

Corresponding author: danielamaladas@gmail.com

Abstract. Friction Welding of Tube to Tube Plate using an External Tool (FWTPET) is also used to fuse materials from steel in this study. This inquiry involves the existence and absence of a backrest of the SA 213 boiler-grading tube to the SA 387 tube base in FWTPET. Taguchi L9 orthogonal array and Analysis of Variance (ANOVA) are used to determine the best tensile strength and hardness from the input parameter. The Optimization Part has been performed using Principal Component Analysis (PCA) in the welding process parameter. Moreover, both the welds' interface that can create defect-free welds was found in surface morphological studies. The tensile and hardness properties were evaluated, and the values were tabled so that optimum parameters could be calculated.

Keyword: FWTPET Process; Boiler Grade; SA 213 and SA 387; PCA, GA, Taguchi, ANOVA

1 Introduction

Among some techniques for joining materials, the methodology for joining solid material has a deep adoration in the metallurgical field. The current condition is the mechanism by which the Solid-State Metal Assembly solution ultimately attaches plates to cylinders for various uses such as vehicles, aerospace, and machinery. FWTPET process is the modified friction stir welding process used to join tube and tube plate perpendicularly. This process produces defect-free welds.This process was discovered in 2006, which is a significant step forward in metal joining technology. In 2008, further disputes were patented [1].

In comparison with the fusion welding process, there are various advantages after the invention of this FWTPET process [2]. The use of compression forces results in this method of connecting metal. Excellent weld joints were formed by friction stir welding [3] without any cracks and porosities of solidification. The effective regulation of process parameters contributes to the improvement of mechanical and metallurgical properties [4]. Daniel Das investigates the interference process, using compression strength as output parameters. The unit's rotational speed, which generates friction [5-8], dramatically contributes to its optimum efficiency. Since the author uses many computational methods, the classification of unique welding process types has taken a comparative approach. Most scholars assume from the FSW technique that the critical parameter is the device's rotational speed. Therefore, this study has been achieved by measuring the tensile parameters and optimizing the adjusted UTM configuration's conventional and non-traditional parameters. The better sample was analyzed, and the same samples underwent metallurgical analysis.

2 Materials and Methods

The study analyses sample welded with a cast-iron block backrest using the FWTPET process. The same work is compared during the FWTPET process without the use of a backrest. This technology is used to help the sample during the soldering process. Seamless samples of the hybrid grade Ferritic and Austenitic alloy are used in this research to be dissimilar in the tube and plate's shape. A stand-alone tungsten carbide gear, a high-temperature resistant material, is produced for joining tube and plate. Figure 1 displays the dimensions of the instrument.During the FWTPET process, a cast-iron block in the form of a cube was used to support the experimental sort. The job piece was attached to the vice rig. Figure 2 displays the diagrammatic view of the backrest. A cube-formed block is holed at the middle for a certain depth of 30 mm and a diameter of 20.5 mm.

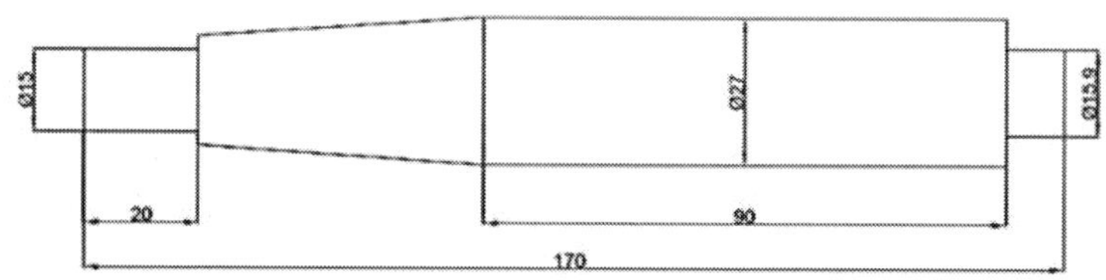

Figure 1 Skeleton Diagram of the FWTPET gear with dimensions

The 0 mm, 1 mm, and 2 mm tube variants are designed for processing purposes. The capital joint's power is determined as the tube projection is the most significant parameter [9-15].

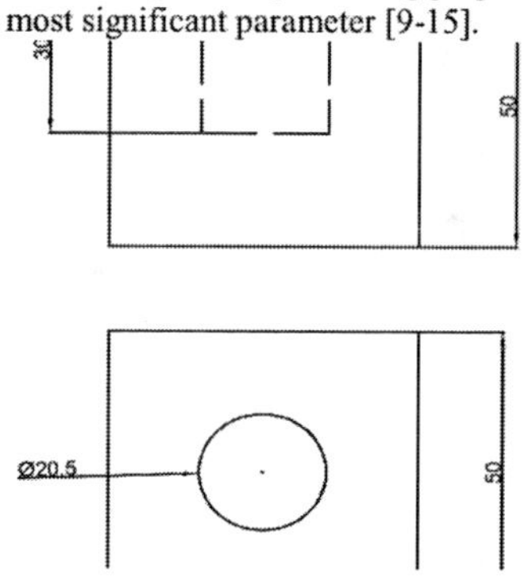

Figure 2 Skeleton diagram of the backrest with dimensions

2.1 Sample Preparation, FWTPET Process, and Microstructure Procedure.

The method FWTPET is a modified Friction Stir process technique. It is a separately updated technique because it was already patentable in 2008. A friction welding machine has a machine vice to carry, and the dimensions shown in Figure 3 are 20.5 mm and 16 mm in the external and internal diameters of the shaft. Besides, the plate measures 50 X 50 and 6 mm wide. The tube and the tube plate are closely fixed perpendicularly. There is a tool pin from Fig 1, which is like a mangrove that supports a knee that restricts the unwanted motion sideways. The pressure is applied by hand; the shoulder rubs the tube plate, which causes friction. The ideal metal bonding was achieved due to extreme plastic deformation. Figure 3 shows that the FWTPET joining strategy uses a different method. Figure 4 indicates that the details of the samples sold using the backrests arrangement [16].

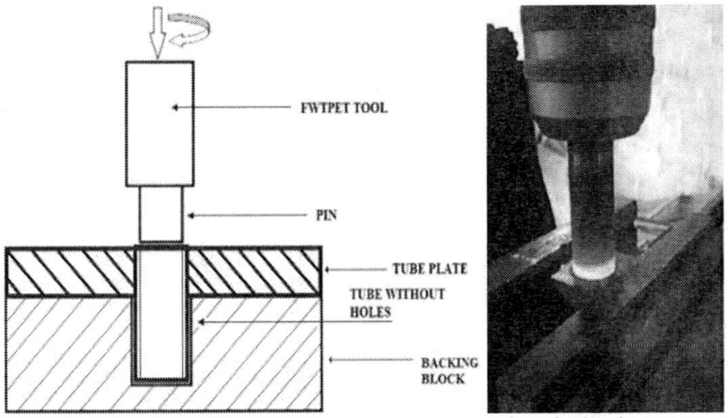

Figure 3 FWTPET Process with samples using the backrest

The samples are segmented into four parts for metallurgical investigation. The structure was determined by introducing the piece into the Nital solution after polishing its surface. Methanol and nitric acid are mixed in desirable amounts with Nital. Using UTM (ASTM E1856 – 13), a tube failure used to evaluate the samples' adhesion strength has been tested with a tensile force. The hardness is tested on the surface of the specimen using Vickers hardness tester (ASTM E92-17). The following values are seen in Table 1 from the exam. Clearance fit is the solid-state metal fitting method that shows the difference between the tube's internal diameter and the pin's outer width in Figure 5.

Figure 4 Welded samples using an arrangement of the backrest and without backrest

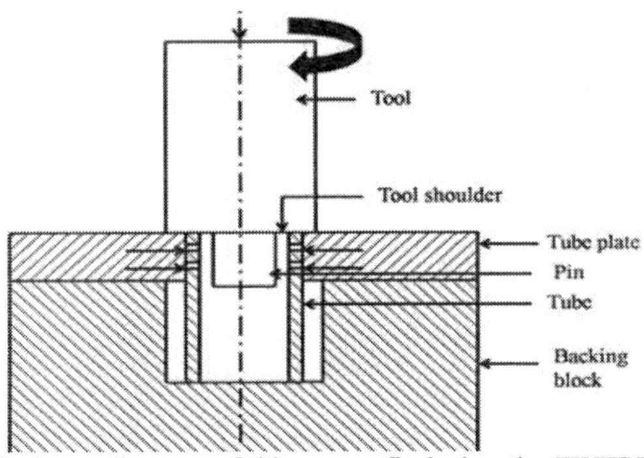

Figure 5 Skeleton diagram of Clearance fit during the FWTPET process.

3 Results and Discussion

3.1 Effects of Tensile Strength and Hardness

Input parameters have welded nine specimens for the orthogonal array. The performance parameter is the intensity of the tractor. Table 1 shows the Tensile Strength validation. The joint capital force generates at 1300 rpm with a depth of 0.5 mm, and a projection of 1 mm by experimental values produces 806.28 MPa with the support of the backrest. The same parameter builds optimum joining strength of 746.5 MPa without the support of the backrest.

Table 1. Tensile Strength validation of FWTPET sample welded using supporting block

Ex.No	Tool Speed (rpm)	Tube Projection (mm)	Depth (mm)	Tensile Strength (MPa)
1	705	0	0.5	597.37
2	705	1	1	550.75
3	705	2	1.5	591.37
4	1005	0	1	665.73
5	1005	1	1.5	726.4
6	1005	2	0.5	709.25
7	1305	0	1.5	719.2
8	**1305**	**1**	**0.5**	**806.28**
9	1305	2	1	726.5

Table 1 and Table 2 show the ranking procedure, in which tube forecasts were a significant parameter for the clearance timing system in generating joint power with and without employing backrest.

Table 2. Tensile Strength validation of FWTPET sample welded without employing supporting block

Ex.No	Tool Speed (rpm)	Tube Projection (mm)	Depth (mm)	Tensile Strength (MPa)
1	705	0	0.5	597.9
2	705	1	1	590.44
3	705	2	1.5	582.56
4	1005	0	1	558.57
5	1005	1	1.5	645.5
6	1005	2	0.5	636.3
7	1305	0	1.5	662.84
8	**1305**	**1**	**0.5**	**746.5**
9	1305	2	1	692.84

Table 3 and Table 4 show the optimum hardness value compared with supporting block and non-supporting block, respectively.

Table 3. Hardness validation of FWTPET sample welded using supporting block

Ex.No	Tool Speed (rpm)	Tube Projection (mm)	Depth (mm)	Tensile Strength (MPa)
1	705	0	0.5	256.25
2	705	1	1	230.6
3	705	2	1.5	242.12
4	1005	0	1	264
5	1005	1	1.5	283.7
6	1005	2	0.5	270
7	1305	0	1.5	279
8	**1305**	**1**	**0.5**	**290**
9	1305	2	1	284

Table 4. Hardness validation of FWTPET sample welded without employing supporting block

Ex.No	Tool Speed (rpm)	Tube Projection (mm)	Depth (mm)	Tensile Strength (MPa)
1	705	0	0.5	257.8
2	705	1	1	240.5
3	705	2	1.5	239.5
4	1005	0	1	236.69
5	1005	1	1.5	259.25
6	1005	2	0.5	258.64
7	1305	0	1.5	260.8
8	**1305**	**1**	**0.5**	**287**
9	1305	2	1	269

3.2 Principal Components Analysis

The tensile strength and hardness levels are validated using principal component analysis (PCA) from this research work. The first principal component value's eigenvalues have been analyzed as 1.9593, and the eigenvalues of the second principal component value are interpreted as 0.0407.

Table 5. Tensile and hardness validation of FWTPET sample welded by employing supporting block using PCA

Actual Value		PCA – Eigen Values Calculated value		MRPI
Tensile Strength (MPa)	Hardness (Hv)	Tensile (MPa)	Hardness (Hv)	
597.37	256.25	298.59	128.09	426.68
550.75	230.6	275.29	115.27	390.56
591.37	242.12	295.60	121.02	416.62
665.73	264	332.76	131.96	464.72
726.4	283.7	363.09	141.81	504.90
709.25	270	354.52	134.96	489.48
719.2	279	359.49	139.46	498.95
806.28	**290**	**403.02**	**144.96**	**547.97**
726.5	284	363.14	141.96	505.10

Table 6. Eigen analysis of the Correlation Matrix

Eigen value	1.9593	0.0407
Proportion	0.980	0.020
Cumulative	0.980	1.000

Table 7. Eigenvectors of the correlation matrix

Variable	PC1	PC2
Tensile Strength (MPa)	0.707	-0.707
Hardness (Hv)	0.707	0.707

The value of proportion for the first and second principal components is 0.980 and 0.020, respectively. Table 5 shows the tensile and hardness validation of the FWTPET sample using PCA with the backrest presence. The Eigen Vector for both Tensile strength and Hardness values are 0.707 each. Table 6 and Table 7 shows the correlation matrix of Eigen Values and Eigen Vectors [17].

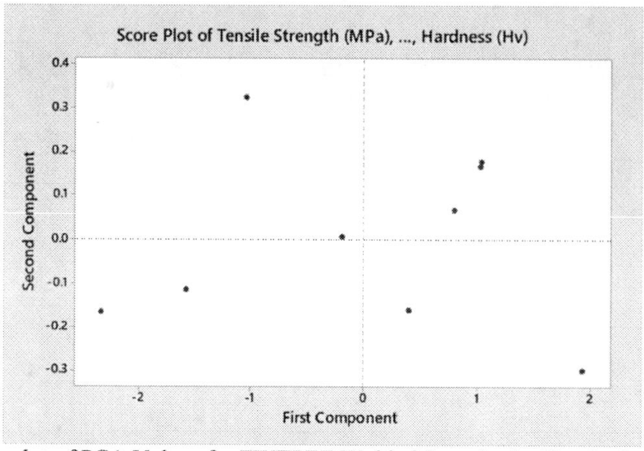

Figure 6. Score plot of PCA Values for FWTPET Welded Samples by Employing backing block

Fig 6 and Fig 7 are the graph plot of scores and scree PCA Values for FWTPET Welded Samples by Employing backing block. From the Score plot, the values are not in the same place, and it scatters to a different location. The exact amount has to be omitted during the time of research. From the scree plot, the first principal component's eigenvalue is higher than the second principal component. The graph plot is plotted between the no of principal components and the eigenvalue [18].

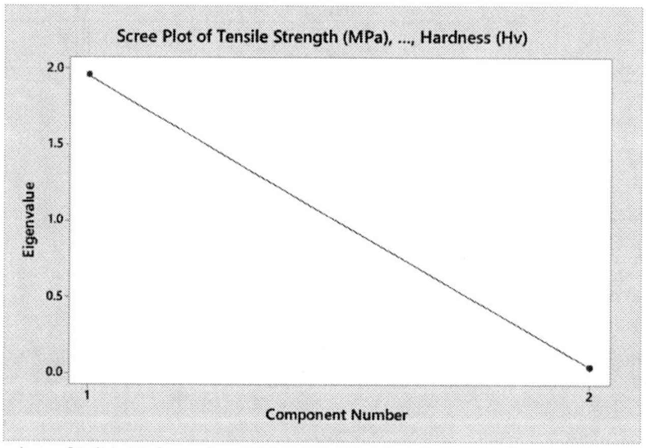

Figure 7 Scree plot of PCA Values for FWTPET Welded Samples by Employing backing block

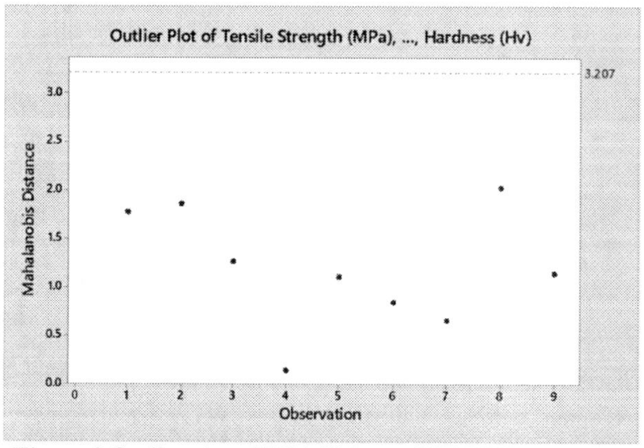

Figure 8 Outlier plot of PCA Values of FWTPET samples.

From Fig 8, it is identified that there is no outlier value. The multi-response performance index is mainly calculated to validate the effects of process parameters and the output using the ranking process. Here the most convincing parameter has been identified at the maximum level of speed (1305 rpm), where are the projection level of the tube is 1 mm and depth has 0.5 mm of the FWTPET sample by the presence of the backing block arrangement [19].

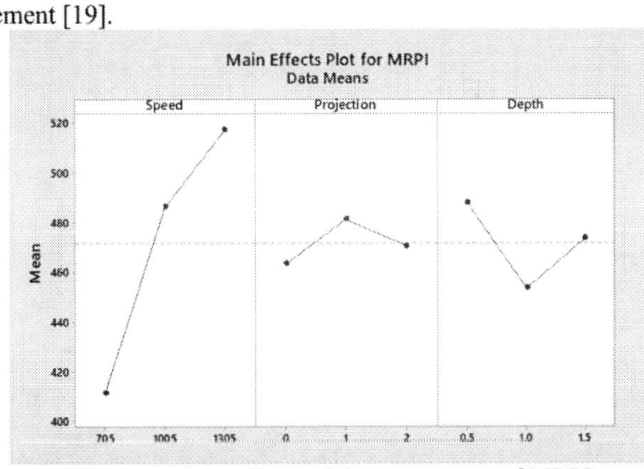

Figure 9 Main effects of the parameters of MRPI

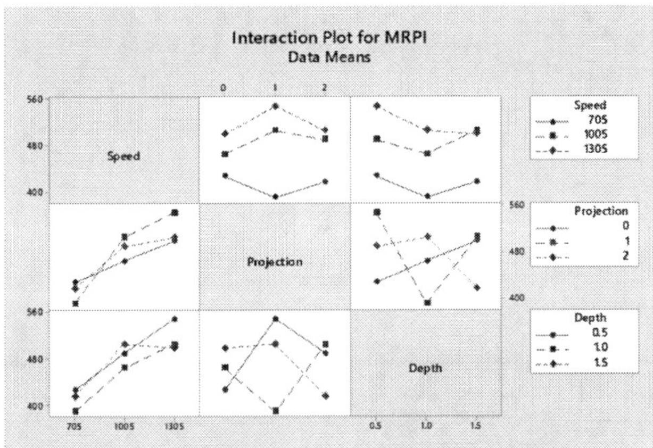

Figure 10 Interaction of parameters of MRPI

Fig 9 and Fig 10 shows the main effect plot of process parameters and interaction plot of the process parameters of the FWTPET samples by the presence of the supporting block. Table 8 shows the tensile and hardness validation of FWTPET samples without supporting block arrangement [20-22]. The first principal component value's Eigenvalues have been analyzed as 1.9593, and the eigenvalues of the second principal component value are interpreted as 0.0407.

Table 8. Tensile and hardness validation of FWTPET sample welded without employing supporting block using PCA

Actual Value		PCA – Eigen Values Calculated value		MRPI
Tensile Strength (MPa)	Hardness (Hv)	Tensile (MPa)	Hardness (Hv)	
597.9	257.8	298.86	128.86	427.72
590.44	240.5	295.13	120.21	415.34
582.56	239.5	291.19	119.71	410.91
558.57	236.69	279.20	118.31	397.51
645.5	259.25	322.65	129.59	452.24
636.3	258.64	318.05	129.28	447.33
662.84	260.8	331.32	130.36	461.68
746.5	**287**	**373.14**	**143.46**	**516.59**
692.84	269	346.32	134.46	480.77

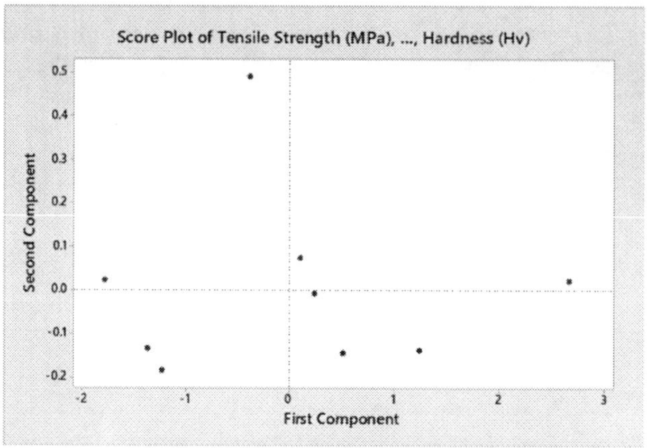

Figure 11. Score plot of PCA Values for FWTPET Welded Samples by the absence of backing block

Fig 11 and Fig 12 are the graph plot of scores and scree PCA Values for FWTPET Welded Samples without employing a backing block. From the Score plot, the values are not in the same place, and it scatters to a different location. The exact amount has to be omitted during the time of research. From the scree plot, the first principal component's eigenvalue is higher than the second principal component. The graph plot is plotted between the no of principal components and the eigenvalue.

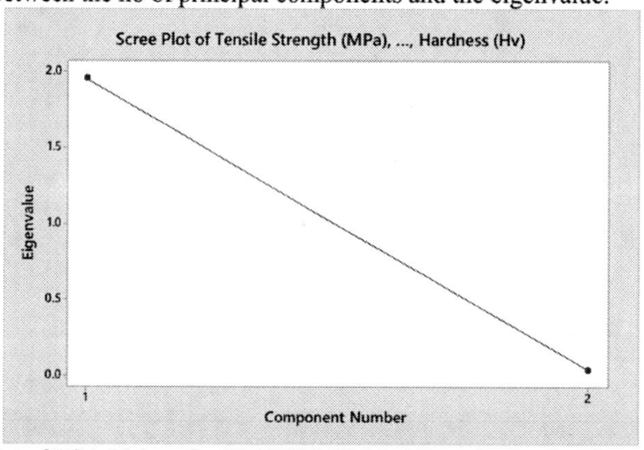

Figure 12. Scree plot of PCA Values for FWTPET Welded Samples by the absence of backing block

From Fig 13, it is identified that there is no outlier value. Here the most convincing parameter has been determined at the maximum level of speed (1305 rpm), where are the projection level of the tube is 1 mm and depth have 0.5 mm of the FWTPET sample by the absence of the backing block arrangement.

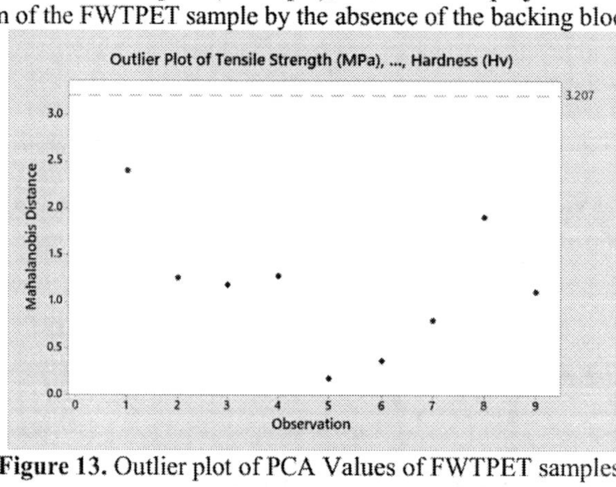

Figure 13. Outlier plot of PCA Values of FWTPET samples.

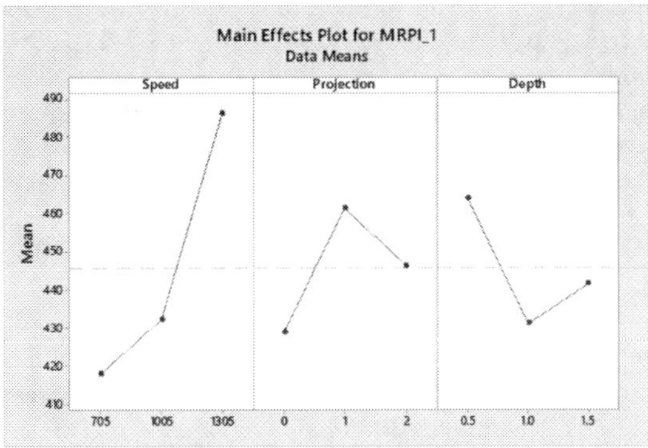

Figure 14. Main effects of the parameters of MRPI (Without Backing block)

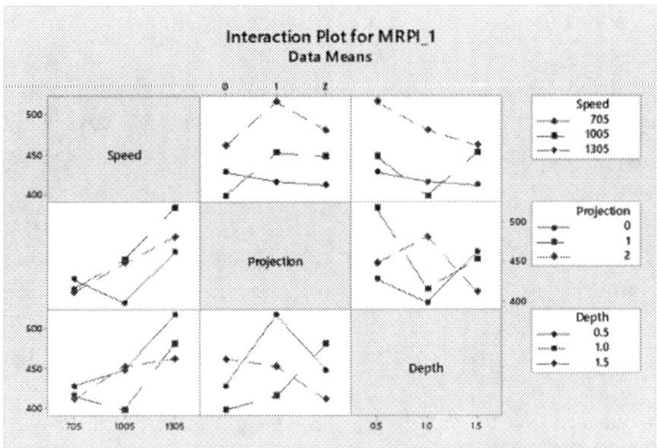

Figure 15. Interaction of parameters of MRPI (Without Backing block)

Fig 14 and Fig 15 shows the main effect plot of process parameters and interaction plot of the process parameters of the FWTPET samples by the presence of supporting block

4 Conclusions

In this research work, the FWTPET samples are welded using a backing block, and the values are compared with the pieces without using a backing block.

- By analyzing the effects of parameters of the models welded using supporting block, the most convincing parameters are 1305 rpm, 0.5 mm depth, and 1 mm tube projection, the output tensile strength is 806.28 MPa, and hardness value are 290 Hv
- By analyzing the effects of the samples welded by the absence of supporting blocks, the most convincing parameters are 1305 rpm, 0.5 mm depth, and 1 mm tube projection; the output tensile strength is 746.5 MPa hardness value is 287 Hv.
- Both Values are validated using Principal Component analysis. The Eigen Vectors of Tensile Strength and Hardness value are 0.707. The first principal component value's eigenvalues have been analyzed as 1.9593 and the eigenvalues of the second principal component value are interpreted as 0.0407 in both cases (With and Without backing block).
- In both cases, the scree plot shows that the first principal components' eigenvalue is greater than the second principal component. From the outlier plot, there are no outlier values.
- By analyzing the MRPI, the main effects plot and interaction show that the optimum tensile and hardness value have been identified using maximum tool rotational speed, optimum tube projection and lower depth of cut.
- The MRPI validation of the sample welded by the presence of the backing block is 547.97, and the MRPI validation of the sample welded by the absence of the backing block is 516

References

[1] Muthukumaran S A process for friction welding tube to a tube sheet or plate by adopting an external tool Indian patent Application No. 189/KOL/06 filed on 07-03-2006, Patent No. 217446, 2006 (granted on 26-03-2008).

[2] Thomas W M, Nicholas E D 1997 Friction-stir welding for the transportation industries *Mater. Des.* **18** 269–73.

[3] Senthil Kumaran S, Muthukumaran S and Vinodh S 2010 Optimization of friction welding of tube to tube plate using an external tool Structural and Multidisciplinary *Optimization.***42** 449-57.

[4] Senthil Kumaran S, Muthukumaran S and Vinodh S Optimization of friction welding of tube to tube plate using an external tool by Taguchi method and genetic algorithm *International Journal of Advanced Manufacturing Technology.* DOI 10.1007/s00170-011-3286-6 (In press).

[5] Kumaran, Senthil S and Daniel Das A 2018 An Examination of Seamless Ferritic tube and Austenitic alloy tube plate joining by the Friction Welding process *Materials Today: Proceedings.***5** 2 8539-46.

[6] Kumaran, Senthil S, and A. Daniel Das 2018 An investigation of Boiler Grade Tube and Tube Plate without block by using friction welding process *Materials Today: Proceedings.* **2** 8567-76.

[7] Kumaran, Senthil S, and Daniel Das A 2018 Friction Welding Joints of SA 213 Tube to SA 387 Tube Plate Boiler Grade Materials by using Clearance and Interference Fit Method *Materials Today: Proceedings.***5** 8557-66.

[8] Sivakumar S, Senthil Kumaran S, Uthayakumar M, and Daniel Das A Garnet and Al-flyash composite under dry sliding conditions *Journal of Composite Materials.* **52** 2281-88.

[9] Daniel Das A., Thirunavukkarasu K 2020 Multi Criteria Decision Making Investigation on Friction Welded Samples of Boiler Grade Materials using Topsis Method *Test Engineering and Management.* **83** 1529-99.

[10] Daniel Das A 2018 Effect of GTAW welding parameters on mechanical properties of aluminium six-series welded samples *Journal of Advanced Research in Dynamical and Control Systems,***11**1112-8.

[11] Daniel Das A 2018 Effect of TIG welding parameters on mechanical properties of Al6063 welded samples *Journal of Advanced Research in Dynamical and Control Systems.***11**1119-25

[12] Karuppusamy, Daniel Das A 2017 Investigations on mechanical properties of squeeze casted Al MMC reinforced with TIC and BN *Journal of Advanced Research in Dynamical and Control Systems.***11** 1086-92

[13] Daniel Das A, Senthilkumaran S 2016 FWTPET Investigation on SA213 Tube to SA387 Tube Plate *Applied Mechanics and Materials.* **852** 355-61.

[14] Das, Daniel A, and Thirunavukkarasu K 2020 Parametric optimization of FWTPET for optimum tensile strength by genetic algorithm approach *Materials Today: Proceedings.* **2** 118-22

[15] Aravind S, and Daniel Das A 2020 An examination on GTAW samples of 7-series aluminium alloy using response surface methodology *Materials Today: Proceedings.* doi.org/10.1016/j.matpr.2020.05.623

[16] Prakash, M, and Daniel Das A 2020 Investigation on effect of FSW parameters of aluminium alloy using Full Factorial Design *Materials Today: Proceedings.* doi.org/10.1016/j.matpr.2020.05.622

[17] Das, Daniel A, and Karuppasamy R 2020 Optimization on hardness values of FSW samples for Al 7–Series & AA 6–Series samples using Taguchi method *Materials Today: Proceedings.* doi.org/10.1016/j.matpr.2020.05.610

[18] Sathishkumar D, and Daniel Das A 2020 Investigations on effect of process parameters on GTAW of aluminium alloy welding using full factorial design technique *Materials Today: Proceedings.* doi.org/10.1016/j.matpr.2020.05.624

Minimization of Performance Error for A Planar Robot By Tolerance Optimization using an Evolutionary Algorithm

Velmurugan S[1] and Sivakumar K[2]

[1&2] Department of Mechanical Engineering, Bannari Amman Institute of Technology, Sathyamangalam – 638 401, Tamilnadu, India

vsvmurugan@gmail.com & sivakumark@bitsathy.ac.in

Abstract

This paper is approached in a different way to consider mixed type of tolerance combinations from the fine (F), medium (M) and coarse (C) type tolerance class applied on the links length of a three-link planar parallel manipulator. The optimal links length and optimal mixed tolerance combination are obtained through Genetic Algorithm (GA) with the minimization of three objectives such as positional error, assembly cost and joint torque. Positional and angular deviations of optimum links length are plotted for the target curve. Finally the variations due to the optimal links length from the true links length is illustrated for a defined trajectory through inverse kinematics.

Keywords: planar robot, genetic algorithm, tolerance analysis

1. Introduction

Flexible tolerance leads the manufacturer to give lesser effort to produce the parts. This will give opportunity to get low cost products. At the same time, customer is not ready to sacrifice the quality on performance by the product. This situation forces the design engineer to optimize the parametric and geometric tolerances for the individual parts to satisfy both customer as well as the manufacturer in terms of their spending cost.

Individual part has its own manufacturing tolerance. During the assembly of parts, the effect of this allocated tolerances impact on the product's performance accuracy. For example, (i) the parametric tolerance ranges of piston, connecting rod, cam, cam shaft, crank, crank shaft and bearings will impact on the swept volume variations [1] (ii) the geometric tolerances of parts of tail stock will impact on the variations on lathe axis [2] (iii) tolerances based on different manufacturing processes like on grinding, turning, welding, etc. will impact on the surface roughness variations on the output product [3] (iv) improper meshing of mating gears due to the parametric and geometric tolerance variations in the shaft, keyway, key, gear tooth [4] (iv) positional error cum workspace variations on planar and spatial mechanisms due to the parametric and geometric tolerances, etc [5]-

[7]. To support the design engineer, more researchers focused on different evolutionary/optimization techniques like genetic algorithm, differntial evolution, ant colony optimization, particle swarm optimization, design of experiments, harmony search, etc. to select optimized tolerance allocations for the individual parts to get effective assembly product with minimum error on its target performance [10].

This paper aims to obtain optimized mixed tolerance combinations with minimum error in the target trajectory by using the famous evolutionary algorithm. The performance variations of obtained optimal links length is illustrated for a line trajectory using inverse kinematics approach.

2. Mixed Tolerance Combinations

A serial three link planar robot with revolute joints is considered for this tolerance analysis. The degrees of freedom of the given planar robot is 3. The following Figure 1 illustrates the robot configuration along with the sample target trajectories such as line and circle trajectories.

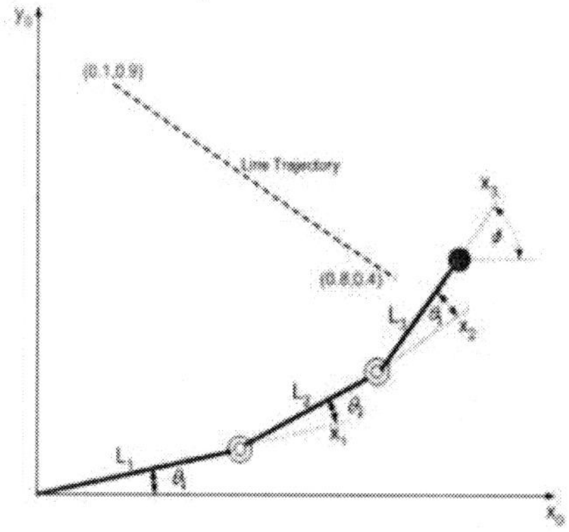

Figure 1. Target trajectories for the Three-link planar robot configuration[8]

The link parameters and the general tolerance for each link are listed in the Table 1. Here the mixing of three types of tolerances like fine (F), medium (M) and Coarse (C) tolerances are considered for analysis for all the three links.

Table 1. Link length, mass and manufacturing tolerance (General Tolerances based on DIN ISO2768)

Link No.	True Link Length (m)	Mass of the True Link (kg)	Fine Tolerance (F) (m)	Medium Tolerance (M) (m)	Coarse Tolerance (C) (m)
L_1	0.4	m_{link1} = 1.0	$\delta L_1 = \pm$ 0.0002	$\delta L_1 = \pm$ 0.0005	$\delta L_1 = \pm$ 0.0012
L_2	0.3	m_{link2} = 0.5	$\delta L_1 = \pm$ 0.0002	$\delta L_1 = \pm$ 0.0005	$\delta L_1 = \pm$ 0.0012
L_3	0.3	m_{link3} = 0.5	$\delta L_1 = \pm$ 0.0002	$\delta L_1 = \pm$ 0.0005	$\delta L_1 = \pm$ 0.0012

The minimum and maximum length of links are listed in Table 2 for the three types of tolerances. The 27 tolerance combinations like FFF, MFF, CFF, FMF, MMF, CMF, FCF, MCF, CCF, FFM, MFM, CFM, FMM, MMM, CMM, FCM, MCM, CCM, FFC, MFC, CFC, FMC, MMC, CMC, FCC, MCC, CCC are possible while considering the mixing up of three types of tolerances and are applied to the links length where a minimum of 127 and a maximum of 15,625 link length combinations possibly to obtain.

Table 2. Minimum and maximum length of each link for the three types of tolerances

Link No.	Min./Max.	Tolerance type		
		Fine (F)	Medium (M)	Coarse (C)

L_1	Min.	0.3998	0.2998	0.2998
	Max.	0.4002	0.3002	0.3002
L_2	Min.	0.3995	0.2995	0.2995
	Max.	0.4005	0.3005	0.3005
L_3	Min.	0.3988	0.2988	0.2988
	Max.	0.4012	0.3012	0.3012

3. Mathematical Modeling & Evolutionary Technique

To minimize the performance error, a combined objective function is to set for the minimization of positional error, assembly cost and joint torque as follows:

3.1 Objective function

The combined objective function (f_c) is defined as follows:
Minimize $f_c = [(w_1z_1 \times N_1) + (w_2z_2 \times N_2) + (w_3z_3 / N_3)]$

Positional Error (z_1) $E_{rms} = \sqrt{\frac{1}{(n+1)}\sum_{i=0}^{n}(E_i^2)}$

Assembly cost (z_2) $a[\exp(-b(L_1 \pm \delta L_1)) + \exp(-b(L_2 \pm \delta L_2)) + \exp(-b(L_3 \pm \delta L_3))]$
(where a = 16.51256 and b = -134.9252)

Joint Torque (z_3) $T = T_1 + T_2 + T_3$

Normalizing factors $N_1 = 1000$
$N_2 = 10$
$N_3 = 10$

Weightage factors $w_1 = 40\%$
$w_2 = 30\%$
$w_3 = 30\%$

The kinematic constraints are

$0° \le \theta i \le 90°$ (i=1,2&3)

$0 \le \dot{\theta}_i \le 4$

$-10 \le \ddot{\theta}_i \le 10$

The dynamic constraints are

$-25 \le \tau_1 \le 45$

$-15 \le \tau_2 \le 20$

$-2 \le \tau_3 \le 5$

The positional error can be obtained by calculating difference between the end effector position for the true links length and tolerated link length. Forward kinematics approach is utilized to find the end effector position (x, y) from the origin.

$$x = L_1 \cos\theta_1 + L_2 \cos(\theta_1 + \theta_2) + L_3 \cos(\theta_1 + \theta_2 + \theta_3))$$

$$y = L_1 \sin\theta_1 + L_2 \sin(\theta_1 + \theta_2) + L_3 \sin(\theta_1 + \theta_2 + \theta_3)$$

The end effector can be assumed to follow the cubic polynomial path. So the orientation of end effector ϕ is as follows:

$\phi = C_0 + C_1 t + C_2 t^2 + C_3 t^3$

The joint torques [9] are explored along with the consideration of manufacturing tolerance variations by the following equations:

For joint 1, the torque T_1 is:

$$
\begin{aligned}
T_1 = {} & \ddot{\theta}_1 \Bigl(I_1 + I_2 + I_3 + M_1\left(\frac{L_1 \pm \delta L_1}{2}\right)^2 + M_2 (L_1 \pm \delta L_1)^2 + M_2\left(\frac{L_2 \pm \delta L_2}{2}\right)^2 + M_3 (L_1 \pm \delta L_1)^2 + M_3 (L_2 \pm \delta L_2)^2 \\
& + M_3\left(\frac{L_3 \pm \delta L_3}{2}\right)^2 + 2M_2 (L_1 \pm \delta L_1)\left(\frac{L_2 \pm \delta L_2}{2}\right)\cos(\theta_2) + 2M_3 (L_1 \pm \delta L_1)(L_2 \pm \delta L_2)\cos(\theta_2) \\
& + 2M_3 (L_2 \pm \delta L_2)\left(\frac{L_3 \pm \delta L_3}{2}\right)\cos(\theta_3) + 2M_3 (L_1 \pm \delta L_1)\left(\frac{L_3 \pm \delta L_3}{2}\right)\cos(\theta_2 + \theta_3) \Bigr) + \\
& \ddot{\theta}_2 \Bigl(I_2 + I_3 + M_2\left(\frac{L_2 \pm \delta L_2}{2}\right)^2 + M_3 (L_2 \pm \delta L_2)^2 + M_3\left(\frac{L_3 \pm \delta L_3}{2}\right)^2 + M_2 (L_1 \pm \delta L_1)\left(\frac{L_2 \pm \delta L_2}{2}\right)\cos(\theta_2) \\
& + M_3 (L_1 \pm \delta L_1)(L_2 \pm \delta L_2)\cos(\theta_2) + 2M_3 (L_2 \pm \delta L_2)\left(\frac{L_3 \pm \delta L_3}{2}\right)\cos(\theta_3) + M_3 (L_1 \pm \delta L_1)\left(\frac{L_3 \pm \delta L_3}{2}\right)\cos(\theta_2 + \theta_3) \Bigr) + \\
& \ddot{\theta}_3 \left(I_3 + M_3\left(\frac{L_3 \pm \delta L_3}{2}\right)^2 + M_3 (L_2 \pm \delta L_2)\left(\frac{L_3 \pm \delta L_3}{2}\right)\cos(\theta_3) + M_3 (L_1 \pm \delta L_1)\left(\frac{L_3 \pm \delta L_3}{2}\right)\cos(\theta_2 + \theta_3) \right) \\
& + \dot{\theta}_1 \dot{\theta}_2 \Bigl(-2M_2 (L_1 \pm \delta L_1)(L_2 \pm \delta L_2)\sin(\theta_2) - 2M_3 (L_1 \pm \delta L_1)(L_2 \pm \delta L_2)\sin(\theta_2) \\
& - 2M_3 (L_1 \pm \delta L_1)\left(\frac{L_3 \pm \delta L_3}{2}\right)\sin(\theta_2 + \theta_3) \Bigr) + \\
& \dot{\theta}_1 \dot{\theta}_3 \left(-2M_3 (L_2 \pm \delta L_2)\left(\frac{L_3 \pm \delta L_3}{2}\right)\sin(\theta_3) - 2M_3 (L_1 \pm \delta L_1)\left(\frac{L_3 \pm \delta L_3}{2}\right)\sin(\theta_2 + \theta_3) \right) \\
& + \dot{\theta}_2 \dot{\theta}_3 \left(-2M_3 (L_2 \pm \delta L_2)\left(\frac{L_3 \pm \delta L_3}{2}\right)\sin(\theta_3) - 2M_3 (L_1 \pm \delta L_1)\left(\frac{L_3 \pm \delta L_3}{2}\right)\sin(\theta_2 + \theta_3) \right) + \\
& \dot{\theta}_2^2 \left(-M_2 (L_1 \pm \delta L_1)\left(\frac{L_2 \pm \delta L_2}{2}\right)\sin(\theta_2) - M_3 (L_1 \pm \delta L_1)(L_2 \pm \delta L_2)\sin(\theta_2) - M_3 (L_1 \pm \delta L_1)\left(\frac{L_3 \pm \delta L_3}{2}\right)\sin(\theta_2 + \theta_3) \right) \\
& + \dot{\theta}_3^2 \left(-M_3 (L_2 \pm \delta L_2)\left(\frac{L_3 \pm \delta L_3}{2}\right)\sin(\theta_3) - M_3 (L_1 \pm \delta L_1)\left(\frac{L_3 \pm \delta L_3}{2}\right)\sin(\theta_2 + \theta_3) \right) + M_1 g\left(\frac{L_1 \pm \delta L_1}{2}\right)\cos(\theta_1) + \\
& M_2 g\left((L_1 \pm \delta L_1)\cos(\theta_1) + \left(\frac{L_2 \pm \delta L_2}{2}\right)\cos(\theta_1 + \theta_2) \right) + M_3 g \Bigl((L_1 \pm \delta L_1)\cos(\theta_1) + \\
& (L_2 \pm \delta L_2)\cos(\theta_1 + \theta_2) + \left(\frac{L_3 \pm \delta L_3}{2}\right)\cos(\theta_1 + \theta_2 + \theta_3) \Bigr)
\end{aligned}
$$

For joint 2, the torque T_2 is:

$$
\begin{aligned}
T_2 = {} & \ddot{\theta}_1 \Bigl(I_2 + I_3 + M_2\left(\frac{L_2 \pm \delta L_2}{2}\right)^2 + M_3 (L_2 \pm \delta L_2)^2 + M_3\left(\frac{L_3 \pm \delta L_3}{2}\right)^2 + M_2 (L_1 \pm \delta L_1)\left(\frac{L_2 \pm \delta L_2}{2}\right)\cos(\theta_2) \\
& + M_3 (L_1 \pm \delta L_1)(L_2 \pm \delta L_2)\cos(\theta_2) + 2M_3 (L_2 \pm \delta L_2)\left(\frac{L_3 \pm \delta L_3}{2}\right)\cos(\theta_3) \\
& + M_3 (L_1 \pm \delta L_1)\left(\frac{L_3 \pm \delta L_3}{2}\right)\cos(\theta_2 + \theta_3) \Bigr) + \\
& \ddot{\theta}_2 \left(I_2 + I_3 + M_2\left(\frac{L_2 \pm \delta L_2}{2}\right)^2 + M_3 (L_2 \pm \delta L_2)^2 + M_3\left(\frac{L_3 \pm \delta L_3}{2}\right)^2 + 2M_3 (L_2 \pm \delta L_2)\left(\frac{L_3 \pm \delta L_3}{2}\right)\cos(\theta_3) \right) + \\
& \ddot{\theta}_3 \left(I_3 + M_3\left(\frac{L_3 \pm \delta L_3}{2}\right)^2 + M_3 (L_2 \pm \delta L_2)\left(\frac{L_3 \pm \delta L_3}{2}\right)\cos(\theta_3) \right) + \dot{\theta}_1 \dot{\theta}_3 \left(-2M_3 (L_2 \pm \delta L_2)\left(\frac{L_3 \pm \delta L_3}{2}\right)\sin(\theta_3) \right) \\
& + \dot{\theta}_2 \dot{\theta}_3 \left(-2M_3 (L_2 \pm \delta L_2)\left(\frac{L_3 \pm \delta L_3}{2}\right)\sin(\theta_3) \right) + \\
& \dot{\theta}_1^2 \left(M_2 (L_1 \pm \delta L_1)\left(\frac{L_2 \pm \delta L_2}{2}\right)\sin(\theta_2) + M_3 (L_1 \pm \delta L_1)(L_2 \pm \delta L_2)\sin(\theta_2) + M_3 (L_1 \pm \delta L_1)\left(\frac{L_3 \pm \delta L_3}{2}\right)\sin(\theta_2 + \theta_3) \right) \\
& + \dot{\theta}_3^2 \left(-M_3 (L_2 \pm \delta L_2)\left(\frac{L_3 \pm \delta L_3}{2}\right)\sin(\theta_3) \right) + M_2 g\left(\left(\frac{L_2 \pm \delta L_2}{2}\right)\cos(\theta_1 + \theta_2) \right) + \\
& M_3 g\left((L_2 \pm \delta L_2)\cos(\theta_1 + \theta_2) + \left(\frac{L_3 \pm \delta L_3}{2}\right)\cos(\theta_1 + \theta_2 + \theta_3) \right)
\end{aligned}
$$

For joint 3, the torque T_3 is:

$$
\begin{aligned}
T_3 = {} & \ddot{\theta}_1 \left(I_3 + M_3\left(\frac{L_3 \pm \delta L_3}{2}\right)^2 + M_3 (L_2 \pm \delta L_2)\left(\frac{L_3 \pm \delta L_3}{2}\right)\cos(\theta_3) + M_3 (L_1 \pm \delta L_1)\left(\frac{L_3 \pm \delta L_3}{2}\right)\cos(\theta_2 + \theta_3) \right) + \\
& \ddot{\theta}_2 \left(I_3 + M_3\left(\frac{L_3 \pm \delta L_3}{2}\right)^2 + M_3 (L_2 \pm \delta L_2)\left(\frac{L_3 \pm \delta L_3}{2}\right)\cos(\theta_3) \right) + \ddot{\theta}_3 \left(I_3 + M_3\left(\frac{L_3 \pm \delta L_3}{2}\right)^2 \right) \\
& + \dot{\theta}_1 \dot{\theta}_2 \left(2M_3 (L_2 \pm \delta L_2)\left(\frac{L_3 \pm \delta L_3}{2}\right)\sin(\theta_3) \right) + \dot{\theta}_1 \dot{\theta}_2 \left(2M_3 (L_2 \pm \delta L_2)\left(\frac{L_3 \pm \delta L_3}{2}\right)\sin(\theta_3) \right) + \\
& \dot{\theta}_1^2 \left(M_3\left(\frac{L_3 \pm \delta L_3}{2}\right)(L_2 \pm \delta L_2)\sin(\theta_3) + M_3 (L_1 \pm \delta L_1)\left(\frac{L_3 \pm \delta L_3}{2}\right)\sin(\theta_2 + \theta_3) \right) \\
& + \dot{\theta}_2^2 \left(M_3 (L_2 \pm \delta L_2)\left(\frac{L_3 \pm \delta L_3}{2}\right)\sin(\theta_3) \right) + M_3 g\left(\left(\frac{L_3 \pm \delta L_3}{2}\right)\cos(\theta_1 + \theta_2 + \theta_3) \right)
\end{aligned}
$$

3.2 Genetic Algorithm (GA)

The GA has its own capability to deal the problems effectively exclusive of inference of connectivity for the primary objective functions. It is a search heuristic stimulated by Charles Darwin's instinctive progression concept. Here the robust genes are opted for reproduction with the intention of producing progeny of the next generation.

The algorithm is written with the population size of 100. Total number of generations are considered as 100. The cross over probability is taken as 1.0000 and the mutation probability is choosen as 0.0500.

4. Results and Discussion

4.1 Forward kinematics

The following figures illustrate the deviations of mixed tolerance links in positional error, assembly cost, individual joint torque, total toruqes and the main objective function (combined errors).

The Figure 2 provides the minimum positional error by the FFC mixed tolerance. If the manufacturer considered only on positional error, then there is a possibility of providing relaxed (coarse) tolerance on the link L_3. At the same time, it is important to provide fine tolerance for the links L_1 and L_2.

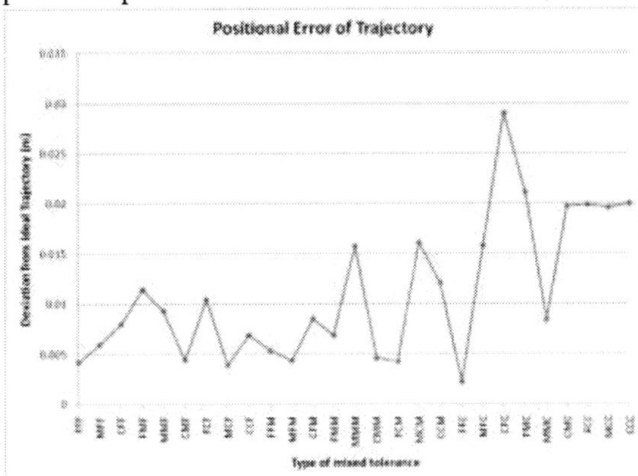

Figure 2. Positional error of the end effector trajectory for various types of mixed tolerances

The Figure 3 shows the minimum assembly cost by the MMF mixed tolerance. Suppose to consider only the assembly cost, it is possibile to provide medium tolerance to the links L_1 and L_2. The link L_3 is forced to be on fine tolerance.

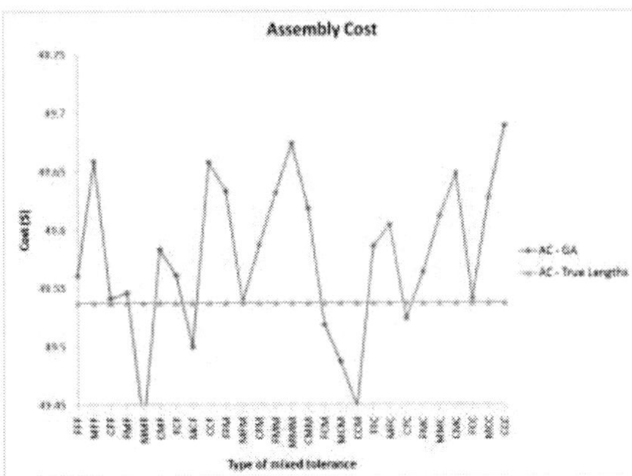

Figure 3. Assembly cost variation for various types of mixed tolerances

The Figures 4-6 shows the individual joint torque errors. During the consideration of total torque error (Figure 7), FCF mixed tolerance provides the minimum error when compared to the other combinations.

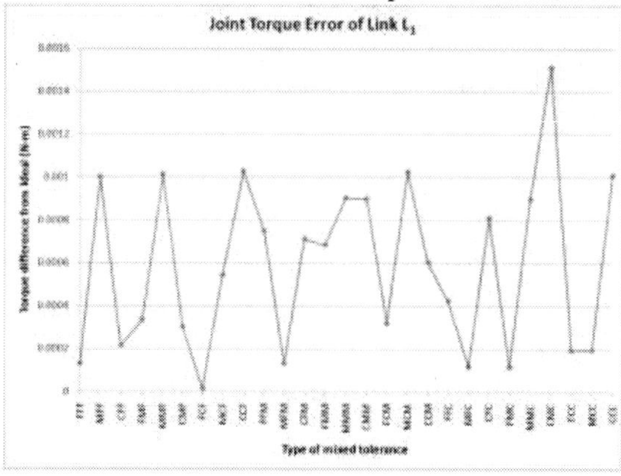

Figure 4. Joint torque error (T_1) of link L_1 for various types of mixed tolerances

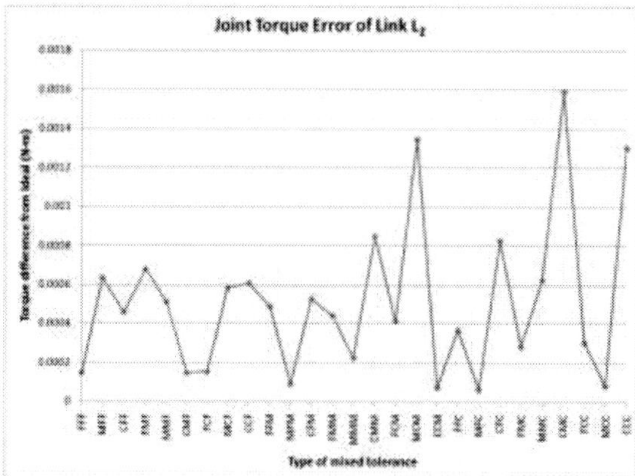

Figure 5. Joint torque error (T_2) of link L_2 for various types of mixed tolerances

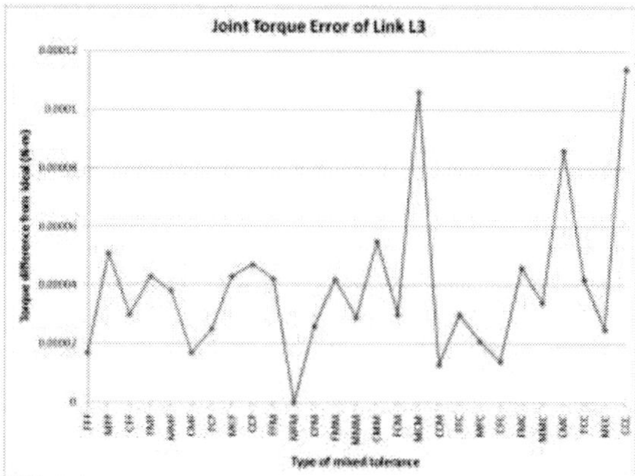

Figure 6. Joint torque error (T_3) of link L_3 for various types of mixed tolerances

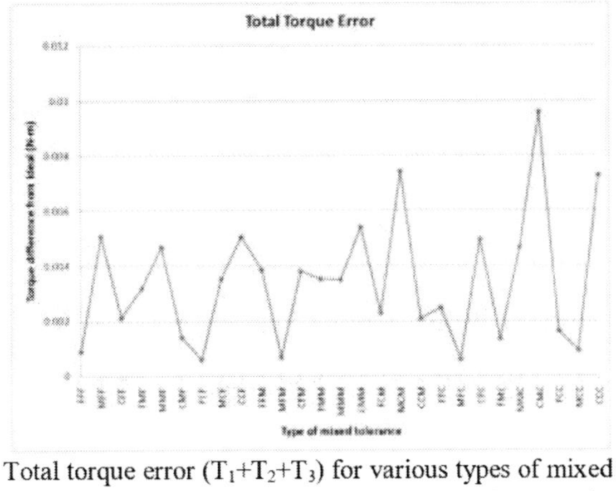

Figure 7. Total torque error (T_1+T_2+T_3) for various types of mixed tolerances

Based on the mathematical modeling for the objective function, the Figure 8 shows the FFC mixed tolerance provides minimum error in all aspects. So, during the consideration of minimum positional error, assembly cost and torque minimization, the FFC combination gives optimum links length.

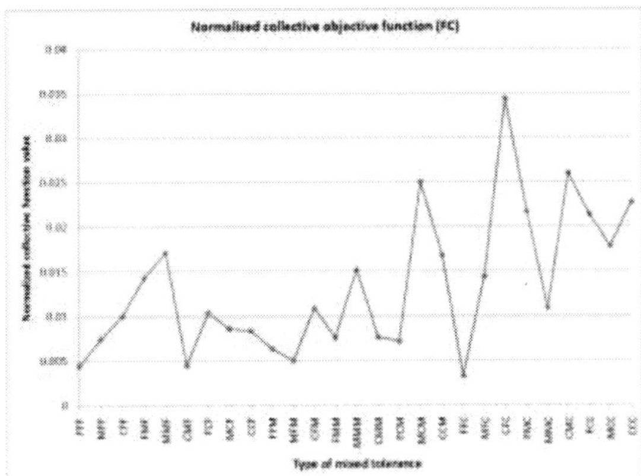

Figure 8. Combined objective function (f_c) variation for various types of mixed tolerances

4.2 Inverse kinematics

The following Figure 9 shows the joint angles of the three links and end effector for the FFC mixed tolerance link lengths using analytical method to follow the given line trajectory.

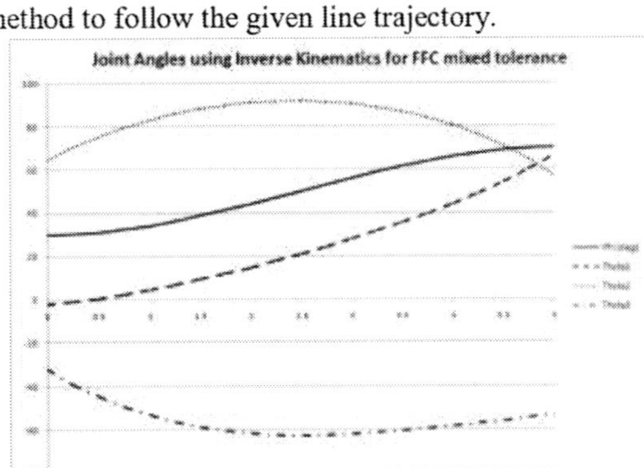

Figure 9. Angular positions variation for true links (Ideal values) vs tolerated links (Actual values) for the FFC mixed tolerance

The simulation is carried out for the line trajectory and the angular deviations of the links are illustrated in the following Figure 10.

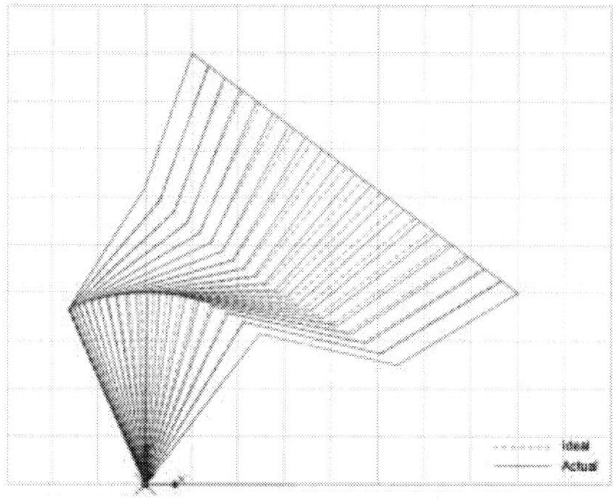

Figure 10. Angular positions variation for true links (Ideal values) vs tolerated links (Actual values) for the FFC mixed tolerance

The following Table 3 listed the angular variation values θ_1, θ_2 and θ_3 for the true links Vs actual (FFC) links while the end effector follows the given line trajectory. The graphical illustration of the same is given in Figure 11.

Table 3. θ_1, θ_2 and θ_3 for the true links Vs actual (FFC) links while following given line trajectory

t	True-Theta1	Actual-Theta1	True-Theta2	Actual-Theta2	True-Theta3	Actual-Theta3
0	51.82948	51.82436	-64.23915	-64.25039	42.40966	42.42603
0.25	57.81719	57.81135	-70.30202	-70.31342	42.77482	42.79207
0.5	63.29893	63.29246	-75.30311	-75.31468	43.12417	43.14221
0.75	68.37784	68.37081	-79.4735	-79.48522	43.52565	43.54442
1	73.11782	73.11028	-82.94699	-82.95887	43.98917	44.0086
1.25	77.56033	77.55233	-85.80154	-85.81357	44.49121	44.51123
1.5	81.73306	81.72468	-88.08125	-88.0934	44.98819	45.00873
1.75	85.65492	85.64622	-89.80921	-89.82147	45.4243	45.44525
2	89.33899	89.33008	-90.99545	-91.00779	45.73646	45.75771
2.25	92.79435	92.78533	-91.6419	-91.65428	45.85755	45.87895
2.5	96.02703	96.018	-91.7454	-91.75778	45.71838	45.73977
2.75	99.04039	99.0315	-91.29917	-91.3115	45.24878	45.27
3	101.83511	101.82647	-90.293	-90.30524	44.37789	44.39876

t	True-Theta1	Actual-Theta1	True-Theta2	Actual-Theta2	True-Theta3	Actual-Theta3
3.25	104.40858	104.40035	-88.7121	-88.7242	43.03351	43.05385
3.5	106.75385	106.74615	-86.53448	-86.5464	41.14063	41.16025
3.75	108.8577	108.85068	-83.72635	-83.73804	38.61864	38.63736
4	110.69742	110.6912	-80.2345	-80.24592	35.37708	35.39471
4.25	112.23529	112.23003	-75.9739	-75.985	31.3086	31.32497
4.5	113.40878	113.40461	-70.80583	-70.81658	26.27705	26.29197
4.75	114.1115	114.10858	-64.49594	-64.50628	20.09444	20.1077
5	114.15125	114.14977	-56.62019	-56.63008	12.46894	12.48031

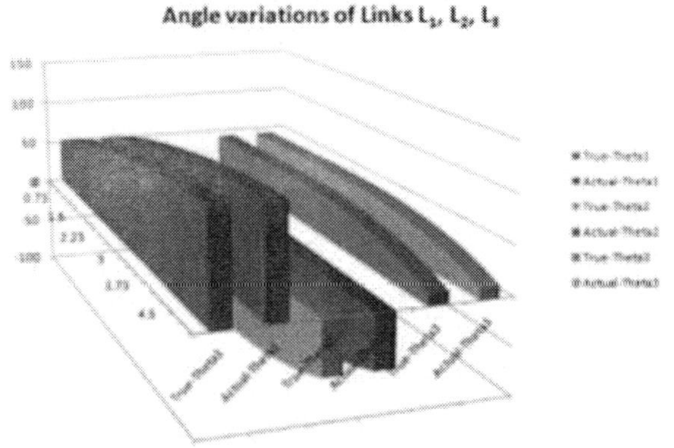

Figure 11. Variations in angles for the true Vs actual links

5. Conclusion

Instead of that applying common tolerance for all the parts, this study creates a way to do the mixed tolerance combination analysis and select optimal tolerance combination. In this three-link mechanism, the link having the end-effector (L_3) of FFC combination provides coarse tolerance. The manufacturing relaxation is provided for this third link and it will leads to reduce the manufacturing cost when compared to the other two links having fine tolerance.

References

[1] Jayaprakash G Sivakumar K and Thilak M 2012 Tolerance design of mechanical assembly using NSGA II and finite element analysis *Journal of mechanical science and technology* **26(10)** 3261-68.

[2] Wallner J Schröcker H P and Hu S M 2005 Tolerances in geometric constraint problems *Reliable computing* **11(3)** 235-51.

[3] Wang P and Liang M 2005 An integrated approach to tolerance synthesis, process selection and machining parameter optimization problems *International journal of production research* **43(11)** 2237-62.

[4] Shen G Xiang D Zhu K Jiang L Shen Y and Li Y 2018 Fatigue failure mechanism of planetary gear train for wind turbine gearbox *Engineering Failure Analysis* **87** 96-110.

[5] Saravanan R Godwin Raja Ebenezer N Ramabalan S Balamurugan C 2012 Evolutionary uncertainty analysis and allocation of joint tolerances in robot manipulators *Journal of Mechatronics and Intelligent Manufacturing*, **1(1)** 1–22.

[6] Ghasemi M H Kashiri N and Dardel M 2012 Time-optimal trajectory planning of robot manipulators in point-to-point motion using an indirect method *Proceedings of the Institution of Mechanical Engineers, Part C: Journal of Mechanical Engineering Science* **226(2)** 473-84.

[7] Nie J Wu X and Yan H 2020 Inverse Kinematics Analysis of Hybrid-Driven Special Five-Bar Mechanism for Shield Machine. *MS&E*, **782(4)** 042043.

[8] Ata A A and Myo T R 2005 Optimal point-to-point trajectory tracking of redundant manipulators using generalized pattern search *International Journal of Advanced Robotic Systems* **2(3)** 24.

[9] Hussain Zakia and Norsinnira Zainul Azlan 2016 Kane's method or dynamic modeling *IEEE International*

Conference on Automatic Control and Intelligent System, **12** 174–79.
[10] Groover M P Weiss M and Nagel R N 1986 Industrial robotics: technology, programming and application *McGraw-Hill Higher Education*.

IOP Conf. Series: Materials Science and Engineering 1059 (2021) 012068 doi:10.1088/1757-899X/1059/1/012068

Thermal Analysis on Regenerative Cooled Liquid Propellant Rocket Thrust Chamber Using ANSYS

Pon.Maheskumar[1], *R.Girimurugan[1], G.Sivaraman[2], S.Purushothaman[3], M.Vairavel[4]

[1]Assistant Professor, Department of Mechanical Engineering, Nandha College of Technology, Perundurai, Tamilnadu, India
[2]Assistant Professor, Department of Mechanical Engineering, Sona College of Technology, Salem, Tamilnadu, India
[3]Assistant Professor, Department of Mechanical Engineering, Mailam Engineering College, Villupuram, Tamilnadu, India
[4]Assistant Professor, Department of Mechanical Engineering, Annapoorana Engineering College, Salem, Tamilnadu, India

E-mail: *dr.r.girimurugan@gmail.com

Abstract In this research work, thermal analysis has done on a rocket engine's thrust chamber with four different shapes like, circular, rectangular with greater breath (b>d), rectangular with greater depth (d>b) and rectangular with greater depth with edges filleted coolant channel models to establish perfect suitable shape for the coolant channels in order to maintain the entire thrust chamber within the melting point of its materials. Finite Element Method (FEM) has adopted to establish the thermal behaviour of the coolant channel's materials such as Nickel, Copper and Zirconium under the specific boundary conditions. Temperature distribution with respect to the nodal distances at thrust chamber's inlet, throat and outlet portions with four different shapes of coolant channel models under Nickel, Copper and Zirconium materials were attained and compared with each other. Finite element results reveals that the successful configuration for generatively cooled rocket thrust chamber must have the coolant passage shape of rectangular cross section with greater depth (b>d).

1 Introduction

Liquid fuelled rocket engine's performance is based on the huge value of thrust for its higher propulsion rate. Higher propulsion rate can be achieved by means of proper mixing, effective burning and regulated ejection of the fuel in the rocket engine's thrust chamber [1]. The reliability of the liquid fuelled rocket engine's thrust chamber is purely based on the exact prediction of its wall temperatures and amount of combustion gases heat flux [2]. Due to the usage of the liquid propellant in a liquid fuelled rocket engine, distribution of the temperature is not even in all wall sides of the combustion chamber. Temperature values are very higher at the nozzle end and very lower at throat portion of the nozzle [3]. Elevated temperature and pressure levels, and thrust chamber materials capable of withstanding high heat fluxes are necessary to meet these requirements [4]. Need of regenerative cooling methods are significantly increased to obtain the superior performance in the liquid fuelled rocket engine's combustion chamber. Analysis of the rocket engine's thermal behaviour

IOP Conf. Series: Materials Science and Engineering 1059 (2021) 012068 doi:10.1088/1757-899X/1059/1/012068

is well defined with related to the coupled heat transfer problem burning gas, combustion chamber wall and the liquid coolant [5, 6]. Life cycle of the combustion chamber can be considerably improved, when the side wall temperature is minimal [7]. In the thermal design of cooling systems for the thrust chambers of rocket engines, reliable estimation of the cool- ant side and the hot gas side heat transfer coefficients is of great importance [8]. In general, cooling systems design for the thrust chamber is playing a vital role during the liquid propellant rocket engine's overall design [9]. Combustion process optimization, combustion chamber's integrity and forecasting of heat flux moving between the combustion gages and the engine wall can be ensured using inclusive thermal analysis [10]. Liquid propellant rocket engine with high pressure thrust chamber reveals the superior performance operation during the regenerative cooling process. It can be used frequently used in huge and small size thrust chambers [11]. The life of the rocket engine is purely depending upon the active and proficient cooling system. The distinctive thermal analysis is much needed to explore the effect of cooling in the thrust chamber to maximize the combustion gases power in a rocket engine [12]. An upright physical perception on the rocket engine mechanism, special effects of geometrical changes calculations on the heat transfer and cooling necessities are the basic needs for the pioneering cooling approaches of the liquid propellant rocket engines [13]. Extreme and active cooling approaches are accomplished in the thrust chamber of the rocket engine by flowing a liquid propellant into the suitable cooling channels [14, 15]. In a high-pressure liquid rocket engine, unique circuit regenerative cooling is extensively used to enhance coolant enthalpy also diminish the thrust wall temperature [16]. In this research work, a distinctive thermal analysis has been conceded on a regenerative cooled liquid propellant rocket thrust chamber to establish perfect suitable shape for the coolant channels in order to maintain the entire thrust chamber within the melting point of its materials.

2 Materials and Methods

In general, thrust chamber wall should be designed to withstand the enormous heat expelled by hot combustion gas, which comes out of combustion chamber. The two important material properties such as thermal conductivity and melting point were considered for this analysis. Based on the above said properties, three materials were chosen to serve the purpose of reducing the temperature along the thickness of thrust chamber wall. At the same time it should be viewed seriously, the temperature should not exceed the melting point of the materials. The three materials namely Copper (Cu), Nickel (Ni) and Zirconium (Zr) were selected in the design of thrust chamber wall and the thermal properties of that materials were considered to carry out the analysis using ANSYS. At room temperature, high electrical and thermal conductivity has observed in Copper material among the pure metal. Copper has been placed between the inner layer Zirconium and outer layer Nickel. High polish is available in Nickel material, because it is a silvery white metal. It be in the right place to the changeover metals, and is rigid and flexible. It is the outermost layer of thrust chamber. Zirconia or Zirconium dioxide (ZrO_2) is a white translucent oxide of Zirconium. Zirconia is an enormously intractable material. The thermal properties of the selected materials such as Copper, Nickel and Zirconium were given in table.1.

Table 1 Physical and thermal properties of rocket thrust chamber wall materials

Rocket thrust chamber wall material	Density (kg/m^3)	Thermal conductivity (W/mK)	Specific heat (J/kgK)	Melting point (K)
Copper	8900	410.00	401.90	1540
Nickel	8890	90.70	456.39	1666
Zirconium	6000	120.75	418.00	2500

Coolant plays an important role in the generatively cooled rocket engine. In general, rocket engine nozzle and other parts are cooled using the liquid nitrogen before mixed with an oxidizer. The different Thrust chamber's wall geometry configurations, which are used in this analysis were given in table.2. Before doing the finite element analysis, calculation of heat transfer coefficient and bulk heat transfer is much needed to establish the heat transfer behaviour between the propellant and the combustion chamber respectively. Though the formula is available, it is very difficult to determine the heat transfer coefficient of combustion gas as its temperature is in terms of 2800 K. At this enormous temperature, the liquid hydrogen dissociates into steam and it is very difficult to find heat transfer coefficient of steam. The heat transfer coefficient and bulk heat transfer for the rocket thrust chamber's inlet, throat and outlet portions were derived from an existing equation and it has given in table.3 respectively. The different physical properties of the liquid hydrogen are shown in table.4.

Table 2 Thrust chamber's wall geometry configurations

Model	Inlet geometry of thrust chamber wall
Model - A	Circular
Model - B	Rectangular with greater breath (b>d)
Model - C	Rectangular with greater depth ((d>b)
Model - D	Rectangular with greater depth with edges filleted

Table 3 Calculated heat transfer coefficient and bulk temperature at various portions in the combustion chamber

Portion of combustion chamber	Heat transfer coefficient (W/m^2K)	Bulk temperature (K)
Inlet	50000	2500
Throat	70000	2400
Outlet	40000	2200

Table 4 Physical properties of liquid Hydrogen

Density (kg/m^3)	Velocity (m/s)	Viscosity (Ns/m^2)	Specific heat (J/kg)
9.46	20	1.3×10^{-5}	7200

3 Finite Element Analysis (FEA) on thrust chamber's half-cooling channel models

An inclusive finite element analysis package (ANSYS) is used to accomplish the entire thermal analysis on the liquid propellant rocket engine's thrust chamber. Initially half-cooling channel section had to be drawn before it was analysed. The drawing part was done in CATIA, a CAD package. Throughout the analysis, the mass flow rate had to be kept constant. The only way to do that while trying out different shapes was to keep the area of the coolant channel constant. Hence, the area of coolant slots was maintained uniform throughout the investigation, irrespective of the shape of it. In this analysis, SHELL 57 element is used to mesh the entire geometry of the half-cooling channel. The complete finite element models, which have meshed with the SHELL 57 elements of the four different thrust chamber's half-cooling channels (Model-A, Model-B, Model-C and Model-D), have depicted in figure.1 (a) to figure.1 (d) respectively. In this investigation it is very important to know the various modes of heat transfer that were involved at the time of analysis. The following heat transfer conditions such as, convection from combustion gases (hot-gases), convection from the coolant, convection from the atmosphere and three-dimensional conduction within the wall were fed during the heat transfer analysis. In this current heat transfer analysis, only half of the cooling section was taken the heat flux across the edge of the coolant channel was restricted by making it zero.

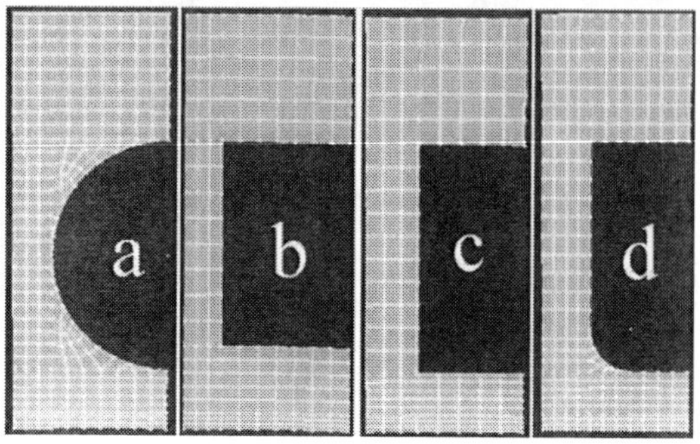

Figure 1. Mesh geometry of rocket thrust chamber half coolant channel inlet for (a) Model-A (b) Model-B (c) Model-C (d) Model-D

In the graphical plot, distributions of temperature with respect to the nodal distance were depicted clearly. It has observed from the graphical plot, distribution of temperature from the gases to the thrust chamber's inlet portion has increased with respect to the increasing nodal distance. From the temperature distribution and nodal distance plot that has represented in figure.7 (a), it has observed that the cooling wall, which has analysed with Nickel material, can withstand the maximum temperature of 1666 K at the thrust chamber's inlet, whereas this maximum temperature value is lower than that of its melting point of 2500 K. The analyses were carried out at inlet, throat and outlet sections of the thrust chamber wall. At each section, four analyses were done to establish the better geometrical shapes (table.2) of coolant passage. The four different geometry of thrust chamber wall was drawn by using CATIA software and the same geometries were imported to ANSYS software to analyse the temperature distribution under the three different cooling channel materials. The four different thrust chamber-cooling wall's geometries like, Model-A, Model-B, Model-C, Model-D with different material properties assigned, and boundary conditions applied configurations were illustrated in figure.2 (a) to 22 (d) correspondingly. After these applied boundary conditions and the suitable material properties on the four different models of the combustion chamber cooling wall's inlet were analysed in ANSYS software by using the solver option to establish the temperature behaviour between the cooling channel walls under the three different materials.

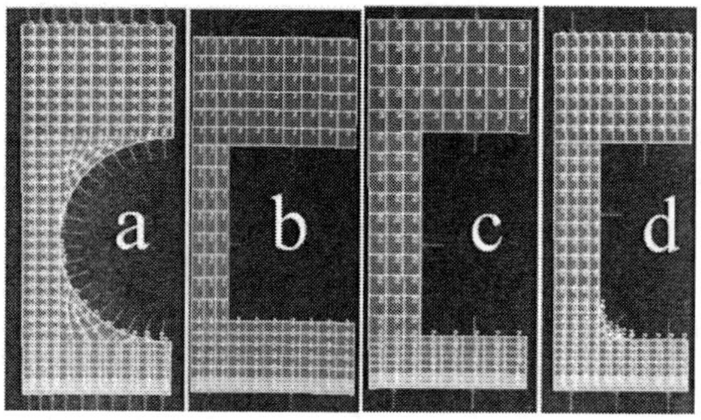

Figure 2. Material properties and boundary conditions applied rocket thrust chamber half coolant channel model's geometry for (a) Model-A (b) Model-B (c) Model-C (d) Model-D

4 Results and discussions

A distinctive heat transfer analysis have been made on an existing rocket combustion chamber's cooling wall at its inlet, throat and outlet portions by changing its geometry and material to expose the suitable shape of cooling wall geometry for the effective heat transfer between the produced gases and the cooling wall. Temperature distribution over the inlet portion of the thrust chamber's cooling wall surfaces for the four different cooling passage geometry by changing the cooling wall's material were illustrated in figure.3 (a) to figure.3 (d) correspondingly.

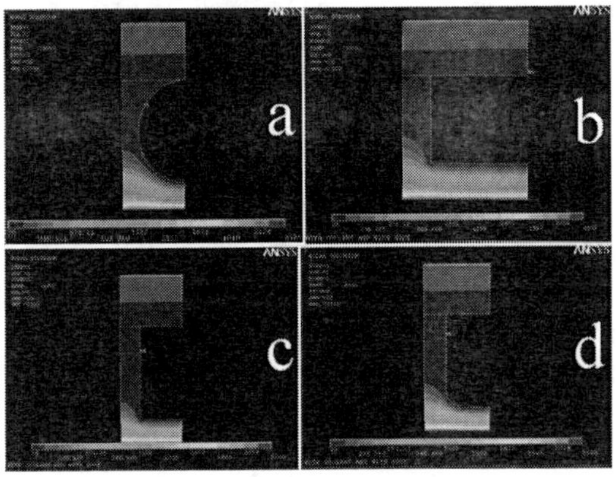

Figure 3. Temperature distribution at rocket thrust chamber's half coolant channel inlet for (a) Model-A (b) Model-B (c) Model-C (d) Model-D

Detailed colour plots, that have attained through heat transfer analysis from ANSYS software for the combustion chamber's cooling passage geometry with four different shapes like, Model-A, Model-b, Model-C and Model-D have depicted in figure. 4 (a) to figure.4 (d) respectively. It has observed from the detailed colour plots temperature distributions were found maximum at the inner surface of the combustion chamber's cooling passage wall.

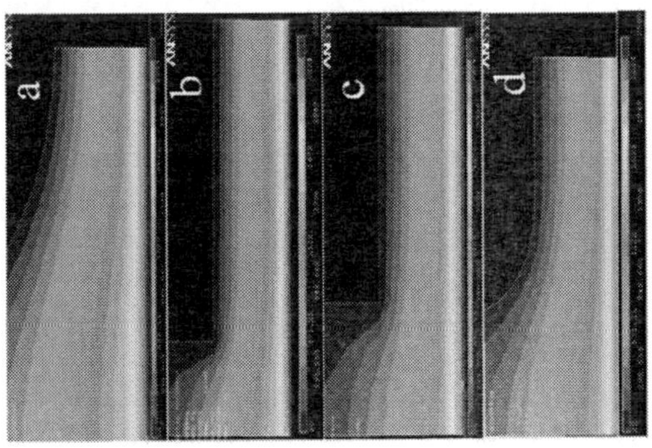

Figure 4. Detailed temperature distribution contour plot for rocket thrust chamber's half coolant channel inlet with (a) Model-A (b) Model-B (c) Model-C (d) Model-D

Conversely, the temperature distribution over the combustion chamber's wall has found minimum with respect to the increasing distance from the gas inlet portion. Similarly, distribution of temperature in a thrust chamber with respect to the nodal distance at coolant channel inlet in model-A for Nickel, Copper and Zirconium materials were shown in figure.5 (a), 5 (b) and 5 (c) respectively.

IOP Conf. Series: Materials Science and Engineering 1059 (2021) 012068 doi:10.1088/1757-899X/1059/1/012068

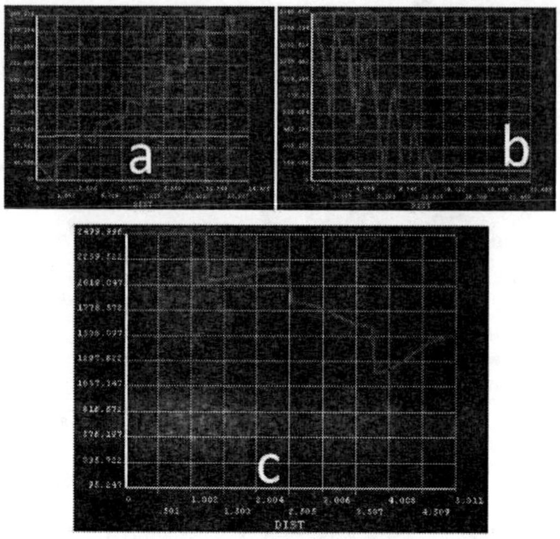

Figure 5. Temperature distribution with respect to the nodal distance at coolant channel inlet in model-A for (a) Nickel (b) Copper (c) Zirconium material

In the graphical plot, distributions of temperature with respect to the nodal distance were depicted clearly. It has observed from the graphical plot, distribution of temperature from the gases to the thrust chamber's inlet portion has increased with respect to the increasing nodal distance. From the temperature distribution and nodal distance plot that has represented in figure.5 (a), it has observed that the cooling wall, which has analysed with Nickel material, can withstand the maximum temperature of 1666 K at the thrust chamber's inlet, whereas this maximum temperature value is lower than that of its melting point of 2500K. From the temperature distribution and nodal distance plot which is represented in figure.5 (b), it has observed that the cooling wall, which has analysed with Copper material, cannot withstand the maximum temperature of 1540 K at the thrust chamber's inlet due to the exceeding its melting point of 1500 K. In the temperature and nodal distance graphical plot for copper material, regions marked by red ellipse are the nodes exceeding the melting point of copper. From the temperature distribution and nodal distance plot that has represented in figure.5 (c), it has observed that the cooling wall, which has analysed with Zirconium material, can withstand the maximum temperature of 2500 K at the thrust chamber's inlet, whereas this maximum temperature value is lower than that of its melting point of 2899 K. Distribution of temperature in a thrust chamber with respect to the nodal distance at coolant channel inlet in model-B for Nickel, Copper and Zirconium materials were shown in figure.6 (a), 6 (b) and 6 (c) correspondingly. In the graphical plot, distributions of temperature with respect to the nodal distance were depicted clearly. It has observed from the graphical plot, distribution of temperature from the gases to the thrust chamber's inlet portion has increased with respect to the increasing nodal distance. From the temperature distribution and nodal distance plot that has represented in figure.6 (a), it has observed that the cooling wall, which has analysed with Nickel material, can withstand the maximum temperature of 1666 K at the thrust chamber's inlet, whereas this maximum temperature value is lower than that of its melting point of 2500 K. From the temperature distribution and nodal distance plot which is represented in figure.6 (b), it has observed that the cooling wall, which has analysed with Copper material, cannot withstand the maximum temperature of 1561 K at the thrust chamber's inlet due to the exceeding its melting point of 1500 K. In the temperature and nodal distance graphical plot for copper material, regions marked by red ellipse are the nodes exceeding the melting point of copper. From the temperature distribution and nodal distance plot that has represented in figure.6 (c), it has observed that the cooling wall, which has analysed with Zirconium material, can withstand the maximum temperature of 2500 K at the thrust chamber's inlet, whereas this maximum temperature value is lower than that of its melting point of 2899 K. Distribution of temperature in a thrust chamber with respect to the nodal distance at coolant channel inlet in model-C for Nickel, Copper and Zirconium materials were shown in figure.7 (a), 7 (b) and 7 (c) correspondingly. In the graphical plot, distributions of temperature with respect to the nodal distance were depicted clearly. It has observed from the graphical plot, distribution of temperature from the gases to the thrust chamber's inlet portion has increased with respect to the increasing nodal distance.

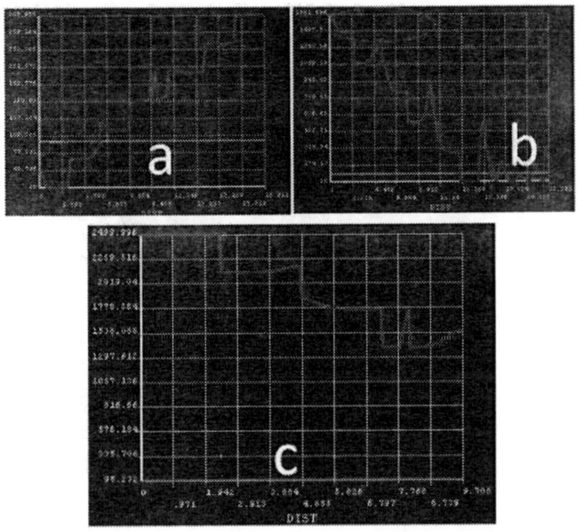

Figure 6. Temperature distribution with respect to the nodal distance at coolant channel inlet in model-B for (a) Nickel (b) Copper (c) Zirconium material

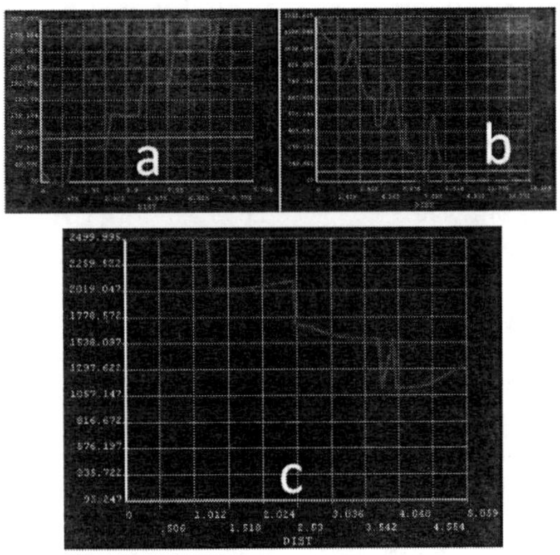

Figure 7. Temperature distribution with respect to the nodal distance at coolant channel inlet in model-C for (a) Nickel (b) Copper (c) Zirconium material

From the temperature distribution and nodal distance plot that has represented in figure.7 (a), it has observed that the cooling wall, which has analysed with Nickel material, can withstand the maximum temperature of 1666 K at the thrust chamber's inlet, whereas this maximum temperature value is lower than that of its melting point of 2500 K. From the temperature distribution and nodal distance plot which is represented in figure.7 (b), it has observed that the cooling wall, which has analysed with Copper material, can withstand the maximum temperature of 1315 K at the thrust chamber's inlet, whereas this maximum temperature value is lower than that of its melting point of 1358 K. From the temperature distribution and nodal distance plot that has represented in figure.7 (c), it has observed that the cooling wall, which has analysed with Zirconium material, can withstand the maximum temperature of 2500 K at the thrust chamber's inlet, whereas this maximum temperature value is lower than that of its melting point of 2899 K. Distribution of temperature in a thrust chamber with respect to the nodal distance at coolant channel inlet in model-D for Nickel, Copper and Zirconium materials were shown in figure.8 (a), 8 (b) and 8 (c) correspondingly.

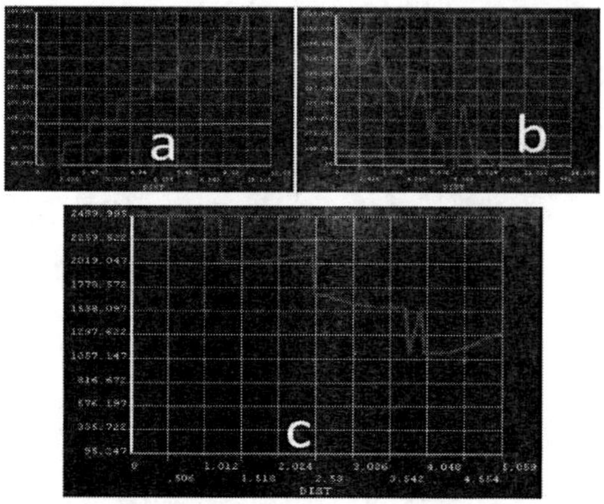

Figure 8. Temperature distribution with respect to the nodal distance at coolant channel inlet in model-D for (a) Nickel (b) Copper (c) Zirconium material

In the graphical plot, distributions of temperature with respect to the nodal distance were depicted clearly. It has observed from the graphical plot, distribution of temperature from the gases to the thrust chamber's inlet portion has increased with respect to the increasing nodal distance. From the temperature distribution and nodal distance plot that has represented in figure.8 (a), it has observed that the cooling wall, which has analysed with Nickel material, can withstand the maximum temperature of 1666 K at the thrust chamber's inlet, whereas this maximum temperature value is lower than that of its melting point of 2500 K. From the temperature distribution and nodal distance plot which is represented in figure.8 (b), it has observed that the cooling wall, which has analysed with Copper material, cannot withstand the maximum temperature of 1415 K at the thrust chamber's inlet due to the exceeding its melting point. From the temperature distribution and nodal distance plot that has represented in figure.8 (c), it has observed that the cooling wall, which has analysed with Zirconium material, can withstand the maximum temperature of 2500 K at the thrust chamber's inlet, whereas this maximum temperature value is lower than that of its melting point of 2899 K. Temperature distribution over the throat and inlet portion of the thrust chamber's cooling wall surfaces for the Model-C by changing the cooling wall's material were illustrated in figure.9 (a) and figure.9 (b) respectively. Temperature distribution over the outlet portion of the thrust chamber's cooling wall surfaces for the Model-C by changing the cooling wall's material were illustrated in figure.9 (b). It has observed from the detailed colour plots, temperature distributions were found maximum at the inner surface of the combustion chamber's cooling passage wall. Conversely, the temperature distribution over the combustion chamber's wall has found minimum with respect to the increasing distance from the throat portion.

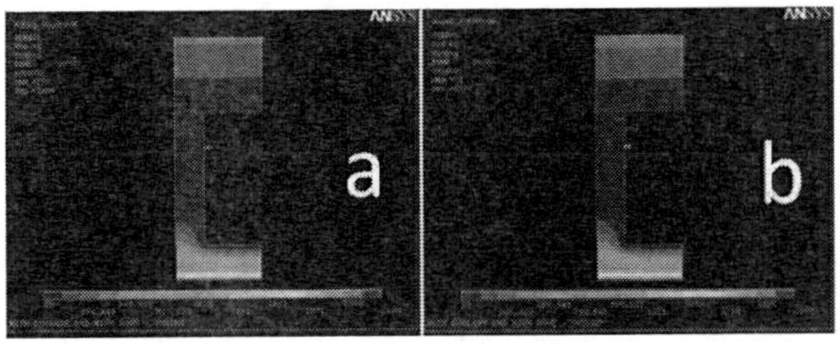

Figure 9. Temperature distribution in rocket thrust chamber's half coolant channel for Model-C at (a) throat portion (b) outlet portion

Detailed colour plots, that have attained through heat transfer analysis from ANSYS software for the combustion chamber's cooling passage geometry at its throat portion for Model-C have depicted in figure. 10 (a) to figure.10 (c) respectively. Similarly, distribution of temperature in a thrust chamber with respect to the nodal distance at coolant channel throat in model-C for Nickel, Copper and Zirconium materials were shown in figure.10 (a), 10 (b) and 10 (c) respectively. In the graphical plot, distributions of temperature with respect to the nodal distance were depicted clearly. It has observed from the graphical plot, distribution of temperature from the gases to the thrust chamber's throat portion has increased with respect to the increasing nodal distance. From the temperature distribution and nodal distance plot that has represented in figure.10 (a), it has observed that the cooling wall, which has analysed with Nickel material, can withstand the maximum temperature of 1666 K at the thrust chamber's throat portion, whereas this maximum temperature value is lower than that of its melting point of 2500 K. From the temperature distribution and nodal distance plot that has represented in figure.10 (b), it has observed that the cooling wall, which has analysed with Copper material, can withstand the maximum temperature of 1269 K at the thrust chamber's throat portion, whereas this maximum temperature value is lower than that of its melting point of 1358 K.

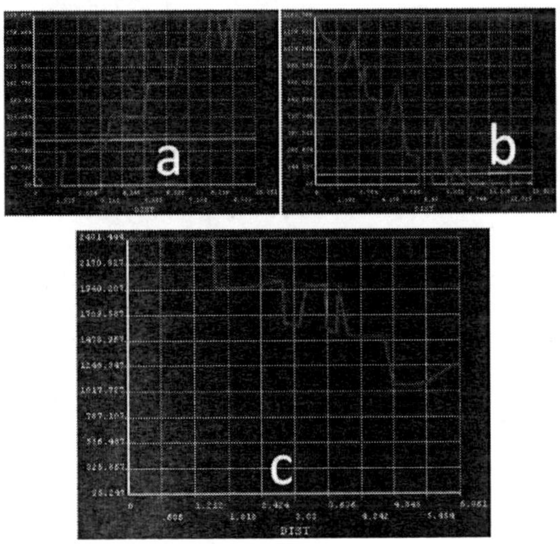

Figure 10. Temperature distribution with respect to the nodal distance at coolant channel throat in model-C for (a) Nickel (b) Copper (c) Zirconium material

From the temperature distribution and nodal distance plot that has represented in figure.10 (c), it has observed that the cooling wall, which has analysed with Zirconium material, can withstand the maximum temperature of 2500 K at the thrust chamber's throat portion, whereas this maximum temperature value is lower than that of its melting point of 2899 K. Detailed colour plots, that have attained through heat transfer analysis from ANSYS software for the combustion chamber's cooling passage geometry at its outlet portion for Model-C have depicted in figure.11 (a) to figure.11 (c) respectively. It has observed from the detailed colour plots, temperature distributions were found maximum at the inner surface of the combustion chamber's cooling passage wall. Conversely, the temperature distribution over the combustion chamber's wall has found minimum with respect to the increasing distance from the outlet portion. Similarly, distribution of temperature in a thrust chamber with respect to the nodal distance at coolant channel outlet portion in model-C for Nickel, Copper and Zirconium materials were shown in figure.11 (a), 11 (b) and 11 (c) respectively. In the graphical plot, distributions of temperature with respect to the nodal distance were depicted clearly. It has observed from the graphical plot, distribution of temperature from the gases to the thrust chamber's outlet portion has increased with respect to the increasing nodal distance.

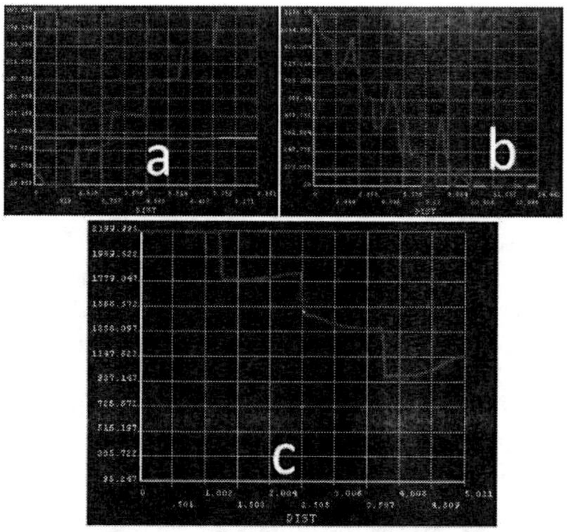

Figure 11. Temperature distribution with respect to the nodal distance at coolant channel outlet in model-C for (a) Nickel (b) Copper (c) Zirconium material

From the temperature distribution and nodal distance plot that has represented in figure.11 (a), it has observed that the cooling wall, which has analysed with Nickel material, can withstand the maximum temperature of 1666 K at the thrust chamber's outlet portion, whereas this maximum temperature value is lower than that of its melting point of 2500 K. From the temperature distribution and nodal distance plot that has represented in figure.11 (b), it has observed that the cooling wall, which has analysed with Copper material, can withstand the maximum temperature of 1269 K at the thrust chamber's outlet portion, whereas this maximum temperature value is lower than that of its melting point of 1358 K. From the temperature distribution and nodal distance plot that has represented in figure.11 (c), it has observed that the cooling wall, which has analysed with Zirconium material, can withstand the maximum temperature of 2500 K at the thrust chamber's outlet portion, whereas this maximum temperature value is lower than that of its melting point of 2899 K.

5 Conclusion

After the series of thermal analysis, which has conducted to optimize the better coolant shapes in a rocket engine thrust chamber, the following conclusions were made. Coolant shapes of circle, rectangle with greater breadth (d>b) and rectangle with filleted edges fail in the copper region due to the inability of coolant to reach the maximum copper region. Later, rectangle shape was considered with greater depth (d>b) so that the coolant could reach the copper region to the maximum extent as only copper was the area of concern analysis. Analysis results showed the successful configuration for generatively cooled rocket thrust chamber must have the coolant passage shape of rectangular cross section with greater depth (b>d). This shape will have the greater influence of coolant to suppress the heat conduction along thrust chamber wall, especially in copper region.

References

[1] Asraff, A K Sheela S Anju Paul, Ansu Mathew and Savithri S 2016 Cyclic stress analysis of a rocket engine thrust chamber using chaboche, voce and creep constitutive models, *Transactions of the Indian Institute of Metals* **69** 495-500.
[2] Pizzarelli M Carapellese S and Francesco Nasuti, 2011 A quasi-2-d model for the prediction of the wall temperature of rocket engine cooling channels, *Numerical Heat Transfer - Part A* **60** 01-24.
[3] Shinichi Toyama High temperature thermal analysis on nozzle of thrust chamber *33rd plasma dynamics and lasers conference* **20-23** May Maui Hawaii.
[4] Vinod K Arya 1992 Nonlinear structural analysis of cylindrical thrust chambers using Viscoplastic models *Journal of Propulsion and Power* **8** 598-604.
[5] Wennerberg J C Anderson W E Haberlen P A Jung H and Merkle C L 2005 Supercritical flows in high aspect ratio cooling channels, *AIAA* Paper **4302**.
[6] Preclik, D Wiedmann D Oechslein W and Kretschmer J 1998 Cryogenic rocket calorimeter chamber experiments and heat transfer simulations, *AIAA* Paper **3440**.
[7] Qiuwang Wang, Feng Wu, Min Zeng and Laiqin Luo 2006 Numerical simulation and optimization on heat transfer and fluid flow in cooling channel of liquid rocket engine thrust chamber Engineering Computations, *International Journal for Computer-Aided Engineering and Software* **23** 907-921.
[8] Sugathan, N Srinivasan K Srinivasa Murthy S 1993 Experiments on heat transfer in a cryogenic engine thrust chamber, *Journal of Propulsion and Power* **9** 240-244.
[9] Martin J Chiaverini, Matthew J Malecki, Arthur Sauer J William H Knuth and Daniel J Gramer 2003 Vortex thrust chamber testing and analysis for O_2-H_2 propulsion applications, *39th AIAA/ASME/SAE/ASEE Joint Propulsion Conference and Exhibit* July **20-23** Huntsville Alabama 01-19.
[10] Jokhakar J and Naraghi M H 2008 A CFD-RTE model for thermal analysis of regeneratively cooled rocket engines, *44th AIAA/ASME/SAE/ASEE Joint Propulsion Conference* Hartford July **21-23**.
[11] Huzel D K and Huang D H 1992 Modern Engineering for Design of Liquid-Propellant Rocket Engines, *AIAA Progress in Aeronautics and Astronautics* **147** 84-104.
[12] Pizzarelli M Nasuti F and Onofri M 2007 A simplified model for the analysis of thermal stratification in cooling channels, *2nd European Conference for Aerospace Sciences*, 1-8.
[13] Naraghi M H Dunn S and Coats D 2004 A model for design and analysis of regeneratively cooled rocket engines, *40th AIAA/ASME/SAE/ASEE Joint Propulsion Conference and Exhibit* **11-14** July, Fort Lauderdale, Florida, 01-13.
[14] Mohammad H Naraghi and Matthieu Foulon 2008 A simple approach for thermal analysis of regenerative cooling of rocket engines, *Proceedings of IMECE2008 ASME.*
[15] Jiawen Song and Bing Sun 2017 Thermal-structural analysis of regeneratively cooled thrust chamber wall in reusable LOX/Methane rocket engines *Chinese Journal of Aeronautics* **30(3)** 1043-53.
[16] *International Mechanical Engineering Congress and Exposition* 2008 31st November Boston Massachusetts USA 01-08.

Performance Assessment of Concrete Incorporating Recycled Coarse Aggregates and Graphene Oxide

T. Balasubramaniam [1], Karthick M[2] and Arun M[3]
[1] Assistant Professor
[2] Post Graduate Student
[3] Associate Professor
[1 & 2] Department of Civil Engineering, Coimbatore Institute of Technology Coimbatore 641014
[3] Department of Civil Engineering, PSG Institute of Technology and Applied Research, Coimbatore 641062

Email ID :Balasubramaniam.t@cit.edu.in[1], kmkarthick4001@gmail.com[2]and
Corresponding Author :arun@psgitech.ac.in[3]

Abstract:The recycled coarse aggregate (RCA) can be utilized in concrete, which leads to produce a sustainable development in construction industry. On the utilization of recycled coarse aggregate minimize the amount of construction waste deposition towards environment. Also contemporary knowledge is agreed chance to progress the recycling material usage. The current research work has to access the effect of recycled aggregate concrete (RAC) containing 30% quantities of RCA with different quantities of reduced graphene oxide and to find the optimal percentage of reduced graphene oxide for RAC. The concrete mixtures formed by replacing the natural coarse aggregates(NCA) with 30% RCA and cement having0%, 0.3%, 0.6% and 0.9% of reduced graphene oxide. The mechanical properties present in the study include compressive strength and split tensile strength each evaluate after 28 days of curing all concrete mixes will be investigated .It provides a vibrant understanding on in what way it can be applied to progress the usage of 30% RCA with reduced graphene oxide content in concrete.

1 Introduction

Because of reliability in terms of strength, durability and economic aspect compared to other construction materials, the rate of production and use of concrete in the construction sector has been increasing in the recent past. On average approximately one percent of concrete was produced per person worldwide each year. Concrete is manufactured by using a huge amount of natural materials such as natural coarse aggregates, etc [1]. It is found that an average amount of 40 billion tons of a natural aggregate has been used annually around the world. The massive consumption rate of natural aggregates contributes to the limited use of these resources, which in many countries around the globe leads to a lack of natural aggregates [2]. Furthermore, natural aggregate production and processing contribute significantly to the emission of dust, noise and greenhouse gases, with a serious negative environmental effect [4]. Therefore, it is necessary for today's society to establish an alternative source of aggregates. In fact, almost all countries worldwide generate significant amounts of building and demolition (C&D) waste each year from the construction industries. The total production of C&D waste worldwide exceeds three billion tons a year. China, India and the USA, which generate approximately two billion tons of waste, contribute greatly to this waste generation. Natural forces, such as earthquakes, cyclones etc., create an enormous volume of solid waste, as large buildings are destroyed.

Consequently, recycling and reusing such waste has been described as a sustainable aggregate for the manufacture of recycled concrete as a fruitful way of mitigating the depletion of natural resources, waste management and environmental problems. RCA (recycled coarse aggregates) are produced from other discarded materials by removing demolished concrete pieces and shoring them to a suitable grading for engineering applications. RCA is an ancient, natural coarse aggregate (NCA) and two-phase material. The recycled aggregate cement (RAC) is generated in a different quantity by substituting natural aggregates for RCA. Compared to virgin aggregates, the main reason for the restricted use of RCA lies in its lower properties [6]. Several inquiries have established the lower characteristics of RCA. It is found that, due to vacuum and micro cracks on the old mortar attached to its surface, the RCA is characterized by high water absorption, low density, weak grading, etc [8]. Several experiments have been carried out to show that concrete RCA systems for structural applications can be used. Several factors like RCA, w / c ratio, strength and source age of concrete, adherence mortar and RCA humidity content have a major impact on the characteristics of RAC [5]. The properties of parent concrete such as the NCA size and the form of crushing measures impact the properties of RCA. Generally speaking, the RCA produced from the high strength and large size parent concrete NCA is of higher quality and therefore produces better RAC efficiency [7,10]. From several prior investigations, the assets of the RAC were not affected when the RCA replacement rate was below 30 percent [3,9]. Nonetheless, substantial deterioration of concrete products was observed at the higher replacement level of RCA. The compression strength (CS), bending strength (FS), elasticity module and tensile dividers have been decreased compared to conventional cement [11,12]. RAC water absorption was also improved over 50% when RCA was included in concrete. However, the durability properties of the RCA were significantly lowered by injecting RCA into concrete at higher level of substitution, such as acid resistance, carbonation, sulphate attack, chloride ion penetration, freezing and thawing [13]. While RAC showed great properties, the materials are also semi-weak with higher tensile strength and lower strain cap [14].
The research will fix RAC's weaknesses in different ways. Since nanotechnology will change the world and particularly cement-based materials, focusing on their nanoscale structure would hopefully give us more information about how we can develop their qualities [15,16]. Nevertheless, the high cost and repetitive generation approach are important considerations requiring exceptional contemplation [17]. Graphene experiments have shown that the structure and the application of graphene-concrete nano-composites. The worked reduced graphene oxide will increase interfacial efficiency, which increases their mechanical characteristics. In this investigation there was an increasing degree of rigidity in the solid examples of reduced graphene oxide (rGO) substituted by concrete with a percentage weight of cement and a good grip between rGO surrounds and cement frames. In fact when analysed with ordinary cement concrete, the rGO mortar had dispersed calcium silicate hydrate (C-S-H) gels between cement and aggregate [19]. The aim of the study is to investigate the effect of reduced graphene oxide in ordinary cement paste to enhance compressive and tensile properties of recycled coarse aggregate concrete.

IOP Conf. Series: Materials Science and Engineering 1059 (2021) 012069 doi:10.1088/1757-899X/1059/1/012069

2 Materials

2.1 Cement
Ordinary portland cement was used in this project. The cement quantity is obtained for testing, and deposited in a state of the laboratory environment. 0.3%, 0.6%, 0.9% reduced Graphene oxide will add with weight of the cement.

2.2 Aggregate

2.2.1 Natural coarse aggregate
Throughout the research experiments, natural aggregate was used according to Indian standard code in concrete mixes with a maximum size of 20 mm.

2.2.2 Recycled coarse aggregate
RCA is obtained from the management of the remaining, unwanted concretes and the concrete that will be destroyed. RCA's proportional size has been used in 20 mm.

Figure 1. Recycled Coarse Aggregate

2.2.3 Fine aggregate
Fine aggregate used in mixtures of concrete made of M sand with an average particle size of 5 mm according to Indian standard code book.

Table 1. Characteristics of aggregate

Characteristics	Fine Aggregate M sand	Natural Coarse Aggregate	Recycled Coarse Aggregate
Specific gravity	2.68	2.65	2.48
Bulk density	1.64	1664	1356
Water absorption	0.5	1.1	5.1

2.3 Reduced graphene oxide
Graphene oxide is a type of graphene that contains oxygen functional groups and has interesting

Figure 2. Reduced Graphene Oxide

properties that may vary from graphene. Such oxidized functional groups are extracted by reducing graphene oxide in order to obtain a graphene sheet. This substance of graphene is called reduced oxide of graphene, sometimes abbreviated to rGO. Graphite oxide, a material made of many layers of graphene oxide, can also be derived from rGO after a process of graphene oxide reduction and then rGO.

Table 2. Parameters of reduced graphene oxide

Parameters	Approximate values
Product Purity	99%
Lateral Dimension	5-9 micrometer
Bulk Density	0.135g/cc
Thickness	0.9-2nm
Surface area	>125m2/g
Number of layers	3-6

Table 3. Chemical composition of reduced graphene oxide

Element	Percentage of Composition
Carbon	>90%
Nitrogen	<5%
Oxygen	<5%

3 Mixture method
A certain volume of reduced graphite oxide powder was dispersed into the water mixture and the polycarboxylate ether (PCE) super plasticizer to prepare rGO solution. The resulting aqueous reduced graphene oxide suspension at a concentration of 0.5 g/L was sonicated. Three mixes where prepared based on the differentiation of reduced graphene oxide % of 0.3, 0.6, 0.9.

Figure 3. rGo inSonication

Figure 4. rGo mixture

4 Mix proportion
The mixing proportion of the concrete shall be maintained in 1:1.8:2.8. 30 per cent of the recycled coarse aggregate has been replaced by natural aggregates according to Indian standard code book. To determine GO's microstructural effect on RCA concrete, GO was blended into RCA concrete based on various proportions, respectively. 0 percent, 0.3 percent, 0.6 percent, 0.9 percent by weight of cement. Cement water (w / c) was selected as 0.45 due to high water absorption in RCA. Mixture was made in mixture machine. A mixture of recycled coarse aggregate, fine aggregate, cement and mixture of water with reduced graphene oxide was poured into mixture machine and rotated for 5 min. Then concrete was poured in specimen mould, after 24 hours specimen has to demould and the specimen let to water curing for 7, 14, 28 days.

4.1 Test method

To determine the compressive strength of recycled coarse aggregate cube specimen size should be in 150x150x150mm are used. Top surface should be smooth and even. Load should be applied gradually in compressive strength machine till the specimen get failed. To determine the split tensile strength of recycled coarse aggregate cylinder specimen size should be in 300mm in height and 150mm in diameter. Test should be carried out at interval of 7 days, 14 days and 28 days.

5 Results and discussion

5.1 Compressive strength

The result has been obtained from compressive strength test. Recycled coarse aggregate with reduced graphene oxide has higher compressive strength when compare with recycled coarse aggregate without reduced graphene oxide. Where reduced graphene oxide increase the improvement in the C-S-H gel could be one of the reason for increasing strength. So RAC with rGo has higher strength when compare with ordinary Portland cement and RAC without rGo. The optimum percentage of recycled coarse aggregate with reduced graphene for compressive strength is at 0.6%.

Figure 5. Compressive Test

Table 4. Compressive strength of cubes with incorporating of 30 % RCA concrete with different percentage of rGO at different ages.

% of rGo with RCA concrete mix	Compressive strength in N/mm^2		
	7 days	14 days	28 days
0%	16	20	25
0.3%	20	24	28
0.6%	25	30	34
0.9%	22	26	31

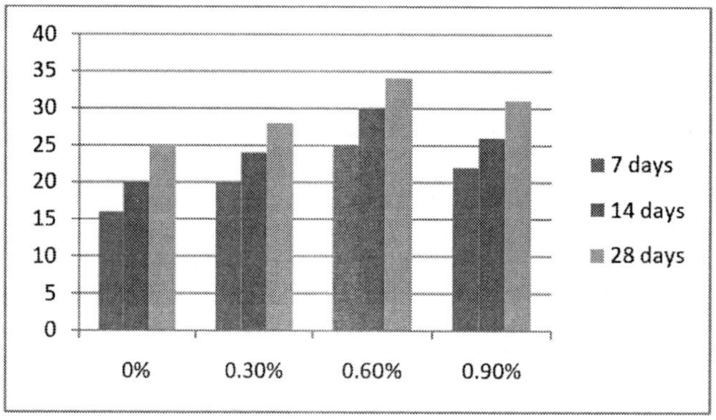

Figure 6 Comparing the compressive strength of cubes with of 30 % RCA concrete with different percentage of rGO.

5.2 Split tensile strength

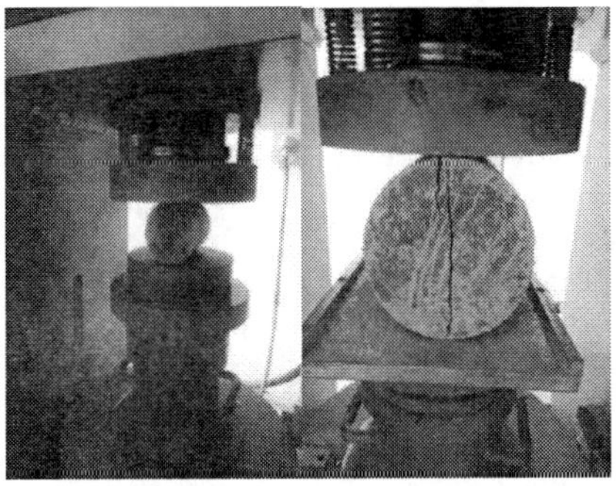

Figure 7. Split tensile strength

The result has been obtained from split tensile strength test. Recycled coarse aggregate with reduced graphene oxide has higher tensile strength when compare with recycled coarse aggregate without reduced graphene oxide. Where reduced graphene oxide is good at tensile strength and increased tensile strength of specimen could be one of the reason for increasing strength. So RAC with rGo has higher strength when compare with ordinary Portland cement and RAC without rGo. The optimum percentage of recycled coarse aggregate with reduced graphene for compressive strength is at 0.6%.

Table 5. Split tensile strength of cylinder with incorporating of 30 % RCA concrete with different percentage of rGO at different ages.

% of rGo with RCA concrete mix	Tensile strength in N/mm^2		
	7 days	14 days	28 days
0%	2.12	2.35	2.93
0.3%	2.39	2.78	3.45
0.6%	2.91	3.53	4.02
0.9%	2.67	3.14	3.89

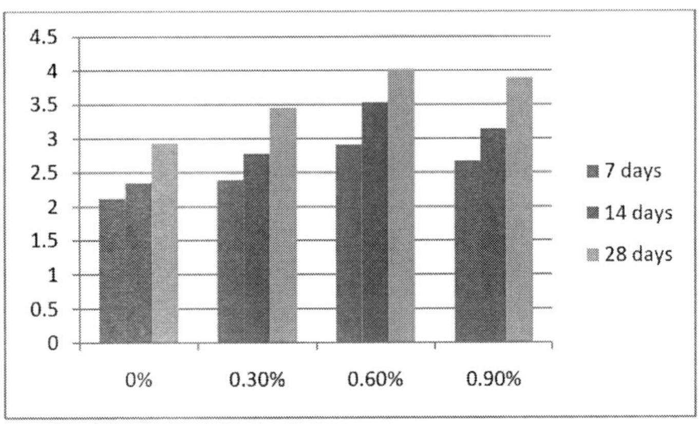

IOP Conf. Series: Materials Science and Engineering 1059 (2021) 012069 doi:10.1088/1757-899X/1059/1/012069

Figure 8 Comparing the split tensile strength of cubes with 30 % RCA concrete with different percentage of rGO.

5.3 Load deflection behaviour

All the beams were tested under two point loading condition. The Monotonic load was applied by using hydraulic jack and to record the load precisely, proving ring was used. The beam was gradually loaded by increasing the load level at an increment of 5kN. The beams were loaded up to failure and the values of load at first crack and at ultimate failure state were noted. As the load level was increased, the observed deflection was increased slightly in each increment of load. But the beam where analysed only by ANSYS software.The value of flexural strength is found to increase for the 0.6 percentage of reduced graphene oxide concrete than compared to ordinary Portland cement concrete. The ultimate load carrying capacity of 0.6% of reduced graphene concrete is 85kN and the load carrying capacity of OPC concrete is 65kN. Hence, the 0.6% of reduced graphene oxide with recycled coarse aggregate concrete comprises higher load carrying capacity than the OPC concrete. The maximum load deflection of beam carried out is 5.05mm.

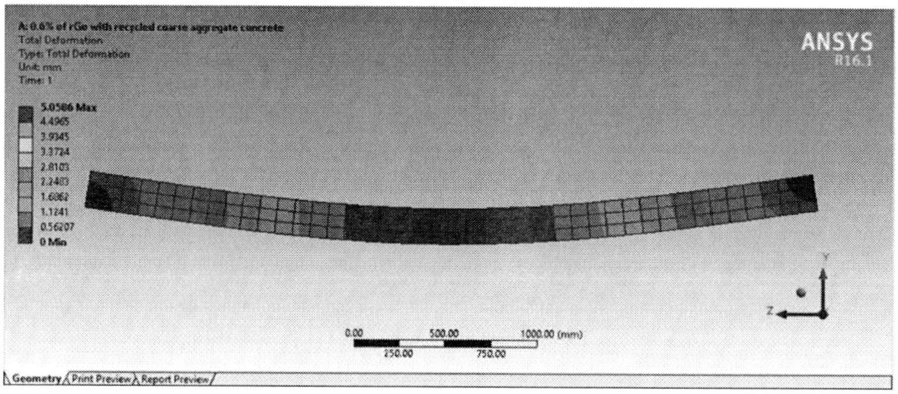

Figure 9. Total deformation of beam

6 Conclusion

In this research the mechanical properties of recycled concrete aggregate with reduced graphene oxide were studied of compressive strength and tensile strength.

The results indicate an increase in strength of compressive and tensile strength of recycled aggregate concrete with addition of reduced graphene oxide with an optimum percentage of 0.6 reduced graphene oxide with cement. Such changes may mainly be due to better interaction in the ITZ between the RA and cement matrix.

Because of the cohesive action of hydration products coupled with go filling effect, interfaces between the RCA and the cement matrix have been enhanced. The strengthened ITZ could increase the performance of RCA cement composites, which could be of important ecological and economic benefits to the construction industry.

With increasing of rGo in mortar the number of different pores decreased by transportation of charges between solid and fluid phases was hindered by the addition of rGo in the C-S-H gel in pores.

Results indicate that by adding rGo in recycled coarse aggregate haws better environmental performance than natural aggregate, which may be additional, and has equivalent mechanical strength. Speed up C&D waste recycling and then foster the sustainable building sector development.

References

[1] Marie, H. Quiasrawi 2012 Closed-loop recycling of recycled concrete aggregates, *J.Clean. Prod.* **37** 243–8.
[2] Tam V W Y, M. Soomro, Evangelista A C J 2018 A review of recycled aggregate inconcrete applications (2000–2017), *Constr. Build. Mater.* **172** 272–92
[3] Mukharjee B B, Barai S V 2014 Influence of nano-silica on the properties of recycledaggregate concrete *Constr. Build. Mater.* **55** 29–37
[4] Akhtar, Sarmah A K 2018 Construction and demolition waste generation andproperties of recycled aggregate concrete: a global perspective, J. Clean. Prod.**186** 262–81
[5] Zhou, Chen Z 2017 Mechanical properties of recycled concrete made withdifferent types of coarse aggregate, Constr. Build. Mater. **134** 497–506
[6] Huda S B, Alam M S 2014 Mechanical behavior of three generations of 100%repeated recycled coarse aggregate concrete, Constr. Build. Mater. **65** 574–82
[7] Committee 551. Removal and reuse of hardened concrete (ACI551R-01). In: American Concrete Institute, editor. Farmington Hills, MI: American ConcreteInstitute; 2001.
[8] Fathifazl G, Abbas A, Razaqpur A G, Isgor O B, Fournier B, Foo S 2009 Newmixture proportioning method for concrete made with coarse recycledconcrete aggregate, *J. Mater. Civ. Eng.* **21** 601–11
[9] Majhi R K, Nayak A N, Mukharjee B B 2018 Development of sustainable concreteusing recycled coarse aggregate and ground granulated blast furnace slag,*Constr. Build. Mater.* **159** 417–30
[10] Kou S C, Poon C S 2015 Effect of the quality of parent concrete on the properties ofhigh performance recycled aggregate concrete *Constr. Build. Mater.* **77** 501–8
[11] Tavakoli M, Soroushian P 1996 Strengths of recycled aggregate concrete madeusing field-demolished concrete as aggregate *ACI Mater. J.* **93** 182–90.
[12] Bravo M, De Brito J, Pontes J, Evangelista L 2015 Mechanical performance of concrete made with aggregates from construction and demolition wasterecycling plants, *J. Clean. Prod.* **99** 59–74
[13] Tam V W Y, Kotrayothar D, Xiao J 2015 Long-term deformation behaviour ofrecycled aggregate concrete, Constr. Build. Mater. **100** 262–72
[14] Kwan W H, Ramli M, Kam K J, Sulieman M Z 2012 Influence of the amount ofrecycled coarse aggregate in concrete design and durability properties, *Constr.Build. Mater.* **26** 565–73
[15] Li J, Xiao H, Zhou Y 2009 Influence of coating recycled aggregate surface withpozzolanic powder on properties of recycled aggregate concrete, *Constr. Build.Mater.* **23** 1287–91
[16] Kou S C, Poon C S, Wan H W 2012 Properties of concrete prepared with low-graderecycled aggregates, *Constr. Build. Mater.* **36** 881–9
[17] Rafiee M A, Rafiee J, Wang Z, Song H, Yu Z-Z, Koratkar N 2009 Enhanced Mechanical Propertiesof Nanocomposites at Low Graphene Content *ACS Nano.* **3(12)** 3884-90.
[18] Kim J, Cote L J, Kim F, Yuan W, Shull K R, Huang J 2010 Graphene Oxide Sheets at Interfaces *JAmChem Soc.* **132(23)** 8180-6
[19] Samuel Chuah, Zhu Pan, Jay G Sanjayan, Chien Ming Wang, and Wen HuiDuan 2014 Nano reinforced cement and concrete composites and new perspective from graphene oxide *Construction and Building Materials* **73** 113–24

IOP Conf. Series: Materials Science and Engineering 1059 (2021) 012070 doi:10.1088/1757-899X/1059/1/012070

Design and fabrication of sanitizer sprinkler robot for covid-19 hospitals

G Sundar raju[1], K Sivakumar[2], A Ramakrishnan[3], D Selvamuthukumaran[4], and E Sakthivel Murugan[5]

[1,3,4,5] Assistant Professor, Bannari Amman Institute of Technology, Sathyamangalam, Erode, Tamil Nadu, India

[2] Professor, Bannari Amman Institute of Technology, Sathyamangalam, Erode, Tamil Nadu, India

E-mail: sunmetly@gmail.com

Abstract. The world is severely affected by the Covid-19 pandemic health-care workers are hostile to work inside of COVID-19 hospital due to social distancing. Supplying essentials food and delivering medicine in hospitals has a challenge. To address the need of the World health organization and the certified sanitizing system is a solution for personnel and surface decontamination. To overcome the problem developed automatic portable sanitizing equipment for spraying sanitization solution. The portable sanitizer unit is attached to the top of the mobile robot. The system integrates a sprinkler mechanism and is used to distribute air and disinfectant fluid mixture. The mobile robot main components consist of a DC motor, Bluetooth module, Arduino, Motor driver, Submergible pump, Sprinkler, Battery, DC Converters are used. The system is capable of sanitizing the floors of hospitals. The application area can include hospital corridors, medical shops, operation theatre, walking pathways, doctor room, testing center, and patient room, etc.

Keywords: Covid-19; Hypochlorite; DC motor; Bluetooth; Arduino.

1. Introduction

The robot can play an important role during the present pandemic situation. It reduces human involvement in all areas starting from a patient check-up to a medicine delivery system [1]. Health organizations were also instructed to maintain social distancing from corona positive found people. It is the biggest challenge for doctors and health care support workers. The government and medical workers are working day and night to protect the people from infection and infected people. During this pandemic situation, many hospitals are facing the biggest issue is a shortage of labourers, and many healthcare people afraid to enter the patient room. Unfortunately, everyday hospitals need to clean their floors for disease prevention and maintain hygiene. Recently many high-risk and high-touch areas, intelligent navigation, and detection systems are used. Further, Robot also used in modern farming to increase productivity. Zant et al. (2018) [2] designed an agricultural robot for greenhouse mildew treatment operations. This system allows the farmers to specify their farm, including fields, roads, and docking stations. In recent days house cleaning robots are famous for the hygiene room cleaning system. Asafa et al. (2018)[3] developed a vacuum cleaner robot for household applications. They fabricate disk type robot attached with

Vaccum and cleaning tools. In the system, it sucks dust via a folding dust bin on top, and a cooling fan is also mounted. The robot navigation via two motor controlled rear sweepers had each driven by 3v DC Motor. This compact type of robot making cleaning is a fully autonomous duty.

The application of robots arisen in all the fields, types such as wheeled mobile robots, legged robots, flying robots, robot vision, artificial intelligence, and so on. Some robots can walk, jump, and fly. Rubio et al. [4] reviewed the mobile robots in the aspects of the concept, framework, and applications. Mobile robots have the ability to move independently, with adequate intelligence to respond and make a decision based on the opinion received from the situation. Normally, any robot fabrication consists of locomotion, perception, cognition, and navigation. Vehicle movement problems are rectified by understanding the mechanism and kinematics, dynamics, and control theory. The sensitivity of the mobile robots depends on signal analysis, computer vision, and sensor technologies. Cognition is the process of analyzing input data from sensors and taking consistent action to achieve the purpose. Navigation systems require knowledge of positioning, starting point, and stopping point. According to the locomotion system, mobile robots can be classified as stationary, Land-based, Air-based, and water-based, etc.

A robot can help a human being in various applications and risky operations in difficult and fickle situations such as fire rescue, anti-terrorism, and removing explosives, etc. Cui et al. (2016) [5] analyzed the wheel /track mobile robot mechanism design and Mobil ability. A flat ground transformable wheel track is used to analyze the mechanical structure of the wheel based robot. In this vehicle structure, the two sides of the driven rod maintain the symmetry of the motion and double four-bar linkage mechanism used in the track expansion and contraction situations.

2. Materials & Methods

The robot is capable of effectively killing microorganisms on the floor, such as fungi, bacteria, and viruses, and has significant effects on harmful microorganisms [6]. The robot cans automatically patrol for sterilization and disinfection in a predetermined area. The sprinkler is equipped with a double-acting pneumatic cylinder, and the purpose of the cylinder is used to pushes the air into the tank [7]. The main components consist of a sprinkler robot, is DC motor, Bluetooth module, Arduino, Motor driver, Submergible pump, Sprinkler, Battery, DC Converters. The frame is made up of a 3mm Galvanized iron sheet with a dimension of 440 mm length and 240 mm width. The frame acts as a base and is used to assemble all the components.

Control system

Lpc2148 microcontroller is used to control the sprinkler unit. It is used to control the mobile robot's mechanical structure and comment to perform the desired objectives. The purpose of the control system is mainly perception, cognition, and action. The perception system provides information about the outside atmosphere. Based on the output data, appropriate commands are sent to actuators, which activate the mechanical structure.

Navigation System

The important unit of a mobile robot is the navigation system. This system has three types path tracker, obstacle detection system, and remotely operated system. The objective is for the robot to move from one place to another place to complete the assigned task. HC-05 is a Bluetooth module that establishes a wireless network, The specification of this module is operating frequency is 2.4GHZ, and the operating voltage is 3.3v why we preferred these modules. These modules require less power, and the execution speed is also very high compared with other processors.

Motion Planning

Path planning is essential for mobile robots to reach the target without any disturbance and collision. Trajectory tracking, path following remotely operated wheel-based robots are an important part of motion planning. Light-based sensors are used in the sprinkler robot.

In specific infrared sensors are used to identify obstacles through distance measurement and vision systems [8].

Wheel System

The choice of wheel types and number of wheels for a mobile robot is strongly linked to the choice of wheel arrangement. In sprinkler robots, four wheels have been selected because these robots are more stable than two or three-wheeled robots. In this type of wheel system, the center of gravity is located inside of the rectangle. The wheel is made up of polyethylene material and which is harder and more durable than many polymeric materials. The wheel geometry designed suitable for high velocity in flat floors.

Battery

A battery is an electrochemical device that converting electrical energy into chemical energy. Normally three types of batteries are used in mobile robots such as Lithium-polymer, Lithium-ion, and Alkaline [9].12V Lithium-ion battery is used in the sprinkler Robot. These batteries provide longer storage capacity and durability.

Tank

The sanitizer tank is made up of poly ethylene material with capacity of 3 litres.

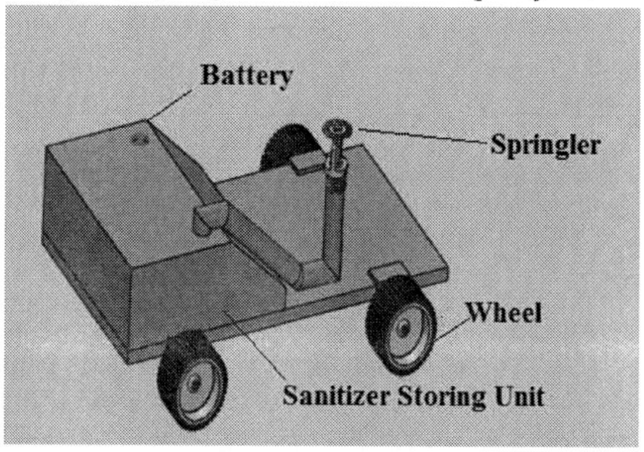

Figure.1 3D Model of Sprinkler Robot

Remote Operating System

Commonly wheel-based robots face some difficulties when traveling in complex environments such as steps and obstacles area [10]. To avoid this problem, vision systems and remote-based wheel operating systems are used in mobile robots. The working principle of the automatic sanitizer sprinkler robot is similar to a Television remote. When pressing the button in the remote, it obeys the input signal through the app. Initially, the input data reaches the Bluetooth module, and then the signal goes to the Arduino module. The Arduino program passes the commands to the motor drive. During the forward and backward stroke of the cylinder, air enters inside and moves the sanitizer outside of the tank.

FORWARD-F

Forward function F button allows robot to move forward when user gives F command in terminal mode.

BACKWARD-B

Backward function B button allows robot to move backward when user gives B command in terminal mode

LEFT-L

Leftward is the function used in Loop block it allows robot to move left side when user gives L command in terminal mode.

RIGHT –R

Rightward is the function used in Loop block it allows robot to move right side when user gives L command in terminal mode.

PUMP START-O

Pump Start is function used in Loop block it allows to start pump when user gives O command in terminal mode

PUMP STOP-G

Pump top is function used in Loop block it allows to stop pump when user gives G command in terminal mode.

STOP-S

ST in terminal mode it will send certain frequency to the Bluetooth module and the robot will stop automatically.

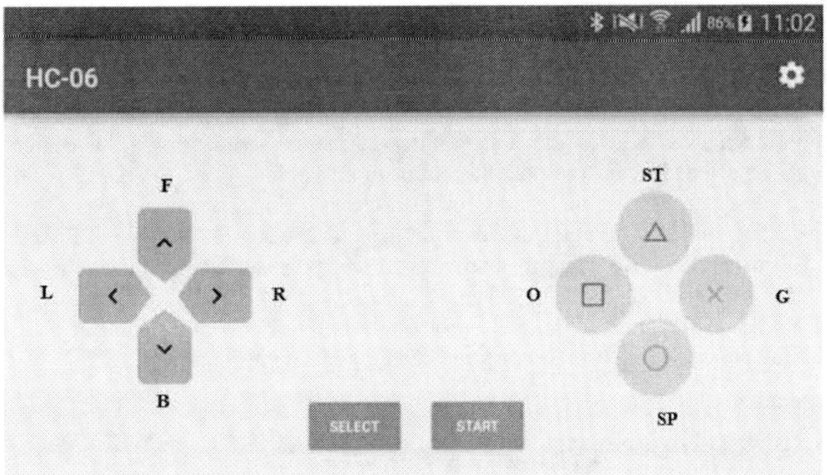

Figure 2. Mobile Controller mode

3. Results and Discussion

I. Design Calculations

(i) Sprinkler flow rate

Theoretical

$$Q=k\sqrt{p}$$

p=20 psi

k=5.6

$$Q=5.6\sqrt{20}$$

Q= 25 GPM

Q=Flow Rate (GPM), P=Operating PSI of head/Outlet

K=K Factor of Head/outlet

(ii) Analysis

For 1 litre, the flow rate of the Sprinkler is 476seconds.

For the project the flow rate of the sprinkler is 200 seconds.

The area of the sanitizer covered is 600mm.

The acquired Flow rate is 17GPM

(iii) Motor Specifications

Speed = 200 RPM,
Voltage = 12V,
Power = 100W
Torque of the motor
Torque = (P X 60) / (2 x 3.14 x N)
Torque = (100 X 60) / (2 x 3.14 x 100)
Torque = 9.554 Nm,
Torque =9.554 x 10^3 Nmm

(iv) Battery life calculation

Robot working hours for one full charge.
Batter capacity = 12v 7Ah (Ampere Hours)
Total device consumption = 520ma (mill ampere)
Battery Life = Battery Capacity in mAh / Load Current in mA
= 7000mAh/520ma
= 13.46 Hours

4. Conclusion

Autonomous mobile sanitizing robot is becoming most useful in COVID-19 hospital environments. It reduces the human intervention in sanitization. The designed system is very compact, so easily can transport this robot to any place. The COVID-19 pandemic presents even more reason to use mobile robots for safe cleaning in quarantine zones. The proposed model is fabricated and tested in a hospital environment. The system is capable of disinfecting an area of up to 100 m^2 per day. By using the autonomous sprinkler system optimizes the disinfecting areas and reduces the wastage of sanitizer. The designed system is capable of sanitizing an area of up to 100 m2 per day. The application area can include hospital corridors, medical shop, operation theatre, walking pathways, doctor room, testing center, and patient room, etc.

Appendix

ARDUINO Programming

```
#include<SoftwareSerial.h>
 int left_p=4;
int left_n=5;
 int right_p=6;
 int right_n=7;
Software Serial my Serial (0, 1); // RX, TX
String data;
int bt Val;
void setup() {
Pin Mode (left_p, OUTPUT);
Pin Mode (left, OUTPUT);
Pin Mode (right, OUTPUT);
 pin Mode(right, OUTPUT);
}
void loop() {
if (Serial.available()>0)
{
int signal = Serial.read(); switch(signal)
```

```
{
case 'F': Forward ();
break;
case 'B': Reverse ();
break;
default: break;
    }
    }
    }
  void Forward(){
  digitalWrite (left_p, HIGH);
  digitalWrite (left_n, LOW);
  digitalWrite(right_p, HIGH);
  digitalWrite (right_n, LOW);
```

References

[1] Joao Rolim and Jose Teixeira 2016 The design and evaluation of travelling gun irrigation systems Enrolador software *Engenharia Agrícola* **36(5)** 917-27.

[2] Asafa T B, Afonja T M, Olaniyan E A and Alade H O 2018 Development of a vacuum cleaner robot *Alexandria engineering journal* **57(4)** 2911-20.

[3] Zant, Chawki, Klement, Nathalie, Bettayeb, Belgacem, Sahnoun, Mohammed and Havard, Vincent 2018 UV-Robot supervision system design and development.

[4] Francisco Rubio, Francisco Valero and Carlos Llopis-Albert 2019 A review of mobile robots: Concepts, methods, Theoretical framework and applications, *International journal of advanced robotic system* **16(2)** 1729881419839596

[5] Dengqi Cui, Xueshan Gao and Wenzeng Guo 2016 Mechanism design and motion ability analysis for wheel/track mobile robot *Advances in Mechanical Engineering* **8(11)** 1–13.

[6] Joao Rolim and Jose Teixeira 2016 The design and evaluation of travelling gun irrigation systems: Enrolador software *Journal of the Brazilian Association of Agricultural Engineering* **36(5)** 917-27

[7] Harishankar S, Sathish Kumar R, Sudharsan K P, Vignesh U and Viveknath T 2014 Solar Powered Smart Irrigation System *Advance in Electronic and Electric Engineering* **4(4)** 341-6

[8] Velásquez-Aguilar J G, Aquino-Roblero F 2015 Hybrid object detection vision-based applied on mobile robot navigation *International Conference on Mechatronics, Electronics and Automotive Engineering* 51-56

[9] Alberto Brunete, Avinash Ranganath, Sergio Segovia, Javier Perez de Frutos, Miguel Hernando and ErnestoGambao 2017 Current trends in reconfigurable modular robots design, *International Journal of Advanced Robotic Systems* **14(3)** 1–21.

[10] Shiroma, N., Chiu, Y.H., Min, Z., Kawabuchi, I. and Matsuno, F., 2006Development and control of a high maneuverability wheeled robot with variable-structure functionality *IEEE/RSJ International Conference on Intelligent Robots and Systems* 4000-4005.

IOP Conf. Series: Materials Science and Engineering 1059 (2021) 012071 doi:10.1088/1757-899X/1059/1/012071

A Novel System for Surgical Gloves Removing and Reusing - IoT Approach

Priyadharsini. K[1] ,Dinesh Kumar J.R[2] , Deepakkumar A[3], Chandran S[4] Ganesh Babu C[5]

[1,2]Assistant Professor, Department of Electronics & Communication Engineering, Sri Krishna College of Engineering & Technology, Coimbatore, Tamil Nadu ,India,
[5]Professor and Head, Department of Electronics & Instrumentation Engineering, Bannari Amman Institute of Technology, Sathyamangalam, Tamil Nadu ,India,

Email : *[1]priyadharsinik@skcet.ac.in bits_babu@yahoo.co.in

Abstract. COVID -19 is a threat to any living species. This has not left even animals. Hence, proper precaution should be taken to avoid the spread of this deadly disease. This affects humans, which cause severe respiratory disease. Glove is the important element to stop spreading. But proper removal of glove is not maintained. Fully automated sensor unit is built for hand sensing and gloves removal unit is designed using microcontroller device to remove the gloves. Mechanical setup with the rod is designed to remove the glove from hand. An output LCD display that display the step by step status of removal process. Gloves contamination is measured using thermal image camera. If the gloves are sensed as reusable then from the smart dust bin it is send to further sanitization process. Else the gloves are completely discarded. We know the COVID -19 stays in the surgical gloves for 8hrs. So the reusable gloves are proceeding for the further processing of sanitation and sterilization. Smart bin is placed to collect the reusable glove.

1 Introduction

World is affected by a dreadful respiratory diseases SARS-CoV-2 (Severe Acute Respiratory Syndrome). It is a complete collateral damage all over the world. Many countries are affected severely economically. All transportation, businesses sectors, schools are remains close till now still no positive improvement from the medical side to subside the disease. All look into the daily update of WHO (World Health Organization) report related to COVID -19. It started to spread from China, Wuhan state (Zhongnan Hospital) where, the first case was reported. Later it spreads rapidly and now it affects almost all the countries. It is started with the mild fever and later it turns out to be a severe respiratory disorder, which even leads to death. The dangerous diseases that have stared already take its count on death. And no outbreak in the medicinal is more concern. With the report of WHO social distancing, maintaining one meter distance between person and wearing mask are made a regular practice. Now government has made it as a mandatory rule in many countries. This has made the impact in such a way if any one violates rules they are punishable. Therefore, this COVID -19 is a threat to all living species. This virus has not left mercy to animals. Even cases are file that COVID - 19 can affect to animals and they too die from respiratory disorder. This threat has rise up to 10 million cases by the end of August 2020. This may turn to double the rate of spread of the diseases if the current scenario is formulated. This cross the stages of community spread and many countries have seen the second wave of COVID – 19. Only this can be controlled, if all stand together and fight against it. Globally many researchers, academics [1,9], national governments agree to make this situation as more prioritized and word in research and development. With this pandemic situation WHO has developed nearly 150 guidance documents on various aspects of preparedness [11]. A weekly survey of reposts is circulated by WHO for the spread of this disease. The Figure 1 and 2, reports the spread of COVID 19 confirm cases world wide as per the report release 207. Through online and in-person training many concern was taking to the people even in remote areas. Through all possible social networking the spread of virus knowledge has to take to all age groups of people. WHO helped almost 80% of the country regional office to take collaborative initiatives for Global Outbreak Alert and Response Network (GOARN). It also helps to translate guidance and strategies into national plans. ACT tools have been launched to measure medical development [13]. The spread scale of this disease is more rapid and it even breakdown the most resilient health systems [2]. This does not have any pattern change and its states are more importance specialty of this disease. It changes its nature based on the geographical nature of the county. It is continuously changing its nature so prediction of anti bodies is delayed. When the Severity is taken then it is estimated as 20% cases are severe or critical, with an increased risk of spreading.

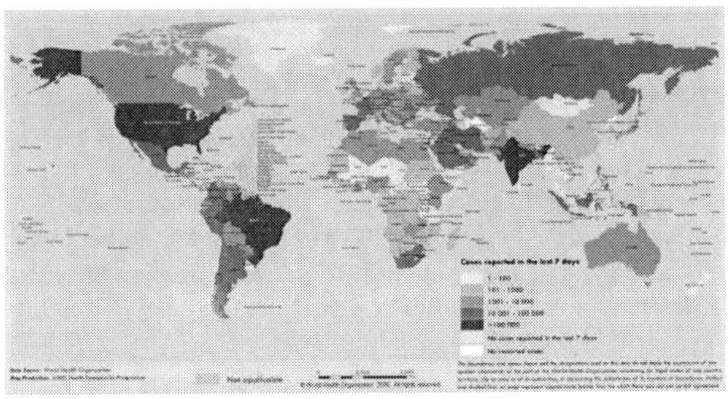

Figure 1. Number of confirmed* COVID-19 by WHO situation report - 207.

The WHO published strategic response and preparedness plan to tackle the disease spread. To limit the harm caused by the disease SPRP has set three objectives. First act of plan is to support countries to plan rapidly establish steps needed financially. It also maintains the correct supply of medicines and equipment evolving epidemiology and risks [3]. It is mandate to give guidance on latest technical training about it. Second, the research and innovation is given top priorities to find cure to SARS-CoV-2. SPRP can help fast track the research and development process and last tackle states that the operational planning during emergency condition.

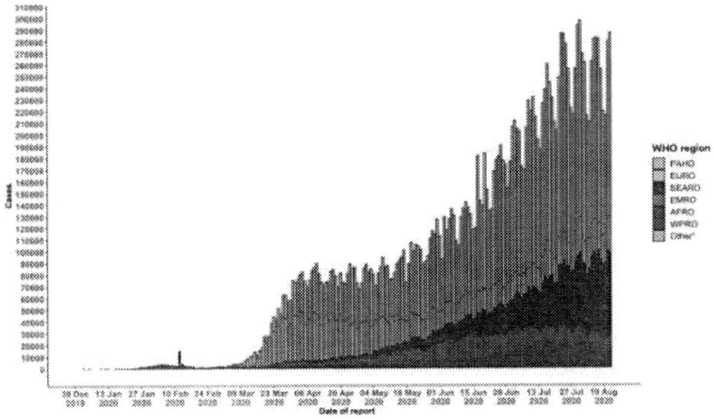

Figure 2. Graphical representation of Number of confirmed* COVID-19 by WHO situation report - 207.

1.1 Coronavirus - Impact

Why is this COVID -19 seen very pandemic?. What are it is effects? We know globally all activities are stopped due to the pandemic outbreak. This COVID – 19 is a virus that causes illness in both human and animals. In case of humans this corona viruses cause several respiratory disorders started from common cold to more severe breathing problem as Middle East Respiratory Syndrome (MERS) and Severe Acute Respiratory Syndrome (SARS). The common symptom found due to the affect of this disease is cold and fever [4]. The COVID – 19 structures was shown in figure 3.

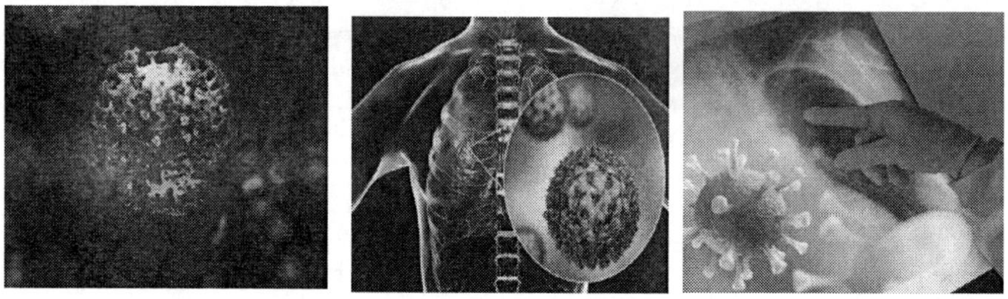

Figure 3. COVID -19 structures and the deposition of virus on the patient lungs causing difficulty in breathing.

The most common symptoms were fever and cold [2]. Some may experience with nasal congestion, runny nose, sore throat or diarrhea. In some cases people get infected by who does not have any of such symptoms. Most people get out of these without any special treatments but older people and those who have problem with high blood pleasure and other heart and lung disease are affected severely [3]. This is spread from person to person, when the infected person sneezes or coughs. Then small droplets of virus from their body are spreads to the environment. The droplets fell to the land of any objects and when any person is touch, the droplets then it spread to healthy person.

Table 1. Life span of SARS-CoV-2 in various metals and house utensils

Surface Type	Life time on Surface
Metals (doorknobs, jewellery, eating utensils)	5 days
Wood	4 days
Plastic	2 -3 days
Stainless Steel	2 -3 days
Cardboard	24 hours
Copper	4 hours
Aluminium	2 -8 hours
Glass	5 days
Ceramic	5 days
Surgical Gloves	8 hours
Paper	5 days

The healthy person if breath the virus or touch the virus where it landed and unknowingly touches his/her mouth, eyes and nose then it goes inside any human body. Therefore, it is advisable to have 1 meter distancing when we talk with a person. This can stay for long time in the landing objects. The table 1 represents the duration of SARS-CoV-2 in various metals and house utensils.

1.2 Protective measures

Be updated with the precaution and information release about COVID-19 outbreak by state and local public health authority. The spread of this disease is started from china but we see it is be sustained in major cities and spreading is slow down. We can reduce it by:

- Regularly cleaning our hands by sanitizers or soap to stop spread of virus into our body through hands.
- When we come across the person with cough and sneezing 3 feet (i.e., 1 meter) distance.
- Our hand is the source of virus that gets inside our body. So, avoid touching eyes, nose and mouth. Because we place hands in many areas where, there may be change of virus particles.
- Look around the people we stay, if any person shows respiratory problem inform it to regional hospital and maintain distance with the person's family members.
- Self isolate if we feel even mild cold and avoid contact with family members.
- Calling in advance the health care facility is more importance so that we could stop spread to other family members.

1.3 Role of Doctors and Health care workers

As we know the world is totally stopped due to this pandemic outbreak there are some people who are in frontline on to stop spreading of this disease and to find cure. Globally people are insisted to stay indoor the frontiers who work for us to stop spread of the disease [5] and they are preferring the smarter ways to reduce the stress like smart farming [6] & expect low power models of computations [7 & 8]. India has received the first case of virus with the travel history from abroad. Doctors and health care worker are working excruciatingly day and night for the welfare of patients even after knowing the severity of the disease. Without the proper cure it makes situation too worse for doctors and other health care worker. Doctors have to attend patients and also they have their own family. So risk of spreading virus to their own family members is too high. The threat of catching the disease and that may spread to their family members they stop being to go to their homes. They also extended their duty hours even they work 18hrs per day [9 & 15]. With the report of doctors from AIIMS who continuously monitor the COVID – 19 patients, they could not get a solution for treatment of this pandemics outbreak. With Personal Productive Equipment (PPE) government insists continuously about the important of the wearing mask and gloves to reduce the chance of spreading of the disease. With the PPE kit doctor face much difficulty and they are in the apathetic situation that wearing of PPE kit more than 7hrs is not possible. But with this continuous work in corona ward, it is meant mandatory to wear mask and gloves. The study says that several doctors and health care persons are affected with SARS-CoV-2 while treating patients. The Indian Council of Medical Research (ICMR) filed a report that the chance of virus to spread to doctors is too high. There are certain procedures to remove gloves and mask [15]. But because of miss leading in the procedure they may have the high risk of getting this pandemic disease. Wearing of PPE equipment is another issue which was shown in figure 4. It has multi layer of protective clothing which profuse sweating. Spectacles and Goggles fog up affects the clear vision. The breathing trouble also there in using face masks and goggles. Lowering the mask and removing the gloves is often done by the doctors due to breathing difficulties [16] and it is more better to biomedical based wearable devices to cause the changes [17 & 18].

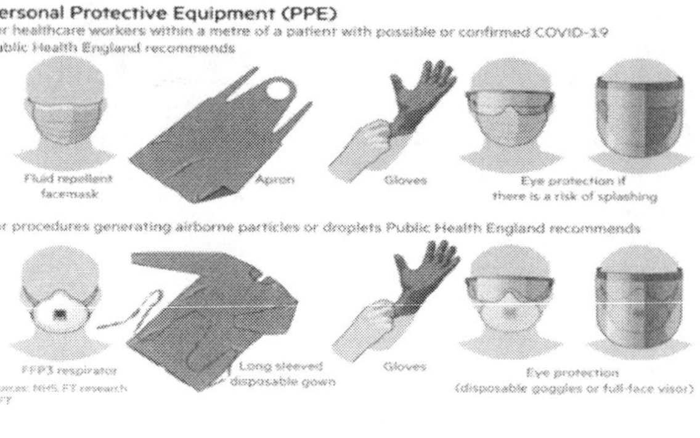

Figure 4. Personal Protective Equipment (PPE)

Hence, to make them more comfort on removing the gloves and mask is regular habitual of doctors. It increases the risk of spreading the disease from patients. There are certain procedure to follow while removing and wearing of gloves and PPE. However, those are not following properly and doctors and health care people are affected by SARS-CoV-2. So, the proposed a system focuses on the ways by which reduces the risk of spread of disease.

2 Proposed Methodology

When the gloves are removed often by the doctor, they do not follow the procedures properly. The dangerous virus stays in gloves for 8hrs. So it is important to follow the procedure to stop spreading of the disease.

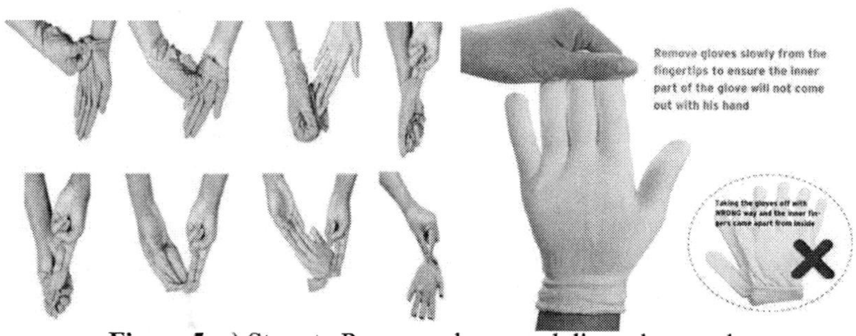

Figure 5. a) Steps to Remove gloves and discard properly
b) Improper way of removing the gloves leads to spread of COVID -19

As shown in Figure 5a and 5b the improper way of removing the mask may leads to chance of spread of disease. Many researchers who test the virus for getting proper vaccine, health care workers who dispose the medical wastes and other medical representatives use these kinds of gloves regularly. But removal procedure is not properly formulated then they may get chance of spreading of virus. There are different types of gloves available in the market. But we study about the surgical gloves which will be used by doctors regularly. With the proximity sensor and image processing technique [5] the complete automated process for gloves removal is done and that reduce the spread of the pandemic virus like SARS-CoV-2.

2.1 Fully automated sensing and Gloves removal block

Fully automated sensor unit is built for hand sensing and gloves removal unit as shown in the Figure 7. A microcontroller that gets input from the sensor and the other unit connect to it and performs the activity of completely automated hand gloves removal. With the image processing the gloves can be separated as reusable and non-reusable. Smart IoT device with training algorithm performs the operation with lower power consumption [19 & 20]. If the gloves are sensed as reusable then from the

IOP Conf. Series: Materials Science and Engineering 1059 (2021) 012071 doi:10.1088/1757-899X/1059/1/012071

smart dust bin, it is send to further sanitization process. Else the gloves are completely discarded. We know the COVID -19 stays in the surgical gloves for 8hrs. So the reusable gloves are proceeding for the further processing of sanitation and sterilization. A movable rod is used to remove the gloves from the hands. They are placed inside the sanitizer block to avoid infections. Ionizing rays are passed to movable rod each time, when one person completes the removal [21]. Ionization is the used to remove the microbiological [7 & 8] contamination in the rod while removing the gloves. High voltage electron beam 10 MeV is passed and gamma rays with radioisotopes Co-60 and Cs-137 are used for removal of microbiological contamination. With the passage of gamma rays that has the capability of killing micro organism. Co-60 is widely used because it can be easily manufactured from natural metals. This gamma radiation sterilization is more for larger scale in commercial manufacturing and also in hospitals [16 & 14].

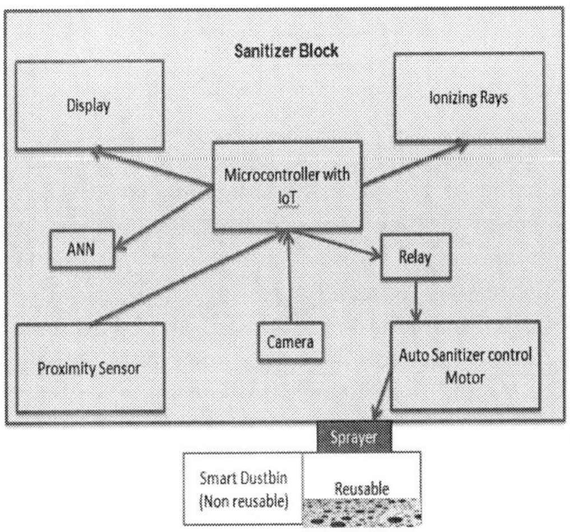

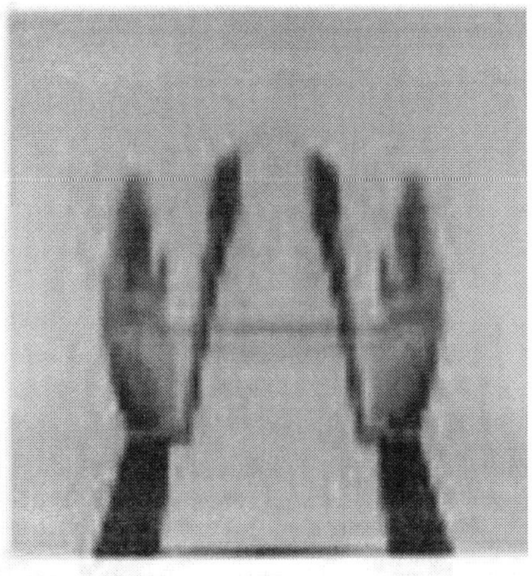

Figure 6. a) Sanitizer block diagram and Circuit implementation for sanitizer block b) Gloves fit in movable rod

This is declared as safe to use by US Center for Disease Control and Prevention, the United Nations and the World Health Organization. The ionization radiation is passed based on the range coated in [1].

2.2) Methodology

The working of the sanitizer block is to sanitize the movable rod each time after removal of gloves. When any health care worker is in need of removal of gloves, then they march downward from the sanitizer block towards the automatic hand sensing unit which, is also placed inside the sanitizer block [3] like a cardiac pacemaker to heart [2]. For the health care worker who tries to remove the gloves an automatic sanitizer block will sense the hand and removes the gloves. This helps them from prevention of virus spreading. Infrared proximity [10 & 6] sensor is used for automatic detection of presence of hands as shown in the Figure 6 a.

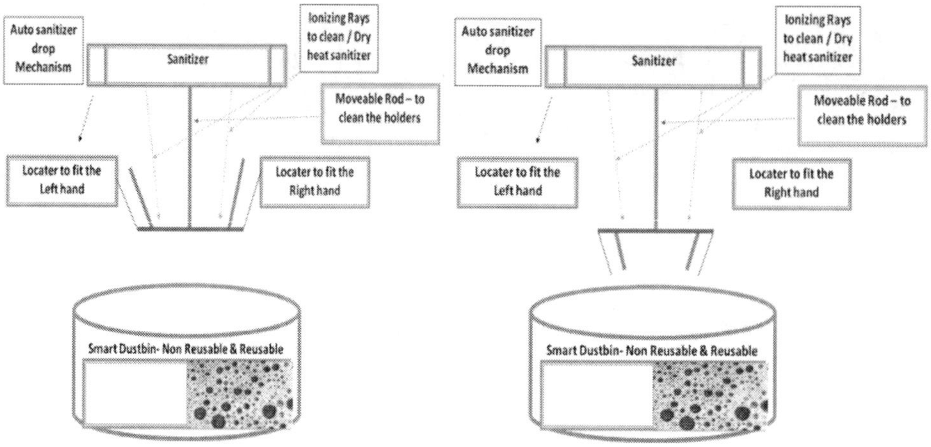

Figure 7. Complete automated mechanism for removal of gloves

When the presence of hand with gloves is detected then the sensor signals are transmitted to the microcontroller. Hence, the movable rod is inducing a start with the solenoid drives. The voltage variation cause the solenoid drives to get movement of pushing and pulling of movable rod. We have contact free environment, which will reduce the spread of virus. This is highly preferable in the current situation of COVID -19. We know that because of frequently wearing and removal of gloves the doctors and other health care worker are not often maintain the follow up procedure for removal [11 & 12]. Hence there may a chance of spreading of infection inside the body through hands. The sanitizer block contains the automatic hand detection unit when the hand with the gloves is place it will sense the type of glove and an rod from the sanitizer block take its path downward to collect the gloves.

The health care worker should fit the hands inside the removable section of rod as shown in the Figure 6 (b). Then a pull down a movement to removes the gloves from the hand. When the gloves removed from the hand, again the IR sensor will detect the completion of process and proceed to next level. In the next step, the camera captures the image of gloves and further processing is done at gloves contamination block. Display section in the Figure 6 displays the hand detection message, completion of glove removal message and the next step of segregation of reusable and non-reusable gloves. Smart IOT device performs the calculation related to the result obtain [4 & 19]. Once the gloves are being removed they are disposed to the smart dust bin for the further process. The gloves can be classified to reusable and non reusable in this process. It is done by smart dust bin which have a colour sensor to measure the quality of the gloves as one parameter. Gloves quality is measured in mile. The figure 8 shows the identification of different surgical glove. In medical field mostly two types of gloves are used which is Nitrate and latex. The first step in the process of reuse of glove is to identify the type of glove. Usually gloves are available in 2 to 15 mile of thickness [15 & 16]. Threshold is set to differentiate the glove as mentioned in the figure 8. If the threshold is maintained between 0.15mm then it should be a type of nitrate glove else it will be identified as latex glove.

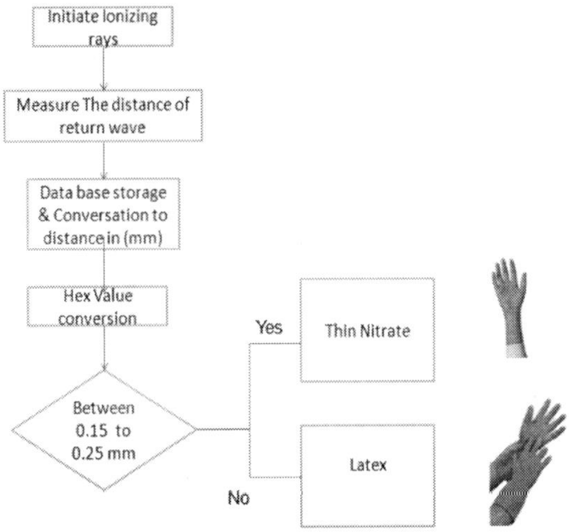

Figure 8. Flow Chart to Identify the type of Material

Gloves contamination is done using thermal image camera. With the white representation of warm, hot and bright object the image is segmented from its surrounding. The blobs in the infrared image are extracted and intensity of the image is adjusted with the mean if original image. During segmentation the edge detection and spoken filtering process are done. The first order statistics is give by,

$$T = k_1 \times \mu + k_2 \times \sigma \tag{1}$$

For low resolution image k1 = k2 = 1 and for high resolution image the value if k1=k2= 1.5 to 2. The values of k1 and k2, are determined manually. "T" value should be maintained greater than 255. Two step processes is done for feature extraction and it needs to be located first [14 & 3]. Then secondly threshold is obtained from region grown around the defected seeds. Threshold for image f should be T = f(i,j), agglomerated of T with n pixels are calculated. The brightness is maintained as multiplication factor of n and T. This is repeated until the boundary of the pixel is reached.

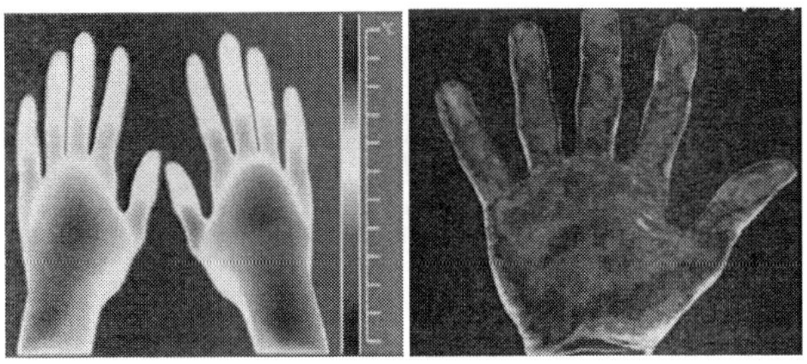

Figure 9. Heat variation on Non – Defective Gloves

Hamadani's based algorithm is not fully automated the centre crossbar is detected by considering 50 pixels for each seed. Olivo's method is an automatic detection with dilated wavelet set [18]. Optimal threshold image is obtained with the maximum level of sum of two class entropies which is represented by histogram. Many algorithms were proposed to get the accurate result. But we use local intensity based neighbouring operation. This has the capability to brighten the dark area and dark the brighter area. And in the second stage this does vies-versa. This technique has an assumption that the defect is present at the brightest area. Consider the pixel f(i,j) with the grey scale representation from f(i-1, j-1) to f(i+1, J+1). This can be simplified as Z. The figure 9 shows heat variation on non-detective gloves.

$$Z = \prod_{k=0}^{8} z_k \qquad (2)$$

IBO and IDO is given in equation 3. Pixels are normalized using maximum pixel value is obtained after each iteration.

$$g(i,j) = Z$$
$$g(i,j) = \begin{cases} \frac{1}{Z} & \text{for } Z \neq 0 \\ 1 & \text{for } Z = 0 \end{cases} \qquad (3)$$

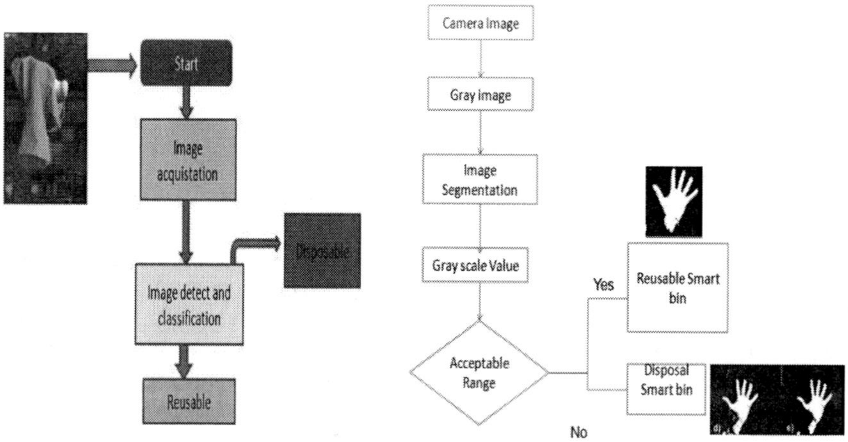

Figure 10. Flowchart for Quality Inspection of Gloves

Pixel got reallocated with the histogram image processing. IBO can be applied repeatedly and IDO is applied only once. In need of further separation IBO is applied again. Mean absolute thresholding is calculated with max and min values obtained. With the obtained results the defective glove is identified. The quality inspection process for the glove was shown in figure 10.

3 Results and Discussion

The value of k1 and k2 determine the threshold results. The high resolution and low resolution image have different impacts. The values have been adjusted for different images as they have variation in the resolution shown in the figure 11 and 12. The histogram plot representation of fault detection is also shown.

Sample Glove 1

Sample glove 2

Sample glove 3

Figure 11 a) Histogram plot of faulty gloves b) Fault detection in gloves

Number of defected glove is analysed. IR analysed the 97% of the defects correctly. 8 out 9 defects are shown perfectly. Reliability of IR sensor camera identifies the fault carefully.

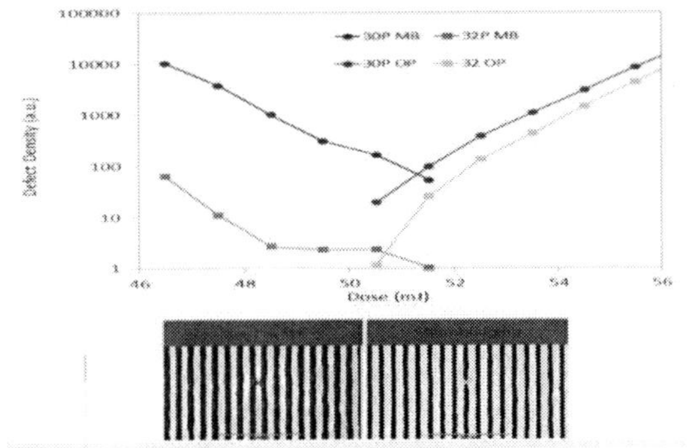

Figure 12. Number of defects

Pareto chart analysis for faulty gloves. This is also known as 80/20 rule. The data set is taken after the pre processing and image segmentation. The value of Pareto gives 20% value with 80% problems. The set of 10 is processed to find the defect percentage. The defective glove is identified from the batch and they are segregated. Identifying different defect using Pareto algorithm is shown in the figure 13. (b).

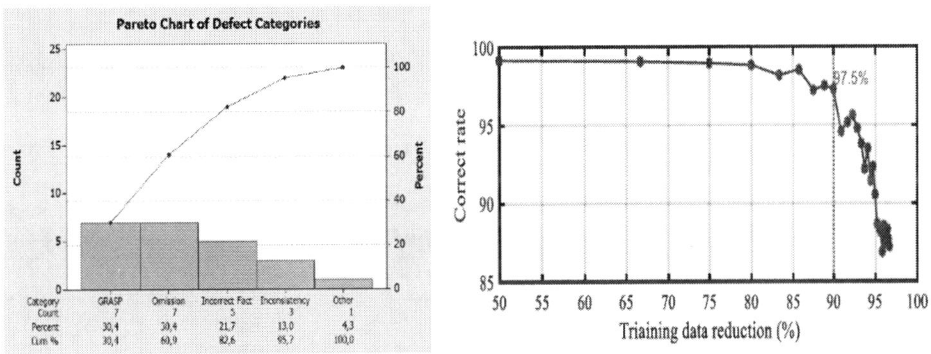

Figure 13. a)Pareto chart for Defect gloves b)Accuracy graph

Accuracy is one important factor to consider which is shown in figure 13b. Surgical glove is used to avoid the harmful germs that get inside our body. So the proper sterilization and checking of defects is done with 97.5 per accuracy that shown in the figure.

4 Conclusion

SARS-CoV-2 a deadly virus which affects all over country and it cause severe respiratory problems. The wearing of gloves by health care person like doctors, medical representatives and medical trash collectors are to be taken care for preventing further spread of dangerous disease. Hence our proposed model provides the proper way of removal of gloves and it uses the smarter technology on classifying them and the information are processed via IOT cloud. The lifespan of that virus is 8hrs in gloves. The associated sensor unit for detection of hands and the movable electronically controlled rod to hold them and smart dustbin to separate them are connected together to enhance the virtual understanding of the proposed model. Henceforth, proper removal of gloves and the contamination leave are connected measured and stored in the cloud for the future purposes also the hardware block is controlled by the microcontroller which gets input from the sensory unit and actuators connect to it. It activity performs the automated hand gloves removal followed by sanitation and gloves classification. Also the reusable of surgical gloves after sterilization is composed using image processing technique. Smart bin is fitted with the entire model to collect the gloves and image processing technique is used for classification of reusable and non reusable glove. The faulty glove and reusable glove can be collected separately using smart bin. Henceforth, a better community can build using this model which focuses the cut down of spreading from the origin of creation.

References

[1] Giorgiana M. Catunescu, Maria Tofana Effects of Ionizing Radiation on Microbiological Contaminants of Foods, Bulletin UASVM Agriculture, 67(2)/2010, Print ISSN 1843-5246; Electronic ISSN 1843-5386.

[2] Priyadharsini, K, Nanthini N, Soundari D V, Manikandan R 2018 Design and implementation of cardiac pacemaker using CMOS technology , ISSN 1943-023X, *Jour of Adv Research in Dynamical & Control Systems*. **10** 1157-61.

[3] Soundari D V, Padmapriya R, Thirumariselvi C, Nanthini N, Priyadharsini K 2019 Detection of breast cancer using machine learning support vector machine algorithm, *Journal of Computational and Theoretical Nanoscience*. **16** 441–4.

[4] Nanthini N, Soundari D V, Priyadharsini K 2018 Accident Detection and Alert system using Arduino *Jour of Adv Research in Dynamical & Control Systems*.**10** 1167-70.

[5] Dinesh Kumar J R, Priyadharsini K, Yazhini V 2020 Performance Metric Interpretation on Handwritten Digit Recognition Using Neural Networks *IJAST* **29(7)** 2639-51.

[6] Priyadharsini K, Yazhini V 2020 A Novel Methodology For Smart Soil Management And Health Assessment System To Support The Contemporary Cultivation Using Ai On Mobile App Environment, *IJAST* 7 **29(7)** 2620-30.

[7] Dinesh Kumar J R, Ganesh Babu C, Balaji V R 2020 Analysis of effectiveness of power on refined numerical models of floating point arithmetic unit for biomedical applications, *IOP Conference Series***764** 012032.

[8] Dinesh Kumar J R and Ganesh Babu C 2019 Performance investigation multiplier for computing and control application *Journal Advance Research in Dynamic and Control System* **13.**

[9] Pollard, E C 1966 Phenomenology of Radiation Effects on Microorganisms. p. 1-34. In: Encyclopedia of Medical Radiology. vol. 2. 2nd ed. ed. A. Zuppinger. Springer. New York. 13.

[10] Radomyski 1994 Elimination of Pathogens of Significance in Food by Low-dose Irradiation: a review *Journal of Food Protection* **57(1)** 73-86.

[11] Roberts, P B 2003 Irradiation of Foods. 3381-3396. In: Caballero. B., L. Trugo, P. Finglas (Eds.). Encyclopedia of Food Sciences and Nutrition (Second Edition).

[12] Molins, R A 2001 Food and Irradiation Principles and Applications. John Wiley & Sons Inc. New York.

[13] Monk, M. L., L. R. Beuchat and M. P. Doyle. 1995 Irradiation Inactivation of Food-borne Microorganisms. *In: J. Food Prot.* **58(2)** 197-200.

[14] Somwang Danchaivijitrmd, Pairoj Suwanasuthi, Rachada Cherdrungsi, Yuwadee Tuntiwattanapibul 2005 A study on powdering of reusable surgical gloves *Journal of the Medical Association of Thailand* **88(10)** 188-91.

[15] Dana M LeMoine, Valerie K Bergdall, and Carrie Freed 2015 Performance Analysis of Exam Gloves Used for Aseptic Rodent Surgery *J Am Asso. Lab Ani. Sci.* **54(3)** 311-6.

[16] Esah Yip,Paul Cacioli 2002 The manufacture of gloves from natural rubber latex *Journal of Allergy and Clinical Immunology* **110(2)** 3-14.

[17] Dinesh Kumar J R, Ganesh Babu C., Balaji V.R., Priyadharsini K., Karthi S.P. 2020 Performance Investigation of Various SRAM Cells for IoT Based Wearable Biomedical Devices. *Inventive Communication and Computational Technologies* **145** 573-88.

[18] Balaji V.R., Sathiya Priya J., Dinesh Kumar J.R., Karthi S.P. 2020 Radial Basis Function Neural Network Based Speech Enhancement System Using SLANTLET Transform Through Hybrid Vector Wiener Filter *Inventive Communication and Computational Technologies* **145** 711-23.

[19] JRD Kumar, N Dakshinavarthini 2015 Analysis and design of double tail dynamic comparator in analog to digital converter *IJPCSC* **7 (2)**106-12.

IOP Conf. Series: Materials Science and Engineering 1059 (2021) 012071 doi:10.1088/1757-899X/1059/1/012071

[20] JRD Kumar, CG Babu 2018 Survey on Consummation of Subsided Power and High Speed Adders For Mixed Signal Processing Applications *IJPAM* **119(12)** 1471-83.
[21] Priyadharsini, K., JR Dinesh Kumar, C. Ganesh Babu, P. Surendiran, S. Sankarshnan, and R. Saranraj 2020 An Experimental Investigation on Communication Interference and Mitigation during Disaster Using Lifi Technology *In 2020 International Conference on Smart Electronics and Communication (ICOSEC)* **20052072** 794-800.

IOP Conf. Series: Materials Science and Engineering 1059 (2021) 012072 doi:10.1088/1757-899X/1059/1/012072

PV Fed Grid Power Control by Discrete CUK Converter and SPWM Inverter Control Techniques

Viji K[1], Chitra K[2] and Lakshmanan M[3]

[1] Assistant Professor, Department of EEE, CMR Institute of Technology, Bengaluru, India.

[2] Associate Professor & Head, Department of EEE, CMR Institute of Technology, Bengaluru, India.

[3] Associate Professor, Department of EEE, CMR Institute of Technology, Bengaluru, India.

E-mail: kvijiperumal@gmail.com

Abstract. In this paper the power supplied to the grid is regulated with the help of Discrete Sliding Mode Controlled (DSMC) CUK converter and Sinusoidal Pulse Width Modulation (SPWM) topology controlled 3-phase Inverter. The CUK converter is used in the input side to Buck or Boost the PV (Photo Voltaic) array voltage to the required value of the 3-phase Inverter. Natural isolation of source and load is obtained in the CUK converter with the help of capacitor energy transfer so that the ripple content is less in both load and source side. These CUK converters are operating in continuous current mode so that regulation of power from the solar panel is easy and efficient compared to Buck-Boost converters. In this paper DSMC CUK converter topology is implemented so that converter is feasible to incorporate with any digital controller with less hardware requirements, it has flexibility in control characteristics and guaranteed stability of highly fluctuating solar power under different climate conditions. The simulation model has battery back-up which is charged during high irradiation time and provides power to load during less irradiation climate conditions. The 3-phase Inverter is controlled by SPWM technique to reduce the switching losses of the Inverter so that the harmonics in the load side is less and it is easy to implement. So that the overall system is efficient, ease of implementation, robust and economical.

1. Introduction

The power generation using renewable energy sources (RES) is in demand due to the growing need of electrical power and the exhaustion of conventional sources including the growing price of those. Solar energy is one of the RES which converts heat energy from the sun into electrical energy which is free of cost, will not pollute the environment and available enrich in nature. In the beginning days the price of the PV array is high with poor system efficiency; but due to energy demand and the research carried out in that related area leads to the development of new control techniques to overcome all the related issues in present days. So there is a need of efficient converter and robust control technique to stabilize the power produced by the PV array all the time irrespective of the climate conditions. Thus CUK converter is used in this application which is operating in continuous current mode so that it is easy to control the output power in all the climate conditions. DSMC technique is implemented to regulate the output power and the stability of the system is ensured by Lyapnov stability criteria [9]. The system is

IOP Conf. Series: Materials Science and Engineering 1059 (2021) 012072 doi:10.1088/1757-899X/1059/1/012072

robust due to the implementation of the Sliding Mode Control technique under unexpected uncertainties. The Perturb and Observe algorithm along with Incremental Conductance method is used to attain Maximum Power Point Tracking (MPPT) of the PV array but the power is fluctuating with increase in losses [1]. The voltage controlled MPPT and current controlled MPPT are implemented in paper [2] and the response was studied. The CUK DC-DC converters are used to transfer the energy produced by the PV module to the load side using MPPT algorithm with the increased energy transfer capability so that the switching losses are reduced and the efficiency increases [3]. So that CUK converter is selected in this paper for the PV fed grid application. Once the required DC power is received from the PV array using CUK converter it has to be efficiently connected to the grid to satisfy the load demand; for that there is a requirement of DC-AC converter called Inverter. The converted AC power from Inverter is stepped up by means of step-up transformer to send it to the transmission line and once again stepped down on the end before connecting it to the grid [4]. In this paper 3-phase Inverter is used to interface the CUK converter with grid and the Inverter is controlled with SPWM technique to reduce the switching losses and harmonics.

1.1. Block Diagram Description

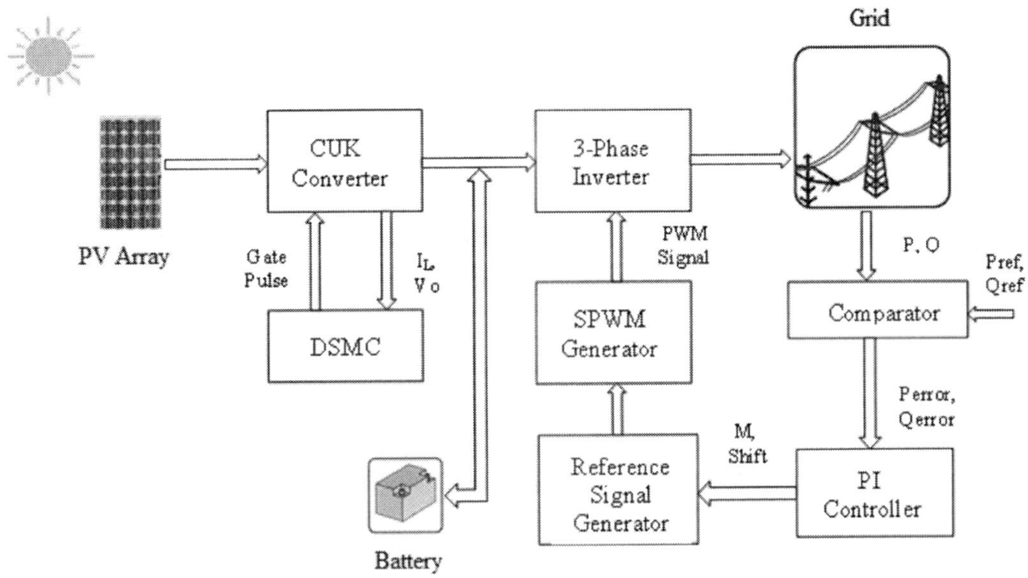

Figure 1. Block Diagram of PV Fed Grid.

The block Diagram of the PV fed Grid is shown in the figure.1. The heat energy of the sun is converted into electrical energy with the help of solar cells which is given as input to the CUK converter. The CUK converter is a type of DC-DC converter which can be step-up or step-down the input voltage according to the load requirement [5-6]. The inductor current and output voltage of the converter is sensed and the controlled with the help of DSMC. Since the controller is discrete it is easy to communicate with any digital devices; the generated gate pulse of the digital controller is given to the gate of the MOSFET. The controlled DC voltage is given as input to the 3-phase Inverter at the same time the DC voltage is charging the battery as well. During the night time and rainy season supply will be taken from the battery.

The 3-phase Inverter is controlled by measuring the real and reactive power of the grid. These values are compared with the reference and the error is obtained. With the help of PI controller, the error values are converted into proportional Modulation Index (M) and shift values. The reference signal generator generates 3-phase reference signal for Sinusoidal Pulse Width Modulation (SPWM) generator with respect to the new values of M and shift. The generated PWM signals are given the driver circuit of the 3-phase Inverter to control the AC voltage output of the Inverter connected to the grid. In such a way the active and reactive power of the grid is controlled.

2. Methodology of Work

2.1. Design of PV Array

The heat energy from sun which is a kind of renewable energy is converted into electrical energy with the help of PV array. It has series and parallel combination of PV module. The equivalent circuit of solar cell is shown in figure 2; it has one current source, one diode and two resistors in series and shunt to represent the losses in the cell. The voltage produced in the cell is given by V and the current flowing through the cell is given by I.

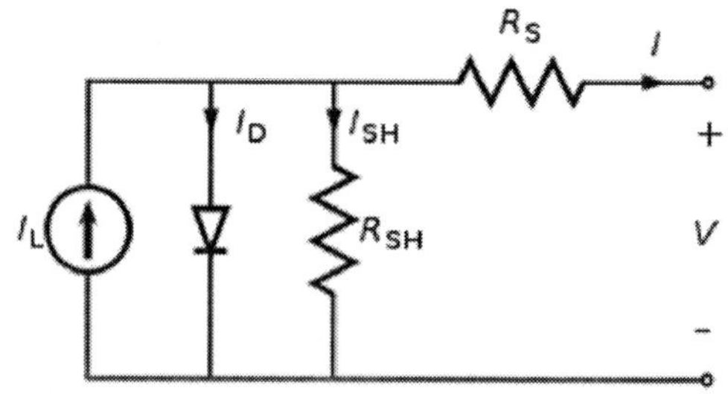

Figure 2. Solar Cell Equivalent Circuit

In this application the PV array of 1STH-215-P type is selected with 4 series modules and 3 parallel strings; the voltage, current and power response of the Array is shown in the figure 3 for 1KW/m^2. This response is obtained by selecting the irradiance of 1000W/m^2 with the temperature of about 25°C. The table.1 gives the selected PV Array Module data.

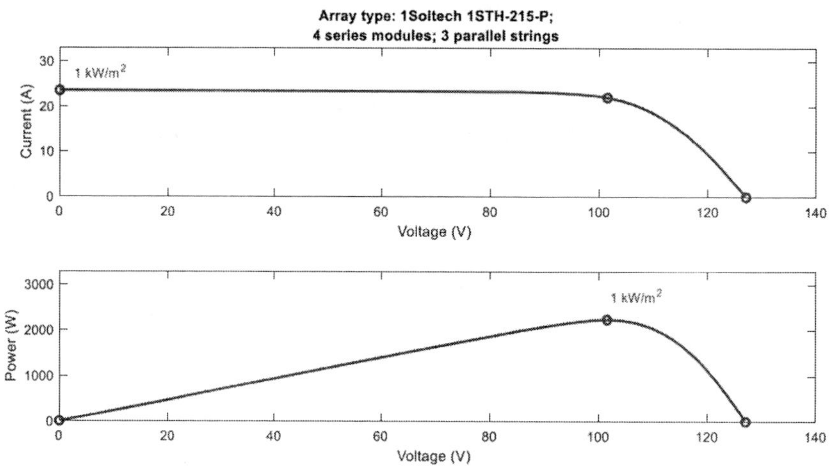

Figure 3. The Voltage, Current and Power Response of PV array Module

Table 1. PV Array Module Data

Parameter	Value
Module Type	1Soltech 1STH-215-P
Maximum Module Power	213W
Open Circuit Voltage Voc	36.3V
Voltage at MPP	29V
Temperature coefficient of Voc	-0.36(%/deg.C)
No of Cells per Module	60
Short-Circuit Current Isc	7.84A
Current at MPP Imp	7.35
Temperature Coefficient of Isc	0.102(%/deg.C)

2.2. Mathematical Modeling of DSMC CUK Converter

In this paper CUK converter is used as DC-DC converter which converts input PV array DC voltage to the required output DC voltage for the Inverter. These converters can step-up or step-down the input voltage similar to Buck-Boost converter with inverted output voltage. The schematic diagram of Cuk converter is illustrated in figure 4. This converter uses an additional capacitor C2 and inductor L2 along with C1 and L1 to buck or boost the input voltage. The inductor L1 is connected in series with the source Vs resembles the Boost Converter and the switch S1 is connected in shut with the source. The load resistance (R) of 150Ω is used which is connected across the capacitor C2 so that the load voltage is the voltage across the capacitor C2. The parameters to be under control is the inductor L1 current iL1 and the output voltage Vo which is nothing but the voltage across the capacitor VC2. The freewheeling diode D is used to charge and discharge the inductor current and capacitor voltage [7-8].

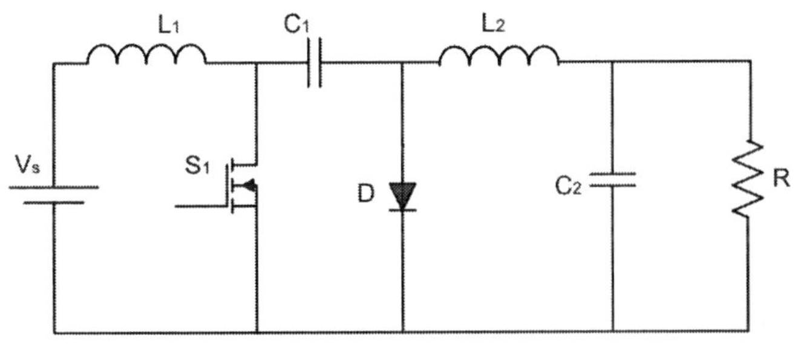

Figure 4. Circuit Diagram of CUK Converter

From the equivalent circuit the switch ON and OFF conditions are considered to obtain the state space model of the system. To ensure the stability the system should track the sliding surface given in equation 1.

The sliding surface is given by

$$S[k] = e_p[k] + \rho_1 e_i[k] + \rho_2 e_v[k] \quad (1)$$

Where ρ_1, ρ_2 are the switching functions of the sliding surface. Current Error is expressed as $e_i = (i_{L1} - i_{ref})$, Voltage Error is expressed as $e_v = (V_{c2} - V_{ref})$ and the power Error is expressed as $e_p = (V_{c2} i_{L1} - V_{ref} i_{ref})$.

The control law is given by

$$u = 1/2\,(1-\text{sgn}(S)) = \{(0 \text{ if } S>0 \text{ else } 1 \text{ if } S<0)\}. \quad (2)$$

The stability is obtained by Lyapnov Stability method. Table.2 shows the parameters used for the CUK Converter for the simulation.

Table 2. Parameters for CUK Converter.

Parameter	Value
Inductor L1	2.01mH
Inductor L2	1.09mH
Capacitor C1	830μF
Capacitor C2	49 μF
Load Resistance R	150Ω
Switch Frequency	40KHz
Sampling Frequency	400KHz

3. Results and Discussion

The 5KW Grid active and reactive power is to be regulated with the power factor of about 0.9. PV Array of Model: 1Soltech 1STH-215-P is used for this simulation with 4 series modules and 3 parallel strings are used to give the voltage of around 100V under the irradiance of 1000 W/m2 at 25°C. MOSFET is used as a switch for CUK converter and the simulation is done with the help of MATLAB/SIMULINK simulation software. Figure 5 Shows the CUK converter response in which 5 (a) shows the PV array response of 100V and 20A current. Figure 5(b) shows the generated PWM signal for the CUK converter using DSMC technique. Figure 5(c) shows the CUK converter output voltage which is boosted of about 130V.

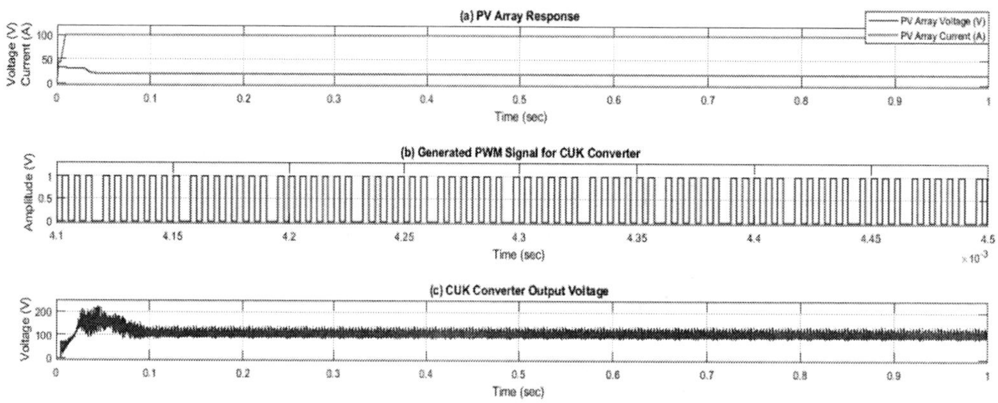

Figure 5. CUK Converter Response (a) PV Array Response (b) Generated PWM Signal for CUK Converter (c) CUK Converter Output Voltage

The output of the CUK converter is given to the 3-phase inverter. Grid voltage and current are sensed active power (P) and reactive power (Q) values are calculated; the calculated values are compared with the references and the error values are obtained. Proportional-Integral (PI) controller is used to control the error values and to generate the modulation Index (M) and the phase shift. These values are given to the 3-phase reference signal generator to produce the Sinusoidal Pulse Width Modulated (SPWM) gate pulse for the 3-phase Inverter [10]. The SPWM signals generated for the 3-phase Inverter switches is shown in Figure 6 (a-f).

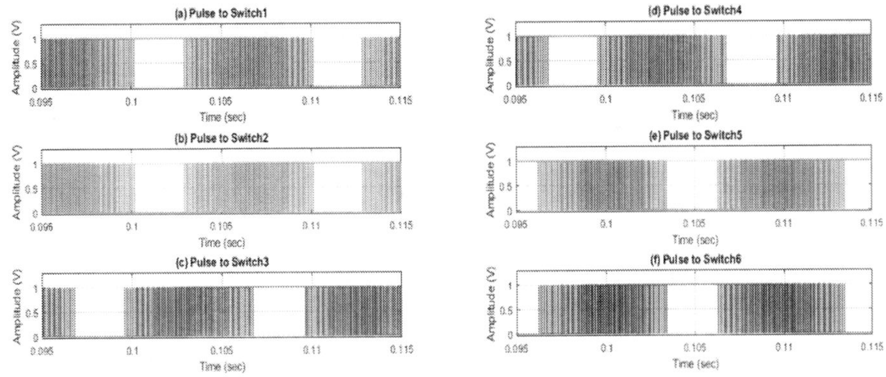

Figure 6. SPWM Signals Generated for 3-Phase Inverter

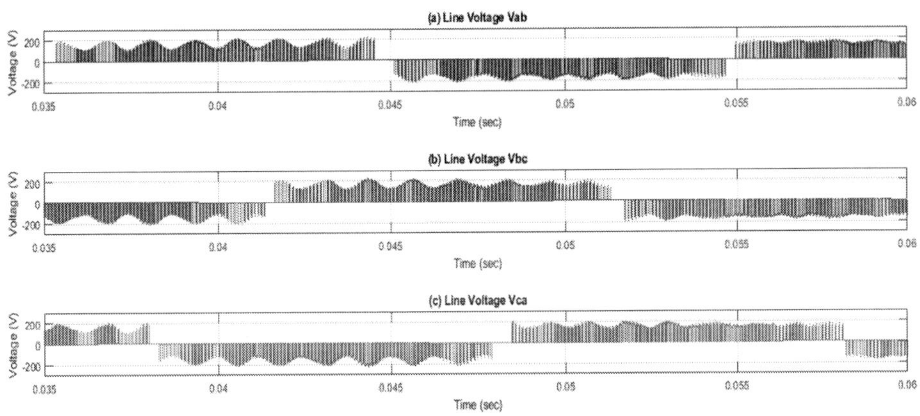

Figure 7. Line Voltage of 3-Phase Inverter

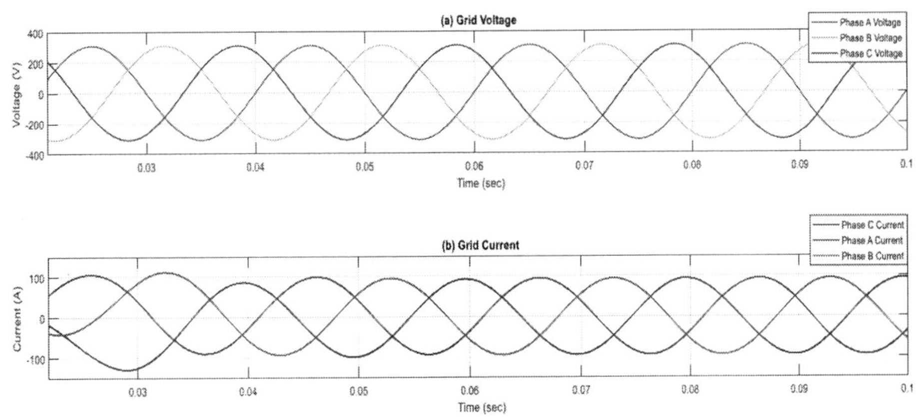

Figure 8. Response of the Grid

The response of the 3-phase Inverter is shown in figure 7 in which 7(a) gives the line voltage V_{ab}, 7(b) gives the line voltage V_{bc} and 7(c) shows the line voltage V_{ca}. The maximum voltage of about ±200V is obtained from the 3-phase Inverter. The voltage and current response of the grid is shown in the figure

8 with the maximum voltage of 300V and maximum current of 100A. The total harmonic distortion of the grid current is less than 5% and it is the accepted range according to IEEE standards.

4. Conclusion

The PV fed Grid power using DSMC CUK converter and SPWM controlled Inverter is simulated using MATLAB/Simulink software and the responses are plotted. The stability and robustness of the system is achieved by DSMC algorithm and the performance of the system is good for the range of switch functions of the sliding surface ρ1 from 150 to 200 ρ2 is from 150 to 300. The Inverter is controlled with the help of SPWM technique which reduces the harmonics of the load current as less than 5% so the controller is more efficient.

References

[1] D. P. Hohm M. E. Ropp,Comparative 2000 Study of Maximum Power Point Tracking Algorithms Using an Experimental, Programmable, Maximum Power Point Tracking Test Bed *IEEE* pp 1699-1702

[2] Peftitsis D et al 2008 A new MPPT method for Photovoltaic generation systems based on Hill Climbing Algorithm *Proc. International Conference on Electrical Machin* 1-5

[3] Azadeh Safari and Saad Mekhilef 2011 Simulation and Hardware implementation of Incremental Conductance MPPT with Direct Control Method Using Cuk Converter *IEEE Transactions on Industrial Electronics* **58**

[4] Syam M S and T.Sreejith Kailas 2013 Grid Connected PV System using Cuk Converter *International Conference on Microelectronics, Communication and Renewable Energy* (ICMiCR-2013)

[5] Jayalaksmi N. S. Et al 2017 Design and Implementation of Single Phase Inverter Based on Cuk Converter for PV System International *Journal Of Renewable Energy Research* **7** 585-591

[6] Piyush Choudhary and Som Nath Mahendra 2016 Feedback control and simulation of DC-DC Cuk converter for solar photovoltaic array *2016 IEEE Uttar Pradesh Section International Conference on Electrical, Computer and Electronics Engineering* (UPCON) Indian Institute of Technology (Banaras Hindu University) Varanasi 591-596

[7] P.Nagalakshmi Kanna and B.Meenakshi 2015 Analysis and Design of DC- DC/AC NonIsolated Cuk Converter using Sliding Mode Controller *IEEE 2015 International Conference on Circuit, Power and Computing Technologies* [ICCPCT]

[8] Sidharth Samantara et al 2015 Modeling and Simulation of Integrated CUK Converter for Grid Connected PV System with EPP MPPT Hybridization 2015 *IEEE Power, Communication and Information Technology Conference* (PCITC)

[9] Viji.K et al 2017 Qualified Analysis of DSMC over SMC for Boost Converter International Journal of Control and Automation (IJCA) **10** 2207-6387

[10] Chitra K et al 2019 Design And Implementation Of Simple Boost Pwm Controlled T-Source Inverter For Solar PV Application International Journal Of Scientific & Technology Research **8** 717-720

IOP Conf. Series: Materials Science and Engineering 1059 (2021) 012073 doi:10.1088/1757-899X/1059/1/012073

Wear and Mechanical Behaviour of Magnesium AZ31 alloy Reinforced with MoS_2 through Friction Stir Processing for Aerospace Applications

Tajdeen A, Kamal Basha K, Roshan Shandres C, Sandeeprajkumar S, Saddam Hussain S and Sanjay R
Department of Mechanical Engineering, Bannari Amman Institute of Technology, Sathyamangalam – 638401, Erode(Dt), Tamil Nadu, India.
tajdeena@bitsathy.ac.in

Abstract. Friction Stir Processing (FSP), a solid-state processing technique employed to modify the microstructure, is based on the principles of friction stir welding. FSP is one of the most promising techniques for the production of surface composites. In this investigation, AZ31 magnesium alloy is reinforced with molybdenum disulfide (MoS_2) through a friction stir processing method. Before and after the fabrication of the surface composites the wear tests were conducted by varying the condition of sliding load (25 N & 35 N), sliding distance (1 km & 2 km), and sliding velocity (1 m/s & 2 m/s). Hardness values were measured by using the Brinell hardness tester. Both hardness and wear tests were conducted as per the ASTM standards. After reinforcement, composite material enhances hardness values. The wear rate also reduces as MoS_2 acts as a solid lubricant material.

1 Introduction

Nowadays, Al and Mg-based Metal Matrix Composites (MMC) are used for the various applications such as aerospace, shipbuilding, defence and automotive applications because of their attractive properties such as good strength and elasticity [1,2]. Surface composites show enhanced characteristics of the material on the surface while retaining several properties of the matrix by using Friction Stir Processing (FSP) [3]. The selection of size and profile of pin is also an important consideration during FSP. For Magnesium AZ31 alloys, Normally Circular pin without and with thread produces the best result in FSP[4].

Friction stir processing and addition of SiC and CNT reinforcement through this route enhance the mechanical and microstructural changes in the material [5]. A single-pass was carried out using a fabricated cylindrical profile FSP tool with operating conditions such as tool rotation speed, travel speed, and downward forces which were 1100 rpm, 60 mm/min, and 6 kN respectively. The hardness of the FSP specimens increased by 40% which contained 50 % SiC + 50 % B_4C which is higher than that of the aluminum matrix alloy [6]. Our present work focuses on the wear and the mechanical properties of magnesium AZ31 alloys reinforced with MoS_2 through FSP.

2 Materials and Methods

2.1 Friction stir processing

FSP can be used to increases the mechanical, wear, corrosion behaviour of the materials. Friction stir processing is a thermomechanical process for improving material properties. AZ31 Mg alloy was used in this process. The dimensions of the plate were 125 x 75 mm with 6 mm thickness. The 18% volume fraction is fixed. In this present study, the molybdenum disulfide (MoS_2) is used as reinforcement. This reinforcement is having excessive reactivity compare to the other bulk materials. The molybdenum disulfide (MoS_2) particles are in the black stable form.

During the drilling, the fixed spindle speed at 750 rpm and fixed feed rate of 20 mm/min was chosen. In the same way, 42 holes were drilled and removed the workpiece from the vertical CNC milling machine.

The rotating non-consumable tool comes in contact with the material surface and starts moving along the seam. The two workpieces get fused into a single workpiece by the solid-state welding process, the joint is by the heat produced by the friction of tool with the workpiece. In this process, the workpiece is held on to a rigid table and the FSP tool is made to rotate over the surface of the material at high speed to generate heat and to join with the base material. Schematic illustration of Friction stir processing system is shown in below Figure 1.

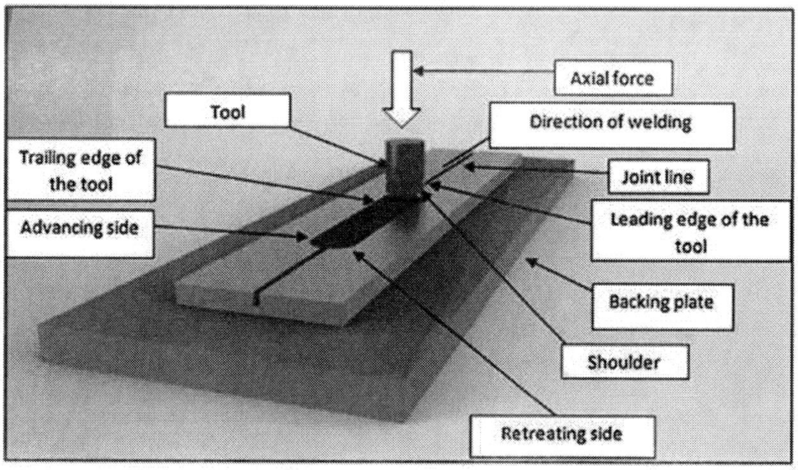

Figure 1. Schematic representation of FSP

The FSP is done mostly by using the groove method, To improve the material distribution, hole method is used instead of the groove method. The drill was done at the center of the AZ31 workpiece with the drill ratio of 18%. The holes are filled with the molybdenum disulphide (MoS_2) and the friction stir processing was done on the surface of the workpiece.

2.2 Composition of AZ31 alloy

In this present study, AZ31 Mg alloys selected as base material. Because, AZ31 Mg alloys having good casting properties, high stiffness to weight ratio, high corrosion resistance, wear resistance, etc. The figure 2 shows various elements in AZ31 Mg alloy and its percentage.

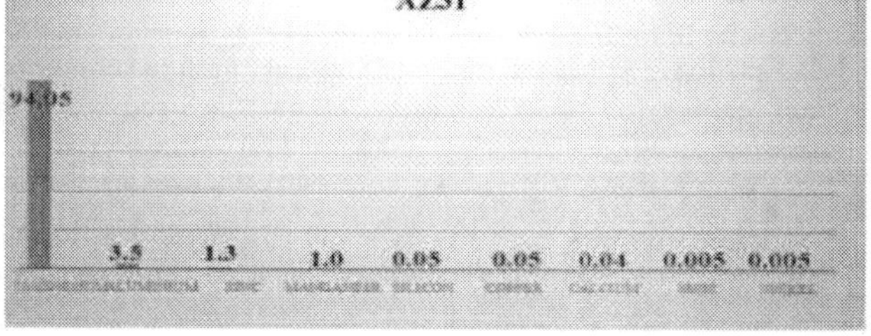

Figure 2. Chemical composition of AZ31 alloy in %

2.3 Properties of AZ31 alloy

AZ31 alloy concerning its properties it has been used in many application such as automobile industries, aerospace, shipbuilding, railway department, pylons and towers etc. The mechanical properties, physical properties of AZ31 alloy is shown in Table 1 & 2 respectively.

Table 1. Mechanical properties of AZ31 alloy

S.NO	PROPERTY	VALUE
1	Tensile Strength (N/mm^2)	260
2	Hardness Number - Brinell (500kg load,10mm ball)	49
3	Shear modulus (N/mm^2)	17000
4	Elastic modulus (N/mm^2)	44800
5	Poisson's ratio	0.35
6	Yield strength-strain 0.200% (N/mm^2)	200

Table 2. Physical properties of AZ31 alloy

PROPERTY	METRIC	IMPERIAL
Density	1.77 g/cm^3	0.0639 lb/in^3

3 Experimental Procedure

3.1 Preparation of Reinforcements:

MoS_2 particles were used as reinforcements. The reinforcement powder has undergone a ball milling process for the required size of particles and also for required percentage. The MoS_2 particle was filled in the hole and the FSP non-pinned tool was used to close the hole(capping). Figure 3 shows the EDAX image of MoS_2 with its chemical compositions.

Figure 3. EDAX image of MoS_2

The chemical composition of MoS_2 is shown below the table 3.

Table 3. Chemical composition of MoS_2

ELEMENT	CONTENT(%)
Molybdenum	59.94
Sulfur	40.06

3.2 Preparation of FSP tool:

The FSP tool is made up of H-13 tool steel material and the shoulder diameter of the tool is 18 mm. The pin diameter and length of the pin were 6 mm and 3.5 mm respectively [7,8]. A cylindrical pin profile was used for the process which is plunged into the capped surface on the plate. FSP tool and FSP setup was shown in figure 4 and figure 5 respectively.

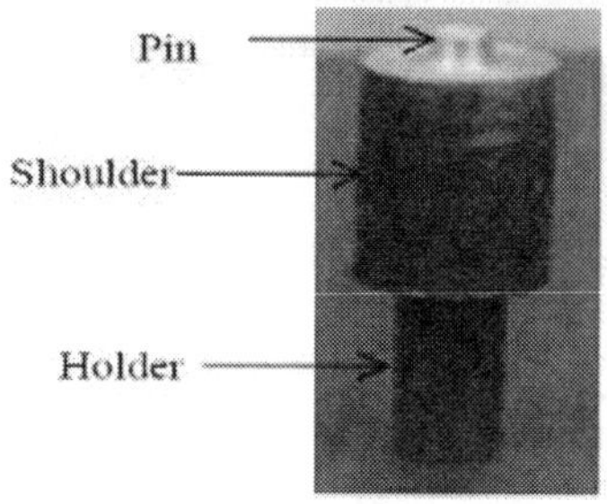

Figure 4. FSP tool

Figure 5. FSP setup

The FSP tool consists of a shoulder and a pin. The primary purpose of the tool is create the localized heating and the plastic material flow in the stir zone. The too shoulder also provides confinement for the heated volume of material. The hardness of the workpiece is based on the tool rotational speed [9, 10]. The hole diameter and depth depends upon the amount of powder to be used to make the metal matrix. The FSP tool dimensions are given in table 4.

Table. 4 FSP tool dimensions

Pin	Values (mm)
Pin length	5.5 (single pass) 3.5 (double pass)
Tool shoulder diameter, D	18
Pin diameter, d	6
Holder diameter	8
Holder length	36
Shoulder length	10

3.3 Experimental work:

After making the hole in the base plate, the hole was filled with reinforcements in the following order hole was filled with 100% of MoS_2. The experimental setup of AZ31 plate with reinforcements is shown in figure 6. Based on pilot study analysis, the process parameters used in the process is given in Table 5. The FSP process is conducted on Vertical Machining Centre.

Table 5. FSP process parameters

Sl. No	Process Parameters	Values
1	Rotational Speed (rpm)	1200
2	Traverse speed (mm/min)	20
3	Axial force (kN)	7
4	Number of passes	1

Figure 6. FSP experimental setup of AZ31 plate

As the process commences, the tool tends to touch the workpiece, the rotating pin comes in contact to the surface of the work piece.

Friction is produced due to which the workpiece gets heated and provides the softening of the matrix. This helps to attain a fine-grained, uniform structure. After the process, the material is removed from the setup and it is subjected to polishing for mechanical testing. The FSP processed plate has been shown in figure 7.

Figure 7. Friction Stir Processed plate

3.4 Brinell Hardness Measurement:

Workpiece was cut for the hardness test in the FSP stir zone. Testing was conducted in the Brinell hardness tester at various places of different indentation. Specimen were cut as per ASTM (ASTM E10) standard for Brinell hardness test. If the material is softer, a smaller force is used and the material is harder, a large force is used. Hardened steel ball is used for indentation.

3.4.1 Procedure for Brinell Hardness Measurement

The following are the procedure for hardness measurement. The specimen is placed on the testing area. To turning the loading lever in a backward direction and to apply the minor load on the specimen surface. The load is maintained at 10 to 15 seconds for a dwell time. After the dwell time is completed, the indenter is removed from the specimen. To repeat the above procedure for different loads and materials.

3.5 Wear Measurement:

When, two solid surfaces are sliding, rolling or impact conditions the wear will occur. wear is one kind of material removal process. The dry sliding wear test was conducted by using the pin on disc wear testing machine. Workpiece was cut for the wear test in the FSP stir zone. Testing was conducted in the pin on disc wear test rig. Specimen were cut as per ASTM G99 standard for the wear test.

3.5.1 Procedure for Pin on Disc Wear Measurement

The following are the procedure to wear measurement. The specimen is placed on the testing holder. The track diameter is selected based on our requirement. The load is added in the loading hanger. The input parameters such as load, time, speed, track diameter are entered. The LVDT sensors are used to measure the displacement. The computerized pin on disc system shows results in the display. To repeat the above procedure for different parameters such as load, speed, track diameter, time. The Pin on disc wear test rig apparatus is shown in below figure 8

Figure 8. Pin on disc wear test rig

4 Results and Discussion

4.1 Effect of reinforcement on the hardness

It was observed that base material AZ31 alloy, friction stir processed with and without reinforcement (MoS_2) shows the increasing trend in hardness because of the grain refinement during FSP and uniform distribution of reinforcements. Figure 9 shows the hardness value variation for base material and FSPed Material.

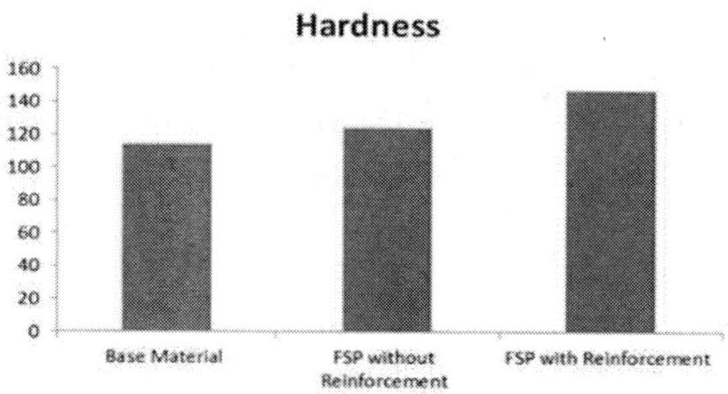

Figure 9. Hardness values

4.2 Effect of Reinforcement on Wear :

It was observed that magnesium AZ31 alloy material (base material) and AZ31+MoS_2 composite material (with reinforcement material) were tested at the same wear condition, results show that the wear rate has decreased due to MoS_2 acting as a dry lubricant and uniform distribution reinforcements reduces direct contact on counter surfaces. Figure 10 shows the wear rate variation for base material and FSPed Material.

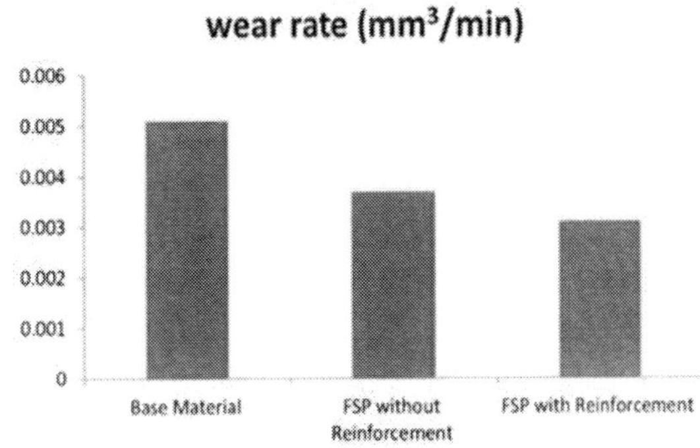

Figure 10. Wear rates

The observed wear values without reinforcement have shown in table 7

Table 6. Wear test values without reinforcement

ROTATIONAL SPEED (rpm)	TRAVEL SPEED (m/min)	SLIDING SPEED (m/s)	LOAD (N)	WEAR (mm^3/min)
900	10	1	25	0.04
		2	35	0.041
1100	20	1	25	0.0046
		2	35	0.0037

From table 6, it was observed that both normal AZ31 material (without reinforcement material) and AZ31+MoS_2 composite material (with reinforcement material) are tested at the same condition. But, the normal AZ31 material (without reinforcement material) gives wear rate of 0.0051 mm^3/min at 1100 rpm rotational speed. The AZ31+MoS_2 composite material (with reinforcement material) gives wear rate of 0.0037 mm^3/min at 1100 rpm rotational speed. Based on the above results observed that the combination of AZ31 and MoS_2 (Molybdenum disulfide) shows higher wear resistance than they act as individuals.

5 Conclusions

The FSP on magnesium AZ31 alloy with reinforcement of MoS_2 particles was successfully carried out and the following results were obtained:

- The Magnesium AZ31 alloy showed the hardness value of 114 BHN. After FSP with a single pass without reinforcement, it showed a hardness value of 124 BHN and after reinforcement, with a single pass, the FSP composite material showed the 147 BHN. Hence the hardness value increased due to the grain refinement and reinforcement.
- The Magnesium AZ31 alloy showed a wear rate of 0.0051 mm^3/min, after FSP with a single pass without reinforcement showed a wear rate of 0.0037 mm^3/min and after reinforcement with a single pass, the FSP composite material showed the value of 0.0031 mm^3/min. So the wear rate decreased due to the grain refinement.

References

[1] Daulat Kumar Sharma, Vivek Patel, Vishvesh Badheka, Krunal Mehta and Gautam Upadhyay 2019 Fabrication of Hybrid Surface Composites AA6061/ (B_4C + MoS_2) via Friction Stir Processing. *Journal of Tribology*, **141** 052201-10.

[2] Morteza Kiani, Imtiaz Gandikota, Masoud Rais-Rohani and Keiichi Motoyama 2014, Design of lightweight magnesium car body structure under crash and vibration constraints *Journal of Magnesium and Alloys*, **2** 99-108

[3] Vipin Sharma, Ujjwal Prakash and B.V. Manoj Kumar 2015 Surface composites by friction stir processing: A review *Journal of Materials Processing Technology* **224** 117-134

[4] Ratna Sunil B, Pradeep Kumar Reddy G, Hemendra Patel and Ravikumar Dumpala 2016, Magnesium based surface metal matrix composites by friction stir processing, *Journal of Magnesium and Alloys*, **04** 52-61

[5] Alavi Nia A and Nourbakhsh S H 2016 Microstructure and Mechanical Properties of AZ31/SiC and AZ31/CNT Composites Produced by Friction Stir Processing, *Transactions of the Indian Institute of Metals*, **69** 1435–1442

IOP Conf. Series: Materials Science and Engineering 1059 (2021) 012073 doi:10.1088/1757-899X/1059/1/012073

[6] Prithivirajan R, Sivagami S M and Narayanasamy P 2015 Characterization of Aluminium Alloy 2024-T351/ different hybrid ratio of SiC and B4C fabricated by Friction Stir Processing. *International Journal of Applied Engineering Research* **10** 444-448
[7] Babu J 2018 Experimental studies on Friction stir processing of AZ31 Magnesium alloy *International Conference of Materials Processing and Characterization*, **5** 4515–4522
[8] Akash Gupta 2015 Effect of Tool rotation speed and feed rate on the formation of tunnel defect in Friction Stir Processing of AZ31 Magnesium alloy, *4th International Conference on Materials Processing and Characterization*, **2** 3463-3470
[9] Elangovan K 2008 Effect of Tool Pin Profile and Rotational Speed on Mechanical Properties of friction stir welded in AA6061 aluminium alloy, *Materials and Manufacturing process* **23** 251-260
[10] Salehi M 2011 Optimization of process parameters for producing AA6061/SiC nanocomposites by Friction Stir Processing, *Trans. Nonferrous Met.Soc.China*, **22** 1055-1063

IOP Conf. Series: Materials Science and Engineering 1059 (2021) 012074 doi:10.1088/1757-899X/1059/1/012074

Development of Areca Nut Climbing And Harvesting Machine With Pesticide

C Sathishkumar, G Dineshkumar, K S Karthikeyan, P Prasanna

[1]Assistant Professor, Department ofMechanical Engineering, Bannari Amman Institute of Technology,
sathishmechace@gmail.com

Abstract. Recent days there were lots of improvement and advancement have been done in the field of agriculture, but the sad thing is the large scale farmers always earns more than the small scale farmers when comparing its ratio. The project focuses on the growth of small scale farmers as equal to the growth of the large scale farmers in the field of agriculture in growing the palm trees. The palm trees usually grow in narrow dimensions and the time taken to harvest the crop nuts will be around 3 to 4 years. Total investment is based on the harvesting and the cost and machines to harvest the crop nuts were of huge amount and the small scale farmers struggle a lot and that leads to loss. Also there is a chance of losing the quantity of crop nuts when harvested from the top of the palm tree when all the nuts will be scattered and is wasted falls down on the ground. Therefore we thought of designing a product that can be operated by the by the farmer itself a user friendly and easy to handle, also we have focused on safe landing of the crop nuts from the top of the palm tree. This product completely works based on the mechanical concepts and designed with mechanical links. It is operated manually now; in the future it will be automated. This product will decrease the cost of harvesting and increase the profit for the small scale farmers to grow their ration as that of the large scale farmers. This will create a revolution in the field of crop harvesting which benefits lots of farmers to come up from their bad time to good time.

1 Introduction

The main purpose of our Project is to help small scale farmers who having land area less than 5 acres, by designing areca nut tree climbing and harvesting machine, we can harvest more nuts according to the previous method, It's a fully manual device, without any external source, Our project work will focus on economy of harvesting operation. It is a time saver at low cost by considering different costly equipment in the market. As the requirement for areca nut is increasing day by day, therefore the target was to create the machine which is affordable to each and every farmer, which is cheaper, efficient and will reduce the total harvesting cost. To achieve this aim, it is decided to follow the following steps
(i) To understand farmer's problem which they are facing about harvesting, for this it is decided to interview the farmers. So as decided we surveyed the farmers who are having land less than 5 acres.
(ii) As the design is based on the requirement and demand for compact efficient and affordable harvester. This demand could be seen only with interaction with farmers of having land less than 5 acres. The most of the farmers were from India. We also found that there is essential to have small scale harvester with each farmers for reducing their harvesting cost.

1.1 Areca Nut

Areca nut known by its biological name Areca catechu is a species of palm tree which mainly grows in tropical Pacific region, south East Asia, Indian subcontinent and parts of eastern Africa. Areca nut known by its biological name Areca catechu is a species of palm tree which mainly grows in tropical Pacific region, south East Asia, Indian subcontinent and parts of eastern Africa. It is a medium-sized tree, growing straight up to 20-meter-tall, with a trunk. diameter varying from 10–15 cm. The leaves are 1.5–2 m long usually present in the top region of the tree. The tree bark is usually narrow and grows vertically, with a medium soft trunk and isn't rough like other tree out barks. The trunk diameter varies constantly from the bottom narrowing towards the top.

The farmers struggle a lot in harvesting the nuts from the top of the trees by the traditional methods, since it is time consuming and lots of requirements were need in order make it work. As the cost for climbing a tree is Rs.25 for a person to climb, also one coconut cost about Rs.8 in every tree. The cost for harvesting the crop or nuts becomes high and thus the small scale farmers suffer a lot in this process since the amount of harvesting is less and the cost to harvest is high when compared both the farmers feel a lack or shortage in the crops harvested.

Figure 1. Harvesting Machine

The cost is comparatively not high for the large scale workers since they have a large crop field to undergo the harvesting process. Since the field is large the need and requirement for the harvesting is high and thus there will be large need and requirement of labors or machines figure 1. to harvest the crop from the top of the tree and thus the cost of the labors and machines and the time it takes to harvest the crop[from the field is comparatively equal to the outcome of the harvested nut crops after selling. Thus ultimately there will not be any loss for the owner and the harvester; here both of them will get profited.

But in the small scale workers the crop field is less in size and the harvester demands a large amount of money in order to harvest the crops from the top of the tree. This leads the small scale workers to end in loss and there won't be any profit to them except the loss and hence the need of a solution here is much needed in order to achieve some profit in the crop harvest. So that the small scale workers will be highly benefited in this process.

The cost and expense after finding the solution to harvest the crops in the small scale farmer's field is comparatively low when compared to large scale farmers with large fields. Though the time consumption will be high in the small scale crop field as only fee crops have to be harvested in the harvesting process. When compared to the large scale field it will take lot of time to harvest the crops by this method hence, machines are much advised to harvest the crops in large scale crop field in order to reduce the time consumed to harvest the crops from the crop field. This is mainly because the time of harvest must not make any changes in the field of profit. The small scale farmers are happy with the solution and thus ultimately they achieve the same profit as the large scale farmers without using harvesting machines and labors.

There is a serious problem in the harvesting of crop nuts from the top of the tree, as there will be lots of crop nuts over the top of the tree at the bottom of the branches. By harvesting the crop in traditional method there is a huge loss of crop nuts when it falls from that height and lands on the ground due to There is a serious problem in the harvesting of crop nuts from the top of the tree, as there will be lots

of crop nuts over the top of the tree at the bottom of the branches. By harvesting the crop in traditional method there is a huge loss of crop nuts when it falls from that height and lands on the ground due to gravitational force the falling rate of crop nuts were increased and at high impact the crops burst out over the ground and spreads over the vast areas on the ground and thus it leads to loss of crops over the field itself.

Tree climber is mainly focused on two units RH and LH .The RH unit create the downward movement of the pedal, through which the steel wire rope is stretched and locks the areca tree. Now the LH unit is lifted up by pulling the handle attached to it to climb and the same process is repeated to reach the required height. To descend the tree the pedal of RH unit is pushed down and the handle of the LH unit is also pulled down alternatively till the bottom of the tree [1]. On the account the safety, reliability, the ease of use which is capable of climbing trees, cutting down coconuts, cleaning the tree tops and spraying pesticides. He designed a system that can be controlled by anyone. The designed proto type responds to human gestures with negligible gap in the response time and hence can be implemented in real time [2]. The first element is to implement land reforms in order to remove institutional bottle necks and the second element was to undertake massive investment in irrigation and other infrastructure in order to update the existing agricultural technology [3]. Tree climbing Robot consists of a four bar linkage and screw mechanism. The mechanical structure is designed to move the structure upwards against the gravitational forces in successive upper body and lower body movements. The result shows that the tree bot can successfully climb the trees with diameter range 15cm to 25cm [4].
a novel tree climbing robot is presented that the maneuverability surpasses the state of the art tree climbing robots .It is composed of a pair of Omni directional tree grippers for holding the robot on a
7 tree surface and a novel 3 DOF continuum manipulator for the manicuring. The locomotion and the workspace of the robot are also discussed [5]. A novel method for climbing and harvesting is proposed.

The design of the hardware setup is discussed and the controlling units are designed. The sharpened the materials selected are appropriate for selected are appropriate for the device to function [6]. Machine works on timely gripping and release of the tree by two metal wire ropes locked to the moving frame. By this design the structure is absorb a load of 100 kg .It is flexible to change the height of the equipment up to 100mm according to the equipment of the user. It has easy maintenance.
This structure will be beneficial to middle class [7]. The mechanical properties in relation to de-husking of the areca nut are found that lateral shear with a rubbing action might be suitable for de-husking the fruit. The moisture content of the dry fruit has to be in the range of 5–6% of areca nut weight [11]. Prior to the properties a device for de-husking dry areca nut fruit was developed comprising of a scissor mechanism, a frame, a platform and a pedal operated lever mechanism [12].

2 Experimental Setup

This model is powered by a human using a rope and pulley mechanism. This rotary motion of the pulley is converted into translational motion, in the opposite direction of gravitational force. The spring fitted along the longitudinal axis of the machine helps in reforming the original structure this allowing the tool to move upwards as the lower end is free moving and the upper end is grabbing the tree. By this, the tool will climb up to the top of the tree and ready to cut down the seeds of the tree, which is handled by a sliding cutter and the cut down part of seeds are collected by two long hands positioned in the inferior part of the machine. The harvesting equipment will help in controlling the yield cost of the small scale farmers, as the cost of it is very much low when compared to the cost of the traditional and conventional machines. The product which we have developed with the raw materials that is available at the chest rates near it makes the small scaled far5mers to fix the damaged parts by themselves. The product that is manufactured by the designer is very much helpful for the small scale workers. The workers requirement will not be necessary with this product since even the owner can make this product to work with this product.

The product with this ability can achieve and solve all the problems that is faced by the small scale farmers, in the process of harvesting the crop nuts from the palm tree. The time period for the cultivation of crop nuts will be about more than 3 to 4 years all those years waiting for that moment of harvesting the crop nuts. It is the time were all the work gets on to profit zone. The Harvesters have the

same idea of making profit in this; it is the places were our product will make more profit for the small scaled labours.

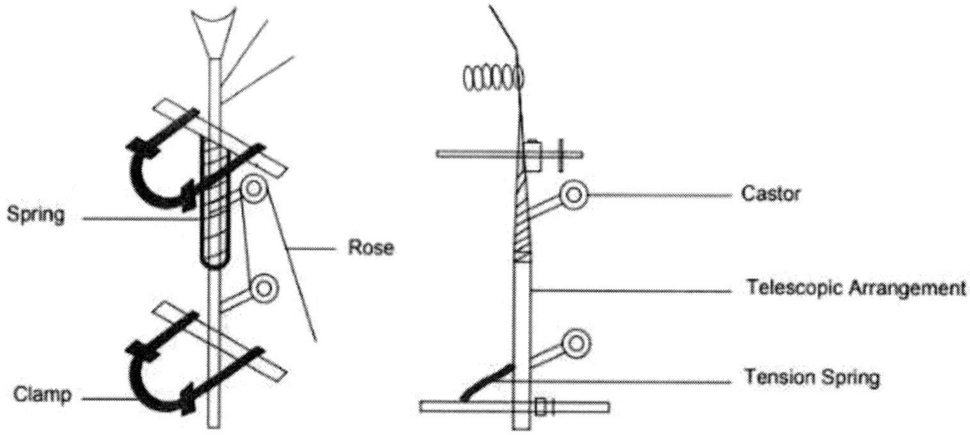

Figure 2. Component Details

The components used in order to fabricate the end product to harvest the crop nuts from the palm tree. The fabrication of the product is done by the raw materials obtained from the products that are available on the market itself. This is mainly helpful for the small scale farmers to operate by their own ways and thus they can easily able to understand the concept and working of the harvesting equipment. The assembly of the components explains the complete apparatus and thus the components figure 2 were fixed exactly based on the design approved by the design engineers.

The harvesting equipment is completely prepared and assembled, the raw materials makes the small scale farmers to easily replace the defective component simpler and easier without any struggle. This makes the equipment user friendly and thus after assembling the testing is done with the experimental setup and the results of the testes were analyzed and evaluated based on the theoretical calculations processed before the start of the project. This result mainly helps in modifying the components and fine tuning the product. The above drawing represents the detailed isometric view of the crop nut harvesting equipment that helps in harvesting the crops from the top of the palm tree o the bottom of the branches. The isometric view gives a detailed explanation about the arrangement of the product apparatus. The mechanical Links and mechanism in the equipment is adjusted by the links in the isometric view. The figure 3. Isometric view gives the clear understanding about the product how the links were attached to the product and the construction details were mentioned clearly in the diagram. The product view says that the vertical construction of the product is required in order to make it work.

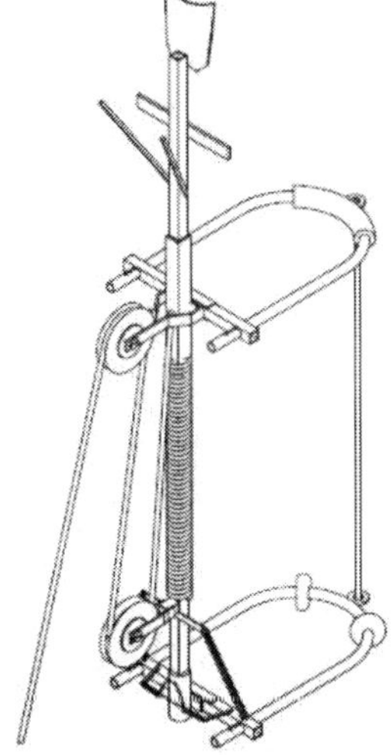

Figure 3. Schematic diagram of Areca Nut Climbing And Harvesting Machine

3 Results and Discussion

The product that is fabricated is mainly useful for the small scale farmers, who suffered a lot from several types of problems in harvesting the crop nuts from the palm tree. The solution for their problem is solved by the product that is developed by designing and fabricating the product with design calculations and test drives. The product is capable of gripping the tree, climbing the tree to the required height. The areca tree climber unit LH/RH is as shown in the figure 4. Complete product helps in reducing the cost of the harvesting process when compared to the other conventional and traditional harvesting methods and machines. The harvesting process takes some time period by this product but the machines will make this work fast and quick. This product is mainly befitted to the small scale farmers, to harvest the crop nuts at low cost itself.

Figure 4. Working model Areca Nut Climbing And Harvesting Machine

The above mentioned diagram is the product which is present in the compression state and thus at that state the spring is compressed by the tension created on the ropes over the pulley from the bottom. The compressed spring holds the potential energy in the compressed state of the spring. The releasing of rope decreases the tension in the pulley and now the potential energy stored in the compressed spring is now turn into expansion stage and thus the potential energy is converted in to kinetic energy and thus creates a movement over the parts of the product.

4 Conclusions

The design and analysis of Areca nut tree climber with pesticide sprayer will be implemented in a comparatively low cost. This will reduce the rate of accident in future. This is more effective technology for climbing Areca nut trees. The aim of our device is achieved. The design of the product is fabricated as per the calculation and the exact efficiency is achieved by the finished product. The loads of the parts as well as the crop nuts were held firmly by the product. The product satisfies the small scaled farmers purpose by providing high profit to them in the harvest process, thus ultimately provides right time to harvest the crop nuts. The yield capacity is perfect so that the harvesting process takes only one time for a single tree as the product can harvest the crop nuts from the palm tree that is to be harvested. The advantage of the areca nut tree climbing is very smoothly without damaging the tree. An unskilled labour can operate the machine safely and efficiently. Climbing the single tree, an average of 15 to 20 trees can be harvested/sprayed. This project concludes that the areca-nut tree climber is a safe, reliable, efficient and which reduces the risk in climbing the areca -nut tree to a good extent.

References

[1] Basavaraja J, Sharana and Somashekar H Hedge 2015 Design and Development of Areca tree Climber *International Journal of research in mechanical mechatronics and Automobile Engineering* 1.2 83-90.

[2] Rahul V, et al. 2015 Semi automated coconut tree climber *IJRET: International Journal of Research in Engineering and Technology* **4**.

[3] DileepEdacheri, PrakashUnakal – Design of a Coconut Tree Climbing Device - in International *Journal of Innovative Research in Science, Engineering and Technology Manufacturing Technology*

[4] Ahmed Salih Mahdi. Al-Zuhairi 2013 Semi automatic tree climber – in *International Journal of Agriculture & Research (IJAR)* – Volume **4** Issue 7.

[5] Nallusamy.V,N.Shrivathsan, Vinoop.U, Satish.M, Yashwanth.N.G 2015 Tree Climbing Robot – *JEST-M*, Vol **4**, Issue 4.

[6] Veerabhadhra - Development of Automated Coconut Harvester Prototype - *International journal of innovative research in science, engineering and technology.*

[7] N S Kannur - Design And Fabrication Of Coconut Tree Climbing Equipment - *International journal of innovative research in science ,engineering and technology.*

[8] Balasubramanian M; Panwar J S. 1986 Mechanical properties of Areca Nut (Areca catechu linn) as related to de-husking. *Journal of Agricultural Engineering (ISAE)*, **23**(1), 82–88.

IOP Conf. Series: Materials Science and Engineering 1059 (2021) 012075 doi:10.1088/1757-899X/1059/1/012075

Optimization of Multi-Criteria Decision Making Analysis for Selection of Suppliers in Supply Chains Using Fuzzy Logic

R Rajiev[1], S Saravanan, C Sasikumar[2], G Sasikumar[3]
[1]Assistant Professor, Department of Mechanical Engineering,Bannari Amman Institute of Technology, Sathyamangalam.
[2]Professor, Department of Mechanical Engineering,Bannari Amman Institute of Technology, Sathyamangalam
[3]Professor, Department of Mechanical Engineering,GMR Institute of Technology, Rajam, Srikakulam.

rajievr@bitsathy.ac.in

Abstract. In today's business environment, suppliers play key roles in making business strategies successful. Since quality, price, and on-time delivery of the products depend on the supplier's ability to support the business strategies, choice of the correct providers assumes a key part in any association since it altogether decreases the unit costs and improves corporate value intensity. Supplier selection endures generally a multi-models dynamic issue which incorporates both subjective and quantitative variables. The main objective of this work proposes coordinating the Analytic Hierarchy Process (AHP) and Fuzzy set hypothesis to create assessment models to decide the best suppliers. The project reported here has attempted to integrate and Fuzzy set theory to determine best suppliers. The Fuzzy set hypothesis situated to the sanity of exposure because of imprecision or dubiousness. The Fuzzy set uses semantic elements to sort out rules and sub-rules weight, with a pair wise examination with the Analytical Hierarchy Process (AHP). The fuzzy AHP choice technique, which decides the loads of emotional judgment for logical assessment structure of the provide determination issue. At long last, a provider choice issue was explained utilizing the suggested model in an industrial cleaning machines manufacturing company.

Keywords: Suppliers, Fuzzy - Analytical Hierarchy Process, Fuzzy System, Linguistic Variable.

1 Introduction

The supplier choice cycle is a multi-models issue, which incorporates both subjective and quantitative elements. Buying orders a huge situation in many associations since bought parts, segments, and supplies commonly speak to 40 to 60 percent of the deals of its final results have suggested that supplier determination is a multi-rules dynamic issue which incorporates assessment of variables [1]. The AHP-based specimen is planned and relevant to a genuine contextual analysis to inspect its attainability in choosing a seller for a telecommunication framework [2]. In this paper, the various levelled dynamic system is introduced where ecological and security guidelines and developing interest for the item quality were additionally considered [3].

It proposed a structure which coordinates the Analytic Hierarchy Process (AHP) and number programming to rate providers presentation with respect to approaching raw materials with regards to supplier board and afterward to dispense periodical purchases [4]. An Outlined mixture strategy, consolidating various strategies into an assessment cycle, so as to choose serious suppliers in a supply chain [5]. This article plot a mixture strategy, which consolidates different methods (i.e., AHP, DEA, and NN) into an assessment cycle, so as to choose serious suppliers in a supply chain. At last, a case of machine device choice is utilized to delineate and approve the proposed approach [6]. Another methodology called "Analytical Hierarchy Process weighted Fuzzy Linear Programming (AHP-FLP)" for supplier determination half breed technique for supplier choice, AHP-FLP is applied to a genuine industry case.

The loads of the different standards, taken as nearby loads from a given judgment framework, are determined Analytical Hierarchy Process (AHP) that are likewise considered as the loads of the fuzzy linear programming model have conferred [7]. The literature, overviewed demonstrated that a precise methodology and assessment technique utilizing AHP and fuzzy-AHP strategies which use the progressive choice structure considering all the connected components can be created and an additionally created numerical model can likewise be planned. It is seen that Analytical Hierarchy Process (AHP) is as a rule dominatingly utilized in the provider determination issues. AHP is utilized as an independent instrument in the greater part of the cases. It is a lot of basic to address the issue of complexities emerging out of the coordinated utilizations of AHP and different methods in supplier's choice issues.

From the literature, we proposed a systematic approach and evaluation method using AHP and fuzzy-AHP methods to consider the hierarchical decision structure considering all the related factors and also we developed mathematical model. Among the issues of supply chain management regarding purchasing decision, supplier selection is the most important activity of a firm's purchasing department.

2 Methodology of Supplier Selection using AHP in Consumer Product Manufacturing Industries.

The AHP divides a complex decision problem into a hierarchical system of decision elements. A pair wise comparison matrix of these elements is constructed, and then the normalized principal eigenvector is calculated for the priority vector, which provides a measure of the relative importance (weight) of each element.

As indicated by literature, some supplier determination models are found to shift in various circumstances, and specialists concur that there is nobody most ideal approach to assess and choose suppliers. An investigation did by overviewed purchasers so as to recognize factors they considered in granting agreements to providers. Out of the 23 variables considered, he inferred that quality, conveyance, technical capability, cost, price, and production capacity, reliability, desire for business, geographical location, and reputation, are the ten most important criteria which were extracted [8]. Organizations utilize a wide range of approaches in their assessing measures based upon Brain storming and Likert-Satty's 1-9 scale The hierarchical structure which is required to be formed according to the AHP is shown in fig 1 as shown select the best supplier from the first level of the Hierarchical structure The second level factors such as (i) Aspect, (ii) Consignment, (iii) Technical Capability, (iv) Cost, (v) Price, (vi) Production Capacity, (vii) Reliability, (viii) Desire for Business, (ix) Geographical Location, (x) Reputation. These are the factors to be concluded for the selection of suppliers. At last, at the third level, we have the choice options that should be assessed through these measures in an extraordinary manner.

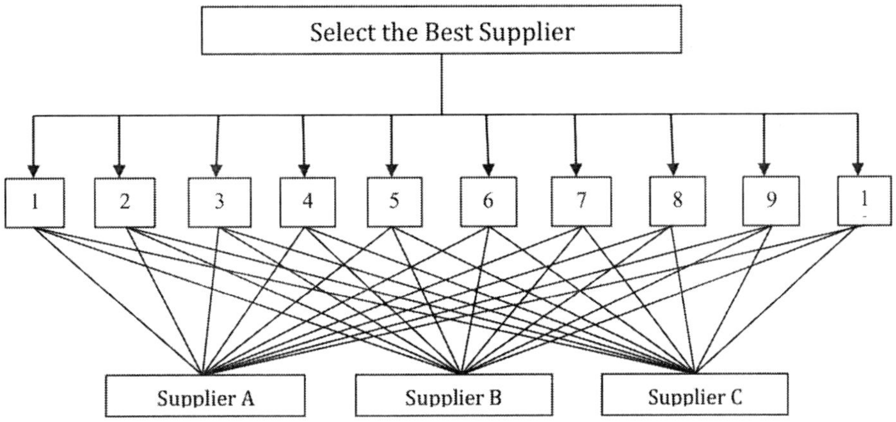

Figure 1. Hierarchical Structure

3 Fuzzy Set Theory

A fuzzy number A, which indicates a fuzzy set A, addresses If it is triangular Fuzzy number. Let M, be a triangular fuzzy number with a triplet (L, M, and U). The membership function can be characterized as relation 1

$$\mu(x) = \begin{cases} L & x = M \\ \frac{x-L}{M-L} & L \leq x \leq M \\ \frac{U-x}{U-M} & M \leq x \leq U \quad \cdots \\ 0 & otherwise \end{cases} \tag{1}$$

In the event that a variable can accept words in normal dialects as its worth, it is known as an alliance variable, where the words are portrayed by fuzzy sets characterized known to mankind of talk in which the variable is characterized. Alliance variables are characterized by (Very bad, Bad, Average, Excellent and Outstanding.

Utilizing fuzzy numbers to demonstrate the overall commitment or effect of every option on a measure, a fuzzy judgment vector is then acquired for every model. The fuzzy judgment matrix A is worked with all the fuzzy judgment vectors.

$$A_{ij} = \begin{bmatrix} G11 & G12 & G13 \ldots\ldots.. G1n \\ G21 & G22 & G23 \ldots\ldots.. G2n \\ G31 & G23 & G33 \ldots\ldots.. G3n \end{bmatrix} \tag{2}$$

A standardization cycle is applied to encourage a coordinating cycle with the weight vector to be determined. Every models in relation 2 is standardized by utilizing relation 3. A fuzzy judgment matrix (A) is gotten as relation subsequent to normalizing.

$$a_{11} = \frac{G11}{\sqrt{\sum_{i=1}^{n}(G1m)}} \tag{3}$$

$$A_{ij} = \begin{bmatrix} a11 & a12 & a13 \ldots\ldots.. a1n \\ a21 & a22 & a23 \ldots\ldots.. a2n \\ a31 & a32 & a33 \ldots\ldots.. a3n \end{bmatrix} \tag{4}$$

Where a_{11} denotes the judgment score of alternative (Ai) with respect to criteria

The weight denotes the overall significance among every standards. It tends to be gotten by applying pair-wise examination of the AHP or quickly decided by a decision maker.

$$Ds = \begin{bmatrix} b11 & b12 & b13 \dots\dots..b1n \\ b21 & b22 & b23 \dots\dots..b2n \\ b31 & b32 & b33 \dots\dots..b3n \end{bmatrix} \quad (5)$$

Equation 5 where a score speaks to a decision maker (Ds) gauges the general significance by utilizing Saaty's. At that point a thorough pairwise comparison matrix (Ds) is worked by coordinating all decision maker's evaluations. Where a far reaching score speaks to the overall significance among every basis with triangular fuzzy numbers.

The significance of every rule is extraordinary. So as to get a weight (w) that relates to explicit models these loads of measures altogether make up a fuzzy weight vector (w) in relation 6.

$$w = (w_1, w_2, w_3 \dots w_n) \quad (6)$$

For deciding the general load between every model, a few decision makers' assessments are incorporated to improve the basic pairwise correlation of the first AHP and structure a more objective fuzzy weight vector. In the judgment matrix, the general judgment scores of alternative (A) are discovered as for models without thinking about the overall load between every basis. Accordingly, the judgment matrix (A) and the weight vector (w) are combined. Every standards weight (w) is taken independently to increase its relating measures and the judgment matrix (H) is appeared in equation 7.

$$H = \begin{bmatrix} w1\otimes a11 & w1\otimes a12 & w1\otimes a13 & \dots & w1\otimes a1n \\ w2\otimes a21 & w2\otimes a22 & w2\otimes a23 & \dots & w2\otimes a2n \\ w3\otimes a31 & w3\otimes a32 & w3\otimes a33 & \dots & w3\otimes a3n \end{bmatrix} \quad (7)$$

$$H = \begin{bmatrix} h11 & h12 & h13 & \dots & h1n \\ h21 & h22 & h23 & \dots & h2n \\ h31 & h32 & h33 & \dots & h3n \end{bmatrix} \quad (8)$$

Where h_{11} signifies the fuzzy exhibition score with alternative (A) compares to measures with triangular fuzzy numbers (L, M and U). By utilizing the relation 8, the fuzzy exhibition scores of every option regarding all measures.

Each fuzzy performance score (h_{11}) is joined α-slice to separately from a span $[h^{\alpha}_{ijl}, \; h^{\alpha}_{ijr}]$

$$h^{\alpha}_{ijl} = Lij + \alpha(Mij - Lij) \quad (9)$$

$$h^{\alpha}_{ijr} = Uij - \alpha(Uij - Mij) \quad (10)$$

h^{α}_{ijr} and h^{α}_{ijl} respectively indicate the right point and left point of the scope of the triangular in the wake of utilizing α is somewhere in the range of 0 and 1. Further, the overall span execution matrix (Hα) with α value can be resolved as equation 11.

$$H^{\alpha} = \begin{bmatrix} [h^{\alpha}_{11l}, h^{\alpha}_{11r}] & [h^{\alpha}_{12l}, h^{\alpha}_{12r}] & \dots & [h^{\alpha}_{1nl}, h^{\alpha}_{1nr}] \\ [h^{\alpha}_{21l}, h^{\alpha}_{21r}] & [h^{\alpha}_{22l}, h^{\alpha}_{22r}] & \dots & [h^{\alpha}_{2nl}, h^{\alpha}_{2nr}] \\ [h^{\alpha}_{31l}, h^{\alpha}_{31r}] & [h^{\alpha}_{32l}, h^{\alpha}_{32r}] & \dots & [h^{\alpha}_{3nl}, h^{\alpha}_{3nr}] \end{bmatrix} \quad (11)$$

The estimation of α is engaged with the decision makers level of certainty as they individually assess elective scores and standard loads for this MCDM problem.

The risk issues typically go with the dynamic cycle, so a risk-index is considered. The risk issue is related with item qualities in the connected market. The specialized intricacy and estimation of the thing are decidedly associated with the level of apparent risk.[9]. The risk index is also applied to be a defuzzifier. Defuzzification is implemented by joining the risk index to produce the final crisp numbers.

The overall crisp performance matrix H^{α}_{β} is calculated in equation The overall crisp exhibition network H^{α}_{β} is determined in equation (12) through equation (13)

$$h^{\alpha}_{ij\beta} = \beta h^{\alpha}_{ijl} + (1-\beta)\, h^{\alpha}_{ijr}, \quad 0\leq\alpha\leq1;\ 0\leq\beta\leq1 \tag{12}$$

$$H^{\alpha}_{\beta} = \begin{bmatrix} h^{\alpha}_{11\beta} & h^{\alpha}_{12\beta} & h^{\alpha}_{13\beta} \cdots & h^{\alpha}_{1n\beta} \\ h^{\alpha}_{21\beta} & h^{\alpha}_{22\beta} & h^{\alpha}_{23\beta} \cdots & h^{\alpha}_{2n\beta} \\ h^{\alpha}_{31\beta} & h^{\alpha}_{32\beta} & h^{\alpha}_{33\beta} \cdots & h^{\alpha}_{3n\beta} \end{bmatrix} \tag{13}$$

Where $h^{\alpha}_{ij\beta}$ signifies the crisp performance score which every other option (Ai) compares to all standards (Cj) under α level of certainty and β hazard level.

In TOPSIS, the ideal arrangement $h^{\alpha+}_{j\beta}$ is characterized as the finest crisp performance score result over all choices on a measure. For example, relation 12 on the opposite the ideal arrangement $h^{\alpha-}_{j\beta}$ is resolved as the most exceedingly awful fresh execution score result over all options on standards as shown in equation 13.

$$h^{\alpha+}_{j\beta} = \{\max h^{\alpha}_{ij\beta}, \quad i = 1,2,\ldots,n \qquad j = 1,2,\ldots.m\} \tag{14}$$

$$h^{\alpha-}_{j\beta} = \{\min h^{\alpha}_{ij\beta}, \quad i = 1,2,\ldots,n \qquad j = 1,2,\ldots.m\} \tag{15}$$

By relation 13 and 14 the ideal arrangement and negative ideal arrangement can be discovered for every basis. Subsequent to deciding the ideal arrangement and negative ideal arrangement, the separation between the ideal arrangement and negative ideal solution for every substitute is in that order to determined by relation 15 and 16 as following

$$D^{\alpha+}_{i\beta} = \sqrt{\textstyle\sum_{i=0}^{n}(h^{\alpha}_{ij\beta} - h^{\alpha+}_{j\beta})^2} \quad i = 1, 2\ldots n \tag{16}$$

$$D^{\alpha-}_{i\beta} = \sqrt{\textstyle\sum_{i=0}^{n}(h^{\alpha}_{ij\beta} - h^{\alpha+}_{j\beta})^2} \quad i = 1, 2\ldots n \tag{17}$$

Where $D^{\alpha+}_{i\beta}$ and $D^{\alpha-}_{i\beta}$ represent the distance between the crisp performance scores

The best option has a more extended separation to the negative ideal arrangement and a more limited separation to the ideal arrangement simultaneously. Thusly, the overall closeness to the ideal solution for every alternative can be detailed as relation 17

$$P^{\alpha}_{i\beta} = \frac{D^{\alpha-}_{i\beta}}{D^{\alpha+}_{i\beta} + D^{\alpha-}_{i\beta}}, \qquad i = 1, 2\ldots n \tag{18}$$

Where $P^{\alpha}_{i\beta}$ indicates a final presentation score that contains the decision makers α degree of certainty about their assessments and perspective of risk level β for every alternative.

4 Result and Discussion

The assessment of Quantitative and Qualitative rules is the initial phase in showing up the fuzzy judgment matrix. In light of the true value (Quantitative information) of every supplier the sub-scores are determined for each substantial measure. The fuzzy judgment matrix for this supplier choice case can be set up by combining all scores.

The scores for every standards are appeared in relation 18 the accompanying matri

$$A = \begin{bmatrix} (5,7,9) & (7,9,9) & (3,5,7) & (3,5,7) & (8,13,18) \\ (7,9,9) & (5,7,9) & (5,7,9) & (5,7,9) & (8,14,18) \\ (3,5,7) & (3,5,7) & (1,3,5) & (3,5,7) & (8,14.5,18) \end{bmatrix} \tag{19}$$

The leftover estimations of standardization were registered utilizing a similar technique. The fuzzy judgment matrix (A) built is shown in relation 19

$$= \begin{bmatrix} (0.34\ 0.56\ 0.98)\ (0.48\ 0.72\ 0.98)\ (0.24\ 0.54\ 1.18)\ (0.22\ 0.50\ 1.06)\ (0.26\ 0.54\ 1.19) \\ (0.48\ 0.72\ 0.98)\ (0.34\ 0.56\ 0.98)\ (0.40\ 0.76\ 1.52)\ (0.37\ 0.70\ 0.37)\ (0.26\ 0.58\ 1.19) \\ (0.20\ 0.40\ 0.76)\ (0.20\ 0.40\ 0.76)\ (0.05\ 0.32\ 0.84)\ (0.22\ 0.50\ 1.06)\ (0.32\ 0.61\ 1.19) \end{bmatrix} \quad (20)$$

The fuzzy weight vector (W) is worked by applying a pair wise correlation of the AHP. Yet, it is difficult to keep away from the decision makers' abstract judgment. Consequently, the cooperative choice with the AHP is utilized to change over into a fuzzy structure. Four decision makers separately analyze relative significance between every model by utilizing Saaty's nine-point scale.

$$W = \{(0.3, 0.43, 0.63)\ (0.16, 0.24, 0.36)\ (0.05,\ 0.07, 0.1)\ (0.12, 0.17, 0.24)\ (0.06, 0.1, 0.14)\} \quad (21)$$

The fuzzy judgment matrix by the comparing fuzzy weight vector utilizing relation 6.

$$H = \begin{bmatrix} (0.10\ 0.24\ 0.61)(0.07\ 0.17\ 0.35)(0.01\ 0.03\ 0.11)(0.02\ 0.08\ 0.25)(0.01\ 0.05\ 0.16) \\ (0.14\ 0.30\ 0.61)(0.05\ 0.13\ 0.35)(0.02\ 0.05\ 0.15)(0.04\ 0.11\ 0.08)(0.01\ 0.05\ 0.16) \\ (0.06\ 0.17\ 0.47)(0.03\ 0.09\ 0.27)(0.02\ 0.02\ 0.08)(0.02\ 0.08\ 0.25)(0.01\ 0.06\ 0.16) \end{bmatrix} \quad (22)$$

The defuzzification process is executed by deciding the span execution matrix with α-cut and considering the risk index. The estimation of α can be fixed by the decision maker's level of certainty when they emotionally assess elective scores and measures loads. The higher worth α esteem communicates the further extent of certainty and the closer to the conceivable estimation of the triangular numbers. In this supplier determination issue, the decision makers set up α = 0.85, to beat vulnerability in their assessments. The stretch $[h_{ijl}^{\alpha}\ \ h_{ijr}^{\alpha}]$ can be sequentially solved using relation 13 and 14. This is shown below.

$$h_{11l}^{0.85} = 0.10 + 0.85\,(0.24 - 0.10) = 0.219$$

$$h_{11r}^{0.85} = 0.16 - 0.85(0.61 - 0.24) = 0.290$$

$$[h_{11l}^{0.85}\ \ h_{11r}^{0.85}] = (0.219\ \ 0.290)$$

Likewise, the remaining stretches were determined like the equations 13 and equation 14 with α = 0.85 built is shown below

$$H^{0.85} = \begin{bmatrix} (0.21\ 0.29) & (0.15\ 0.19) & (0.02\ 0.04) & (0.07\ 0.10) & (0.49\ 0.63) \\ (0.27\ 0.34) & (0.11\ 0.16) & (0.04\ 0.06) & (0.10\ 0.11) & (0.49\ 0.63) \\ (0.15\ 0.21) & (0.08\ 0.11) & (0.01\ 0.01) & (0.07\ 0.10) & (0.56\ 0.69) \end{bmatrix} \quad (23)$$

The potential dynamic risk problems envelop the supplier determination issues inside a supply chain. In the proposed approach, the decision makers of key buying change hazard record β alongside buying diverse raw materials attributes in the connected market.

$$h_{11,0.2}^{0.85} = 0.2 * 0.219 + 0.85 * 0.290 = 0.27$$

$$h_{21,0.2}^{0.85} = 0.2 * 0.27 + 0.85 * 0.34 = 0.32$$

$$h_{31,0.2}^{0.85} = 0.2 * 0.15 + 0.85 * 0.215 = 0.19$$

$$H_{0.2}^{0.85} = \begin{bmatrix} 0.27 & 0.18 & 0.03 & 0.09 & 0.60 \\ 0.32 & 0.15 & 0.05 & 0.11 & 0.60 \\ 0.19 & 0.10 & 0.01 & 0.09 & 0.66 \end{bmatrix} \quad (24)$$

To adjust these crisp exhibition scores and execute final positioning, TOPSIS (Techniques of request inclination by similitude to ideal framework) strategy is utilized [10]. The final presentation score $(P_{1,0.2}^{0.85})$

speaks to the general closeness to the ideal arrangement. The final exhibition score with α = 0.85 and β = 0.2 are understood utilizing relation 17.

$$P^{0.85}_{1,0.2} = \frac{D^{0.85-}_{1,0.2}}{D^{0.85-}_{1,0.2} + D^{0.85-}_{1,0.2}} = 0.521$$

$$P^{0.85}_{2,0.2} = \frac{D^{0.85-}_{2,0.2}}{D^{0.85-}_{2,0.2} + D^{0.85-}_{2,0.2}} = 0.624$$

$$P^{0.85}_{3,0.2} = \frac{D^{0.85-}_{3,0.2}}{D^{0.85-}_{3,0.2} + D^{0.85-}_{3,0.2}} = 0.287$$

The execution of the proposed model dependent on Fuzzy AHP for selection suppliers of manufacturing industry has been portrayed. The calculation procedure is detailed understand the use of the proposed model.

5 Conclusion

From the result proposed it is evident that Fuzzy-AHP method can be applied to any supplier selection problem of a manufacturing company. In the two cases, the choices came to by utilizing the model as harmonized with those acquired by utilizing the prior supplier choice cycle. This may be because of the set number of suppliers and choice standards. When there are more number of suppliers and determination models, the current strategy won't be feasible because of the tedious calculations involved and the possibilities of mathematical errors. Whereas, in the Fuzzy-AHP model, quite a few standards for suppliers determination and alternatives can be incorporated. This helps the decision makers to assess the quality and shortcoming of suppliers systems by contrasting them with deference with the rules. The proposed system can likewise decrease the time and exertion in dynamic huge.

References

[1] Heung-Suk Hwang, Chiung Moon, Chun-Ling Chuang and Meng-Jong Goan 2005 Supplier Selection and Planning Model Using AHP *International Journal of the Information Systems for Logistics and Management* **1(1)** 47-53.

[2] Maggie C Y and Rao tummala V M 2001 An application of the AHP in vendor selection of a telecommunications system *International Journal of Management Science* **29** 171-82.

[3] Krisztina G 2005 Multi criteria decision making in process integration *Journal of cleaner production* **13** 475-83.

[4] Jing-Rung Yu and Chao-Chia Tsai 2008 A decision framework for supplier rating and purchase allocation: A case in the semiconductor industry *Computers & Industrial Engineering* **55** 634–46.

[5] Sung Ho Haa and Ramayya Krishna 2008 A hybrid approach to supplier selection for the maintenance of a competitive supply chain *Expert Systems with Applications* **34** 1303–11.

[6] Orlando Duran and Jose Aguilo 2008 Computer-aided machine-tool selection based on a Fuzzy-AHP approach *Experts system with applications* **34** 1787-984.

[7] Mehmet Sevkli and Selim Zaim 2008 Hybrid analytical hierarchy process model for supplier selection *Industrial Management and Data Systems* **108** 122-42.

[8] Dickson G W 1966 An Analysis of Vendor Selection Systems and Decision *Journal of Purchasing* **2** 5-17.

[9] Zsidisin G A 2003 Ground Definition of Supply Risk *Journal of Purchasing and Supply Management* **9** 217-24.

[10] Hwang, C L and Yoon, K. 1981. Multiple attribute decision making: a state of the art survey *Lecture Notes in Economics and Mathematical Systems* **186(1)**

AUTHOR INDEX